WASTE DISPOSAL BY LANDFILL

PROCEEDINGS OF THE SYMPOSIUM GREEN '93 – GEOTECHNICS RELATED TO THE ENVIRONMENT / BOLTON / UNITED KINGDOM / 28 JUNE – 1 JULY 1993

Waste Disposal by Landfill

Edited by

R.W. Sarsby

Bolton Institute of Higher Education, United Kingdom

A.A. BALKEMA / ROTTERDAM / BROOKFIELD / 1995

All articles were refereed before publication; the refereeing did not extend to typographical errors or to graphic quality.

The GREEN'93 symposium was developed with the assistance of a Senior Advisory Committee comprising:

The texts of the various papers in this volume were set individually by typists under the supervision of each of the authors concerned.

Published by
A.A. Balkema, P.O. Box 1675, 3000 BR Rotterdam, Netherlands (Fax: +31.10.413.5947)
A.A. Balkema Publishers, Old Post Road, Brookfield, VT 05036, USA (Fax: 802.276.3837)

ISBN 90 5410 356 6

Printed in the Netherlands

Waste Disposal by Landfill – GREEN'93, Sarsby (ed.) © 1995 Balkema, Rotterdam, ISBN 90 5410 356 6

Table of contents

2 *Basic principles*

4 *Pollutant control*

Waste Disposal by Landfill – GREEN'93, Sarsby (ed.) © 1995 Balkema, Rotterdam, ISBN 90 5410 356 6

Foreword

Public awareness of environmental issues has increased greatly in recent years and the growing body of environmental legislation and regulation reflects both the need, and the desire, to exercise control of pollution to protect mankind's environment. Topics such as rain forest depletion, fossil fuel burning and global warming, disposal of high level nuclear waste, and consequential effects on the environment, have received wide publicity and public attention has been focussed on these as the major environmental issues for our times. The reality is different and more mundane. As the world population increases and there is greater urbanisation the major environmental problems for mankind will be a shortage of 'clean' water, 'clean' land and 'clean' air.

The construction industry is supposed to make the best use of earth's natural resources for the benefit of mankind. Through the centuries the industry has made major contributions to the fight against disease and the improvement of living standards. All works of construction interact with the environment in some way, be it accidentally, intentionally, or as an inevitable consequence of the works, and the interaction may be beneficial or harmful. Much of the interaction between construction work and the environment is related to engineering of the ground since this immediately involves the soil and ground water. The term 'Environmental Geotechnology' has been coined to describe the technology which is a synthesis of geotechnical engineering and environmental considerations. Examples of topics in this area are: Engineered waste disposal by landfill, investigation and remediation of contaminated land, use of industrial waste materials (Pulverised Fuel Ash, Minestone, etc.) as bulk fill, noise and ground vibrations from piling and drilling operations. The GREEN symposia (Geotechnics RElated to the ENvironment) are intended to provide a regular, primarily European, forum for the interchange of ideas on aspects of Environmental Geotechnology. The symposia will be held at four-yearly intervals and each meeting will address a particular aspect of Environmental Geotechnology.

The idea of the GREEN symposia developed at a breakfast discussion in May 1989 between several European delegates to the second International Conference on Environmental Geotechnology in Shanghai. The group felt that there was a need to encourage European engineers/scientists/environmentalists to meet periodically. Everybody agreed that the theme of the first symposium had to be 'engineered waste disposal by landfill' because of the rapidity with which the waste disposal industry and the associated technology was developing, and the high geotechnical content of the topic. The dream has been realised and these proceedings are the outcome. However the initial idea would have come to nothing without the input of a host of people, too numerous to mention individually, but to whom the organising committee would like to say 'Thank You'. We would particularly like to thank the session chairs, general reporters, session secretaries and the Spa Road 'crew' for their patience, diligence and hardwork and for ensuring that the symposium ran smoothly.

Finally I would like to express my deepest gratitude to Irene, Louise, Drew and Bernard, for their limitless enthusiasm, energy and sheer grit; without your help there would have been no proceedings to produce.

R.W. Sarsby
Chairman of the GREEN'93 organising committee

Waste Disposal by Landfill – GREEN'93, Sarsby (ed.) © 1995 Balkema, Rotterdam, ISBN 90 5410 356 6

Organisation

Symposium Directorate

R W Sarsby

Chairman

B A Leach

Secretary

Chairs and Reporters

D E Daniel

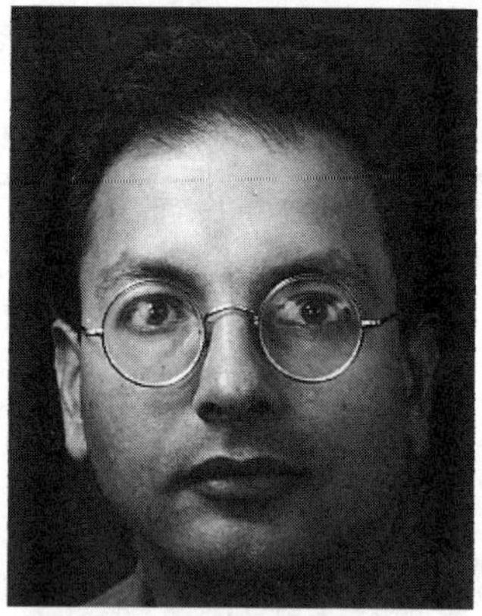

M S de Silva

H Y Fang

S A Jefferis

H L Jessberger

N A Kalteziotis

M Manassero

P J McKendry

J Saarela

C Sagaseta

Y Wasti

W Wolski

VIP lectures

Waste Disposal by Landfill – GREEN'93, Sarsby (ed.) © *1995 Balkema, Rotterdam, ISBN 90 5410 356 6*

The European Community and the environment

G.Titley
Member of the European Parliament for Greater Manchester West, UK

Waste disposal is a particularly sensitive issue. My purpose is to try to set out a broad european perspective on environmental protection.

Many people would argue that environmental matters can only be dealt with on a global, or at least an international, level and therefore the environment is very much a european issue. It has not always been accepted that the environment is an issue for european policy making, in fact the Treaty of Rome, which established the European Community, did not mention the environment because the community was conceived as purely an economic union. However the treaty does set out, as one of the tasks of the union, an improvement in the quality of life of its citizens.

The 1960's saw a growing awareness of environmental matters and in 1972 the Paris Summit [(1)] set out, for the first time, the beginning of a european policy on the environment which led to the first of a series of environmental action programmes. This early policy was very modest but by the 1980's there was a demand for a much more dynamic european environmental policy, partly because of public concern about pollution, but more importantly because the European Community set out to establish the Single Market. The logic of the Single Market means that there has to be a 'completely flat playing field' - no member state would be able to go it alone. Equally it was believed that the Single Market would lead to greater economic activity and that activity would lead to even greater pollution. Consequently in the Single European Act of 1987 [(2)] environmental action was placed very firmly on the european agenda, and that action was based on certain basic principles namely; prevention, the polluter pays, subsidiarity, industrial cooperation, and integration into other areas. Since then there has been a great increase in european environmental legislation. Now the body of European Community environmental policy consists of more than 200 directives which range from protecting the habitat of migrating birds to setting detailed standards for the quality of air and water.

The Maastricht Treaty [(3)] contains a number of provisions which will bring care for the environment into a new dimension. Among the principles it sets out in article 2 it refers to; the promotion of a harmonious and balanced development of economic activities, sustainable and non-inflationary growth respecting the environment. It goes on to provide that the activities of the European Community shall include a policy, rather than as now actions, in the sphere of the environment. It also stiplulates that this policy shall aim at a high level of protection and be based on the precautionary principle. Moreover it is required that environmental protection requirements are integrated into the definition and implementation of other European Community policies. Furthermore the agreed changes in the European Community's decision-making procedures, providing for majority voting in the Council of Ministers on more environmental issues and the strengthening of the role of the European Parliament in this sector through the co-decision procedure, should improve the efficiency and quality of the legislative process in the future. However it would be wrong to suggest that all is well with European Community environmental policy - the opposite is in fact true. The European Community is actually going through a crisis in relation to the environment.

The sheer weight of European Community legislation, in recent years, has created a number of difficulties and the priority (in my view) of the European Community has got to be how to resolve those difficulties before they cripple policy-making in this field rather than worrying about extension of its powers.

First among these difficulties is the question of 'subsidiarity' which is the latest 'euro-buzz-word'. In essence this simply means taking legislative action at the most appropriate level, ie international, national, regional or local. It is a principle, of course, that has always been behind environmental policy. Most people would recognise that environmental matters should be dealt with by european legislation even if not dealt with on a global level. So in theory subsidiarity ought to cause no problems whatsoever in this field. However at least one government has called into question the necessity and the appropriateness of much of the European Community's environmental policy, arguing that under subsidiarity many existing measures should be abandoned and the issues they deal with should be dealt with at a national level. Clearly we must not get bogged down in unnecessary legislation and we must constantly re-appraise the appropriateness of the existing legislation. There must also be a debate about which level of government is the most appropriate one for action in any field. However what has to be clear is that the principle of subsidiarity was never meant to simply mean re-nationalising policy areas. Such an approach to the environment would actually be very damaging, it would undermine the Single Market, and seriously weaken the fight against pollution. Of course it is true that the strategy and objectives of the European Community have to be met by action, not simply at Community level alone but on a basis of sharing of responsibility at all levels of society including Governments, Regional and Local Authorities, non-governmental organisations, financial institutions, companies and indeed individual citizens. That is why it is important that those governments which claim strong support for subsidiarity should first of all ensure that they practice it within their own countries and then we can get a coherent approach on the environment running through all levels of government and society. In my view it is absolutely vital for the future direction of European Community environmental policy that a clear statement of the precise nature and operation of subsidiarity is agreed upon by the Council of Ministers very soon otherwise chaos is going to result.

The second area we have to address is that of enforcement and implementation. European directives are actually very flexible and allow member states to ignore and misinterpret them either at the stage of transposing them into national law or at enforcement level. The number of cases currently being brought before the European Court of Justice for non-compliance illustrates this point. This just brings environmental legislation into disrepute. We have to make enforcement and implementation a priority and to do that the European Community needs several things. Firstly there should be reliable and compatible information on member states environmental policies - in order to have that we need to get the European Environmental Agency up-and-running as soon as possible. Secondly there have to be tougher sanctions for non-compliance. Thirdly there has to be greater openness of information at member-state level, so that it is easier for the public and non-governmental organisations to monitor the implementation of directives. All of the policies in the world are actually meaningless unless they are implemented.

The final issue to be resolved is in the integration of environmental considerations into other European Community policy areas. It is nonsense, for example, for the Community to stress action on the environment while at the same time european regional funds are used for projects which damage the environment. For instance - how can the quality of drinking water be improved without changing certain means of agricultural production, and how can CO_2 emissions be brought under control without changes in both energy and transport policies. Time and time again integration is stated as an objective of policy but nobody appears to want to do anything about it when it comes to actual legislation.

So the foregoing items are the general frameworks and issues facing european environmental policy at the moment. I would now like to turn to the specific question of waste management. European Community waste management is based on two concepts; firstly environmental protection and secondly the need to make sure there are not major distortions of trade and competition or the fragmentation of the Single Market within the Community. The priorities of European Community environmental policy as it affects waste starts off with the principle that waste should be prevented, secondly it should be re-used and thirdly it should be disposed of. The waste management strategy which was developed in 1989 [(4)] laid out five strategic guidelines for Community waste management - firstly prevention, secondly recycling and re-use, thirdly high standards for final disposal, fourthly high standards for transport, and fifthly remedial action for contaminated sites. And so we have currently on the table a number of important pieces of legislation on waste for example; framework directives on waste (which set out basic guidelines and definitions), standards for new and existing municipal waste incinerators, a directive on the trans-frontier shipment of hazardous waste, directive on the disposal of waste oils, a resolution on waste management which incorporates the principle of self-sufficiency in waste management and the proximity principle, directive on waste from the titanium dioxide industry (which was revised in 1992) etc.. We also have a number of proposals in progress; the shipment of waste into/out of/within the European Community, the landfill directive (which has gone through its first reading), the incineration of hazardous waste (which had its first reading in March 1993), the packaging and packaging waste directive (currently being examined), primary waste streams (a proposal is expected shortly).

I would like to amplify my previous points by referring to the packaging and waste directive because it is a current source of significant controversy. The first question to be addressed is why a directive on packaging and packaging waste is needed. The answer is several fold. Firstly the European Community produces over two billion tonnes of waste each year of which 15 million tonnes is packaging waste and it is growing rapidly. Only 18% of this is recycled - most is landfilled. Hence resources are depleted through constant use of new materials and energy is wasted. Further, competition is distorted by different management régimes. We have a shortage of landfill sites, added to which the phenomenon known as NIMBY (Not In My Back Yard) is growing as a factor in politics because nobody wants to live next to a landfill site. There are also fears about landfill sites, no matter how well-managed they are, about long-term contamination (particularly of ground-water). It is all those factors together which make a european directive necessary.

The European Community has tried in the past to deal with this matter because the current directive owes its origins to a directive on containers for liquid foods for human consumption which was developed in 1985. That frankly did not work properly, and therefore member states started to develop a variety of different approaches, which has forced the European Community to abandon the sectoral approach to get an overall approach to packaging waste.

Two of the initiatives which have caused problems are worth considering. Firstly (in Denmark) all containers for beer and soft drinks must be returned, and marketed in approved re-usable containers, with an absolute ban imposed on beverage cans. That was challenged in the European Court of Justice as a technical barrier to trade. The Court's decision was extremely interesting because it upheld Denmark's provisions on the grounds of environmental protection so long as there was no European Community legislation which offered the same or higher degree of protection. The second initiative is the German one (commonly known as the Töpfer decree [(5)] of June 1991) which sets specific targets for recycling and waste disposal. By 1995, 85% of all packaging wastes must be collected and of that collected 90% of glass tinplate and aluminium and 80% of other materials must be recycled. That law has created many problems, largely because it failed to address the lack of recycling facilities available and the lack of markets for recycled goods. Hence we have had german waste and german recycled goods being dumped on other markets and there has been a distortion of the market leading to, for example, local paper recycling plants being put at a price disadvantage. Clearly a European Community - wide approach had to be agreed if the Single Market was not to disintegrate. So in July 1993 the Commission brought forward a final proposal with the aim being to provide a framework for establishment of packaging and package waste management programmes at Community level, not only to protect the environment but to stop the disintegration of the Single Market. The objectives of the proposal are to reduce the overall impact of packaging waste on the environment by reducing the quantity of waste going into landfill, and to encourage a harmonising framework for all existing and future measures to avoid barriers to the free movement of goods. That proposal is in its early stages, for the moment it sets out high standards for recovery, recycling and reduction of landfill. Member states are to reduce the landfill of packaging to just 10% within 10 years of the adoption of the directive. Ninety percent of packaging waste must be recovered for the purposes of waste management - 60% is to be recycled, the remaining 30% may be incinerated with energy recovery. All packaging will be covered by the proposal from blister packs to wooden pallets, from packaging for pharmaceutical products to containers for hazardous waste. It almost goes without saying that this proposal has proved extremely controversial, and appears to have upset industry, environmentalists and consumers and some governments have also been hostile. The proposal had its first reading in the European Parliament in June 1993, and during its consideration in committee there were no less than 367 amendments with nearly 100 plenary amendments. The directive was approved at its first reading although the Parliament is asking that 60% of waste should be recovered within 5 years and asking for the 'Polluter-pays' principle to be imposed on third-country importers, with an exception for pharmaceutical products and a reduction in the content of dangerous substances used in packaging such as lead, calcium and mercury. Clearly however this debate is going to run for a long time.

In conclusion I would say that European Community environmental policy is designed to ensure that there is common action taken on the environment. That is much easier said than done when countries wish to travel at different speeds, in response to different problems, and in accordance with different national traditions. What is very clear though is that unless we get to grips with these issues, and unless we can resolve these issues for the benefit of the environment without at the same time disadvantaging industry, the long-term costs of failure are going to be far greater than the short-term costs of change.

1 Paris Summit (1972) - Meeting of Council of European Community Leaders to establish policy guidelines.

2 Single European Act (1987) - Amendment to Treaty of Rome, article 130, (environment).

3 Maastricht treaty (1992) - Treaty of European Union.

4 Waste Framework Directive 75/442EC - Amended to 91/156/EEC.

5 Töpfer decree (1991) - German federal law governing packaging and packaging waste.

Waste Disposal by Landfill – GREEN'93, Sarsby (ed.) © 1995 Balkema, Rotterdam, ISBN 90 5410 356 6

United Kingdom policy on waste management

P. Hinchcliffe
Waste Technical Division, Department of the Environment, London, UK

Waste management is a busy area just now. Legislation is being developed in the United Kingdom, in the European Community and internationally. I will focus this article on those developments and initiatives which relate mainly to landfill.

The basis behind the UK's waste management policy is one with which most people should be reasonably familiar because there is international acceptance of this hierarchy of waste management options :

(i) One should do everything possible to minimise the production of waste in the first place.
(ii) Those materials which can be re-cycled, or re-used in some way, or from which energy can be taken, should then be removed from the waste stream. There are a whole range of different possibilities (the French use the word 'revalorisation' which sums it all up).
(iii) On the bottom rung of the hierarchy is environmentally - sound final disposal.

After waste has been minimised, after we have 'revalorised' as much of it as possible, there still remains an enormous amount left. In the UK this is roughly 400 million tonnes a year. Much of the waste arises from agriculture and minerals extraction. The remainder, which is about one-third of the total, amounts to some 140 million tonnes and is termed 'controlled wastes'. In extant UK legislation 'controlled wastes' has a special meaning viz. they are the wastes which are not agricultural, nor from mineral extraction processes but are basically industrial/ commercial/ municipal/and other-related wastes - those wastes for which there is not an obvious, built-in disposal route. Therefore one has to take particular steps to dispose of them in an environmentally-sound way. At the moment about 90% of these 'controlled wastes' go to landfill and it is self-evident that landfill is a key final disposal route to which we must pay particular attention. For the wastes which do not go to landfill, the next most important route is incineration and the remainder go to special treatment processes of various descriptions.

Under the European Community Framework directive[1] there is a requirement that landfill sites (and waste management processes generally) should be licensed. In UK law this has been a requirement for almost 20 years - it was contained within the Control of Pollution Act[2] which introduced the concept of site licensing. Site licensing gives a regulatory authority the ability to impose on a waste management facility conditions which are aimed at reducing to a minimum environmental pollution and the possibility of harm to human health. This has worked quite well and, despite the fact that there are currently over 3000 operating licensed landfill sites in the UK and landfilling has been in operation for a great many years, the number of problems that have arisen is rather small. Nevertheless problems have arisen and the Environmental Protection Act (EPA)[3] was introduced in 1990 to try to correct some of those problems. Part 2 of the Environmental Protection Act deals specifically with waste management, and it introduces a set of new features :

(i) There is clear separation of regulation and waste disposal functions whereas previously Local Authorities were both disposers and regulators. Under Part 2 of the EPA this cannot now occur, either because the Local Authorities have set up at-arm's-length waste disposal companies (which are effectively acting in the private sector), or because the large private sector companies are developing and moving into the area of landfill in a private capacity. This has quite a benefit for waste management in the UK, not only from the point of view of the separation of poacher and gamekeeper roles but also because of The Duty of Care;

(ii) The Duty of Care places a responsibility on waste producers for the care of their refuse long beyond the time when they had actual possession of it. It is no longer acceptable for a waste producer to hand his refuse over to anybody with a truck and say "please take it away and do not tell me where it is going". A waste producer will be responsible for the safe management of his waste from the time that he produces it to the time that it is safely disposed of. This means that the private sector waste management companies cannot afford to get a poor reputation for disposing of waste because otherwise they will lose business. People who convey waste must be properly authorised and registered with their local authorities so that a track can be kept of what happens to the refuse. Hence there is a commercial pressure which will drive up environmental standards and this can be seen to be happening in UK at the present time.

(iii) The Environmental Protection Act imposes a new Waste Management Licensing system. New regulations will serve both to bring into effect Part 2 of the Environmental Protection Act and to implement the European Community Framework Directive on waste. It is quite a difficult job technically to marry the British and Community legislation, and this has caused delays. Thus the industry is currently in a state of flux.

(iv) An essential feature of the new Waste Management Licensing system is that it introduces the concept of 'fit and proper persons'. A company or person cannot be granted a waste management licence until they have demonstrated to the Local Authority that they are competent to hold such a licence.

(v) Under the Control of Pollution Act it was possible, (or even a requirement), to impose conditions on a site licence. A similar situation will exist under the Waste Management Licensing system, but now much more clear and coherent guidance will be given on what the Department of the Environment expects to see in these licences. This direction will be based on the twenty years of experience of site licensing gained under the Control of Pollution Act. Consequently the technical guidance contains a strong emphasis on the need to carry out proper site assessment and proper monitoring (before operations start at a landfill site, whilst operations are continuing and during the long post-closure period of a landfill site).

(vi) A new feature of waste management licensing, which was not seen under the Control of Pollution Act, is the surrender of licences. It will no longer be possible for a waste management operator simply to surrender his licence when he has completed his tipping operations. A waste regulation authority will be prohibited from accepting such a surrender unless they are satisfied that the landfill site (or whatever the facility) is no longer likely to cause environmental pollution or harm to human health. The problem then is to decide whether in the future a site is likely or not to cause harm to health or environmental pollution. Consequently the Department of the Environment has been engaged in drawing up guidelines, specifically aimed at landfill sites, which both regulators and operators can use to determine whether or not a site has reached a reasonably-stable state and so conforms with the preceding criterion. This has led to debate with industry and other interested parties and a consensus view is now appearing. Certain key parameters, (in terms of a landfill site), can now be identified and one is the very long time period needed. Once operations have stopped at a site processes will continue there for decades. The consequence is that in most cases the surrender of site licences might not be accepted for several decades after operations at a site have stopped. In some special cases the time period might be very much shorter, in a number of cases it may be longer.

To return to the concept of a 'fit and proper person'. There are three limbs to the test for a 'fit and proper person' and these relate to any person responsible for a waste management facility:-

First of all there is the question of whether an applicant has any convictions for breaking laws concerning environmental pollution and if so, how often and with what consequences.

Secondly the local authority will want to see that the waste management company is not 'a company of straw', i.e. it has reasonable expectations of financial solvency during the after-care period (probably decades, in the case of landfills). It is impossible to provide absolute guarantees of long-term financial stability but the regulatory authority will need to see that at least the intent is there and that the means exist to provide stability over the operations period of the landfill site.

Finally there is technical competence. A training and advisory board has been developed in conjunction with the waste management industries and a package of technical competence qualifications has been developed. This includes certificates of technical competence relating to different aspects of waste management operations. An operator must have the competence certificates relevant to the licence requested or operation being undertaken.

Several other areas related to landfill are currently under consideration.

The government is trying to minimise unnecessary regulation wherever possible and the Department of the Environment is actively investigating alternative ways to secure environmental improvements. One way which has attracted significant attention is the use of economic instruments, in particular a landfill levy. The question to be considered is what would be the effect of imposing a levy, or a tax, of between £2 to £10 per tonne of waste going to landfill. The ideal outcome is that such an economic instrument would drive waste management up the hierarchy mentioned previously, i.e. move away from final disposal, up towards 'revalorisation' and possibly encouraging waste minimization. This however is a slow process and would take quite a large levy. Initial suggestions are that a landfill levy might push more waste away from landfill towards incineration in the first stage - but is this a good or bad thing. It is suspected that it is good since there is an existing substantial imbalance between the amount of waste going to landfill and the amount going to incineration. With modern incineration plants, which have high standards of emission control and which can recover energy from the waste being burnt, it is an attractive proposition to pass more waste down this route. However incineration has its own environmental disbenefits.

Landfill gas is primarily thought of in terms of its potential to suffocate people, blow up houses and destroy vegetation. Now the methane content of landfill is being viewed as an important contributor to global warming. Although much of the gas can be recovered, and in many landfill sites it is converted into energy, much of it still escapes. This might be one of the driving forces behind the government's attitude towards landfill and the need for a landfill levy.

Another main area of concern is the question of containment of landfill sites. The Department of the Environment is now moving towards a policy of total containment of landfill sites. It is acknowledged that 'dilute and disperse' (which has been the main form of landfill system in the UK for many years) has not worked perfectly. Nevertheless it has worked sufficiently well to recognise that with tight regulation, better monitoring, better site investigation and risk assessment then it should be possible to establish policy which allows both for 'dilute and disperse' to continue and for improvements in containment technology and engineering.

Finally, the disposal of hazardous waste is a matter of heated debate in the context of the proposed landfill directive. In the UK co-disposal of municipal refuse with industrial wastes (particularly hazardous ones) has been seen as an important way of dealing with the latter materials. The powerful biological reactions in a municipal landfill site can destroy and neutralize many wastes. When this is done properly in a well-defined, scientific way we have a very useful way of treating substances which otherwise would be quite difficult to dispose of. At the present time it is not acceptable to ban, or progressively phase-out, co-disposal even though drafts of the European Directive on Landfilling have suggested that might be a policy objective for other member states.

1 Waste Framework Directive 85/442 EC, European Community.

2 Control of Pollution Act (1974)
Part 1 - Waste Disposal, HMSO, London.

3 Environmental Protect Act (1990),
Part 2 - Waste on Land, HMSO, London.

Waste Disposal by Landfill – GREEN'93, Sarsby (ed.) © 1995 Balkema, Rotterdam, ISBN 90 5410 356 6

An overview of the technology of waste disposal by landfill in the United Kingdom

N. Robinson
Robinson Fletcher Environmental Consultants Ltd, Wigan, UK

INTRODUCTION

According to the press the state of waste disposal in the UK is very poor. Whilst there is always a very large potential to improve, I and many professionals in the industry believe that this bad press is incorrect and misleads the public at large. This leads to undue concern by the public when there are proposals for new, and usually large, landfills nearby. It is true that there are still some poor landfill sites in use, as is the case in most developed countries, but the vast majority of new landfill is of a high standard and the rate of improvement is very high.

According to published data the UK produces about 515 Million Tonnes of waste per year (Table 1). About 300 million Tonnes are outside the Control of Pollution Act as they are agricultural, mining, radioactive and explosive wastes. Controlled Wastes comprise some 215 M Tonnes which are incinerated, recycled, treated but mainly landfilled. The amounts of waste are increasing and the constituents of the wastes have changed and will continue to do so.

Landfill is therefore the predominant method of waste disposal in the UK today and there is no reason to suppose that there will be a major change in the near future despite platitudes and some move towards recycling. It is therefore of strategic importance to the country and to industry to provide adequate landfill capacity, to a suitably high standard to provide environmental protection yet to keep industry competitive.

Standards, and therefore costs, are rising but quantum leaps in standards produce imbalance in the landfill industry and economic uncertainty for the customer. Cost of landfilling waste has increased dramatically over the past ten years or so in response to the increase in the cost of technology in landfilling. Going fast are the days of the poorly controlled "council tip" and "cowboy operators". Today the minimum investment for preparing an single landfill cell is in the order of £ 0.25 to 1.0 million on engineering costs dependent on complexity, risk and size, without infrastructure or restoration costs. Unfortunately the public at large do not appreciate the efforts that now go into engineering all new sites.

The new Waste Management Licensing regime should enable the better controlled, generally larger companies, to produce high quality sites. However, as landfill sites take several years to develop from inception to restoration there are still many poor quality sites in existence which do not help the responsible operators, because they provide unfair competition, and in particular affect the image of landfill. Unfortunately the public is not aware of the change in standards and cannot appreciate the difference that is so obvious to those in the industry. Indeed uninformed, erroneous and/or mischievous misrepresentation by the media and some environmentalists does not help the image of an industry which is changing fast.

The cost of landfill as a disposal method has been rising but is still very low by the standards of many developed countries. This is due to a variety of factors including;

Table 1 Estimated total annual waste arisings in UK : by Sector

Sector	Annual arisings (million tonnes)	Date of estimate	Status [1]	Source	Percentage of total arisings
Agriculture	250	1990	NC	DOE	49
Mining and quarrying					
- colliery and slate	50	1990	NC	DOE	10
- china clay	27	1990	NC	DOE	5
- quarrying[2]	30	1989/90	NC	QI[6]	6
Sewage sludge[3]	1	1990/1	PC	MAFF/WSA[7]	-
Dredged spoils[4]	21	1989	NC	MAFF	4
Household	20	1990	C	DOE	4
Commercial	15	1990	C	DOE	3
Demolition and construction[5]	32	1990	C	DOE	6
Industrial					
- blast furnace and steel slag	6	1990	C	DOE	1
- power station ash	13	1990	C	DOE	3
- other	50	1990	C	DOE	10
Total	515				

1 NC = non-controlled wastes;
PC = controlled wastes only when landfilled or incinerated, but not otherwise;
C = controlled wastes.

2 Excludes wastes from open-cast coal mining.

3 Dry weight.

4 Dry weight. The data are for all UK waters, ie external and internal waters.

5 Includes road plannings.

6 Quarries Inspectorate.

7 Water Services Association.

- poor standards in the past and some sites/operators are still operating at low overheads,

- a surfeit of holes in the ground and the planning strategy regarding them as the best location for engineering landfill sites,

- a past lack of consistency in enforcement -small operators and Local Authorities are often not prosecuted whereas large national and regional companies are hit hard,

- lack of adequate funding for the regulation authorities,

- lack of consistency of neighbouring Waste Regulation Authority (WRA) and National Rivers Authority (NRA) requirements and operations which has permitted unnecessary movement of wastes to the areas with lower standards and therefore the cheapest air-space,

- subsidy by local government in the past of some public sector sites,

- some very efficient very large sites operated by the larger companies which have economy of scale.

High standards of enforcement are in the interests of the environment, the landfill industry, and for all the high calibre operators who are investing in expensive high technology sites. High standards will;

- improve standards overall
- reduce risks
- improve the public image
- drive out of existence the low-overhead inadequate sites and companies
- improve the profitability of all sites
- put everyone on an even playing field.

DESIGN HISTORY OF LANDFILL

To appreciate the current state of landfill technology it is necessary to have an appreciation of the historical development of landfill design. There was very little control of landfill during the 1950's and 60's. In the early 1970's in the run up to the implementation of Control of Pollution Act[1] the River Boards were becoming concerned about pollution of groundwater and the Government started one of the most intensive investigations into the formation, migration and attenuation of leachate from existing landfills. This was concluded with the publication of "The Co-operative Programme of Research on the Behaviour of Hazardous Wastes in Landfill Sites"[2] which gave early insight into attenuation processes, movement of leachate through the unsaturated zone and led to the widespread acceptance of the "dilute and disperse" (presently called the "dilute and attenuate") philosophy. The pros and cons of this philosophy are summarised in figure 1. Protection of groundwater was the prime concern and two philosophies have evolved since that time;

- Containment
- Dilute and Disperse.

Containment at the time could be classed as "casual containment" because it relied upon natural, in-situ barriers. The UK, along with many, but not all, northern European countries have basically advantageous geological conditions for the area has a significant cover of low permeability clays left by the glaciations. There are also large areas of outcropping low permeability rocks or mudrocks eg Palaeozoic mudrocks (Cambrian mudstones, Coal Measures), Mesozoic clays (Keuper marls, Oxford Clay, London Clay etc.). Many of these materials have been worked for generations for brick clays, refractory clays, pottery clays etc. and the Industrial Revolution and the subsequent period has left a huge number of holes in the ground which planners generally wish to see restored.

Such worked out quarries were looked on as a major resource for landfill as the low permeability would act to minimise flow of leachate to the groundwater. Minimal investigation was carried out, possibly because it was recognised that they did not represent total containment but increased the dilute and disperse effect but probably many people did consider that they provided total containment. However, all these basically low permeability

FOR

- all competing forms of treatment or disposal are dilute and disperse (containment and incineration)
- it is sustainable
- possible to monitor

AGAINST

- parameters can be difficult to assess accurately
- monitoring takes a very long time
- remedial action may be difficult or unrealistic
- more designers able to understand and implement containment
- policy makers generally lack background to understand theories and risks
- containment is an easier concept for the public
- lack of confidence of the public in scientists' predictions
- politically expedient to support a concept understood by the public at large

Figure 1 Factors relating to Dilute and Disperse approach

materials have inhomogenities of some degree caused by the variable depositional environment or due to subsequent weathering processes.

"Dilute and disperse" was the start of conscious design in landfill in the UK. Concepts that were researched and developed included:

- Water Balance -
  Analysis of the water balance in the landfill and its environs

- Capping -
  To reduce the through flow of incident rainfall

- Attenuation -
  Processes researched to reduce the effect on groundwater including attenuation blankets and Co-disposal.

CO-DISPOSAL

Co-disposal as a philosophy has the support of many eminent landfill scientists and engineers even today when it is obviously becoming a pariah of a design concept. Co-disposal is a major topic in the European Community. There is much disagreement about its use but most is apparently caused by wide variations in the use of the terms co-disposal and mono-disposal in different countries.

Co-disposal as practised in the UK is defined as: The intentional disposition of industrial or difficult wastes with household waste in a manner which utilises the attenuation processes in the landfill site to minimise the impact of the industrial or difficult waste on the environment.

Co-disposal is a technique which fosters a change in the physical or chemical nature of the industrial or difficult waste so that it is not retained in its original form with its potential to create environmental pollution.

Typical conditions in a landfill biomass are:

- long hydraulic retention
- high temperatures
- high metabolism of organic compounds
- high Eh (reducing conditions)
- pH is neutral

Typical processes are:

- absorption
- filtration
- dilution
- dispersion
- adsorption - physisorption and chemisorption
- ion exchange
- chemical reaction - especially precipitation and formation of chemical complexes.

The landfill site containing household waste is a biochemical reactor. It can be demonstrated that the leachate from a co-disposal site is no different from that from a landfill site containing only household waste.

CAPPING

Capping was the first "engineering" activity carried out on early landfills in the UK. With the widespread use of low permeability holes in the ground during the 1960's and 1970's the Regional River Authorities became aware that some were filling up with incident rain and overtopping with disastrous results to local watercourses. High leachate levels also increased flow through the bottom of the sites, at some sites into aquifers. The obvious solution was to require the capping of such sites with a low permeability material - usually clay. This was introduced during the late 1970's and early 1980's along with rudimentary leachate control measures including:

- leachate chimneys (manholes)
- maximum permissible leachate levels - eg 1 metre above base
- construction of cellular facilities based on water balance

FLEXIBLE MEMBRANE LINERS

Flexible Membrane Liners (FML's), also known as geomembranes, appeared on the UK scene during the early 1970's. They were marketed as virtually the panacea for the protection of our water resources. Certainly butyl rubber and High Density Polyethylene (HDPE) were so marketed and probably others. A few cells were constructed in highly permeable strata where there was no clay and with primitive welding techniques, no seam testing, no subgrade specification and no liner protection. Not surprisingly they leaked :-

- Early 1970's butyl rubber liner (1973/74) Chippenham
- Mid 1970's HDPE (1977) Scarborough
- Various attempts with thin gauge LDPE.

The well publicised stress-cracking of the very high density sheet lining of the lagoons at Shar Chesma mining complex in Iran which was caused primarily by high altitude, high UV and high temperature variation, did not assist in their widespread acceptance either.

Quality Control and subsequent construction Quality Assurance (CQA) by an independent third party has been the saviour of FML lined sites in the UK and this started around the mid 1980's.

MINERAL LINERS

Mineral liner development has progressed significantly over the last 7 to 8 years also. The first regulator-produced Specification(3) for clay liners was North West Water (now the NRA North West Region) who produced a first edition in 1986 but some consultants had previously produced contract specific specifications.

This Specification basically requires three phases of works:

- laboratory testing of sampled clay
- field compaction trial
- testing of the compacted clay

The main criteria are:

- minimum 1000 mm thickness
- permeability less than 1 x 10^{-9} m/sec
- proven range of moisture content and dry density
- no greater than 5% voids ratio
- minimum 10% clay particles
- consistent plasticity indices
- earthworks in accordance with DTp guidelines

This was a splendid document for a first attempt, it was practical but also had additional benefits;

- it focused people's attention on the correct parameters for clay earthworks
- operators had to seriously plan sites
- it started the move towards continuously improving design and engineering
- operators had to use consultants who understood earthworks
- it was modified utilising comments from the consultants, operators and other regulators who were implementing it.

This specification was last revised in 1989 and further revisions have been avoided because it was recognised that nationally agreed specifications should be produced. This has not stopped it being adopted and/or parts of it utilised by many other NRA regions.

It is interesting to note that this specification is not vastly dissimilar from other national guidelines formed at about the same time.

COMPOSITE LINERS

The use of composite liners in the UK appears to date from the mid 1980's. One of the first was a clay/FML/clay composite on a site in Greater Manchester. It was designed in good faith but there were design flaws which caused a slip failure which in retrospect was inevitable. Despite severe movement of the waste mass, the liner did not apparently shear as monitoring of underdrainage has yet to show any sign of leachate contamination.

Today the HDPE/Clay composite is the most common lining system in northern regions of the UK for sites accepting leachate producing wastes. There is regional variation, some due to geology and some due to differing philosophy.

A current major development in the UK is the widescale acceptance of Bentonite Enhanced Soils (BES) where clay is of poor or variable standard.

Multiple liners are being installed where there is high risk, underdrainage is common and leak detection systems (beneath the composite rather than the North American practice of beneath the primary FML) are being installed.

Until recently much of the design was intuitive using accepted, probably over-conservative, slope and standard earthworks procedures, whether they were totally pertinent or not. Many standard methods cannot be utilised as more sophisticated multi-layer geotextile and geomembrane slopes are designed.

The need for calculated design has been recognised by the engineers involved to; maximise the design, make the operators more competitive, and to avoid litigation in case of failure.

There were, and still are, no universally-accepted methods of design although the DoE have a working party Report on Design of Landfills which is due to be published later this year, but this will only be for guidance.

PROCESSING NATURAL MINERAL LINERS

Processing of mineral liners started with reworking natural clay deposits but it was primitive breaking up the fabric of the natural deposited clay. After the introduction of the NRA Specification and the involvement of soils and geotechnical engineers it was inevitable that further refinements would be introduced to improve marginal materials and to permit other, otherwise unsuitable materials, for use in mineral liners.

With the advent of the containment philosophy there were many sites which did not have suitable clay but required containment. The cost of importing clay was regularly uneconomic in those days when landfill gate prices were much lower than they are today. Bentonite Enhanced Soils (BES), then perceived as Bentonite Enhanced Sand, was introduced to provide a containment liner in such situations. The standard of workmanship was, by today's standards, poor. The bentonite was spread and mixed using local agricultural contractors and there was insufficient testing of any sort. The industry and the regulators were not impressed by these early attempts and the use of this method for lining, as opposed to capping where it is still used, was severely curtailed. Needless to say, market forces prevailed and the technique evolved to a much more controlled method of mix-in-place technology with vastly better Quality Control.

With the advent of Construction Quality Assurance (CQA) the regulators and quality engineers recognised that to pass BES for use in a liner required much more sophisticated QA and QC procedures. This has produced much interest in mix-in-plant technology which permits sophisticated computer control of continuous or batch mixing with full records. Combining this traditional laboratory testing has produced an upsurge in interest in this technique.

A large amount of research work is being carried out or has recently been carried out including; work to verify the performance of different types of bentonite (Na enriched Ca-bentonite has been used as in most parts of Europe), the effect of polymer treatment, the effects of leachate, long term effects.

BES is expensive per cubic metre but the regulators have recognised the vast improvement in quality available using mix-in-plant technology and they permit lower thicknesses of bentonite sand (typically 300 mm thickness with a permeability of 1×10^{-10} m/sec) than for conventional clay liners.

It is recognised that a thin layer of well controlled liner material has the following benefits;

- permits more waste in the site
- is quicker to lay as there is less volume
- less risk of delay which can affect the start of filling operations
- is often less susceptible to shrinkage and wetting
- the "man-made" nature reduces the risk of variation

Processing also permits the mixing of bentonite with crushed mudrocks or other base materials to provide a suitable material. The technology has been used extensively in Germany and, whilst crushing mudrocks has been carried out here, it is believed that bentonite addition has not yet been used except in trials. Several large contracts for crushed mudrock liners are planned. Other processing has been tried out, some successfully, including the removal of oversized stone from boulder clay.

SUMMARY

Our Environmental Protection Act is amongst the best legislation in Europe and although the implementation of some aspects have been delayed this is because the will of government to implement it has outstripped the practicalities of getting realistic enforceable details and mechanisms. It is only hoped that this initiative continues.

In conclusion it must be stressed that site specific design, as generally used in the UK, makes for high quality, low risk, cost effective landfills. That is what industry, the environment and the public deserve. The technology of landfill engineering has advanced rapidly in recent years to provide the understanding, methods and materials to facilitate such designs.

1 Control of Pollution Act (1974), Part 1 - Waste Disposal, HMSO, London.

2 Cooperative Programme of Research on the Behaviour of Hazardous Wastes in Landfill Sites - Final Report (1978), Department of the Environment, HMSO, London.

3 Earthworks for landfill sites (1986), North West Water Authority, Warrington.

Keynote lectures

Waste Disposal by Landfill – GREEN'93, Sarsby (ed.) © 1995 Balkema, Rotterdam, ISBN 90 5410 356 6

Engineering waste disposal (geotechnics of landfill design and remedial works)

Hans L. Jessberger
Institute of Soil Mechanics and Foundation Engineering, Ruhr-University Bochum, Germany

Mario Manassero
I.G. Ingegneria Geotecnica, Torino, Italy

Bertrand Soyez
Laboratoire Central des Ponts et Chaussées, Paris, France

Andrew Street
MRM Partnership, Bristol, UK

ABSTRACT: The European Technical Committee (ETC 8) prepared recommendations on "Landfill Design and Remedial Works". In this paper recommendations are discussed, dealing with:

- liner deformations and drainage systems
- geotextiles and geomembranes
- cut-off walls
- contaminant migration assessment

1. GENERAL

The design and construction of landfills and also remedial works for abandoned landfills and contaminated land are mainly based on geotechnics. Therefore it is an important task of environmental geotechnology to establish specific principles for landfill design and remedial works. Actual experiences show an increasing need for guidelines on the geotechnical aspects of landfill design and remedial works.

The European Technical Committee No. 8 (ETC 8), founded at the ICSMFE 1989 in Rio de Janeiro aims at providing the engineering practice with recommendations for its work. The members of ETC 8 are at present: Ch. Schlüchter (CH), B. Soyez (F), H. L. Jessberger (G), M. Manassero (I), P. Ruardi (NL), and A. Street (UK). The work of this Committee comprises the major geotechnical aspects in landfill design and remedial works which are believed to be decisive like:

- General design principles including multi-barrier concepts
- Composite sealing systems
- Stability and settlement analysis of the waste body
- Deformability of mineral seals
- Contaminant migration
- Construction methods
- Quality assurance

Therefore several recommendations have been worked out, which can be subdivided into the following groups:

- Recommendations on the Site Assessment
- Recommendations on the Principles of Design
- Recommendations on Suitability Testing
- Recommendations on Construction Methods
- Recommendations on Quality Assurance

A first draft version of recommendations prepared by ETC 8 has already been published in 1991 and a second edition containing further recommendations is to be published in 1993 [1]. It is called "Geotechnics of Landfill Design and Remedial Works - Technical Recommendations GLR" and partly based on the Recommendations of the Committee on Geotechnics of Landfill Design and Remedial Works of the German Geotechnical Society [2].

In this paper design principles of mineral sealing layers and basal drainage systems, application of geotextiles and geomembranes, construction of cut-off walls and assessment of contaminant migration will be discussed.

2. RECOMMENDATIONS ON LINER DEFORMATIONS AND DRAINAGE SYSTEMS

To guarantee the long-term barrier function of a landfill various design principles have to be followed in the planning and construction phase. Regarding the function of landfill lining systems it is important to assess the deformation of mineral sealing layers (R 2-11) and to design an appropriate leachate drainage system (R 2-12 and A 2-12).

2.1 ASSESSMENT OF MINERAL SEALING LAYER DEFORMATION

A mineral sealing layer should not be subject to settlement to such an extent that the barrier function is compromised by deformation as these can cause cracking or shearing mechanisms (Fig. 1). Deformation induced shear mechanisms without open cracks do not result in a significant increase in leakage rates through the liner [3]. But the development of open cracks has to be avoided as these might lead to an unacceptable increase of leakage rates. Open cracks develop when the tensile strength of the material is exceeded. In order to assess if a mineral sealing deformation results in an unacceptable development of open cracks a specific assessment procedure should be conducted. Input data for this procedure are the extent of deformation and the stress-strain behaviour of the sealing material, including its properties under tension.

The first step of the assessment procedure is to compare the induced deformation and the deformation capacity of the mineral material in terms of tensile strains according to Table 1. ϵ_{zq} represents the maximum tensile strain at the onset of cracking measured in direct tension tests [3], whereas ϵ_{RF} is the maximum edge fibre strain of the deformed mineral liner given by the settlement curve. If this quotient of strains given in Table 1 exceeds a recommended value of 2.0 the sealing material is considered capable of withstanding induced deformations without cracking. Overburden loads can be considered in this validation step by comparing the deformation induced edge fibre strain ϵ_{RF} with the overburden induced transverse strain. If the transverse strain exceeds the edge fibre strain, cracking of the mineral sealing layer is prevented.

If such strain conditions are not fulfilled, stress conditions excluding bending stresses should be taken into consideration in the second step of the assessment procedure. Unacceptable cracking is avoided when the overburden load σ_0 acting on the mineral sealing layer exceeds the required load (see Table 1). Here a safety factor as used in slope stability analysis should be introduced (e.g. 1.4). In case that σ_0 does not exceed the required load sufficiently a further criteria is the consideration of bending stresses according to Figure 2. If the existing material cohesion ex.c' is less than the required cohesion req.c', the shear resistance is actually too low to maintain equilibrium. At this stage shear mechanisms have already occured and hence no open cracks develop.

2.2 DESIGN OF LEACHATE DRAINAGE SYSTEMS

A leachate drainage system forms part of a

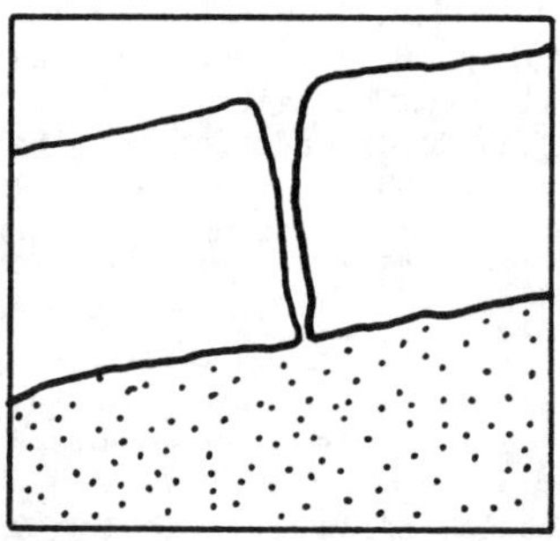

a)

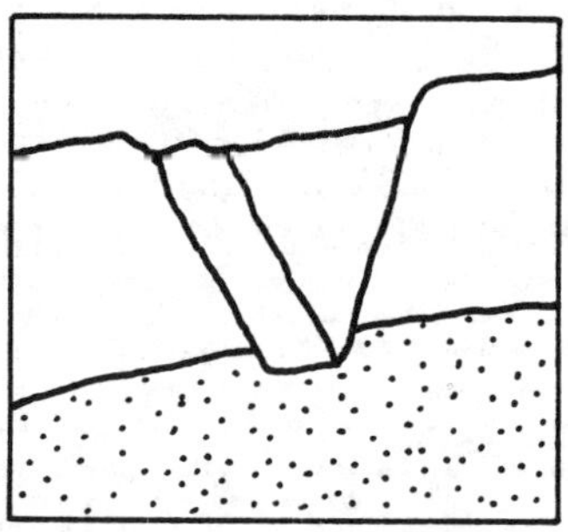

b)

Fig. 1: a) Cracking and b) shearing mechanism caused by settlement induced deformation in mineral sealing [3] composite lining system and normally comprises a drainage blanket, a protective layer, drainage pipes, and collection and monitoring shafts. The design of any drainage system has to follow specific national regulations and to consider all risks that a specific site may pose to the environment. In general leachate needs to be controlled:

- To prevent liquid levels rising to such an extent that they can spill over and cause uncontrolled polution to ditches, drains, watercourses etc.
- To reduce the potential for seepage out of the landfill through the sides or the base either by exploiting weaknesses in the liner or by flow through its matrix
- To minimise the chemical interaction between leachate and the liner
- To influence the processes leading to the formation of landfill gas and chemical and biological stabilisation of the landfill
- To ensure the stability of the waste body in the case of above ground landfills.

Reliable leachate drainage and collection can only be achieved by appropriately engineered systems. Therefore the leachate drainage system should be described as part of the site infrastructure in the planning documentation. The overall objective and the consequent design criteria should be stated. For certain requirements it will be necessary to include design calculations and proofs for structural stability (e.g. slip resistance, stability of drainage pipes) and hydraulic efficiency of the drainage system components.

Any calculations should be approved by the responsible certifying engineer. Leachate drainage systems should be included in the construction quality control programme.

As general design requirements leachate collection systems should

- be designed to collect the anticipated volumes of leachate generated. This will vary during the life of the site and can be estimated by a water balance calculation
- be able to drain down leachate within the landfill such that the leachate head acting on the base of the landfill is less than any specified level
- be strong enough to withstand physical damage due to the loading imposed by the waste, and any compaction equipment working over the system
- be able to accomodate the predicted settlement at the base of the landfill under the applied loads and foundation conditions
- be resistant to chemical attack in the corrosive environment of the landfill
- be capable of inspection and maintenance until such time as the system is no longer required to function and
- incorporate a suitable level of "redundancy" in the design to take account of failure of parts of the system and provide alternative drainage routes to collection points.

In addition specific influences on the functioning of the leachate drainage systems like leachate quality, temperatures at the landfill base and leachate viscosity have to be considered appropriately.

Table 1: Assessment procedure [3]

Content	Formulae
STEP 1 deformation and material properties	$\dfrac{\epsilon_{zq}}{\epsilon_{RF}} > \eta_v \ (> 2{,}0)$
STEP 2 vertical stresses σ_o check the deformation (shear mechanism is acceptable)	$\sigma_o > 2 \cdot c' \cdot \mathrm{TAN}\ (45° + \Phi'/2)$ req. c' > ex. c'

In order to satisfy these requirements the following features are suggested as minimum guidance for design features:

a) A leachate collection system should extend over the entire base of a landfill and, if below ground, should extend up its sloping side walls.

b) Underdrainage with granular materials should be at least 300 mm thick and should have a hydraulic conductivity of at least $1 \cdot 10^{-3}$ m/s (based on German requirements [2]).

c) The cell base should be profiled to have sufficient gradient to promote efficient drainage to the leachate collection point. Due regard should be given to potential settlement of the landfill base. Minimum values ranging from 3 % to 0.1 % are stipu-

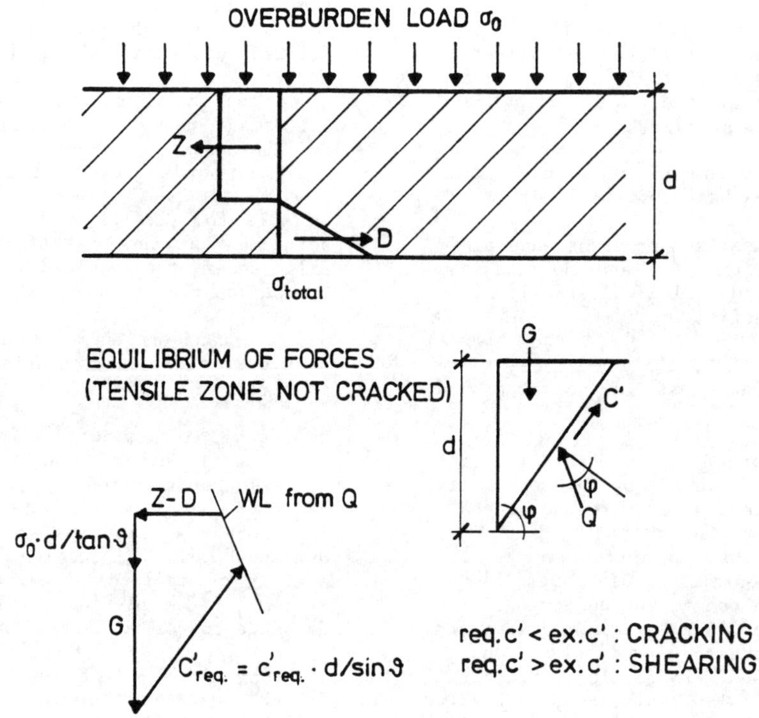

a) overburden pressure $\sigma_o = 0$:
$$\text{req. } c' = 1/2 \cdot \gamma \cdot d \cdot (F_1 - 2/3 \cdot F_2) + q_z \cdot (2/3 - 1/12 \cdot R_E) \cdot F_2$$

b) overburden pressure $\sigma_o > 0$:
$$\text{req. } c' = (1/2 \cdot \gamma \cdot d + \sigma_o) \cdot F_1 - \sigma_o \cdot (1/3 - 1/12 \cdot R_E) \cdot F_2 + q_z \cdot (2/3 - 1/12 \cdot R_E) \cdot F_2$$

with: γ = specific weight of mineral sealing material [kN/m³]
d = mineral layer thickness [m]
σ_o = overburden pressure [kPa]
q_z = uniaxial tensile strength of mineral sealing [kPa]
R_E = stiffness ratio: tangential stiffness from uniaxial compression tests / tangential stiffness from direct tension tests [-]
$F_{1,2}$ = parameter [-], see right

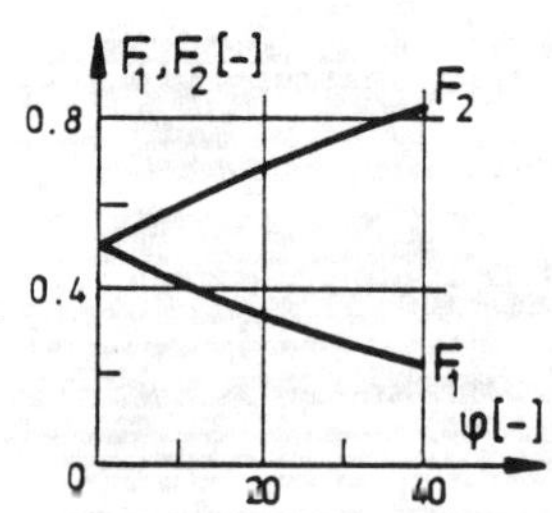

Fig. 2: Bending tensile stresses in the deformed mineral sealing layer (tensile zone not cracked) [3]

lated by different national regulations, depending on the topographical and hydrogeological setting of the landfill and the design objectives. An example of such a cell base according to German requirements is included in A 2-12 of the Technical Recommendations GLR. Figure 3 shows the plan view.

d) The granular material of the drainage layer should be prewashed to remove fines.

e) Limestone or other calcareous materials should not be used.

f) A network of perforated collection pipes should be laid within the drainage media with continuous (self cleansing) gradients towards the leachate collection point(s) and should be capable of being inspected and cleaned out.

g) Leachate removal from a common collection point should be capable of continuous and automatic functioning. Sump emptying should be automatically activated as leachate accumulates to avoid a build-up above the prescribed level/head.

h) Facilities to monitor leachate levels relative to the drainage media, which are remote from the collection and removal points.

i) The system should be examined regularly and cleaned out accordingly.

j) Construction documentation should be retained.

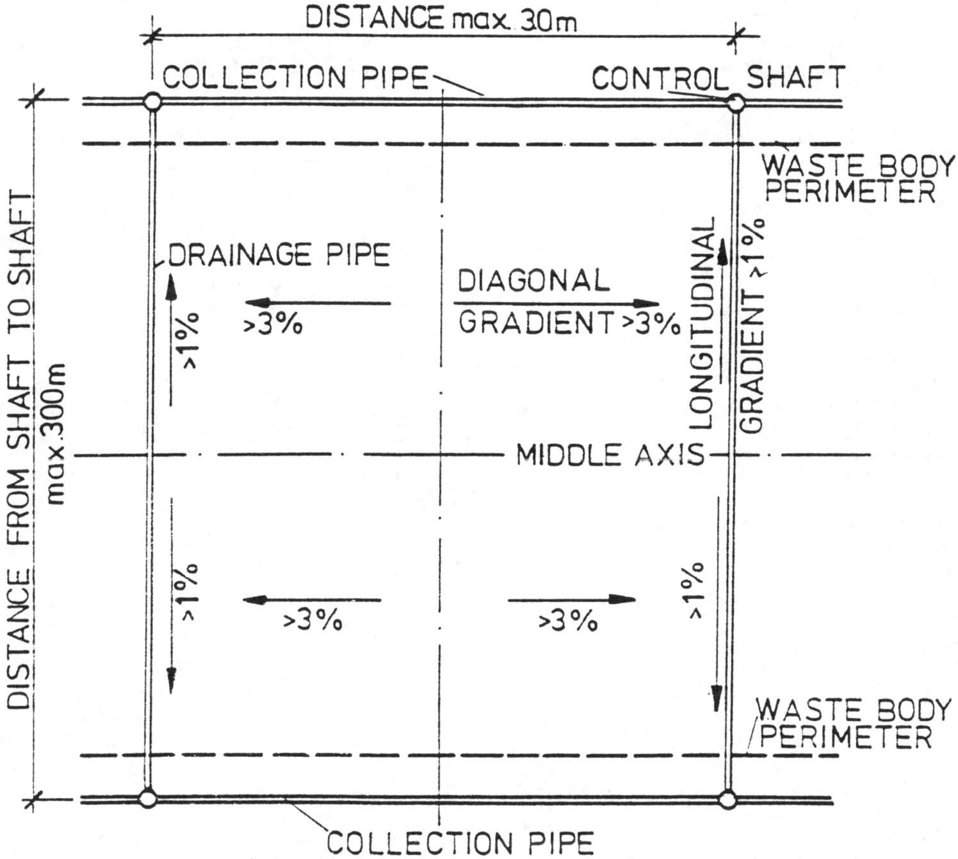

Fig. 3: Basal Drainage System (plan view) according to [2]

3. RECOMMENDATIONS ON GEOTEXTILES AND GEOMEMBRANES

3.1 GENERAL

Geotextiles, geomembranes and related products are currently used as structural components of landfills in numerous countries. Due to the lack of documents or associated specifications dealing specifically with landfill applications of these geosynthetic products the Technical Recommendations "GLR" also include recommendations on geotextiles and related products (geogrids, geonets, ...) and on geomembranes to provide the user with additional guidance.

Two technical committees, CEN TC 189 and CEN TC 254, are currently working on the use of these products. Besides, additional informations are provided by the work of national groups (e.b. [2]). In the GLR-Recommendations already adopted CEN decisions have been incorporated in the main part of the text as far as possible. Additional country-specific proposals will be found under "Appendix status".

3.2 GEOTEXTILES AND RELATED PRODUCTS

Geotextiles and related products (geogrids, geonets ...) are dealt with in the following GLR-Recommendations:

R 2-9 Principles for the Use of Geotextiles and Related Products
R 3-9 Suitability Tests for Geotextiles and Related Products
R 5-5 Quality Assurance for Geotextiles
A 2-9 Geotextiles in Lining Systems (French Proposal)
A 5-5 French Certification Procedure for Geotextiles

The main point in recommendation R 2-9 is the analysis of the functions of geotextiles in lining systems:

- Drainage of fluids (liquids and/or gases)
- Filtration in drainage systems
- Protection of geomembranes and other liners
- Reinforcement (soils or waste).

To fulfil the design requirements, geotextiles and related products have to be carefully selected. There is a need to guarantee the satisfactory long term performance of the products, especially at the base of a landfill where no access for maintenance and remediation can be provided once waste is deposited.

Table 2 is listing the suitability tests that are either necessary (N) of applicable (A), from a French point of view, for the selection of products (or combinations of products), for each of four functions indicated hereabove. This proposal is a landfill related modification of the CEN TC 189 WG 1 N48E official document issued for more general applications of geotextiles and related products in lining systems.

In recommendation R 3-9 testing procedures that can be adopted to perform the tests listed in Table 2 are summarized.

Recommendation R 5-5 deals with quality assurance for geotextiles in landfill applications; the proposed quality assurance procedure classically covers:

- The manufacture of the product (product quality)
- Its acceptance as a possible component of a landfill;
- Quality assurance on site.

It must be pointed out that the recommendations clearly make a distinction between:

- "Certified" geotextiles for which the performance characteristics are certified by an external independant party and

- "Non-certified" geotextiles.

As an example, Appendix A 5-5 of the GLR-Recommendations describes the French certification procedure for geotextiles; this procedure is fully effective in France.

3.3 GEOMEMBRANES

The recommendations R 2-10, R 3-10 and R 5-6 are dedicated to geomembranes either synthetic or bituminous. Geosynthetic clay liners have not been introduced in the recommendations because they can be considered as relatively new materials for which more experience is needed before any extensive use in landfill applications.

Recommendation R 2-10 "Principles for the use of geomembranes" is completed by Appendix A 2-10 which lists the tests, also (N) or (A), that, from a French point

Table 2: Geotextiles in lining systems of waste disposals

This table modified from CEN TC 189 WG 1 N48E corresponds to a French proposal to present the tests to be specified in regard of the wanted function of the geotextile.
(N) Results of the corresponding test to given in any case for all the projects
(A) Results of the corresponding test to be given if required according to the specifities of the projects
(1) Geomembrane hydraulic performance to be tested after performing the test
(2) Tested in association with the associated geomembrane

Function	Filtration	Reinforcement	Drainage	Protection
performance	long term filter stability static puncture resistance resistance to chemical degradation (A)	friction (inclined plane) (A) in soil creep test (A) in soil tensile test (A) resistance to chemical degradation (A)	long term drainage (biological clogging) (A) resistance to chemi-cal degradation (A)	long term puncture resistance (A) (1,2) restistance to chemical degradation (A) burst test (A)
index tests	opening size (N) permittivity (N) resistance to chemical degradation (A) thickness at specified pres. (N)	tensile test (N) friction (direct shear test) (N) tensile for joints or seams (A) tensile creep behaviour (A)	transmissivity (N) water penetration resistance (A)	static puncture test(see Gmb) (N) (1,2)
installation test	tensile test (N) construction damage (N) thermal oxidation (A) temperature (junction of gmb) (A) tensile for joints or seams (A) dynamic perforation test (A) identification of roll (N) mass per unit area (A) thickness at specified pres. (A)	tensile test (N) construction damage (N) dynamic perforation test (N) thermal oxidation (A) identification of roll (N) thickness at specified pres. (A)	tensile test (N) construction damage (N) identification of roll (N) mass per unit area (A) thickness at speci-fied pres. (A)	tensile test (N) construction damage (N) (2) thermal oxidation (A) dynamic perforation test (A) identification of roll (N) mass per unit area (A) thickness at specified pres. (A)

Table 3: Tests for geomembranes in landfill lining systems (derived from CEN TC 189 WG 1 N 48 E)

(N) necessary: Results of the corresponding test have to be given in any case
(A) applicable: Results of corresponding test have to be given if required according to the specifities of the project
(1) Fluid, i.e. liquids and gases
(2) Wheathering, chemical, biological, permanent elongation
(3) Strain stress behaviour; static & cyclic
(4) Tested on the system geotextiles and geomembranes

Function	Fluid Barriers (1)
Performance tests	chemical gas permeation (A) (2) long term liquid "permeability" (N) resistance to chemical degradation (2) friction properties (anchorgae, slope, ...) (N) (4) tensile test (performance) (A) (3) tensile test on joints (performance) (N) hydrostatic puncture resistance (A) (4) static puncture resistance (A) (4) burst creep/relaxation behaviour (A) bursting (large diameter) (A) (4) resistance to chemical degradation , biological, UV, Oz, water, roots, tested on gmb and joints (A)
Index tests	water vapour transmission (A) thickness at specified pressure (N) stress crack (A) carbon black (content, dispersion) (A) tensile test (performance) (N)
Installation tests	tensile test (performance) (N) impact test : gmb. joints (A) (4) construction damage (N) (4) straightness flatness (A) low T° folding (N) high T° stability (N) coefficient of linear expansion (A) identification of roll (N) mass per unit area (A) thickness at specified pressure (A) joints : peel test (A), and tensile shear test (A), vacuum box (N), sparkle test (A), dual seams test (A), air lance test (A)

of view, have to or can be done for a proper selection of geomembranes as fluid barriers. Table 3, where possible testing procedures for the suitability tests are listed, is also a modification of the more general CEN document, where other applications of geomembranes are considered.

In R 5-6 of the GLR-Recommendations quality assurance for geomembranes in landfill applications is considered. Similar to geotextiles the quality assurance of geomembranes comprises:

- Product quality
- Acceptance of the product as a possible landfill component
- Quality assurance on site

Reference is made to a German certification procedure of geomembranes [4].

4. RECOMMENDATIONS ON CUT-OFF WALLS

Cut-off wall construction is addressed in the following sections of Geotechnics of Landfill Design and Remedial Works Technical Recommendations "GLR":

R 3-2 Suitability Testing for Mineral Sealing Compounds
R 4-1 Cut-off Wall Construction
R 5-3 Quality Assurance of Vertical Cut-off Walls
A 4-1 Geomembranes for Composite Cut-off Walls

Cut-offs are used for a variety of purposes. They may be installed as part of the site preparatory works, or as remedial works subsequent to landfilling. They generally consist of physical, passive barriers, but active pumped systems may also be used, alone or in combination with passive systems.

4.2 APPLICATIONS

The objectives of cut-off systems are likely to include one or more of those listed below. In any event it is essential to define the fluid to be intercepted, i.e. clean or dirty (contaminated) water or leachate, gas or non-aqueous liquids:

i) Interception up gradient of the site
ii) Interception down gradient of the site
iii) Barrier - i.e. impoundment of the fluid, with drainage only by diffusion through the surrounding external ground
iv) Drainage
  - For control of fluid level
  - For control of fluid level, also enabling separation of fluids into "clean" and "dirty" for discharge or treatment respectively
v) Stability enhancement
  - By lowering groundwater levels
vi) Groundwater lowering
  - For leachate control
  - To permit excavation of soils from within a site

4.3 CUT-OFF WALL CONSTRUCTION

Cut-off walls can be constructed using a range of foundation engineering techniques. A distinction should be made between cut-off walls constructed by excavating the soil and those installed by soil displacement. A summary of the principal methods available is shown in Figure 4. A brief description of the various techniques available is included in Section R 4-1 of the GLR-Recommendations [1].

In certain circumstances (for instance, where high concentrations of toxic pollution are present, or where there is a significant risk of gas migration into the subsoil) a geomembrane may be introduced into diaphragm walls in order to further reduce permeability and increase the chemical resistance and durability of the cut-off barrier.

A discussion of the use of geomembranes in cut-off walls is presented in Appendix A 4-1 of the GLR Recommendations and this includes comment on suitability testing and quality assurance aspects. It should be noted that the principal geomembrane material used is High Density Polyethylene (HDPE).

In most cases cut-off walls are designed to key into a continuous low permeability strata of sufficient thickness and at such a depth so as to ensure an adequate seal is achieved. Defining the keying depth is an important part of the design process and requires detailed site investigation to define ground conditions.

4.4 SUITABILITY TESTING

Suitability testing for sealing compounds used in the construction of vertical cut-off systems is addressed in Section R 3-2 of the GLR-Recommendations. This aspect is covered in some detail, and reflects the care that must be taken to ensure that the mix used is appropriate for the particular application.

Aspects covered in R 3-2 include:

- Composition and manufacture of sealing compounds
- Preparation of samples
- Testing the characteristics of fresh samples
- Workability and hardening behaviour tests
- Strength and stress deformation behaviour
- Permeability testing

It should be noted that in many cases the cut-off wall will be installed in very aggressive environments where leachates and high concentrations of toxic pollution are present. It is therefore essential to demonstrate the suitability of a sealing

Principle	Cut-off wall system	Plan view	Soils	Material
Excavation of soil in place and installation of a sealing material	Diaphragm wall one-phase method	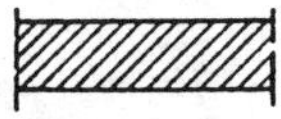	Limited application in the case of peat/humic acid	Bentonite-cement suspension with/without filler
	Diaphragm wall two-phase method	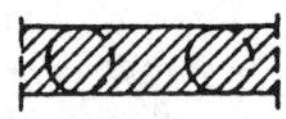	As above	Bentonite suspension, natural concrete
	Composite diaphragm wall	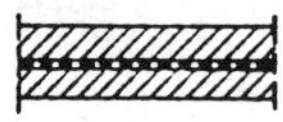	As above, only in the one-phase method	Bentonite-cement suspension, geomembrane (eg HDPE, sheet piles, etc.)
	Intersecting pile wall	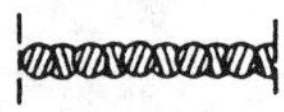	No restrictions with cased borings	Natural concrete, concrete
Displacement of soil in place and installation of a sealing material	Thin wall	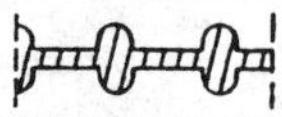	Suitable for driving or vibrating in	Bentonite-cement susp., with filler
	Sheet pile wall	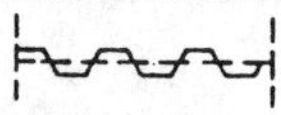		Steel
Reduction of permeability of soil in place	Grout curtain		Injectable soils	Cement, clay-cement susp., silica gels
	Jet grouting curtains		Also in very fine grained soils	Bentonite-cement susp. with/without filler
	Freezing wall	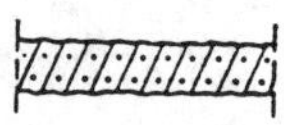		Liquid nitrogen, freezing plant

Fig. 4: Cut-off Wall Systems

compound using representative leachate samples or test liquids.

4.5 QUALITY ASSURANCE

Recommendations on the Quality Assurance procedures which should be adopted for vertical cut-off walls are presented in Section R 5-3 of the GLR-Recommendations. The recommendations are presented in three main sections related to the type of construction:

- One-phase diaphragm walls
- Two-phase Diaphragm walls
- Thin walls

Tables are presented giving detailed recommendations on the type and frequency of testing which should be adopted.

5. RECOMMENDATIONS ON CONTAMINANT MIGRATION ASSESSMENT

In planning new landfills and the rehabilitation of abandoned and contaminated sites it is necessary to assess contaminant migration and the effectiveness of hydraulic and/or structural groundwater remediation schemes.

Many theoretical approaches are available today for modelling both groundwater flow and contaminant migration. The rage of the methods extends from closed form analytical solutions with very simple boundary conditions to complex numerical methods.

The modelling of groundwater flow in a saturated porous media is highly advanced in the field of hydrogeological applications like water resources management.

$$\frac{\partial(\theta\cdot R\cdot c)}{\partial_t} = \frac{\partial}{\partial_x}\left[(\underbrace{\theta\cdot d_o\cdot}_{1} + \underbrace{\alpha\cdot/v_s/)}_{2}\ \frac{\partial_c}{\partial_x} - \underbrace{v_s\cdot c}_{3}\right] - \underbrace{\mu\cdot\theta\cdot c}_{4} \qquad (1)$$

[1] diffusion
[2] dispersion
[3] advection
[4] degradation

$$R = 1 + \underbrace{\frac{\rho_d \cdot k_d}{\theta}}_{\text{sorption}} \qquad (2)$$

c = Concentration of the material dissolved in water (kg/m^3)
d_o = Diffusion coefficient of the water-dissolved material in free water (m^2/s)
$v_s = k_f \cdot i/\theta$ = average seepage velocity in x- direction (m/s)
i = Hydraulic gradient (-)
k_d = Coefficient of the linear equilibrium sorption isotherms (distribution coefficient) (m^3/kg)
k_f = Hydraulic conductivity (permeability coefficient) (m/s)
R = Retardation coefficient (-)
t = time (s)
x = Distance in the direction of transport (m)
$\alpha = d.V_s$ = Dynamic dispersivity (m^2/s)
δ = Dispersion length (m)
γ = Impedance factor (-)
θ = Volumetric water content (m^3/m^3)
μ_F = Degradation constant for 1st order material degradation processes in the solid part of the porous media (grains) (1/s)
μ_W = Degradation constant for 1st order material degradation processes in the solution (1/s)
ρ_d = Dry density of the sealing material (kg/m^3)

Such models are based on flow parameters such as permeability ratio and storage coefficient, which require correlation against long-term groundwater level records from observation boreholes. In some circumstances, small-scale inhomogenities and variations in the permeability do not show up in the observation borehole hydrographs and are therefore difficult to calibrate. However, they are of prime importance indicating flow velocity - and thus for convective transport.

On the other hand, contaminant migration modelling techniques are limited to persistent (conservative) substances, i.e. those which do not undergo any reactions or adsorption in the saturated zone. The development of simulation models for reactive substances is still in progress.

Basic factors of the contaminant migration process in a mineral sealing layer are as follows:

Advection -
passive transport of dissolved contaminants in pore water by reference to the average pore water velocity (Darcy)

Dispersion -
geometric diffusion of concentration areas as a result of different flow velocities

Diffusion -
physical equalization process as a result of Brownian movement. Materials dissolved in water pass from areas of high to low concentration

Sorption -
retardation of materials dissolved in

water during transport through the sealing layer, e.g. molecular filtration, ion exchange

Materials dissolved in water can be attenuated by conversion and degradation processes. For the quantitative description of material transport in a mineral sealing layer, the physical regularity of the processes is formulated mathematically and balanced in accordance with the mass conservation law. The following partial differential equation [5], known as the contaminant migration equation, is commonly used to describe the contaminant migration process.

The equations cannot be used unless representative transport parameters are determined for landfill site conditions. More detailed information on the transport parameters is given in Appendix A 6-1 of the GLR Recommendations [2].

When considering the use of theoretical models for contaminant migration it is important to stress the differences, with respect to groundwater flow modelling in porous media, used for the water resource management:

- Contaminant migration plumes and/or risk areas around landfills are usually much smaller than regional aquifers investigated for water supplies. Therefore scale effects with reference to smaller subsoil volumes must be taken into account where local subsoil variations are important.

- Water resource management usually requires a reliable estimation of the global flow rate through wells or vertical sections of the aquifer. This result comes from an integration and averaging process that smooths the effects due to local soil variation. Conversely, contaminant migration assessment usually requires a concentration distribution of the pollutants in ground water at any time. This scenario is very sensitive to local variations of subsoil characteristics (i.e. seepage velocity through different permeability sublayers), therefore in the latter case a detailed knowledge of the subsoil macro-structure is necessary to obtain reliable indications.

- As far as the investigation for input parameters assessment is concerned, it is important to observe that open stand pipe piezometer measurements are able to give a good evaluation of the average permeability and transmissivity of the aquifer. On the other hand they are not able to detect local variation in permeability due to finder sublayers. More sophisticated monitoring tools are necessary to check these features (i.e. "Casagrande" type piezometers, piezocone tests etc.).

In conclusion if one neglects the chemical and biochemical phenomena, the problem of establishing a contaminant migration model does not lie in the theoretical description of the phenomena or in the solution of the corresponding equations, but rather in the availability of suitable data of the subsoil structure.

The choice between sophisticated and simple theoretical models for contaminant migration and groundwater flow depends on the geometry of the problem; good results are strongly linked to an accurate choice. In principle, two-dimensional horizontal models can be used when referring to flow and migration problems where the horizontal range is very much greater than the relevant aquifer depth.

In the case of waterproofing measures and/or the combination of extraction and waterproofing measures, a three-dimensional scheme is generally necessary - and only in special cases axi-symmetric or two-dimensional vertical models are possible.

The use of theoretical models, within the context of the contaminant migration phenomena, is useful for a deterministic prediction only where an adequate density of data is available, on the other hand simplified calculations carried out with assumed parameter variations can also yield valuable information for sensitivity and preliminary studies for the selection of sites for waste disposal.

Finally, it is important to underline that when dealing with conservative input parameters it is important to carefully check that they are really conservative when referring to the specific answers the designer is looking for.

Even if the theoretical models can be useful for the comparision of different alternatives and trends, the complete definition and the success of a remedial solution for a contaminated area is a very complex combination of aspects including chemical, biochemical, subsoil and aquifer physical parameters. Therefore only experience, monitoring and possible modifications during the activity, can provide the effectiveness of the final result.

6. CONCLUSIONS

The text deals with the Technical Recommendations "Geotechnics of Landfill Design and Remedial Works" (GLR) prepared by the European Technical Committee No. 8 (ETC 8) of ISSMFE. The 2nd edition of these recommendations will be published in 1993 and covers additional important geotechnical aspects, e.g.:

- Design principles of mineral sealing layers and basal drainage systems
- Application of geotextiles and geomembranes
- Construction of cut-off walls
- Assessment of contaminant migration

In this paper major aspects of the according recommendations have been presented showing the progress of the work of ETC 8.

7. LITERATURE

[1] Geotechnics of Landfills and Contaminated Land - Technical Recommendations "GLC" (German Geotechnical Society for ISSMFE). Ernst & Sohn, Berlin, 1991. 2nd Edition (GLR), 1993 (to be published)

[2] Empfehlungen des Arbeitskreises "Geotechnik der Deponien und Altlasten" - GDA (Deutsche Gesellschaft für Erd- und Grundbau). Ernst & Sohn, Berlin, 1990. 2nd Edition, 1993 (to be published)

[3] Scherbeck, R., Jessberger. H.L.: Zur Bewertung der Verformbarkeit mineralischer Abdichtungsschichten. Bautechnik, Heft 9, 1992, pp 497-506

[4] Richtlinie für die Zulassung von Kunststoffdichtungsbahnen als Bestandteil einer Kombinationsdichtung für Siedlungs- und Sonderabfalldeponien sowie für Abdichtungen von Altlasten. Bundesanstalt für Materialprüfung und -forschung. Berlin, 1992

[5] van Genuchten, M.Th., Wagenet, R.J.: Two site/two region models for pesticide transport and degradation: Theoretical development and analytical solutions. Soil Sci. Am. J. 1989, 53, pp. 1303-1310.

Waste Disposal by Landfill – GREEN'93, Sarsby (ed.) © 1995 Balkema, Rotterdam, ISBN 90 5410 356 6

Risk assessment of engineered containment landfill designs

P.J. McKendry
WS Atkins Consultants Ltd, Epsom, UK

ABSTRACT: The traditional approach to landfill in the UK has been based upon the "dilute and disperse" principle for the control of leachate. Recent EC legislation has caused this principle to be replaced by that of engineered "containment", which has in turn allowed the application of well-established geotechnical/civil engineering techniques to the design and construction of landfill sites. This more rigorous, engineering approach to landfill design and construction has, in turn, enabled the methodology of quantitative risk assessment to be applied. The benefits of this process are applicable to both designer/operator and regulator alike, allowing designs to be evaluated for their environmental risk under a range of "what if" scenarios.

1 INTRODUCTION

The traditional approach to landfill design in geology containing permeable strata has been that of "dilute and disperse". With the formation of the NRA (National Rivers Authority) as a regulatory body to protect controlled waters (ie surface and groundwaters) and the introduction of the EC Directive on groundwater, the net effect has been to disallow the "dilute and disperse" philosophy for future site developments. Instead this "dilute and disperse" approach has been replaced with the engineered "containment" approach. In this context engineered containment includes landfill designs using both natural and synthetic materials, such as clay, geomembranes, geotextiles, bentonite enhanced soils and mats etc.

It should be noted that whilst the containment principle applies equally to landfill gas, the bias in this paper is towards leachate, as it potentially has a longer term, adverse environmental impact compared to landfill gas.

The move towards engineered containment is welcomed by many in the waste management industry, as it establishes landfill as an engineering discipline, which sadly has hitherto been lacking in some areas. This is not to say that engineered containment is an inherently better approach to landfill than that of "dilute and disperse", as one set of problems is exchanged for another. With a containment site the problem becomes that of collecting and disposing of the leachate generated, an issue which has yet to be both fully appreciated and fully solved. It has been shown that under the right conditions, a properly operated "dilute and disperse" landfill is a satisfactory means of waste disposal, having minimal, adverse, environmental impact. However the reproducibility of such properly operated sites cannot be guaranteed due to the sensitive nature of the process to operational parameters, whereas engineered "containment" landfills, by definition, are a consistent and more robust product, which can be constructed to a performance specification.

Of importance also is the degree of "containment", an aspect which has largely been ignored on a site specific basis by virtue of the zoning approach adopted by the NRA to defining aquifers. One of the benefits of applying risk assessment to containment designs is that an appropriate degree of protection can be designed-in for a specific site, rather than relying upon a standard specification which may be unnecessary and excessive for certain sites.

It is within this framework that the need for an objective assessment of engineered containment designs is required, if unnecessary and unwarranted capital is not be to expended by waste management operators due to subjective judgements on the part of the regulatory authorities. This paper develops the background to risk assessment, establishing the methodology and developing the application to engineered containment landfills but does not itself attempt to define the background to risk assessment in a formal manner, for which there are numerous erudite references. A subsequent paper will provide a case-study comparison of some typical containment systems and analyse the relative cost/benefit of these designs.

2 RISK AND RISK ASSESSMENT

What is risk? Risk can be defined as the likelihood or expected frequency of a specified adverse consequence. The term risk assessment (RA) is a much misused concept and is often confused with a simpler, non-quantitative concept referred to as a hazard assessment. Thus a hazard is something which may occur but it is the combination of that consequence with the likelihood or probability of it actually occurring that is the basis of the RA approach. For example, the probability of failure of a component may be high but the consequences insignificant. In contrast, the probability of failure may be low but the consequences catastrophic. Such situations are particularly relevant in the aerospace and nuclear industries where human life may be at risk and the development of the approach to quantitative risk assessment has largely derived from work in these industries.

There are several key stages to RA, the purpose of which is to address the following issues:

i. What can go wrong?
  - Hazard Identification
ii. How likely is it to go wrong?
  - Hazard Analysis
iii. What would happen if it did go wrong?
  - Consequence Analysis
iv. What are the associated risks?
  - Risk Determination
v. Are the risks acceptable and can they be reduced?
  - Risk Evaluation

Apart from the emotive aspects associated with risk being quoted in terms of the likelihood of a death due to a particular event (which must not be underestimated in influencing public opinion), people often do not have a feel for what is regarded as an "acceptable" or normal level of risk. Actuarial tables can be used to indicate the mathematical probability of risk of death associated with everyday events, about which most of us would scarcely think twice before undertaking. Table 1 gives some examples or risks, expressed as deaths per million per year, for various activities.

Table 1. Examples of individual risk

Causes	Risk per million per year
All causes	11,900
Cancer	2,800
All violent causes	396
Road accidents	100
Accidents in private homes	93
Fire or flame	15
Drowning	6
Gas incident	1.8
Lightning	0.1
Deep-sea fishing	880
Construction	92
Rock climbing	8,000

The differentiation here is the degree of choice applicable to particular events and public perception of the normality of such activities. Thus the risk of dying from cancer is 2800 people per year per million persons, an event generally regarded as a serious concern, while for rock climbing it is 8000 and yet people still do the latter, perhaps because it is a voluntary action and perceived, subjectively, to be a satisfactory level of risk. Similarly with a gas incident with a risk of 1.8 compared with a fatal road accident at 100. Both events receive substantial media attention yet the perceived opinion, judging by the publicity given to gas incidents, is that it is a far worse event, perhaps because one is a voluntary action, ie driving a car, whilst the other is imposed.

In the context of engineered landfill containment designs, it is hoped that subjective opinions can be avoided by assessing the risks on an objective basis, taking due account of specific circumstances relevant to that location. It is this need for objective, rather than emotive or subjective decision-making that RA can fulfil.

Whilst the RA process and methodology is well-established, its application to particular situations is usually hindered by the lack of suitable data. This is especially true of the materials and construction techniques used in engineered containment sites and it is at this juncture that empirical or professional judgements may be required, in lieu of actual data. Such judgements need not devalue the output of the RA process, as they can be assigned conservative values and indeed can be changed as part of the subsequent sensitivity analysis, to see the influence of particular aspects on the containment design. This interactive approach is one of the benefits of RA, allowing the "what if" scenarios to be fully explored and evaluated at the design stage. Such an interactive approach is useful in the context of comparing the cost/benefits of various alternative, engineered containment designs and made easy with the application of the approach to a PC-based software package.

3 DATA AVAILABILITY

As noted above, there are 5 key stages in the RA process, all of which require data, some of which will be documented and some based upon professional judgement.

Data on the chemical and physical properties of the natural and synthetic materials used in engineered containment designs is available from a variety of sources. One of the complaints made against the use of high density polyethylene (HDPE) geomembranes in such designs is the apparent lack of field data on the environmental performance of such membranes, in the context of a landfill. Whilst many landfill geomembranes have been installed, some over 20-25 years ago, there is a lack of reliable, scientifically documented data, both about the nature of the membrane used and its method of installation and any subsequent monitoring of performance. It is believed that the longest, documented example is a site in Germany, with a current installed life of some 15 years: the results to date do not show any deterioration in properties of the HDPE membrane. Despite this paucity of field data, the application of HDPE as a material in severe chemical environments is well documented in the chemical process and gas industries and it is from these industries that much of the present data on HDPE has been derived. Therefore the validity and applicability of these materials data, in the context of landfill, should neither be underestimated nor undervalued. A literature review of the use and performance of geosynthetic materials in waste management and related applications yielded many references, from which useful data was gleaned for inputting into the RA process. A variety of data-bases exist, providing information on a range of material properties and user experience. The British Library for example offers a number of such data-bases, while the contract research organisation RAPRA, which has previously undertaken work in this field for geomembrane manufacturers, holds 3 useful data-bases, PLASCAMS (for general properties), CHEMRES (for chemical resistance) and the US-based COMPENDEX PLUS data-base. A search of these sources indicated some 260 references in total, of which 40 yielded information and data suitable for RA purposes.

Key references are given at the end of this paper and it is interesting to highlight some of the main features from these references associated with HDPE membranes and the potential for leachate leakage:

1. Leakage through an HDPE membrane can occur either by permeation through the intact material and/or by defects (ie holes) in the material.

2. Permeation is a function of membrane thickness, liquid composition and pressure differential across the membrane. Thus for HDPE, solvent vapour transmission

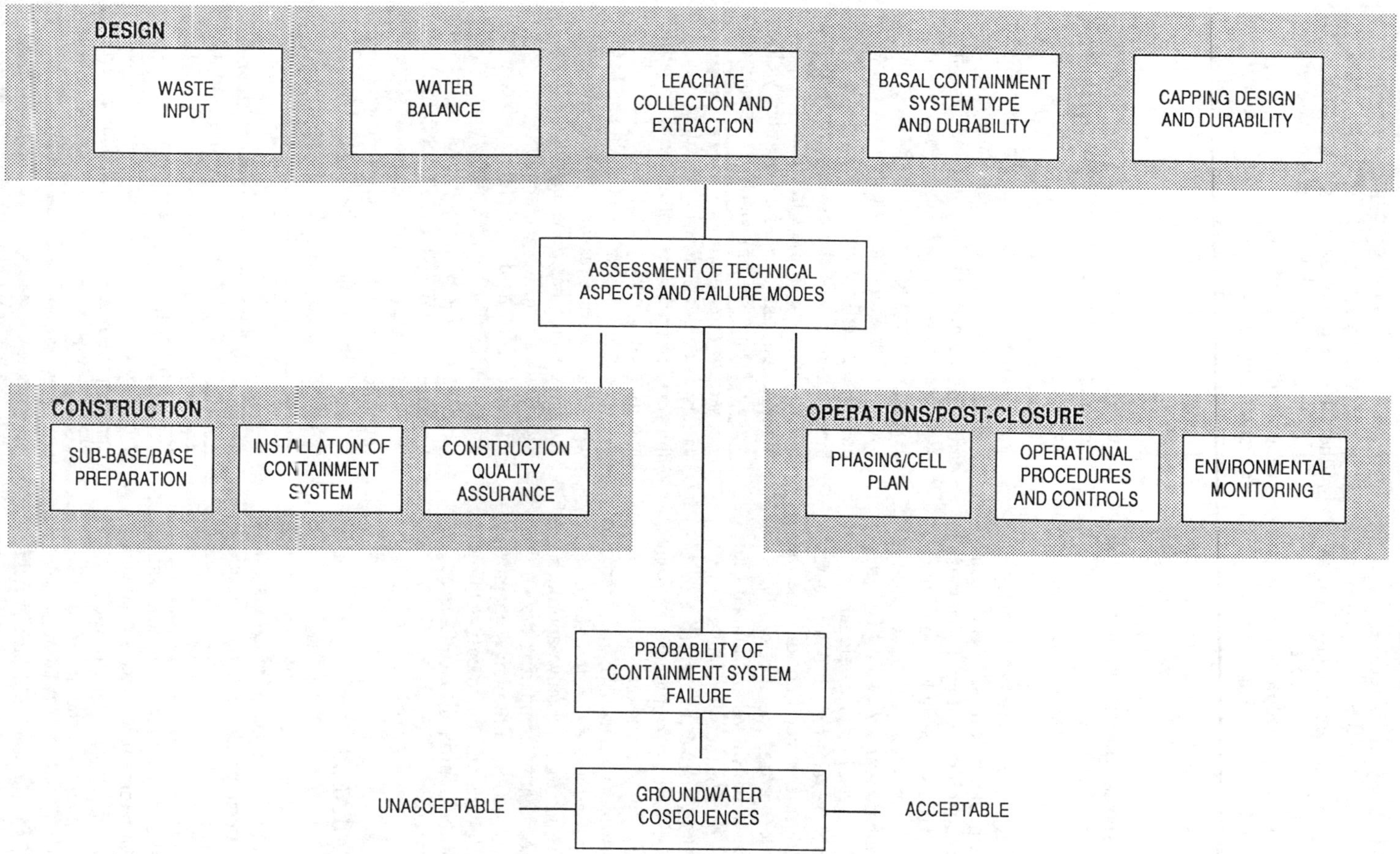

FIGURE 1 - **Key Elements**

rates are 1-3 orders of magnitude greater than those of water, which is estimated at $1x10^{-14}$ m/s.

3. Liner defects can be defined as pinholes (generally produced during manufacture and with a diameter < membrane thickness), holes (produced during construction and typically 1-10mm) and cracks (the product of manufacture and/or construction, caused by excessive tensile stresses acting on a crack initiation site in the membrane).

4. The rate of flow through these defects is dependent upon the head of leachate acting on the membrane, the hydraulic conductivity of the underlying material/layer and the number and size of defects.

5. At the design stage, a number of factors have an effect on the potential leakage rate, including basal floor slopes, leachate head limit, membrane thickness, size/type of leachate collection system and the nature of the lining system configuration (ie underlying clay liner or witness/leak detection layer, type and thickness of protective layers etc).

6. The construction phase of the containment system is probably the most crucial stage in determining subsequent system performance: construction quality assurance, CQA, is essential therefore if the full design performance of the system is to be achieved. For example, with minimum CQA, up to 75 holes/ha have been reported, compared with good CQA at 2-3 holes/ha. Two-thirds of these holes are associated with welding/seaming and it has been shown that with good CQA that seam defects can be reduced from 1 per 10m of field seam to 1 per 300m. Similarly with the preparation of the sub-base where inadequate specification and CQA could lead to the presence of large or angular aggregates coming into contact with the geomembrane lever.

7. During the operational phase, careless use of heavy plant at the start of a new cell can severely distort the underlying membrane, unless adequate protection cover is used: this distortion can directly induce defects or cause initiation of cracks which subsequently propagate during later operations or post-closure. Similarly the placement of the first lift of waste should be carefully controlled to minimise the potential for sharp/bulky objects which could penetrate the containment system: the use of only household or dustcart waste for this first lift is therefore a sensible precautionary measure.

The value of these data-bases depends largely upon the requirements of the searcher but the common element in all of these information sources is the relative lack of field-trials and documented case-studies. This suggests that the waste management industry have not foreseen the legislative trends which have lead, in a relatively short space of time, to the current position in the UK whereby the acceptable "dilute and disperse" philosophy of only 5 years ago is no longer allowable and the "quantum" leap to engineered containment sites has occurred.

4 HAZARD IDENTIFICATION AND ANALYSIS

The hazards posed to an engineered containment landfill may arise during the design, construction, operation or post-closure phases. The possible consequences which could arise due to these hazards will vary with time for the same event, due to the changing circumstances within the wastes undergoing stabilisation. Examples of possible aspects which could lead to an adverse environmental impact are given below:

Design:
- inadequate base slope to promote leachate drainage to sumps.
- thickness of geomembrane.
- leachate collection system, including design of sumps.
- incorrect leachate head.

Construction:
- inadequate preparation of sub-base, leading to settlement.
- poor quality control over materials used and their storage.
- penetration of liner containment system.
- failure of welded geomembrane seams.
- inadequate/absence of QA procedures.

Operations:
- poor operational procedures leading to damage to the liner system.

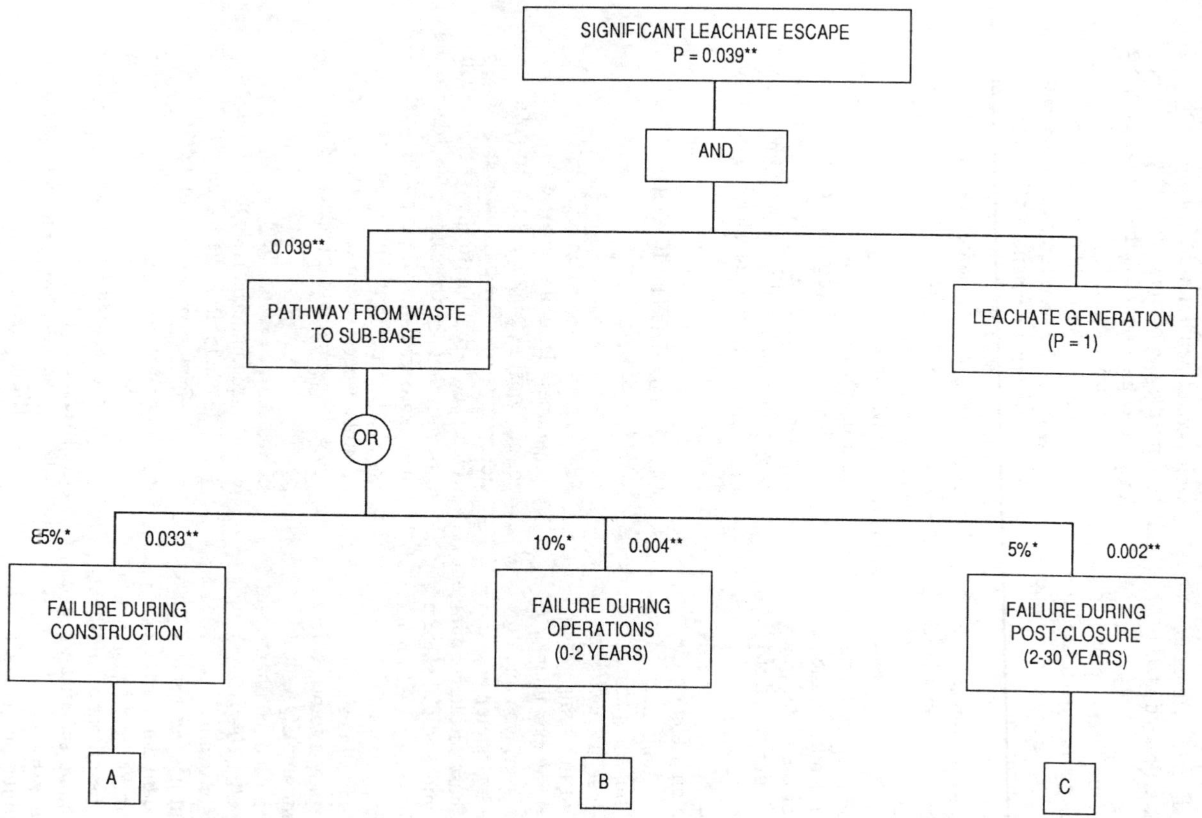

** Probability of adverse event
* % Contribution of likely failure modes

FIGURE 2- **Basic Fault Tree for Engineered Containment System**

- failure of leachate drains due to poor site practices.
- excessive rainfall entry.
- excess leachate head on the liner through poor site management.

Post-Closure:

- failure of the capping system due to human/natural actions.
- collapse of leachate collection system.
- harmful long-term effects of inappropriate wastes.

This is not an exhaustive listing but rather an indication of the type of hazards or event that may arise at various stages. The routes by which these hazardous events could occur are examined using fault trees, which are then quantified to give the likelihood, or probability, of such events occurring.

5 CONSEQUENCE ANALYSIS

The hazards described above need to be evaluated to determine what would happen if they occurred. Depending upon the design of the containment system and the surrounding hydrogeology/ hydrology/geology, many individual consequences could be deduced. However in simple terms the single most important consequence in nearly all cases is the breaching of the containment system by whatever means, leading to the escape of leachate or landfill gas. To illustrate this consider the following examples:

1. Within the first layer of waste placed on the completed basal containment system, skip-waste is inadvertently deposited and compacted. Unseen objects, such as broken reinforced concrete with the reinforcement bars protruding, angle-iron, lengths of timber etc are pushed through the protective layer and penetrate the engineered containment system. The severity of any subsequent damage will depend on the type of containment system and its design, such as the thickness of protective layers etc.

2. A poorly designed/installed leachate collection system blocks, due to clogging and settlement of the sub-base, leading to an excess head of leachate on the HDPE liner and subsequent tensile failure of the liner. In the case of a clay liner, the clay may fail in shear due to excessive settlement of the sub-base, or the volume of leachate passing through the clay liner is increased, due to the increased head of leachate.

3. Inadequate design of the foundation of a leachate sump leads to undue tensile forces in the underlying HDPE liner, causing failure and the subsequent escape of leachate. Similarly, the poor installation of a leachate pipe through the lining system is inadequately sealed and allows the escape of leachate along the length of the pipe.

4. Damage to the sides of the containment system during landfill operations causes a breach in the integrity of the system and allows the escape of landfill gas into the surrounding porous strata.

The magnitude of any adverse environmental impact arising from these consequences can be deduced once some assumptions are made. It should be stressed that whilst any assumption can be made, it is more beneficial that these be realistic rather than theoretical. For example, assuming that the worst case for a synthetic liner is that it "instantaneously" disappears is not really a practical option and the adverse pollution impacts that derive from this are not really worthy of consideration. Instead, a reasoned scenario based upon whatever empirical data or field experience is available is more likely to provide a useful case-study and add to the overall understanding of what could, realistically, occur.

The output from the RA of the engineered containment system may in fact be only the first stage in an overall RA study, which could encompass a groundwater modelling study for a particular site. From the various failure scenarios examined for the engineered containment RA phase of the work, the resultant leachate leakage rates and contaminant levels would become the input data to the groundwater RA study; hence the importance of ensuring that the failure scenarios examined are realistic. This is illustrated in Figure 1, which shows the outline structure of the key elements involved and how the output from the containment system RA provides the input data to a subsequent groundwater modelling study.

The level of technical input to such studies, including both man-hours by the client or consultant and the costs of any field-work associated with installing observation and

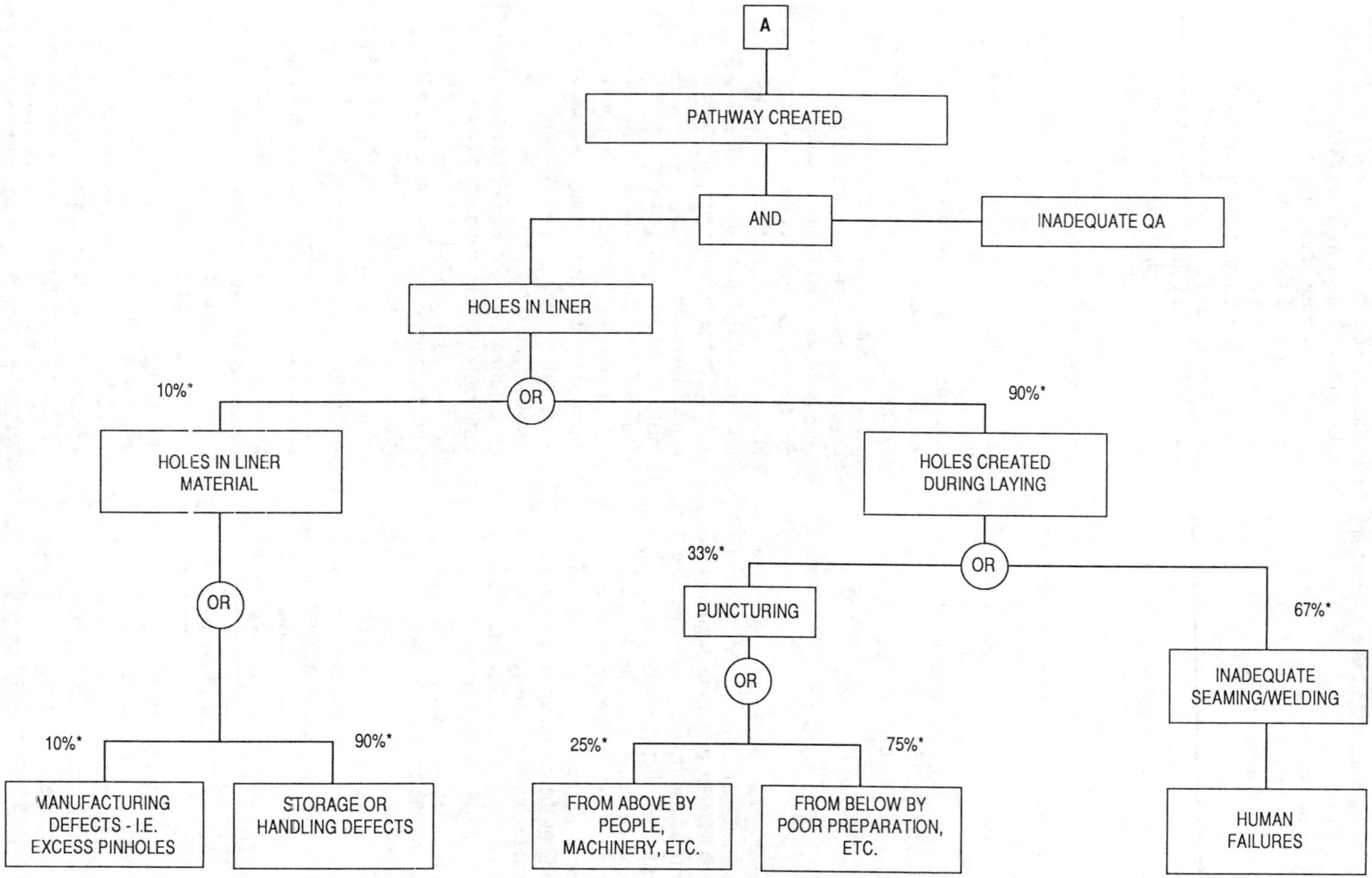

* % Contribution of likely failure modes

FIGURE 3 - **Possible Failure modes During Construction (GEOMEMBRANE)** pumping-wells for example, should not be underestimated. It is with these cost implications in mind that the regulatory authorities should act responsibly and not insist that unnecessary or impractical failure scenarios be examined.

6 RISK DETERMINATION AND EVALUATION

The purpose of this stage of the overall process is to determine the likelihood of leakage from the site causing unacceptable contamination, be that leachate into potable water supplies, landfill gas into surrounding strata etc. The estimation of the likelihood of the failure of the containment system is determined using fault trees. Fault trees are a diagrammatic representation of the following series of questions and answers:

Q1. What conditions are necessary to produce a significant leachate escape?

A1. A pathway from the waste to the sub-base and leachate generation.

Q2. What would cause a pathway to exist?

A2. Failure of the containment system during construction, operations or post-closure.

Figure 2 shows a basic fault tree which links hazards (eg failure during construction) via a pathway to produce a consequence, namely leachate escape. It should be noted that this adverse consequence requires not only the generation of leachate (which is taken for granted in this analysis) but also the presence of a suitable pathway. Taking this process one step back on the construction branch of the tree, figure 3 shows the possible mechanisms by which failure of the containment system can occur. Once fault trees have been developed for all possible situations, they can be quantified, using data of the type referred to in section 3 on Data Availability. The approach to quantification is to work through each fault tree and to ascribe the relative probability of the various events on the basis of the available data, producing an overall probability for the significant escape of leachate. At this stage the validity of the data used may be questioned and tested, via a sensitivity analysis which would allow various interested bodies to assess the final outcome, or result, using other data.

Thus figure 2 also contains the assigned probability for each of the elements involved, producing the net result that the calculated probability of a significant leachate escape 0.039, ie 1 in 25. Once such a system has been established for a particular containment design, other designs can be similarly evaluated for their probabilities of failure. A comparative RA can then be undertaken for a given set of assumptions and using cost estimates for the construction of the various containment systems, a relative cost-benefit can be established.

The benefits of this process to both the landfill operator and the waste regulatory authorities are significant, especially if the process is made interactive using a suitable PC-based model. Thus various alternative containment deign configurations can be (relatively) easily examined and quantified. By running a sensitivity analysis, the effects of changing the ascribed probabilities of particular events/hazards of interest can be determined and comparative cost-benefit values produced.

7 DISCUSSION

The basic principles of quantified risk assessment are well established and have been used in the chemical and nuclear industries for many years. The process represents a logical method of evaluating the probability of a given adverse event, based upon the identification and analysis of known or possible hazards, their consequences and the associated risk of that consequence occurring. In the preparation of the logic pathways, or fault trees, linking cause (hazards) and effect (consequences), a degree of professional judgement is inevitably involved in identifying potential hazards and evaluating the likely consequences. The availability of suitable data to allow quantification of the subsequent fault trees is therefore key to the RA process.

Whilst such data is readily available for the chemical and nuclear industries, this is not the case for the waste management industry, primarily because there has not perceived to have been the need. The historic situation in the UK with respect to waste disposal by landfill has been a reliance upon the principle of "dilute and disperse", allowing the by-products of waste stabilisation to be dealt with naturally by a combination of biological and physical processes. In the past 2-3 years however this philosophy has been overtaken by the engineered "containment" approach to landfill, a direct consequence of EC legislation. The waste management industry has been remiss in not foreseeing this significant shift in policy and has been forced into accepting possibly much higher standards than are necessary, by virtue of the lack of documented evidence to disprove the basis of the need for higher standards. In an attempt to redress this situation and to apply some objectivity to determining landfill containment requirements on a site specific basis, it is suggested that the application of formalised RA to landfill designs is appropriate. The apparent lack of data referred to previously, to allow RA to be satisfactorily undertaken is, upon closer examination, not necessarily the case. The combination of an examination of the literature and the judicious application of professional judgement is sufficient to enable RA to be applied in a meaningful and useful manner. In addition, application of the process to a PC-based model allows modifications to the assumptions made to be easily and rapidly evaluated, providing a useful tool to both designer and regulator alike. With the increasing availability of materials data, documented case studies and field-data, RA of containment designs could, in due course, become an accepted part of the design process, providing a degree of confidence to all concerned of the suitability of the chosen system.

Using the output of a RA of an engineered containment design as input data for subsequent groundwater modelling studies should enable landfill to remain as an environmentally acceptable and important option within a total waste management strategy.

In a future paper, the cost/benefit of a range of containment designs will be examined, using the RA approach outlined above.

The opinions and views expressed in this paper are those of the author.

REFERENCES

Geomembrane Role for Nigerian Dam, Civil Engineering, May 1987, pp24-26.

Hessel J, et al (1988): Long Term Stability of waste Landfill Sealing Sheet made from Polyethylene, Kunstoffe Vol 78 #2 (Feb 1988), pp155-60.

Leaversuch R D, (1989): Geosynthetics Prove to be Vital Organs in Sanitary Landfill, Modern Plastics International, Vol 19 #4 (April 1989), pp54-56.

Muenk G, et al (1989): Lining of Waste Landfills, Kunstoffe German Plastics, Vol 79 #4 (April 1989) pp31-36.

Lawson C R, (1990): Geosynthetics, in Polymers and Polymer Composites in Construction (Ed: L C Hollaway), London, Thomas Telford (for ICE), pp206-45.

Koerner R M, (1989) (ed): Durability and Aging of Geosynthetics, London and New York, Elsevier Applied Science.

Walton J C & Sagar B, (1990): Aspect of Fluid Flow through Small Flaws in Membrane Liners, Environmental Science and Technology, Vol 24 #6 (June 1990), pp920-924.

Khanbilvardi R M & Ahmed S, (1988): Leachate Flow through Landfills, proceedings of the Conference: "Critical Water issues and Computer Applications" held in Norfolk, VA, USA on 1-3 June 1988.

Bonaparte R and Gross B, (1990): Field Behaviour of Double-liner Systems, Waste Containment Systems: Construction, Regulation and Performance, ASCE, Geotechnical Publication No 26, November 1990.

Haxo H E, et al (1984): Permeability of Polymeric Membrane Lining Materials, International Conference on Geomembranes Proceedings, Denver, Colorado, June 20, 1984.

NUS Corporation, (1987 - for US EPA): Bottom Liner Performance in Double-lined Landfills and Surface Impoundments, dated April 1987.

GeoServices Inc, (1988):
Evaluation of the Rate of leakage Through a Composite Liner, report dated 8 July 1988.

Giroud J P, (of Geoservices Inc) (1984): Impermeability: The Myth and a Rational Approach, International Conference on Geomembranes Proceedings, Denver, Colorado, June 20, 1984.

Peggs I A and Carlson D S, (1988): Stress Cracking of Polyethylene Geomembranes: Field Experience, Proceedings of "Durability and Aging of Geosynthetics" (ed: R M Koerner), Philadelphia, December 1988.

Bonaparte R, et al (1989): Rates of Leakage Through Landfill liners, Geosynthetics Conference Proceedings, San Diego, California, 21 February 1989.

Haxo H E & Haxo P D, (1988): Environmental Conditions Encountered by Geosynthetics in Waste Containment Applications, Proceedings of "Durability and Aging of Geosynthetics", Philadelphia, December 1988.

Parkinson C D, The Permeability of Landfill Liners to Leachate.

Hall D H, (of Golder Associates): The Role of Construction Quality Assurance in the Installation of Geomembrane Liners.

Halse Y H, et al (1988): Laboratory Evaluation of Stress Cracking in HDPE Geomembrane Seams, Proceedings of "Durability and Aging of Geosynthetics", Philadelphia, December 1988.

Waste Disposal by Landfill – GREEN'93, Sarsby (ed.) © *1995 Balkema, Rotterdam, ISBN 90 5410 356 6*

Engineering behaviour of urban refuse, compaction control and slope stability analysis of landfill

H.Y. Fang
Lehigh University, Bethlehem, Pa., USA

ABSTRACT: This paper covers three aspects concerning urban refuse. The first is the basic characteristics of urban refuse (garbage) including composition, classification and engineering behavior such as volume change, hydraulic conductivity, bearing capacity and strength characteristics. The second aspect is the decomposition processes of refuse in the landfill including causes and mechanisms. Finally, the third topic is the construction of landfill sites with special emphasis on compaction control during disposal process and slope stability of the landfill in both newly established slopes and existing abandoned landfill sites.

1 INTRODUCTION

1.1 Characteristics of Urban Refuse (Garbage)

Waste disposal material consists of anything that cannot be further used or recycled economically. Its composition varies from community to community and from country to country as well as from season to season. The density varies from 50 pcf to 400 pcf (800 kg/m^3 to 6400 kg/m^3) depending on the amount of metal and construction debris it contains. Table 1 gives the average composition of solid waste according to the U.S. national survey results and data from the cities of New York and Osaka (Japan). Table 2 presents some typical industrial wastes generated in the Commonwealth of Pennsylvania. Eventually these wastes will be dumped into a landfill. Needless to say, some industrial wastes are highly hazardous and toxic.

Waste tires are one of the most important environmental and economic concerns as landfill. The US Environmental Protection Agency (USEPA) estimated that there are approximately 280 million waste tires discarded annually in the United States alone. Out of this, only a small quantity of tires are recovered for recycling

Table 1 Comparison of composition of urban refuse (garbage)

Types	New York City (1968)	U.S. National (1973)	Osaka, Japan (1980)
Paper	58.8	51.6	37.1
Food Wastes	9.2	19.3	13.2
Metal	7.6	10.2	5.5
Glass	8.6	9.9	12.3
Wood	2.5	3.0	2.5
Textiles	0.8	2.7	4.0
Rubber, Leather	0.8	1.9	0.3
Plastic	0.8	1.4	15.2
Garden Wastes	10.1	-	2.5

Data from Sowers (1968, 1973); Fang, Slutter and Steubben (1976); Yamamura (1983)

Table 2 Some industrial wastes generated in the State of Pennsylvania

Sources	Type of Wastes	Approximate Composition
Food & Food Products		
Food Additives Meat Grain Mills	Trimming wastes Sludges Residues	Organics & Acids
Textile Products Cotton, Wool, Synthetics	Sludges	Acids, Alkalis, Metal Salts, Solvents
Paper & Paper Products	Sludges, Pulping sludges	Sulfates, Organics, Soaps, Mercaptans
Soaps, Detergents	Sludges	Surfactants, Polyphosphates, Aluminium-copper-oxides
Chemical, Fertilizers	Sludges	Sulfuric Acids, Organic phosphorus, Copper sulfate, Mercury arsenates
Paints, Varnishes	Sludges	Metal salts, Liquid toxics
Petroleum Refining	Sludges Fly ash	Acids, Hydrocarbons, Metallic salts
Leather & Products	Sludges	Chrome salts, Oils, Dyes, Organic acids
Metals & Fabricated Metals	Sludges Slag Slimes & Tailings	Sulfur, Chlorides, Phenols, Oils, Chrome, Alkalis, Acids. Metallic salts

purposes (5.8% in 1988). Most waste tires are dumped in landfill space. The scrap tyre piles are growing each year and pose two major threats to the public; a fire hazard and a health hazard.

Table 3 presents a preliminary classification of urban refuse - there appear to be two groups of urban refuse, i.e. the degradable materials and relatively non-degradable materials.

1.2 Identification of Potentially Incompatible Wastes

Many wastes, when mixed with other wastes or other materials can produce effects that are harmful to human health and the environment, such as:

- Heat or Pressure;
- Fire or Explosion;
- Violent chemical reaction;
- Toxic dust, Mists, Fumes or Gases;
- Flammable fumes or Gases

A detailed list of these items has been prepared by USEPA (1990).

1.3 Recycling and Energy Recovery from Waste

Recently, in the USA a successful program of recycling and energy recovery from waste has been developed. Aluminium cans, plastic containers and used papers have been recycled and used again. Heavy metals such as lead (Pb), mercury (Hg), zinc (Zn) trapped in the ground in landfill areas have been recovered using electrokinetic and electromagnetic processes as part of soil decontamination and energy recovery programs. Therefore, the composition of landfill material has changed since this program began.

1.4 Engineering Problems

There are two basic problems of landfill design which need engineering knowledge:

(i) During Waste Disposal Period

Problems involved include compaction, drainage systems, stability of landfill slopes and liners or hydraulic barriers.

(ii) Utilization of Wastes or Landfill Sites

The first type of utilization uses waste as conventional construction material, such as scrap waste tires as lightweight concrete aggregates. The second type is utilization of old abandoned landfill sites for housing projects, parks, highway routes etc.. In such cases, the engineering behavior characteristics of wastes must be known. Overall systems concerning landfill construction and control are presented in Figure 1.

1.5 Scope and purpose of the paper

- Close examination of the engineering behavior of urban refuse (garbage) including hydraulic conductivity, bearing capacity, strength characteristics.
- Explanation of two basic design factors for compaction control and slope stability analysis of landfill.

2 ENGINEERING BEHAVIOR OF URBAN REFUSE (GARBAGE)

2.1 Introduction

In general, the engineering properties of urban refuse can be grouped into two categories:

(i) Fresh garbage (freshly dumped active landfill) with little or no decomposition.

(ii) Decomposed garbage (old/abandoned landfill).

As the engineering properties change with time, the composition of refuse as indicated in Tables 1 and 2 is highly non-homogeneous. Therefore, the engineering behavior of refuse (garbage) is also nonuniform.

The engineering behavior of above mentioned garbage type depends on:

(a) The composition of the garbage itself, as indicated in Table 3.

(b) The condition of test specimens, such as;

TABLE 3 PROPOSED ENGINEERING CLASSIFICATION OF URBAN REFUSE (GARBAGE)

Categries			Types	Descriptions	Engineering Behavior
Fresh Garbage	Fast Degradable	A-I	Kitchen	Vegetable food wastes; Meat scraps; Fried fats; Citrus peelings	Unit Weight Varies Bad Odors Combustion Potentials
		A-II	Garden Trash	Wood chips & bark ; Tree leaves, & roots; Flowers & garden plants; Lawn grass; Evergreens & shrub trimmings.	
		A-III	Dead Animals & Manure	Dead animals, birds; Manure; Oils & paints	
		A-IV	Paper & Paper Products	Cardboard; Newspaper; Mail; Magazines; Paper food cartons; Tissue paper; Wax paper, Plastic coated paper; Brown bags.	
	Slow Degradable	B-I	Textiles, Toys, Rugs	Textiles; Rugs; Cloths; Shoes	
		B-II	Glass, Ceramics	Soft drink & beer bottles; Broken glass & dishes; Velse	
		B-III	Plastic, Rubbers, Leathers	Plastic bags; Rubber tires; Rugs; Leather products.	
	Relative Non-Degradable	C-I	Metals, Appliance	Various metals; Electrical appliance; Abandoned vehicle parts.	Unit Wt.=200-400 pcf
		C-II	Demolition & construction materials	Broken stones, bricks, walls, concrete & asphalt pavements; Sand-gravels; Mortor.	Unit Wt.= 110-150 pcf
		C-III	Soils	Soils; Dust, Ash, Vacuum cleaner wastes	Unit Wt.= 100-135 pcf
Decomp. Garbage	D-I		Partial Decomposed Garbage	Organic soil contained some glass or metal articles.	Unit Wt.=85-120pcf
	D-II		Decomposed Garbage	Organic soils contain some sand	Unit Wt.=100-135pcf

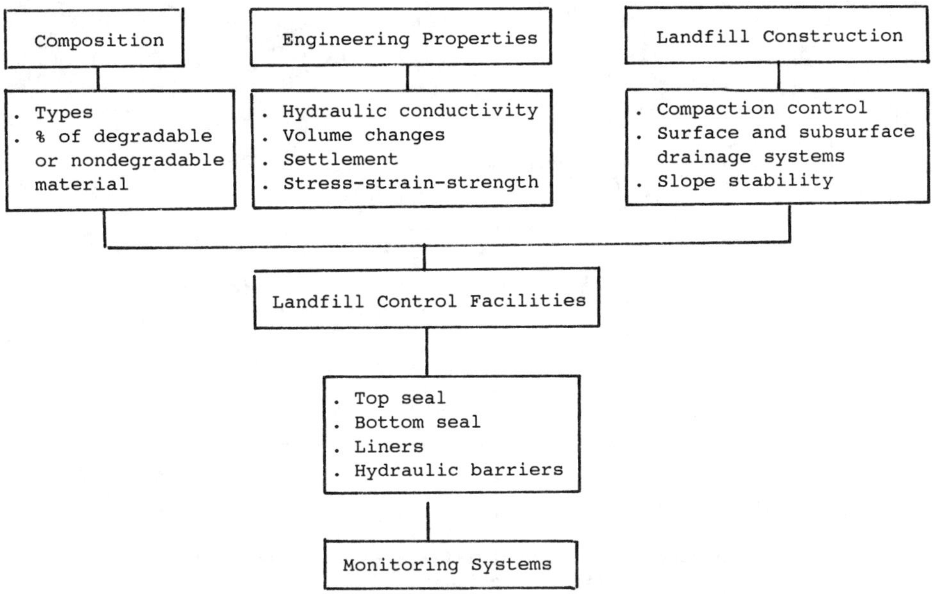

Figure 1 Some major elements to be considered in landfill system

- compacted or loose;
- saturated or unsaturated;
- sizes and shapes;
- laboratory prepared or in-situ condition, etc.

Decomposed garbage is generally referred to as an organic soil. The properties of these types of soil must be determined in-situ, on location, and in the laboratory, in a state which is comparable to the intended engineering use.

2.2 Odor and Color

Odor and color of garbage or ground soil or soil-water near the landfill site can be used to predict the engineering characteristics of contaminated soil or soil-garbage mixtures as reported by Fang (1993a). The characteristics of soil related to its color under normal environmental conditions have been discussed be ASTM (D2488-83). However, if the ground soil is contaminated, these colors will change. In most cases, these colors are shiny with multi-colored hues.

Odor also can be used to identify many naturally existing gases, including chlorine, hydrogen chloride, hydrogen sulfide, sulfur dioxide, and ammonia. However, some highly poisonous gases exist in nature, or as by-products of manufacturing, which are colorless and odorless such as carbon monoxide, methane and radon gas. Figure 2 shows the gas distribution near a landfill as reported by National Research Council (1977). Testing for odor in water has been standardized by ASTM (D1292-80).

2.3 Volume Changes and Cracking Patterns

Cracking patterns and soil color are closely related. They are both affected by contaminated soil/water, especially by organic acid such as acetic acid, aniline, carbon tetrachloride. Significantly different volumetric changes between non-polluted and polluted soil samples have been observed. Therefore, ground surface cracking patterns are a useful tool for evaluation of engineering behavior of contaminated ground soil near landfill sites (Fang, 1993a).

2.4 Capillarity and Hydraulic Conductivity

The capillarity and hydraulic conductivity behavior of refuse itself as well as contaminated ground soil near a landfill site, are directly related to the stability and effectiveness of liners or hydraulic barriers. These elements are used to

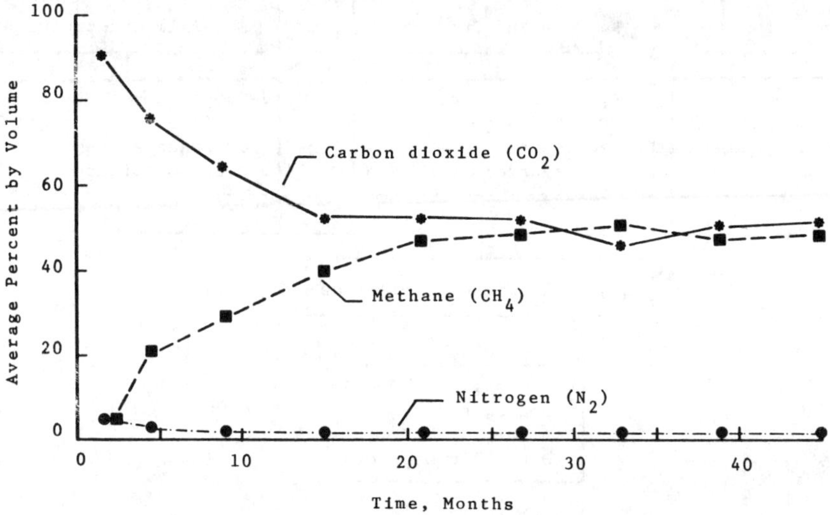

Figure 2 Gas distribution near the landfill (After NRC, 1977)

prevent the contaminated pore fluid (leachate) seeping from a landfill into an aquifer or surrounding areas.
Figures 3 and 4 show the particle size versus capillarity height and hydraulic conductivity. It can be seen that the smaller the particle size the greater is the variability of test results especially for polluted water because pore fluid plays an important role in these tests. Also, flow through a porous medium following Darcy's law is commonly used in geotechnical engineering, however, this law can not be effectively applied to micro-particle soil, especially for colloidal systems. Comprehensive studies on the relationship between the hydraulic conductivity determined in the laboratory and field for a compacted clay layer (relating to sanitary landfills and hazardous waste sites) have been conducted by USEPA (Rogowski, 1990). Other mechanical and hydraulic properties of refuse are given by Jessberger and Kochel (1993), Sanchez-Alciturri, (1993a) and others.

2.5 Settlement Prediction

A schematic diagram illustrating the settlement potential versus time in a landfill area is presented in Figure 5. Terzaghi's one-dimensional consolidation theory covers only a part of this settlement behavior, because the theory is based on loading, i.e. mechanical energy. In landfill areas, the decomposition process involves multi-media energy fields including biological and physico-chemical processes. Further discussion on landfill decomposition process will be presented later. Current practice of settlement prediction is still based on the Terzaghi theory with modifications to overcome the environmental effects on future settlement.

Sowers (1973) suggested that continuing settlement of sanitary landfill is analogous to secondary compression of soil. The settlement with respect to time and depth of fill was described by Sowers as follows:

$$\Delta H = \alpha \frac{H}{1+e} \log\left(\frac{t_2}{t_1}\right) \qquad (1)$$

ΔH = total settlement
α = coefficient which depends on field condition
α = 0.09e (for conditions favorable to decomposition)
α = 0.03e (for unfavorable conditions)
H = fill depth
t = time
e = initial void ratio

Yen and Scanlon (1975) suggested that settlement rate can be computed by the following equation:

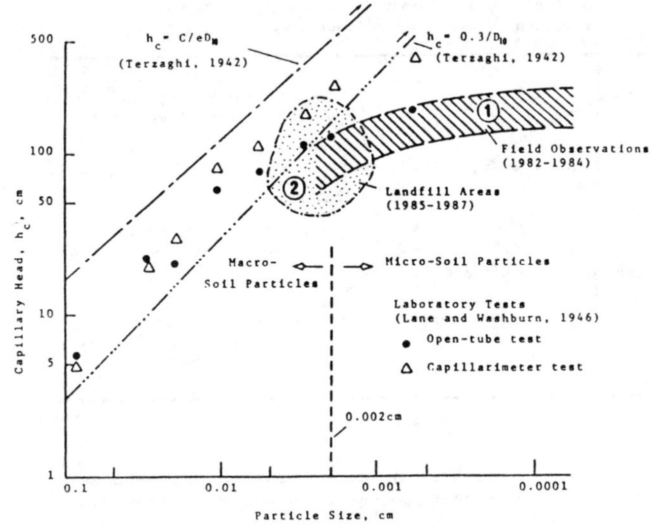

Figure 3 Characteristics of capillary head near the landfill site (After Fang, 1989)

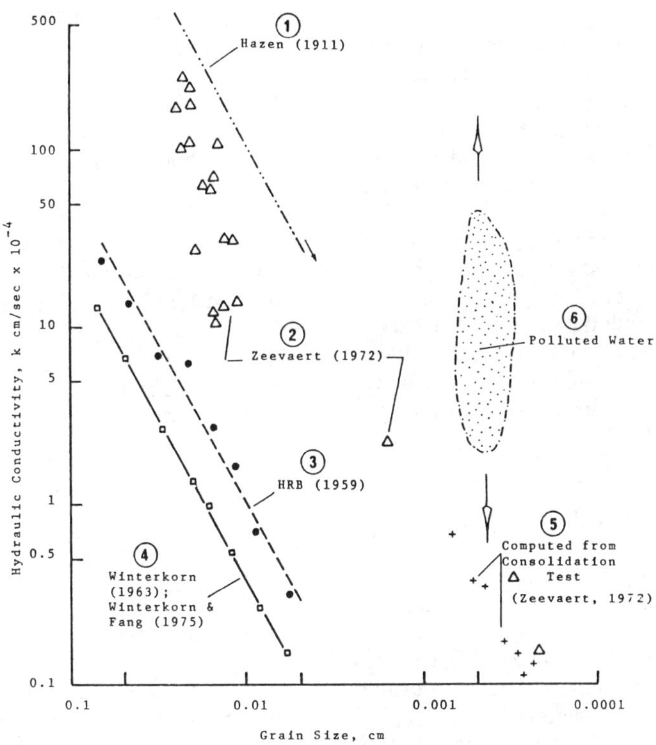

Figure 4 Effects of soil particle size on hydraulic conductivity

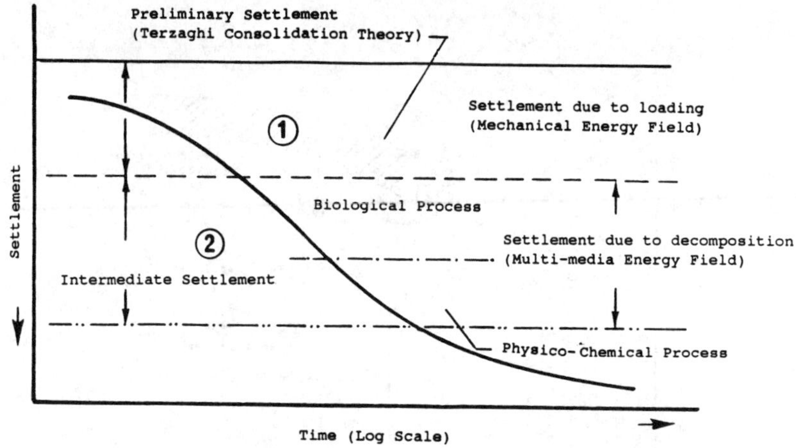

Figure 5 Schematic diagram illustrating settlement versus time for a degradable material

$$m = \frac{\alpha}{1+e} \frac{H}{t-1} \log(t) \quad (2)$$

m = settlement rate in feet per month
t = time elapsed in months
Other notations in Eq. (2) are the same as defined in Eq. (1).

Due to the nature of the problem, the settlement analysis of landfill cannot be solved by mathematical equations. Semi-empirical methods such as equations (1) and (2) may be the best approach at the present time. Yen and Scanlon (1975) suggested that precompaction of landfill can eliminate some of the initial settlement. The rate of settlement in landfill site is an important factor and numerous investigators have reported on various aspects; modelling (Neely and Artz, 1972), laboratory (Andersland and Al-Khafaji, 1985), field observations (Merz and Stone, 1962, Chang and Hannon, 1976), and case studies (Byrne, et al. 1992, Sanchez-Alciturri, et al. 1993a).

2.6 Stress-Strain-Strength Characteristics

2.6.1 Shear Strength Parameters of Garbage.

To use a traditional geotechnical engineering approach for the estimation or prediction of the engineering behavior of material e.g. bearing capacity or slope stability, it is necessary to have the shear strength parameters c and ϕ of the material. Unfortunately, it is difficult to determine c and ϕ for garbage by a routine laboratory approach. However, a short cut method has been developed for estimation of these parameters for compacted garbage (Fang, Slutter and Koerner, 1977). A brief description of the method is as follows:

(a) Procedure

The method is based on the modified Mohr-Coulomb failure envelope as shown in Figure 6. In order to establish the failure envelope, three pieces of information are required:

Point A: Determined by tensile test.

Point B: Determined by unconfined compression test (ASTM D2166-72).

Point C: Determined by tensile/cohesion ratio (ξ).

$$\xi = \frac{\sigma_t}{c} \quad (3)$$

$$c = \frac{\sigma_t}{\xi} \quad (4)$$

(b) Test Specimens

The test specimens are compacted natural waste disposal blocks (bales). Each bale was 4' x 4' x 5' (1.2m x 1.1m x 1.5m) in size, formed with three-dimensional baling forces

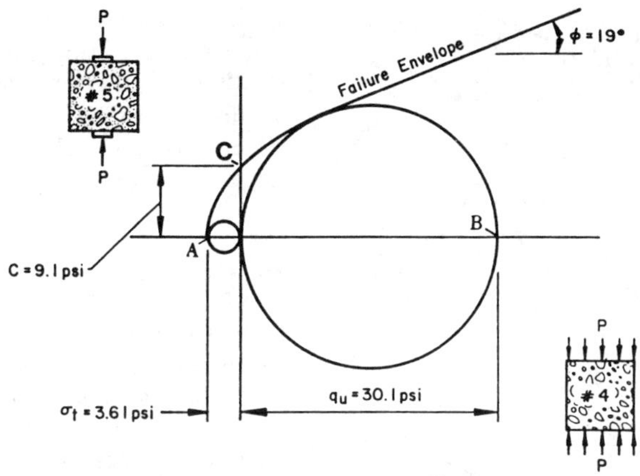

Figure 6 Procedures for estimation of cohesion (c) and internal friction angle (ϕ) for a compacted urban refuse (After Fang, Slutter and Koerner, 1977)

varying from 1300 psi to 5300 psi. (8.97 to 36.57 MN/m^2). The material composition for each bale is residential garbage and commercial waste with the average components based on national survey results which are shown in Table 1 as reported by the American Hoist and Derrick Co. (1974).

(c) Tensile/Cohesion Ratio (ξ)

Considerable information is available regarding the value of (ξ). For soil, the value is a function of the plasticity index (ASTM D424-71) and varies from 0.34 to 0.44 as reported. For rocks, the average value is 0.5. The ratio used for compacted garbage 0.4 which is the average value of all previous published data.

(d) Computation

If all three pieces of information are known, then the failure envelope shown in Figure 6 can be established. From Figure 6, the friction angle (ϕ) can be computed as:

$$\phi = 2\tan^{-1}\left(\frac{q_u}{c}\right) - \frac{\pi}{4}, \quad where\ \phi < \frac{\pi}{2} \qquad (5)$$

The friction angle ϕ, is computed either from equation (5) or by a graphical method from Figure 6.

For the large, irregular size of test specimen such as baled waste material conventional test methods cannot be satisfactorily used, however this suggested method may serve that purpose. It must be noted for highly non-uniform garbage, all test results obtained are used only as a reference

2.6.2 Stress-Strain Relationship of Compacted Soil-Garbage Mixture

Laboratory prepared samples of soil-garbage mixtures were used for this study. The composition of this garbage was based on the national survey indicated in Table 1. The tensile strength was used to indicate the stress-strain relationship. The test samples were molded and compacted in a standard Proctor mold. Standard AASHTO procedures were followed (ASTM D698-70).

Soil-garbage mixes were made with different amounts of soil, i.e., 100%, 94%, 84% and 25% by weight. As the amount of soil added to the garbage increased so did the strength of the mixture indicating that garbage needs soil to fill the various sizes of voids in order to increase the strength and bearing capacity. It was also observed that there existed more than one failure peak in the stress-strain relationship curves indicating the typical phenomenon of non-uniform composition in the soil-garbage mixtures. For further discussion on this study see Fang, Slutter and Stuebben (1976).

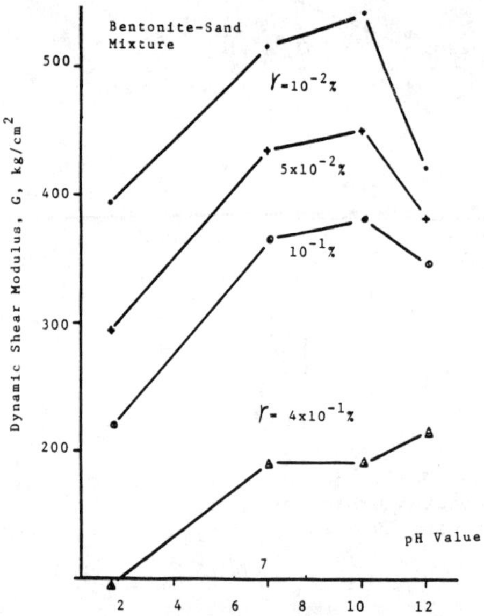

Figure 7 Effect of pore fluid (as reflected by pH value) on dynamic shear modulus of laboratory prepared landfill liner material (After Du, Mikroudis and Fang, 1986)

2.6.3 Stress-Strain Relationship of Compacted Soil-Scrap Waste Tire Mixture.

As indicated in a previous section, large amounts of waste tire are dumped into landfills. To reduce environmental hazards and to save landfill space, waste tires have been considered for conventional construction use. The stress-strain relationship of compacted soil-scrap waste tire mixture is similar to the other soil-garbage mixture as reported by Fang, et al. (1992) and Lyons, et al. (1993).

2.7 Dynamic Shear Modulus

Recent research results indicate that when acid concentration increases, as reflected in the pH value, the dynamic shear modulus of laboratory prepared liner specimens (sand-bentonite mixtures) decreases as shown in Figure 7. In general, acid is used to stabilize soil as it is found to increase its shear strength when measured in unconfined compression test or other shear tests using static loading. However, dynamic load destroys the bond created between the soil particles by the acid as reflected in Figure 7. In other words, dynamic loading can cause more damage than static loading in landfill areas. Therefore, precautions must be taken when landfill sites fall in an earthquake zone or when they are subjected to any other dynamic loading.

2.8 Summary of Engineering Behavior of Refuse (Garbage)

(a) Color, odor and cracking patterns of ground soil near the landfill sites can be used to identify or characterize some degree of engineering behavior of refuse.

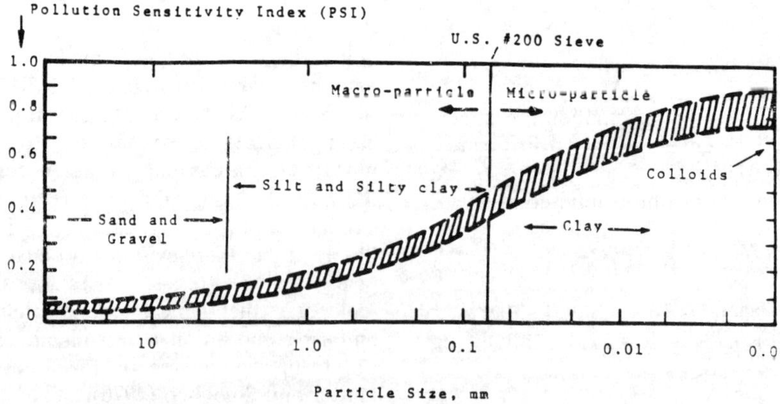

Figure 8 Pollution sensitivity index (PSI) relating to soil particle size

TABLE 4 CLASSIFICATION OF CONTAMINATED SOIL AS REFLECTED BY POLLUTION SENSITIVITY INDEX AND SOIL PARTICLE SURFACE AREA
(After Fang, 1993b)

PSI	Soil Type	Size, cm	Surface Area cm^2/cm^3	Sensitivity
0 - 2	Gravel	0.2 - 8.0	15- 0.125	Very Low
2 - 4	Sand	0.005 - 0.2	500- 15	Low
4 - 6	Silt	0.0005 - 0.005	6,000- 600	Medium
6 - 8	Clay	0.0001 - 0.0005	30,000- 6,000	High
8 - 10	Colloids	< 0.0001	> 30,000	Very High

(b) Darcy's law cannot be applied effectively to the evaluation of flow movement through polluted or unsaturated fine-grained soil near landfill sites. The reason for this is that flow movement is a mass transport phenomenon, not a hydraulic conductivity one, as commonly used in geotechnical engineering.

(c) Settlement analysis of landfill cannot be solved by mathematical equations due to the nature or composition of garbage. Semi-empirical methods may be the best approach at the present time. As suggested by Yen and Scanlon (1975) precompaction of landfill can eliminate some of the initial settlement.

(d) Garbage itself is an unstable material. Its strength is low and its stress-strain relationship is irregular. Without proper stabilization or ground improvement techniques being applied, garbage itself cannot take large loads.

(e) In a landfill site, dynamic loading can cause more damage than static loading, because the bonding energy created by acidity is destroyed by excitation of dynamic loading.

(f) Garbage will decompose as time goes on, and in most cases, converts into organic or inorganic soils. Figure 8 and Table 4 give some idea of the sensitivity of soil types which will interact with environments. In examining Table 4, the smaller soil particle has more surface area per volume and more chance for soil to interact with environment.

(g) Shear strength, hydraulic conductivity and settlement are affected by time. In general, the shear strength increases and hydraulic conductivity and settlement decrease as garbage ages.

3 DECOMPOSITION PROCESS AND REACTIONS/PHENOMENA PRODUCED DURING THE PROCESS

3.1 General Discussion

Decomposition is a natural process. Due to various environmental factors any material in a given location will after a certain time gradually change its form and properties. Figure 9 shows the concept of geomorphism of refuse. In examining Figure 9, there are basically four stages of decomposition. The mechanical alteration is considered a short-term process while the others are long-term. The chemical processes include hydrolysis, hydration, carbonation, oxidation and solution. These processes involve principally dissolving processes or deterioration of minerals.

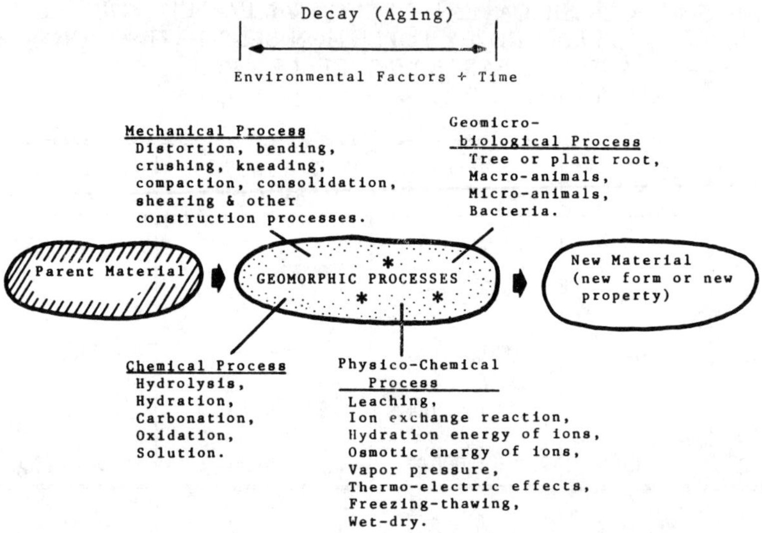

Figure 9 Decomposition processes of urban refuse (garbage)

On the other hand, during the decomposition process in the landfill area, the ground temperature increases. When temperatures are high (a minimum temperature of 80°F/27°C), retention times are long and sulfate concentration is appreciable. Therefore, some additional reactions and/or phenomena can be expected near the landfill site during the decomposition processes.

3.2 Decomposition Processes

(a) Mechanical Process

The mechanical process itself is not included in the decomposition process. However, a large amount of settlement in landfill areas is contributed by mechanical loads such as surcharge loads, snow, rainwater, etc. Since the refuse is nonhomogeneous and contains organic matter, the settlement in the landfill area is also non-uniformly distributed. Because of this phenomenon, the top seal of a landfill cover frequently breaks down.

(b) Chemical and Physico-Chemical Decomposition Processes

Leaching and ion exchange reactions affect material properties significantly. Material property changes occur when rainwater or drainage processes remove some solubles, such as clay minerals, by leaching. During this natural occurrence, common mineral elements such as calcium (Ca), magnesium (Mg), nitrogen (N), potassium (K) and phosphorous (P) may be removed. This natural process in turn reduces the cementing of the soil matrix or changes the ion concentration within soil-water system, resulting in significant changes in the properties of material.

The ion exchange reaction results in the soil structure changing from a dispersed to flocculated matrix or vice versa. This change can significantly affect the hydraulic conductivity, shear strength and compressibility of materials.

(c) Bacteria Decomposition Process

Most bacteria exist at a shallow depth from ground surface, especially in the A-horizon (Fang 1993b). Disturbed soil has more bacteria than undisturbed because the air (oxygen) enables more bacteria to grow. Most bacteria exist in an aerobic condition which exists above the groundwater table. The rate and degree of bacterial activity depends on the availability of micro organic

TABLE 5 RANGE OF COMPOSITION OF LEACHATE FROM LANDFILL

Component	Range
pH	3.7 - 8.5
Hardness, $CaCO_3$	200 -7,600
Alkalinity, $CaCO_3$	720 -9,500
Ca	240 -2,400
Mg	64 410
Na	85 -3,800
K	28 -1,700
Fe	0.15 -1,640
Ferrous Ion	8 - 9
Chloride	50 -2,400
Sulfate	20 - 750
Phosphate	0.5 - 130
Organic-N	3.0 - 490
NH_4-N	0.3 - 480
BOD	22,000 -30,000
COD	800 -50,000
Zn	0.02 130
Ni	0.15 0.9
Suspended Solids	13 -27,000

Data from Fang and Evans (1988), Evans (1991) and Others.

types, nutrients, substrate composition and temperature. Landfill sites provide ideal environment for bacteria to grow.

An individual bacterium acts as a vehicle which carries ions and speeds-up the ion exchange process in the soil-water system. Consequently, changes in soil-water structure or the characteristics of the pore fluid occur through decomposition of soil mineral constituents. For example, bacteria can change hydrogen sulfide (H_2S) to sulfuric acid (H_2SO_4) in a landfill area. For further discussion on bacteria attack on landfill see Fang (1987, 1993b).

Figure 10 illustrates the geomorphic process for ten-year garbage, as reported by Witmer, Volk and Naik (1984) and Witmer (1987), which is reflected in the grain size distribution curve.

3.3 Pore Fluid Produced and Its Movement Characteristics

Pore fluid produced during the decomposition process in the landfill generally is referred to as leachate (Evans, 1991). The properties of leachate change from community to community, as well as from season to season. Table 5 shows national survey results and ranges. Table 6 shows the composition of leachates from central Pennsylvania. Leachates can be grouped into inorganic and organic pore fluids:

(a) Aqueous inorganics have water as the solvent and the solute is mostly inorganic.

(b) Aqueous organics are those in which water is the solvent and solutes are predominantly organic. The organic part covers organic wastes and organic fluids.

TABLE 6 PHYSICO-CHEMICAL PROPERTIES OF LANDFILL LEACHATE FROM CENTRAL PENNSYLVANIA (After Fang and Evans, 1988)

Parameter	Leachate Concentration, mg/ℓ
Biochemical Oxygen Demand (BOD)	7,300
Chemical Oxygen Demand (COD)	11,000
Total Organic Carbon (TOC)	3,900
Chloride	898
Magnesium	300
Sulfate	400
Total Organic Nitrogen	220
Manganese	48
Aluminum	2.1
Total Iron	278
Nitrite	0.21
Nitrate	0.89
Fluoride	0.41
Oils & Greases	43
Color	4,957 color units

(c) Organic acids are those fluids which react with bases and include a proton donor. In organic bases, any organic fluid is capable of accepting a proton to become an ionized cation.

Flow through or around an area of landfill is complex. The flow is low velocity, unsteady, non-uniform or sometimes discontinuous and contains toxic or hazardous substances in solid or gaseous forms. There are at least six possible causes which may affect the movement from one place to another as summarized in Table 7. In all cases, the stress or flow movement are small in magnitude. However, the accumulations of these small amounts can be sufficient to cause groundwater pollution thereby affecting the stability of all waste control systems.

3.4 Reactions/Phenomena Produced During the Decomposition Processes

(a) Acidity

Most pore fluids produced from landfill are acidic as discussed in previous sections. These acids seep through landfill into the liner, regardless of the types of liners used. The polluted pore fluid interacts with ground soil or any other material at various stages. All soils contain H^+ and OH^- ions. Higher H^+ concentrations are due to exchangeable H ions on the soil particles and/or the presence of mineral or organic acids, eg H_2SO_4 from oxidation of pyrites, H_2CO_3, and other organic acids from the decomposition of vegetation.

Acidity in landfill varies with the season and it affects many geostructural members because of its corrosive effect on metals and other construction materials as it pollutes the groundwater aquifer. On the other hand, acidity is sometimes desirable because of its

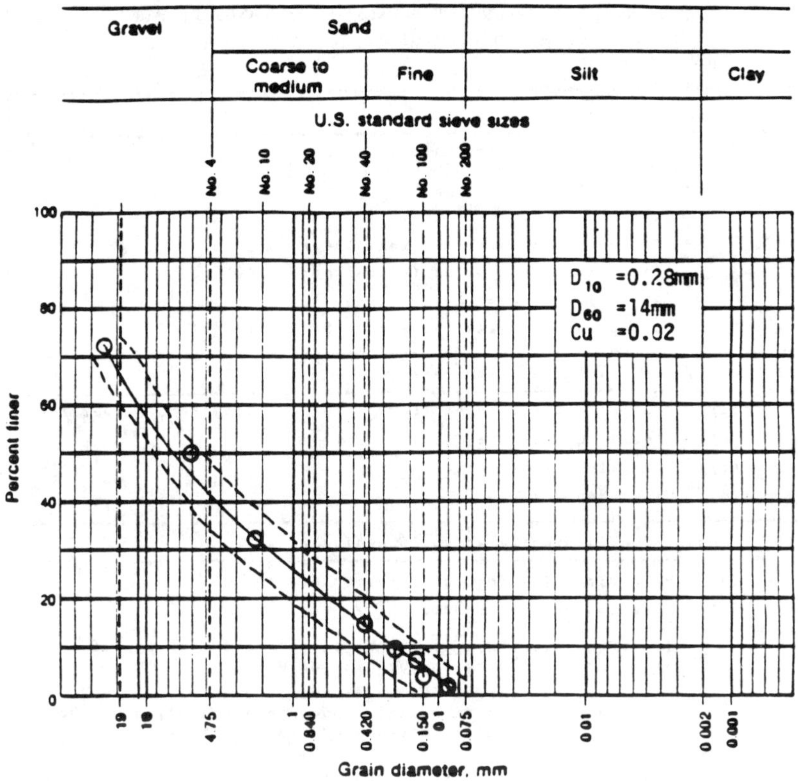

Figure 10 Grain size distribution for 10-year old garbage (After Witmer, Volk and Naik, 1984)

catalytic effect on certain reactions employed in soil stabilization (Winterkorn and Pamukcu, 1991).

(b) Environmental Pressures

Additional forces or stresses are developed within the soil-water system during the decomposition process in the landfill site. These additional forces are referred to as environmental stresses (Fang, 1987). There are at least four types of stresses existing, in addition to the external loads (Table 7).

(c) Other Reactions and Phenomena

When temperature increases in the landfill site the cumulative settlement also increases as indicated in Figure 11. When two or more wastes interact they may create additional reactions as discussed in Section 1.2. In landfill areas underground combustion or fire is frequently reported.

In the lower portion of a landfill site, there are not only higher temperatures but also higher hazardous/toxic residues, consequently higher bacterial activity and higher ion exchange reactions. These chemical erosions will cause instability of landfill.

4 COMPACTION OF LANDFILL

4.1 General Discussion

Among the factors relating to the stability and safety of landfill, the compaction control during the garbage disposal process and the slope stability of landfill during and after construction are the most important from an engineering view

TABLE 7 MASS TRANSPORT PHENOMENA IN LANDFILL SITE
(Based on Fang, 1987)

I.	Types of Water	
	(a)	Free Water (Gravity Water)
	(b)	Environmental Water
		(1) Double-Layer Water
		(2) Hydrated Water (Eucken, 1940)
		(3) Hydratation Water (Winterkorn, 1942)
		(4) Osmotic Water
		(5) Oriented Water (at air-water interface)
II.	Environmental Forces (Stresses)	
	(a)	Heat of Wetting Force (force produced from heat of wetting)
	(b)	Kinetic Dispersive Force
	(c)	Electric-Motive Force
	(d)	Water-Vapor Force
III.	Causes of Moisture Migration (Creeping Flow)	
	(a)	Mechanical (Hydrostatic) Potentials
	(b)	Physico-Chemical Causes
		(1) Hydration Energy: due to the hydration energy of ions, related to the heat of wetting;
		(2) Osmotic Energy: due to the osmotic energy of ions either held in a kind of Donnan equilibrium on the solid particle surfaces or free in the aqueous solution;
		(3) Capillary Potential: due to the surface tension of water and the size and geometry of the soil pores;
		(4) Electric Potentials (Electroosmosis);
		(5) Thermal Potentials (Thermoosmosis);
		(6) Vapor Pressure Potentials.

point. Further consideration of these two aspects is included together with suggested methods for the improvement of the effectiveness of the stability and landfill sites.

When waste (garbage) is delivered daily by truck and dumped into a landfill site - this is the "old fashioned" approach to waste disposal. In some cases this garbage is spread to make thin layer, mixed with some earth material and compacted by conventional compaction equipment. This is done to cover up an unattractive landfill site or to minimize odor or to prevent animal and bird vandalism. There is no standard rule or regulation on how garbage should be dumped or compacted at the present time. In this section some compaction control in landfill areas is suggested and past experiences are reviewed.

4.2 Compaction Process

Compaction or densification is a simple and low-cost mechanical process for all types of construction. The main purpose of this process is to change loose material into a denser state for the following reasons:

- To reduce future settlement
- To increase bearing capacity
- To increase shear strength
- To reduce hydraulic conductivity

In a landfill area when the proper compaction is applied it can also reduce the potential fire hazard. The smaller the amount of air that is trapped in the landfill the smaller is the potential for combustion of the waste.

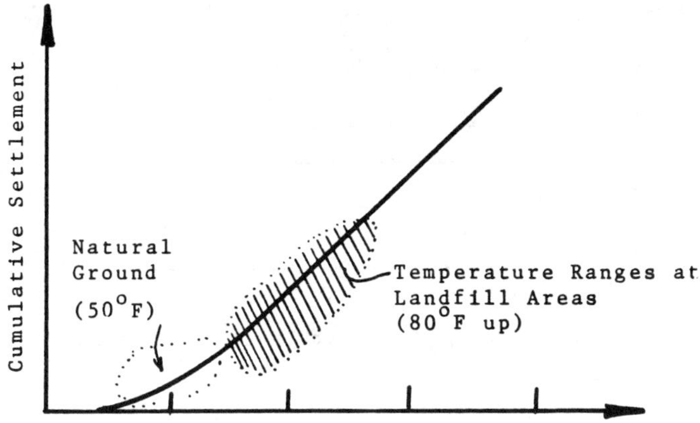

Figure 11 Cumulative settlement versus temperature trends due to decomposition of organic solids (Based on Wardwell, Charlie and Doxtader, 1982)

4.3 A review of Compaction Experience - Sanitary Landfill Test Sections

Several test sections of landfill have been carried out by the states of New York (Moore and McGrath, 1970) and California in early 1970 (Chang and Hannon, 1976). The test sections performed in California are the most comprehensive. This test site was located on the proposed Route 52 in San Diego County, California. This is within the City of San Diego's Sanitary Fill accessible from Mercury Street north of Clairemont Mesa Boulevard. Refuse materials, including glass, wood, kitchen garbage, paper, rags, metals, etc. were dumped in this area between 1964 and 1967, and covered with a thin layer of earth. The area receives only about 10 inches rainfall yearly and little decomposition has taken place, except for food or vegetation waste. Based on this experience the following benefits were obtained:

(a) A 50-ton roller can be used to compress a foundation material consisting of dry or unsaturated sanitary landfill to reduce total primary settlement. Maximum prerolling settlement was achieved after 25 passes of the 50-ton roller. Eighty-five percent of the maximum prerolling settlement developed by 10 passes of the 50-ton roller.

(b) Surcharge loading is of major benefit in accelerating consolidation of sanitary landfill. Settlement rate accelerated substantially when the maximum surcharge height of 10 feet (3m) was reached. Fifty percent of the average total surcharge settlement occurred prior to completion of the surcharge embankment and eighty percent had occurred 30 days after completion of the embankment.

4.4 Surface Compaction Process and Suggestions

The process of surface compaction plays an important role in stabilizing the landfill. However it requires planning during the waste disposal process period and the following measures are suggested:

(a) Garbage comes in all types and it cannot be uniformly distributed in the whole landfill. However, within limits, it can be distributed uniformly within a layer (Figure 12a).

(2) With non-uniform spreading the heavier items should be dumped close to the center of landfill for the purpose of controlling the stability of the fill (Figure 12b) - avoid dumping them around the edges of the landfill (Figure 12c).

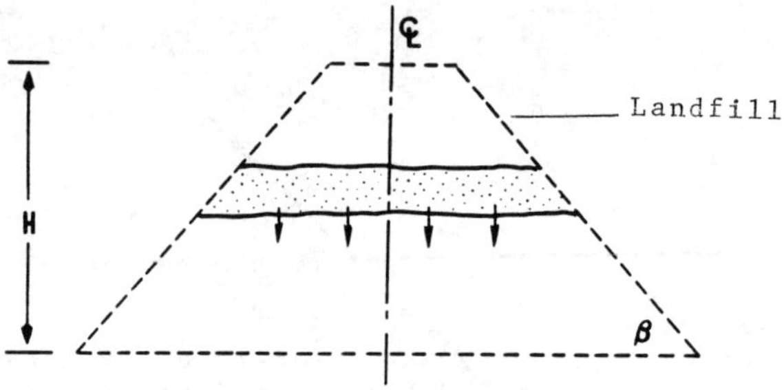

(a) Dumping Uniformly: Equal Settlement - Stable Slope

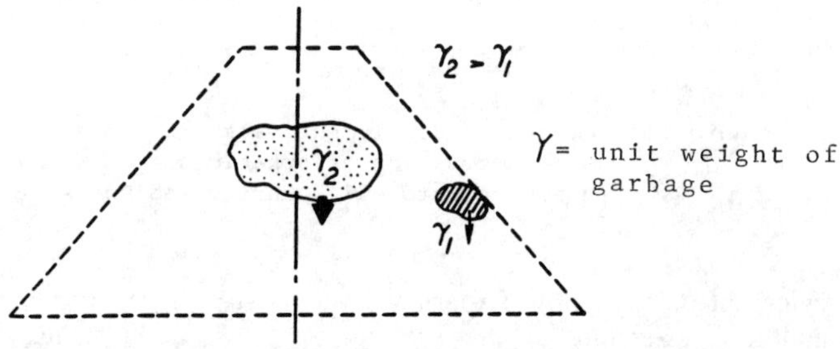

(b) Dumping Irregularly: (Case I)
Differential Settlement-Relatively Stable Slope

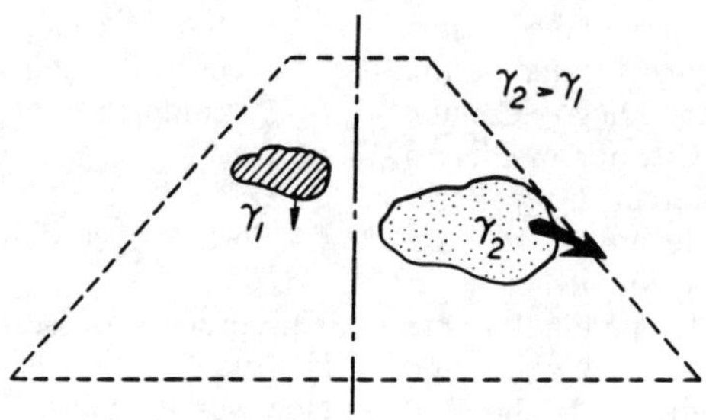

(c) Dumping Irregularly: (Case II)
Differential Settlement-Unstable Slope

Figure 12 Effect of refuse (garbage) dumping procedure on stability of landfill slopes

(3) Spread newly dumped garbage as thinly as possible. Since garbage is an unstable material (as discussed in section 2), it requires stabilizing to reduce excessive settlement/subsidence. Mix locally-available soil with the garbage or add flyash, lime etc., as discussed by Witmer, et al. (1984), to act as a stabilizing material.

(4) Heavy rolling as achieved by a 50-ton roller can be used. The weight of the roller is related to the thickness of landfill layers. Theoretically, the heavier the roller the better the landfill stability. Use

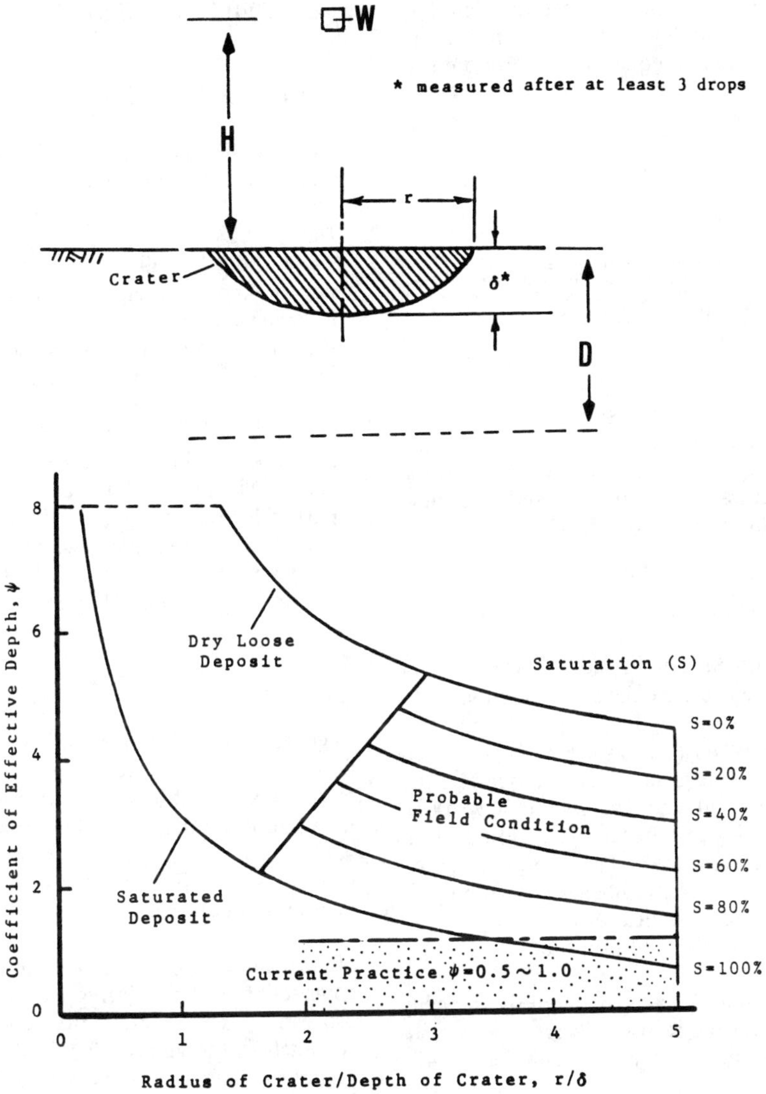

Figure 13 Estimation of coefficient of effective depth (ψ) by dynamic consolidation (Based on Fang and Ellis, 1983)

of dynamic consolidation (Section 4.5) to control the compaction in landfill areas has recently come into practice.

4.5 Deep Compaction on Landfill Sites

Conventional compaction techniques used in landfill areas sometimes give ineffective results for compacting highly nonuniform materials. Recently, dynamic densification (dynamic consolidation) has been used in this situation with good results.

The effective depth or the depth of influence is important for landfill compaction. The effective depth (D) is defined as the depth that the input energy can reach travels into the ground from the surface as illustrated in Figure 13. The weight or pounder, which has specified weight (W) is dropped freely from a predetermined height (H), and affects the landfill material below the ground surface. A simple equation is proposed as follows:

$$D = \psi \sqrt{W H} \qquad (6)$$

D = effective depth or depth of influence, m
ψ = environmental coefficient (due to degree of saturation of the landfill material and compaction characteristics as indicated in Figure 13.
W = weight of pounder, metric tons
H = height of free drop, m.

Where the number of drops (δ) is at least three, the coefficient of effective depth (D) is a function of the shape of the crater (r) produced and the degree of saturation (S) of the landfill ground surface as shown in Figure 13.

4.6 Compaction Control

In general compaction in landfill areas is not really controlled because the major intention is to dispose of the wastes with a simple attitude of "who cares". When this waste land is converted into useful land at a later date, frequently, the regret is expressed that "we should have done that at an earlier date because we did not do it properly for current use". Therefore, during the waste disposal process advanced planning is needed to avoid complications at a later date. Two basic approaches are proposed:

(i) Even though the urban refuse is highly nonhomogeneous there can be improvements as discussed in Section 4.3.

(ii) Compaction control is suggested but proper records must be kept of the composition of the landfill.

Application of random inspection and a sampling plan, as described by Hilf (1991), and as commonly used in geotechnical engineering should also be used in landfill areas to reduce the frequency of sampling and permit the estimation of the in-place performance of a landfill site.

5 LANDFILL SLOPE STABILITY ANALYSIS

5.1 Characteristics of Landfill Slopes

Slope stability analysis in landfill relates to two cases i.e. fill and cut conditions.

Case (i) : During the waste disposal process, the slope at the landfill site is in the fill stage. We need to know how high the garbage landfill can be made without slope failure. In this situation the critical height (H) of the slope is the most important factor. (Figure 12a).

Case (ii) : When old or abandoned landfill sites are to be developed for other purposes such as commercial parking lots, recreation parks, or as part of highway routes, etc, excavation into the landfill is sometimes necessary for construction purposes. In such a case, it is the slope angle (β) that is important (Figure 12a).

5.2 Failure Modes of Landfill Slopes

In general, landfill slope failure is similar to slope failure in earth or rock slopes apart from the exceptions caused by highly nonhomogeneous materials and nonuniform decomposition process and the types of failure also include falls, topples, slips and slides. Falls and topples are due to lack of cohesion between loose refuse pieces.

Slope failure potential is directly related to the compaction control during the waste disposal process. The better the control of the compaction, the less is the risk of slope failures. Figure 14 illustrates some typical landfill slope failures:-

(a) Differential decomposition due to weathering (Figure 14a).

(b) Lack of cohesion between two or more refuse materials (Figure 14b).

(c) Hydrostatic or environmental force acting on decomposed or loosened refuse pieces (Figure 14c).

(d) Differential decomposition and differential settlement causing slope cracks and falls (Figure 14d).

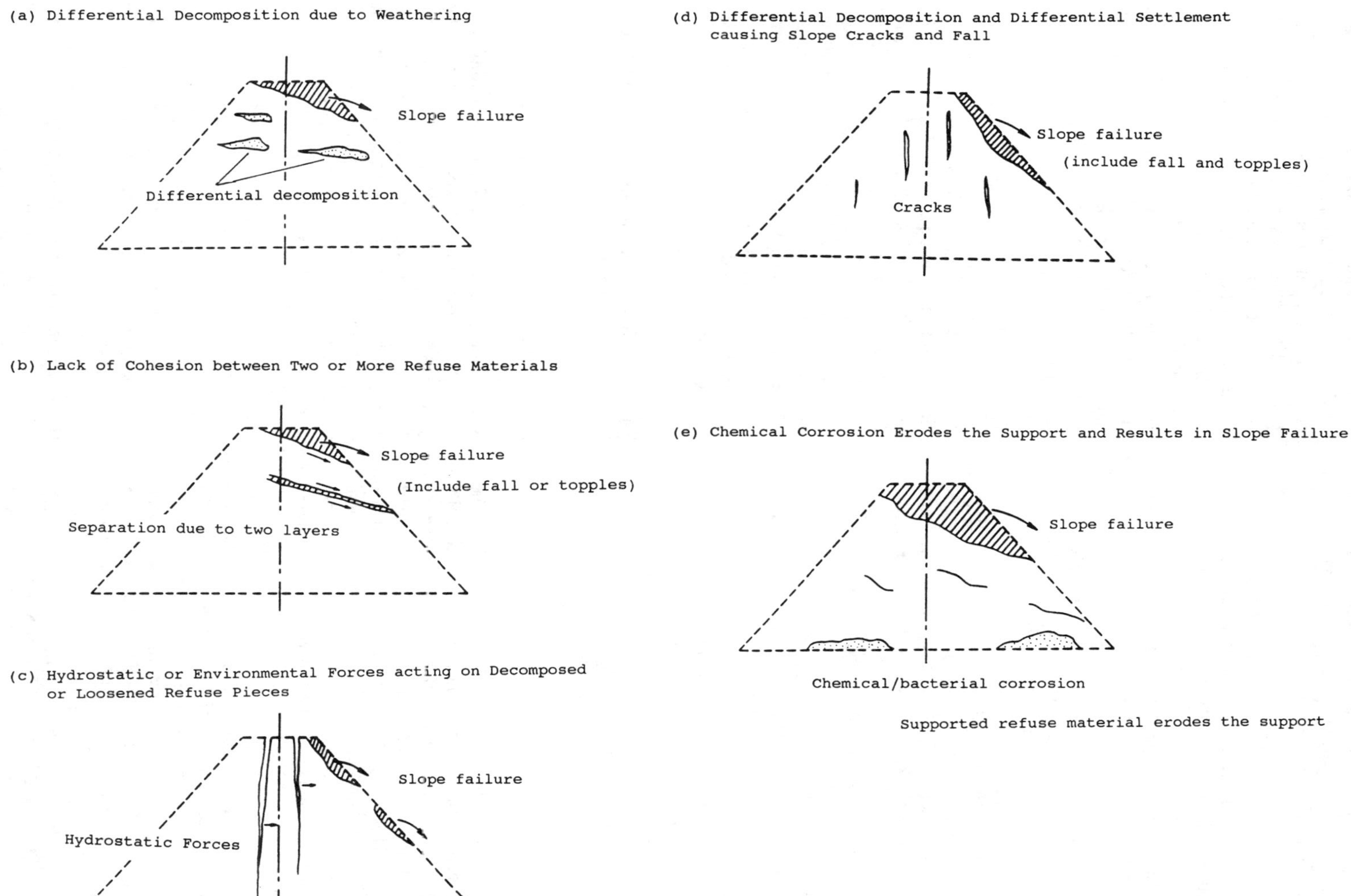

Figure 14 Failure Modes of Landfill Slopes

(e) Chemical corrosion of some supported refuse material erodes the support and results in failure (Figure 14e).

5.3 Problems and Precautions Needed For Analyzing Slope Stability in Landfill Areas

Characteristics of garbage which lead to slope stability problems include:

- Highly nonhomogeneous materials.
- Nonuniform decomposition process.
- Nonuniform compaction within the layer or lift.
- Nonuniform rate of settlement (high differential settlement).
- Garbage itself is an unstable material and cannot take both vertical and horizontal loads.
- Porewater pressures are varied.

Based on the nature of refuse and the characteristics of landfill sites three basic problems must be evaluated:

- Complex material composition.
- Complex geometry of landfill site.
- Complex and adverse environment.

Precautions should be exercised in the following areas when Analyzing Slope Stability Problems

(i) Selection of Strength Parameters

Due to complex material composition the selection of strength parameters is difficult. Commonly these parameters are determined by unconfined compression, direct or triaxial shear tests. However, laboratory experiments indicate that results obtained from these conventional tests are too conservative because they are not sensitive enough to interact with contaminated pore fluids or leachate (Fang, 1991).

Laboratory experiments indicate that tensile test results are more sensitive to the environment than other standard shear tests commonly used in the geotechnical laboratory. It is suggested that with conventional shear tests values are generally overestimated and certain reductions must be applied for practical uses.

(ii) Static and Dynamic Shear Strengths

Most contaminated landfill pore fluids are acidic. In general the acid may help to increase the shear strength from a soil stabilization view point. For years these shear tests have been determined by unconfined or direct shear tests while under static load. Recent findings for contaminated soil, as reflected by the pH value, indicate that as the acid concentration increases (pH decreases) the dynamic shear modulus also decreases as indicated in Figure 7. Therefore in selecting strength parameters it is important to consider whether the landfill site is located in an earthquake region.

(iii) Geometry of Landfill Sites

Landfill sites come in various sizes, shapes and depths and with the unit weight (density) of the refuse varying from 40 pcf to 400 pcf (640 kg/m^3 to 6400 kg/m^3 as indicated in Section 1.1 and Tables 1 and 2. These complex variations of unit weight in a landfill site create problems of slope stability. Computer aided design (CAD) and knowledge-based expert systems have been applied to these problems with limited success and some of the problems are still unanswered.

5.4 Slope Stability Analysis In Landfill Site-A Genetic Approach

(1) Genetic Diagnosis -Approach and Procedure

Refuse (garbage) in a landfill site is complex as discussed in previous sections. It cannot be generalized as a conventional geotechnical problem. Therefore a genetic diagnosis approach is proposed which may assist in understanding or solving landfill slope stability problems. A overall system is presented in Figure 15. The proposed procedures are as follows:

(a) Identification and characterization of the refuse as indicated in Section 1 and Table 3.

(b) Evaluation of the nature of the refuse and its engineering properties as discussed in Section 2.

(c) Examining the degree of sensitivity towards environment and potential decomposition as discussed in Section 3.

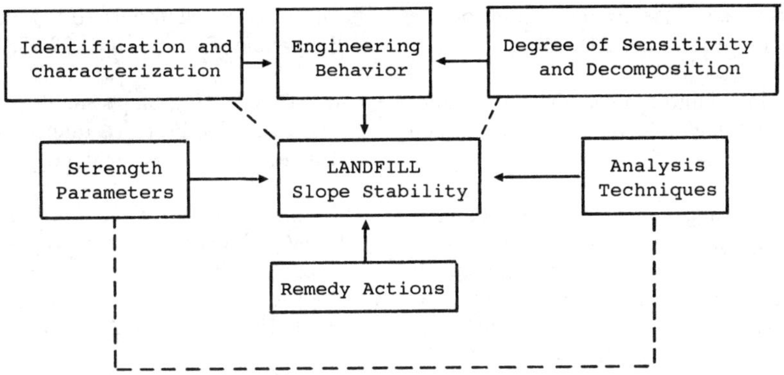

Figure 15 Stability of slope of landfill - genetic diagnosis approach

(d) If the landfill site is newly established, proper records must be kept for the purpose of future land use. Items to record include waste sources, types, amounts, compaction characteristics etc.

(e) If it is an old/abandoned landfill site, and the necessary records are no longer available, the boring logs should be obtained.

(f) Other necessary information includes the age of the refuse deposit, thickness/depth, local weather conditions, groundwater table, etc.

(2) Compaction Control

The stability of a landfill site is directly related to the compaction control during the disposal period. Detailed discussion has been presented in Section 4.2.

(3) Selection of Strength Parameters

As discussed in Sections 2 and 4, landfill is composed of different materials and is subject to seepage which results in a very small effective cohesion (c). The complications that arise are due to complex geometry and material conditions (Oweis and Khera, 1986).

(4) Ground improvement or soil stabilization

Various low-cost energy saving ground improvement techniques are available for repairing or strengthening landfill slopes. Such methods include sand piles, stone columns, lime columns as well as bamboo-lime composite piles (Witmer et al. 1984). Solidified residue from municipal solid waste is also important as discussed by Jackson (1990). This approach also can reduce possible combustion in the landfill sites.

(5) Slope Stability Analysis (Including Earthquake Considerations)

The numerous computer programs that have been developed for slope stability analysis for landfill sites include Chen (1986), Dvirnoff and Munion (1986). Those programs which consider seismic analysis include Anderson et al. (1992) and Seed and Bonaparte (1992). Computer programs or knowledge-based expert systems may assist in solving the problems, however, experience is urgently needed for making such decisions.

6 SUMMARY AND CONCLUSIONS

(1) Garbage itself is an unstable material. Its strength is low and its stress-strain relationship is irregular. Without proper stabilization or ground improvement techniques being applied, garbage itself cannot take large loads both vertically or horizontally.

(2) Surface compaction is an important factor for controlling the stability of landfill slopes. Proper compaction control is needed during the waste disposal period. Fresh garbage dumped into a landfill site should be well spread within the layer, if not, the heavier items (Figure 12b)

should be dumped close to the center of the landfill, and not around the edges.

(3) Landfill slope failure modes that form under various environmental conditions have been explained (Figure 14).

(4) Dynamic loading can cause more damage than static loading, because the bonding energy created by acidity is destroyed by excitation under dynamic loads. Therefore, precautions must be taken when landfill sites are in an earthquake zone or if any other dynamic loading may be applied (Figure 7).

(5) Refuse in the landfill area is complex. It cannot be generalized as a conventional geotechnical problem, therefore, a genetic diagnosis of landfill refuse may enhance our understanding. A detailed approach and procedure has been outlined.

(6) Computer aided design or knowledge-based expert system can be of assistance in solving slope stability analysis in landfill areas. However, human experience based on genetic diagnosis is more effective.

REFERENCES

American Hoist and Derrick Co. (1974), Seminar on High Compression Bailing of Solid Waste, American Solid Waste Systems, St. Paul, Minnesota.

Andersland, O.B. and Al-Khafaji, A.W.N. (1985), Soil volume change due to decomposition of organic solids, ASTM January F1.

Anderson, D.G., Hushmand, B. and Martin, G.R. (1992), Seismic response of landfill slopes, Proc. ASCE Stability and Performance of Slopes and Embankments-II, July.

Byrne, R.J., Kendall, R.J. and Brown, S. (1992), Cause and mechanism of failure, Ketteman Hills Landfill B-19, Unit IA, Proc. ASCE Stability and Performance of Slopes and Embankments-II, July.

Chang, J.C. and Hannon, J.B. (1976), Settlement performance of two test highway embankments on sanitary landfill, New Horizons in Construction Materials, Envo Publishing Co. Bethlehem, PA, v. 1, pp. 139-158.

Chen, R.H. (1986), Slope stability analysis of a waste landfill, Proc. lst. Intern. Symposium on Environmental Geotechnology, v. 1, pp. 37-42.

Du, B.L., Mikroudis, G.K. and Fang, H.Y. (1986), Effect of pore fluid pH on the dynamic shear modulus of clay, Hazardous and Industrial Solid Waste Testing and Disposal, ASTM STP 933, pp. 226-239.

Dvirnoff, A.H. and Munion, D.W. (1986), Stability failure of a sanitary landfill, Proc. 1st Intern. Symposium on Environmental Geotechnology, v. 1, pp. 25-35.

EPA (1990), EPA Report on Definition, Potentially Incompatible Wastes and Hazardous Waste Determination.

Evans, J.C. (1991), Geotechnics of hazardous waste control systems, Chapter 20, Foundation Engineering Handbook 2nd ed. Van Nostrand Reinhold Co. NY, pp. 250-777.

Fang, H.Y., Slutter, R.C. and Steubben, G.A. (1976), Stress-strain characteristics of compacted waste disposal materials, New Horizons in Construction Materials, Envo Publishing Co. Bethlehem, PA, v. 1, pp. 127-138.

Fang, H.Y., Slutter, R.C. and Koerner, R.M. (1977), Load bearing capacity of compacted disposal materials, Proc. of the Speciality Session of Geotechnical Engineering and Environmental Control, 9th Intern. Conf. on Soil Mechanics and Foundation Engineering, Tokyo, Japan, v.1., pp. 265-278.

Fang, H.Y. and Ellis, G.W. (1983), Laboratory Study of Ground Response to Dynamic Densification, Fritz Engineering Laboratory Report no. 462.6, Department of Civil Engineering, Lehigh University, March.

Fang, H.Y. (1987), Mass transport phenomena and stability of waste control systems, Proc. 1st Intern. Symposium on Environmental Geotechnology, v.2, pp. 340-342.

Fang, H.Y. and Evans, J.C. (1988), Long-term permeability tests using leachate on a compacted clayey liner material, ASTM, STP 963, pp. 397-404.

Fang, H.Y. (1991), Sensitivity of soil to environment, Proc. 1st Canadian Conf. on Environmental Geotechnics, Montreal, Canada, May.

Fang, H.Y., Hitchens, D. and Hontz, D. (1992), Use of scrap rubber tires for construction materials, Proc. 24th Mid-Atlantic Industrial Waste Conf., pp. 263-275.

Fang, H.Y. (1993a), Identification and characterization of contaminated ground soil for engineering purposes, Proc. Intern. Conf. on Environment and Geotechnics, Paris, April.

Fang, H.Y. (1993b), Bacteria and tree root attack landfill liners, Green '93, An Intern. Symposium on Geotechnics related to the Environment, Waste Disposal by Landfill, Bolton, U.K., June.

Hilf, J.W. (1991), Compacted fill, Chapter 8, Foundation Engineering Handbook, 2nd Ed. Van Nostrand Reinhold Co. NY, pp. 249-316.

Jackson, D.R. (1990), Evaluation of solidified residue from municipal solid waste combustors, U.S. EPA Report no. EPA/600/S2-89/018, February.

Jessberger, H.L. and Kochel, R. (1993), Determination and assessment of the mechanical properties of waste materials, Green '93, An Intern. Symposium on Geotechnics related to the Environment, Waste Disposal by Landfill, Bolton, U.K., June.

Lyons, J.C. Monticello, D.A. and Hitchens, D.G. (1993), Recovery of scrap rubber tires from landfill for construction uses, Green '93, An Intern. Symposium on Geotechnics related to the Environment, Waste Disposal by Landfill, Bolton, U.K. June.

Martin, C.C. and Hu, Z.X. (1987), The geomorphic processes of soil/rock with examples, Proc. 1st. Intern. Symposium on Environmental Geotechnology, v. 2, pp. 390-393.

Merz, R.C. and Stone, R. (1962), Landfill settlement rates, Public Works, v. 93.

Mitchell, R.A. and Mitchell, J.K. (1992), Stability evaluation of waste landfill, Proc. ASCE Stability and Performance of Slopes and Embankments-II, July.

Moore, L.H. and McGrath, M.E. (1970), Highway construction on refuse landfills, Highway Focus, v. 2, no. 5, pp. 11-26.

Naik, D. (1986), Effect of temperature and pore fluid on shear characteristics of clay, Proc. 1st Intern. Symposium on Environmental Geotechnology, v.1, pp. 382-390.

NRC (1977), Methane Generation from Human, Animal and Agricultural Wastes, National Research Council, Washington, D.C.

Neely, G.A. and Artz, N.S. (1972), Model sanitary landfill in Kansas City, Civil Engineering, ASCE, v. 42, no. 10, pp. 72-76.

Oweis, I.S. and Khara, R. (1986), Criteria for geotechnical construction on sanitary landfills, Proc. 1st Intern. Symposium on Environmental Geotechnology, v. 1, pp. 205-222.

Rogowski, A.S. (1990), Relationship of Laboratory and Field Determined Hydraulic Conductivity in Compacted Clay Layer, U.S. EPA Report no. EPA/600/2-90/025.

Sanchez-Alciturri, J.M., Palma, J., Sagaseta, C. and Canizal, J. (1993a), Mechanical properties of wastes in a sanitary landfill, Green '93, An Intern. Symposium on Geotechnics related to the Environment, Waste Disposal by Landfill, Bolton, U.K., June.

Sanchez-Alciturri, J.M., Palma, J., Sagaseta, C. and Canizal, J. (1993b), Three years of deformation monitoring at Meruelo landfill, Green '93, An Intern. Symposium on Geotechnics related to the Environment, Waste Disposal by Landfill, Bolton, U.K., June.

Seed, R.B. and Bonaparte, R. (1992), Seismic analysis and design of waste fills : current practice, Proc. ASCE Stability and Performance of Slopes and Embankments-II, July.

Sinacori, J.W. (1951), Compaction of unstable material with heavy pheumatic tired roller, Highway Research Board Bulletin no. 42.

Sowers, G.F. (1968), Foundation problems in sanitary landfill, Journal of the Sanitary Engineering Div. Proc. ASCE, v. 94, no. 1, pp. 102-116.

Sowers, G.F. (1973), Settlement of waste disposal fills, Proc. 9th Intern. Conf. Soil Mechanics and Foundation Engineering, Moscow, v. 4, pp. 207-210.

Turner, C.F. and McCreery, J.W. (1981), The Chemistry of Fire and Hazardous Materials, Allyn and Bacon, Inc., Boston.

Wardwell, R.E., Charlie, W.A. and Doxtader, K.A. (1982), Test method for determining the potential for decomposition in organic soils, ASTM STP 820, pp. 218-229.

Winterkorn, H.F. and Pamukcu, S. (1991), Soil stabilization and grouting, Chapter 9, Foundation Engineering Handbook, 2nd ed. Van Nostrand Reinhold Co. NY, pp. 317-378.

Witmer, K.A., Volk, J. and Naik, D. (1984), Low-cost ground improvement technique for landfill area, Proc. Low-cost and Energy Saving Construction Materials, pp. 497-516.

Witmer, K.A. (1987),, Geomorphic process : from fresh garbage to organic soil, Proc. 1st Intern. Symposium on Environmental Geotechnology, v. 2, p. 289.

Yamamura, K. (1983), Current status of waste management in Japan, Waste Management and Research, v. 1, pp. 1-15.

Yen, B.C. and Scanlon, B. (1975), Sanitary landfill settlement rates, Journal of the Geotechnical Engineering Div. Proc. ASCE, v. 101, GT5, pp. 475-487.

Waste Disposal by Landfill – GREEN'93, Sarsby (ed.) © 1995 Balkema, Rotterdam, ISBN 90 5410 356 6

Pollution prevention in landfills using engineered final covers

D.E. Daniel
University of Texas, Austin, Tex., USA

ABSTRACT: The problem of designing a final cover system for a solid waste landfill can be an enormously challenging task. The most difficult problem is to provide a barrier to infiltration of water. The materials that have traditionally been considered for the barrier layer within final covers are, in decreasing order of popularity, compacted clay (compacted mineral liner), geomembranes, and geosynthetic clay liners (GCLs). Data are presented to suggest that compacted clay is often a poor choice of materials. Unless the compacted clay is buried under a very thick layer of protective soil or covered with a geomembrane, the compacted clay is likely to desiccate and lose its low hydraulic conductivity. Differential settlement of compacted clay from uneven compression of underlying waste is almost certain to produce cracks within the clay. Geomembranes do not suffer as much from these problems, and geosynthetic clay liners are much better able to resist damage from freeze-thaw, desiccation, and differential settlement than compacted clay. It is suggested that designers of landfill covers should make more use of geomembranes and GCLs and less use of compacted clay. However, given the large settlement of waste, particularly in the first few years after the landfill is closed, it seems undesirable to try to construct a final, engineered cover on a fresh, unstable mass of waste. A suggested design approach is: (1) initially to place a temporary, relatively permeable cover on the waste and collect leachate for several years, treating the landfill as an active bioreactor; and (2) after significant biodegration of the waste has occurred and the waste mass has become relatively stable, to place a final, engineered cover over the waste. By stabilizing the waste prior to placement of a final cover, the probability of the cover performing as designed is maximized.

1 INTRODUCTION

Final cover systems are a critical component in the overall process of managing liquid and gas movement into and out of landfills. The main objectives in designing a final cover system are to separate the buried waste or contaminated material from the surface environment, to restrict infiltration of water into the waste, and (in some cases) to control release of gas from the waste. If the objective is prevention of pollution to ground water, then one of the obvious strategies is to try to minimize the amount of liquid entering a landfill. If there is no leachate, there is little or no potential for pollution of ground water.

Although much as been written about landfill liners, relatively little has been written about landfill covers. The purpose of this paper is to explore the design considerations for engineered landfill covers.

In succeeding sections, the basic components of an engineered landfill cover will be identified, the materials that are often employed will be discussed, and some of the problems with current designs will be mentioned. Finally, a possible framework for improved implementation of engineered landfill covers will be suggested.

2 BACKGROUND

Cover systems may be called upon to perform one or more of the following functions:

- Raise the elevation in low-lying areas
- Control runoff of precipitation
- Separate the waste from plants, animals, insects, and humans
- Prevent migration of perched leachate out of waste on side slopes
- Limit infiltration of precipitation into the waste
- Control release of gas from the waste.

The U.S. Environmental Protection Agency (EPA) considers keeping water out of the contaminated material to be the prime element in the final cover (EPA, 1989).

The cover must perform the intended functions for extended periods of time. The "design life" of a cover depends primarily upon the nature of the waste, the hydrogeology of the site, the length of time that maintenance of the cover will be provided, and the likelihood of funds being available for repair of the cover. Unfortunately, too many designers are forced to design a final cover system for an actively decomposing and deforming body of waste; a better approach for such wastes might be to construct a temporary cover and then wait until substantial decomposition of the waste has occurred before attempting to construct a final cover system. Unfortunately, this latter approach may require the passage of many years before most of the decomposition is complete; special financial arrangements, such as setting closure funds aside in a publicly-controlled account, would be required because the responsible party may not be viable when the time finally comes to construct the cover. At present, final covers are usually constructed shortly after the tipping of waste is complete in any particular cell within the landfill.

The design of the final cover is complicated by factors that include:

- Temperature extremes, possibly including freeze/thaw to significant depths
- Cyclic wetting and drying
- Penetration by plant roots, burrowing animals, worms, and insects
- Total and differential settlement caused by compression of the underlying waste or foundation soil
- Down-slope slippage or creep
- Vehicular movements on haul roads traversing the cover
- Wind or water erosion
- Deformations caused by earthquakes in seismically-active areas
- Long-term moisture changes caused by water movement into or out of the underlying waste

Because of these rather difficult conditions, the design of a cover system can be very challenging. It is often more difficult to provide an effective hydraulic barrier layer in a cover system than in a liner system because the cover system is challenged in a larger variety of ways by mostly unknown and unquantifiable stresses that do not act on liner systems buried deep beneath the waste.

Most engineered cover systems in the U.S. are composed of multiple components. As shown in Fig. 1, the components of a cover system can be grouped into five categories: (1) surface layer; (2) protection layer; (3) drainage layer; (4) barrier layer, and (5) gas collection/foundation layer. Not all components are needed for all final covers. For example, a drainage layer would normally not be needed for a landfill located in an arid region. Similarly, a gas collection layer may be required at some landfills but not others. Further, some layers may be combined -- for instance, the surface layer and protection layer may be combined into a single layer of soil over the surface of the landfill.

The materials typically used for an engineered final cover system are also shown in Fig. 1. The materials include natural earth materials, geosynthetics, waste materials, blends of natural materials, or blends of earth and geosynthetic materials, e.g., geosynthetic clay liners.

3 COMPONENTS OF COVER SYSTEMS

3.1 Surface Layer

Soil. The material that is by far the most commonly used material for the surface layer is soil. Soils have the advantages of being available locally and supporting growth of vegetation, which helps to minimize erosion and promote transpiration of water back to the atmosphere. Vegetation also creates a leaf cover above the soil to reduce rainfall impact and decrease wind velocity (Lutton et al., 1979). Procedures for vegetating cover soils are reviewed by Gilman et al. (1985). The main drawbacks to a surficial layer of soil are problems in initiating vegetative growth (particularly during some parts of the year) and vulnerability to wind or water erosion if vegetative growth is not suitably thick.

Erosion can be controlled by placing a geosynthetic erosion control material at the surface. Such materials are commonly used on bare slopes next to highways and hold the soil in place until vegetation can become established.

At some sites, cobbles have been used. Cobbles are highly resistant to erosion from wind and water. A problem with cobbles is that they allow water to enter the cobbles but do little to promote evapotranspiration of water back to the atmosphere. Thus, a surficial layer of cobbles may allow far greater infiltration of water into the cover than soil.

Paving material has occasionally been used for the surface layer, but paving materials are not normally recommended for permanent final covers because of problems with cracking of the paving material as a result of differential settlement.

Erosion of the surface layer of soil has been common at many landfill sites. The author has never been to a landfill site and not seen some evidence of erosion in the surface layer. Gullies extending to a depth of approximately 100 to 200 mm are very common at many sites. Swope (1975) studied 24 landfill covers in the U.S. and found that 33% had slight erosion, 40% moderate erosion, and more than 20% had severe erosion. Johnson and Urie (1985) report that erosion is made more severe from the installation of hydraulic barriers (which, without a drainage layer, can produced soaked soil with large amounts of runoff after heavy rains) and vehicles driving on the surface of the cover. Many designers underestimate the problem of erosion of the surface layer prior to establishment of a mature vegetative cover.

Additionally, the problem of stability of the cover soil must be addressed. A number of failures of covers were reported by Boschuk (1991). Most of these failures were attributed to fundamental errors made in assessing the stability of the cover, or in several instances, lack of an analysis of the stability of the cover. Geogrid or geotextile reinforcement layers can be incorporated in the cover to increase factors of safety against slope instability to acceptable levels.

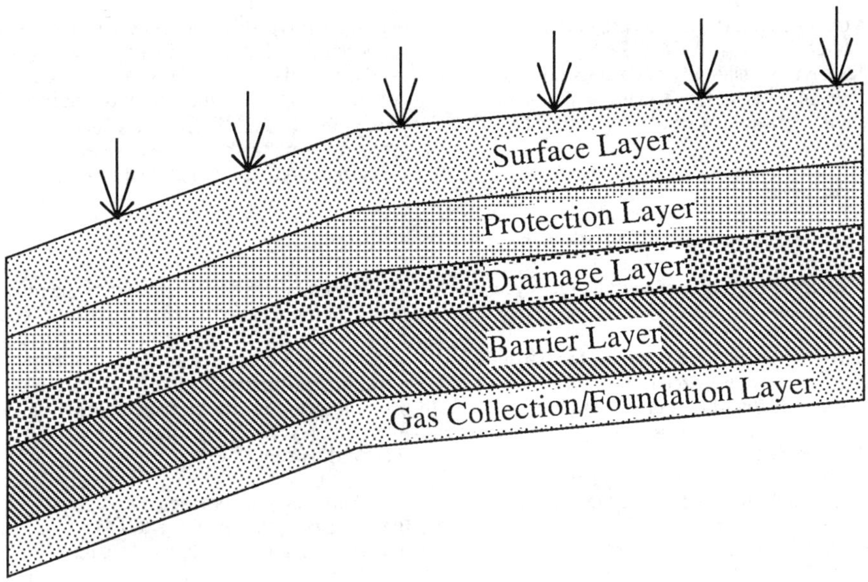

Layer	Description of Layer	Typical Materials
1	Surface Layer	Topsoil; Geosynthetic Erosion Control Layer; Cobbles
2	Protection Layer	Soil; Recycled or Reused Waste Material; Cobbles
3	Drainage Layer	Sand or Gravel; Geonet or Geocomposite
4	Barrier Layer	Compacted Clay; Geomembrane; Geosynthetic Clay Liner; Waste Material
5	Gas Collection Layer and/or Foundation Layer	Sand or Gravel; Soil; Geonet or Geotextile; Recycled or Reused Waste Material

Figure 1. The Five Possible Components of an Engineered Final Cover System.

3.2 Protection Layer

A protection layer (which will generally be some type of local soil) may serve several functions:

- To store water that has infiltrated into the cover until the water is later removed by evapotranspiration
- To physically separate the waste from burrowing animals and plant roots
- To minimize the possibility of human intrusion into the waste
- To protect underlying layers in the cover system from excessive wetting/drying (which could cause cracking)
- To protect underlying layers in the cover system from freezing (which could also cause cracking).

Soil is the material that is most commonly used for the protection layer. Many designers lump the surface and protection layers together and call the combination the "cover soil." Soil can be very effective in storing water for later evapotranspiration and provides protection against freezing of underlying materials if the thickness of the protection layer is sufficiently large. Maps of maximum frost penetration depths are given in various publications, e.g., EPA (1989).

Cobbles have occasionally been considered as a barrier to plant roots and burrowing animals but are not commonly used for solid waste landfills.

3.3 Drainage Layer

A drainage layer is sometimes placed below the protective layer and above the barrier layer (Fig. 1). There are three reasons why a drainage layer might be desirable:

1. To reduce the head of water on the barrier layer, which minimizes infiltration
2. To drain the overlying protection layer, which increases the water-storage capacity of the protection layer
3. To reduce pore water pressures in the cover soils, which improves slope stability.

The materials used for drainage layers are:

- Sand or gravel with either a soil filter or a geotextile filter
- A thick geotextile, which serves as both a drain and a filter
- A geonet with a geotextile filter/separator
- A geocomposite drain, which consists of a polymeric drainage core and an overlapping geotextile filter/separator.

Selection of the material type is usually based on economics; all the materials listed above will function adequately if properly designed. If air space has an economic value (e.g., for a new waste containment unit), then thin, geosynthetic materials rather than a thicker layer of sand or gravel will usually prove to be the most economical material.

The designer is cautioned to be careful about two details. First, water must discharge freely from the drainage layer at the base of the cover, e.g., Fig. 2. If the outlet plugs or is of inadequate capacity, the toe of the slope will become saturated, develop excess pore water pressure, and potentially become unstable. Drainage pipes that commonly surround a site at its low elevations must have adequate capacity and cannot be allowed to plug. Conversely, the designer must beware of the possibility that the drainage layer may drain overlying soils so well that vegetation will not grow during parts of the year without irrigation. This problem is minimized by selecting surface and protective soils with adequate water-holding capacity to maintain vegetative growth.

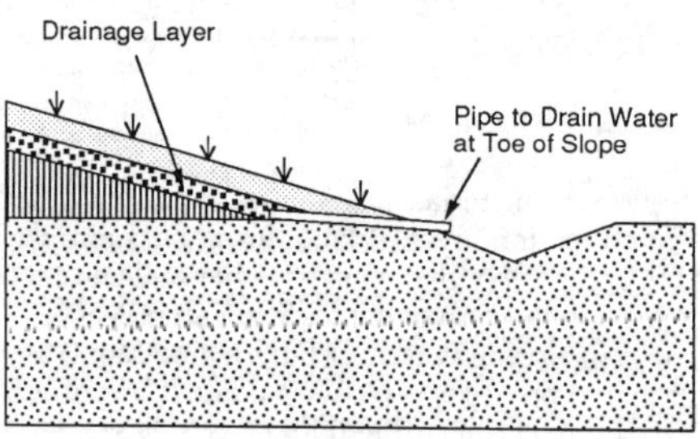

Figure 2. Toe Drain.

3.4 Barrier Layer

The barrier layer is often viewed as the most critical component of an engineered final cover system. The barrier layer minimizes percolation of water through the cover system directly by impeding infiltration through it and indirectly by promoting storage or drainage of water in the overlying layers, where water is eventually removed by runoff, evapotranspiration, or internal drainage.

Historically, a compacted clay liner with a thickness of approximately 300 to 600 mm and a hydraulic conductivity that is less than or equal to 1 x 10^{-9} m/s has been the requirement in the U.S. Typically, the compacted clay liner has been used without a geomembrane liner and has been covered with a relatively thin (150 - 600 mm) layer of cover soil.

It is unfortunate that compacted clay has been so widely and indiscriminately used in cover systems because there are several problems which make the long-term performance of a compacted clay liner questionable. The problems with clay liners used in cover systems include:

- Clay liners are difficult to compact properly on a soft foundation (i.e., waste)
- Compacted clay will tend to desiccate from above and/or below and crack unless protected adequately
- Compacted clay is vulnerable to damage from freezing and must be protected from freezing by a suitably thick layer of cover soil
- Differential settlement of underlying compressible waste will crack compacted clay if tensile strains in the clay become excessive
- Compacted clay liners are difficult to repair if they are damaged.

Subsequent sections will consider the barrier layer in much greater detail and explore the issue of compacted clay barriers used in cover systems in much greater detail.

3.5 Gas Collection/Foundation Layer

The purpose of a gas collection/foundation layer is to collect gas for processing or discharge (gas collection layer) and/or serve as a foundation for the overlying barrier layer. A foundation layer would almost always be constructed of soil or, if suitable materials are available, select waste material. Gas flows from the gas collection layer to *vents*. Flow of gas may occur passively under the pressure gradient generated naturally or actively through assistance from a vacuum pump at the surface. A gas collection layer is only necessary for capping wastes that can produce gas or volatiles that must be released in a controlled manner.

The gas collection layer must have a high in-plane gas flow rate and must not become plugged with fine-grained materials. The usual materials are:

- Sand or gravel with a soil filter
- A thick geotextile which serves as both drain and filter
- Geonet drain and a geotextile filter.

Any of these materials will function well if properly designed.

4 PROBLEMS WITH COMPACTED CLAY IN ENGINEERED FINAL COVER SYSTEMS

4.1 Field Test Plots

Some designers have assumed that a compacted clay liner can be protected from desiccation with a thin layer of cover soil. Montgomery and Parsons (1989) describe field experiments in which three test plots were constructed and monitored for 3 years. A 1.2-m-thick layer of compacted clay was covered with either 150 mm or 450 mm of topsoil at a temperate site in the northern U.S. After 3 years, excavations were made into the compacted clay. The condition of the two plots with 150 or 450 mm of top soil was about the same. The upper 200 to 250 mm of clay was weathered and blocky, cracks up to 12 mm wide extended up to 1 m into the clay, roots penetrated up to 250 mm into the clay in a continuous mat, and some roots extended up to 750 mm into the clay. Clearly, neither 150 mm nor 450 mm of topsoil was enough to protect the clay adequately.

Corser and Cranston (1991) describe test plots in which a layer of compacted clay was covered with either (1) a 600 mm layer of topsoil, (2) an unprotected geomembrane, or (3) a geomembrane overlain by 450 mm of topsoil. The site was located in a relatively arid part of California. In less than a year, significant drying and cracking occurred in the plots with soil cover alone and the geomembrane cover alone, but no significant desiccation occurred in the clay covered with both a geomembrane and soil. However, the tests were short-term tests, and long-term drying may have eventually occurred even with the geomembrane and soil overburden. Nevertheless, the studies of Montgomery and Parsons (1989) and Corser and Cranston (1991), taken collectively, illustrate that the best and perhaps only practical way to protect a compacted clay liner from desiccation from the surface is to cover the clay liner with both a geomembrane and a layer of cover soil. To provide less protection is inappropriate if the designer's intention is for the compacted clay liner to remain intact and not crack. A geomembrane may not be necessary if an extremely thick layer of cover soil is used, but the layer would have to be so thick that for most projects soil alone would be impractical. Extremely thick layers of protective soil are only practical for unusual wastes, e.g., radioactive wastes.

4.2 Desiccation

Although designers have sometimes considered desiccation from above in the design of clay liners in cover systems, it appears that few designers have given careful thought to the potential for the clay liner to dry from below. Compacted clay liners (CCL's) are usually placed at water contents wet of optimum and with soil water potentials of 0 to -1 bar. The equilibrium relative humidity of the air phase in soils with water potentials > -1 bar is in excess of 99.5%. If the relative humidity of the gas phase in the underlying material (e.g., soil, final daily cover, gas collection layer, or the waste itself) is less than 95.%, the clay liner will eventually dry and, depending on the amount of drying, possibly crack. Even landfill gas may possess the capacity to dry a CCL, and dry soils or waste materials certainly possess that capability. The issue of desiccation from below has not been studied in landfills.

4.3 Differential Settlement

LaGatta has summarized available data on the tensile strains that clays undergo when they begin to fail in tension (i.e., crack), and these data are shown in Table 1. The published data indicate that most compacted soils cannot withstand tensile strains greater than 0.1 to 1.0 % without cracking. The question, then, is how likely it is that tensile strains greater than 0.1 to 1.0% will develop in final covers over landfills.

It is convenient to define *distortion* as the differential settlement Δ that occurs over a distance L (see Fig. 3). Distortion in a cover stretches the barrier layer and, as a result, produces tensile strains in the cover components. Murphy and Gilbert (1985) computed the relationship between

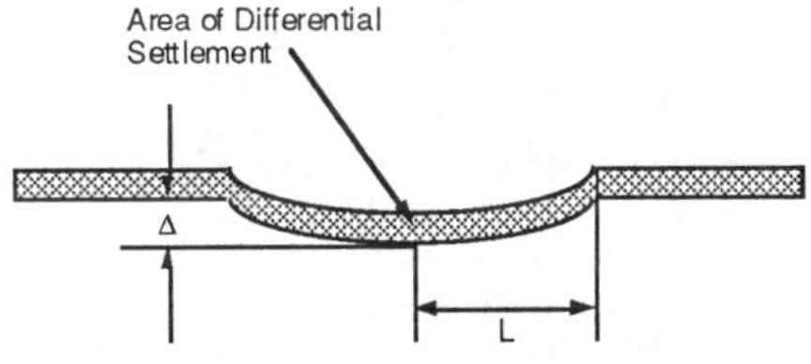

Figure 3. Definition of Distortion, Δ/L.

Table 1. Published Data on the Tensile Strain at Failure for Clay Soils (from LaGatta, 1993).

Reference	Type or Source of Soil	Water Content (%)	Plasticity Index (%)	Failure Tensile Strain
Tschebotarioff and DePhilippe (1953)	Natural Clayey Soil	19.9	7	0.80%
	Bentonite (Montmorillonite)	101	487	3.4%
	Illite	31.5	34	0.84%
	Kaolinite	37.6	38	0.16%
Leonards and Narain (1963)	Portland Dam Portland, CO	16.3	8	0.14%
	Rector Creek Dam Napa, CA	19.8	16	0.16%
	Woodcrest Dam Riverside, CA	10.2	Non-plastic	0.18%
	Shell Oil Dam Ventura, CA	11.2	Non-plastic	0.07%
	Willard Test Dam Embankment, UT	16.4	11	0.20%

distortion and tensile strain for cover components. The maximum distortion Δ/L associated with a tensile strain of 0.1% to 1% is 0.05 to 0.1.

Jessberger and Stone (1991) performed centrifuge tests on 35- to 45-mm-thick prototype clay liners that would be used for cover systems. A "trap door" beneath the liner was deformed to produce settlement like that which would be expected in a landfill. Flow rates through the soil were measured as a function of the distortion in the liner. It was found that the flow rate through the clay remained low until the distortion reached $\Delta/L = 0.1$, at which point the flow rate increased dramatically. Tension cracks were noted in later examinations of the material. Thus, the experiments reported by Jessberger and Stone support the findings from Murphy and Gilbert (1985), i.e., CCL's in cover systems cannot withstand Δ/L distortions greater than approximately 0.05 to 0.1 without cracking.

What does this level of distortion mean in practical terms? Suppose that one observes a circular crater in a landfill cover with a diameter of 5 m. What is the maximum settlement at the center of the crater before significant cracking would be expected in the liner? The horizontal distance from the edge to center of the crater, L, is 2.5 m, and 0.05 to 0.1 times L yields a maximum allowable settlement (Δ) of 0.125 to 0.25 m. It is the authors' experience that many, if not all, covers for municipal solid waste landfills have areas with distortions of this magnitude or larger. In such cases, the author would argue that it is pointless to waste time and money compacting an extremely impermeable layer of compacted clay if the layer is destined to suffer tension cracks and lose its initially low hydraulic conductivity.

In contrast to the sensitivity of CCL's to cracking caused by differential settlement, geomembranes can withstand comparatively large tensile strains, even when stressed three dimensionally. Tensile stains of 5 to 20% are no problem for many modern geomembrane materials. The maximum tensile strain that a geomembrane can withstand is at least an order of magnitude greater than for a compacted clay. Indeed, a geomembrane suffers from few of the problems experienced by compacted clay in a cover system; the geomembrane can withstand large differential settlement and is not vulnerable to damage from desiccation or freeze-thaw. A single geomembrane liner is a lot more sensible liner component for most municipal solid waste cover systems than a single compacted clay liner.

5. GEOSYNTHETIC CLAY LINERS

5.1 Materials

Geosynthetic clay liners (GCLs) are thin "blankets" of bentonite clay attached to one or more geosynthetic materials (e.g., geotextile or geomembrane). Bentonite is a unique clay mineral with very high swelling potential and water absorption capacity.

Geosynthetic clay liners are manufactured by laying down a layer of dry bentonite, approximately 5-mm thick, on a geosynthetic material and attaching the bentonite to the geosynthetic material. Two general configurations are currently employed in commercial processes: bentonite sandwiched between two geotextiles (Fig. 4a) or bentonite glued to a geomembrane (Fig. 4b). The primary purpose of the geosynthetic component or components is to hold the bentonite together in a uniform layer and permit transportation and installation of the material without losing bentonite or altering the thickness of the bentonite. However, the geosynthetic components may serve other important purposes, as well, such as adding tensile or shear strength to the material.

The bentonite component of a manufactured GCL is essentially dry, and there are open voids between bentonite granules in the manufactured material. When the bentonite is hydrated with water, the bentonite swells and the voids between bentonite granules close. The swelling action of bentonite is crucial to attainment of low permeability.

Geosynthetic clay liners contain approximately 4.9 kg/m^2 of bentonite that has a hydraulic conductivity (coefficient of permeability) of approximately 1×10^{-11} m/s or less. Continuous gravity percolation under unit hydraulic gradient through a material with a hydraulic conductivity of 1×10^{-11} m/s would result in an infiltration rate of 0.3 mm per year, or approximately 30 mm every 100 years. For landfill covers, an intact GCL may be considered essentially impermeable to water.

Geosynthetic clay liners were first manufactured in the early 1980's and were initially used for foundation water proofing and for sealing water retention structures. Geosynthetic clay liners were first used for landfill liners in 1986. Since 1986, geosynthetic clay liners have been used for a variety of lining applications and also in several final cover systems for hazardous wastes, radioactive wastes, and non-hazardous solid wastes.

Four geosynthetic clay liners are currently manufactured: Bentofix®, Bentomat®, Claymax®, and Gundseal®. The GCLs fall into the broad categories shown in Fig. 4 as follows:

- Bentonite sandwiched between two geotextiles: Bentofix®, Bentomat®, and Claymax®
- Bentonite mixed with an adhesive and glued to a geomembrane: Gundseal®.

The GCLs are sketched in Fig. 5. Bentofix® and Bentomat® consist of bentonite sandwiched between a woven and non-woven geotextile that are needle-punched together. Claymax® 200R consists of bentonite mixed with glue and sandwiched between two woven geotextiles. Claymax® 500SP consists of bentonite mixed with glue and sandwiched between two woven geotextiles that are sewn together. The purpose of stitching the two geotextiles together is to provide additional internal reinforcement and greater shear strength. With all the geotextile-encased GCLs, special geotextiles can be selected to "custom design" the GCL to a particular application. Gundseal® is made by mixing bentonite with an adhesive and attaching the bentonite layer to a polyethylene geomembrane. Gundseal® can be supplied with high density polyethylene (HDPE) or very low density polyethylene (VLDPE), and the geomembrane can be either smooth or textured.

All GCLs are manufactured in panels with widths of approximately 4 to 5 m and lengths of

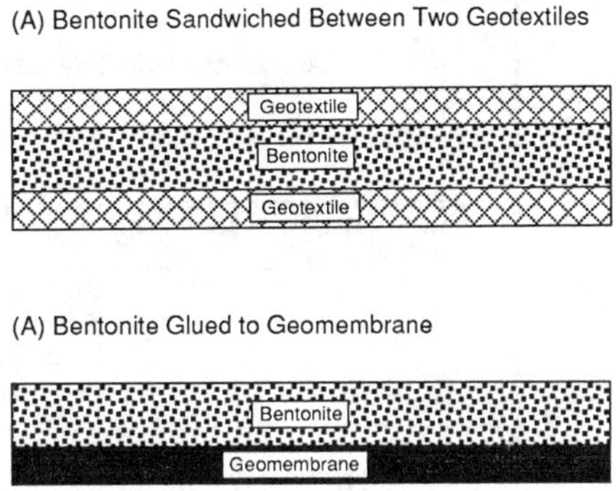

Figure 4. Four Types of GGLs.

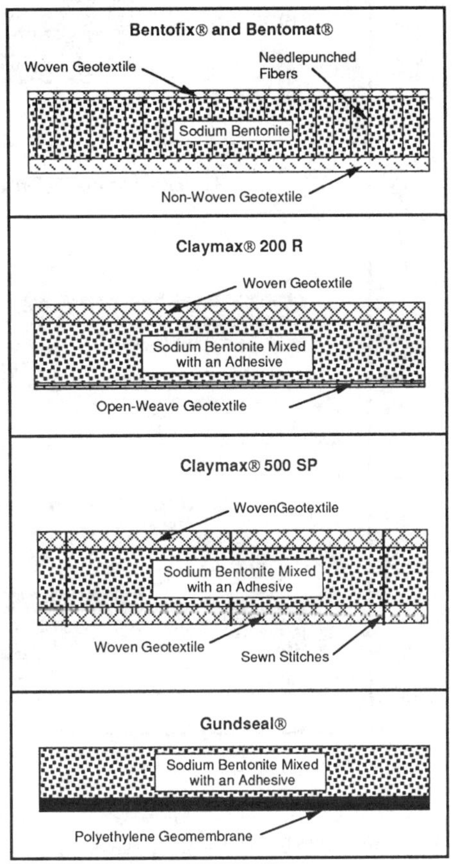

Figure 5. Commercially-Produced GCLs.

approximately 25 to 60 m. The panels are placed on rolls at the factory and are unrolled at the time of installation. The mass of the roll varies, depending on size and materials, from about 600 to 2,000 kg.

The panels are typically overlapped 75 to 300

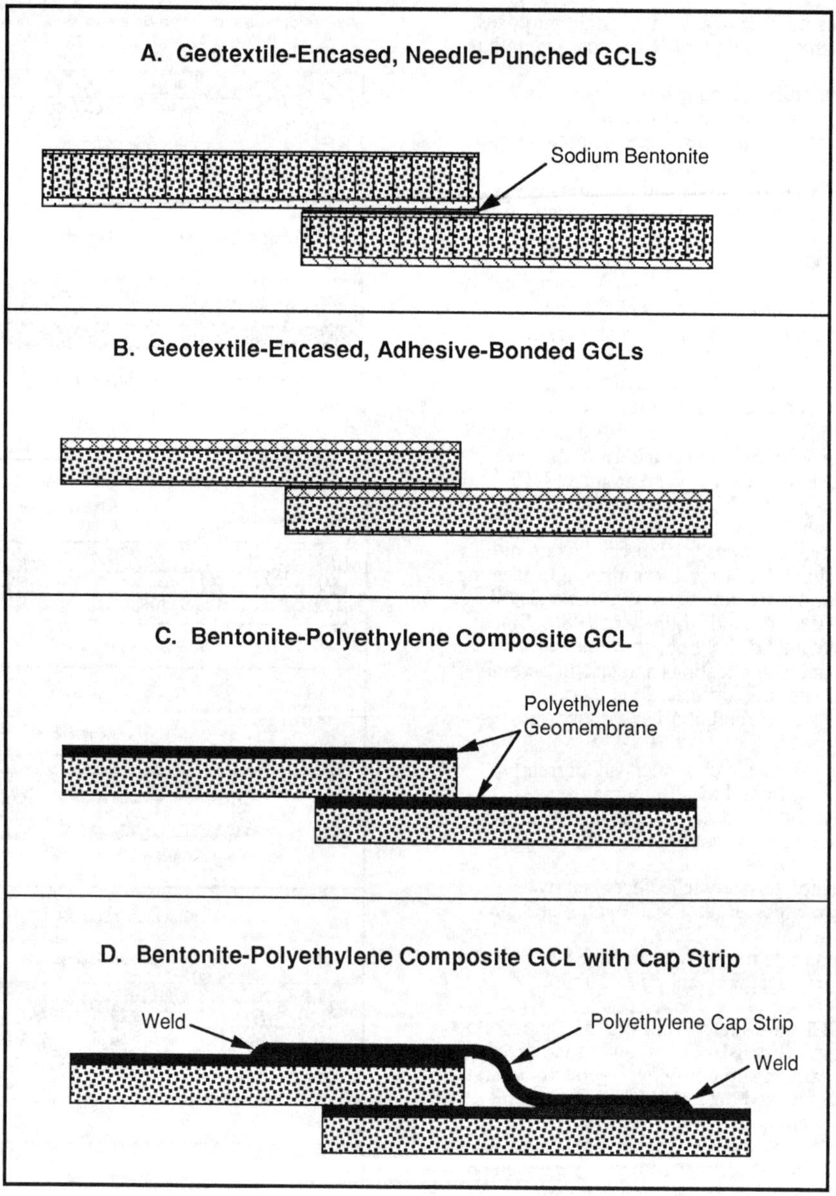

Figure 6. Overlapped Sections.

mm and are said to be "self sealing" at the overlaps. A sketch of the overlapped zones is shown in Fig. 6. With geotextile-encased, needle-punched GCLs, sodium bentonite is placed along the overlap (Fig. 6a) at a rate of approximately 0.8 kg/m. The bentonite penetrates the pores of the geotextiles and is said by the manufacturers to cause the materials to self seam when the bentonite hydrates. With geotextile-encased, adhesive-bonded GCLs, no additional bentonite is needed (Fig. 6b). The material is said to self seal upon hydration at the overlaps through expansion and "oozing" of bentonite out through the openings of the geotextile in the overlap area.

With GCLs containing a geomembrane, the GCL can be placed with the bentonite facing upward or, as shown in Fig. 6c and 6d, downward. If the GCL will be used by itself as a composite geomembrane-clay liner, the geomembrane would face upward. If a separate geomembrane is to be placed on the GCL, the bentonite would face upward. The material is said to be self sealing at overlaps with no need for any mechanical seam at the overlap (Fig. 6c). However, if one wants to form a continuous geomembrane out of the geomembrane component of the GCL, a cap strip can be welded over the overlap (Fig. 6d).

5.2 Uses in Engineered Final Covers

Geosynthetic clay liners can be used in final cover systems in several ways, as shown in Fig. 7. One choice (Fig. 7a) is to use the GCL by itself as a barrier to water infiltration. The GCL would be buried below a layer of protective soil. As indicated earlier, the bentonite component is expected to be essentially impermeable to water after it is has been hydrated, assuming that the GCL withstands the potentially damaging effects of wet-dry cycles and differential settlement (discussed later). One possible problem with using a GCL by itself as a barrier layer is that the dry bentonite is initially highly permeable to landfill gas -- the bentonite would have to absorb water, hydrate, and swell before the bentonite becomes an effective barrier to gas migration, and the bentonite could not be allowed to dry out because the bentonite would again become permeable to landfill gas. At extremely arid sites, there may not be adequate water available to hydrate the bentonite to the extent that is necessary in order for the GCL to have a low permeability to gas. However, for those GCLs that contain a geomembrane, the geomembrane itself provides a barrier to gas migration. In addition, a barrier to gas migration within the final cover may or may not be a design consideration, depending on site-specific considerations.

The second potential use of a geosynthetic clay liner in a final cover system is in conjunction with a geomembrane (Fig. 7b) to form a composite geomembrane/GCL liner. The composite could either be formed by using a GCL that contains a geomembrane or by separately constructing a geomembrane on top of a GCL. By placing clay under the geomembrane, the clay serves to seal off any imperfections in the geomembrane, e.g., pinholes or defects in seams, and to help in providing an extremely effective composite barrier to infiltration of water. The geomembrane would protect the underlying GCL from wet-dry cycles and would serve as a gas barrier for those periods when the bentonite component of the GCL is relatively dry. The main advantages of a separately-constructed geomembrane are that a separate polyethylene geomembrane liner could be seamed with the most advanced welding equipment available, which is microprocessor-controlled, dual-track, hot wedge welding equipment, or that some other type of geomembrane besides polyethylene could be used, if desired. If a bentonite-polyethylene composite GCL is used and the polyethylene components are to be seamed at overlaps, a cap strip is typically placed over the overlapped region and the edges of the cap strip are welded with fillet extrusion welding apparatus (Fig. 6d). However, because water flow through non-welded seams is expected to be negligible, the author encourages designers not to use cap strips over overlapped panels unless there is a good reason to do so.

A third option is to sandwich the GCL between two geomembranes (Fig. 7c). One or both geomembranes would be separately installed, depending upon the GCL material employed. The advantage of this design is that even less percolation of water through the barrier would occur. In fact, the bentonite component would become wetted only around minor imperfections in a geomembrane or its seams, where the bentonite would serve to seal off the leakage through the imperfection. This type of design approach, with a triple-composite liner, has rarely (if ever) been used for final covers over solid waste landfills and would be considered an extreme design for those facilities requiring extraordinary protection from water percolation or gas migration through the final cover.

A fourth option is to place the GCL on top of a low-permeability, compacted soil liner (Fig. 7d), possibly with a geomembrane placed on top of the GCL (Fig. 7d). This design adds redundancy of materials and enables one to provide a very high degree of protection in the final cover system. In such cases, the GCL may replace part of a conventional compacted clay liner, or the low-permeability soil component may have a hydraulic conductivity that is greater than the usual 1×10^{-9} m/s (i.e., use of the GCL lessens the need for extremely low permeability in the underlying soil barrier layer).

A fifth option is to place the GCL on top of a low-permeability, re-used waste material (Fig. 7e), possibly with a geomembrane placed on top of the GCL (Fig. 7e). This design adds redundancy of materials and enables one to make productive use of waste materials. An example of a waste material that might be considered is paper industry sludges (Maltby and Eppstein, 1993).

5.3 Engineering Properties

Hydraulic Conductivity. In general, the hydraulic conductivity of the bentonite component of GCLs varies between about 1×10^{-12} and 1×10^{-10} m/s, depending on the confining stress. The higher the compressive stress, the lower the hydraulic conductivity. There are some differences between the hydraulic conductivities of the various GCLs, but, except for bentonite-geomembrane composite GCLs (for which the geomembrane will significantly reduce the overall hydraulic conductivity), the differences do not appear to be very large. The available data are summarized by Schubert (1987), Daniel and Estornell (1990),

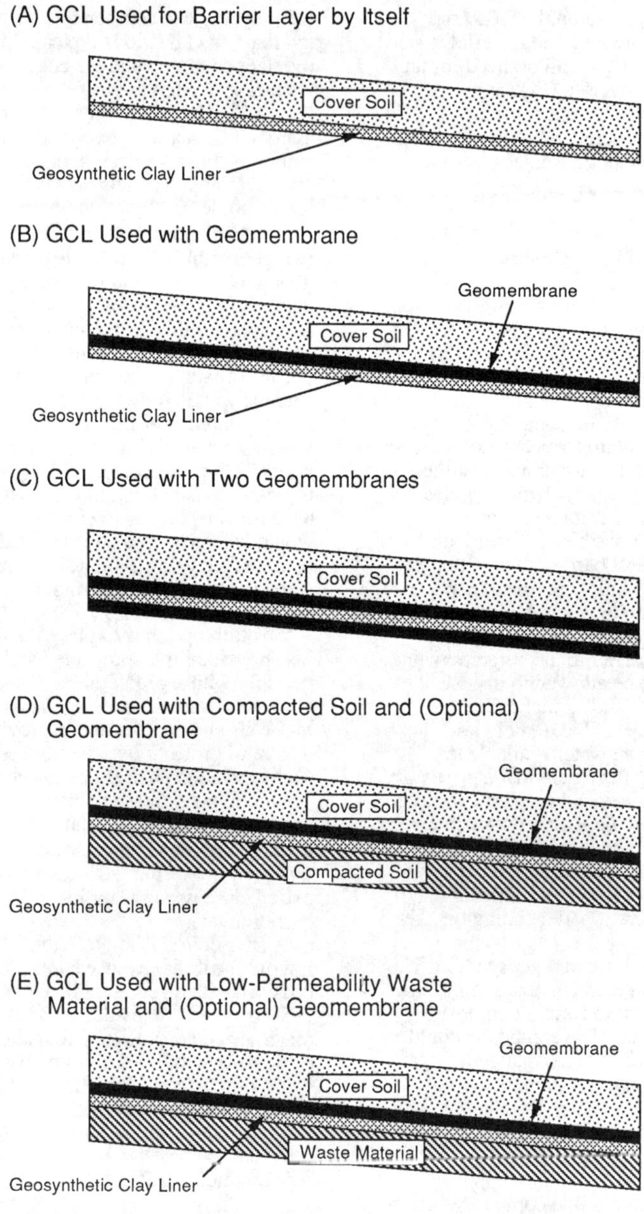

Figure 7. Possible Uses of Geosynthetic Clay Liners for Barrier Layer in Final Cover System.

Scheu et al. (1990), Daniel (1991), Eith et al. (1991), Shan and Daniel (1991), Estornell and Daniel (1992), Grube (1992), Daniel et al. (1993), and Daniel and Boardman (1993).

For a final cover system, a confining stress on the order of 10 to 30 kPa is a reasonable range. Laboratory hydraulic conductivity tests performed on backpressure-saturated test specimens in flexible-wall permeameters indicate that the hydraulic conductivity of the bentonite component of GCLs in this range of compressive stress is approximately 1 to 4 x 10^{-11} m/s. Estornell and Daniel (1992) measured the hydraulic conductivity of GCLs in large tanks. The tests were specifically set up to simulate conditions of low overburden stress that are typical of final cover systems and to test very large specimens with overlaps. Of the 10 tests for which hydraulic conductivities were measured, the average value was 4.6 x 10^{-11} m/s (normal averaging) or 2.2 x 10^{-11} m/s (logarithmic averaging). Based on all the data, a reasonable assumption is that a GCL can be supplied with a hydraulic conductivity for a landfill cover application less than 1 to 5 x 10^{-11} m/s.

Studies of the hydraulic properties of overlapped seams performed by Estornell and Daniel (1992) indicate that the overlapped seams in GCLs self seam in the manner described by the manufacturers. For geotextile-encased, needle-punched GCLs with additional bentonite along the overlap, the bentonite appears to swell upon hydration and plug voids in the geotextiles present in the overlap. For the geotextile-encased, adhesive-bonded GCLs that have been tested, the bentonite within the GCL appears to ooze out through the openings in the geotextile and to allow the material to self seal. For bentonite-geomembrane composite GCLs, the bentonite swells upon hydration, seals at the bentonite-polyethylene interface, and effects self-seaming at the overlap. Thus, based on the available data, it is reasonable to assume that with proper quality control in the field, seams can be installed that will self-seal.

Strength. The internal shear strength of GCLs has been determined by the manufacturers and various organizations and testing laboratories. "Internal shear strength" refers to the strength of the material when sheared through the mid-plane of the bentonite. The author and his students at the University of Texas have performed a number of direct shear and tilt table tests (Shan and Daniel, 1991; and Daniel et al., 1993).

Dry bentonite is much stronger than water-saturated bentonite. For dry GCLs or slightly damp GCLs, the angle of internal friction (even for the materials that are not internally reinforced) is approximately 35°. For unreinforced GCLs, the angle of internal friction drops to about 10° for fully saturated bentonite.

For those GCLs that are needle-punched or sewn together, the internal reinforcement of the GCL makes the material's internal shear strength much less sensitive to the strength of the bentonite contained between the attached geotextiles. However, the reader is cautioned that for landfill covers, the GCL may be exposed to prolonged shearing stresses for periods of years, decades, or even centuries, and that the long-term shearing resistance should be carefully considered.

"Interfacial shear strength" refers to the shearing strength between two adjacent components of a liner or cover system. The GCL may be placed against soil, a geomembrane, or a geotextile. Because the range of possible materials at an interface is unlimited, the actual interfacial shearing properties are usually determined on a project-specific basis. It is the author's experience that the internal shear strength will often govern the design because, with proper selection of materials, relatively high interfacial strengths can usually be obtained.

Durability. Shan and Daniel (1991) studied the effects of punctures on a geotextile-encased, adhesive-bonded GCL. The manufacturers of other GCL products have developed similar data for their particular products. The effects of punctures on the hydraulic conductivity of the GCL were studied by drilling or cutting circular holes into the dry GCL, setting the punctured GCL up in flexible-wall permeameters, and permeating the GCL slowly until steady flow was achieved. Results are summarized in the following table:

Diameter of Puncture	Hydraulic Conductivity (m/s)
No Punctures	2 x 10^{-11}
12 mm	3 x 10^{-11}
25 mm	5 x 10^{-11}
75 mm	> 1 x 10^{-6}

Small (≤ 25-mm-diameter) punctures made in the dry material self-sealed upon hydration of the bentonite. These tests illustrate the self-healing capability of bentonite. Each particular GCL has a different capacity to self-heal punctures. However, all GCLs are capable of self healing small punctures in the dry GCL when the bentonite is hydrated. It should be emphasized that these tests were performed under carefully controlled conditions in which no material other than bentonite was allowed to fill the puncture. In the field, other materials may fill large punctures. Although GCLs have some capability to self-seal if punctured, there are clearly limitations in the size of puncture that could self seal in the field.

Concern has been expressed that the bentonite component of a GCL may swell when hydrated but may later dry out, shrink, crack, and lose its impermeability. Shan and Daniel (1991) investigated the healing capability of one geotextile-encased, adhesive-bonded GCL that was subject to wet-dry cycles. Samples of the GCL were permeated in a flexible-wall permeameter, removed from the permeameter, and allowed to air dry with a small vertical stress applied to the specimens. All specimens exhibited severe cracking upon drying. The specimens were then set back up in a flexible-wall permeameter, slowly rehydrated, and then repermeated. There was no change in hydraulic conductivity from the initial value of 2 x 10^{-11} m/s, even after three wet/dry cycles. These tests reinforce the fully reversible shrink/swell nature of bentonite and suggest that any desiccation cracks will self-heal when the bentonite is hydrated.

In research recently completed at the University of Texas (Boardman, 1993), large samples of GCLs (with and without overlaps) were buried under 600 mm of gravel and subjected to a wet-dry cycle that simulates severe conditions that might occur in a final cover for a landfill. The GCLs were set up in the tanks, hydrated with water until a steady hydraulic conductivity was measured, and then severely desiccated by draining away the water on top of the GCL and circulating heated air into the gravel that was placed over the GCL. The heated air caused severe desiccation cracking in the GCLs. However, when the GCLs were rehydrated, the bentonite quickly swelled and the hydraulic conductivity eventually returned to the original, extremely low value. Thus, it appears from the available data that GCLs have an excellent capacity to self seal from desiccation-induced cracking. Geosynthetic clay liners probably possess much greater ability to self seal than conventional compacted clay liners.

Compacted clay liners are known to be vulnerable to damage from freezing (Kim and Daniel, 1992). When water in soil freezes, the water expands, and when the water thaws, the water contracts. This expansion and contraction causes small cracks to appear in the soil and causes other alterations in the soil structure that tend to increase hydraulic conductivity.

Shan and Daniel (1991) subjected a geotextile-encased, adhesive-bonded GCL to freeze/thaw. A test specimen was set up in a flexible-wall permeameter, hydrated with water, and permeated until a steady hydraulic conductivity was obtained. Then the specimen was removed from the flexible-wall permeameter and subjected to five freeze/thaw cycles at constant water content. The specimen was repermeated, and it was found that the hydraulic conductivity did not change. Similar results have been obtained by commercial testing laboratories for other GCL products. Available data indicate that the high shrink-swell capability of bentonite gives bentonite the ability to self-heal if any alteration occurs from freeze/thaw cycles. Geosynthetic clay liners appear to have a much better capacity to remain undamaged after freeze-thaw than conventional compacted clay liners.

5.4 Comparison of GCLs to Compacted Clay Liners

Many regulatory agencies have traditionally required a low-permeability, compacted clay liner (or the equivalent) as the primary hydraulic barrier within landfill covers. The thickness of a compacted clay liner typically ranges from 300 to 600 mm and the maximum allowable hydraulic conductivity is typically 10^{-9} m/s. If one wishes to substitute a GCL for a compacted clay liner, one must usually demonstrate that the GCL will be equivalent in terms of meeting performance objectives.

One should not really think of a geosynthetic clay liner as being equivalent to a compacted clay liner. Indeed, it is not possible that a thin layer of bentonite could ever be equivalent to a much thicker layer of compacted clay in all respects. The critical issue is whether substitution of an alternative material such as a GCL for the more traditional compacted clay liner in a landfill cover will meet or exceed the performance objectives of the compacted clay liner. If the GCL will meet or exceed the performance objectives, then it should be considered that equivalency has been established.

Some of the differences between compacted clay liners and geosynthetic clay liners are listed in Table 2. On many projects, there are numerous practical advantages that might arise from use of a GCL rather than a compacted clay liner.

5.5 Equivalency of GCLs in Meeting Performance Criteria.

If a GCL is to be used rather than a compacted clay liner, the GCL must meet or exceed the stated or implied performance objectives for the compacted clay liner. Equivalency can be assess in three areas:

1. Hydraulic Issues
2. Physical/Mechanical Issues
3. Construction Issues (Including Quality Assurance Issues).

<u>Hydraulic Issues</u>. Hydraulic issues are the easiest to quantify. The suggested criteria, which are discussed separately, include steady water flux, time to initiate release of water from the base of the liner ("breakout time"), production of consolidation water, and air permeability.

<u>1. Steady Flux of Water</u>

Water flux is defined as the quantity of flow across a unit area in a unit time. For a barrier in a final cover system, water flux is equal to the rate of percolation of water through the barrier layer.

Water flux is usually analyzed based on the long-term, steady state water flux. The flux of water (v) through an individual layer of porous material is defined from Darcy's law as:

$$v = k \frac{H + T}{T} \qquad (1)$$

where k is the hydraulic conductivity, H is the depth of liquid ponded on the liner, and T is the thickness of the liner. The water pressure on the base of the liner is assumed to be atmospheric pressure in Eq. 1.

Equation 1 is applicable only for flow through the bentonite component of a GCL; if the GCL contains a geomembrane, water flux will be controlled by water vapor diffusion through the geomembrane component. The geomembrane component, if present, should be considered in the equivalency analysis and in computation of water flux. The simplest way to do this is to adjust the hydraulic conductivity of the GCL to reflect the presence of a geomembrane. Also, Eq. 1 applies to a CCL or GCL liner alone and not to composite liners involving one or more separate geomembrane components. Composite action with

Table 2. Differences Between GCLs and Compacted Clay Liners.

Characteristic	Geosynthetic Clay Liner	Compacted Clay Liner
Materials	Bentonite, Adhesives, Geotextiles, and Geomembranes	Native Soils or Blend of Soil and Bentonite
Thickness	Approximately 12 mm; Consumes Very Little Landfill Volume	Typically 300 to 600 mm; Consumes More Landfill Volume
Hydraulic Conductivity	≤ 1 to 5×10^{-11} m/s	$\leq 1 \times 10^{-9}$ m/s
Speed and Ease of Construction	Rapid, Simple Installation	Slow, Complicated Construction
Ease of Quality Assurance (QA)	Relatively Simple, Straight-Forward, Common-Sense Procedures	Complex QA Procedures Requiring Highly Skilled and Knowledgeable People
Vulnerability to Damage During Construction from Desiccation and Freeze-Thaw	GCLs Are Essentially Dry; GCLs Cannot Desiccate during Construction; Not Particularly Vulnerable to Damage from Freeze-Thaw	Compacted Clay Liners Are Nearly Saturated; Can Desiccate during Construction; Vulnerable to Damage from Freeze-Thaw
Vulnerability to Damage from Puncture	Thin GCL Is Vulnerable to Puncture	Thick Compacted Clay Liner Cannot Be Punctured Accidentally
Vulnerability to Damage from Differential Settlement	Can Withstand Much Greater Differential Settlement than Compacted Clay Liner	Cannot Withstand Much Differential Settlement Without Cracking
Availability of Materials	Materials Easily Shipped to Any Site	Suitable Materials Not Available at All Sites
Cost	Reasonably Low, Highly Predictable Cost that Does Not Vary Much from Project to Project	Highly Variable -- Depends Greatly on Characteristics of Locally Available Soils
Ease of Repair	Easy to Repair with Patch Placed Over Problem Area	Very Difficult to Repair; Must Mobilize Heavy Earth-Moving Equipment If Large Area Requires Repair
Experience	Limited Due to Newness	Has Been Used for Many Years
Regulatory Approval	Not Explicitly Allowed In Most Regulations -- Owner Must Gain Approval on the Basis of Equivalenc in Meeting Performance Objectives	Compacted Clay Liners Are Usually Required by Regulatory Agencies

a geomembrane is considered later.

The flux ratio for water, F_w, is defined as the flux through the GCL divided by the flux through the compacted clay liner (CCL):

$$F_w = v_{GCL} / v_{CCL} \tag{2}$$

or:

$$F_w = \frac{k_{GCL}}{k_{CCL}} \frac{T_{CCL}}{T_{GCL}} \frac{H + T_{GCL}}{H + T_{CCL}} \tag{3}$$

If the flux ratio is ≤ 1, then the GCL is equivalent to the CCL in terms of steady water flux. For example, for a situation with H = 1 ft (0.3 m) and a GCL with $k_{GCL} = 1 \times 10^{-11}$ m/s, $T_{GCL} = 7$ mm, and a compacted clay liner (CCL) with $k_{CCL} = 1 \times 10^{-9}$ m/s and $T_{CCL} = 0.6$ m, then F_w from Eq. 3 equals 0.3, which means that there would be less water percolation through the GCL than a compacted clay liner -- equivalency is established for these conditions.

Alternatively, one can assume that water flux through the GCL is equal to the water flux through a CCL (i.e., $F_w = 1$):

$$v_{GCL} = v_{CCL} \tag{4}$$

and compute the required hydraulic conductivity of the GCL by substitution in Eq. 4:

$$k_{GCL} \frac{H + T_{GCL}}{T_{GCL}} = k_{CCL} \frac{H + T_{CCL}}{T_{CCL}} \tag{5}$$

to obtain:

$$(k_{GCL})_{Required} = k_{CCL} \frac{T_{GCL}}{T_{CCL}} \frac{H + T_{CCL}}{H + T_{GCL}} \quad (6)$$

Equation 6 may be used to determined the hydraulic conductivity of the GCL necessary to establish equivalency. So long as the job specifications require that the actual hydraulic conductivity be less than the value computed from Eq. 6, equivalency in terms of steady water flux is theoretically guaranteed. The required hydraulic conductivity of the compacted clay liner (k_{CCL}) is almost universally established as 1 x 10^{-9} m/s by regulatory agencies in the U.S. The thickness of GCLs (T_{GCL}) varies from product to product, but is typically about 7 mm after hydration at low overburden stress. The head of liquid on the barrier layer is expected to be low in a final cover system; evapotranspiration and the nature of rainfall events makes the buildup of head on the barrier layer much less likely in final covers than in landfill liners. For illustrative purposes, three values of head of water (H) on the CCL or GCL are assumed: 0, 25 mm, and 300 mm. The required hydraulic conductivity of the GCL for equivalent performance to a compacted clay liner in terms of steady flux of water through the liner is computed as follows:

For a 300-mm-thick compacted clay liner:

- $(k_{GCL})_{Required}$ = 1 x 10^{-9} m/s for a negligibly small head of water on the liner
- $(k_{GCL})_{Required}$ = 2 x 10^{-10} m/s for a water head of 25 mm on the liner
- $(k_{GCL})_{Required}$ = 4 x 10^{-11} m/s for a water head of 300 mm on the liner

For a 600-mm-thick compacted clay liner:

- $(k_{GCL})_{Required}$ = 1 x 10^{-9} m/s for a negligibly small head of water on the liner
- $(k_{GCL})_{Required}$ = 2 x 10^{-10} m/s for a water head of 25 mm on the liner
- $(k_{GCL})_{Required}$ = 3 x 10^{-11} m/s for a water head of 300 mm on the liner

As discussed earlier, the hydraulic conductivity of the bentonite component of commercially-produced GCLs is typically $\leq$ 1 to 5 x 10^{-11} m/s. Thus, it is clear that equivalency of a GCL to a CCL, in terms of the amount of water that passes through a GCL under conditions of steady seepage, can be established for most, if not all, landfill covers.

The idea of combining GCLs with native soils (e.g., Fig. 7d) is appealing because of the redundancy that the combination provides and the fact that a relatively low-permeability, native soil material is backing up the GCL.

A composite liner consists of a geomembrane placed in contact with a low-permeability soil. A geomembrane/GCL composite may be considered as an alternate to a geomembrane/CCL composite. If so, flow through the composite should be analyzed. The rate of flow through a flaw in a geomembrane in a composite liner depends on the size of the flaw, the hydraulic conductivity of the underlying clay component, the hydraulic gradient across the clay component, the hydraulic contact between the geomembrane and the clay component, and the presence of a geomembrane within the GCL. No equations have been published for explicit purpose of computing flow rates through a defect in a geomembrane component of a geomembrane/GCL composite liner. The presence of a geotextile between the geomembrane and bentonite may influence overall performance. This is a topic of current research. However, it is likely that equivalency can be demonstrated with reasonable assurance for some or all GCLs that are used with geomembranes to form composite liners.

2. Time to Initiate Discharge of Water from Base of Liner ("Breakout Time")

Geosynthetic clay liners and compacted clay liners are initially unsaturated with water. Geosynthetic clay liners contain essentially dry bentonite, but compacted clay liners are often very close to saturation at the time of construction. When liquid first enters the upper surface of an unsaturated liner, no liquid discharges from the base of the liner until the liner absorbs enough water to reach field capacity at the base.

The time to discharge water from the base of the liner is difficult to analyze in a simple way. For CCLs, the time depends greatly upon the hydraulic conductivity, initial water content, tendency to swell, and rate of water infiltration into the top of the liner. For GCLs, the time to initiate discharge of water from the base is usually fairly short (a few weeks) if the liner is continuously flooded with water or may be extremely long if water is slowly absorbed by the bentonite. For GCLs that contain a geomembrane, the time may be much greater. A comparison of time to initiate discharge of water from the base of the liner would have to be performed on a site and product specific basis.

In general, it is not believed that breakout time should be an important issue in an equivalency assessment. Other factors seem far more important.

3. Production of Consolidation Water

Application of load to a compacted clay liner tends to squeeze water out of the clay. If this were to occur in a cover, the water might eventually become leachate. Dry GCLs have no capacity to produce consolidation water loading upon loading. In general, the GCL should be viewed as superior to a CCL in terms of minimizing production of consolidation water. However, because the applied loads in final covers are so small, the entire issue of production of consolidation water is usually moot for final covers. This issue is far more important for clay liners located above leak detection layers in bottom liner systems for landfills.

4. Air Permeability

The permeability of a barrier layer to gas may be very important if the barrier layer is expected to restrict the movement of gas through the cover. For porous materials, the air permeability is extremely sensitive to the water content of the soil. Dry materials are highly permeable to air, but water-saturated porous materials are practically impermeable to air.

Compacted clay liners are compacted at a water content that is wet of optimum. Any air present in the CCL tends to be present as isolated bubbles and not in continuous channels. Thus, the air permeability of CCLs tends to be very low. The air permeability of GCLs depends greatly on whether or not a geomembrane is present and how much moisture has been absorbed by the bentonite. The air permeability is high for dry bentonite that is sandwiched between two geotextiles. For GCLs that contain a geomembrane, the geomembrane dominates the material's air permeability and gives it a very low permeability to air. Equivalency in terms of air permeability probably can be demonstrated for GCLs that contain a geomembrane or for GCLs that are sufficiently hydrated to attain a low permeability to air. The bentonite in the GCL can be forced to hydrate quickly either by placing the GCL in contact with a moist soil or by applying water to the overlying soil after the GCL is placed and covered. Laboratory tests indicate that absorption of water by the bentonite occurs within a few weeks (Daniel et al., 1993) -- the hydration of the bentonite can be forced to occur if air permeability is a critical issue.

Physical/Mechanical Issues. The physical/ mechanical issues that might be considered in an equivalency analysis include freeze/thaw effects, wet/dry effects, response to total settlement, response to differential settlement, stability on slopes, vulnerability to erosion, and bearing capacity.

1. Freeze/Thaw Resistance

Compacted clay liners are known to be vulnerable to large increases in hydraulic conductivity from freeze/thaw (e.g., Kim and Daniel, 1992, and the references therein), although compacted soil-bentonite mixtures may not be as vulnerable to damage. As discussed earlier, limited laboratory data indicate that GCLs do not undergo increases in hydraulic conductivity as a result of freeze/thaw. Thus, from the available data, GCLs appear to be superior to CCLs in terms of freeze/thaw resistance.

2. Wet/Dry Effects

Wetting and drying of CCLs and GCLs can cause either type of clay liner to swell or shrink. The main concern with clay liners is that desiccation can lead to cracking and to an increase in hydraulic conductivity.

As discussed earlier, available laboratory data indicate that desiccation of wet GCLs does cause cracking, but rehydration of the GCL causes the bentonite to swell and the material to self heal. Thus, GCLs appear to be superior to CCLs in terms of ability to self-heal if the material is wetted, dried, and then rewetted.

3. Response to Total Settlement

Total settlement refers to block-like settlement without significant bending or distortion. It is believed that GCLs and CCLs would both respond similarly to total settlement and that neither would be damaged if there is no bending or distortion.

4. Response to Differential Settlement

LaGatta (1992) studied the effects of differential settlement on the hydraulic conductivity of GCLs. LaGatta placed a water-filled bladder in a "false bottom" located beneath the GCL. The GCL was placed over the bladder and was then covered with 600 mm of gravel to simulate cover material. The GCL was flooded with 300 mm of water, and water draining out the bottom of the experimental apparatus was collected for 2 to 4 months, until the flow rate became steady. Then the bladder was incrementally deflated to produce differential settlement. Boardman (1993) performed similar tests but subjected dry (rather than hydrated) GCLs to differential settlement; the GCLs were hydrated and permeated after the distortion took place in the dry material. The extreme differential settlement caused by the deflated bladders did not produce large increases in hydraulic conductivity for most of the GCLs tested.

Distortion (defined in Fig. 3) produces tension, which can lead to cracking. It appears from LaGatta's and Boardman's tests that many GCLs can withstand large distortion (Δ/L up to 0.5) and tensile strain (up to 10 to 15%) without undergoing significant increases in hydraulic conductivity. This finding is in sharp contrast to the results for compacted clay, which are summarized in Table 1. Normal compacted clay materials cannot withstand tensile strains greater than approximately 0.1 to 1.0% without failing (cracking). Pure bentonite, on the other hand, is reported to have a tensile strain at failure of 3.4%, but LaGatta measured much greater tensile strains without cracking in many GCLs, probably due to the beneficial reinforcing effects from the geotextiles or geomembrane in the GCLs. In any case, the available data indicate that GCLs can withstand much greater tensile deformation than normal compacted soils without cracking, which is a very favorable characteristic for final covers. Geosynthetic clay liners are considered to be superior to compacted clay liners in terms of resistance to damage from differential settlement.

5. Stability on Slopes

The shear strength of CCLs varies widely, depending on materials, water content, and compaction conditions. In stability analyses, one often must consider not only internal shear failure but interfacial shear with an adjacent layer, e.g., a geomembrane. No general statement can be made about equivalency of a GCL to a CCL in terms of shear strength because the assessment depends on specific materials, the degree to which the bentonite can wet, slope angle, and other site-specific conditions.

6. Vulnerability to Erosion

Erosion resistance may be of concern in final covers if inadequate cover soil is present. With a well-designed and properly maintained cover system, the barrier layer should never be subjected to forces of erosion after the construction phase is over and equivalency should not be an issue. In some cases, however, there may be insufficient cover soil to guarantee that the barrier layer will not be exposed. Because of the presence of erosion-resistant geosynthetic materials in GCLs, most GCLs can potentially be more resistant to erosion than CCLs. However, if the clay liner is exposed to erosive forces, the bentonite may be washed out of some GCL materials. Thus, equivalency depends upon the specific materials being considered. For many sites, erosion will not be of any concern, e.g., for a GCL underlying a geomembrane or a cover with adequate cover soil.

7. Bearing Capacity

A clay liner must have adequate bearing capacity to support loads, e.g., wheel loads from construction or maintenance equipment. The clay liner must not thin or pump clay into adjacent layers under static or dynamic (e.g., traffic) loads.

Hydrated bentonite is not as strong as most materials used in constructing CCLs. However, under most circumstances, both a GCL and a CCL will provide adequate foundation bearing capacity, particularly if the GCL or CCL is buried under sufficient soil overburden. Equivalency is heavily dependent upon site-specific conditions.

Construction Issues. The construction issues that might be considered in an equivalency analysis include puncture resistance, effect of subgrade condition on constructability, ease of placement, speed of construction, availability of materials, requirements for water, air pollution effects, weather constraints, and quality assurance requirements.

1. Puncture Resistance

Geosynthetic clay liners are thin and, like all thin liner materials, are vulnerable to damage from accidental puncture during or after construction. Thick CCLs cannot be accidentally punctured. Some GCLs have the capability to self-seal around certain punctures, e.g., penetration of the GCL with a sharp object such as a nail. The swelling capacity of bentonite gives GCLs this self-healing capability. Of greater concern than penetration of the GCL by an object after construction is accidental puncture during construction. For example, if the blade of a bulldozer accidentally punctures the GCL during spreading of cover material, the GCL would probably not self seal at the puncture.

The puncture resistance of GCLs will generally not be equivalent to that of CCLs. However, this does not mean that a GCL cannot meet or exceed the performance objectives of a compacted clay liner. Quality assurance and quality control procedures can be established and implemented to make the probability of puncture during construction extremely low. In final covers, one or two accidental punctures would probably not have a major impact on the overall performance of the barrier layer. In a bottom liner system subjected to a continuous head of liquid, a different conclusion might be drawn about the significance of undetected and unrepaired damage to a GCL from puncture. Ultimately, site-specific conditions and quality assurance procedures will be critical in dealing with the issue of puncture and in establishing equivalency of a GCL to a CCL for a particular project.

2. Effect of Subgrade Condition

Compacted clay liners are constructed with heavy equipment. If the subgrade is compressible (e.g, solid waste), the GCL, which can be installed with lightweight equipment, will be easier to construct. On the other hand, stones and rocks can puncture a GCL but not a CCL; if the subgrade contains stones or rocks, the integrity of the GCL may be compromised. Also, in order for the overlapped seams in a GCL to self seal properly, the overlapped panels must be placed on a reasonably smooth and even subgrade. Thus, equivalency of a GCL to a CCL in terms of the effect of subgrade depends on the condition of the subgrade and will have to be evaluated on a site-specific basis.

3. Ease of Placement or Construction

A GCL will generally be easier to place than a CCL, except under rainy condition -- both GCLs and CCLs are difficult or impossible to construct in heavy rain. In general, GCLs are superior to CCLs in terms of ease of construction because GCLs can be installed in more varied conitions.

4. Speed of Construction

Geosynthetic clay liners can be placed much more quickly than CCLs. Geosynthetic clay liners are superior to compacted clay liners in terms of speed of construction.

5. Availability of Materials

Suitable clays for construction of a CCL may or may not be available locally, depending on the site. Because GCLs are a manufactured material, they are readily available and can be shipped to a site quickly. The cost of shipment is usually not a large percentage of the total cost of a GCL. Thus, GCLs will always be at least equivalent to CCLs in terms of availability of materials and will be superior to CCLs at sites lacking local sources of suitable clay.

6. Requirements for Water

Construction water is necessary for many compacted clay soils, which must usually be placed at a moisture content wet of optimum to achieve the desired low hydraulic conductivity. The total amount of water required to moisten a clay liner can be very large. For example, if a 0.6-m-thick compacted clay liner were to be constructed over a 4-ha site, and the water content of the soil had to be increased 5% to achieve the required moisture conditions, the total amount of water used would be approximately 2.3×10^6 L. In arid regions, this water may represent a valuable resource, and in some remote locations, it may be very expensive to provide the water.

Geosynthetic clay liners do not require water for construction and are superior in this regard to CCLs.

7. Weather Constraints

Compacted clay liners are difficult to construct when soils are wet, heavy precipitation is occurring, the weather is extremely dry (clay desiccates), the soil is frozen, or the temperature is below freezing. Geosynthetic clay liners are difficult to construct during precipitation. Weather constraints generally favor GCLs.

Some, if not all, GCLs must be covered before they hydrate. If a geomembrane will be placed over the GCL, the GCL must be covered almost immediately with the geomembrane. Additional weather constraints, e.g., wind speed, may apply to the geomembrane and, indirectly, influence the GCL. The fact that many GCLs must be covered before they are hydrated can be a significant weather constraint for GCLs. However, CCLs have weather constraints, too: CCLs must not be allowed to freeze or desiccate, and wet weather often brings construction of compacted clay liners to a halt. GCLs cannot desiccate during construction because they are dry, and dry GCLs are unaffected by freezing temperatures.

Equivalency in terms of weather constraints must be considered on a site-specific basis, but weather constraints generally favor GCLs over CCLs.

8. Ease of Quality Assurance

The proper construction of a low-permeability, compacted clay liner is a very challenging task. Careful control must exist over materials, moisture conditions, clod size, maximum particle size, surface preparation for a lift of soil, lift thickness, compaction coverage and energy, and protection of each completed lift. Comparatively, quality assurance (QA) requirements are much less extensive for GCLs compared to CCLs, but no less critical. In general, while QA for a compacted clay liner requires a number of relatively sophisticated tests and points of control by very experienced and capable personnel, QA for GCLs is more nearly the application of common sense. Far fewer things can go wrong with the installation of a GCL compared to placement and compaction of a CCL. However, testing procedures and observational techniques are well established for CCLs but are not for GCLs. Many people are working to establish testing methods for GCLs. While it would appear that GCLs are superior to CCLs in terms of ease of quality control, more work needs to be done to establish standard test methods for GCLs.

Summary of Equivalency Issues. Table 3 summarizes the preceding discussion of equivalency. Equivalency can be demonstrated generically in many categories. In several areas, geosynthetic clay liners (GCLs) are clearly superior to compacted clay liners. However, in one category, equivalency probably cannot be demonstrated: thin GCLs do not have the same resistance to puncture as much thicker compacted clay liners. Although thin GCLs can be punctured during construction, careful QA should be capable of addressing this potential problem. Further, for final covers, an occasional small puncture may be of little consequence. Indeed, puncture is probably of much greater concern for a bottom liner than a final cover. Also, if puncture is of concern, a layer of relatively low permeability soil or waste material may be placed below the GCL to provide a back-up should puncture occur at an isolated location. In any case, the GCL enjoys several important advantages over a compacted clay liner which may more than offset greater vulnerability to puncture.

As suggested by Table 3, many equivalency issues depend on the GCL product and the particular conditions unique to a given site. Equivalency will have to be evaluated on a case-by-case basis. The most important site-specific

Table 3 - Potential Equivalency Issues.

Category	Criterion for Evaluation	Equivalency of GCL to CCL: GCL Is Probably Superior	GCL Is Probably Equivalent	GCL Is Probably Not Equivalent	Equivalency Depends on Site or Product
Hydraulic Issues	Steady Flux of Water		X		
	Breakout Time of Water				X
	Production of Consolidation Water	X			
	Permeability to Gas				X
Physical/ Mechanical Issues	Freeze-Thaw	X			
	Wet-Dry	X			
	Total Settlement		X		
	Differential Settlement	X			
	Slope Stability				X
	Erosion				X
	Bearing Capacity				X
Construction Issues	Puncture Resistance			X	
	Subgrade Condition				X
	Ease of Placement	X			
	Speed of Construction	X			
	Availability of Materials	X			
	Requirements for Water	X			
	Weather Constraints				X
	Ease of Quality Assurance	X			

issues are likely to be permeability to gas and slope stability. It may be difficult to provide adequate factors of safety against slope failure on relatively steeply sloping final covers that contain GCLs, but designers have a variety of reinforcement materials (such as geogrids) available for use, if necessary.

6 SEQUENCING OF FINAL COVER

Common practice with landfilling is to construct the final engineered cover as soon as a cell has been filled. This practice is less than ideal in two important respects:

1. The waste is bound to undergo substantial differential settlement, which will change the drainage patterns on the surface of the cover and generate distortion and tension within the various components of the cover. Some degree of cracking within the cover is probably inevitable if the cover is placed over fresh waste.

2. A relatively impermeable cover system impedes percolation of meteoric water into the waste and thus minimizes or eliminates production of leachate. In many ways, it would be better if leachate were produced and collected in the landfill and the landfill were operated initially as an active biological reactor. Biological decay leads to stabilization of the waste.

It is suggested that a potentially better choice at many landfill sites would be to place a temporary and rather permeable cover over the waste. Meteoric water would percolate through the cover and into the underlying waste, which would promote biodragation of the waste. Leachate and gas would be collected. After a period of time (typically a few years), the waste would be far more stable than initially. At this point, when much of the biodegradation that will occur has occurred, a final, engineered cover may be placed over the waste. Because the underlying waste is relatively stable, the cover has a far greater probability of performing well than a cover placed over fresh waste.

Why has this approach not been tried more often? There are probably many reasons, one of which is financial. The owner of a landfill is expected to pay for construction of the final cover. If the construction of the cover is delayed, the necessary money may not be available. However, this problem is easily resolved by requiring that the landfill owner place the necessary funds for closure into a fund that is controlled by public officials.

7 CONCLUSIONS

This paper has summarized the basic principles employed for design of final cover systems for landfills. The most important conclusions relate to the barrier layer within the final cover. The main conclusions are as follows:

- Compacted clay liners by themselves are not the best type of material to use for a barrier layer in landfill covers because the clay is vulnerable to damage from wet-dry cycles, freeze-thaw at some sites, and differential settlement
- Geomembranes or geosynthetic clay liners are better overall choices for many sites because these materials can better resist wet-dry cycles, freeze-thaw conditions, and differential settlement
- Geosynthetic clay liners have numerous advantages over more traditional compacted clay liners for landfill covers, but also have some disadvantages, as well -- the material of first choice should be determined on a site-specific basis
- Through application of rational performance criteria, the equivalency of a geosynthetic clay liner to a compacted clay liner can be assessed

The differential settlement that a decomposing mass of waste causes in the final cover system alters surface drainage and creates distortion and frequently cracking in the cover materials. A sensible alternative is to place an interim, permeable cover over the waste, wait several years for decomposition and settlement to occur, and than place the final engineered cover over the relatively stable waste mass.

8 REFERENCES

Boardman, B.T. 1993. "The Potential Use of Geosynthetic Clay Liners as Final Covers in Arid Regions," M.S. Thesis, University of Texas, Austin, Texas, 109 p.

Boschuk, J. 1991. "Landfill Caps - An Engineering Perspective," *Geotechnical Fabrics Report*, Vol. 9, No. 4, pp. 23-34.

Cline, J.F., et al. 1982. "Biobarriers Used in Shallow Burial Ground Stabilization," *Nuclear Technology*, Vol. 58, pp. 150-153.

Corser, P. and Cranston, M. 1991. "Observations on Long-Term Performance of Composite Clay Liners and Caps," *Proceedings*, Geosynthetics Design and Performance, Vancouver Geotechnical Society, Vancouver, British Columbia.

Daniel, D.E. 1991. "Geosynthetic Clay Liners," *Geotechnical News*, Vol. 9, No. 4, p. 28-33.

Daniel, D.E. 1993. "Clay Liners," in *Geotechnical Practice for Waste Disposal,* D. E. Daniel (Ed.), Chapman and Hall, London, in press.

Daniel, D.E., and Estornell, P.M. 1990. "Compilation of Information on Alternative Barriers for Liner and Cover Systems," U. S. Environmental Protection Agency, EPA 600/2-91/002, Cincinnati, Ohio, 81 p.

Daniel, D.E., Shan, H.Y., and Anderson, J. 1993. "Effects of Partial Wetting on Strength and Hydrocarbon Permeability of a Geosynthetic Clay Liner," *Proceedings, Geosynthetics '93*, International Fabrics Association International, Vol. 3, pp. 1483-1496..

Daniel, D.E., and Boardman, B.T. 1993. "Report on Workshop on Geosynthetic Clay Liners," U.S. Environmental Protection Agency, Cincinnati, Ohio, in review.

Eith, A.W., Boschuk, J., and Koerner, R.M. 1991. "Prefabricated Bentonite Clay Liners," *Landfill Closures: Geosynthetics, Interface Friction, and New Developments*, R.M. Koerner (Ed.), Elsevier Applied Science, New York, pp. 193-218.

Estornell, P. and Daniel, D.E. 1992. "Hydraulic Conductivity of Three Geosynthetic Clay Liners," *Journal of Geotechnical Engineering*, Vol. 118, No. 10, pp. 1592-1606.

Gilman, E.F., Flower, F.B., and I.D. Leone (1985), "Standardized Procedures for Planting Vegetation on Completed Sanitary Landfills," *Waste Management and Research*, Vol. 3, pp. 65-80.

Grube, W.E. 1992. "Geosynthetic Liners Offer Cover Option," *Environmental Protection*, May, pp. 29-33.

Kim, W. H., and Daniel, D.E. 1992. "Effects of Freezing on the Hydraulic Conductivity of a Compacted Clay," *Journal of Geotechnical Engineering*, Vol. 118, No.7, pp. 1083-1097.

LaGatta, M. D. 1992. "Hydraulic Conductivity Tests on Geosynthetic Clay Liners Subjected to Differential Settlement," M.S. Thesis, University of Texas, Austin, Texas.

Leonards, G.A., and J. Narain (1963), "Flexibility of Clay and Cracking of Earth Dams," *Journal of the Soil Mechanics and Foundations Division*, ASCE, Vol. 89, No. SM2, pp. 47-98.

Jessberger, H.L., and Stone, K. 1991. "Subsidence Effects on Clay Barriers," *Geotechnique*, Vol. 41, No. 2, pp./ 185-194.

Johnson, D.I, and Urie, D.H. 1985. "Landfill Caps: Long-Term Investments in Need of Attention," *Waste Management and Research*, Vol. 3, pp. 143-148.

Maltby, V., and Eppstein , L.K. 1993. "A Field-Scale Study of the Use of Paper Industry Sludges as Hydraulic Barriers in Landfill Cover Systems," *Hydraulic Conductivity and Waste Contaminant Transport in Soils, ASTM STP 1142*, D. E. Daniel and S. J. Trautwein (Eds.), American Society for Testing and Materials, Philadelphia, in press.

Montgomery, R.J., and Parsons, L.J. 1989. "The Omega Hills Final Cap Test Plot Study: Three Year Data Summary," Presented at the 1989 Annual Meeting of the National Solid Waste Management Association, Washington, DC.

Murphy, W.L., and Gilbert, P.A. 1985. "Settlement and Cap Subsidence of Hazardous Waste Landfills," U.S. Environmental Protection Agency, EPA/600/2-85-035, Cincinnati, Ohio.

Scheu, C., Johannben, K., and Saathoff, F. 1990. "Non-Woven Bentonite Fabrics - A New Fibre Reinforced Mineral Liner System," *Geotextiles, Geomembranes and Related Products*, D. Hoet (Ed.), Balkema Publishing, Rotterdam, pp. 467-472.

Schubert, W.R. 1987. "Bentonite Matting in Composite Lining Systems," *Geotechnical Practice for Waste Disposal '87*, American Society of Civil Engineers, New York, pp. 784-796.

Shan, H.Y., and Daniel, D.E. 1991. "Results of Laboratory Tests on a Geotextile/Bentonite Liner Material," *Proceedings, Geosynthetics '91*, Industrial Fabrics Association International, St. Paul, MN, Vol. 2, pp. 517-535.

Swope, G.L. 1975. "Revegetation of Landfill Sites," M.S. Thesis, Pennsylvania State University, State Park, Pennsylvania.

Tschebotarioff, G.P., and DePhillippe, A.A. 1953. "The Tensile Strength of Disturbed and Recompacted Soils," *Proceedings*, Third International Conference on Soil Mechanics and Foundation Engineering, Switzerland, Vol. 1, pp. 207-210.

U.S. Environmental Protection Agency. 1983. "Standardized Procedures for Planting Vegetation on Completed Sanitary Landfills," EPA600/2-83-055, Cincinnati, Ohio.

U.S. Environmental Protection Agency. 1989. "Technical Guidance Document, Final Caps on Hazardous Waste Landfills and Surface Impoundments," EPA/530-SW-89-047, Washington, DC.

Waste Disposal by Landfill – GREEN'93, Sarsby (ed.) © 1995 Balkema, Rotterdam, ISBN 90 5410 356 6

Old landfills: Perception and Remediation of problem sites

S.A.Jefferis
Golder Associates Ltd, Maidenhead, UK

ABSTRACT: Old landfills are all around us and have been since man first started to live in settled communities rather than as a nomad. In the past they have aroused little interest except perhaps to archaeologists. However, over the last few decades the environment has become a progressively more important issue. Old landfills may be visually unpleasant and have poor vegetation cover. They will also generate gas and leachate which may impact the adjacent environment though in general monitoring data will be few and the impact may be difficult to assess. Before designing remedial systems it is necessary to understand the risks that they pose and the mechanics of pollution migration from them. and to apply this information so as to prevent today's new landfills becoming tomorrow's problem landfills. At the present time slurry trench walls are widely used for the prevention of leachate and particularly gas migration from old landfills and the technique and some of the associated problems are reviewed.

1 INTRODUCTION

Individuals may be regarded as having an in built environmental regulator which sets their sensitivity to environmental issues. At one extreme there is pure emotion. Any damage to the environment is unacceptable. This position could be labelled the BNAANA (build nothing at all near anyone) syndrome. A milder form of the syndrome is NIMBY (not in my backyard). At the other extreme is the purely technical view. The environment is a scientific and engineering problem and must be analysed and dealt with in these terms. The problem is that the environment is very difficult to analyse and on major issues we often risk scoring 'own goals'. Thus with hindsight the expansion in the use of unleaded petrol in the UK may not have brought the environmental benefits it first appeared to offer. Also recycling of low value high volume materials such as paper could prove to be environmentally negative. Indeed the cynical might say that such recycling has gained popularity only because anyone can be do it and it does not affect lifestyle in the way that say reduction in car usage would.

Whatever standpoint one takes it is clear that there are votes in the environment - the public want environmental improvement. Though sadly one must allow that many of those who currently enjoy relatively the best environment can be the most vociferous in demanding a better environment - and often for themselves rather than for the environmentally disadvantaged. The technologist could regard environmental awareness as something to be encouraged as good for business - and indeed it is. However, we must be careful to avoid the own goals and to ensure that real benefits are to be achieved - and of course especially in the field of landfill design.

We must also consider financial as well as technical issues. Good landfill practice may be a major priority in developed countries. In developing countries the situation may be very different. Where money is minimal the improvement of landfill practice and the remediation of old landfills may not be the most urgent priorities. We must be careful not to export our own priorities when exporting technology and more importantly we must be prepared to seek low technology as well as high technology landfill designs and remedial measures

Social issues also must be considered in all parts of the waste chain from producer to landfill. For example in lesser developed parts of the World landfills may become quarries from which resources may be recovered and a living albeit at a miserable level extracted - especially if recycling is encouraged and at the landfill rather than higher up the waste chain. As a result communities will develop on and adjacent to landfills. These communities will be seriously at risk not only from disease but also from physical instability of the landfill and fire etc. Especially as in low technology landfills the waste may be simply end-tipped from a high point eg a hillside to the valley below. As a result the waste may take up a very steep and unsafe angle of repose. Slips may occur as a result of fires, natural degradation of the waste or as a result of rainwater ingress in wet seasons. If a major slip occurs sufficient methane may be released to cause an explosion - as appears to have happened when the waste mass on the Umraniye landfill, near Istanbul landfill slipped and engulfed a number of houses with the loss of 39 lives. Explosions on landfills are extremely rare but this Turkish incident regrettably shows that they can occur. Could they also occur if an earthquake caused a major gas release from a landfill?

2 THE SITE

A common factor for most old landfills is that little is known about them. There may be few records of filling and no groundwater or gas monitoring etc. Even their extent may be poorly defined. Some old municipal waste landfills may be easily identified by a walk over survey. The soil cover may be thin there may be no trees as the soil cover is insufficient or the methane levels are lethal to deeper rooting species. Evanescent ponds may occur and the electrical conductivity of the water may be high suggesting the presence of leachate species.

However, other old landfills may be less obvious. Love Canal, the toxic waste dump that was a major trigger of environmental legislation in the USA was not immediately obviously a hazardous waste dump.

Aerial photography is one of the most powerful procedures for the identification of unmapped old landfills and may be able to give information on landfill practice, waste type and even on occasion the source of the waste though the earliest data available will be from around the late 1930s/early 1940s.

3 SOME OF THE PROBLEMS

The problems include:

a) Waste is chemically and mechanically very complex.

b) Data on the wastes that have been put into any landfill usually will be scarce.

c) Landfills will settle and we lack well developed geotechnical models for waste.

d) Many old landfills will have been sited for convenience or because holes were available rather than because of favourable geology and hydrogeology.

e) Around any conurbation there may be many old landfills. When first established these landfills may have been reasonably remote from the conurbation. However, subsequent development may have occurred very close to them.

f) The landfill degradation process is slow. Landfills may continue to produce gas and leachate for decades.

g) The leachate generated from landfills is complex and will vary with age.

h) Leachate may be migrating from the landfill.

i) Landfill gas may be migrating from the landfill both horizontally and vertically,

j) Today we are unsure of the reason for putting the waste into a landfill. Is a landfill to be a tomb for toxic materials or a bioreactor? Will today's practice will be the seed of tomorrow's problems?

Many of the above problems stem not so much from engineering issues but the from the complex system that a landfill represents. Landfills are enormous structures and even quite minor design errors may produce major problems. When we introduce design changes we hope that there will be immediate benefits. However, there also may be problems and it may be quite some time before the consequences are manifest.

For example the move from simple end-tipping of waste to compaction and daily cover layers lead to more concentrated leachates and the reduction in vertical gas escape and hence enhanced lateral gas migration. In contrast the issue of clay-leachate interaction - cracking, increase in permeability etc. that seemed very significant just a few years now seems rather less important particularly as we know more about the chemistry of leachate and the design of composite systems. Though there were problems on the learning curve with membrane barriers. Initially a few clay liners were simply replaced by membrane liners and as these proved impossible to lay without minor imperfections the results were some spectacular leaks. We now have a much better understanding of liner systems and can design composite systems which address the strengths and weaknesses of the different elements such as HDPE and clay that may be used in a barrier. As a result we can produce systems which are more tolerant of and effective than any of the individual elements.

There also have been problems with the mechanics of clay/liner interaction and major slips occurred (Byrne et al, 1990). New design procedures are now available and are continually under development especially for steep sided landfills to enable maximise use of the landfill area.

As regards today's designs possible causes for concern are:

a) There could be desiccation of clay liners under HDPE as a result of the self heating within the landfill or as a result of solar heating prior to placement of waste.

b) Drainage of landfills can be a problem. Mineral and microbiological fouling of drains occurs - and may do so very rapidly. Regulations may stipulate that only a small head of leachate is permitted within a landfill but can we keep drains unclogged for 60 years - or even 30 years? Remedial drainage measures will be expensive.

3.1 Leachate level

A potential issue for future contention can be identified in Waste Management Paper 26A, 'Landfill completion' which develops the completion condition set out in the Environmental Protection Act 1990 (in the UK completion is the time at which a landfill licence may be surrendered to the waste regulation or other authority) and requires that 'the landfill has stabilised physically, chemically and biologically to such a degree that the undisturbed contents are unlikely to cause pollution of the environment or harm human health'. It also states that where a site is dry: 'it may be necessary to analyse the residual solid wastes to determine if there is a potential for further gas generation'. However, the National Rivers Authority, the UK regulator for the water environment is likely to require that the maximum head of leachate on the liner of a landfill is of order 1 metre and thus much of the mass of the landfill could be effectively dry - unless leachate recirculation is employed - and this is currently more of a research concept than a proven technique. As the rate of microbiological activity is very sensitive to the moisture content of a waste it is clear that in a dry site the potential for such gas generation will persist for a very substantial time so that completion as defined above may not be achievable within any reasonable time-scale.

Thus to one regulatory authority a 'dry site' represents a continuing risk but to another it is the ideal to be sought after. How do we resolve the conflict? Quantitative risk assessment is currently very much the vogue - and may help us but we should be prepared to change our thinking if we do adopt it. It may radically twist the focus of our attention. Research is like mining we tend to follow a good vein but we may forget that the un-researched areas can be where the greatest reduction in risk can be achieved for the least expenditure.

4 THE LANDFILL SITUATION

The local situation of a landfill always will be a major issue. This will include the geology and hydrogeology (old landfills are unlikely to be lined) also nuisance, odour, visual impact, and the general quality of the site. Many old sites may have developed only a thin veneer of soil cover. Thus vegetation may be sparse and trees non existent. Leachate springs may appear in wet seasons and disappear in dry seasons to leave patches of discoloured soil. Though in passing it may be noted that this colour may be due to iron and relatively harmless though the other species associated with it could be less so.

Old landfills could be used for the grazing of cattle, rabbits may burrow into it and vermin live on it. All of these may have health implications but the pathways may be very difficult to analyse - though probabilistic risk based procedures certainly will help.

4.1 Risk assessment

Risk assessment can be a very useful management tool but if we adopt risk based strategies we must learn to understand their strengths and weaknesses and also to 'sell' them. A risk in the form of extra lifetime cancers may calculate as 1 in 10^4 or 10^5 rather than the 1 in 10^6 or 10^7 that we could all feel is comfortably negligible (without any risk awareness training). The protestor in the street (or the public at large) may not be able to understand leachate concentrations in say mg/litre of chlorinated solvents but everyone has some concept of risk and the same scenario if stated as an extra life-time cancer risk of 1 in 10^4 may prove difficult to 'sell'.

At the present time we can produce good probabilistic assessments of the technical issues involved in complex structures such as landfill (eg gas and leachate migration, groundwater contamination etc.). However, the available toxicological models to link the effects of contamination to human or environmental damage can produce a very wide scatter of results. For example in the Guidelines for the Evaluation of Chemicals for Carcinogenicity (1991) the following statement is made about mathematical models for the evaluation of toxic effects of chemicals: 'These models may give an impression of precision which cannot be justified from the approximations and assumptions on which they are based'. The guidelines support a broadly based approach to the assessment of putative carcinogens 'which uses all the available data and which draws on expertise and information from a wide range of medical and scientific opinion'. These words may be very difficult to apply when we are trying to assess the effects of chemical doses which may be several orders of magnitude below those for which dose - response data exist. Indeed such words may seem quite unhelpful though it must be accepted that they are the considered opinion of a group of experts and may not be lightly discarded.

Even though we may not have numerical precision at the present time toxicological modelling can help with decisions on what to change and where to focus our attention both in the field and in the laboratory.

4.2 Landfill modelling

Let us now consider the interior of a landfill. We have gas generation, leachate generation and migration etc. - processes that can be defined theoretically and quantified. In conventional mathematical analysis the modeller may vary the boundary conditions such as heads, fluxes, etc. so as to minimise the impact of the landfill on the surrounding environment. Today the engineer or technologist is still responsible for some of the modelling and the design of landfills but regulators may control the boundary conditions such as liner permeability and leachate head. Before long regulators may also prescribe how the modelling and design should be undertaken - a situation which the USEPA has in part achieved.

This may inhibit the best design using the current best science. It also begs the question of liability. Regulators should beware. Models will develop and environmental sensitivities may migrate to new pastures. Those who design or influence design must accept responsibility for their designs. If a regulator prescribes fundamental design parameters, modelling procedures or boundary conditions he may become responsible for the design and there is an enormous potential to get things wrong within landfills because of their sheer size. If we are to design landfills by codes of practice then those codes must be very thorough and all embracing.

Of course anyone accepting liability in the environmental arena may hope to escape that liability because actions could be time barred (it may take a substantial time for today's landfills to become 'old and problem landfills' and of course we aim to ensure that none of them will). To cite an example from another arena: two decades ago fire authorities required the inclusion of asbestos in a building for fire control. Today health authorities require its removal. Who is liable? Who will pay? Will landfill generate the same kinds of problems? The answer seems very likely to be yes if we continue to have a multiplicity of regulatory authorities influencing design.

5 POLLUTION MIGRATION

For many old landfills the main problems are with the migration of leachate and gas from the site. It is therefore important to examine the driving potentials that may cause migration both from lined and unlined sites. Figure 1 shows some of the potentials that may induce flow from/to a landfill site.

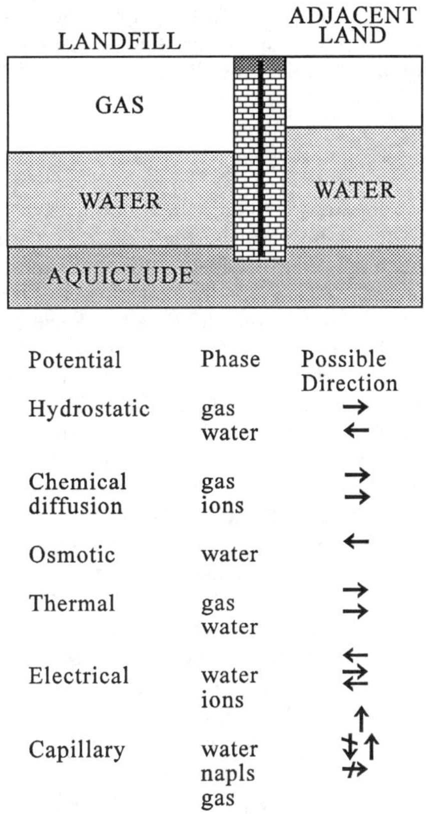

Figure 1: Flow Potentials

5.1 Hydrostatic pressure

Flow under a hydrostatic pressure is the simplest migration mechanism. However, even this may not be simple. There will be a tendency for gas to be driven from a landfill site but there may be a tendency for inward flow of water if the internal water level is below the outside level. Thus gas may be going one way and liquid the other and interfacial conditions may be interesting.

5.2 Chemical diffusion

Chemical diffusion also will be important and generally as the landfill should be the area of highest concentration the diffusive flux of gas and chemicals in solution will be from the landfill to the 'outside world'. In theory this flux could be minimised if there were an inward flow of water. In practice with composite barriers this can be difficult to achieve especially if real barriers are considered with their quota of small but real defects.

5.3 Osmosis

Osmosis also will occur. That is there will be movement of water from less chemically concentrated solutions to more concentrated solutions and thus generally from outside the landfill to within. In multi-barrier systems osmosis could lead to the build up of water in or desiccation of clay adjacent to HDPE. The timescale for any landfill will be so long that even such minor flow potentials could ultimately have significant effects. Unfortunately data on landfill liners seems rather scarce despite the fact that many of those involved in membrane research are concerned with maximising flow through membranes for reverse osmosis systems etc.

However, there is little apparent cross-fertilisation between the reverse osmosis research community and the landfill research community.

5.4 Temperature

The temperature within a landfill may be markedly greater than that of the natural surroundings. This will lead to thermally induced migration. Clay under an intact HDPE membrane could be dried and cracked by thermally induced water movement. If the membrane were to fail later and leachate start to escape would the underlying clay swell to close the cracks? Low swelling clays placed at high water contents (to ensure low initial permeabilities) may not. In contrast a bentonite-sand mix might well swell sufficiently as the scale of any cracking may be limited to the sand grain size and thus easily healed.

Of course low temperatures also must be considered. Freeze thaw cycles may damage the permeability of clay liners and capping layers prior to placement of respectively waste or soil cover. Problems may be particularly severe if clays are subject to concentrated pollutant solutions. Chemicals in solution will depress the freezing point of water but the effect of even strong leachates may be quite modest. On exposure to freezing conditions ice will begin to form (at a temperature slightly below the freezing point of water) and the solution will become more concentrated. As more water freezes the solution will become progressively more concentrated and the freezing point of the remaining solution will drop until the solution becomes so concentrated that a eutectic mixture freezes. On thawing the reverse process will occur and the clay will again be subject to a strong chemical solution. Thus a simple freeze-thaw cycle may expose a clay to very strong chemical solutions and the clay structure is likely to be adversely affected. In general strong chemical solutions are likely to promote face to face aggregation of the clay particles and increase in permeability. This structure may not be lost on reduction in the pore water salt concentration. Drying and re-wetting will produce similar effects.

5.5 Electrical potentials

Electrical phenomena also must be considered. If methane is being generated in a landfill then the local redox potential must be of order -400 millivolts. The HDPE may act as an electrical insulator and thus the electrical regime within a landfill will be complex. The author is not aware of any analyses of the effects of self potentials within landfills or indeed in any natural or contaminated soils on pollution migration (though the potentials have been used to help with the location of holes in liners) and to stabilise landslips (though not in landfills - at least not by design). It would seem that -400 millivolts is a significant potential as regards electro-osmotic flow and could counteract quite significant hydraulic heads.

5.6 Redox potential

The construction of a landfill may markedly change the local redox environment. The covering of a large area of land with a liner (irrespective of the subsequent placement of waste) will prevent oxygen ingress and thus reduce the redox potential of the underlying soils. This may lead to the solution of iron and manganese naturally present in the soil and the conversion of sulphates to sulphides etc. This effect may be exacerbated by the escape of leachate which will exert a biochemical oxygen demand and further reduce the redox potential but it is important to realise that the solution of iron and manganese etc. does not confirm leakage. A chemical analysis particularly for species which are largely unaffected by redox eg chloride may enable natural chemical changes induced by change of redox potential to be separated from leachate escape.

5.7 Capillary processes

Surface tension effects must be considered whenever flow involves more than one phase eg gas/liquid interfaces or water/non-aqueous liquids (floating or dense).

Soil pore size is a crucial factor when considering capillary forces. Fine soils may arrest the downward migration of non-wetting liquids such as chlorinated solvents but enable substantial capillary rise of wetting liquids. It will also control the pressure needed to displace water and allow gas flow in a soil. Waste management Paper 27 suggests that a 1 metre thickness of 10^{-9} m/s permeability clay may be suitable for the control of landfill gas and indeed it may be entirely satisfactory when the clay is water saturated as the gas pressure required to force water from the capillaries in a such a clay will be large (several bars). Thus for a saturated clay gas migration may be possible only in solution or as a result of degradation of organics which have passed through the barrier in solution. However, a warning is necessary about drying. If a clay barrier dries so that water is no longer a continuous phase in the pore system then it will become gas permeable. Data on gas permeability of clays are few and thus design of remedial gas barriers is difficult unless drying can be avoided or a membrane is used.

Capillary migration also can be important in cap design to ensure that toxic chemicals are not drawn up to the surface in periods of drought.

5.8 Effective stress

Although not a pollution driving force effective stress can be a very effective inhibiting mechanism. In some soils increase in the effective stress may dramatically reduce the permeability and not just by a factor of 2 or 3 times as might be suggested by an analysis based on a the Carman-Kozeny equation but by several orders of magnitude. High effective stress may also much reduce the impact of chemical and other clay - environment interactions. Compatibility tests based on forcing a few pore volumes of leachate through a clay under high gradients and thus at high effective stresses may fail to identify effects which occur at lower stresses. As already noted clay/chemical interactions are less feared than they were. However, testing still may be necessary and when this is carried out it should be done at an effective stress and flow gradient appropriate to that in-situ - though such tests may take a very extended time.

6 SETTLEMENT

Settlement prediction is undoubtedly improving. Models are being developed and calibrated by back-analysis. However, it should be remembered that landfills will be heterogeneous and large voids may exist within the waste for example slabs, beams etc from building waste may trap large voids. Collapse of voids within a landfill may lead to sharp differential settlements over small distances which can be damaging to capping layers. Settlements over larger distances may affect the surface profile and lead to problems of surface water ponding etc.

A further point to be considered is settlement of the ground. The author recently saw a new hospital unit built on a landfill. The building was founded on piles to prevent settlement and appeared in perfect condition. Unfortunately the designers apparently had not allowed for the settlement of the ground around the building. The ground had settled so much that it had been necessary to extend the rainwater pipes by some 150 mm so that they connected to the gulleys in the ground. More seriously the drainage system appeared to be being damaged by down-drag as possibly were the other service connections: gas, electricity and water.

The settlement of the ground might have been limited by dynamic compaction but this most probably would have re-initiated gas generation and the whole building phase would have had to be delayed for several years as indeed it seems it should have been as tolerance of ground movements was not built into the design.

7 LANDFILL GAS

Gas migration is clearly a major issue where it may put buildings at risk but how often does it pass unnoticed when the result is merely the death of a hedge-row or patches of stunted growth in a farm field? It should also be remembered that methane may be oxidised by bacteria and methane oxidising bacteria may flourish on the walls of fissures venting methane. Local ground temperatures may be substantially elevated, certainly up to 50°C. Such temperatures will be lethal to plants but lesser temperature increases may actually lead to more luxuriant plant growth.

Gas extraction and flaring can be an effective method for controlling methane escape. But a landfill may also contain significant quantities of chlorinated solvents (total emissions possibly in the region of several tonnes per year) and these will not be properly destroyed in some flare systems or in internal combustion engines if power is generated from the gas. The potential for the generation of dibenzofurans and dioxins during the cooling of flare gases also needs to be considered.

Thus in a risk analysis all volatile emissions including those from power generation plant and gas flaring need to be considered and not just methane and carbon dioxide.

There is a further problem in that when utilising the gas for power generation a rich mixture is needed, but to control gas migration the ideal will be draw relatively large quantities of gas towards the venting points. So there is an immediate conflict between use of gas and control of gas because the former requires high concentrations whilst for the latter a net influx of air into the landfill is wanted.

8 LEACHATE

The ability to remove leachate from a site can be of prime importance but may be a problem in the long-term. Pumps, pipework, drainage channels and geotextiles etc. may become fouled with microbiological slimes and mineral precipitates. For example landfill leachates may contain high levels of dissolved carbon dioxide. If some of this is allowed to de-gas for example by flow into a well at atmospheric pressure the carbonate-bicarbonate equilibrium may be disturbed and carbonates deposited. Such deposition can occur very rapidly and may be difficult to control. The microbiological slimes may contain consortia of many different bacterial including powerful blocking bacteria such as the iron bacteria gallionella. The volume of slime may be reduced if conditions can be kept anaerobic. Bacterial slimes may be very resistant to erosion by water flow and drain cleaning systems should take account of the nature and dynamics of slimes in so far as this is understood. Research is in progress to develop predictive models for slime generation but much remains to be done.

9 CAPPING

Old landfills often will need to be capped as they are unlikely to be lined and the net rainfall may all escape as leachate Clearly excavation and disturbance of the waste should be kept to a minimum during capping operations but some re-grading may be essential if reasonable slopes are to be achieved so as to ensure satisfactory surface drainage and/or to control erosion from steeper areas. Until vegetation is established erosion of caps is a major problem, even with shallow slopes. The cap has to be sloped to ensure run-off but erosion from unvegetated areas may be severe even at slopes as shallow as 5%. Furthermore a cap may shed large volumes of water and the resulting run-off may need to be assessed for its potential to affect the adjacent environment.

The weight of materials added as a cap will induce some settlement and as a result may lead to a temporary increase in the leachate level within a landfill and thus an increase in the rate of flow of leachate from the landfill. The compaction may also re-initiate gas production in sites where it has reduced to low levels.

At the present time caps are designed to prevent water ingress to landfills. For this very low overall permeability are required. The effective rainfall on a UK landfill might be equivalent 10^{-9} to 10^{-8} m/s. The gradient across the barrier element of a cap may be of order unity and thus in simple terms even a clay of permeability of order 10^{-9} m/s may allow significant proportion of the net rainfall to enter the landfill.

Capping itself is a subject of debate. Should we continue to exclude rainfall or should we permit it so as to increase the moisture content of the landfill and increase the rate of degradation? Of course overtopping would have to be prevented and other regulatory control levels respected and the effects of leachate escape from the site investigated. Leachate collection and removal also would be necessary. Such removal would contribute to the clean-up of the site but would increase running costs though it should reduce the time to completion - especially if leachate recirculation etc were employed and thus overall costs might be reduced.

10 CONTROL OF GAS AND LEACHATE MIGRATION

For existing landfills which are leaking the escape of leachate and gas can be controlled by slurry trench cut-off walls. These are now widely used in the UK and are advancing around landfills and other contaminated sites at a rate of perhaps 1 to 4 km/month. They therefore will be considered in some detail.

In the process a trench is excavated in the ground and during this excavation a slurry is pumped into the trench so as to keep it filled with the slurry. The slurry generally will be a mixture of bentonite clay, ground granulated blast furnace slag, ordinary Portland cement and water. During excavation the slurry has to support the sides of the excavation and it then has to set to form the barrier or support a membrane barrier. This type of wall is known in Europe as a single phase wall. Two phase walls also may be employed but are seldom used except for deep walls or other situations where the excavation time would be so extended that a single phase cement-bentonite slurry would have set or become unworkable. For two phase walls the excavation is carried out under a bentonite-water slurry and once it is complete the hole is backfilled with a cement-bentonite-aggregate plastic concrete or a soil-bentonite mix. The latter type of backfill is widely used in the USA but rare in Europe.

In the UK the set properties of single phase cement-bentonite slurries present a dilemma because of local specification requirements. The permeability of the wall must be $<10^{-9}$ m/sec (which is reasonable) and the preferred strength is around 100 to 500 kN/m^2. The reason for such a low strength is the requirement that the strain at failure should be not less than 5%. The ideal is therefore a plastic material but the slurry wall material is cement-based. Cement-based materials in the form of concrete typically have a strain at failure of around 0.3%. The fundamental problem is that to achieve a high strain at failure in a cement-based material it must be confined so that it is in a plastic state. The confining pressure has to be comparable with the strength of the material. In a slurry trench the available confining pressure will be dictated by the thrust the ground exerts on the material, and this will be small except at depth in a deep wall or unless there is an unusually thick cap over the wall. Since in general the confining pressure will be small to obtain the 5% failure strain the strength of the material must be kept very low so that the confining pressure : strength ratio is high. This makes it extremely difficult to meet the permeability requirement of 10^{-9} m/sec. To achieve low permeability the slurry mix should have high clay and cement contents. For high strain at failure the mix should have a minimal amount of cement. It is the author's opinion that the 5% failure strain requirement needs to be reconsidered when slurry trench walls are used to contain landfills. Whenever possible the trench should be installed in stable ground so that large strains are unlikely. If a slurry trench must go through or very close to a landfill and strains are expected to be large it may be necessary to include a membrane in the trench. The prime requirements for slurry trench cut-off materials should be low permeability and high durability and these should not be compromised by the need to develop high strain at failure which is always going to be unsatisfactory in a cement-based material. Rather we should design out the requirement for high failure strain. It would seem that in Europe strain at failure is not a specified parameter but a deformation modulus criterion is used. If such a parameter is to be used the test conditions (drained/undrained/rate of strain) must be specified as the modulus is sensitive to these conditions and also the strain level at which it is to be measured.

10.1 Mix proportions

Typical slurry mixes contain 25 to 60 kg of bentonite with 100 to 350 kg of cement per 1000 kg of water, that is a mix which comprises, by volume, perhaps 90% water and 10% solids. The range of useful bentonite contents will depend on the source of the bentonite. The 60 kg upper limit is typical for the bentonites currently used in the UK. At higher levels the slurry may become too thick. Thus the selection of the bentonite content must be based on the properties of the bentonite to be used. Blast furnace slag makes a very substantial difference to the properties of the material. Typically it is used to replace up to 80% of the cement content and at these levels it can produce a substantially more impermeable material than a mix with the same total cementitious content but containing only ordinary

Portland cement. However, the mixes also tend to be much stronger and if replacement levels much over 80% are used the material may become rather too brittle. Pulverised fuel ash (pfa) does not behave in the same way as slag. It has a different mineralogy and behaves differently in the mix. If pfa is used at replacement levels over about 30% the resulting mix may be very slow setting. However, it may be used not as a replacement material but as an additional material to provide extra fines to the mix without developing the excessively strong mixes that might result if extra cement were added. Extra fines can be useful in systems that may be subject to drying as the extra solids will reduce cracking.

10.2 Durability

Mix designs are generally controlled by the permeability, strength and strain requirements stated in the specification. However, durability is also very important - especially for leachate barriers. Four general points may be made:

a) Cement-bentonite materials may behave quite differently under confined and unconfined conditions. This is important because in the ground the material will be confined and any expansive forces resulting from chemical reactions will be resisted. For example sulphate attack is expansive and can lead to substantial damage under unconfined conditions. Whereas under confined conditions there may be some softening but the material may remain intact and retain some useful impermeability.

b) If chemical attack produces expansive reactions or expansive crystal growth (eg sulphate attack) then stronger cement bentonite materials if unconfined may be more obviously damaged than weaker materials.

c) If the attacking liquid is likely to cause solution of the cement-bentonite material or some phases within it then a high solid content mix is to be preferred to a low solid content mix.

d) Pfa mixes also can exhibit markedly better sulphate resistance than slag mixes though the benefit may be sensitive to the confining pressure.

10.3 Rate of reaction

The data shown in Figure 2 relate to a clay cement material subjected to permeation with an unusually concentrated magnesium chloride solution (30 g/litre of magnesium ion). Initially there is a period of reducing permeability followed by a period of relatively constant permeability. Had the test been stopped at 45 days the material might have been accepted as entirely satisfactory whereas at 100 days the material might be rejected as showing a rapidly increasing permeability. However, the permeability then levels off at a tolerable if rather high value and remains unchanged to about 600 days (and very probably much beyond this).

To understand this type of behaviour it is necessary to consider the mechanics of permeability change. If a material is permeated with a reactive solution a reaction-front will propagate through it. Cement -bentonite materials are reactive to magnesium ion because of the lime from the cement hydration. If the reaction causes a change of permeability then as the reaction-front moves through the material so the permeability will change. Thus the permeability of the material in the zone through which the permeating liquid has passed and fully

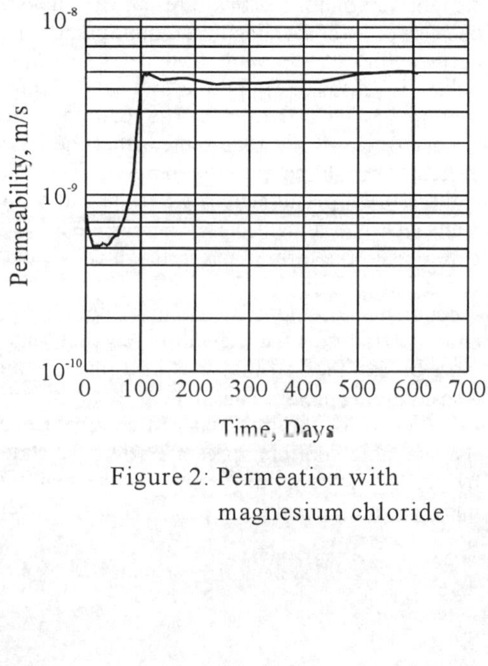

Figure 2: Permeation with magnesium chloride

reacted will be different from that of the remaining unreacted material. Thus there will be a zone of reacted permeability followed by a zone of unreacted permeability.

The volume of permeant needed to move the reaction-front forward will be that required to fill the reacted pore volume plus the fluid necessary to carry the chemicals used in the reaction. The permeability measured in a standard laboratory procedure is calculated from the flow through the sample and the total head drop across the sample (including both reacted and unreacted zones) and it will not possible to separate this into contributions from the 'reacted' and 'unreacted' zones until the liquid has permeated completely through or unless the sample is cut up and the reacted and unreacted portions tested separately - if they can be identified.

A fuller theory for flow with a reaction front is developed in Jefferis (1992 and 1993a). Until the reaction-front has passed completely through the sample the final permeability cannot be predicted. The problem is to know when this has occurred, especially as there may be a family of reaction-fronts travelling through a material as a result of different reactions between the material and the permeating fluid (perhaps a complex landfill leachate).

It is becoming common practice to require that proving trials are carried out on cut-off mixes whereby they are permeated with contaminated groundwater or landfill leachate. Such tests are entirely reasonable but it must be appreciated that it may take a very substantial time for any deleterious effects be manifested. Short term tests of a few days or weeks duration may give false confidence in the material. For example consider the effects of permeation with an imaginary but particularly damaging solution that completely destroys the impermeability of all material with which it has reacted. If this solution was set to permeate through a sample by the time the reaction front had reached the middle, the first half of the sample would be of effectively infinite permeability. However, the remainder would still have its original permeability and thus the overall permeability would have increased by only a factor of 2 - an amount that could easily pass unnoticed if the second half of the sample had in the meantime undergone a modest time dependent decrease in permeability. Such a reduction easily could occur as the permeation time to destroy half a sample of permeability $<10^{-9}$ m/s is likely to be order months or even years.

There is of course always a temptation to increase the rate of advance of the reaction front by using a high test gradient. However, this will also require a high confining pressure on the sample. If the confining pressure on the sample during a test is not comparable to that which will be sustained in the ground the test results also may not be comparable. In general high/excessive confining pressures will reduce the effects of both expansive and leaching type attack and may so compress the sample that there is more limited permeability change than might be seen at lower confining pressures.

If samples are permeated with landfill leachate it must be remembered that the leachate may not be chemically stable and that methane generation may occur and disturb or reverse the permeation flow. Also the high pH and calcium ion content of the slurry material may lead to the immediate precipitation of calcium carbonate and other materials on the inflow face of the sample so that its permeability is very markedly reduced and thus test times become very long. This is a useful feature of the material as it improves its performance though it does make testing very lengthy.

In terms of chemical systems, water itself is probably one of the most aggressive fluids because of its dissolving power. Though interestingly the permeability of cement-clay mixes drops with time when they are subjected to sustained permeation with water. Figure 3 shows a test on a typical bentonite cement material and it can be seen that there is a decrease of two orders of magnitude in 200 days of permeation.

From the graph it can be seen that log (permeability) declines almost linearly with log(time). However, it should not be assumed that this reduction will continue for all time. It will cease when the reaction front for the permeating water (distilled but saturated with carbon dioxide under atmospheric conditions) has passed through the sample (see the discussion on Figure 2). Thus for cut-off wall tests the age of the specimen at the start of the test and the duration of the test can be important parameters.

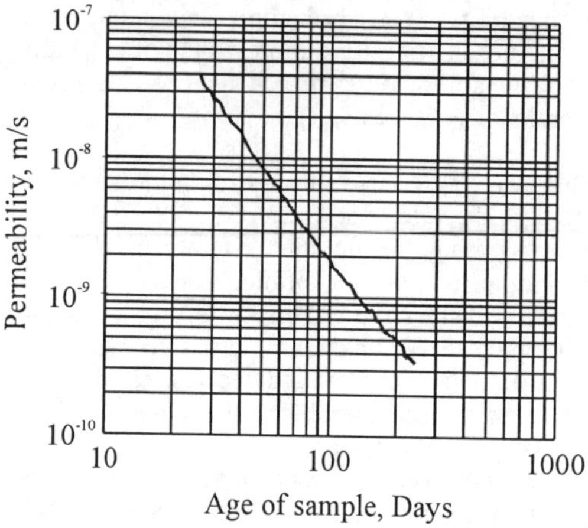

Figure 3: Permeation with water

Finally it should be noted that the chemicals in a leachate that may be the most harmful to human health or the environment may have little effect on cement-bentonite materials and vice versa. This may be because they are present only at low concentration or because they are not physically or chemically reactive with cement-bentonite systems.

10.4 Drying

Because the slurry mix is approximately 90% water drying shrinkage can present a problem. Cement-bentonite slurries should not be used in isolation where there is a possibility of drying or where there is a need for gas control. The material should be capped and ideally a clay layer should be compacted along the line of the wall prior to the works and the wall then excavated through it and the top of the trench then plugged with further rolled clay to form a continuous clay layer across the trench and perhaps to 0.6 to 1 m on either side. The cap will also increase the effective stress on top of the wall and this will improve the mechanical performance and durability of the near surface material.

10.5 Gas pressure

Landfill gas pressure can pose problems during the excavation phase of a wall. Consider a slurry trench wall which is being excavated near a landfill. The density of the slurry will be about 1.1 to 1.2 Mg/m^3 but the pressure of the gas at depth may higher than the hydrostatic pressure of the slurry so that gas is forced up through the slurry before it sets. This action can destroy the integrity of the wall. If high gas pressures exist and persist in the adjacent ground after the end of excavation it can be very difficult to prevent damage to the integrity of the slurry. Serious consideration should be given to including a membrane in walls excavated near landfills or peats which are still evolving gas.

11 VIBRATED BEAM WALLS

Barrier walls may be formed by a number of other techniques (Jefferis, 1993b) including the vibrated beam process. In this process an 'H' pile is vibrated into the ground and slurry is injected at the toe of the beam during both insertion and subsequent extraction. The slurry injected during insertion lubricates the beam penetration and that during excavation forms the barrier and should be injected at a rate such that the slot created by the beam is kept filled with slurry. As slurry penetration/filtration into the ground may continue after the beam is extracted it is good practice to drive the beam through a small slurry filled header trench so that extra slurry always is available to top up the barrier slot. Vibrated beam walls may be only of order 50 mm thick and thus may have rather poor chemical resistance. However, developments are in progress to enable HDPE membranes to be placed by the vibrated beam process and at least one membrane wall has been installed in this way.

12 BARRIERS UNDER LANDFILLS

Jet cutting utilising a high-pressure water/grout jet which can be rotated, at depth, to produce a series of interpenetrating discs of grout can be used to install a base under existing landfill. However, such rotary jet cutting is an extremely labour intensive way of installing a sealing layer beneath a landfill.

An alternative way is to use inclined drilling under the landfill. The drills are inclined so as to pass under the landfill and a trench is dug on the side of the landfill remote from the drilling rig so that the exit point of the drill string can be seen. After drilling a high pressure jet cutting tool is inserted. The tool has two grout injection points designed to cut and inject grout to form a rather flat 'V' shaped slot (the tool is not rotated). The tool is then run in an adjacent hole and the aim is that the 'V' shaped slots from the two holes should cross so as to create a 'W' shaped section. The process then repeated so as to form a continuous corrugated trough beneath the landfill. The 'V' shaped cut is used because if the jets merely cut a horizontally, the slots from adjacent drill holes might pass above or below each other and not interlink to form a continuous barrier layer. The technique has been proven in trials (Wheeler, 1993) and for these the grout was an interesting and unusual emulsion of a wax (derived from a particular type coal) in water.

13 TESTING BARRIER WALLS

To date no coring procedure has been developed which has enabled undisturbed samples to be taken from hardened slurry in a wall. Coring almost invariably so damages the material that tests for mechanical properties and permeability give meaningless results. If a wall is to be cored the damage inflicted by the coring operation on the in-situ material should be considered. Great care also should be exercised when coring in or adjacent to a wall to ensure that the wall will not be hydrofractured by any fluid pressures employed.

Because coring of the wall is impracticable most wall materials are proved by laboratory tests on samples taken as liquid slurry taken from the trench and cast into cylindrical moulds (generally plastics pipes 100 mm in diameter by of order 450 mm long).

Testing procedures to actually prove the in-situ performance of walls are very much needed. One way is to form a test box by excavating a short length of wall parallel to the main wall and linking it to the main wall with cross walls. A pumped well is then installed within the box and piezometers installed within the box and in the adjacent ground. Such boxes have been installed on a few sites in the U.K. In general the in-situ permeability has been found to be in reasonable agreement with that measured in the laboratory though it must be allowed that there are many uncertainties in the box test. The effective porosity of the soil in the box (and thus the water storage) may be uncertain and flow through the base of the box may be significant but difficult to assess. If it is required to prove a wall to 10^{-9} m/s permeability then the test duration may have to be several months and the pump(s) within the box operated to maintain a constant water level and the flow quantity recorded as a function of time.

A new technique which holds much promise is as follows: on one side of the wall an injection well is installed and on the other side an observation well. Water is pumped into the injection well at a fixed flow rate for a fixed time and the resulting pressure in the well recorded. The flow is then stopped and the pressure decay monitored. The pressure-time response at the observation well is also recorded The resultant pressure profiles are back analysed to estimate the permeability of the wall. Analysis shows that with this procedure it should be possible easily to detect sections of a wall of permeability $>10^{-9}$ m/s. If observation and injection wells are spaced at intervals of order 40 m along a wall it should be possible to isolate leaks or poor quality sections of the wall to a precision of about 15 m. Further details of the procedure and the analytical method are given in Eiben and Jefferis (1994).

14 CONCLUSIONS

There are very many old landfills and work is in progress to remediate a number of these sites. Major problems relate to the general lack of monitoring information, visual impact, poor soil cover and gas and leachate migration Slurry trench walls are being widely employed for gas and leachate control (most frequently for gas). Landfills are also being capped to reduce leachate generation and encourage re-vegetation. Over the next generation landfill designs will change as will waste management practice. Of this we can be sure. What we cannot be sure about is where the changes will occur. Changes may be driven by societal issues, recycling/incineration etc. or because of improved landfill design procedures or developments in landfill practice. We must be prepared to adapt not only the design of new landfills but also the remediation of the old to this ever changing scene.

We must also use old landfills to obtain data on the long term effect of leachate on the mechanical properties of soils such as permeability and stress-strain behaviour. Landfill archaeology is well established within landfills. We must now move outside the landfill and look how the surroundings have been affected.

During the 1960s and 1970s many earth dams were built and it was normal practice for a substantial amount of instrumentation to be included in a dam not only to confirm the safety of the structure during and after construction but also to enable design procedures to be refined. When landfills are constructed or remediated monitoring also ought to be included so that barrier systems etc. can be proven at full scale. Happily such monitoring is a part of some new landfills but unfortunately it is rather rarer in remedial works.

In the UK we have a national soft clay site for large scale geotechnical research. In 1990 the author put forward the idea of a national landfill test site so that the landfill research in different disciplines including microbiology, chemistry and engineering could be integrated. The idea seems to be gathering some momentum but landfill work is still unnecessarily fragmented and it value often reduced because of lack of control data.

15 REFERENCES

Byrne, R.J., Kendall, J. and Brown, S. 1990. Cause and mechanism of failure, Kettleman Hills Landfill B-19, Phase 1a.

Department of the Environment 1994. Waste Management Paper 26A, Landfill Completion, HMSO, London.

Department of the Environment 1989. Waste Management Paper 27, Landfill Gas, HMSO, London, 1989.

Department of Health 1991. Committee on Carcinogenicity of Chemicals in Food, Consumer Products and the Environment, Guidelines for the Evaluation of Chemicals for Carcinogenicity, Report on Health and Social Subjects, HMSO, London.

Eiben, T. and Jefferis, S.A. 1994. Implementation of interactive numerical modelling and hydraulic testing in the design and quality control of low permeability barrier systems, Geol Soc Engineering Group Meeting on the measurement of low permeability in soils and rocks, University of Surrey, November.

Jefferis, S.A. 1992. Contaminant - grout interaction, ASCE Specialty conference, Grouting, Soil improvement and Geosynthetics, New Orleans, February.

Jefferis, S.A. 1993. Design and performance of vertical barriers to pollution migration, in Contaminated Soil, editors - Arendt, F., Annokkee, G.J., Bosman, R. and van den Brink, W.J., Kluwer Academic Publishers.

Jefferis, S.A. 1993. Vertical Barriers to gas and liquid migration, in Contaminated Land Problems and Solutions, editor, Cairney, T., Blackie Academic and Professional Publishers, Glasgow.

Wheeler, P. 1993. Sealing Power, Ground Engineering, Vol. 26 No. 8, October.

Waste Disposal by Landfill – GREEN'93, Sarsby (ed.) © 1995 Balkema, Rotterdam, ISBN 90 5410 356 6

Investigation of closed sanitary-landfill sites for rehabilitation

J. Saarela
National Board of Waters and the Environment, Finland

Abstract: Waste disposal and management are in many ways in focus in most industrial countries and so also in Finland. Wastes are placed or treated in approximately 500 landfills, in 80 industrial waste ponds, in 20 sludge ponds situated in the sanitary landfills and in 2 incinerators. In recent years there have been started many research projects and measures to minimize waste quantities and their environmental effects for example, dam safety legislation, which also concerns sludge ponds, estimation of environmental impacts of hazardous wastes in landfills, inventory of chemically contaminated soils sites and inventory of closed sanitary landfills for rehabilitation.

1 INTRODUCTION

Waste disposal and waste management are nowadays in many ways in focus in most industrial countries and so also in Finland. Wastes are placed or treated approximately in 500 landfills, in 80 industrial waste ponds in 20 sludge ponds and in 2 incinerators. In recent years there has been started many research projects to minimize waste quantities and their environmental effects for example dam safety legislation, which concerns also sludge ponds, estimation of environmental impacts of hazardous wastes in landfills, inventory of chemically contaminated soils sites and inventory of closed sanitary landfills for rehabilitation.

The yearly quantity of municipal waste is 3.1 million tonnes from which 2.5 million tonnes is transported to disposal sites and 0.6 million tonnes to recycling (Ministry of the Environment 1988).

2 WASTE PONDS

Waste pond project included the following points (Saarela 1986, 1988, 1990):

1) the inventory of waste ponds
2) the hazard potential caused by

Table 1. The waste ponds of different industries

Type of industry	Number
Mining industry	18
Forest indutry	21
Ash lagoons of power stations	5
Metal industry	5
Potato "	6
Sugar "	5
Consruction industry	4
Fodder industry	4
Oil waste pond	1
Chemical industry	3
Textile "	1
Leather "	3
Waste plaster	1
Waste lye	2
Dredging lagoons	2
Agricultural	1
Sludge ponds in sanitary landfills	20
	100

different types of dammed wastes
3) the structural requirements
 a) for new waste dams
 b) for improvement of the old waste ponds
4) the safety inspection of waste ponds

In the industry there are 80 waste ponds and in the municipal sanitary landfills there are 20 waste ponds. The waste ponds of the different industries are shown in the table 1.

2.1 Collapses of waste ponds

In the years 1975 - 1985 there have been 6 collapses of the waste ponds of the industry and other problems, e.g. leakages of the lagoons through the bottom 5 cases. There have been 13 cases of collapse of the waste ponds in the municipal sanitary landfills.

In the most cases the cause of failure has been the lack of the geotechnical knowledge in the planning, constuction or in the raising of the dams or mistakes in the operation of lagoons. Reasons for failures are in appencixes 1 and 2.

2.2 The inspections of the waste ponds

The waste ponds structures are classified into four categories in the following way:

- P dams are those which in the case of an accident will endanger life of health, or cause serious damage to the environment or property.
- N dams are those which do not belong to categories P, O or T
- O dams are those which cause only minimal danger.
- T dams are temporary structures as defined in the law.

The Dam Safety Codes of Practice include some technical recommendations concerning the stability, the drainage system, the height of the dam (including frost protection), the slope protection and the permitted location and size of trees.

The safety surveillance programme includes:

- regular inspections every 5 years by a competent expert;
- yearly inspections in the intermediate years, by maintenance personnel; and,
- surveillance between inspections according to the programme defined in the basic inspections.

The responsibilities of different parties for dam safety are as follows:

- the owner is responsible for the safety of a particular dam (the dam must be designed by a competent and experienced person: and,
- the National Board of Waters and the Environment must supervise the fulfilling of all reguirements of the dam safety legislation.

3 INVENTORY OF CHEMICALLY CONTAMINATED SOILS

Chemically contaminated soils were mapped in the years 1988 and 1989 with the following results: (Puolanne 1992)

- There are 10 000 - 20 000 chemically contaminated soil sites.
- 5 % sites are considerable contaminated
- 20 % of sites are situated on usable groundwater areas
- 250 sites need urgent treatment contaminated sites are mainly:
- landfills
- waste handling places
- saws
- impregnation plants
- service stations
- repair shops

4 INCINERATORS

There are two incinerators and the number will be same for many years in the future because for example in the Helsinki region the political decision makers did not approve the new incinerator in the year 1990 though it was long and well planned by communal officials.

5 SANITARY LANDFILLS

In our organization there are many wide projects concerning sanitary landfills: Main projects have been:

- environmental risks of hazardous
- wastes in landfills
- Chemically contaminated soil sites
- geotechnical properties of tailing dams and wastes
- waste analysis methods,
- procedures for the assessment of landfill disposability of wastes, and
- waste research programming
- surface capping and rehabilitation of closed sanitary landfills

5.1 Environmental risks of hazardous wastes in landfills

The environmental impacts and risks caused by hazardous wastes disposed in Finnish landfills were assessed on the basis of a 5-yr field study of 43 landfills. The experimental investigations were concentrated on the chemical quality and toxicity of surface water runoff from municipal solid waste landfills. Additionally, measurements have been made of the quality of waste deposits, landfill gas, ground water and soil.

The following matters were researched:

- contaminant concentrations in landfilled solid wastes
- contaminant concentrations in interstitial water of landfill bodies
- contaminant concentrations in pore gas of municipal landfill bodies
- contaminant concentrations in landfill leachate runoff
- estimates on waterborne emissions of some toxicants in runoff from the studied landfills
- contaminant concentrations in ground water near landfills

The environmental risks of hazardous wastes in landfills were estimated to be small in relation to other chemical risks but locally significant, depending e.g. on migration and attenuation in wastes (Asmuth et al. 1990, Assmuth et al. 1991).

Ettala (1987, 1988, 1989) has researched in many projects vegetation and hydrological aspects in landfill management in Finnish cold climate conditions. According to his results by removal of the snow cover and establishment of short-rotation tree plantations at a sanitary landfill infiltration and leachate discharge can be diminished even while the site is in operation. Evapotranspiration from the Salix aquatica stand during the growing season exceeded the mean annual precipitation.

6 INVENTORY OF OLD SITES

6.1 Introduction

National Board of Waters and the Environment started this project because about 300 sanitary landfills will be closed up to the end of this century and there are no practical directives or code of practice on how to cover and rehabilitate sanitary landfills after their use.

The project has included among other things the following points:

1) Inventory of closed sanitary landfills in Finland (157 landfills)
2) Inventory of geotechnical structures of covers of closed sanitary landfills
3) Inventory of landscaping of closed sanitary landfills
4) Development 6 different types of the cover structures of sanitary landfills.
5) Development of water-balance model

In the inventory the following points were investigated: the starting and closing year of the sanitary landfill, the type of waste, the height and the area of the landfill, the material of the cover structure, the afteruse, the landscaping, the quality control of the leachate waters and the environmental consequences of the sanitary landfill.

6.2 RESULTS OF THE INVENTORY

6.2.1 Number of the landfills

The inventory was done in the region of the 13 different water and environmental districts. The number of the landfills varied between 4-30 in the different regions. It was seen specially that in the communities of eastern Finland there were many small village sanitary landfills. The total number of the landfills was 157.

6.2.2 Age of the landfills

51 landfills were older than 20 years. 28 landfills were older than 15 years but not over 20 years. 13 landfills were older than 10 years but not over 15 years and 7 were younger than 10 years. However the age of all landfills was not known.

6.2.3 Type of wastes

Most of the landfills were for the household and industrial wastes (77). However almost the same number of landfills were only for household wastes (74). 6 landfills were for industrial wastes only.

6.2.4 Height and the area of the landfills

The height of the most landfills was under 5 meters (89). 35 landfills were 5-10 meters. 19 landfills were between 10-20 meters and 14 were over 20 meters.

The area of the most landfills was under 0.5 hectars (66). 45 landfills were 0.5-2 hectars. 23 were between 2-5 hectars, 8 landfills were between 5-10 hectars and 15 landfills were over 10 hectars.

6.2.5 Material of the covers

Most of the landfills were covered with waste soils (103). Sand or silt were on 13 landfills. Gravel and moraine together were on 5 landfills, and clay on 9 landfills. Other types of cover structures were on 9 landfills for example asphalt, peat and chips. Layer depths varied between 0.5-20 m. Vegetation layers were between 0.1-0.3 m.

The cover structures were in most cases very defective, for example hydraulic barrier, the vegetation layers and gas control.

6.2.6 Afteruse

Most of the landfills had no afteruse (124). However for example the most developed afteruse was in Helsinki city region, where there are modern and complex recreation areas on the closed landfills. Other ways of afteruse were for example store areas of the industry, compost areas and motor sport areas.

6.2.7 Landscaping

Most of the landfills were not landscaped (108). The landscaping was made on 12 landfills wholly and partially on 21 landfills. The landscaped landfills were near to the big cities. The landscaping in general was also connected to some kind of afteruse.

6.2.8 Quality control of the leac hate waters

Quality control of the leachate waters was not uniform due to the great differencies between the landfills for example their location, area and waste content. The leachate waters of 61 landfills are or have been under control and the number of the water samples varied from one sample during the whole life of the landfill to the 4 yearly samples. The control results of the 31 landfills were discussed in the inventory.

6.2.9 Environmental effects

Two of the landfills have caused the groundwater contamination. 7 landfills have caused changes in the grounwaters but there is not yet serious contamination. However the situation needs to be controlled.

In 16 cases the effects of the landfill leachate have been observed in the ground or surface waters but they are not serious due to the small quantity of the contamination, location of the landfill or due to the other greater contamination sources.

It was seen that the most serious consequences of the closed landfills were those when the leachate waters cause the contamination of groundwaters.

When the landfill is situated on the groundwater area and the leachate is observed, causing yearly contamination values in the groundwaters, then one must think of even the removal of the landfill from that area. In the inventory there was one case where one has planned the removal of the landfill but that has not yet been done.

In all Finland there are on the groundwater areas 157 landfills. It has been estimated that 71 have had some kind of effect on the quality of the groundwater.

Figure 7 shows the landfills where the inventory was made.

7 REHABILITATION AND TYPES OF THE COVER STRUCTURES

In principle the rehabilitation of the closed landfill can be done in the following ways:

- cover it,
- remove it, for example because it contaminate the groundwaters
- inject or build cut wall around the landfill to prevent the contamination groundwaters

However the most common rehabilitation method is to cover the landfill properly and in this connection there is discussed only cover structures because it is the most common way of the rehabilitation.

7.1 Requirements for the cover structures of closed sanitary landfills

The main problem is how they should be covered so that the leachate is minimised and how they should be vegetated after their use. Sanitary landfill differs considerably from the mine waste or build rubble area because for example their settlement can usually be controlled by compaction by earth-moving machinery during the period when the tip is being built up. Sanitary landfill, however, suffers considerable settlement by the decomposition of organic contents over a long period of time.

The other problem is gas generation in relation to the engineering of landfill reclamation with building and vegetation.

A cover at a landfill site is an earthwork construction designed to exclude surface wastes from contact with disposed wastes, to act as a barrier against human or animal contact with the wastes and to prevent release of vapors to the environment. Covers are preventive measures used in most site remediation plans. They are seldom used alone but are part of an integrated concept of waste containment.

A cover may perform a number of functions, as listed in Table 1. Along with functions are listed necessary attributes, which the cover must have to accomplish these functions. Attributes pertain mainly to the cover's own durability and permanence. The purpose of good design is to provide these attrtibutes while assuring the performance of the desired function.

While all of the functions are important, that of preventing percolation is predominant, because contamination of surface- and groundwater supplies is the most serious threat from uncontrolled hazardous waste sites.

7.2 MATERIALS USED IN THE COVER SYSTEMS

Soils can be used in most cover systems. They are usually are available on or near the site and are relatively cheap. Not all local soils are suitable for all cover purposes, accordingly either im-

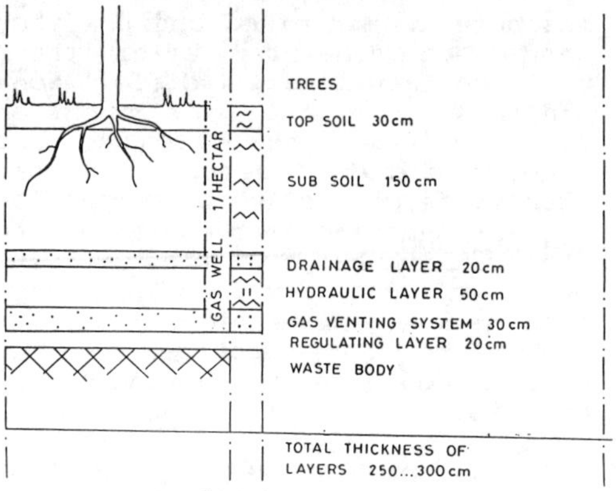

Figure 1. Cover structure type 1.

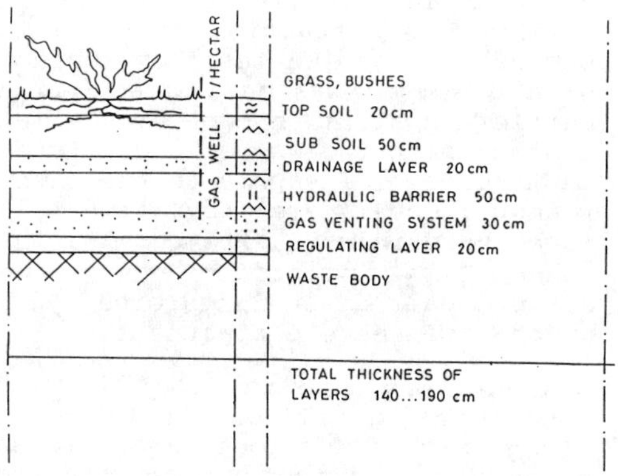

Figure 2. Cover structure type 2.

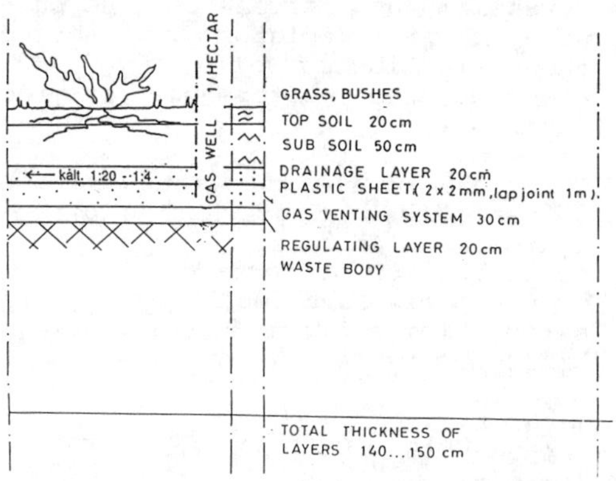

Figure 3. Cover structure type 3.

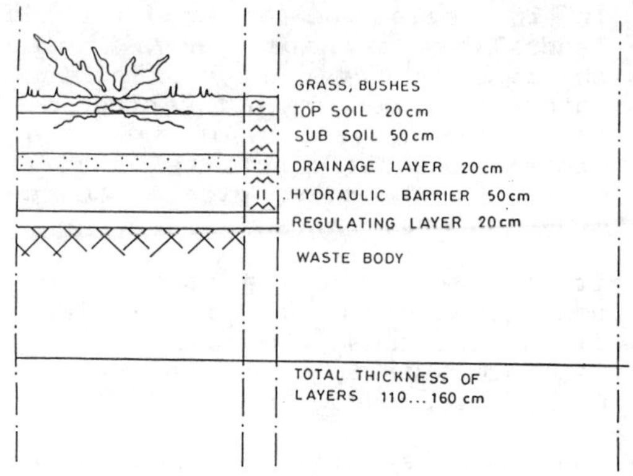

Figure 4. Cover structure type 4.

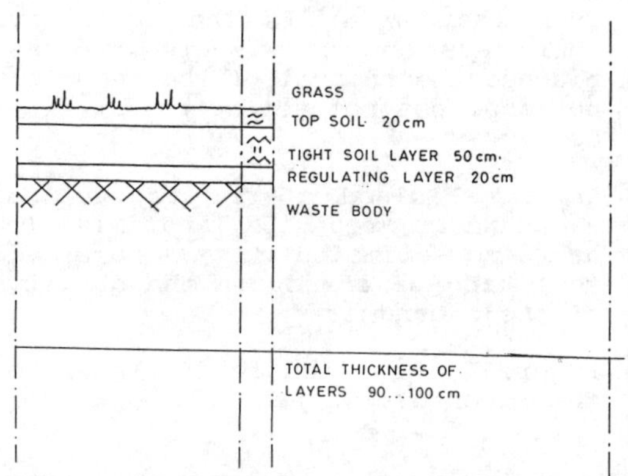

Figure 5. Cover structure type 5.

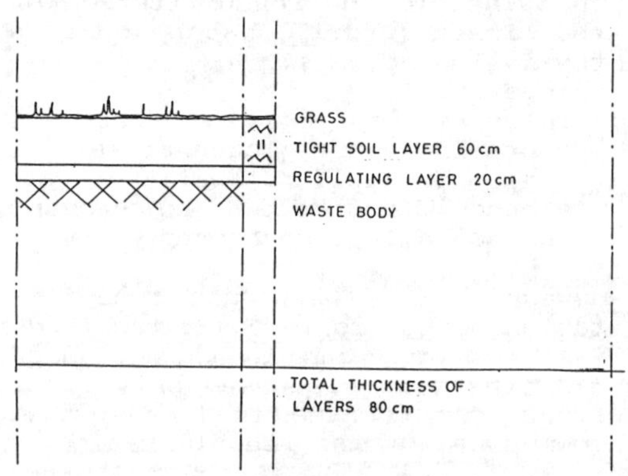

Figure 6. Cover structure type 6.

Table 1. Cover functions and attributes (McAneny and Hatheway 1982).

Functions
Prevention or Minimization of Percolation
Promotion of Aesthetics
Suppression of Vectors
Containment of Gases
Suppression of Fire Danger
Prevention of Blowing Litter of Dust
Promotion of Site Reclamation
Necessary Attributes
Water Erosion Resistance
Wind Erosion Resistance
Stability Against Slumping and Cracking
Stability Against Slope Failures
Resistance to Cold-Weather Distress
Resistance to Disruption by Animals or Plants

ported soils or non-soil materials may have to be employed. The cover designer needs to consider soils both from the engineering and the agricultural points of view.

Permeability is an important property of cover soils since the cover system is largely intended to retard the inflow of water. When a soil is unsaturated, its permeability is less than when it is saturated. Out of a large number of possible flow channels through the soil, only a fraction is utilized. Unsaturated flow is important because the soils in a cover system will be completely saturated for only a small portion of the time, if ever.

7.3 STRUCTURE OF COVER SYSTEMS

In the inventory of closed sanitary landfills it was seen that the cover structures of closed sanitary landfills do not come up to the requirements of modern landfill technology. It was also seen that their size and location vary from each other for example from the sanitary landfill of a small village in the outlying district to the great landfills of big cities.

For this reason different sanitary landfills must be covered in different ways. Sanitary landfills of big cities require totally different cover structures than small village landfills and also their costs are different. So one must select different requirements for different types of sanitary landfills.

For this reason in this project there has been developed the types of the cover structures for the Finnish code of practice on how to cover closed sanitary landfill. Also their costs have been calculated and there has also been developed simulation models for calculating water balance and leaching production, which go through these cover structures. However in this connection there are mainly discussed the types of structure and their costs.

In general the cover of landfill is a multi-layer system and any system must have a surface layer. Other layers, include the following:

1) A hydraulic barrier
2) A gas-control layer intercepts gases evolved by the wastes and leads them to the atmosphere via venting mechamisms
3) A regulating layer (a foundation or buffer layer) isolates the hydraulic barrier from the wastes and also serves as a strong base to support the rest of the system
4) A drainage layer intercepts downward-percolating water and conveys it laterally out of the system
5) Filter layers, serve to separate fine from coarse materials and prevent clogging of coarse interstices by fine particles.

In this project there were developed 6 different types of cover structures, which vary very much from each other. In the following they are discussed in detail.

Structure type 1 is the most superior one, which comes up to the full requirements of modern landfill technology (fig 1). It is used when there is the intention to build on the closed sanitary landfill for example a modern and complex recreation area. It can also be used on the hazardous waste landfills. The total thickness of the cover system varies between 2.5-3.0 m and it can be used for planting trees on the sanitary landfills.

Structure type 2 has same structure layers as type 1 but they are thinner (fig. 2). In structure type 3 the hydraulic barrier is replaced by plastic sheet. In this special case the quality of plastic sheet does not need to be very good because this type is for the recirculation of leaching waters (fig. 3). The structure type 4 has the same standard but there is not the gas ventilating layer, because in many landfills one builds a separate system with its own wells and pipes for gas collecting. These three types fulfill minimum requirements for health and environmental protection when there is the intention to use the sanitary landfill for example for the recreation. These types are not suitable for hazardous wastes.

Types 5 and 6 represent simple and light cover structures (fig. 5 and 6). They are used for example on the old landfills where environmental risks are already decreasing. These light structures are also sufficient for small landfills which are situated far from the settlement when there is no need for activity purposes for example for recreation. For their structural simplicity they are not sufficient for preventing leaching water nor for controlling gas discharge.

8 BUILDING COSTS OF COVER SYSTEMS

Costs of cover systems consist of construction and building material costs. Costs of soil material are calculated so that the transport distance is about 10 km. If the transport distance increases for example to 20 km, it increases construction costs 20 % in that structure where all soil materials must be bought and transported to landfill. Costs are calculated in the year 1990.

Costs calculations are made for two alternative in the following way:

1) Cover structures can be built of waste and excess soils free of charge of soils, which are transported to the landfill site.
2) All soil materials must be bought and transported to the construction site.

The building costs of the covers built of waste and excess materials are 40-80 % of the costs of the bought materials. The cheapest cover type costs 37 000 US $ and the most expensive one costs 403 000 US $ (table 2).

Table 2. Construction costs of the cover systems in the year 1990.

Cover Type	Waste and excess soils can be used free of charge 1 000 US $	All materials must be bought and transported 1 000 US $
1	257	403
2	225	307
3	251	300
4	151	233
5	66	114
6	37	92

9 DEVELOPMENT OF WATER-BALANCE MODEL BY FINITE ELEMENT METHOD FOR THE FINAL COVER DESIGN

For the final cover design we are doing the water balance model by finite element method.

In the following there are some facts about the model. The model needs for example the following data about the sanitary landfill:
- permeability of cover
- thickness of cover
- gradient of cover
- vegetation on the cover etc.

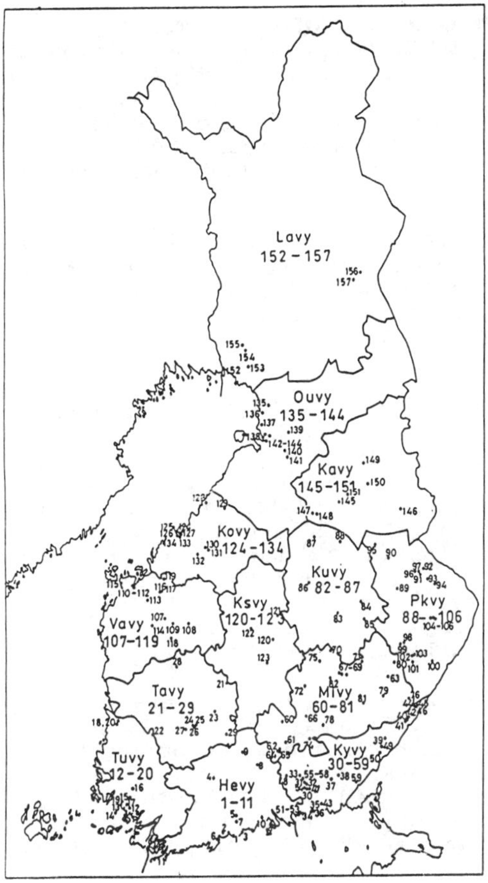

Figure 7. Inventory of closed sanitary landfills. The numbers refer to landfills.

In the development and testing of the model following research results were obtained. Tests were made mainly on three large landfills near Helsinki.

- weight sounding of the sanitary landfills was very difficult due to the inhomogeneous waste material for example building rubble containing concrete pieces.
- water table was 5-10 m from the cover of landfill. The total height of landfills was between 10-20m
- unit weight was at a depth of 20 m 17...19 kN m^{-3} and at the surface 12 kN m^{-3}
- from the resistance of rub drilling it was estimated that the angle of friction was at a depth of 20 m approximately 42...45° and near the surface 32...36°
- frost depth was between 40-90 cm. It is less than in natural soils
- conductivity was between 95-2020 µs cm^{-1}. The value is low and wastes contain little salt and ionized compounds
- pH of wastes was 5,16-9,15. The value is high containing alkaline compounds
- bearing capacity of landfills was 10-25 Mpa. The value is low and for example the bearing capacity of sport field is approximately 50 Mpa
- permeability of covers was $10^{-5.49}$-$10^{-8,91}$ m s^{-1}
- methane measurements showed that in the research points, where methane comes to the surface there is no vegetation and there is also very high temperature at those points maximum 46°C
- nutrient analysis of surface soils showed that there was metal content in some samples but high metal content of soils were not so important in the failure of vegetation as for example poor nutrient content of soil.
- there was not higher gamma and radon content in the soil samples taken on the surface of sanitary landfills as in the surrounding.

REFERENCES

ASSMUTH, T., STRANDBERG, T., POUTANEN, H., MELANEN, M., PENTTILÄ, S. and KALEVI, K. (1990) "Environmental impacts of hazardous wastes in landfills". Helsinki, National Board of Waters and the Environment. Publ. Water and Environment Administration - Series A 67. 211 p. (In Finnish, with English summary.) ISBN 951-47-4281-8. ISSN 9786-9592.

ASSMUTH, T., POUTANEN, H., STRANDBERG, T. and MELANEN, M. (1991). Occurrence, attenuation and toxicological significance of hazardous chemicals in uncontrolled landfills - codisposal risks reconsidered. Proceedings Sardinia 91, Third International landfill Symposium S. Margherita di Pula, Cagliari, Italy; 14-18 October 1991. CISA, Environmental Sanitary Engineering Centre, Cagliari, Italy.

ETTALA, M. 1987. Kaatopaikkavesi kastelun vaikutus biomassatuotokseen ja haihduntaan jätepenkereellä. Lic. thes. Techical Univ. of Helsinki, Dep. of Civil Engineering.

ETTALA, M. 1988. Short-rotation tree plantations and hydrological aspects in landfill management. Thesis for the degree of Doctor of Technology. Helsinki University of Technology. Otaniemi. Finland.

ETTALA, M. 1989. Vegetation and hydrological aspects in landfill management. Proceedings of ISWA international symposium on process, technology and environmental impact of sanitary landfill. Gagliary, Sardinia. Oct. 1989.

BARNES, F.J. and RODGERS, J.C. 1988. Evaluation of Hydrologic Models in the Design of Stable Landfill Covers. EPA/600/2-08/048.

FENN, D.G., HANLEY, K.J. and DEGEARE, T.V., "Use of the Water Balance Method for Predicting Leachate Generation from Solid Waste Disposal Sites, "EPA/530/SW-168, U.S. EPA, Cincinnati, OH, 1975.

JESSBERGER, H.L., KLOS, U. (1992). Geotechnical aspects of design and construction of waste containment facilities. Proceedings of the Mediterranean Conference on Environmental Geotechnology, Cesme, Turkey 25-27 May 1992. ISBN 90 5410 0559.

LECHNER, P. 1987. The Austrian Guidelines for Sanitary Landfills. Proceedings of International Symposium on process, technology and environmental impact of sanitary landfill. Cagliari, Sardinia, 19th-23rd October 1987.

LUTTON, R.J., REGAN, G.L. and JONES, L.W., "Design and Construction of Covers for Solid Waste Landfills, "EPA-600/2 79 165, Municipal Environmental Research Laboratory, U.S. EPA, Cincinnati, OH, 1979, 250 pp.

MCANENY, C.C. & HATHEWAY, A.W. 1985. Design and Construction of Covers for Uncontrolled Landfill Sites. The 6th National Conference on Management of Uncontrolled Hazardous Waste Sites. November 4-6, 1985. Washington, DC.

MINISTRY OF ENVIRONMENT (1988). Development Programme on Municipal Waste Management 2000. Cin Finnish, ISBN 951-47-4766-6, ISSN 0788-5903.

PUOLANNE, J. (1992). Chemically contaminated soils. (in Finnish). INSKO 28-27.8.1992.

SAARELA, J. 1986. Tailings dams, waste ponds and the dam legislation in Finland. Paper presented at the First International Symposium on Environmental Geotechnology, April 21-23, 1986, Allentown, USA 7p.

SAARELA, J. 1988. Operation and Surveillance of tailings and other waste dams according to the Finnish Code of Practice. Paper presented at ECC Seminar on Dam Safety, Rovaniemi, Finland, August 29 - September 2, 1988.

SAARELA, J. 1990. Geotechnical pro perties and environmental effects of tailings. Thesis of Lic. Tech. (in Finland) Helsinki Technical University, SF.

SAARELA, J. (1992). Kaatopaikan pinnan loppukäsittely ja sen simulointimalli. Väliraportti. Vesi- ja ympäristöhallitus. (Surface structures of closed sanitary landfills and their simulation model: in Finnish; not published) National Board of Waters and the Environment.

SAARELA, J. (1993). Lopetettujen kaatopaikkojen inventoinnin tulokset. (Inventory of closed sanitary landfills: in Finnish; not published) National Board of Waters and the Environment.

SAARELA, J. 1991. Landscaping aspects of sanitary landfills in Finland. Third International symposium 14 - 18 October 1991. Sardinia -91.

VIATEK Oy, (1991). Käytöstä poistuvien kaatopaikkojen pintarakenteiden kustannukset. Esiselvitys. (Costs of surface structures of closed sanitary landfills: in Finnish).

APPENDIX 1. The reasons for failures and other problems of waste ponds in the industry.

Helsinki water and environment district

- Leakage of a waste pond of the plastic industry.
  The reason: unsufficient grout at the bottom
- The leakage of dams of sugar industry. The reason: inadequate dam structures and soft ground.

Tampere water and environment district

- Collapse of a waste pond of the paper industry. The reason: waste pulp layer in the dam, which caused the failure.
- Collapse of a dam of the leather industry. The reason: insufficient structural strength of the dam.

Kuopio water and environment district

- Collapse and overflow of mining industry tailing dam.
- The reason: the iced moraine masses in the heightening of the dam and the sludge pipe too long without moving.

North-Karelia water and environment district

- Collapse of the tailing dam of the mining industry.
  The reason: the rising of the water level caused by icing of the water route and the unsuitable masses in the structures of the dam.

Central-Finland water and environment district

- Collapse of a bark dam of the paper industry.
  The reason: the insufficient structural strength of bark as dam building material.
- The leakage of a waste pond through dam structures in the paper industry.
  The reason: insufficient tight layers in the structures of the dam.

Kokkola water and environment district

- Leakage of a tailing dam through the foundation.
  The reason: insufficient ground investigations and grouting in the building of the dam.
- Cracking on a tailing dam and the rising of ground near the dam.

Oulu water and environment district

- Collapse of a dam of the construction material industry.
- The reason: leaving the surface masses under the dam.

APPENDIX 2. The reasons for failures of dams for waste ponds in the sanitary landfills.

Helsinki water and environment district

- One failure.
- The reason: overtopping

Tampere water and environment district

- 3 failures.
- The reason: insufficient strength of the dams.

Kymi water and environment district

- 2 failures.
- The reason: dams were raised in the winter with iced masses and in the spring when the ice melted the dam collapsed. Hydraulic failure.

Central-Finland water and environment district

- 2 failures.
- The reason: insufficient strength of dams, leakage of the dam.

Oulu water and environment district

- one failure.
  The reason: the raising of the dam with iced masses and the filling of the pond with the melting water from the surrounding area in the spring.

1 Current international practice

Waste Disposal by Landfill – GREEN'93, Sarsby (ed.) © 1995 Balkema, Rotterdam, ISBN 90 5410 356 6

QC/QA program for the construction of a municipal waste landfill

F. Belfiore
Browning-Ferris Industries Italia, Milan, Italy

F. Magri
Milan, Italy

ABSTRACT: A comprehensive and advanced QC/QA program has been carried out for the construction of a new municipal waste landfill in Northern Italy. Due to the lack of reference in the national legislation, the QA plan had been agreed and developed together with the public control authority in such a way to cover at least all the legal and permit technical requirements relevant to the landfill construction.

1. INTRODUCTION

Quality Assurance can generally be defined as a systematic pattern and organization of work designed to provide adequate confidence in the performance of items or services to meet contractual and regulatory requirements.

Quality Control, on the other hand, is a set of actions which provide a means to measure and regulate the characteristics of an item or service according to plans and specifications.

QC/QA programs are an essential feature at all stages of landfill development, including design, construction, operation and monitoring of a site.

The various aspects involved at each stage have very strong implications on all the others, therefore a consistent quality level must be guaranteed in order to optimize the overall landfill performance.

Within this framework, the paper describe a possible approach to the definition and implementattion of a reasonable and effective Construction Quality Assurance plan.

2. SITE DESCRIPTION

The landfill site is located about 80 km south of Milan in a typical alluvial plain area which had been developed as the result of a sand quarry operation.

The total property area is 6 ha, of which 5 ha are used for the landfill and 1 ha for the facilities.

The planned average height of the fill is 6 m, with a resulting total airspace of 300.000 m^3.

The landfill is divided into two adjoining cells, surrounded by outer earth embankments 4 m high, having separate lining and leachate collection systems. The outer berm, along the field drainage canal, was reinforced with geogrids and toe retaining blocks.

A schematic plan view of the site is shown in Figure 1. The typical section of the earth embankments, with and without reinforcement, is shown in Figure 2.

To insure a secure and proper distance between the bottom of the cells and the groundwater table, the floor of the quarry had to be raised by an average of 2 m, using selected backfill material.

The lining system consists of 1 m of compacted clay, with a minimum required hydraulic conductivity of 10^{-7} cm/s, topped by a double HDPE geomembrane, with an interposed leak detection system. This system consists of a sand cushion layer with a grid of electrodes, placed to help locate leak points in the geomembranes.

A leachate collection system,

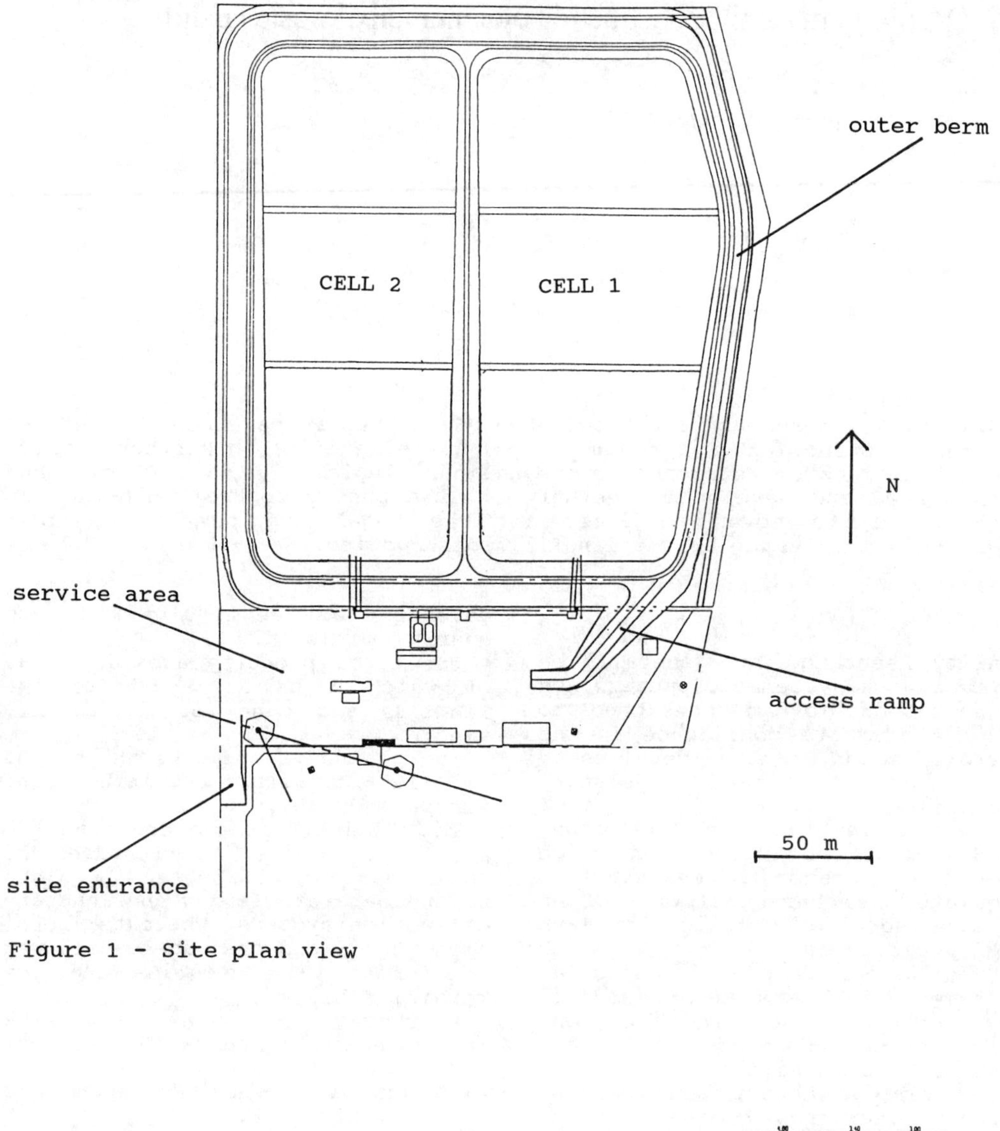

Figure 1 - Site plan view

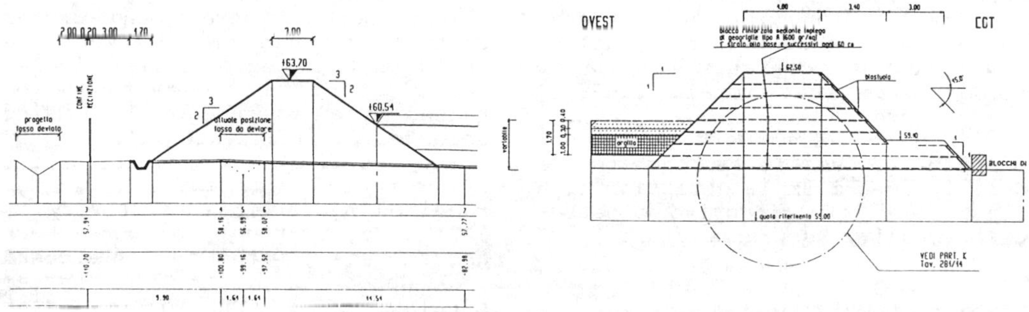

Figure 2 - Typical cross section of the berms

consisting of HDPE pipes in a sand layer, was placed above the lining system to reduce any potential hydraulic head on it and allow for the rapid removal of liquids as they are produced.

3. CONSTRUCTION QA ORGANIZATION

The basic objective of the QA organization is to verify, document and certify that, at the end of construction, the landfill meets or exceeds the design criteria and the legal requirements.

Due to the relatively small size of the project, an effort was made to simplify and optimize the structure of QA.

The parties involved in the program were:

- QA Manager.
- QA Assistant.
- On site soil laboratory.
- External soil laboratory.
- Certification Commission.

The responsibility of managing the QA plan was given to the "Direttore Lavori" (Director of Works), whose figure is defined by Italian law for public works in order to follow and check particularly the financial and bookkeeping aspects for construction and, at a much lower degree, the assurance of quality.

In this case, the QA Manager was assigned the following responsibilities:

- Review the initial documentation.
- Co-ordinate and review the results of the planned quality tests and analyses.
- Inspect construction operations.
- Implement remedial actions to correct non-conformances with the plan specifications.
- Prepare state of progress reports, in connections with the terms of payments agreed with the contractors.
- Prepare final documentation.

In developing the program, the QA Manager has been supported by a QA Assistant, a qualified engineer, to perform the following activities:

- Review QC documentation from the suppliers.
- Check acceptance, construction, sampling and testing practices and procedures.
- Survey quantity of materials and size of components.
- Provide the QA Manager with the monitoring documentation.

The on site and external soil laboratories were used respectively to check the materials and placing procedures in real time, and to certify the proper completion of the single components of the earth structures, as defined in the QA plan.

The Certification Commission, which represents the Public Authority, performs the following duties:

- Inspect and review the general construction process.
- Review the documentation prepared by the QA Manager.
- Check certain characteristics of the components or of the whole structure by means of indipendent laboratories and consultants, if necessary.
- Issue the conformity certification, in order to start the operation phase of the landfill.

4. CQA DETAILED PLAN

The CQA has basically developed and integrated the original design specifications on the natural subgrade, the foundation backfill, the embankments material, the clay liner, all the geosynthetics used, including the HDPE geomembranes, the leachate management system and all the other service facilities installed.

The details of tests on soils, including reference test methods and frequencies, are shown in Table 1, while the QA plan for geomembranes is outlined in Table 2.

On all the other supplies and works, the following preconstruction acceptance and performance assessment procedures were implemented:

- Review manufacturer's data certification and specification conformity of geotextiles and HDPE pipings.
- Review backup documentation and perform operability tests on the leachate collection and storage system, the leak detection system, the truck weighing and wheel washing systems.
- Supervise installation and perform operability tests on all

Table 1. Tests on soils.

Item	Reference Test Method	Minimum Frequency
NATURAL SUBGRADE		
- plate load bearing	SNV 670317 (Swiss Code)	3000 m^2
FOUNDATION FILL		
- grain size distribution	ASTM D422	4000 m^3
- Atterberg limits	ASTM D4318	4000 m^3
- moisture-density relation (Proctor test)	ASTM D698	4000 m^3
- in-situ density	ASTM D1556	500 m^3
- plate load bearing	SNV 670317	3000 m^2
OUTER EMBANKMENTS		
- grain size distribution	ASTM D422	2000 m^3
- Atterberg limits	ASTM D4318	2000 m^3
- moisture-density relation	ASTM D698	4000 m^3
- in-situ density	ASTM D1556	up to 3m high: 500 m^2 above 3m : 200 m^2
- geogrid tensile strength	manufacturer QC	each roll
CLAY LINER		
Preconstruction testing:		
- grain size distribution	ASTM D422	2500 m^3
- Atterberg limits	ASTM D4318	2500 m^3
- natural moisture content	ASTM D2216	500 m^3
- organic content	ASTM D2974	2500 m^3
- moisture-density relation	ASTM D698/1557	5000 m^3
- laboratory permeability of undisturbed samples	ASTM various	2500 m^3
- laboratory permeability of re-compacted samples	ASTM various	2500 m^3
Field testing:		
- in-situ density	ASTM D1556	500 m^3
- moisture content	ASTM D2216	500 m^3
- laboratory permeability of undisturbed samples from finished lift	ASTM various	2000 m^2
- field permeability		20000 m^2
- plate load bearing	SNV 670317	1000 m^2
DRAINAGE LAYERS		
- grain size distribution	ASTM D422	1000 m^3

Table 2. Tests on geomembranes.

Item	Reference Test Method	Minimum Frequency
PRECONSTRUCTION TESTING		
- review raw material data certification		
- review manufacturer's internal QC documentation		
- review and approve installation plan		
- review and approve welding machine type and procedure		
WELD NONDESTRUCTIVE TESTING		
- ultrasonic survey		100 %
- surface visual inspection		100 %
WELD DESTRUCTIVE TESTING		
- shear behaviour	ASTM D638	300 m
- "peeling" behaviour	ASTM D638	300 m

the components of the environmental monitoring system.

All these activities have been fully documented, together with the corrective actions taken, in order to allow the final certification of the landfill.

5. SPECIAL Q/A TESTS

In addition to the above mentioned tests, special in-situ tests were conducted to assess the actual performance of the lining system, including a Sealed Double Ring Infiltrometer test (Daniel & Trautwein 1986), in Cell 1, and a Boutwell Permeameter test (Boutwell & Derick 1986), in Cell 2, on compacted clay and an electrical method of defect location of the geomembranes.

The field values of hydraulic conductivity of the clay, k, were also compared with a number of laboratory values, measured on undisturbed samples taken from the finished liner.

The SDRI test result was k=1.80E-07 cm/s, versus an average value of k=5.23E-09 cm/s measured by oedometer tests.

The Boutwell test result was k=6.50E-08 cm/s (vertical permeability), versus an average value of k=1.06E-08 cm/s in the laboratory.

These results confirm that usually the value of k measured with an oedometer test is underestimated with respect to the field value, which, in turn, can be taken as more representative of the actual clay liner performance.

As a further guarantee of the geomembrane liners performance, an electrical method was also used, measuring the resistivity field along the landfill floor. The principle of this survey is based on the big difference in resistivity between the polimeric membrane and the underlying and overlying materials (sand in this case).

After placement of the upper leachate collection layer, the cell floor was sectioned into a grid of squares, 1 m by 1 m, which were individually surveyed.

In order to verify the accuracy and precision of the method, two sample holes were made on purpose, which were later detected by the survey.

Some results of the survey are shown in Table 3.

It can be seen that the degree of accuracy is very high and that this method represents an improvement of the overall degree of reliability of the lining system.

Table 3. Typical result of the electrical survey on Cell 1.

Discontinuity no.	Line no.	Distance m	Discontinuity Description
1	46.5	2.0	- Hole, 2 mm diameter
2	62.5	2.0	- Sharp cut, 5 mm long
3	108.5	11.0	- Hole on a seam, 2mm dia.
4	177.0	75.0	- Sharp cut, 2 mm long
5	163.0	75.0	- Sharp cut, 1 mm long
6	7.0	46.0	- Hole on a seam, 2 mm dia.

Total number of discontinuities: 6.

These QA methods, results and certified repairs are important to all parties concerned because create those documents necessary to end the construction phase of the project and begin the operation phase.

It is also thought very important that the parties responsible for operational aspects, as well as those parties responsible for closure aspects, attend or have the opportunity to verify the tests and review all the tests results. In this way key aspects of future liability can be more correctly defined.

6. ECONOMICAL ASPECTS OF CQA

An interesting aspect of a CQA program is the cost involved.

For the landfill described herein, costs were tracked during the design and development of the QC/QA plan and maintained throughout the project.

Consultant fees, laboratory fees, third party or internal labour and administration costs were documented not only to control costs on this project but also to create a useful set of data for estimating and controlling future projects.

In this case, a cost of about 2.5 % of the overall value of the contracts has been calculated, all inclusive.

7. FINAL REMARKS

A sound landfill construction Quality Assurance program is a key factor to guarantee that the performance of such a facility is satisfactory from a technical, operational and environmental point view.

In addition to that, Quality Assurance must be consistently applied throughout the entire life of a landfill, from planning to post-closure.

The application of such a program to the construction of the landfill described in this paper has created a foundation to deal with and overcome the initial opposition of the public to the project, thus establishing a new, promising point of view over the general issue of waste disposal.

REFERENCES

Boutwell, G.P. & Derick, R.K. (1986). Groundwater protection for sanitary landfill in the saturated zone. Waste Technology, National Solid Waste Man. Ass., Chicago.

Daniel, D.E. & Trautwein, S.J. (1986). Field permeability test for earthen liners. In S.P. Clemence (ed.), Use of In Situ Tests in Geotechnical Engineering, p.146-160. Blacksburg, Va., ASCE.

Waste Disposal by Landfill – GREEN'93, Sarsby (ed.) © 1995 Balkema, Rotterdam, ISBN 90 5410 356 6

Waste disposal by landfill – German landfill lining systems

David J. Bishop
Naue Fasertechnik GmbH & Co. KG, Lübbecke, Germany

Geoffrey Carter
Netlon Ltd, Blackburn, UK

ABSTRACT: This paper describes the landfill lining systems as laid down in the German regulations.

1 GERMAN LANDFILL LINING SYSTEM DEVELOPMENT

As in other countries, in the past, Germany rubbish was usually disposed of in convenient holes covered with soil and forgotten. In 1972 legislation moved the responsibility for waste disposal from small local authorities to large district councils which developed centralised controlled landfills.

Initially German landfills were lined only with clay. However this was not always sufficient. Tests carried out on clay at the bottom of a 50 year old landfill in Germany showed that heavy metals had only penetrated a few centimeters into the clay. But on another site traces of contaminants from the landfill were found 200 m into the natural clay base. The reason for this is that some clays have natural cracks and fissures allowing the leachate to escape [1].

Clays cannot cope with tensile forces, they are frost susceptible and are dependent upon the weather for successful installation. However they exhibit an interesting capability. Chemicals become "absorbed" by the clay through, for example, ionic exchange.

Because of the problems with clays, geomembranes alone were also used in landfills. They too were not always sufficient for a number of reasons. Some chemicals, e.g., chlorinated hydrocarbons, can penetrate geomembranes by diffusion. Geomembranes are susceptible to damage by puncturing and the welded seams can be a source of problems. In the past the quality control of the welded seams was not as good as it is now.

In view of the advantages and disadvantages of clay and geomembrane lining systems it wasdecided that the best solution would be to combine the two.

2 GERMAN REGULATIONS

2.1 Federal and state law

Germany consists now of 16 states. Legislation comprises of both Federal and State Laws, the Federal Laws having priority. Many states use laws developed in other states and there are organisations involving many states but which are not regarded as federal ones. An example of this is LAGA - Landes-Arbeitsgemeinschaft Abfall (State Working Group Waste).

2.2 Federal laws

There are two main federal regulations effecting landfill lining systems. These are:

2.2.1 TA Abfall

Technische Anforderung Abfall - Technical regulations - Waste. This has been law since 1990 and deals with the disposal of hazardous waste.

For the basal seal:

a) Below the landfill liner there should be

at least 3 m of soil with a k-value $\leq 1 \times 10^{-7}$ m/s and this should have at least 10% clay (by weight). The water table should be 1 m below the base of the landfill.

b) The lower element of the combination lining system is 1500 mm of clay compacted in 250 mm lifts with permeability $k \leq 5 \times 10^{-10}$ m/s ($i = 30$, laboratory values).

c) Installed above this clay and in intimate contact with it is the HDPE geomembrane which must be a minimum 2.5 mm thick with specific characteristics. After settlements the geomembrane should be installed with a cross fall of $\geq 3\%$ and a longitudial fall of $\geq 1\%$.

d) The geomembrane should be adequately protected. This is specified in the certification for the geomembrane.

e) The leachate drainage layer should be at least 300 mm thick with a permeability, $k \geq 1 \times 10^{-3}$ m/s over the long term.

For the landfill cover:

f) The cover system should be so selected that leaks can be localised and repaired for the duration of the after care period.

g) As a cover levelling layer at least 500 mm of non cohesive soil should be used. If gas is present and cannot be collected in this layer an additional gas drainage layer $t \geq 300$ mm should be installed. The carbonate content should be 10% or less by mass.

h) The liner should be a combination of geomembrane and clay. The clay layer should have a thickness, t, > 500 mm with a permeability $k \leq 5 \times 10^{-10}$ m/s ($i = 30$, laboratory values) and the geomembrane $t \geq 2.5$ mm. After the expected settlements the cover should have a fall $\geq 5\%$. It must be shown that the drainage immediately above the geomembrane system is capable of removing the expected rain water. Such a layer is usually min. 300 mm thick. This should be covered with at least 1 m of suitable soil for cultivation.

2.2.2 TA Siedlungsabfall (Draft)

Technische Anforderung Siedlungsabfall (Draft) - Technical Regulations Municipal Waste (Draft). Although not yet finally law the main elements are established and current municipal landfill designs follow these requirements. Two landfill classes are designated, Class I for low grade waste and Class II for middle grade i.e. municipal waste. TA Abfall (see para 2.2.1 above) is for hazardous (Class III) waste.

For the basal seal:

- Class I (low grade waste)

a) The liner should be min. 500 mm clay $k \leq 5 \times 10^{-10}$ m/s ($i = 30$)

b) A suitable protection layer should separate this from a leachate collection layer with a minimum thickness 300 mm and with a permeability, k, $\geq 1 \times 10^{-3}$ m/s

- Class II (middle grade waste i.e. municipal waste)

c) This has a lining as described in para. 2.2.1 a-e except that 750 mm instead of 1500 mm of clay should be used.

For the landfill cover:

- Class I

a) A levelling layer as in para. 2.2.1 g is required. Gas should not be present in this class of landfill.

b) The top liner should be a 500 mm layer of clay constructed in two lifts, $k \leq 5 \times 10^{-10}$ m/s ($i = 30$).

c) The drainage layer and cultivation layers should be as in para. 2.2.1 h.

- Class II

d) This should be designed as in para. 2.2.1 g and h.

A recent amendment to the draft is that

equivalent lining systems can be accepted. This is to ensure that lining systems develop with the state of the art.

2.3 LAGA M3

The States Working Group Waste (LAGA) brought out some guidelines in the same year that TA Abfall became law, known as LAGA M3. This required 750 mm clay with $k \leq 5 \times 10^{-10}$ m/s and a 2.5 mm HDPE geomembrane. The leachate drainage layer was specified as 500 mm of gravel with a grain size range of 16-32 mm.

2.3.1 Geomembrane protection layer

This was specified as 1200 g/m² needle-punched geotextile together with 150 mm of low carbonate 0/8 mm grit or sand.

2.4 State regulations

Most state regulations refer to the geological requirements for landfill sites, reflecting the wide variety of geological conditions found in Germany.

2.4.1 Lower Saxony (Niedersachsen)

The older regulations in this state required 750 mm clay $k \leq 5 \times 10^{-10}$ m/s but with 2.0 mm HDPE geomembrane. The leachate collection layer was the same as LAGA M3. The protection layer was a sandwich of two 400 g/m² needle-punched geotextile with 100 mm sand in between.

2.4.2 North Rhine Westphalia (Nordrhein-Westfalen)

This state is the most industrialised state in Germany and it was the first to develop regulations concerning landfills. The NRW authorities laid down regulations for testing the chemical resistance of geomembranes. This involved immersing the geomembrane in 12 different concentrated chemical environments (one of them leachate and another landfill gas) for 3 months. The tensile strength should not have diminished more than 25% of its original value. Only specific types of HDPE geomembranes can meet this requirement.

2.5 Quality assurance

Both TA Abfall and TA Siedlungsabfall lay down specific criteria for the quality assurance of the landfill lining system. These require that a quality assurance plan to be carried out by an independent third party on all aspects of the works and materials, is established before the lining system is constructed. This is additional to those controls to be carried out by the contractor, the suppliers and the authorities.

3 THE CLAY COMPONENT FOR BASAL AND COVER LINING SYSTEM

3.1 The chosen soil must fulfil the following requirements:

a. > 20% by mass < 2 μ m (DIN 18123) and not be susceptible to suffusion
b. The clay amount and type shall be capable of absorbing water to min. 10 % of its mass (Enslin Neff Test).
c. Soils containing wood, roots and other foreign matter is not to be used. The organic content $\leq$ 5% by mass (DIN 18128). The carbonate content $\leq$ 15% by mass (Scheiber).
d. The soil shall be capable of plastic deformation in keeping with the expected settlement.
e. It should be homogeneous and have the same water content throughout.
f. For each lift of soil the dry density shall be > 95.5 of Proctor maximum.
g. The moisture content must remain above the Proctor optimum.

3.2 Field trials

The ability of the contractor to achieve the required values on site should be established in field trials which do not form part of the final lining system. Field and laboratory tests are carried out and the placing and compacting procedures to achieve the required results are established. Usually a mixed-in-plant system is used to ensure a homogeneous clay layer.

3.3 Construction

Each 250 mm thick layer of clay is sampled and tested in at least three different places (or each 1,000 m²) for in-site density, permeability and moisture content. Laboratory classification and density tests are carried out on every fourth sample and at least once per construction day.

4 GEOMEMBRANES

4.1 Specification for HDPE geomembranes in Germany

Only High Density Polyethylene (HDPE) Membranes with a minimum thickness of 2.5 mm are used as base liners in landfills. The regulations in Germany, TA Abfall and TA Siedlungsabfall (draft) require that the HDPE geomembranes to be used in landfill lining systems are licensed by, for example, BAM (Bundesanstalt für Materialforschung und -prüfung) in Berlin. Only materials which hold such a license are allowed to be used in Germany and extensive testing is carried out, including detailed description of the raw materials used, before this licence is granted. Blown or ring extruded geomembranes do not meet the tight requirements.

The preferred method of welding geomembranes in Germany is the double fusion seam with an air channel in between. This type of seam can be tested by pressurising the air channel. Only licensed operators are allowed to carry out the geomembrane installation.

5 PROTECTION OF GEOMEMBRANES

5.1 Protection Philosophy

Protection of the impervious layer of a lining system is of utmost importance in environmental engineering. There is no point in installing a geomembrane in a lining system if it is going to be punctured by stones etc. placed against it. Any protection layer should last at least one day longer than the geomembrane.

The selection of a protection layer for a geomembrane is based on the allowable strain which the geomembrane can tolerate. In Germany, based on experience gained with HDPE pipes, the strain in an HDPE geomembrane is limited to 6% strain. With a safety factor of 2 this strain limit is reduced to 3%, of this 3%, 2.75% is allowed for large area deformation leaving 0.25% for local deformation. This local deformation can be either due to installation forces e.g. dozers spreading a drainage layer on the protected geomembrane [2], or by the long term vertical stress imposed by the landfill through the drainage layer onto the geomembrane. Apparatus now exist to test a lining system under pressure, at the temperatures expected in a base lining system of a municipal landfill, immersed in leachate. Such tests are presently being carried out at the BAM (Bundesanstalt für Materialforschung und -prüfung) in Berlin and at other institutions.

5.2 Drainage layer selection

Landfill operators often have problems with the blocking of drainage pipes at the base of the landfills [2]. This blocking did not only occur in the pipes but also in the drainage layer itself as was clearly demonstrated by excavations carried out in an experimental area of the landfill site Geldern Pont [3]. Laboratory tests and experience on site have shown that a drainage blanket of coarse grained stone (16-32 mm) allow the leachate to travel fast to the drainage pipes, reducing blockage. The use of fine material, such as sand, can cause problems due to chemical and biological activities in the leachate cementing the sand grains together to form a hard layer through which the leachate cannot drain. The resulting increase in hydraulic head imposes additional strains upon the lining system.

The recommended 16/32 mm stones (gravel or crushed rock) are good for leachate collection but put the underlying geomembrane under stress. The geomembrane must be protected against this drainage layer both during installation and under the long term loading from the weight of refuse at the temperatures expected at the base under the influence of aggressive leachate.

This protection can be achieved in a number of ways:

- 150 mm sand layer,
- needle punched non-woven geotextile layer,
- bentonite filled geosynthetic clay liner,
- combination of geotextile and sand.

5.3 Sand protection layer

A layer of 150 mm of sand with grain sized 2 mm or less has to be placed on the geomembrane probably by hand, to prevent damage due to earth moving plants. The sand layer can also become displaced when the 300 mm drainage layer is installed. Placing a needle-punched geotextile between can alleviate the problem.

Sands have a low friction angle with the geomembrane and cannot normally be installed on slopes steeper than 1:3 (18°) without reinforcement. The sand has a k-value of about 1×10^{-5} m/s which means that the leachate moves slowly through the sand. The leachate is therefore in contact with the geomembrane for a longer time, which allows the ions of hazardous chemicals such as dioxins and chlorinated hydrocarbons to migrate through the geomembrane. The migration rate is dependant both on the hydraulic head of the leachate and the concentration of these substances in the leachate. A thicker sand layer means a larger hydraulic head.

Additionally, as discussed in para 5.2 above, the leachate can cause encrustations to form within the sand, which is then capable of damaging the geomembrane.

5.4 Needle-punched non-woven geotextile layer

Thick needle-punched non-woven geotextiles have been used extensively in Germany to protect geomembranes against damage. The thickness and therefore the weight of the needle punched non-woven geotextile is dependent upon:

a. The grain size of the drainage layer and
b. the weight of waste expected within the landfill.

Field trials carried out by Dr Knipschild have shown that

$tg > ds$ where tg = thickness of geotextile
and ds = particle grain size of the drainage medium

and that for 16/32 mm size drainage medium a minimum of 2000 g/m² are needed. Needle punched geotextiles incorporating a woven one provide even better protection than needle punched ones by themselves.

The selection of the polymer type for the protective geotextiles is of great importance. The chemical tests detailed in para 2.4.2 above have shown that the best resistance against chemical attack is provided by HDPE fibres, followed by PP (polypropylene). PES (polyester) fibres have shown themselves not to be suitable for use within a landfill.

Geotextiles are easy to handle and large areas can be covered using manual labour only therefore reducing the chance of geomembrane damage by machines.

The geotextile layer is much thinner and more permeable than a sand layer therefore the leachate hydraulic head is reduced. Their use also allows more waste to be placed into the landfill.

5.5 Bentonite Geosynthetic Clay Liners (GCLs)

Geosynthetic Clay Liners (GCLs) consist of a layer of bentonite sandwiched between two layers of geotextiles. One example is Bentofix® which consists of bentonite powder between two thick needle-punched geotextiles. An additional needle-punching step during the manufacturing process joins the geotextiles together through the bentonite powder to

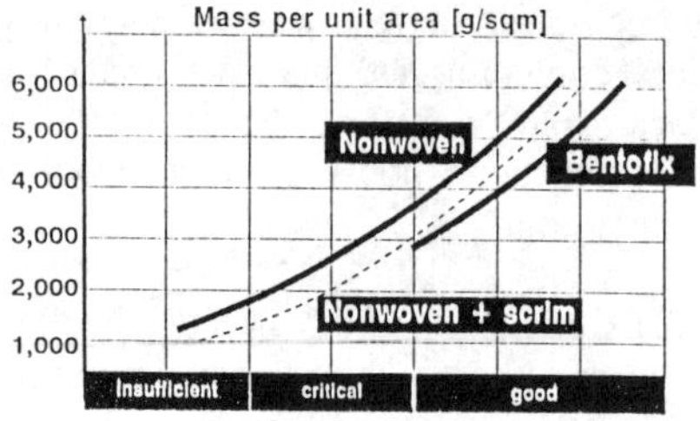

Fig.1 Efficiency of protection layers on lining systems by using granular material 16/32 mm

provide shear resistance across the whole of the bentonite layer even when hydrated. When the bentonite becomes wet it swells to produce a material with a permeability of less than 8 x 10^{-11} m/s.

This type of GCL has been found, both by field trials [2] and in laboratory tests to provide a very effective protection layer between the geomembrane and the drainage layer. The results of tests to determine the efficiency of protection layers on lining systems are shown in Fig.1.

An additional benefit of using a GCL protection layer is that it acts as an additional seal and can self-heal around small penetrations and any defects in the lining system.

The same consideration should be given to polymer type as described in para 5.4. If the landfill contains hazardous or municipal waste then the geotextiles on either side at the bentonite layer should be HDPE.

An example of a GCL being used in this way is Bentofix® Type C at the Oldenwald landfill in Germany.

6 APPLICATION OF GEOSYNTHETIC CLAY LINERS (GCLs) WITHIN GERMAN LANDFILL LINING SYSTEMS

The use of certain GCLs as a protection to the geomembrane has been covered in para 5.5 above. However there have been a number of other applications within German landfill practice which are described below.

6.1 GCLs in landfill base seals

In Germany, GCLs are used to supplement the clay layer beneath a geomembrane.

6.1.1 Geological barrier

If the required geological barrier of a soil of thickness $\geq$ 3 m, k $\leq$ 1 x 10^{-7} m/s (see para 2) does not exist at a particular site, the existing soil strata can be improved by the installation of a GCL.

6.1.2 GCLs as primary liners

GCLs are used on their own as the primary lining system for water retention and also for the containment of wastes with low grade contaminants.

6.2 GCLs in the landfill cover

The two main objectives of a landfill covering are to prevent the ingress of rainwater into the waste and to prevent the egress of gasses from the waste.

Some points to consider when designing the landfill cover with these objectives in mind are as follows:

a) EFFECTIVENESS The cover seal should be as effective as the base liner or leachate generation and hence, treatment, will continue a much longer period.
b) VOLUME By keeping the thickness of the cover to a minimum, more waste can be placed in the landfill for a given final height.
c) SETTLEMENT The surface of the waste will settle due to voids left after the filling process or to the decomposition and collapse of materials within the waste. Any cover system must be able to accommodate these movements without detrimental effects.

Bearing these points in mind the reasons that GCLs are used as an alternative to compacted clay liners for landfill covers in Germany are as follows:

6.2.1 Effectiveness of GCL

The US Army Engineering Corps programme HELP (Hydraulical Evaluation of Landfill Programme) was used to evaluate the effectiveness of the GCL Bentofix® D (which is manufactured from HDPE fibres for the lower geotextile and PP fibres for the upper geotextile). Using the meteorological information corresponding to German conditions and giving details of the fauna, the programme showed that a 750 mm thick compacted clay layer would allow the passage of 31 mm of rainfall per year whilst the Bentofix® GCL would only allow 3 mm/m² [4].

6.2.2 Thickness and weight

In its swollen state a GCL has an effective thickness of approximately 10 mm depending upon the load above it. There is therefore a significant reduction in the capping liner thickness and hence an increase in space available for waste if the compacted clay layer is replaced by a GCL. In addition the GCL weighs only about 10 kg/m² compared with approximately 1600 kg/m² for a clay layer. This weight difference may make a significant difference to the settlement profile.

By using geosynthetic drainage layers on either side of the GCL a complete cover lining system of thickness approximately 75 mm and weight approximately 13 kg can be provided.

6.2.3 Settlement considerations

GCLs, particularly those such as Bentofix® which are manufactured from heavy needle-punched geotextiles and therefore are able to withstand very high elongations without loss of bentonite, can stretch considerably without suffering significant changes in permeability.Bentofix® has been stretched in the laboratory to 40% uniaxially without any significant gain in permeability. It is therefore able to withstand the substantial settlements which may occur in the underlying landfill without significant loss in effectiveness.

6.2.4 GCLs as gas barriers

The Bentofix® GCL has been tested to check its impermeability to methane and petrol fumes at various degrees of hydration and with different gas pressures. Even with twice the gas pressure expected from a landfill and only 50% moisture content the Bentofix® was found to be effectively impermeable to these gasses [5]. Bentofix® has been used as a vertical gas barrier to prevent gas escaping from the side of an old landfill [6].

6.2.5 GCLs compared to Compacted Clay Liners

When compared to the testing and control required for a clay layer (see para. 3), GCLs provide a simple but technically sound alternative. Other aspects of comparison between Compacted Clay Liners (CCLs) and GCLs can be seen in Fig. 2.

6.2.6 GCLs compared to bentonite soils

Soil bentonite mixtures are sometimes used instead of a clay layer especially in areas where clay is not available. Work carried out on leachate - bentonite soils interaction have shown that the micro fissures develop in the mixture increasing the leachate permeability [7]. Bentonite swells when in contact with water. Therefore when such mixtures are tested using water, high impermeability values are measured. However, the bentonite swelling decreases when it is in contact with leachate and the pores within the soil bentonite mixture which were filled with swollen bentonite become less so filled.

The pores do not change size between the soil particles, only the volume of the hydrated bentonite. Micro fissures develop. GCLs - especially those made with powdered bentonite instead of granular bentonite - do not fill voids between soil particles but remain as a

COMPARISON OF CCLs AND GCLs

COMPACTED CLAY LINER (CCL)	BENTONITE MAT (GCL)
Field constructed, high variation	Industrial manufactured
Site-specific data on soil needed	Data immediately available
Often requires test pad	Repeated field testing not needed
Thick (0,75 - 1,5 m)	Thin (0,01 m)
Dry weather installation	Not weather dependent installation
Compaction necessary	Unroll and place without compaction
Heavy installtion equipment necessary	Light construction equipment
Slow installation (approx. 1,000 sqm/day)	Fast installation (approx. 10,000 sqm/day)
Performance depends on installation	Easy handling
Impossible to puncture	Self sealing for small punctures
Difficult to repair	Easy to repair
Takes up space, rubble replacement	Needs less space
High installation costs	Low installation costs
Low tensile strength	High tensile strength
Erosion susceptible	Bentonite filter stable reinforced
Can desiccate and crack	No cracking
Filter stable geotextiles needed	No extra geotextiles needed
Hard to connect to other constructions	Easy to connect to other constructions

Fig. 2 Comparison of CCLs and GCLs

swollen homogeneous layer between two geotextiles. As leachate reduces the swelling, the external pressure from the cover layer (of drainage stones or waste, sludge, etc.) simply reduces the thickness but maintains its homogeneous state and very low k-values without micro fissures developing. Thus GCLs are better liners against leachate than bentonite soil mixtures.

7 CONCLUSIONS

7.1 Landfill lining system design in Germany has advanced considerably. The basal seal consists of a combination of thick highly engineered clay with a high quality 2.5 mm HDPE membrane. To ensure the effectiveness of the liner the hydraulic head is kept to a minimum using high permeability leachate drainage blankets constructed from coarse gravel or crushed stone, usually of 16/32 mm size. To ensure that the geomembrane is not damaged, extensively tested and effective protection layers are used.

7.2 In many areas of the lining system geosynthetic clay layers (GCLs) are used. In the base seal they provide ideal protection to geomembranes and can be used to improve geological barriers below the geomembrane. They have many advantages over compacted clay liners in the covering of landfills and contaminated land. They are also used alone as a base liner to retain low contaminated wastes.

REFERENCES

[1] Incrustration Processes in Drainage Systems of Sanitary Landfills. Brune, Ramke, Collins and Hanert, Proceedings Sardinia 91, Third International Landfill Symposium.

[2] The Combination of Mineral Base Sealing, Protective Non-woven and Filter Layer such as in the Combination Sealing for the Residue Landfill Hopferstadt. Biedermann, Würzburg, 1989.

[3] Shadenanalyse von Deponiebasis-abdichtungssysteme aus Kunststoffdichtungsbahnen. Abschlußbericht - Unwelt-forschungsplan des Bundesministers für Unwelt, Naturschutz und Reaktor-sicherheit. Düllmann/Eisele, 1989, FE-Vorhaben 103-02-225-1989.

[4] Vergleichende Untersuchung zur Hydraulischen Wirksamkeit der Bento fix® Typ D Matte in Oberflächendich-tungssystemen von Deponien, UMTEC, Mai 1990.

[5} Untersuchungsbericht - Gasdurchlässig-keit von Bentofix® Dichtungsmatten gegenüber Methangas und Benzin-dämpfen, ICP, August 1992.

[6] Tech Note 2 - Vertical Methane Barrier Incorporating Bentofix® A. Kent v. Maubeuge, 1991.

[7] Geo-Environmental Engineering - Short Course on Environmental Engineering. Institute of Engineers Malaysia. Yong, Kuala Lumpur, 1992.

[8] Geokunststoffe in Dichtungssytemen - Laboruntersuchungen zum Verhalten von Geotextilien und Kunststoffdichtungs-bahnen. Saathoff, Heft 72 (1991) Mitteilungen des Franzius-Instituts der Universität Hannover.

Waste Disposal by Landfill – GREEN'93, Sarsby (ed.) © 1995 Balkema, Rotterdam, ISBN 90 5410 356 6

Section 61 – Another regulatory nightmare

Bill Darbyshire & Alan D'Arcy
Waste Regulation Section, Lancashire County Council, UK

ABSTRACT: Section 61 of the Environmental Protection Act (EPA) requires Waste Regulation Authorities to inspect land within their areas for sites which could cause pollution to the environment. As a result of this and because of duties under other legislation a review of the information held on former landfill sites was undertaken. A methodology was developed, both to improve existing information and assess the potential scope of the problem.

This paper examines the background to the problem, and the development and application of low-tech methodologies with a view to assessing the risks associated with these former sites.

INTRODUCTION

It is not known how many of the 500 plus closed landfill sites so far identified in Lancashire pose a real environmental hazard. Lack of accurate information limits our ability to assess the hazards, and any risks that they present. Before this potential was recognised, many waste disposal sites were used that may now present a serious hazard to people and the environment, and it is becoming increasingly apparent that even sites older than thirty years are still presenting a potential risk.

Awareness of their potential impact has prompted the need for a site assessment and investigation programme to quantify the extent of the problem within the County.

Work carried out over the last year in the Waste Regulation Authority (WRA), and by a student at the University of Central Lancashire indicates that perhaps three to four times the number of sites originally thought to exist are awaiting identification Carter (1992).

In examining the background to the problem this paper considers the legislative framework, and the stance adopted by Lancashire WRA.

This is followed by a detailed consideration of one possible approach to identifying and categorising sites.

Brief case histories will be used for illustration as appropriate.

THE BACKGROUND

Legislative framework

Section 61 of the 1990 EPA requires Waste Regulation Authorities to inspect their area "to detect whether any land is in such a condition, by reason of the relevant matters affecting the land, that it may cause pollution of the environment or harm to human health". The "relevant matters" are "the concentration or accumulation in, and emission or discharge from, the land of noxious gases or noxious liquids caused by deposits of controlled waste in the land".

It is expected that Section 61 will have implications for county councils, both in terms of extending current powers to control wastes already deposited on land and in terms of finance and manpower.

Although expected at the beginning of April this year Section 61 has not yet been enacted - this is limiting the work being undertaken by many Waste Regulation Authorities on former landfill sites. In Lancashire, the intention is to continue looking at former sites in the way described in this paper. This intention is supported by a number of the "Waste Issues" identified in the Green Audit of Lancashire carried out by the County Council. LCC (1990).

The Town and Country Planning Act General Development Order 1988 (GDO) requires planning authorities to consult with the Waste Regulation Authority on the likely risks from landfills within a 250 meter radius of a proposed development. This legislation resulted in a list of 500 plus sites which is used to formulate consultation replies to district councils.

There is also other guidance, such as Waste Management Paper 27, DoE (1991); Circular 21/87 "Development of Contaminated Land", DoE (1987); Circular 17/89 "Landfill Sites: Development Control", DoE (1989); Building Research Establishment reports, BRE (1991); The Building Regulations, DoE (1991); various British Standards, BSI (1981, 1988, 1991) and publications of ICRCL (1983, 1988).

It is not the intention of Lancashire County Council to create a contaminated land register of the type required under the (recently abandoned) section 143 of the 1990 EPA; but to ensure that places where wastes have been deposited are identified, recorded and evaluated.

Scope of the problem

In 1988 Lancashire WRA circulated the 14 Local councils within the county for information on landfills. This exercise resulted in the creation of a list with 500 entries dealing with sites ranging from large filled quarries producing significant levels and volumes of gas to small filled ponds with little capacity to produce gas.

During the summer of 1992 Lancashire WRA, in collaboration with the University of Central Lancashire, undertook a research project to assess the information collected during 1988.

This work was done by comparing Ordnance Survey 1:2500 plans from the 1960's and the 1990's identifying all areas where waste had been tipped. The information collected was verified by studying aerial photographs and doing walkover studies. The results of the survey suggested that, from the single district studied, a ratio of 3 unrecorded to 1 recorded site could be expected throughout the county. This has led to two subsequent studies being initiated, the early results of which are tending to confirm the results from the initial study. There are, in addition, problems of definition, eg: what constitutes a landfill – are, for instance, factory demolition sites landfill?

Marginal Land

There is continuous and increasing pressure for land to be made available for development; time and money is spent on forward planning to ensure that a supply of land is available to meet the anticipated demand. There is also the potentially conflicting pressure to preserve "green-belt" and areas of open space for public amenity.

Developments regularly approach, and in some cases are directly on, former landfill sites. Where it can be shown that the former site presents little or no risk to the development, then, from this point of view, the development could be permitted. However, opinions may differ as to whether the risk is acceptably low - a recent planning appeal on a gassing site in Leicester considered this - see case study one.

In Lancashire we have recently been advising a District Council on a similar proposal. We were concerned about the long-term integrity of gas control measures, especially where the development is conventional housing. We think that even the removal of permitted development rights will not guarantee that a house-holder will not build a conservatory, or install drainage for their garden, or drill a hole through the kitchen floor puncturing any gas-proof membrane.

Again, considering each site on its merits is essential, as is a detailed site investigation to provide the necessary information for decision making. Developers should perhaps recognise that other forms of development may be more appropriate than housing - in Lancashire there are supermarket-type retail outlets built on former landfills; ensuring the long-term integrity of gas control measures is easier with such buildings because of the corporate management controls possible.

Case Study One

Barkby Thorpe Road, Leicester

The site was identified for housing on the local plan, and outline planning permission was granted in March 1989 - this envisaged that a further application would be made regarding proposals to remove the gassing waste.

The further application, made in March 1989, was referred to the County Council. Doubts as to whether the operation could be carried out in an environmentally acceptable manner led to this being refused. An appeal was lodged, but withdrawn after advice from a consultant on an alternative means of developing the site.

Amongst other things, the new proposals involved moving gassing waste on site to one area, use of a cut-off wall to prevent gas migration, and different types of development across the site. The planning applications were refused and appeals were lodged dealing with a number of issues; here, we are only interested with those related to landfill gas.

The inspector identified that the main issue was whether the proposed measures can be expected to ensure the safety of the developments, and would they remain effective for the foreseeable future?

The inspector recognised that: "the general tenor of advice is that sites known to be producing landfill gas are best avoided for all forms of hard development."

The need for adequate information and site investigation was also identified.

The inspector's report clearly states that each site must be considered on its merits, and that: "if the characteristics of the site and the measures proposed can minimise risk to an acceptable degree for the foreseeable future, then the Council cannot, in my opinion, continue to resist the development of housing on a gassing site as a matter of principle."

The inspector recognised that a situation of no risk is not attainable, but that measures should preclude risk as far as possible.

The appeal was allowed, subject to certain conditions. DoE (1992).

THE METHODOLOGY DEVELOPED BY LANCASHIRE COUNTY COUNCIL

General

Lancashire Waste Regulation Authority has attempted to develop a methodology to assess the hazards that former landfill sites present. The developmental work is continuing on a low-tech, hands-on approach which it is hoped will be applicable to other types of contaminated land problems.

It is felt that the heart of the system should be an accurate catalogue of sites, compiled by a small team, that is updated and expanded as new sites are licensed and previously unrecognised sites identified. The system should also be sufficiently flexible to allow sites to be removed from the catalogue if it is found that a site no longer presents a risk or to allow mistakes to be rectified. To minimise risk to people (and the environment) strict planning and building controls on developments near such sites are needed. New developments should not be undertaken without an assessment of the likely hazards and risks being undertaken.

Why In House ?

A decision was made at an early stage to develop the methodologies and do the work within the County Council. This decision was made on the basis of cost and efficiency. An option was available to employ consultants to undertake certain parts of the survey, but it was felt that the need for the continuous updating of the catalogue required a dedicated team. A small team was created within the Scientific Support Unit of the Waste Regulation Section, and as part of its remit was given the task of reviewing the data collected in 1988. The team plays a number of other roles within the section, including supporting the Licensing Unit and providing scientific advice to other departments at County and District level.

In Lancashire, our starting point was the database created after the 1988 General Development Order, this involved written responses from the 14 District councils giving details of site name, location (National Grid Reference) and other data available at the time. The information collected during this exercise was found to be variable due to the subjective nature of the replies.

The Method

In 1991 Lancashire WRA recognised that a revue of the information held was required if our consultation responses to District Councils were to be of use. A structured but flexible approach was chosen with the dual aims of stream-lining and improving the system.

The methodology has been reviewed and changed throughout the last 12 months, we now use a four-stage approach - which could be applied to any contaminated land problem. Within Lancashire the method has primarily been used to improve responses to enquiries; this will provide a sound basis for a catalogue of sites which will be required by section 61.

The four stages are:

1) creating a catalogue of landfills
2) initial assessment of each site
3) desk study
4) low-cost site investigation

Stage 1 Creating a Catalogue of Landfills

There are a large number of potential sources for information relating to landfill sites, those we have found particularly useful include:-

i) Local Councils-Planning, Environmental Health, Landscaping/Reclamation Departments
ii) County Councils - Waste Regulation Authorities, Libraries and Archives, Planning, Property Services Departments
iii) Ministry of Defence
iv) Department of the Environment
v) Ordnance and Geological Surveys
vi) British Coal
vii) Mines and Quarries Inspectorate
viii) Private companies such as British Gas, British Waterways, ICI, and British Railways.

A combination of methods was used to obtain information, such as: direct letter writing to the above, comparison of Ordnance Survey maps and aerial photographs. Personal visits to the above were found essential; for example, visits to the County Records Office and Planning Department were needed to access old Ordnance Survey maps and aerial photographs, particularly for the work carried out during 1991.

Stage 2 Initial Assessment of Each Site

After the initial trawl for information it was decided to visit each site, this involves the following steps:

(i) site walkover - essential to assess: geography, geomorphology, pattern of development, topography and ease of access.
(ii) verification of data held on file - data acquired should be checked for accuracy; in our experience, information provided on site location has often proved to be inaccurate.
(iii) Photography - photo's have been found to be a useful tool in giving an impression of the site to other people, and as a memory aid.
(iv) Closed site report - A detailed report of any site visit, including maps, diagrams and photo's is appended to the site file. The report also details any particular action the visiting officer feels is necessary
(v) Site assessment - the information is reviewed and a decision taken whether to proceed with further site investigation or accept the site as low risk.

The method itself was reviewed, initially at monthly team meetings to ensure that it was providing the information wanted. This review proceedure is continuing periodically.

Stage 3 Desk study

If a site is considered a potential risk a desk study is undertaken to allow the subsequent site investigation to be targeted, the main steps include:

(i) Historical maps - comparing maps of different ages for individual sites.
(ii) Aerial photo's - these give information on site location, boundaries and depth of fill.
(iii) Planning records - provide useful information on site location, waste types and depth of fill.
(iv) Archive records - besides Ordnance Survey maps, etc. these hold other useful information such as local

government committee reports, company records, etc.

(v) Site assessment - as in stage 2 above.

Stage 4 Low-Cost Site Investigation

On completion of the desk study a site investigation may be planned to collect data on landfill gas production and waste types and depths. These investigations are normally completed in two days and involve 2-3 people.

(i) Access - permission from the landowner and access for vehicles is checked prior to any work being undertaken.

(ii) Services - the owners of underground services (eg. gas, water, electricity, sewers) are contacted for detailed layout diagrams.

(iii) Spike tests - Although only able to penetrate to between 1 and 1.5 metres, they are proving useful in identifying areas for further investigation. See case study two.

(iv) Hand augers - this simple manual excavation technique is useful as it allows the installation of temporary monitoring points and sampling of the wastes.

(v) Trial pits - A useful technique on areas of rough land, involving the excavation of pits or trenches; these give a clear indication of the types of waste present on site. Pits can be backfilled to include a simple gas and/or leachate monitoring point.

(vi) Boreholes - Normally used to provide points for long-term monitoring, establishing the depth of fill and for gas migration and groundwater studies.

(vii) Monitoring - Monitoring is carried out, starting on a weekly basis for methane, carbon dioxide, oxygen, barometric pressure and groundwater/ leachate. Although not yet common in the UK gas flows and pressures are useful additions to this regime. The monitoring frequency needs regularly reviewing. Chemical analysis of gases, wastes, leachate, surface and groundwater provides essential information for characterising the site.

(viii) Site assesment - as in stage 2 above

Case study two illustrates the methodology in practice.

Problems encountered and short cuts to good data

Initial attempts at prioritising sites and workloads revolved around an attempt to correlate existing information and apply a numerical risk assessment procedure, HSE (1989), D'Arcy (1991).

We found it extremely difficult to use a scoring system, the principal problems being: lack of information in key areas such as site-specific geology, presence of gas migration pathways and presence of groundwater; we also found that staff scored differently to each other. Scoring was no better than asking staff to prioritise a list without assigning scores, discussion between team members tended to resolve any difficulties encountered and ensured consistency. We were also concerned that use of a scoring system could lead to a false sense of security in the results, because of the pseudo-scientific method used.

Hazard and risk assessment techniques may be of more use on operational or recently closed sites where detailed information on key areas such as liner design, waste intake and environmental monitoring is available. As a predictive tool on older sites, before the site investigation stage, they seem to be of very limited use.

There were a number of problems relating to obtaining information from private sources where individual companies expressed concern that liabilities could accrue if information on landfills was entered onto the public record.

We found it was important to maintain a standard approach within the team giving well defined aims and criteria, (even though this may be stating the obvious) as there was a great temptation to be side tracked by issues which were not central to the programme.

The accumulation of data led to a rapid demand for file space, over and above that initially predicted, the option of a computer based Geographical Information System (GIS) was assessed, and although it is able to hold much of the catalogue the need for paper files was not removed.

The basis of the success of this methodology within the county is its simplicity and flexibility, enabling sufficient data to be collected on a large number of sites to allow an assessment of hazard to be made.

Case Study Two

Links Road, Blackpool

As part of the continuing review of information on old sites it was found that one particular site was partially overlain by residential housing. A decision was made to initiate the second and third stages of the methodology above in order to make a detailed assessment.

The desk study confirmed the original indications and a site investigation was undertaken. The initial work undertaken was an impact searcher bar (spike) survey based on a ten metre grid. Portable gas analysers were used to take readings of methane, carbon dioxide and oxygen levels to build a picture of the subsurface gas regime. This technique highlighted a number of areas where significant gas concentrations (1 - 5% v/v methane) were noted. These areas were then chosen for hand auger holes to identify waste types and depths - and to allow the installation of temporary gas monitoring points. Gas concentrations recorded in the auger holes were found to be up to ten times those noted in the spike holes.

The results of the survey confirmed that the nearby domestic housing could be at risk and house to house gas monitoring was carried out with the local Environmental Health Department. The results of this survey found that landfill gas was not present within the properties and it was decided to monitor the site for a number of months to establish the gas regime. It is likely that permanent monitoring points will be installed and further data collected prior to a final decision being made as to whether remedial works will be required.

This work highlighted the need for a flexible system which allows potential problems to be dealt with at any stage of the method.

CONCLUSION

It is apparent that implementation of section 61 of the Environmental Protection Act, 1990 has been delayed. However, the methodology developed in Lancashire is being applied, and a programme of site assessment and investigation developed. The improvements in information are allowing a far better assessment of the likely impact of sites on proposed developments and the environment.

Without a structured approach to the problem time would have been wasted, and the quality of information gathered diluted. We believe that the methodology could act as a model for authorities embarking on this type of work on both landfill sites and land potentially subject to other types of contamination.

REFERENCES

BSI (British Standards Institution)
BS1377 1991 *Methods of test for Soils for civil engineering purposes* (9 parts)
BS5930 1981 *Code of practice for site investigations.*
DD175 1988 *Code of practice for the identification of potentially contaminated land and its investigation.*

BRE (Building Research Establishment) 1991 *Construction of new buildings on gas contaminated land.*

Carter, P. *An investigation of closed landfills in the Preston area* (Unpublished).

Commission of the European communities 1988 PAGIS - *Performance Assessment of Geological Isolation Systems for Radioactive Waste. Summary. Commission of the European Communities. 1988.*

Crowhurst, D.1987 *Measurement of gas emissions from contaminated land.* BRE

D'Arcy 1991 *Investigation into and developement of an assessment proceedure for use with sites contaminated by the landfilling of wastes* (Unpublished).

DoE (Department of the Environment) 1991 The Building Regulations

DoE 1991 Waste Management Paper No. 27 *Landfill Gas*

DoE 1987 Circular 21/87 *Development of contaminated land*

DoE 1989 Circular 17/89 *Landfill Sites Development Control*

DoE 1992 *Barratt (East Midlands) Ltd v Leicester City Council* (appeal notice)

Health and Safety Executive 1989 *Quantified risk assessment: its input into decision making*

ICRCL (Interdepartmental committee on the reclamation of contaminated land) *1988 Notes on the redevelopment of landfill sites 17/78*

ICRCL 1983 *Guidance on assessment and redevelopment of contaminated land 59/83*

Lancashire County Council 1990 *Lancashire a green audit.*

Naylor, J.A.; Rowland, C.D.; Young, C.P.; Barber, C. 1978 *The Investigation Of Landfill Sites. Technical Report.* TR91. Water Research Centre. October 1978.

Waste Disposal by Landfill – GREEN'93, Sarsby (ed.) © 1995 Balkema, Rotterdam, ISBN 90 5410 356 6

Landfill cover technology within the United Kingdom, the rest of Europe and the United States

A. Forster
MRM Partnership, Bristol, UK

ABSTRACT: To protect human health and the environment the waste management industry must identify appropriate technology for the safe long term disposal of solid wastes. Effective control of landfilled waste currently attempts to isolate it from the surrounding environs.

This paper summarises landfill cover technology currently in use, or being developed, within the United Kingdom, the rest of Europe and the United States. With particular reference to final covers over landfills, the legislative background and technical guidance documentation is outlined as it related to waste management practices, methods of placement and quality control procedures. Current operational practices are reviewed along with an evaluation of state of the art research projects.

1. INTRODUCTION

The risk to human health and the environment caused by landfilling has become an important contemporary issue. It is recognised that any water entering a landfill may percolate through wastes containing chemical contaminants, which may in turn generate a corresponding quantity of leachate. For many sites precipitation is the predominant source of water ingress. It is therefore essential to design and engineer an effective long term barrier for the surface as well as the base and sides.

The nature of landfill design varies, both within and between countries, depending on waste management strategies and practices, public concern and political will. Factors commonly considered include waste settlement/stabilisation, gas migration and minimising infiltration, although mitigation measures for liner desiccation, plant and animal intrusion, chemical attack (leachate and gas), slope stability and afteruse should also be assessed. A summary of waste management policy within the European Community (EC) and the United States is given below. This paper summarises a more detailed review undertaken by MRM Partnership on behalf of the Department of the Environment (Ref 1).

2. LEGISLATION

Landfill operation, design and aftercare is primarily dependant on national and regional disposal policy. National legislation often transfers detailed policy formulation to the regional level. Within Belgium, France, United Kingdom and the Republic of Ireland this is still largely the case. As an example, Figure 1 indicates the noticeable regional variation in Belgian landfill policy, with particular reference to final covers.

Until recently, a similar decentralised framework existed for most EC Member State countries and in the United States. In the past, Germany for example, although giving guidance at a Federal level, allowed regional

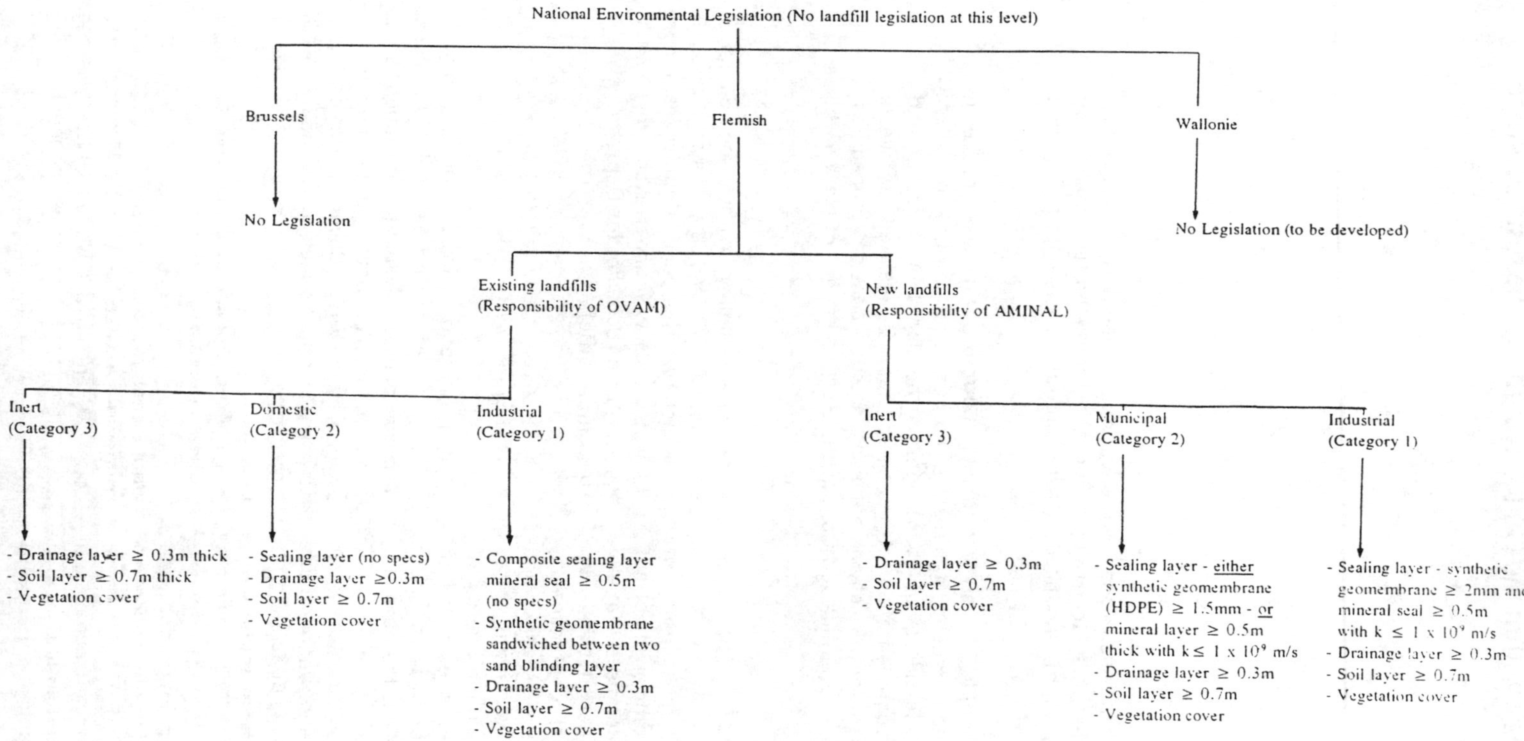

FIGURE 1 : EXISTING AND PROPOSED FINAL COVER LANDFILL REQUIREMENTS WITHIN BELGIUM

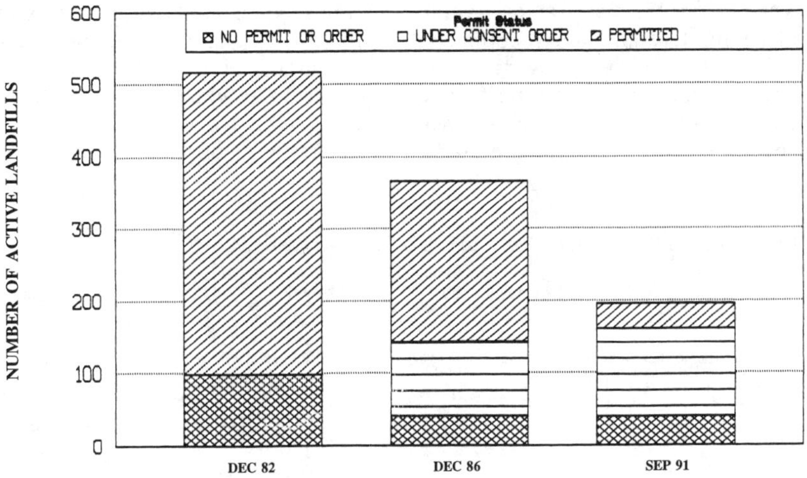

FIGURE 2 : NEW YORK STATE HAZARDOUS LANDFILL PERMIT STATUS, DEC 1982-SEPT 1991

governments (or Laenders) to develop landfill policy and design specifications. Standardisation of landfilling design is now being implemented through the introduction of TA Abfall notes (Waste Technical Notes) which, when approved, impose national minimum requirements. A similar trend has occurred within the United States which, prior to the introduction of minimum requirements under the Resource, Conservation and Recovery Act, allowed final cover specifications to be developed at the state level.

The effect of implementing national minimum standards can be dramatic. An example of this is given in Figure 2, which shows the number of permitted hazardous waste landfills within just one state (New York) between 1982 (when legislation was introduced) and 1991.

Divergence on the method of policy implementation can also be generalised into two categories. Countries such as the Netherlands, Germany, United States, and Italy, introduce final cover requirements into their legal framework and therefore (in theory) ensure that minimum standards are achieved nationally. Other countries chose to identify the function of a final cover within a legal framework and provide supporting reference documents which outline the means to achieve the stated function. The United Kingdom, Denmark, Republic of Ireland and France are included within this category.

3. DESIGN

With the exception of the Netherlands, Republic of Ireland and the United Kingdom, the nature of landfill restoration is determined by waste site type. Waste classification can be broadly grouped into four types; inert, municiple (sanitary or compostable), low strength industrial (non-hazardous) and high strength industrial (hazardous) wastes. Landfill classification generally divides into inert sites, municiple and low strength industrial sites, and high strength industrial/hazardous/co-disposal sites. For each site type variation in design specifications or recommendations can be enormous both between and within each country.

For inert landfill sites, final covers can in theory comprise a full multi-layered cover including a composite sealing layer (eg, within the Netherlands), although in practice a less stringent requirement is more likely to be imposed. Germany specifies a basal support and drainage layer overlain by a sealing component, drainage profile and vegetation support layer. The Flemish region within Belgium and Denmark require and recommend (respectively) only a capillary barrier layer beneath a vegetation support layer, no sealing component being incorporated (Figure 1 and 3). In Italy no special requirements are made.

France, the United Kingdom, the United States and the Republic of Ireland make no direct reference to final covers over inert waste sites within their legislative or policy framework.

Requirements and recommendations for covers

a) (i) Municiple and Industrial (non hazardous) Landfills (Agricultural use)

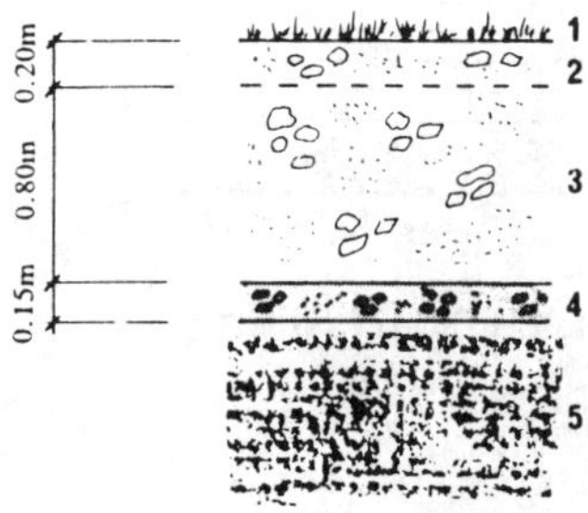

Component Name	Function	Min. Requirement
1. Vegetation Layer	1. - Enhance stability and restoration	1. - Mandatory (non specific)
2. Topsoil/Compost	2. - Provide media for vegetation - Enhance runoff	2. - Min thickness ≥ 0.20m - Slope < 10%
3. Subsoil	3. - Barrier to waste	3. - Min thickness ≥ 0.80m
4. Root Barrier	4. - Prevent root penetration - Prevent leachate leakage	4. - Free draining gravel material - Min thickness ≥ 0.15m
5. Waste Body		

(ii) Inert Landfills (Agricultural use)

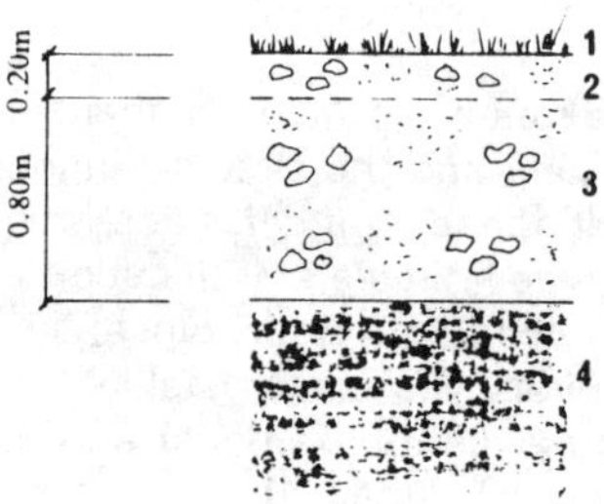

Component Name	Function	Min. Requirement
1. Vegetation Layer	1. - Enhance stability and restoration	1. - Mandatory (non specific)
2. Topsoil/Compost	2. - Provide media for vegetation - Enhance runoff	2. - Min thickness ≥ 0.20m - Slope ≤ 10%
3. Subsoil Layer	3. - Barrier to waste	3. - Min thickness ≥ 0.80m
4. Waste Body		

(b) All Landfills except Hazardous Wastes (Non-agricultural use)

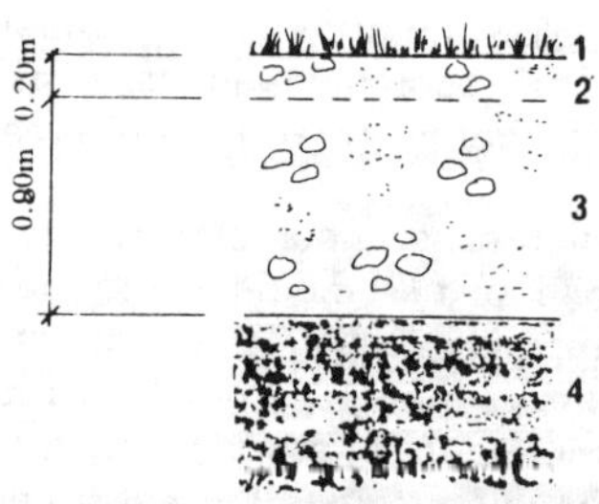

Component Name	Function	Min. Requirement
1. Vegetation Layer	1. - Enhance stability and restoration	1. - Mandatory (non specific)
2. Topsoil/Compost	2. - Provide media for vegetation	2. - Min thickness ≥ 0.20m
3. Subsoil Layer	3. - Barrier to waste	3. - Min thickness ≥ 0.80m
4. Waste Body		

FIGURE 3 : FINAL COVER RECOMMENDATIONS FOR LANDFILL RESTORATION WITHIN DENMARK

over municiple and/or non-hazardous waste sites are more stringent than those above, although again considerable variation exists. All countries with the exception of Italy and the United States specify or recommend a drainage layer in addition to a vegetated soil profile placed above a sealing layer. A composite sealing system will be required by both Germany (no less than 0.5m thick) and the Netherlands (between 0.3-0.5m thick) in future legislation, although present legislation within these two countries is non-specific. Mineral seals are more commonly recommended for sites of this type, ranging in thickness from a minimum of 0.1m in Italy (future legislation) to 1.0m in France, the United Kingdom and the Republic of Ireland. Hydraulic conductivity varies depending on the parameters used but maximum values generally range between 5 x 10^{-10} m/s (in Germany) to 1 x 10^{-9} m/s, elsewhere. Infiltration rates as opposed to hydraulic conductivity values are specified in Italy (Table 1) and the Netherlands and a hydraulic gradient (i) of 30 is prescribed for mineral seal testing within Germany (Figure 4).

Implicit reference to covers over high strength industrial waste sites are made by most countries with the exception of the United Kingdom, the Republic of Ireland, the Netherlands and Denmark. Requirements or recommendations are more stringent than for municiple waste sites, including hydraulic threshold values being set for the drainage layer above the mineral seal (within France, Germany and the United States). Composite sealing systems are advocated within the Flemish region in Belgium, France, Germany and the United States. In Italy a 0.5m thick mineral layer is required.

Construction quality control for final cover placement is a concept in landfill engineering that has recently received much attention. Guidance at the national level on material testing procedures, placement techniques and monitoring protocols is available within

c) Low Permeability Cover

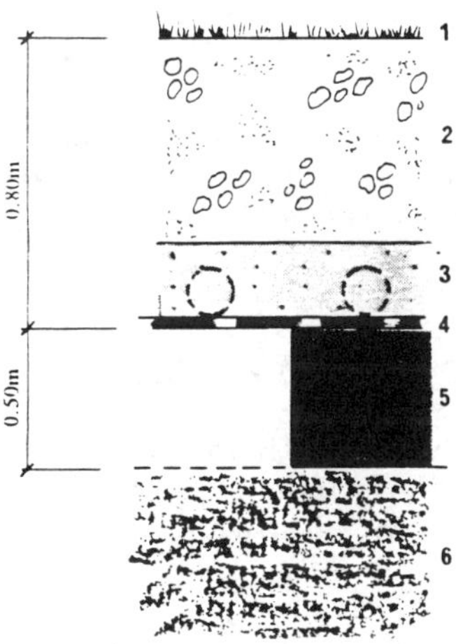

Component Name	Function	Min. Requirement
1. Vegetation	1. - Enhance stability and restoration	1. - Mandatory (non specific)
2. Topsoil/Compost	2. - Support media for growth	2. - Min thickness ≥ 80cm
3. Drainage Layer (with pipe network)	3. - Remove infiltrating water	3. - Min thickness ≥ 0.30m - Pipe spacing ≤ 0.20m (see DIF 436)
4. Sealing Layer	4. - Minimise infiltration	4. - Can either be: - (a) Polymeric Membrane Liner (for specs see DS/R 466) - (b) Natural sealing layer. Thickness between 0.30m and 0.50m Hyd. Cond. ≤ 1 x 10^{-10}m/s Clay content ≥ 10% Plastic Index ≥ 4%
5. Gas/Regulating Layer	5. - Provide foundation for overlying layers - Gas control layer	5. - Min thickness 0m for noncompressible waste: 0.20m for nearly noncompressible: 0.50m for compressible - Highly permeable
6. Waste Body		

FIGURE 3 (cont'd) : FINAL COVER RECOMMENDATIONS FOR LANDFILL RESTORATION WITHIN DENMARK

TABLE 1 : Existing and proposed future final cover requirements for all landfill types within Italy

Landfill Type	Existing Requirements[1]	Future Requirements[2,3]
CLASS I : **Municiple Waste Sites**	Cover must incorporate surface water and gas collection. Leachate gas and groundwater monitoring systems must also be included. No detailed specifications exist.	(a) ≥ 0.1m (b) ≥ 0.5 year (c) ≥ 1m (for agricultural use)
CLASS IIa : **Inert Waste Sites**	No specifications. Requirements set at the planning stage.	No specifications. Requirements set at the planning stage.
CLASS IIb : **Hazardous Waste Sites (Low Strength)**	Same as Class I.	(a) ≥ 0.2m (b) ≥ 1 year (c) ≥ 1m
CLASS IIc : **Hazardous Waste Sites (High Strength)**	Cover must include a compacted mineral layer (no thickness or K value given) and an organic soil layer (approximately 1m thick).	(a) ≥ 0.5m (b) 5 years (c) ≥ 2m
CLASS III : **Special Encapsulated Waste Sites**	No specifications or recommendations exist. No sites at present within Italy.	No specifications or requirements exist. No sites at present within Italy.

1 DPR 915/82 Enacted by Law 15 of 13 September 1984

2 Decree of the Ministry of the Environment. Provisional Draft Version 31 Jan 1989.

3 (a) Impermeable Layer Thickness, (b) Migration time for percolation to occur, (c) Organic Soil Layer Thickness.

Denmark, the Netherlands, the United Kingdom, Germany and the United States. Within this, in an attempt to identify the performance of as placed materials selected, test field sites are recommended in Germany the Netherlands and the United States

The focus of most, if not all, technical guidance or legislation is primarily on properties and engineering details of the sealing component. With perhaps the exception of the Netherlands, testing protocols, placement methods and specifications for the final cover liner generally derive from work originally intended to develop knowledge of basal liner systems. As a result, the effects of thermal variation with time, waste and cover settlement, gas permeation and desiccation on overall cover integrity have generally not been fully assessed. No country (with the exception of the Netherlands and the United Kingdom) specifies anything other than thickness for the soil component overlying the drainage (where present) and sealing layers (Figure 5 and Ref 1).

4. FUTURE LEGISLATIVE CHANGES

Much of the requirements and recommendations discussed earlier in this paper have been developed within the last 3 to 5 years in response to growing public concern and the adoption of recent EC Directives by Member State countries. Up until the mid 1980's, most countries made only general reference to final cover restoration measures and with the exception of the United States (for hazardous sites only) none had made reference to maximum permeability values, acceptable infiltration rates or construction quality control procedures. The legacy of previous, and in

a) Class I (Inert Wastes) sites

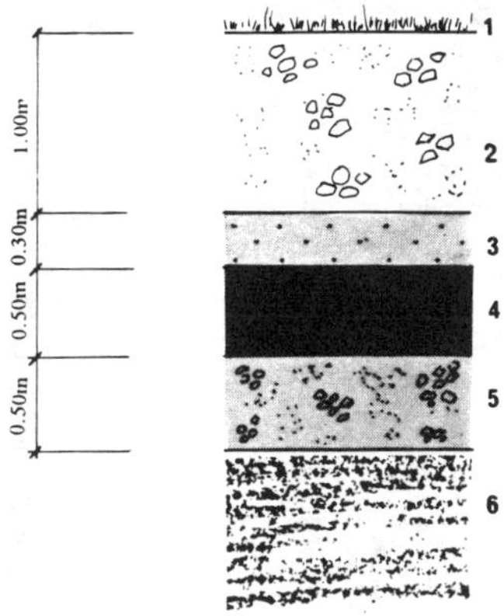

Component Name		Recommendations/specifications	
1.	Vegetation Layer		
2.	Reclamation Layer	2.	Thickness ≥ 1.0m
3.	Drainage Blanket	3.	Thickness ≥ 0.3m - Hydraulic conductivity (k) ≥ 1×10^{-3}m/s
4.	Mineral Sealing Layer	4.	Thickness ≥ 0.5m - Hydraulic conductivity (k) ≥ 5×10^{-10}m/s when i = 30
5.	Compensation (drainage) Layer	5.	Thickness ≥ 0.5m - $CaCO_3$ content ≤ 10% by weight
6.	Waste		

b) Class II (Municipal Solid Wastes) sites.

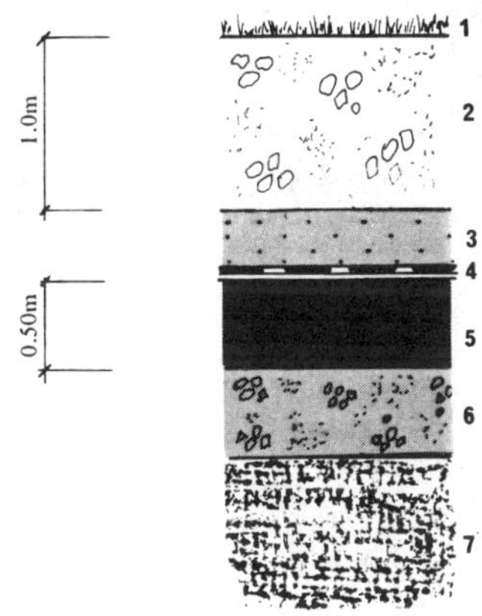

Component Name		Recommendations/Specifications	
1.	Vegetation Layer		
2.	Reclamation Layer	2.	Thickness ≥ 1.0m
3.	Drainage Blanket	3.	Thickness ≥ 0.3m - Hydraulic conductivity (k) ≥ 1 x 10^{-3}m/s
4.	Geomembrane Barrier	4.	Thickness ≥ 2.5m - Approved geomembrane only
5.	Mineral Sealing Layer	5.	Thickness ≥ 0.5m - $CaCO_3$ content ≤ 10% by weight
6.	Compensation (drainage) Layer		
7.	Waste		

FIGURE 4 : RECOMMENDED MINIMUM COVER REQUIREMENTS FOR MUNICIPLE (CLASS II) AND INERT (CLASS I) WASTE SITES IN GERMANY

some countries current, poor control over landfill design can be judged by the costly remediation programmes many countries have now had to adopt.

Future legislative changes for most countries would suggest a move towards a total containment strategy. A noticeable exception to this is Denmark which actively promotes a practice of "controlled leaching", although this is in part due to particular geographical conditions.

5. RESEARCH

A number of contemporary research projects attempt to assess the effectiveness of restoration covers, including composite, mineral, sand bentonite and capillary barrier layers. Much of this work has been reviewed by the Department of the Environment (Ref 1, 2 and 3) and the United States Environmental Protection Agency (Ref 4).

The focus of research work to date has been in identifying material properties, placement specifications and long term performance of the sealing component within a final cover

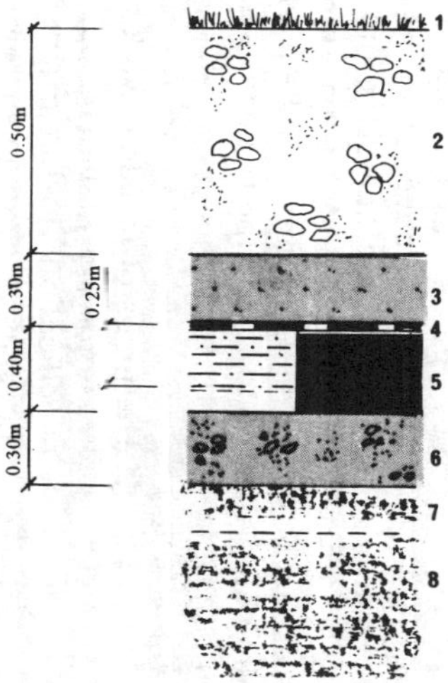

Component Name	Function	Min Recommended Requirement	Suitability Tests/Placement Criteria
1. Vegetation Layer	1. - Prevent erosion of topsoil - Maintain the stratified structure	1. - 100-125 kg/ha seeding mixture of BTK 1100 (Ref. G7) - Seeding completed by September	
2. Topsoil Layer	2. - Provide media for vegetation growth and suitable aftercare	2. - Minimum available water capacity 150mm for grass 200mm for deeper rooting vegt - Minimum thickness of 0.80m for grass 1.00m for deeper rooting vegetation - $K \geq 0.5$ m/d* - Slopes $\leq$ 1:3	2. - Soil classification - Particle size distribution - Soil Chemistry - Soil moisture characteristics - Stability - Permeability estimate
3. Drainage Layer	3. - Discharge surplus rainwater - Barrier to root growth/animals	3. - Porosity $\leq$ 40% - Dry density $\geq$ 1600 kg/m^3 - Min thickness 0.30m - $K \geq 2.9 \times 10^{-5}$ m/s* - Loam fraction (ie, <63 μm) $\leq$ 5% - Pipe network optional	3. - Soil classification - Particle size distribution - Soil chemistry - Density - Stability - Permeability estimate
4. Geomembrane	4/5 - Minimise infiltration - Prevent methane emissions	4. - Physically and chemically resistant - Intimate contact between geomembrane and mineral layer - $\geq$ 2mm thick polyethelene geomembrane, preferably profiled and HDPE - Biaxial and uniaxial elongation of $\leq$ 5% and 20% respectively	
5. Mineral Sealing Layer		5. - $\geq$ 0.25m for a sand/bentonite mix layer - $\geq$ 0.40m thick for a natural clay layer - sand mechanical properties uniformity coefficient approximately 10, density $\geq$ 1700 kg/m^3 - $K \leq$ 20mm/a, when i=5 - $\geq$ 5% bentonite wt/wt	5. - Soil classification - Particle size distribution - Clay mineral analysis - Soil Chemistry - Density - Plasticity - Permeability (Falling Head) - Stability - Conformance tests (geomembrane)
6. Support Layer	6. - Blinding layer - Foundation for above layers - Gas pathway	6. - Grain size < 3mm or if - geotextile used $\leq$5-6mm - Pipe network optional - Minimum thickness 0.30m	6. - Soil classification - Density - Bearing capacity - Particle size distribution - Permeability estimate - Density - Stability
7. Basal Layer (Top Refuse Layer)	7. - Foundation to final cover - Methane extraction	7. - Grain size <0.15m - Compaction - Slope between $\geq$ 1:30 $\geq$ x $\leq$ 1:2½ - Subgrade quality sand - Minimum thickness 0.20m	7. - Soil classification - Stability
8. Waste Body			

* Estimates only

FIGURE 5 : FUTURE REQUIREMENT FOR FINAL COVER SYSTEMS OVER ALL LANDFILL TYPES IN THE NETHERLANDS

profile. In summary, there is mounting evidence to suggest that mineral seals are particularly susceptible to deterioration in overall integrity even after only a few years. Geomembranes however do not provide the solution in their own right, and composite sealing layers are generally recognised as the best available technology. The importance of appropriate quality control procedures before and during placement of the restoration cover was highlighted for two reasons; lack of background data to properly assess liner performance; and the indication that many liner deformities occurred directly as a result of poor placement methods.

Other layers within the final cover are essential to its long term integrity, acting as a barrier and as a media for reducing infiltration into the underlying waste. Component layers, such as a drainage layer, have proven to be highly effective in removing infiltrating water in a number of studies within Europe. Adequate soil thickness and vegetation cover has also been highlighted as essential to overall long term cover integrity and would suggest that at least 1 metre of soil (including the drainage layer), possibly up to 1.5m, should be placed above the sealing layer.

Placement recommendations or specifications for the layers overlying the liner were recognised as essential to avoid root penetration (where necessary), avoid preferential flow pathways, and enhance vegetation growth.

Final cover gradients are recommended in most countries as it is seen to enhance surface water runoff. Gradients, where specified, range from a minimum post settlement value of 3%, within France (municiple sites), Germany (municiple sites), the Netherlands (future legislative requirements), the Republic of Ireland and the United States, to an acceptable maximum of 40% within the Netherlands. An average gradient of 5% is generally acceptable. Research however, suggest that only a small proportion of precipitation is removed as surface runoff, although attenuation mechanisms within the cover profile (ie, sub-surface flow) are greatly affected.

The effect of thermal processes, settlement and gas migration from the waste on final cover integrity is an area requiring much attention. A limited amount of work on these subjects have been conducted which suggests that their influence could be significant.

6. CONCLUSIONS

From the above review a number of conclusions can be drawn.

Legislation and guidance on final covers over landfills can vary considerably both within and between countries, depending on waste management strategies and practices, climate and geographical differences, public concern and political will.

Technical guidance has only been formulated, for the most part, within the last 3 to 5 years and has seen an increasing trend towards the development of national minimum standards. Within the countries reviewed new, or soon to be implemented, policy tends to favour total containment as a waste management strategy. A noticeable exception to this is Denmark where a "controlled leaching" strategy is actively promoted.

Design requirements, where present, generally vary according to the perceived pollution threat (ie, waste site type). In most countries a multi-layered final cover is recommended. The focus of attention is primarily on engineering properties of the sealing component, however, imposing more stringent (ie, other than thickness) requirements for other layers within the restoration cover could see an improvement of overall cap performance in the long term.

Careful preliminary design, material selection, placement specifications and construction quality control is essential in achieving the design requirements discussed above and assessing overall cover performance. This can only be achieved through the implementation of a comprehensive quality assurance programme. Often, however, the function of the final cover is all that is stated, with little or no guidance given on how this may be achieved.

Other factors specific to landfill covers, such as thermal variation, settlement, gas permeation, plant and animal intrusion, have received little attention and should be the subject of continued research.

It is encouraging to see policy for final covers over landfill within Europe and the United States undoubtedly improving. Continuation of this trend will only be achieved, however, through widespread discussion and agreement between operators, design engineers, regulating bodies and protection agencies.

ACKNOWLEDGEMENTS

This paper is published with the permission of the Department of the Environment. The views expressed therein are those solely of the author and should not be considered as those of his or any other organisation.

REFERENCES

1. FORSTER, A (1993). Landfill capping review. An assessment of current practices and performance within the United Kingdom, the rest of Europe and the United States. Department of the Environment Contract No. PECD 7/10/251. March 1993.

2. KNOX, K (1991). Water balance methods and their application to landfill in the United Kingdom. Department of the Environment Contract No. CWM 031/91. December 1991.

3. DOE (1992). Investigation into the use of capillary barriers as landfill caps. Phase 1 : Desk Study. Department of the Environment Contract. April 1992.

4. USEPA (1992). Design and construction of RCRA/CERCLA final covers. Seminar publication. EPA/625/4-91/025. May 1991.

Waste Disposal by Landfill – GREEN'93, Sarsby (ed.) © 1995 Balkema, Rotterdam, ISBN 90 5410 356 6

Leak location survey of a soil covered geomembrane at a landfill site in the UK

D.L.Laine
Leak Location Services Inc., San Antonio, Tex., USA

N.G.Mosley
Hydrotechnica Ltd, Shrewsbury, UK

ABSTRACT: An electrical leak location survey was conducted in a newly constructed phase at a municipal solid waste landfill site located in the United Kingdom. The new phase was lined with an HDPE geomembrane. An adjacent phase of the landfill, constructed five years earlier and also lined with an HDPE geomembrane, was known to have developed a leachate leakage problem after placement of waste. The site owners wanted a test method which could locate leaks in the HDPE liner after final construction of the new phase. The test method was required to be performed as a "pre-service inspection".

The principal of the electrical leak location method is to place a voltage across the geomembrane liner and then to detect areas where electrical current flows through leaks in the liner. The results of the electrical leak location survey indicate that both small "pin hole" size leaks and large cuts are detected and easily located.

INTRODUCTION

Geomembrane liners are made of synthetic plastic panels that are welded together in the field to form one large continuous sheet. They are most commonly used as low permeability barriers to prevent contaminated liquids from percolating into the surrounding environment. Testing and inspection of geomembrane installations for leaks is a very important aspect of liner performance.

The electrical leak location survey is performed after all phases of the liner installation are complete. Therefore, the method is particularly valid because the liner is tested under load, and after the liner has been exposed to possible damage from the construction process. In addition, in a landfill where the geomembrane is covered with a protective soil layer the liner may be damaged during emplacement of the soil. Therefore, an electrical leak location survey of the soil-covered geomembrane is a highly effective method of ensuring that any leaks which exist in the liner will be located and repaired prior to commissioning.

Electrical leak location surveys have been available as a commercial service since 1985. During this time 169 surface impoundments and 17 landfill sites have been investigated. The method is used to locate leaks in the geomembrane liners of single-lined and double-lined facilities covered with water or a protective soil cover. The surveyed facilities have ranged in size from 9 metre (30 foot) diameter above ground tanks to surface impoundments of approximately 6 hectares (15 acres) in size.

The first application of the electrical leak location survey technique in the UK took place in February 1993 during construction of Phase 2 at the Craigmore Landfill site near Belfast in Northern Ireland. The site is owned and operated by Antrim Borough Council.

In 1987-88, Phase 1 of the Craigmore Landfill was designed and developed with a single HDPE geomembrane liner, this being one of the earlier applications of flexible membrane liners in the UK. Due to the experience of leachate containment problems in Phase 1, Antrim Borough Council were keen to ensure that similar problems did not arise in Phase 2. Consequently, Hydrotechnica, the appointed landfill design consultants, was requested to incorporate additional measures into the design and construction exercise to ensure the liner integrity in Phase 2.

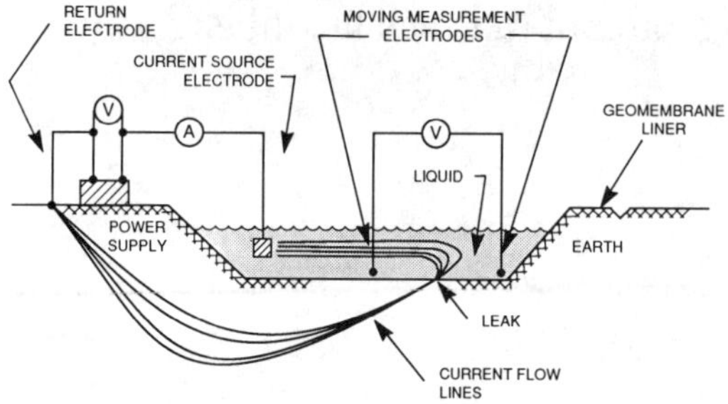

Figure 1 Diagram of Electrical Leak Location Method. This method is used on single or double lined geomembranes covered with water or protective soil.

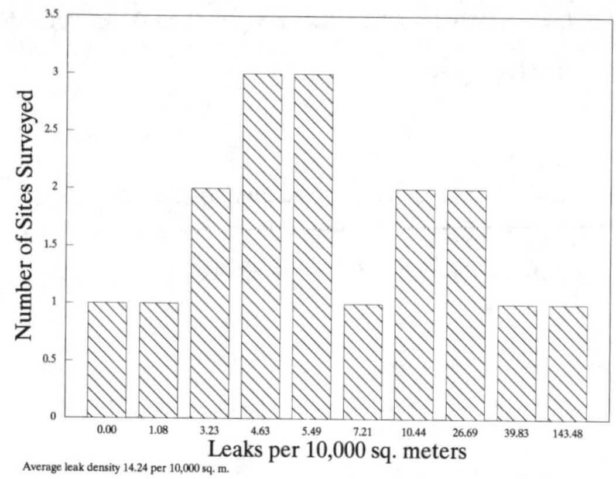

Figure 2 Histogram of Electrical Leak Location Survey Data for Geomembranes Covered with Water

DESCRIPTION OF THE TECHNIQUE

Figure 1 shows a general diagram of the electrical leak location method which illustrates the basic system technique. The method is to connect an electrical power supply to electrodes above and below the liner and then detect areas of localised electrical current flow through leaks in the otherwise insulating liner. When no leaks are present, the voltage impressed across the liner produces a very low current flow and a relatively uniform voltage distribution in the material above the liner. If the liner has a leak, electrical current flows through the leak causing a localised anomaly in the potential gradient. Leaks are located by measuring potential gradients in the material above the liner searching for localised areas of relatively high electrical potential gradients. The electrical leak location method is used in liquid impoundments, and for pre-service inspection of solid waste landfills with soil cover.

With the proper implementation of equipment and survey procedures the electrical leak location method is a very sensitive and accurate technique. The measured amplitude of the leak signal is proportional to the amount of electrical current flowing through the leak. Therefore, leak location surveys should be conducted with the maximum practical safe impressed voltage. To increase the leak detection reliability the detector electronics must be optimised for maximum sensitivity. In addition, extraneous electrical conductors such as metal pipes and pump wiring that provide an electrical conduction path through or around the geomembrane liner should be eliminated or insulated.

GENERAL SURVEY RESULTS

To date the majority of electrical leak location surveys have been conducted in the United States. However, surveys have been conducted in Japan, Saudi Arabia, Germany, Italy, Canada, and the United Kingdom. Figure 2 shows a histogram of the survey data for 169 sites where the geomembrane was covered with water. The results of this data indicate that an average of 14 leaks per 10 000 square metres of geomembrane tested were detected when an electrical leak location survey was conducted. In many cases a rigorous quality assurance/quality control programme was followed during liner installation.

Figure 3 shows leak location survey results for 17 landfill sites where the geomembrane was covered with up to 75 cm (29.5 in) of protective soil. The results indicate an average of 22 leaks per 10 000 square metres, where again rigorous quality assurance/quality control was applied. The technique was able to locate leaks as small as 0.12 centimetre-diameter (0.05 inch-dia) under a soil depth of 75 centimetres. In addition, several large leaks seen as cuts or holes caused during the construction process, were also located.

LINER DESIGN FOR CRAIGMORE SITE

In line with current good practice in the UK it was agreed to develop full containment in Phase 2 of the Craigmore Landfill site by means of a

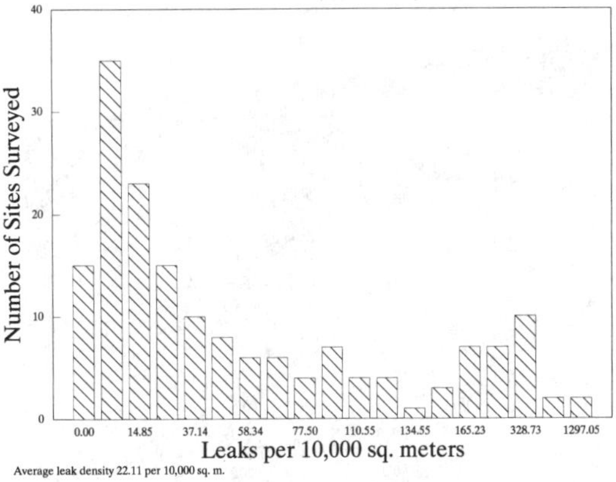

Figure 3 Histogram of Electrical Leak Location Survey Data for Geomembranes Covered with a Protective Soil

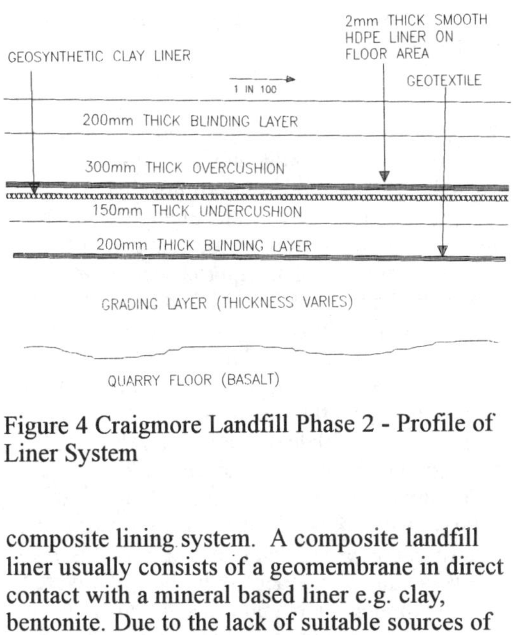

Figure 4 Craigmore Landfill Phase 2 - Profile of Liner System

composite lining system. A composite landfill liner usually consists of a geomembrane in direct contact with a mineral based liner e.g. clay, bentonite. Due to the lack of suitable sources of clay in the region generally, and in particular close to the site, it was decided to use a bentonite based liner. A geosynthetic clay liner (geotextile bentonite mat) was finally selected for a number of reasons:

- very low permeability once hydrated
- ease of deployment and quality control
- economic compared to alternatives.

The alternative of a bentonite enhanced sand layer (BES) was seriously considered but in this instance was not selected due to:

- anticipated construction difficulties in the relatively high precipitation levels in Northern Ireland
- the lack of experienced BES contractors in Northern Ireland
- cost of mobilising an experienced contractor to Northern Ireland

The constructed liner system at Craigmore Landfill Phase 2 comprises a 2 mm HDPE geomembrane overlying a "Bentomat" geosynthetic clay liner. "Bentomat" is the trade name used by Volclay Ltd of the UK for its geosynthetic clay liner product.

A typical profile of the selected liner system is shown in Figure 4.

TECHNOLOGY ASSESSMENT

As part of the landfill design brief Hydrotechnica undertook a review of several systems which were purported to enable the detection and location of leaks in landfill liners, and also collect escaped leachate. The exercise resulted in the selection of the electrical leak location technique which was offered by Leak Location Services Inc based in Texas, USA. The method was considered to be the most promising of the reviewed technologies to ensure the integrity of the as-built liner.

Whilst the principle of the technique appeared to be sound there were some issues which required to be satisfactorily resolved before the technology could be recommended. The issues centred around the ability of the surrounding soils, and incorporated geotextiles and a geosynthetic clay liner, to transmit electricity. These issues were resolved based on the following points:

- Electric current is transmitted through soils via the soil moisture. The surrounding soils (above and below the liner) must be "moist" but are not required to be saturated. The moisture required to achieve adequate compaction is sufficient to transmit an electric current. Usually precipitation onto the cover soil layer provides sufficient moisture which percolates down to the liner. However, a thin saturated zone (fully wetted geomembrane) in the cover soil is not required.

- Sufficient soil moisture must be present to sufficiently wet any geotextile layers in the soils. Previous experience had not shown "dry" geotextiles to be a problem.

- The non-hydrated bentonite in geosynthetic clay liners contains a residual moisture content of some 15% to 17%, and such clays are highly conductive. These conditions were considered sufficient to transmit an electric current and simple laboratory trials on fresh samples of such liners confirmed this point. In a field situation the geosynthetic clay liner is exposed to soil moisture and begins to hydrate, which enhances the conductivity.

CONSTRUCTION QUALITY ASSURANCE AND CONTROL

The construction contract for Craigmore Landfill Phase 2 was let with a strict set of Quality Assurance/Quality Control (QA/QC) procedures built into the contract. The procedures governed the handling, deployment, seaming, testing and repair of all geosynthetic materials used on the contract. The QA/QC procedures were based primarily on those which are routinely applied in the USA and were strictly enforced by Hydrotechnica's QA/QC Inspector. The QA/QC Inspector oversaw the complete installation of the landfill lining system and all activities were fully documented and approved.

Both non-destructive and destructive tests were performed on the welded geomembrane seams and all repairs. Non-destructive testing consisted of air pressure tests on double wedge fusion welds, spark testing of extrusion welds and repairs, and visual inspection.

The electrical leak location survey was specified in the contract to take place after the HDPE geomembrane had been fully installed and covered with a 300 mm layer of protective soil. In this case the soil cover was a uniformly graded sand material. The survey was seen very much as a final level of Quality Assurance (a pre-service inspection) after the liner had been lightly loaded with cover soil and trafficked by construction plant.

ELECTRICAL LEAK LOCATION SURVEY

The electrical leak location survey work was performed on the 300 mm thick sand layer covering the floor area of the landfill. In addition, the survey lines were extended approximately five to ten metres up the side slopes which were covered with the sand material.

Figure 5 Leak Location Survey - Data Collection with Soil Cover

The landfill is divided into four separate leachate collection cells by means of an HDPE partition welded to the floor of the primary liner. The HDPE partitions extend approximately one metre up the side slope. The soil located on the side slope directly above the partition was removed to electrically isolate each of the cells. In addition, the soil on the top of the side slope was removed around the full perimeter of the landfill to electrically isolate the sand from the soil located outside of the landfill.

The survey lines were spaced one metre apart with potential data collected on fifty centimetre centres along each survey line. Figure 5 shows survey data being collected on the soil covered geomembrane. One hundred and ninety-seven survey lines oriented in a north-south direction and varying in length from approximately 30 metres to 80 metres were surveyed in the landfill. Survey data was plotted and analysed for data quality and completeness in the field. Leaks indicated in the data were investigated and if verified the sand cover was removed to expose the leak in the HDPE liner. Figure 6 shows representative plots of the data collected during the survey.

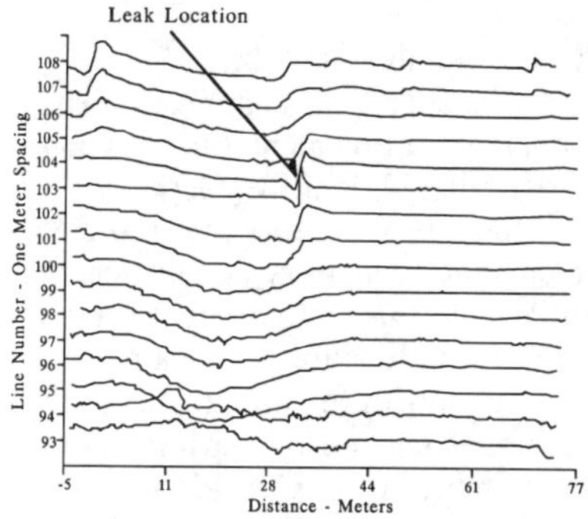

Figure 6 Waterfall Plot of Electrical Leak Location Field Data Collected Over 30 cm of a Protective Soil Cover. Leak Location at Position 30 in line 103.

RESULTS

Ten leaks were detected and verified in the primary 2 mm HDPE liner. All leaks located in the primary liner were uncovered, repaired, and re-tested. Figure 7 shows the general location of leaks which were located and the following notes describe the leaks located in the primary liner:

Leak 1 - Located in extrusion fillet weld on top right side of pipe boot. The leak was repaired and re-tested. No additional leakage was found.

Leak 2 - Two centimetre cut in liner material approximately 15 centimetres northeast of patch R2. Leak was repaired and re-tested with no additional leakage located.

Leak 3 - Leak on fillet weld between panels N2 and B2 approximately one metre south of panel N1. Leak was repaired and re-tested with no additional leakage located.

Leak 4 - Leak in tee weld between panels B14 and S15/S16. This leak was visually confirmed with water flowing from the leak. The leak was repaired and re-tested with no additional leakage located.

Leak 5 - Leak in hand fillet weld on north side of patch R31/D5. Water was observed flowing from the leak. The leak was repaired and re-tested with no additional leak located.

Leak 6 - Leak located on north side of patch R44 in hand fillet weld. The leak was repaired and re-tested with no additional leakage located.

Leak 7 - Leak located in fillet weld on panel S24 near toe seam. The leak was repaired and re-tested with no additional leakage located.

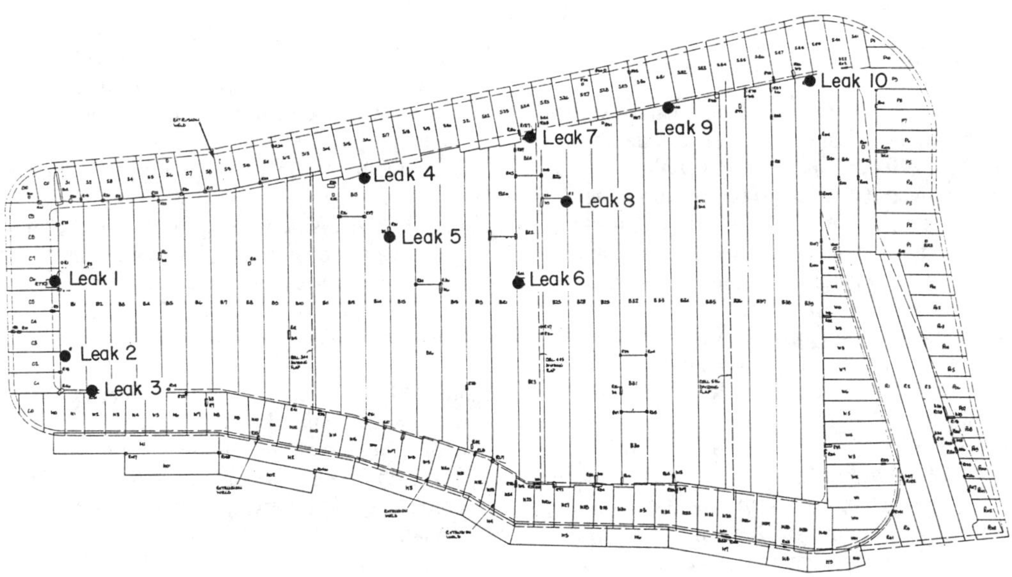

Figure 7 Craigmore Landfill Phase 2 - Geomembrane Panel Layout Showing the General Location of the Leaks Located in the Primary Liner

Leak 8 - Leak located on the north side of patch R53. The leak was patched and re-tested with no additional leakage located.

Leak 9 - Tee weld of panels S31/S32 and B33/B34. The leak was repaired and re-tested with no additional leakage located.

Leak 10 - Fifteen centimetre cut in liner located in panel B39 near toe of side slope. The cut was patched and re-tested with no additional leaks detected.

In addition to locating the leaks in the primary liner the electrical current injected into each of the landfill cells was measured before and after the liner was repaired. Because the HDPE liner is an excellent electrical insulator the electrical current injected into each cell is a measurement of current flow paths around or through the liner material. This is a relative measurement and is not used to quantify the number or size of leaks in the primary liner. If there are no leaks in the liner material and the soil or water located above the liner is not in contact with the soil or water located below the liner then with an input of 350 volts the resultant current flow should be less than approximately 1 milli-amp. This assumes a liner floor area of approximately 10 000 square metres with a fairly high soil or water resistivity.

CONCLUSIONS

The results of the electrical leak location survey at the Craigmore Landfill site clearly demonstrated the effectiveness of the technique in pinpointing a variety of imperfections in the geomembrane installation. As might be expected, the majority of leaks were located in seamed or welded joints but significant holes were also identified in geomembrane panels which would have had serious implications for leachate escape. The significant holes were judged to have resulted from damage during deployment and covering with the protective soil layer.

Including the side slopes, an area of some 12 000-14 000 m^2 of liner was surveyed in three to four days. Data was often processed within a few hours on site to provide preliminary (and almost real-time) information so that leaks could be verified, uncovered and then repaired and re-tested to prove a satisfactory repair. All data was processed overnight as a final check so that the full results were available next day.

Given the relative speed, efficiency and demonstrated effectiveness of this technique it is considered important that all landfill operators, designers, installers, QA/QC inspectors and landfill licensing and regulatory authorities become familiar with this technology. Based on the experience of leak location surveys performed in many countries to date, it is clear that most geomembranes leak despite stringent QA/QC protocols in many cases. Considering the long-term liabilities associated with containment landfills it is in the owners best interest to apply this technique at the construction stage in all landfill developments relying on geomembranes for containment.

ACKNOWLEDGEMENTS

Mr John Quinn, Chief Technical Officer with Antrim Borough Council is gratefully acknowledged for allowing this case study to be published. In addition Environment Service, DOE Northern Ireland are acknowledged for providing financial support, via the European Regional Development Fund under the Intereg Program, which enabled the survey to be conducted.

REFERENCES

Darilek, G.T., Laine, D.L, Parra, J.O. 1989. The Electrical Leak Location Method for Geomembrane Liners-Development and Applications, Industrial Fabrics Association International Geosynthetics '89 Conference, San Diego, CA.

Laine, D.L, and Miklas, M.P. 1989. Detection and Location of Leaks in Geomembrane Liners Using an Electrical Method: Case Histories, Superfund '89, 10th National Conference, Washington, DC.

Laine, D.L., Darilek, G.T. 1993. Locating Leaks in Geomembrane Liners of Landfills Covered with a Protective Soil. Industrial Fabrics Association International Geosynthetics '93 Conference, Vancouver, British Columbia, Canada.

Waste Disposal by Landfill – GREEN'93, Sarsby (ed.) © 1995 Balkema, Rotterdam, ISBN 90 5410 356 6

Main tools for water resources protection

J.A. Plainer
University of Zimbabwe, Harare, Zimbabwe

ABSTRACT: Some of human activities influence negatively quantity and quality of water. Some other influence natural conditions of water cycle and thus the quantity and quality of water as well. Water resources protection has to be a complex system of protection and each of individual tools should be an effective part of it. There are several main tools applicable for water resources protection. This paper deals with the general description of water resources protection and underlines the importance of underground waters. It brings examples of water resources protection and application of some tools in the field of water resources protection.

1 THE EARTH'S WATER CYCLE

The Earth's Water Cycle can be described as a process beginning with evaporation from water surfaces, particularly from oceans, soil and transpiration from plants into the air. The resulting water vapour is transported by air masses which move under influence of the sun radiation and daily rotation about the Earth's axis. Due to changes of pressure and temperature in the atmosphere water vapour can condense and create clouds what can result in precipitation.

Precipitation, which falls on the land surface, is dispersed in several ways. The greater part is temporarily retained in soil near the place where it fell and its substantial portion is ultimately returned into the atmosphere by evaporation and transpiration. A portion of precipitation finds its way over the land surface to rivers, lakes, seas and oceans. The remaining part of precipitation infiltrates and penetrates into soil to become a part of underground waters. Under influence of gravity both surface and ground waters move toward lower elevations and may eventually reach oceans and the water cycle can be considered as to start once again.

1.1 Basic Spaces of the Water Cycle

This simplified concept of the Earth's Water Cycle enables to define four basic phases of the water cycle:

Evapotranspiration, Precipitation, Surface Waters and Subsurface Waters and to distinguish three main spaces where water is distributed:

- Atmosphere, what is a mixture of gases (air surrounding the Earth). Its lower part is named troposphere of an altitude up to 8-20 km above the Earth. It contains water vapour and precipitation,
- Lithosphere, what is the Earth's crust of the thickness about 15 km. Water fills pores and holes of the lithosphere,
- Hydrosphere includes oceans, seas, lakes, swamps and the ice and snow fields in high mountains and in the polar zones.

2 Water Resources as a part of the Water Cycle

Water Resources have been defined as a part of the total volume of water on the Earth which can be used to satisfied human water needs. Therefore they must fulfil some conditions the main of them have been as follows:

- they must have an acceptable quality. From this point of view saline waters (water in oceans and inland saline lakes and seas) are generally not convenient for human needs (with the exception of ship transport, cooling purposes, etc.),
- they must be available for an acceptable price. From this point of view they must be, for example, in an acceptable

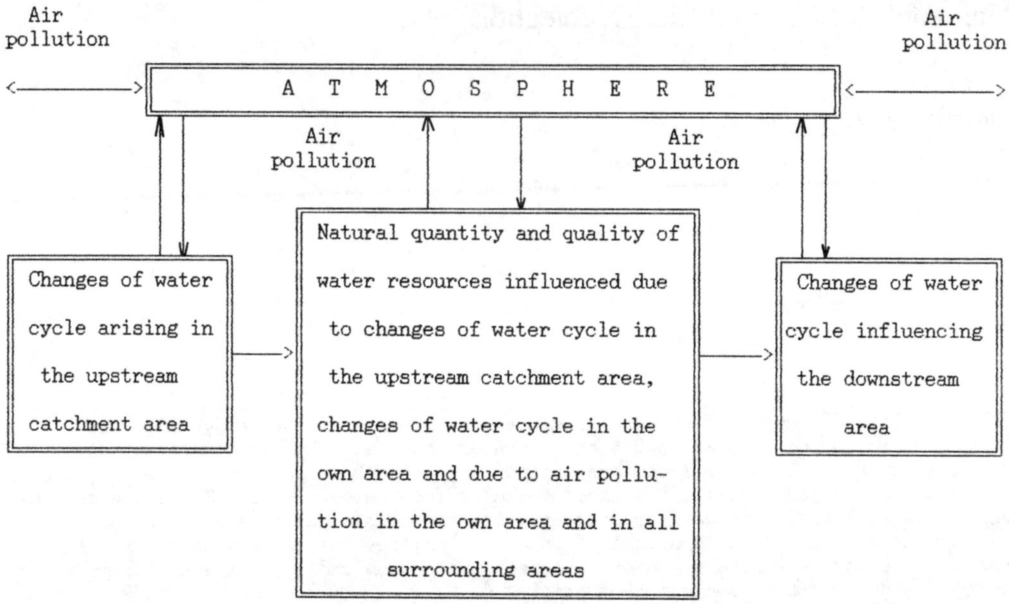

Figure 1: Water resources and system of their changes

distance from the places of consumption to diminish necessary costs for their transportation. This means, that water in polar zones (ice and snow with good quality of water) are not convenient for human needs, etc.,

- they must be technically easily controlled and managed. From this point of view the great amount of water existing as soil moisture and water in the atmosphere cannot be taken into account as water resources.

These conditions show that only very small part from the total volume of water on the Earth can be used for human needs. Water resources are created first of all from water in rivers, in fresh lakes and underground waters. Their amount is estimated to be approx. one half of one percent of the total amount of water on the Earth.

Another important feature of the water cycle is that only a small part of the total amount of water on the Earth takes part in the regularly circulated water on the Earth (approx. 0.04 % of the total volume of water on the Earth). Water resources belong to this regularly recycling water what increases danger of their detorioration due to their repeated utilizations.

On the other hand water resources are on natural conditions inexhaustible because they renew regularly within the Earth´s water cycle.

Above mentioned shows that even if water resources are a negligible part of the total amount of water on the Earth that there are many ways and much space where they can be influenced negatively.

This starts with the air pollution which can deteriorate quality of precipitation, any human activity which influences negatively quality and quantity of water in rivers and lakes, infiltration, percolation, runoff, streamflow and groundwater flow, etc., generally the natural hydrological conditions.

3 MAIN NATURAL HYDROLOGICAL CONDITIONS INFLUENCING WATER RESOURCES

Water resources in individual regions are influenced with a large set of natural factors. The main of them have been as follows:

- precipitation and evaporation which have the basic influence on the amount of water resources in individual regions,
- geological condition and morphology of landscape influencing distribution of precipitation into runoff, underground water and evaporation,
- extent and kinds of vegetation which influence the ratio of evaporation and

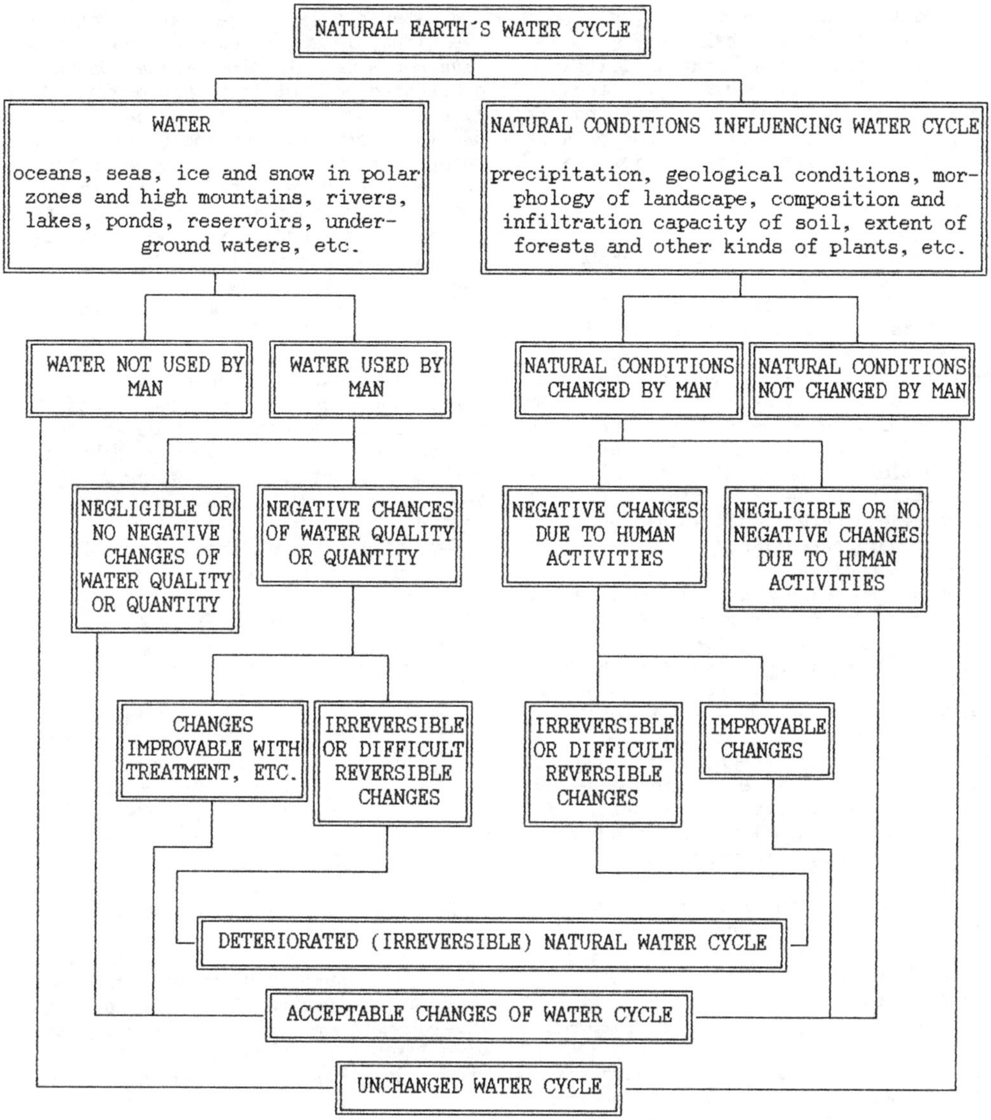

Figure 2.: Changes of the Earth´s water cycle due to human activities

transpiration, runoff and underground waters,

- composition and hydrological properties of soil influencing again the ratio of runoff and underground water flows and first of all formation of underground waters in aquifers.

Some of these natural factors can be influenced with the human activities in a relatively small extent, e.g.: amount of precipitation, basic geological conditions and morphology of landscape (with the exception when a very small region is scrutinized where any even small change of it can be significant).

Some other natural conditions of the water cycle can be influenced with the human activities in such an extent that these changes can result in changes of quality and quantity of water resources, e.g.:

- extent and kinds of vegetation,
- hydrological conditions (river network, amount of lakes, swamps, etc.),

- composition and properties of soil, particularly of agricultural land.

This review shows that human activities which influence the second group of the natural hydrological conditions must be taken into account as dominant when principles of water resources protection are prepared.

4 MAIN KINDS OF WATER DEMANDS AND NEEDS

The quality and quantity of water resources is influenced first of all due to using of water resources for human water demands and needs. These can be divided into two main groups:

1. Using of water directly in rivers, lakes, ponds, etc. This using includes:
   - navigation,
   - recreation and water sports,
   - fishing and water poultry farming,
   - hydropower generation,
   - evacuation of waste waters, i.e., waste waters from industry, agriculture and settlements, less or more treated, even fully untreated, are discharged into rivers, lakes, seas and oceans and in this way water in them is used as a transport medium for replacement of waste waters.

It is obvious that this last kind of using of water brings the highest danger for deterioration of water resources generally and their quality particularly. In water resources protection this using of water plays the most important role.

2. Abstraction of water from rivers, lakes, ponds, etc., and from underground waters for using water for:
   - drinking purposes,
   - other household supply,
   - other industrial needs,
   - irrigation of agriculture land, towns, etc.

All these water demands reduce quantity of water resources in the place of abstraction. Some of above mentioned using of water result in smaller or greater amount of evaporated water (cooling water in industry, evapotranspiration in agriculture, etc.) and the abstracted water never returns into the local water resources what represents losses of water resources in the region.

The remaining part of abstracted water is less or more deteriorated due to using and it becomes waste waters which are discharged into rivers, lakes and ponds, etc., like described above.

Very important feature of all these kinds of water demands is that their amount varies or can vary in a very great extent. For example, two litres of drinking water per capita per day, sometimes even less, is sufficient for man to survive for a certain time period. On the other hand consumption of some hundreds of litres of drinking water per capita per day is a very frequent amount of drinking water consumption in households in the greater cities in many countries and in some cases consumption over one thousand litters of drinking water per capita per day in luxurious hotels and holiday resorts can be recorded.

However, it is not simple to state that the low consumption is always better. It can result in hygiene problems in the case of drinking water, lower production in agriculture and high or higher costs of measures for reducing water demand in industry, etc. This question should be solve with help of balancing of water demands and available water resources in the region.

Some of the possibilities how to save water in industry show figures 3, 4 and 5:

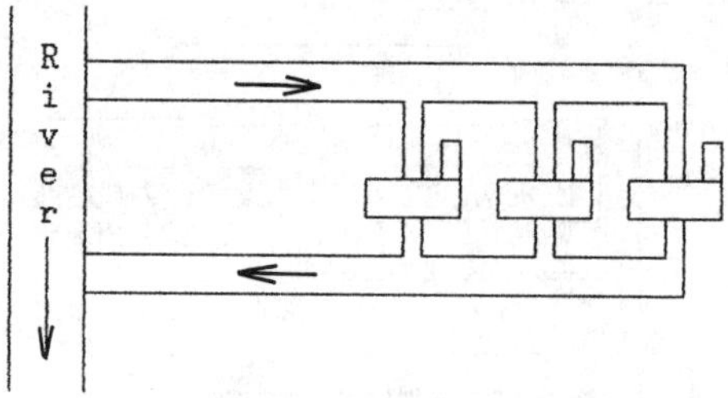

Figure 3: Throw flow supply system

a) The through flow system means that water is abstracted from a source (e.g., a river) and then used separately for individual kinds of production or stages of production. Water once used is after ne-

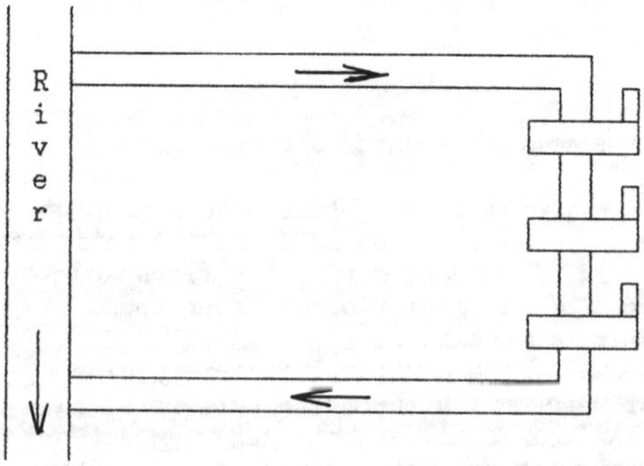

Figure 4: Cascade supply system

cessary treatment discharged into the river.

b) The cascade system means that water is abstracted from the river then used for production or stage of production which calls for the highest quality of water. After the first using water is delivered to the next kind of production or stage of production with a lower requirements for the quality of water (e.g., cooling water) and if possible water is then delivered to the further kind of production or stage of production with the lowest requirements for the water quality (water for washing, for hydraulic transport of wastes, etc.). After necessary treatment water is discharged into the river. It is quite obvious that the total water demand can be diminished up to one third in comparison with the through flow system.

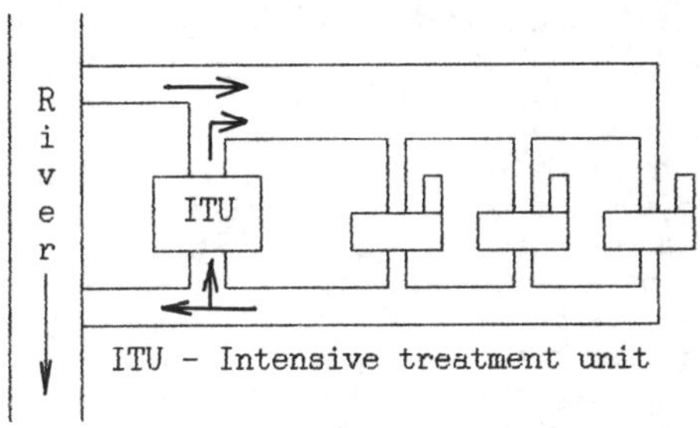

Figure 5: Partly recycling supply system

c) The partly recycling system means that water is abstracted from the river, used for some productions or stages of production and after treatment a part of treated water returns into the factory to be used once again for some kinds of productions or their stages. Remaining part of treated water is discharged into the river. It is obvious that the total water demand can be reduced significantly.

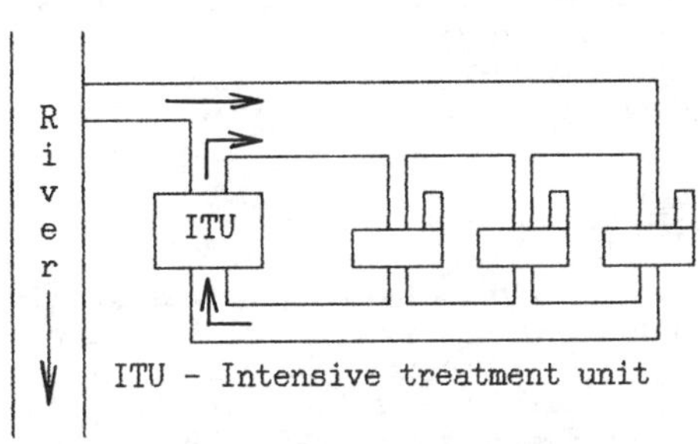

Figure 6: Fully recycling supply system

d) The fully recycling system means that water after using in production is treated intensively and then used again for the same production. The amount of abstracted water from the river is reduced to the amount which covers losses due to evaporation of water in the factory.

5 TYPICAL KINDS OF CHANGES OF NATURAL WATER CYCLE

Figure 2 shows how the natural water cycle can be influenced by human activities. It also shows three possible resulting states of the water cycle due to these activities:

1. The water cycle remains unchanged because the extent of human activities is very small and their influence on the water cycle is negligible.
2. If the extent of human activities is significant, however, their influence on the water cycle cannot be simply classified as negative we can speak about cultural changes of the water cycle. This group of human activities includes, for example, measures against flooding, improvement of infiltration capacity of soil, measures against water-born diseases (drainage of towns, swamps, etc.), storage of water in artificial reservoirs and ponds for water supply in dry seasons, etc.
3. However, nowadays there are many regions all over the world where the impact on water resources due to human activities is harmful and must be classified as undesirable changes of the water cycle and these cases should be avoided. In the worst cases these changes are, however, irreversible.

6 TARGETS OF WATER RESOURCES PROTECTION

Any change of the natural water cycle results in changes of the natural ecosystems and they may result in the total annihilation of some forms of life. It is a question whether the prosperity of man is the only criterion for decision whether the extent of the change of the water cycle is acceptable or not. On the other hand, a return to the natural conditions existing some hundreds or thousands of years ago is an unreal demand or a vision only. This can also be summarised as the problem of the number of the human population on the Earth and its economical and cultural standard of living. It can be assumed that this problem cannot be answered satisfactorily today or in the near future and its solution will therefore not be available to solve the problem of protection of water resources fully.

On the other hand, each country and perhaps each region all over the world face the necessity to solve problems of water resources protection today, without delay. From this point of view, we must accept the rule that the cultural changes of water cycle, described above, are the only acceptable or available target for the water resources protection in individual regions.

7 MAIN KINDS OF WATER RESOURCES

Water resources can be divided into two groups:

- surface waters, like rivers, lakes, ponds and reservoirs, and
- underground waters.

The velocity of motion of underground waters is low, even extremely low. However, earlier or later the greatest part of them drain into rivers and so they become the only source of water in rivers in the periods between rains. Protection of underground waters has therefore the highest priority. This first of all calls for protection of existing infiltration capacity of soil or even for its improvement and, furthermore, for protection of the aquifers against pollution.

8 INFILTRATION CAPACITY OF SOIL

There are two most important groups of human activities that can result in negative changes of infiltration capacity of soil and deterioration of aquifers: agriculture and forestry because they cover the greatest part of land surface. Different kinds of plants result in different infiltration capacity of soil. It is known that one of the best infiltration capacity of soil is in forests. Therefore, deforestation in any region, not only in Amazon catchment area, must be designated as the most dangerous human activity influencing negatively water resources.

It is less known that infiltration of water into arable land is lower than infiltration into grass or meadows because arable land very often enables faster runoff and thus lower ratio of infiltrated water. The acceptable ratio between arable land and grass and meadows should be carefully scrutinized and controlled in individual regions. Unfortunatoly, the change of grass and meadows into arable land, particularly in developing countries, is very frequent case. This activity is very often accompanied with drainage of agricultural land. It should be applied very carefully because it increases velocity of underground waters and diminishes their storage in the aquifers for dry seasons. If it is necessary to apply drainage it should be completed with an adequate storage of drained water, for example with artificial storage like in ponds or small reservoirs located downstream the drainage system.

Another less known fact is that using of heavy machines in forestry and particularly in agriculture leads to an artificial compression of upper parts of soil what again significantly diminishes infiltration capacity of soil. No or minimum effort was done in this field: to prohibit or restrict application of heavy machinery in forestry and agriculture.

Decreasing of infiltration capacity of soil in settlements becomes also a serious problem. Even if the extent of settlements is significantly lower than the extent of agricultural land or forests the infiltration capacity of soil is reduced in settlements dramatically, nearly to zero due to impervious roofs of buildings and tarred roads and pavements. Even all car parks have been mostly fully tarred. The total area of car parks in settlements is relatively hight and even if there are good methods how to cover them with semipervious surfaces (concrete bricks or panels with holes) they are not applied in practice perhaps that they are not too much known or that they call for greater maintenance.

Generally said, agriculture, forestry and many types of structures can significantly diminish natural recharging of underground water resources.

9 POLLUTION OF WATER RESOURCES

The human activities which pollute water resources can be again divided into two main groups:

- human activities their exact location is difficult to define,
- human activities acting in quite exactly defined places.

9.1 Air pollution

The first group includes, for example, air pollution. Even if some of its sources are simply defined factories or other structures air pollution as the whole is a composition of many sources of pollution, like exhausted gases from combustion machines, fires and many other accidents and damages. Polluted matters contained in the air become a part of precipitation and reach the landscape and deteriorate water resources in a large area.

9.2 Transport and storage of dangerous and poisoned matters

Another of human activities which belong to the first group of pollution are transport and storage of dangerous and poisoned matters, particularly of those matters which are transferred in a great amount and for a long distances like crude oil, petrol, etc. Leakage of pipelines, ships and lorries, tanks and other facilities for their storage, their accidents and collapses lead to the entrance of these dangerous and poisoned matters into surface and underground waters.

The detection of pollution of underground waters due to leakages of pipelines, tanks and other storage facilities is very difficult because it cannot be predicted where it could be reliably detected because nearly an infinite amount of detectors would have to be applied and enormous number of employees would have to read them and evaluate results to obtain a perfect detection system.

Very often pollution is detected too late after it has reached underground waters. In these cases the further using of underground water for human needs, particularly for drinking water supply, is then no more possible because even very low amount of these matters is not acceptable for drinking water purposes.

Even if the source of pollution is later defined natural or artificial purification of underground waters calls for tens or hundreds of years due to low velocity and recharging of underground water. Sometimes this pollution must be classified as irreversible and water resource can be used for human activities no more.

9.3 Discharging of waste waters

The second group of pollution includes discharging untreated or insufficiently treated waste waters from industry, communal treatment plants, etc., into rivers and lakes, seas and oceans and sometimes even into soil.

The importance and extent of individual cases of this pollution vary in a very large scale.

It is obvious that tools for water resources protection should be adequate to the importance and scale of water resources pollution and to the kinds of sources of pollution so that the effect of water resources protection should be the highest.

10 MAIN TOOLS FOR WATER RESOURCES PROTECTION

There are three main groups of tools for protection of water resources:

- payments,
- other legal measures and
- enlightenment.

Payments include taxes, fees, charges and similar measures.

Other legal measures include acts, regulations, directions, standards, licences, etc.

These two groups can be accompanied with fines and even with punishment.

Enlightenment has a very broad application starting from conferences, seminars, meetings, books, papers, articles in magazines and newspapers, leaflets, TV and radio broadcasting, etc.

There are some main principles how to apply these individual kinds of water resources protection:

10.1 Payments

The most effective way how to diminish water consumption and amount of discharged waste waters and emitted polluted gasses from locally simply defined places is the direct payment for these activities. However, it can be only applied when metering of them is possible or available.

For example: payment for drinking water in a settlement from a public drinking water network can be organized like a direct monthly payment for an amount of drinking water delivered into an individual household and measured monthly with a consumption meter its readings are approachable both to supplying company (city council) and to the tenants so that they have good opportunity to check their consumption.

To avoid employees in supplying companies for necessary readings of consumption and to save their salaries (even to save money to tenants for installation of consumption meters) payment for drinking water can be included into taxes. This is done in that way that tenants in houses with better equipment pay more than the other groups of tenants or population.

There is another possibility. The blocks of flats have a single consumption meter

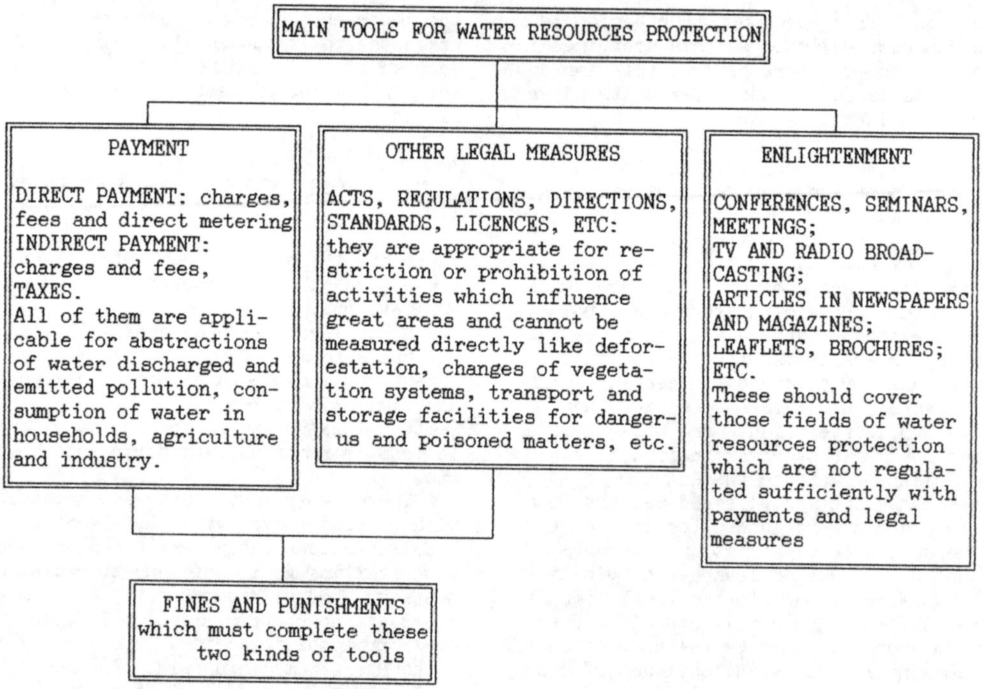

Figure 7: Review of main tools for water resources protection

for the whole building, payment for individual flat´s consumption is defined from the total drinking water consumption in the building and is divided between individual flats according to the number of people living in them.

In my opinion the direct measurement in individual flats and direct (monthly) payment is the best solution. It diminishes water consumption most radically and thus it results in the lowest demand for water resources. Furthermore, the lower amount of used water for washing, flushing, etc., diminishes the amount of waste waters and subsequently the deterioration of water resources. It also reduces requirements for intensification of new drinking water resources: new reservoirs, wells, boreholes, treatment plants (these produce waste water again), etc.

Some years ago I took part in the preparation of the Master Plan of Colombo, the capitol of Sri Lanka. Payment for drinking water was there organized like one of the above mentioned cases. Householders were divided into three main groups: flats with drinking water connections and with connections for waste waters disposal (these flats were named "fully plumbed"), flats with drinking water connections only ("partly plumbed") and inhabitants who had street stands with drinking water available only. Payment for these services was combined with payment for other kinds of communal services (solid waste disposal, cleaning of streets, etc.) and was paid like a part of taxes once a year. Survey of water consumption in individual parts of the town showed that the highest consumption of drinking water per capita was in the suburbs with street stands. In the residential areas with good or high standard houses the drinking water consumption was 150-300 l/cap/day, in the city with hotels and modern blocks of flats it was approx. 600 l/cap/day an in the suburbs with street stands it overreached 700 l/cap/day ! (Literature gives consumption for these cases as 20- 60 l/cap/day). Taps of street stands were damaged or open 24 hours per day. Payment for drinking water was fixed, there was no interest for saving drinking water. The recommendation of Master Plan was to introduce reliable measuring of drinking water and to pay for used amount of drinking water.

In Czechoslovakia in past decades payment for drinking water was organized with another of above mentioned ways. The greatest part of new blocks of flats was equipped with a single consumption meter for the whole block, even for a set of these blocks. Payment was divided between individual flats according to the number of inhabitants in the flat. Very often it was difficult to defined number of inhabitants in the flats (births, deaths, moves, etc.). Monthly deposits with an annual balance were paid for drinking water consumption. Furthermore, the price of drinking water was subsidised and so it was no wonder that the drinking water consumption was approx. 400-800 l/cap/day. After the "velvet revolution" in 1989 subsidising was stopped and tenants themselves started to install drinking water consumption meters in individual flats to introduce a true way for payment for water consumption. Even if the change of the system has not been finished yet and the exact results have not been available it is already known that the drinking water consumption dropped significantly and the building up of new planned drinking water reservoirs and treatment plants could be postponed.

This method, i.e., direct measurement and payment can be applied for many other cases, not only for drinking water. Any abstraction of water from rivers, lakes, from underground waters, discharging of waste waters into rivers, lakes, seas or oceans, emitting of gases from factories, thermal power stations and other similar sources of air pollution. Even in these cases it can be the most effective method for protection of water resources.

Sometimes a problem can arise how to define fees per unit of emitted gases or discharged waste waters because in these cases the concentration of dangerous and poisoned matters must be taken into account. If the fees are low the effect could be low as well. However, nowadays many standards have been available in literature. The method of direct metering and payment can be completed with granting of locally acceptable limits of emitted and discharged pollution and with fines in the case that these limits are not kept.

10.2 Other legal measures

The second group of human activities described above, which includes sources of pollution their places of action cannot be simply defined (air pollution, transport and storage of dangerous and poisoned matters) and controlled with direct measurement and payment, calls for other kinds of tools for protection of water resources, i.e., for other legal measures.

The main way of these legal measures is restriction or full prohibition of such a human activity which can deteriorate water resources. Acts, regulations, directions, standards, licences, etc., can contain list of activities which are restricted and prohibited. It is done in that way that some activities can be prohibited generally in the whole country or they could be prohibited in some areas only and restricted in a certain extent in other regions. In this case we speak about protected zones, areas, belts, etc.

For example, if the country is supplied with drinking water from underground waters and no other source is available protection of underground waters must be high and effective. The aquifers and their recharging areas must be fully defined and proclaimed as protected areas. Storage and transport in pipelines of crude oil, petrol and other similar matters should be prohibited. Building up of chemical factories producing dangerous and poisoned matters should be avoided as well. Similarly, application of some kinds of pesticides and herbicides in agriculture, etc., etc.

However, it is simple to prohibit everything but the great extent of prohibitions can lead to degeneration of the region because it become poor due to low agricultural, industrial and other production, high rate of unemployment, etc. At the same time even very broad list of prohibited activities has not to lead to an efficient water resources protection if this list is not accompanied with a good knowledge of the whole natural, economical, social and political state of the region, of the country. This is an important feature of application of the other legal measures for water resources protection. List of prohibitions which would be necessary in tropical zones cannot be applied in polar zones. There are different conditions for application of restrictions and prohibitions in developed and in developing countries. Legal system is different even in neighbouring countries what calls for modification of methods used. This means that (good or even excellent) acts, regulations, directions, standards, licences, etc. introduced into practice in a certain country and doing there well cannot be translated and simply and fully introduced into practice in another country. This makes the application of legal tools for protection of water resources a little difficult.

Another thing which is important for preparation of good legal measures is that they must be based on narrow cooperation between lawyers and experts from all fields of water resources protection like hydrologists, geologists, civil engineers, etc. This cooperation has not been educated at the universities and it has not been taught and developed systematically in practice. Experts must know and say what is necessary for water resources protection and lawyers must know how these requirements can be implemented into bills, regulations and directions. Very often they must persuade experts that their proposals or ideas are in contradiction with general legal principles in the country. Sometimes, on the other hand, the lawyers must accept new ideas and look for new ways how to legalise them.

Some years ago I had good luck that I took part in preparation of the Water Act in Czechoslovakia and in preparation of several regulations and directions. I share an office with a very good and experienced lawyer for many years. The first stage of our cooperation called for trying to understand each other well. However, very quickly we found out that law and engineering has very narrow linkage. Both of them call for exact and precious knowledge and expressing of it.

Law uses for this definitions and engineering equations. Generally said definitions and equations are the same or at least very similar way of thinking. This finding can, perhaps, simplify cooperation even between other lawyers and engineries and help to improve protection of water resources.

10.3 Enlightenment

The last group of tools for protection of water resources is enlightenment. As it was described above it covers very wide area. It starts with discussions about new findings and ideas at learned conferences, seminars and workshops, their publishing, etc. Sometimes problems have to be remembered to politicians on meetings and even demonstrations. In this field the possibility of enlightenment are sometimes overestimated. It is hardly to believe that a set of demonstrations is sufficient for enlightenment of individual people, the government or even the whole society once for ever. On the other hand even a simple advice in newspaper, magazines, TV and radio broadcasting can bring very good results. For example, in last year there was an extra dry period in southern parts of Africa and water resources became insufficient. Many extraordinary legal and other measures were introduced into practice to improve drinking water supply in countries in question. But enlightenment contributed to solve this problem as well. Any good invention and idea how to save even very small amount of water were broadly published and as survey showed even immediately applied in households, factories, schools, institutions, etc.

Generally said, enlightenment covers the gap not sufficiently solved with the remaining two kinds of tools - payments and other legal measures. Sometimes it even could accelerate these two main methods. Even if payments and other legal methods should be preferred in any society the possibilities of enlightenment should be also exploited fully for protection of water resources.

REFERENCES

Vancura, J. & coauthors 1974. Water Management in Czech Socialist Republic. State Publishing House of Agriculture Literature, Praha.

Plainer, J.A. 1977. Balances of Water Resources and Needs. State Publishing House of Agricultural Literature, Praha.

Plainer, J.A. 1983. Utilization and Protection of Water Resources. State Publishing House of Agriculture Literature, Praha.

Volker, A. & Henry, J.C. 1988. Side Effects of Water Resources Management. IAHS Publication No. 172, Paris.

Plainer, J.A. 1988. Analysis of Influences of the Refined Oil Storage on Water Resources in Maiduguri Region. Final Research Report, University of Maiduguri, Nigeria.

Plainer, J.A. 1991. Protection of Water Resources. Proceedings from the International Conference on Environmental Pollution, Lisabon.

Waste Disposal by Landfill – GREEN'93, Sarsby (ed.) © 1995 Balkema, Rotterdam, ISBN 90 5410 356 6

Waste disposal by landfill in Britain: Problems, solutions, and the way forward

H.A.B. Potter & R.N. Yong
Geotechnical Research Centre, McGill University, Montréal, Que., Canada

ABSTRACT: Landfill is the most widely used method of waste disposal in Britain. The potential for leachate from these sites to cause groundwater pollution is becoming increasingly recognised. Waste regulation authorities seldom have sufficient expertise pertaining to geotechnical practices or hydrogeology to protect groundwater. Scarce resources often prevent the National Rivers Authority from providing assistance concerning landfills. The trend toward containment sites is welcomed, however the British landfill industry has much to learn from North America. The limitations of liners are not well understood in the UK. Field testing of the hydraulic conductivity, and investigation into the significance of diffusive transport is discussed. Research into leachate-liner compatibility is crucial to provide confidence in the long term integrity of liners. The hydrogeological environment around landfill sites is emphasised. Finally attention is drawn to the necessity for academic research into landfill practice plus greater national guidance on regulation to improve the current level of British waste disposal.

1 INTRODUCTION

Landfill is the primary route for the disposal of solid wastes in the UK (Taylor & McLean, 1992). This historically inexpensive method of waste management has traditionally been undertaken in old quarries, former open cast mining areas, and abandoned railway cuttings. Thus landfilling has been seen as a method of restoring these seemingly useless pieces of land to the public for recreation or other development.

Landfills have traditionally been designed on the basis that the surrounding soil and groundwater have the ability to dilute, disperse or attenuate any leachate formed. Thus landfills were often deliberately sited in high permeability strata such as sandstone since the effect of mixing the leachate with a large volume of groundwater was thought to be sufficient to reduce the concentration of any pollutant to a "safe" level. The realisation that the landfilling of wastes can pollute the surrounding groundwater in conjunction with the implementation of the EC "Directive on the protection of groundwater against pollution by certain dangerous substances" (EC, 1980) has resulted in considerably greater emphasis being placed on controlling the outflow of polluting leachate from landfill sites.

The United States Environmental Protection Agency (USEPA) estimate that there are 75,000 sanitary landfills in the USA, and that over 75% of these have caused groundwater pollution (Lee & Jones, 1991). Meanwhile representatives of the UK Department of the Environment (DoE) maintain a belief that landfill leachate has never caused the loss of any public drinking water supply (Robinson & Gronow, 1992). This implies that the DoE do not believe that landfills pose a significant threat to groundwater.

Local government controlled waste regulation authorities (WRAs) are responsible for issuing licences for waste management facilities including landfill sites. The National Rivers Authority (NRA) are statutory consultees for waste management licence applications since they have been charged with the protection of water resources. These authorities have responded to the potential pollution of groundwater by requiring low permeability liners to be constructed at new landfill sites.

This paper will discuss current geotechnical practice concerning landfills in Britain, and will draw on North American experience to propose ways in which landfill engineering can be improved in Britain.

2 REGULATORY FRAMEWORK

In 1976, the Control of Pollution Act 1974 (COPA) came into effect as the first piece of legislation specifically designed to control the management of waste in Britain. COPA provided for licenses to be issued by waste disposal authorities with the intention of preventing any activities which could cause a hazard to the environment. Part II of the Environmental Protection Act 1990 (EPAct) will largely supersede the waste management provisions of COPA once it is fully implemented by the Secretary of State for the Environment.

A critical provision of EPAct when compared to COPA, is the requirement for a landfill operator to be issued with a "Certificate of Completion" by the licensing WRA before the waste management licence may be surrendered. Once the Certificate of Completion has been issued, responsibility for the landfill site, including the cost of any remedial works, passes to the WRA. This passage of responsibility requires WRAs to be confident that the site in question is not likely to cause pollution. It is worth noting that expenditure on remediating hazardous waste facilities in the USA is currently in excess of US$10 billion per year (Bredehoeft, 1992), and that the expense of remediation for municipal landfills is expected by workers in this area to be even greater (Lee & Jones, 1991).

There is currently little provision made in Britain for the expense of cleaning up land and groundwater contaminated by waste disposal activities. The difficulties and protests that have arisen as a result of the register of contaminated land proposed in the EPAct emphasise the need to ensure that pollution is avoided. It has been estimated that the remediation of contaminated land in Britain could cost up to £30 billion (Pearce, 1992).

Article 17 of the proposed EC "Directive on the landfill of waste" (EC, 1991) stipulates that a financial guarantee shall be provided by a landfill operator as a condition of the licence being granted. Such guarantees have already been required by certain WRAs in the form of insurance bonds, and Section 74 of the EPAct allows for refusal of a licence if it is believed by the WRA that the applicant has made insufficient financial provisions. Furthermore, article 18 of the proposed EC directive requires the establishment of a "Landfill Aftercare Fund" analogous to the US "Superfund". This will be created by levying a charge from landfill operators on the basis of the type and amount of waste that they accept, and will be in addition to the financial guarantee. However until these measures are enacted, and given the low cost of landfilling in Britain, it is uncertain who will be liable for the cost of remedial action. It is therefore recommended that WRAs begin to require that operators of both new and existing landfill sites obtain insurance or make other financial provision as a safeguard against the expense of future remedial measures.

3 "CONTAINMENT" SITES

The recognition by regulatory authorities of the need to avoid pollution has resulted in a move toward landfills being designed according to the "containment" principle wherein liner barriers composed of a low permeability soil or an artificial geomembrane are emplaced. Leachate originating from the dissolution products of material in the waste pile is therefore restricted from movement out of the waste pile and can thus be collected, extracted, treated and disposed of by suitable means. This approach has been inspired by regulations introduced many years ago in the USA and reinforced by the proposed EC "Directive on the landfill of waste" (EC, 1991).

3.1 Compacted soil liners

The current trend in the UK is to require that compacted soil liners be one metre in thickness, with a maximum hydraulic conductivity of $1x10^{-9}$m/s. This liner thickness is likely to be increased in the future as a result of the EC landfill directive which requires 3 metres of soil at $1x10^{-9}$m/s or equivalent. The hydraulic conductivity requirement is currently evaluated by laboratory testing of either re-compacted or "undisturbed" samples of the proposed liner. Whilst WRAs encourage these tests to be conducted according to BS1377 (1990), this is not a mandatory requirement. The dangers of relying on samples that have been re-compacted in the laboratory, often after the removal of large particles, are obvious since this is likely to give an underestimate of the true hydraulic conductivity. The more advanced WRAs will already only consider tests on "undisturbed" samples to be acceptable indicators of the hydraulic conductivity of the liner. These undisturbed samples are commonly known as "U100" tests since the sampling diameter is 100mm. The small diameter of each test requires that many samples be extruded if they are to be representative of the entire liner where small preferential flow paths may exist. These flow paths are the result of such factors as micro-fissures, macropores, desiccation cracks, heterogeneities in the source soil, and poor QA/QC during the construction of the liner.

The hydraulic conductivity requirement is usually combined with licence conditions that stipulate a maximum leachate head of 1 metre, to give an expected breakthrough time at the base of the liner of about 30 years (according to Darcy's Law). The leachate is assumed to be non-polluting after this time although the scientific basis for this assumption is extremely questionable. Furthermore, since actual leachate chemistry is considerably different from the "distilled" water used for hydraulic conductivity tests, it is relevant to ask how this impacts on the "real" hydraulic conductivity of the barrier material.

Unfortunately there is a growing body of research which suggests that leachate will travel through liners in the field in a considerably shorter period of time (Anderson et al, 1991; Brown & Daniel, 1991; Daniel, 1984; Panno et al, 1991; Rogowski, 1990; Sai & Anderson, 1991; Thomas & Brown, 1992). These studies have found that laboratory values for the hydraulic conductivity are consistently one to three orders of magnitude lower than those found under field conditions (Sai & Anderson, 1991). Other research conducted on behalf of the USEPA has resulted in the conclusion that a 1 metre thick soil liner, compacted to a hydraulic conductivity of $1x10^{-9}$m/s, could show leachate breakthrough in a period of months rather than the predicted 30 years (Thomas & Brown, 1992). This reduced breakthrough time is thought to be due to the presence of preferential flow paths that are not detected by laboratory tests. It is therefore essential for WRAs to begin requiring that the hydraulic conductivity of compacted soil liners be evaluated in the field using in-situ tests. The different methods of in-situ hydraulic conductivity testing (e.g. infiltrometers) have been reviewed by Sai & Anderson (1991). In addition, the use of lysimeters emplaced within the liner should be considered to assist in evaluating the effects of diffusion. The development of a leachate transport database for UK conditions will be an important step in improving the protection of groundwater from landfill sites.

Partly as a result of the relatively recent advent of compacted soil liners in Britain, and partly due to an incomplete understanding of the transport of liquids through these liners, current landfill liner guidelines emphasise the possible relationship between density, moisture content and hydraulic conductivity (NRA, 1989; WYWRA, 1991; Seymour, 1992). In addition, using the moisture content and density as construction control parameters avoids having to delay work on the liner whilst awaiting the results of the hydraulic conductivity tests. Several studies conducted on behalf of the USEPA however, have concluded that there is insufficient correlation between the density and moisture content of a compacted soil liner and its hydraulic conductivity to give this relationship validity (Rogowski, 1990; Anderson et al, 1991; Thomas & Brown, 1992).

3.2 Diffusive transport

An additional flaw in the reliance on hydraulic conductivity as the performance specification for compacted soil liners, is the failure to recognise the importance of diffusive transport in low permeability materials. It is generally understood by researchers in North America that at the hydraulic conductivity typically found in clay liners, diffusion becomes the controlling mechanism of solute transport (Goodall & Quigley, 1977; Shackelford & Daniel, 1991).

Waste regulation authorities have only begun to realise the importance of low hydraulic conductivity liners within the last few years, and so it is perhaps understandable that the impact of diffusive transport has yet to be fully recognised. Studies at the Keele Valley Landfill near Toronto, Ontario have clearly demonstrated the effect of diffusion in transporting leachate constituents through a compacted clay liner. This liner was constructed to have a maximum hydraulic conductivity, with domestic waste leachate as the permeant, of $1x10^{-10}$m/s (an order of magnitude less than that typically specified in Britain) and yet, after less than five years, chloride has been detected 900mm into the liner (King et al, 1991). Predictions using advective transport alone suggest that this solute should have moved just 100mm into the liner during this time period. This and other similar case studies on real liners as opposed to laboratory studies, prove the necessity of considering diffusive transport mechanisms when evaluating the movement of leachate through a compacted soil liner.

3.3 Leachate-liner compatibility

One of the difficulties caused by the

scarcity of qualified personnel in Britain is that compacted soil liners are viewed as essentially inert barriers whose behaviour will not be affected by changes in the composition of the permeating fluid. The impact of landfill leachate on the hydraulic conductivity and stability of soil liners is rarely if ever evaluated by either developers or regulators in Britain. The need for compatibility testing has been recognised for many years in North America. Bowders et al (1986) have proposed a decision tree for determining the types of tests that are needed to adequately evaluate the compatibility of a landfill leachate with a soil liner. Leachate-liner compatibility testing is therefore commonly undertaken in North America as a means of providing some assurance that the liner material will not undergo significant deterioration, at least under the test conditions. The impact that leachate type solutions have on the permeability of clay liners is well documented (Mitchell & Madsen, 1987; Brown, 1988) but this has not been recognised by WRAs. This limited understanding of the mechanisms by which leachate is transported through low permeability soil liners has resulted in the overconfidence in the ability of the hydraulic conductivity to predict whether a landfill site will cause pollution.

3.4 Flexible membrane liners (FMLs)

Artificial geomembranes or flexible membrane liners have only recently been utilised in Britain for landfill sites. They are manufactured from a variety of different organic materials such as high-density polyethylene (HDPE), polyvinyl chloride (PVC), and chlorinated polyethylene (CPE). These liners have been seen as the panacea for containment sites due to their extremely low hydraulic conductivity. The enthusiasm with which they have been embraced is partially the result of a poor understanding of the difficulties associated with their use, in addition to the doubts that have been raised over natural soil liners. Whilst the necessity of implementing a rigorous QA/QC program has been recognised by WRAs, it has proved hard to overcome the differences between theoretical proposals and actual field practice. The first practical experience that many regulatory authorities have of FMLs occurs when they are asked to approve and supervise their use in a new landfill. It is difficult for Waste Regulation Inspectors to critically assess the design or construction of an FML if this is their first introduction to the material.

The long term ability of FMLs to prevent the outflow of leachate is questionable. FMLs are expected to provide an effective barrier for approximately 25 years (Rowe, 1991) but their continued ability to retain leachate after this time is unknown. In addition, landfill leachate commonly contains organic fluids that can interact with the organic material that is used to manufacture FMLs. Chemical reactions due to contact with dilute organic or inorganic solutions can significantly affect the behaviour of FMLs (Bellen et al, 1987). These chemical reactions can alter the permeability and other properties (e.g. tensile strength) of the FML which could compromise the integrity of the material as a landfill liner Telles et al, 1988). Research into the compatibility of FMLs and municipal solid waste (MSW) leachate has been undertaken in North America and, based on currently available data, a recent report conducted for the USEPA concluded that FMLs should not be given automatic approval for use as MSW landfill liners (Haxo, 1991). This conclusion suggests that research into the compatibility of flexible membrane liners and MSW leachate should be undertaken in Britain, and that the results should be disseminated, with recommendations, to WRAs and the NRA.

Flexible membrane liners are typically only 10-25mm thick but several tens of square metres wide, hence they are very liable to being damaged. If punctured, these materials possess no self-healing ability (unlike soil liners) and so substantial leakage of leachate can occur. FMLs should not be used as the sole barrier to leachate but instead should be combined with a soil liner or similar. The importance of direct contact between the FML and soil components of such a composite liner system in order to minimise leakage appears to be recognised in Britain (Seymour, 1992), however the effect of the hydraulic conductivity of the soil component on the leakage rate does not seem to be fully appreciated.

Research conducted for the USEPA suggests that the rate of leakage through FMLs is primarily dependent upon the size and shape of the hole, the hydraulic head, and the hydraulic conductivity of the underlying soil (Brown et al, 1987). For example, the USEPA study found that for a leachate head of 0.5 metres, the leakage rate through a hole with a diameter of only 0.08cm will increase from 66 litres/year to 540 litres/years as the hydraulic conductivity of the underlying soil rises from $3.4x10^{-9}$ m/s to $3.4x10^{-8}$m/s (Brown et al, 1987).

It is imperative that information such as this is communicated to WRAs and the NRA since these bodies are responsible for permitting the construction of landfills using flexible membrane liners. FMLs should be considered as suitable for providing an extra factor of safety for a landfill rather than having the ability to function as the sole liner component.

4 HYDROGEOLOGICAL CONSIDERATIONS

Whilst the move towards containment sites is a distinct improvement over the "dilute and disperse" philosophy, there appears to be a lack of understanding within Britain as to the limitations of this method of landfill design. In North America, where containment sites have been mandated for several years, it is recognised that even "state of the art" engineering design and construction are not sufficient to guarantee that a landfill site will not cause pollution (LeGrand & Rosen, 1992). The impossibility of guaranteeing the integrity of liners, and the high expense and difficulty of remediating contaminated groundwater, require that consideration is given to the hydrogeological conditions underlying proposed landfill sites. This is especially important since leachate and gas are expected to be produced for many decades, particularly as low permeability cover materials are more widely used (Ham, 1993).

The draft groundwater protection policy of the National Rivers Authority (NRA, 1991) recognises the potential for leakage by excluding the construction of any landfill site within a 50 day travel period of a supply borehole (Seymour, 1992). However, notwithstanding this draft policy, new British landfills are still being permitted in locations such as limestone quarries with high water tables where any leakage is extremely likely to cause pollution of the aquifer. One of the reasons for the continued approval of hydrogeologically unsuitable sites is due to the division of responsibility for landfills between the WRAs and the NRA.

WRAs receive applications for new landfill sites or extensions to existing sites and, whilst the NRA are consulted concerning any application for a waste disposal licence and their recommendations included as licence conditions, the responsibility for supervising the actual design and construction of the landfill site and liner remains with the WRAs. The NRA will generally have a greater depth of experience on hydrogeological matters, and so could provide valuable input to the WRAs. Unfortunately, in spite of their draft policy, the NRA often regard groundwater pollution from landfill sites as being of little importance compared to pollution from agricultural practices (e.g. fertilisers, pesticides) and industrial discharges to surface waters. Therefore the WRAs are left to ensure that groundwater pollution does not occur.

Landfill sites should not be located within vulnerable hydrogeological environments (e.g. permeable limestones and sandstones) particularly where these strata communicate directly with aquifers from which abstraction occurs. Instead landfills should be sited where there is a substantial depth of clayey materials (e.g. thick clay soils or massive shales) which have some ability to attenuate leachate. The USEPA have developed a method for comparing the suitability of different sites based on the relative potential for groundwater pollution at each prospective site (Aller et al, 1987). This method evaluates Depth to water, Recharge (net), Aquifer media, Soil media, Topography (slope), Impact of vadose (unsaturated) zone, and Conductivity (hydraulic) of the aquifer, and is known by the acronym, DRASTIC. However, DRASTIC presupposes that landfill sites will be evaluated based on technical considerations rather than simply the availability of, for example, an old quarry. The current trend toward larger landfills will result in each site having a greater environmental impact. The risk of these potentially negative impacts could be effectively reduced by assessing the suitability of new landfills at a regional level where due consideration could be given to hydrogeological conditions. Any progress away from the current fragmented method of waste regulation (i.e. regional WRAs or the proposed national Environment Agency) is therefore to be encouraged from the perspective of environmental protection.

In order to encourage increased consideration of the hydrogeological environment around a proposed landfill, greater emphasis should be placed on formal training concerning landfill site assessment for WRA staff. This training should be supplemented by requiring landfill developers to predict the likely impact of any leakage of leachate or gas from the proposed landfill. Such predictions will include modelling of potential leachate plumes into the surrounding land. In addition, WRAs need to ensure that the impact of a site can be identified by requiring complete site investigations into the background condition and variation of groundwater and soils.

5 CONCLUDING REMARKS - THE WAY FORWARD

Since the responsibility for the day-to-day supervision and permitting of landfill construction and operation rests with the WRAs, it is essential that these bodies are given sufficient resources to ensure that they can adequately fulfil their role in environmental protection. Similarly, the WRAs must be consulted during the development of new guidelines, and in identifying areas in which further research is needed. Unfortunately, the Department of the Environment do not seem to have recognised this necessity and so the majority of their current research efforts into landfill engineering design appear to be in conjunction with the NRA (Robinson & Gronow, 1992).

The guidelines that are currently being used by WRAs have been adapted from those produced by the North West Region of the NRA (NRA, 1989; WYWRA, 1991). Whilst these guidelines have improved the construction of landfill liners, they would substantially benefit from the incorporation of research that has been undertaken both in North America (e.g. USEPA reports) and in the UK (e.g. research by the British Geological Survey, universities etc). A substantial amount of literature has been produced in North America concerning the design, construction and monitoring of landfill sites, and many of these documents would be of immediate benefit to WRAs (e.g. USEPA, 1989).

Many WRAs simply do not have the resources to dedicate employees to devising research programs or keeping up with research carried out by universities or research institutes. Until the proposed national Environment Agency is in place, the Department of the Environment and HM Inspectorate of Pollution must take greater responsibility for providing support and guidance to WRAs on issues relating to landfills.

Research programs in conjunction with universities and other research institutes need to be implemented to encourage a better understanding of matters such as: leachate-liner compatibility; the long term behaviour of FMLs and compacted soil liners; the recycling of leachate for waste stabilisation; the importance of diffusion in leachate transport.

The UK appears to be several years behind North America in terms of environmental protection and unless we can learn from their experiences we may be doomed to follow the same path which will lead to significant pollution of our valuable groundwater resources with prospects for remediation slim and exceedingly expensive. Much of the research undertaken in North America could be used to improve the current state of guidelines and practice within Britain if its existence was more widely disseminated by such means as incorporation into guidelines (e.g. Waste Management Papers) with statutory weight. Whilst it is essential for the UK to learn from its own experiences, it should be possible to use the North American experience to avoid many potential pitfalls. By selective utilisation of the research published, WRAs, with the assistance of the DoE and the NRA, will be able to bring the levels of landfill practice in Britain to a far higher level than currently exists.

REFERENCES

Aller, L., T. Bennett, J.H. Lehr, R.J. Petty, & G. Hackett 1987. *DRASTIC: A standardized system for evaluating ground water pollution potential using hydrogeologic settings*. EPA/600/2-87/035. Cincinnati.

Anderson, D.C., M.J. Lupo, J.A. Rehage, J.O. Sai, R.L. Shiver, R.C. Speake, K.W. Bellen, G., R. Corry, & M.L. Thomas 1987. *Development of chemical compatibility criteria for assessing flexible membrane liners*. EPA/600/S2-87/067. Cincinnati.

Bredehoeft, J. 1992. Much contaminated ground water can't be cleaned up. *Ground Water* 30: 834-835.

Brown, K.W. 1988. *Review and evaluation of the influence of chemicals on the conductivity of soil clays*. EPA/600/S2-88/016. Cincinnati.

Brown, K.W. & D. Daniel 1991. *Factors controlling minimum soil liner thickness*. EPA/600/2-91/008. Cincinnati.

Brown, K.W., J.C. Thomas, R.L. Lytton, P. Jayawickrama, & S.C. Bahrt 1987. *Quantification of leak rates through holes in landfill liners*. EPA/600/S2-87/062. Cincinnati.

BS1377, 1990. *Methods of Test for Soils for Civil Engineering Purposes*. British Standards Institute, London.

Control of Pollution Act, 1974. HMSO, London.

Daniel, D.E. 1984. Predicting hydraulic conductivity of clay liners. *Journal of Geotechnical Engineering* 110: 285-300.

Environmental Protection Act, 1990. HMSO, London.

EC, 1980. Council of European Communities. Directive on the protection of groundwater from certain dangerous substances (80/68/EEC). *Official Journal*, L20: 43-47.

EC, 1991. Council of European Communities. *Directive on the landfill of waste*

(draft). (Com(91)102 final - SYN 335).
Goodall, D.C., & R.M. Quigley 1977. Pollutant migration from two sanitary landfill sites near Sarnia, Ontario. *Canadian Geotechnical Journal* 14: 223-236.
Ham, R.K. 1993. Overview and implications of U.S. sanitary landfill practice. *Air & Waste* 43: 187-190.
Haxo, H.E. 1991. *Compatibility of flexible membrane liners and municipal solid waste leachates*. EPA/600/S2-91/040. Cincinnati.
King, K.S., R.M. Quigley, F. Fernandez, D.W. Reades, & A. Bacopoulos 1991. Performance monitoring of hydraulic conductivity and diffusion at the Keele Valley landfill, Maple, Ontario. *Proc. 1st Canadian Conference on Environmental Geotechnics:* 49-56. Montréal.
Lee, G.F. & R.A. Jones 1991. Landfills and ground-water quality. *Ground Water* 29: 482-486.
LeGrand, H.E., & L. Rosen 1992. Common sense in Ground-water protection and management in the United States. *Ground Water* 30: 867-872.
Mitchell, J.K., & F.T. Madsen 1987. Chemical effects on clay hydraulic conductivity. *Geotechnical practice for waste disposal '87, ASCE Geotechnical Special Publication No. 13:* 87-116.
National Rivers Authority (North West Region) 1989. *Earthworks on Landfill Sites*. NRA, Warrington.
National Rivers Authority 1991. *Policy and Practice for the Protection of Groundwater (draft)*. NRA, Bristol.
Panno, S.V., B.L. Herzog, K. Cartwright, K.R. Rehfeldt, I.G. Krapac & B.R. Hensel 1991. Field-scale investigation of infiltration into a compacted soil liner. *Ground Water* 29: 914-921.
Pearce, F. 1992. Worst first on list of contaminated land. *New Scientist* 1833: 5.
Robinson, H., & J. Gronow 1992. Groundwater protection in the UK: Assessment of the landfill leachate source-term. *Journal of the Institution of Water & Environmental Management* 6: 229-236.
Rogowski, A.S. 1990. *Relationship of laboratory- and field-determined hydraulic conductivity in compacted clay layer*. EPA/600/S2-90/025. Cincinnati.
Rowe, R.K. 1991. Some considerations in the design of barrier systems. *Proc. 1st Canadian Conference on Environmental Geotechnics:* 157-164. Montréal.
Sai, J.O., & D.C. Anderson 1991. *State-of-the-art field hydraulic conductivity testing of compacted soils*. EPA/600/S2-91/022. Cincinnati.
Seymour, K.J. 1992. Landfill lining for leachate containment. *Journal of the Institution of Water & Environmental Management* 6: 389-396.
Shackelford, C.D., & D.E. Daniel 1991. Diffusion in saturated soil. I: Background. *Journal of Geotechnical Engineering* 117:467-484.
Taylor, M.R.G., & R.A.N. McLean 1992. Overview of clean-up methods for contaminated sites. *Journal of the Institution of Water & Environmental Management* 6: 408-417.
Telles, R.W., S.L. Unger, H.R. Lubowitz, & H.K. Howard 1988. *Technical considerations for de minimis pollutant transport through polymeric liners*. EPA/600/S2-88/042. Cincinnati.
Thomas, J.C., & K.W. Brown 1992. Depth variations in hydraulic conductivity within a single lift of compacted clay. *Water, Air, & Soil Pollution* 65:371-380.
USEPA 1989. *Requirements for hazardous waste landfill design, construction and closure*. EPA/625/4-89/022. Cincinnati.
West Yorkshire Waste Regulation 1991. *Natural Landfill Liners: Internal guidance for waste regulatory staff*. (Unpublished).

Waste Disposal by Landfill – GREEN'93, Sarsby (ed.) © 1995 Balkema, Rotterdam, ISBN 90 5410 356 6

Inventory of old sanitary landfills and their rehabilitation

J. Saarela
National Board of Waters and the Environment, Finland

ABSTRACT: Aftercare of sanitary landfills is in many ways the environmental problem. One of these is how to cover and rehabilitate the old landfill sites. For this purpose there was made the inventory of old sanitary landfill sites and it was seen that the aftercare and the covers of old closed landfill sites does not fulfill the requirements of the modern landfill technology and the environmental reguirements. There were also developed the types of the cover structures for the rehabilitation of the closed landfill sites.

1 INTRODUCTION

Sanitary landfills and waste managament are nowadays in the focus in the most industrial countries and so also in Finland, where National Board of Waters and the Environment has started many projects concerning sanitary landfills. One of these is how to cover closed sanitary landfill sites after their use because about 600 sanitary landfills will be closed up to the end of this century and there are no practical directives or code of practice how to cover and rehabilitate sanitary landfills after their use.

The project has included among other things the following points:

1) Inventory of closed sanitary landfills in Finland (157 landfills)
2) Inventory of geotechnical structures of covers of closed sanitary landfills
3) Inventory of landscaping of closed sanitary landfills
4) Development 6 different types of the cover structures of sanitary landfills.

In the inventory the following points were investigated: the starting and closing year of the sanitary landfill, the type of the waste, the height and the area of the landfill, the material of the cover structure, the afteruse, the landscaping, the quality control of the leachate waters and the environmental consequences of the sanitary landfill.

2 RESULTS OF THE INVENTORY

2.1 Number of the landfills

The inventory was done in the region of the 13 different water and environmental districts. The number of the landfills varied between 4-30 in the different regions. It was seen specially that in the communities of eastern Finland there were many small village sanitary landfills. The total number of the landfills was 157.

2.2 Age of the landfills

51 landfills were older than 20 years. 28 landfills were older than 15 years but not over 20 years. 13 landfills were older than 10 years but not over 15 years and 7 were younger than 10

years. However the age of all landfills was not known.

2.3 Type of wastes

Most of the landfills were for the household and industrial wastes (77). However almost the same number of landfills were only for household wastes (74). 6 landfills were for industrial wastes only.

2.4 Height and the area of the landfills

The height of the most landfills were under 5 meters (89). The height of 35 landfills were 5-10 meters. 19 landfills were between 10-20 meters and 14 were over 20 meters.

The area of the most landfills were under 0.5 hectars (66). The area of 45 landfills were 0.5-2 hectars. 23 were between 2-5 hectars, 8 landfills were between 5-10 hectars and 15 landfills were over 10 hectars.

2.5 Material of the covers

The most of the landfills were covered with waste soils (103). Sand or silt were on the 13 landfills. Gravel and moraine together were on 5 landfills, clay on the 9 landfills. Other types of cover structures were on 9 landfills for example asphalt, peat and chips. Layer depts varied between 0.5-20 m. Vegetation layers were between 0.1-0.3 m. The cover structures were in most cases very defective, for example the vegetation layers and gas control.

2.6 Afteruse

The most of the landfills had no afteruse (124). However for example the most developed afteruse was in Helsinki city region, where there were modern and complex recreation areas on the closed landfills. Other ways of afteruse were for example store areas of the industry, compost areas and motor sport areas.

2.7 Landscaping

The most of the landfills were not landscaped (108). The landscaping was made on 12 landfills wholly and partially on 21 landfills. The landscaped landfills were near to the big cities. The landscaping in general was also connected to some kind of afteruse.

2.8 Quality control of the leachate waters

Quality control of the leachate waters were not uniform due to the great differencies between the landfills for example their location, area and waste content. The leachate waters of 61 landfills are or have been under control and the number of the water samples varied from one sample during the whole life of the landfill to the 4 yearly samples. The control results of the 31 landfills were discussed in the inventory.

2.9 Environmental effects

Two of the landfills have caused the groundwater contamination. 7 landfills have caused changes in the grounwaters but there is not yet serious contamination. However the situation needs to be controled.

16 cases the effects of the landfill leachate have been observed in the ground or surface waters but they are not serious due to the small quantity of the contamination, location of the landfill or due to the other greater contamination sources.

It was seen that the most serious consequences of the closed landfills were those when the leachate waters cause the contamination of groundwaters.

When the landfill is situated on the groundwater area and the leachate is observed yearly upgoing contamination values in the groundwaters then one must think even the removal of the landfill from that area. In the inventory there was one case where one has planned the removal of the landfill but that has not yet been done.

In all in Finland there are on the groundwater areas 157 landfills. It has been estimated that 71 have had some kind of effects to the quality of the groundwaters.

Figure 7 shows the landfills were the inventory was made.

3 REHABILITATION AND TYPES OF THE COVER STRUCTURES

In principle the rehabilitation of the closed landfill can be done in the following ways
- one cover it,
- one remove it, for example because it contaminate the groundwaters
- one inject or build cut wall around the landfill to prevent the contamination groundwaters

However the most common rehabilitation way is to cover the landfill properly and in this connection there is discussed only cover structures because it is the most common way of the rehabilitation.

3.1 Requirements for the cover structures of closed sanitary landfills

The main problem is how they should be covered so that the leachate is minimum and how they should vegetated after their use. Sanitary landfill differs considerably from the mine waste or build rubble area because for example their settlement can usually be controlled by compaction by earthmoving machinery during the period when the tip is being built up. Sanitary landfill, however, suffers considerable settlement by the composition of the organic contents over a long period of time.

The other problem is gas generation in relation to the engineering of landfill reclamation with building and vegetation.

A cover at a landfill site is an earthwork construction designed to exclude surface wastes from contact with disposed wastes, to act as a barrier against human or animal contact with the wastes and to prevent release of vapors to the environment. Covers are preventive measures used in most site remediation plans. They are seldom used alone but are part of an integrated concept of waste containment.

A cover may perform a number of functions, as listed in Table 1. Along with functions are listed necessary attributes, which the cover must have to accomplish these functions. Attributes pertain mainly to the cover's own durability and permanence. The purpose of good design is to provide these attrtibutes while assuring the performance of the desired function.

While all of the functions are im-

Table 1. Cover functions and attributes (McAneny and Hatheway 1982).

Functions
Prevention or Minimization of Percolation
Promotion of Aesthetics
Suppression of Vectors
Containment of Gases
Suppression of Fire Danger
Prevention of Blowing Litter of Dust
Promotion of Site Reclamation
Necessary Attributes
Water Erosion Resistance
Wind Erosion Resistance
Stability Against Slumping and Cracking
Stability Against Slope Failures
Resistance to Cold-Weather Distress
Resistance to Disruption by Animals or Plants

portant, that of preventing percolation is predominant, because contamination of surface- and groundwater supplies is the most serious threat of uncontrolled hazardous waste sites.

3.2 MATERIALS

Soils will be used in most cover systems. They usually are available on or near the site and are realtively cheap. Not all local soils are suitable for all cover purposes, accordingly either imported soils or non-soil materials may have to be employed. The cover designer needs to consider soils both from the engineering and the agricultural points of view.

Permeability is an important property of cover soils since the cover system is largely intended to retard the inflow of water. When a soil is unsaturated, its permeability is less than when it is saturated. Out of a large number of possible flow channels through the soil, only a fraction is utilized. Unsaturated flow is important because the soils in a cover system will be completely saturated for only a small portion of the time, if ever.

4 STRUCTURE OF COVER SYSTEMS

In the inventory of closed sanitary landfills it was seen that the cover structures of closed sanitary landfills does not come up to the requirements of modern landfill technology. It was also seen that their size and location vary much from each other for example from the sanitary landfill of a small village in the outlying district to the great landfills of big cities.

For this reason different sanitary landfills must be covered in different ways. Sanitary landfills of big cities require totally different cover structures than small village landfills and also their costs are different. So one must select different requirements for different types of sanitary landfills.

For this reason in this project there has been developed the types of the cover structures for Finnish code of practice how to cover closed sanitary landfill. Also their costs have been calculated and there has also been developed simulation models for calculating water balance and leaching production, which go through these cover structures. However in this connection there are only discussed the types of the structures and their costs.

In general the cover of landfill is a multi-layer system and any system must have a surface layer. Other layers, which there can be are the following:

1) A gas-control layer intercepts gases evolved by the wastes and leads them to the atmosphere via venting mechamisms
2) A regulating layer (a foundation or buffer layer) isolates the hydraulic barrier from the wastes and also serves as a strong base to support the rest of the system
3) A drainage layer intercepts downward-percolating water and conveys it laterally out of the system
4) Filter layers, serve to separate fine from coarse materials and prevent clogging of coarse interstices by fine particles.

In this project there were developed 6 different types of cover structures, which vary very much from each other. In the following they are discussed in detail.

Structure type 1 is the most superior one, which come up to the all reguirements of modern landfill technology (fig 1). It is used when there is the purpose to build on the closed sanitary landfill for example a modern and complex recreation area. It can also used on the hazardous waste landfills. The total thickness of the cover system vary between 2.5-3.0 m and it can be used for planting trees on the sanitary landfills.

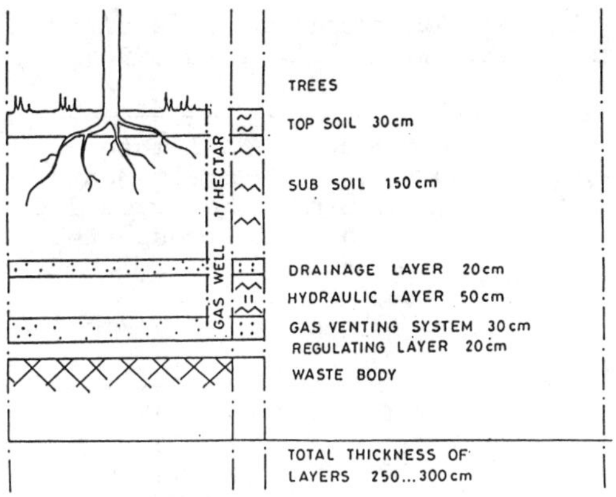

Figure 1. Cover structure type 1.

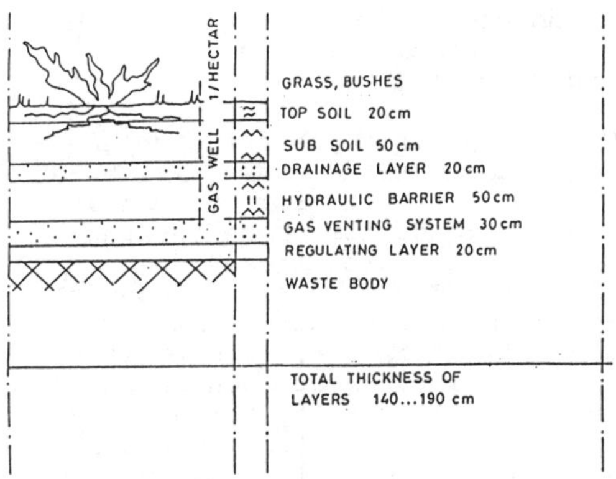

Figure 2. Cover structure type 2.

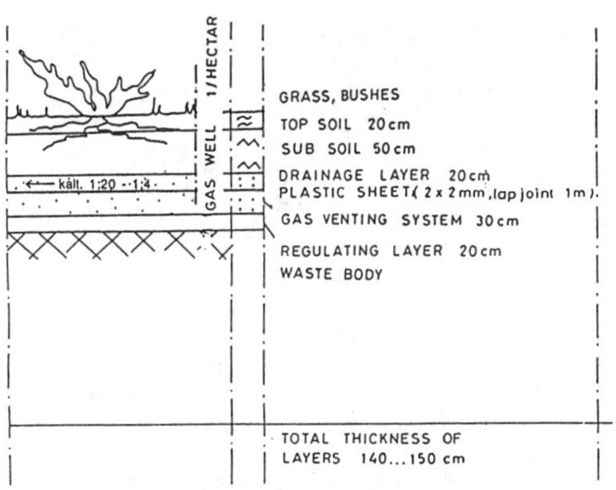

Figure 3. Cover structure type 3.

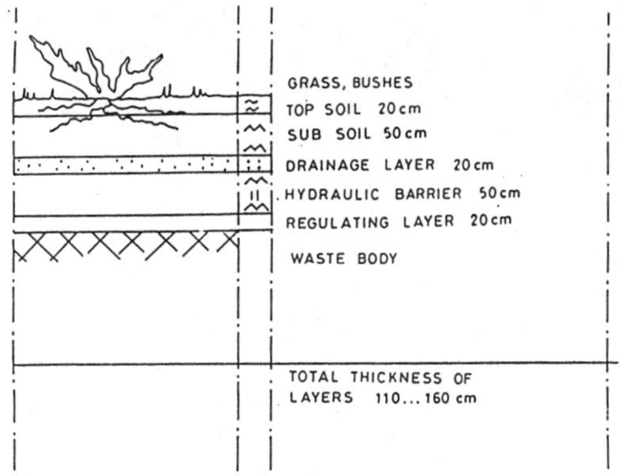

Figure 4. Cover structure type 4.

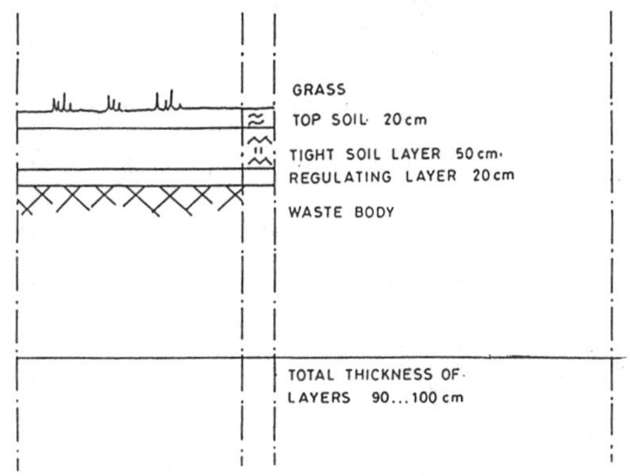

Figure 5. Cover structure type 5.

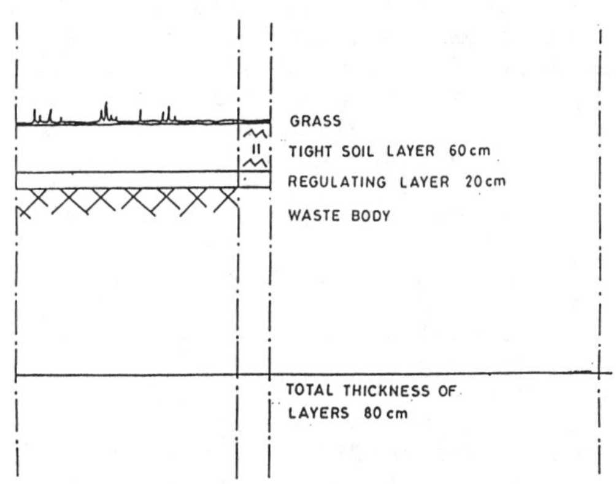

Figure 6. Cover structure type 6.

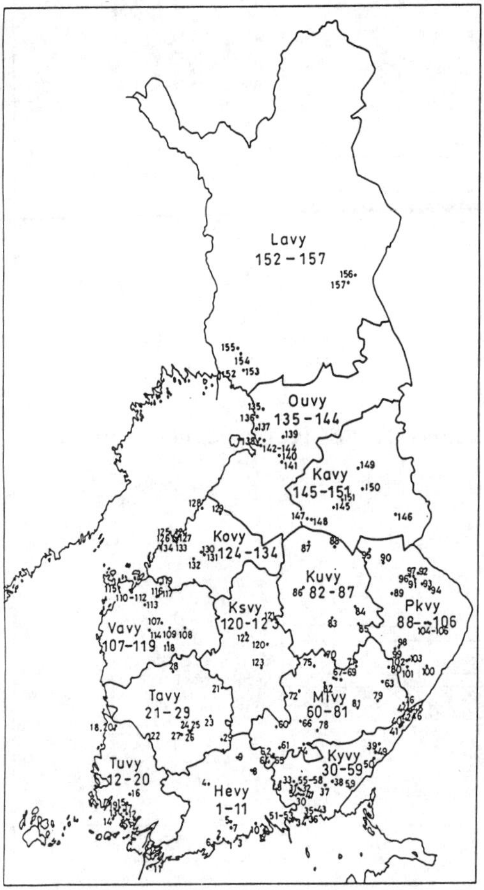

Figure 7. Inventory of closed sanitary landfills. The numbers refer to landfills.

Structure type 2 has same structure layers as type 1 but they are thinner (fig. 2). In structure type 3 the hydraulic barrier is replaced by plastic sheet. In this special case the quality of plastic sheet does not need to be very good because this type is for the recirculation of leaching waters (fig. 3). The structure type 4 has the same standard but there is not the gas ventilating layer, because in many landfill one build separate system with its own wells and pipes for gas collecting. These three types fulfill minimum requirements for health and environment protection when there is the meaning to take the sanitary landfill for example for the recreation use after their use. These types are not suitable for hazardous wastes.

Table 2. Construction costs of the cover systems in the year 1990.

Cover Type	Waste and excess soils can be used free of charge 1 000 Fmk	All materials must be buyed and transported 1 000 Fmk
1	1 440	2 260
2	1 260	1 720
3	1 410	1 680
4	850	1 310
5	370	640
6	210	520

Types 5 and 6 represent simple and light cover structures (fig. 5 and 6). They are used for example on the old landfills where environmental risks are already decreasing. These light structures are also sufficient for small landfills which are situated far from the settlement when there is no need for activity purposes for example for recreation. For their structural simplicity they are not sufficient for preventing leaching water and for controlling gas discharge.

5 BUILDING COSTS OF COVER SYSTEMS

Costs of cover systems consist construction and building material costs. Costs of soil material are calculated so that the transport distance is about 10 km. If the transport distance increases for example to 20 km, it increases construction costs 20 % in that alternative where all soil materials must be bought and transported to landfill. Costs are calculated in the year 1990.

Costs calculations are made for two alternative in the following way:

1) Cover structures can be built of waste and excess soils free of charge of soils, which are transported to the landfill site.
2) All soil materials must be bought and transported to the construction site.

The building costs of the covers built of waste and excess materials are 40-80 % of the costs of the bought materials. The cheapest cover type costs 210 000 Fmk and the most expenxive one costs 2 260 000 Fmk (table 2).

REFERENCES

BARNES, F.J. and RODGERS, J.C. 1988. Evaluation of Hydrologic Models in the Design of Stable Landfill Covers. EPA/600/2-08/048.

FENN, D.G., HANLEY, K.J. and DEGEARE, T.V., "Use of the Water Balance Method for Predicting Leachate Generation from Solid Waste Disposal Sites, "EPA/530/SW-168, U.S. EPA, Cincinnati, OH, 1975.

JESSBERGER, H.L., KLOS, U. (1992). Geotechnical aspects of design and construction of waste containment facilities. Proceedings of the Mediterranean Conference on Environmental Geotechnology, Cesme, Turkey 25-27 May 1992. ISBN 90 5410 0559.

LECHNER, P. 1987. The Austrian Guidelines for Sanitary Landfills. Proceedings of International Symposium on process, technology and environmental impact of sanitary landfill. Cagliari, Sardinia, 19th-23rd October 1987.

LUTTON, R.J., REGAN, G.L. and JONES, L.W., "Design and Construction of Covers for Solid Waste Landfills, "EPA-600/2-79-165, Municipal Environmental Research Laboratory, U.S. EPA, Cincinnati, OH, 1979, 250 pp.

MCANENY, C.C. & HATHEWAY, A.W. 1985. Design and Construction of Covers for Uncontrolled Landfill Sites. The 6th National Conference on Management of Uncontrolled Hazardous Waste Sites. November 4-6, 1985. Washington, DC.

SAARELA, J. (1992). Kaatopaikan pinnan loppukäsittely ja sen simulointimalli. Väliraportti. Vesi- ja ympäristöhallitus. (Surface structures of closed sanitary landfills and their simulation model: in Finnish; not published) National Board of Waters and the Environment.

SAARELA, J. (1993). Lopetettujen kaatopaikkojen inventoinnin tulokset. (Inventory of closed sanitary landfills: in Finnish; not published) National Board of Waters and the Environment.

SAARELA, J. 1991. Landscaping aspects of sanitary landfills in Finland. Third International symposium 14 - 18 October 1991. Sardinia -91.

VIATEK Oy, (1991). Käytöstä poistuvien kaatopaikkojen pintarakenteiden kustannukset. Esiselvitys. (Costs of surface structures of closed sanitary landfills: in Finnish).

Waste Disposal by Landfill – GREEN'93, Sarsby (ed.) © 1995 Balkema, Rotterdam, ISBN 90 5410 356 6

Permeability measurements in the triaxial apparatus

R.W. Sarsby
Bolton Institute of Higher Education, UK

I.S. Heggie
Cumbria County Council, Penrith, UK

S. Rainford
Waste Management UK Ltd, Warrington, UK

ABSTRACT: Comparative Permeability measurements have been made on clay specimens using compaction permeameters (falling head test) and the triaxial apparatus (constant and falling head tests). The moulding moisture content of the soils used was in the preferred range for producing a low-permeability end-state, i.e. well wet of optimum.

MEASUREMENT OF PERMEABILITY

Compacted clay is used extensively to form clay liners to contain, either on their own or in association with man-made membranes, refuse in engineered waste disposal sites. Such clay needs to be placed as to have a permeability not greater than 10^{-6} mm/sec.

Traditionally the permeability of slow-draining, fine-grained soils has been measured using the falling head test. This test method was originally devised as a means of accurately measuring the small flow-rate of water passing through low-permeability soils.

The common test apparatus is the compaction permeameter. This consists of a rigid cylindrical body (into which the soil is compacted as per a compaction test) clamped between a perforated plate and a solid top cap. A water head is applied to the sample via the top cap and water issues from the lower end of the soil specimen via the perforated plate. Two major criticisms have been levelled at this apparatus for measuring permeability:

1) The potential for 'tracking' of water down between the soil grains and the rigid inside of the cell body. If the soil being compacted is stiff then intimate contact may not be achieved between the soil and the cell body over the full depth and voids may even be present.

2) There is no facility for saturating the soil sample. It is not possible to pressurise the water exiting from the sample (to dissolve trapped air and hence produce saturation) because of the design of the apparatus. Permeability measurements can be seriously affected by air in the voids.

Consequently there has been a move towards promoting the use of the triaxial apparatus for permeability measurements. In this apparatus a compacted specimen is set up in a similar way to a conventional sample but with drainage connections at top and bottom. No deviator stress is applied to the sample. Water is forced through the sample under a constant head and the resultant flow rate is measured (preferably at both the inlet and outlet drainage points). Samples are saturated, and air is kept in solution in the drainage lines, by applying a back pressure to the drains.

The major advantages of the triaxial test arrangement are:

1) The sample is surrounded by a flexible membrane which fits intimately to the sides of the soil specimen because of the surrounding cell pressure. The membrane will follow the surface of the specimen even if it contains gouges and voids. Thus the potential for 'tracking' should be eliminated.

2) Full saturation of the soil specimen can be achieved by maintaining a high pressure in the pore water so that air is dissolved.

Notwithstanding the preceding benefits of using the triaxial apparatus there are several other important points to bear in mind:

a) The triaxial apparatus is more complex and more costly than the compaction permeameter so that with the latter equipment many tests may be performed for the price of a few triaxial tests. It must be remembered that compacted soil samples are formed using a crude, variable process where the number of variables, e.g. size of the soil lumps, depth of each layer, precise drop height of the rammer, etc. means that identical samples will not be produced each time. Therefore a range of possible permeability values will always exist for such a formation process.

b) The flow rates are very small (even with high hydraulic gradients) and the accuracy of the measuring devices is not high unless very expensive equipment is used. Measuring instruments are often affected by temperature changes. Both inlet and outlet flow rates should be measured to ensure that steady-state seepage conditions have been attained - however the British Standard permits the use of only one flow-measuring device (if a high degree of saturation has been achieved and confirmed by measurement of the pore pressure parameter B).

c) It is important to recognise that the permeability measurements relate to compacted clay for lining purposes, i.e. the soil will be compacted wet of optimum (by at least 2% moisture content) to ensure low permeability. Hence the soil will not be composed of very stiff lumps and the compaction should not only knead the lumps together but it should also force the soil into intimate contact with the sides of the compaction mould. This will be particularly true for modified (heavy) compaction of plastic clays with moulding moisture contents around optimum plus 3/4%.

d) On the wet side of optimum the samples rapidly approach the zero air voids line as the moulding water content increases - particularly when heavy compaction is used. Hence after compaction the 'wet soil' samples will already be close to saturation and the presence of very small quantities of air in the voids does not significantly affect the permeability value.

e) All three pressures, i.e. inlet, outlet and confining, must be maintained very accurately throughout a test or effective stress changes will occur and these will give rise to consolidation or swelling of the soil and hence false volume changes.

TEST PROGRAMME

Comparative permeability measurements were made on two soils using compaction permeameters and triaxial cells. All soil samples were formed using modified (heavy) compaction as this is appropriate to current earthworks standards - the two test soils have been used to form compacted clay liners at landfill sites in North West England. The soil properties are given in table 1.

Several 'identical' samples of soil A were compacted, at natural moisture content, in compaction permeameters. The resultant permeability, values were in the range 0.1×10^{-6} to 0.4×10^{-6} mm/sec. The values for constant head triaxial tests (on 'identical' samples again) were dependent on the difference between the confining stress and the inlet seepage pressure as shown in figure 1. The permeability value increased rapidly as this pressure difference decreased, presumably due to loss of intimate contact between the membrane and the surface of the test specimen. Whilst British Standard 1377 recommends that the confining pressure should be greater than the flow inlet pressure it does not specify by how much - the Authors' data suggest that the difference needs to be at least 150 kN/m^2. With this latter condition the measured permeability was in the range 0.1×10^{-6} to 0.3×10^{-6} mm/sec i.e. very similar to the compaction permeameter values, although overall the triaxial values indicated a slightly lower permeability than the compaction permeameters.

Table 1 Soil properties

Soil	A	B
Liquid Limit	27	24
Plastic Limit	15	Non-plastic
Clay Fraction	8	4
Proctor optimum - Moisture Content (%) Dry Density (Mg/m^3)	 10.4 2.00	 14.1 1.87
Heavy optimum - Moisture Content (%) Dry Density (Mg/m^3)	 8.5 2.12	 11.8 1.96
Specific Gravity	2.70	2.64
Classification	Low plasticity clay	Low plasticity silt
Natural Moisture Content (%)	13.5	14.3

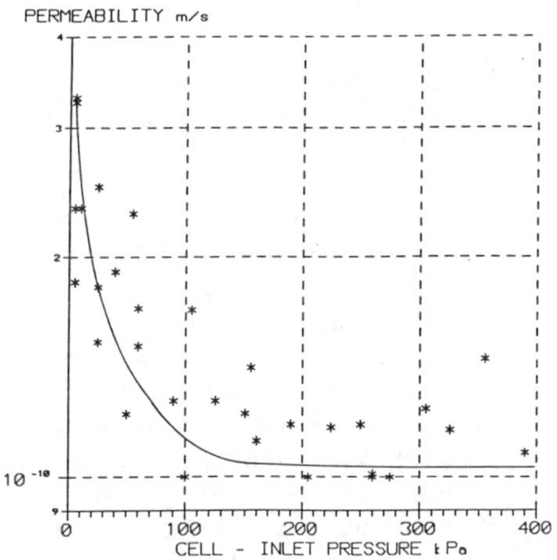

Figure 1 Triaxial permeability data - Soil A

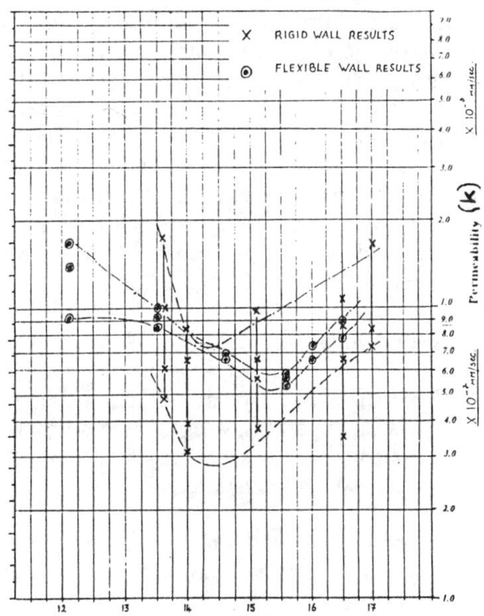

Figure 2 Permeability data - Soil B

The range of permeability values obtained from soil B, using compaction permeameters and triaxial apparatus, is shown in figure 2. As with soil A there is good agreement between the two types of test for the moisture content range relevant to compacted clay liners, i.e. greater than optimum moisture content + 2%, approximately. The difference between the tests on soil A and those on soil B is that in the latter case the permeabilities were measured by conducting falling head tests on triaxial samples.

FALLING HEAD TRIAXIAL APPARATUS

This apparatus and test procedure were devised in an attempt to combine the best features from the triaxial and compaction permeameters. The requirements were:

a) Side-wall tracking should be prevented by enclosing the sample in a flexible membrane.

b) Back pressurisation of the pore fluid should be possible.

c) Inlet and outlet flow rates should be measured.

d) Accurate measurement of small flow rates required.

e) Inlet, outlet and confining pressures should be accurately controlled.

f) The apparatus should be as cheap as possible.

The approach adopted was simply to take a conventional triaxial cell and apply all pressures using old-style 'mercury - pot' pressure systems. Small bore burettes are attached to the uppermost mercury-pots which apply the back pressure (on the inlet and outlet sides) and flow rates are measured by recording the water levels in these burettes. The cell pressure is maintained using the conventional mercury pot system. However the back pressure systems utilise mercury pots set at fixed heights to eliminate errors caused by unequal spring stiffnesses. The apparatus is illustrated in figure 3.

The test procedure is:

a) Set up a compacted soil sample, in the triaxial apparatus, with top and bottom drainage in the usual way and apply the required confining stress.

b) Connect the inlet and outlet backpressure systems together to allow them to equalise.

c) Apply the confining stress and back-pressures to the sample and allow it to saturate.

d) When saturation is complete re-balance the inlet and outlet backpressures but this time ensure that the burettes (on top of the upper mercury-pots) are full to the top with water.

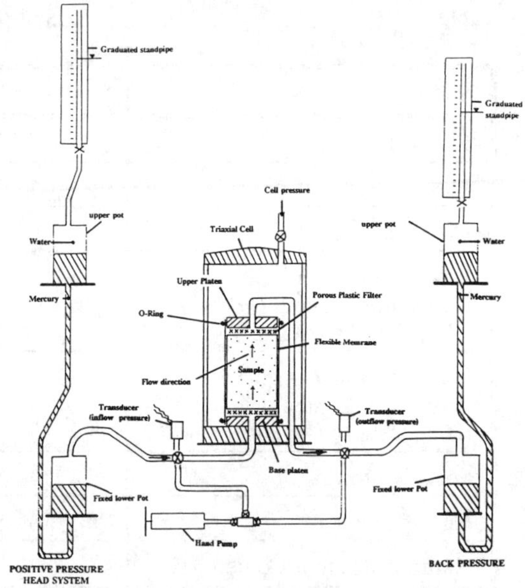

Figure 3 Falling head triaxial apparatus

e) Isolate the two backpressure systems from one another and record the water levels in the two burettes.

f) To start a falling head test the water level in the burette on the outlet backpressure side is dropped to the bottom of the burette. At this stage the head difference across the sample is equal to the amount by which the water level has been dropped - typically this would be approximately 1m as per the compaction permeameter test.

g) With time water will flow from one side of the sample to the other until there is no head across the sample.

The permeability is obtained in the usual way for a falling head test, i.e. by plotting log (Initial Head/Head) against time. The only complication to the test is that at any time after the start of the test the head difference across the sample is affected by the movement of the mercury in the mercury-pot backpressure systems, i.e. the head is not simply the difference between water levels in the burettes. The correction to the head is easily calculated once the system has been calibrated :-

The cross-sectional area of the four mercury

pots in the backpressure system is A_1, A_2, A_3 and A_4.

If the cross-sectional area of a burette is a, then when the water level changes by h the volumetric flow is ah.

Thus the mercury levels change by

$$\left(\frac{ah}{A_1}\right), \left(\frac{ah}{A_2}\right), \left(\frac{ah}{A_3}\right) \text{ and } \left(\frac{ah}{A_4}\right).$$

Thus the total change in water head (in the flow line), Δh, due to the mercury movements is

$$\frac{\rho_{\text{mercury}}}{\rho_{\text{water}}} \times ah \left(\frac{1}{A_1} + \frac{1}{A_2} + \frac{1}{A_3} + \frac{1}{A_4}\right).$$

Since the diameter of the mercury-pots is considerably larger than the glass burettes small errors in A_1, A_2, A_3 or A_4 have negligible effect on the head correction. For the Authors' apparatus the correction was given by

$$\Delta h = \frac{h}{66.2}$$

CONCLUDING REMARKS

Test data indicate that compaction permeameters and triaxial apparatus give very similar values of permeability for soils compacted with moisture contents appropriate for clay liner compaction.

A simple falling head triaxial apparatus has been developed - the resultant permeability data were in close agreement with values measures conventionally. The apparatus utilises low-cost, accurate methods to maintain pressures and to measure flow rates.

REFERENCE

BS1377 (1990) : Soils for Engineering purposes, Part 6, Consolidation and permeability tests, HMSO, London.

Waste Disposal by Landfill – GREEN'93, Sarsby (ed.) © 1995 Balkema, Rotterdam, ISBN 90 5410 356 6

Landfill engineering – UK practice and guidance in a European context

K.J. Seymour
National Rivers Authority (NW Region), UK

A. Street
MRM Partnership, Bristol, UK

ABSTRACT: The move away from reliance on the 'dilute and disperse' principle of site operation and 'natural' containment towards engineered containment of landfills has been a relatively recent trend in the UK. This has been influenced by experience, research and most significantly by new UK and European legislation and policies such as the NRA's "Policy and Practice for the Protection of Groundwater". It has also led to the preparation of a number of guidance documents on engineering standards and practices. These include one currently being drafted by a DoE funded working group on "Good Practice for Landfill Engineering", which complements the work of the ISSFME's ETC8 in the wider European context.

The use of composite mineral/FML liners is rapidly becoming the normal minimum requirement where groundwaters are at risk, with the use of additional (multiple) lines in particularly sensitive situations.

Future developments are likely to include vertical liners and more effective leachate drainage systems.

1 INTRODUCTION

The move towards engineered containment of landfills has been a relatively recent trend in the UK. Accordingly, there has been a general lack of detailed guidance on landfill engineering available specifically for UK situations.

In contrast, engineered containment has been mandatory in certain European companies, principally Germany, and also in the United States over a much longer period, albeit principally for "hazardous" wastes. To date therefore, the majority of research and guidance on landfill design, construction and performance has been based on overseas experience.

However, this imbalance is rapidly being redressed, with current UK best practice and standards being comparable with, if not better than, most in Europe and the United States.

2 BACKGROUND/HISTORY OF UK LANDFILL PRACTICE

Prior to the 1970's there was little control over the UK waste industry. Public outcry over indiscriminate dumping of toxic wastes led firstly to the introduction of the Deposit of Poisonous Wastes Act in 1972 followed by the 1974 Control of Pollution Act. These statutes brought in the requirements for prior approval and authorisation to dispose of waste to landfill.

Research carried out by the Department of the Environment in the early 1970's in and around existing landfill sites concluded that it was generally acceptable to operate on the principle of 'dilute and disperse'. This relied upon the premise that leachate escaping from the base of a landfill would be rendered safe by attenuation as it percolated through the waste and unsaturated zone and subsequently diluted within the saturated zone (DoE 1978).

Under such a regime the only landfill "engineering" involved would be perhaps low permeability capping to minimise leachate production and the installation of boreholes around the site to monitor the migration of the pollution plume. Even as recently as 1986 the philosophy was endorsed by Waste Management Paper No. 26 (DoE 1986)

However, by the mid 1980's the effectiveness and acceptability of the 'dilute and disperse' philosophy was called into question particularly by certain Water Authorities. This followed research which proved that the degree of attenuation in the unsaturated zone varied tremendously and was minimal in certain types of strata (Harris & Lowe 1984). This led to a move towards containment of leachate within the landfill rather than allowing it to escape into the aquatic environment. Implementation of the EC Groundwater Directive (Anon 1980) requiring the protection of all usable groundwaters, and also the emergent dangers associated with uncontrolled migration of landfill gas have accelerated this move.

Landfill sites located in natural low permeability environments have become scarce. Furthermore, many sites which were originally considered to provide natural containment have been found to be subject to either water ingress or leachate escape.

These factors, combined with new European and UK legislation (Section 3 below), have resulted in engineered containment landfills being the norm rather than the exception. Initially these involved single liners, usually re-compacted mineral seals, although geomembranes were used in applications where there was no source of natural low permeability material conveniently available. However, in line with overseas practices, the use of composite mineral/geomembrane liner systems is rapidly becoming the minimum standard required where groundwaters or other "controlled waters" are at risk.

As regards standards of liner design and installation, these have improved dramatically in recent years. When engineered liners were first stipulated in site licence conditions, both regulators and operators alike had no experience of civil engineering construction practices and standards. However, the introduction of sound civil engineering/geotechnical practices into the landfill industry has resulted in very rapid evolution of both ideas and standards. Nowadays proper geotechnical design, rigorous material testing, and on-site construction quality control involving full time independent supervision and certification of the engineering works are the usual prerequisites of schemes being accepted by the Waste Regulation Authorities (WRA's).

The liner system is recognised as being just one element of the whole landfill design, which will also include leachate minimisation, collection, removal and treatment measures.

Cut-off walls and similar vertical barriers are being used increasingly in the UK to contain leachate and gas associated with both new and old landfill sites, as well as in the wider context of remediation of contaminated land sites. The incorporation of geomembranes within the conventional cement/bentonite mix is a relatively recent development which is likely to assume greater importance in the future.

3 CURRENT LEGISLATION AND INFLUENCES

The Environmental Protection Act 1990, the implementation of which is eagerly awaited by regulators, will have a significant impact on the UK waste disposal industry's approach to landfill design and management.

At present, under the 1974 Control of Pollution Act, an operator can surrender his licence at any time, after which the licence conditions become null and void, effectively removing his environmental liabilities.

In contrast, under the Environmental Protection Act the site operator will be required to obtain a Certificate of Completion from the Waste Regulation Authority (WRA) before he can relinquish his site licence. In order to achieve this he will need to demonstrate that the site no longer poses a risk to the public and the local environment. This will involve establishing that the waste has degraded to such an extent that the rate and composition of landfill gas emissions and the quality of leachate produced by the waste body meet specified criteria. To achieve these objectives landfills will need to be engineered to very high standards to provide effective leachate containment and management during the operational and post closure monitoring phases. It is recognised that in most cases this could last decades, if not centuries.

Similarly, the insurance market has started to exert a major influence on site engineering and management standards. To secure cover against environmental damage liability operators will have to persuade the underwriters that their landfill presents an acceptable risk by virtue of the site's integrity.

The National Rivers Authority's Policy and Practice for the Protection of Groundwater issued in December, 1992, (NRA 1992) will also greatly influence the location and design of landfill sites. It sets out a national framework for protecting both groundwater resources in general and sources in particular from potentially polluting activities, including landfill. The policy considers resources in terms of major, minor and non-aquifers. Source protection zones are defined based on contaminant/water travel times, as follows:-

Zone 1 : 50 day travel time or 50 metres (whichever is the greater)
Zone 2 : 400 day travel time or 25% of source catchment (whichever is greater)
Zone 3 : Total source catchment area

A series of matrices have been drawn up which consider the acceptability of the specific polluting activity in relation to its hydrogeological setting and proximity to sources.

In the case of landfills (Figure 1) the pollution potential of the waste dictates its acceptability in any particular situation. Engineered containment and operational safeguards, including leachate management, will be pre-requisites for sites located on aquifers (major and minor) and within Source

SOURCE PROTECTION			
SITE TYPE	I INNER ZONE	II OUTER ZONE	III CATCHMENT ZONE
1.High pollution potential (landfills accepting domestic, commercial and industrial waste either individually or on a co-disposal basis)	Not acceptable	Not acceptable	Only acceptable with engineered containment and operational safeguards
2. Medium pollution potential (landfills accepting construction, demolition industry wastes and similar	Not acceptable	Acceptable subject to evaluation on a case by case basis and adequate operational safeguards	Acceptable subject to evaluation on a case by case basis and adequate operational safeguards
3. Low pollution potential (landfills accepting inert,uncontaminated waste)	Not normally acceptable	Acceptable only with adequate operational safeguards	Acceptable only with adequate operational safeguards
RESOURCE PROTECTION			
SITE TYPE	MAJOR AQUIFER	MINOR AQUIFER	NON- AQUIFER
1.High pollution potential (landfills accepting domestic, commercial and industrial waste either individually or on a co-disposal basis)	Only acceptable with engineered containment and operational safeguards	Only acceptable with engineered containment and operational safeguards	Acceptable only with adequate operational safeguards. Engineering measures may be necessary in order to protect surface waters
2. Medium pollution potential (landfills accepting construction, demolition industry wastes and similar	Acceptable only with adequate operational safeguards	Acceptable only with adequate operational safeguards	Acceptable only with adequate operational safeguards
3. Low pollution potential (landfills accepting inert, uncontaminated waste)	Acceptable	Acceptable	Acceptable

FIGURE 1.

National Rivers Authority 'Policy and Practice for the Protection of Groundwater'

Landfill Acceptability Matrix

Protection Zone 3, which receive leachate forming wastes. However, recognising the inability to guarantee the long term performance of any liner system, regardless of its degree of sophistication, the policy states that no landfill activities will normally be accepted in Zone 1 ie. close to a source.

This approach is consistent with the draft EC Landfill Directive which, when implemented, appears likely to make containment mandatory for leachate forming wastes.

4. EXISTING UK GUIDANCE DOCUMENTS ON LANDFILL ENGINEERING

Although Waste Management Paper No. 26 refers briefly to engineered containment, it lacks any detailed guidance on geotechnical aspects of design and construction.

A review of lining practice in the north-west of England by the North West Waste Disposal Officers Sub-Group led to the publication in 1988 of their "Guidelines on the Use of Landfill Liners" (NWWDO 1989). This included sections on liner selection, the installation of clay, bentonite and flexible membrane liners (FML's), cut-off walls, contract arrangements, site supervision and monitoring. It recognised the dependence of the success of the lining operation on weather conditions.

One of the first documents prepared in the UK dealing specifically with mineral liners was North West Water Authority's "Earthworks on Landfill Sites", originally drafted in 1986. This has subsequently been revised, with the latest version issued by the NRA (North West Region), dated July 1989 (NRA 1989). The document set out minimum requirements in terms of mineral liner performance and provided guidance on material testing and emplacement. It used the relationship between permeability and density, moisture content and soils index parameters to form the basis for liner design and material control during emplacement. This approach is consistent with North American and European practice. Appropriate British Standard soils laboratory test methods were used to encourage widespread applicability and acceptability by both the civil engineering and waste disposal industries. It also focused on the need for the construction works to be designed and supervised by a suitably qualified and experienced geotechnical engineer.

Over the past seven years it has been adopted and adapted by regulators and operators in many parts of the country. For example, in 1991 West Yorkshire Waste Management, in conjunction with NRA Yorkshire Region, issued guidance documents on landfill liners (WYWM 1991). These set out minimum requirements in terms of site licensing, and as such may be considered mandatory in that region.

The National Association of Waste Disposal Contractors published its own "Guidelines on Landfill Liners" (NAWDAC 1991) in September 1991. These are more comprehensive and up-to-date than the earlier NWWDO's guidelines, and reflect an operator's rather than regulator's viewpoint.

The North West Waste Disposal Officers group has recently prepared a comprehensive guide to Leachate Management NWWDO (1991). It includes reference to geotechnical aspects of leachate drainage system design, based on a literature review of North American and German experience.

5. ETC 8

In 1989 a European Technical Committee (ETC 8) was set up under the auspices of the International Society of Soil Mechanics and Foundation Engineering (ISSFME) to consider "geotechnical aspects of landfill design and remedial works". The committee comprised invited geotechnical practitioners and academics, under the Chairmanship of Prof. H.L. Jesseberger, with members from Germany, France, Italy, the Netherlands, Switzerland and the UK. The UK representative was initially K. Seymour of the NRA, and subsequently A. Street of MRM Partnership.

The committee's work has been based on a translation of recommendations of a working group of the German Geotechnical Society, which have been revised to make them applicable in a wider European context. ETC 8's first (draft) document, entitled "Geotechnics of Landfill and Contaminated Land - Technical Recommendations" was presented to the European Conference on Soil Mechanics and Foundations Engineering in Florence, in May 1991 (ISSFME 1991).

It includes a list of cross-referenced national standards relating to the specific tops covered in the document.

From feedback since publication, the title is seen as being somewhat misleading in that the document does not specifically address contaminated land issues. Therefore, at the committee's latest workshop held in Switzerland in March, 1992, it was agreed to amend the title to "Geotechnical Aspects of Landfill Design and Remedial Works". The revised report is due to be issued during 1993.

The document covers a wide range of issues relating to landfill design and remediation and is divided into six principal sections as follows:-

1. Recommendations on Site Assessment - dealing with the approach to site investigation and techniques for sampling groundwater, soil and air.
2. Recommendations on the Principles of Design - this section addresses, inter alia, liner system design, capping systems, use of geotextiles and geomembranes (FML's), leachate drainage, behaviour of waste materials and health and safety matters.
3. Recommendations on Suitability Testing - guidance is presented on testing regimes for natural and synthetic materials used in capping, lining and vertical cut-off systems. Particular emphasis is placed on the need to undertake field trials and to assess the impact of leachates on the long term performance of materials used in containment systems.
4. Recommendations on Construction Methods - this principally addresses the construction of vertical cut-off walls and outlines the various engineering techniques available.
5. Recommendations on Quality Assurance - considerable emphasis is placed on the importance of adopting a Quality Assurance regime for all aspects of landfill design and construction.
6. Recommendations on Hydraulic Measures - present guidance on the assessment of contaminant migration.

It is intended that the document will be further expanded over the next year or so and further modification made in order to produce a document which is relevant to the broader European context. It is important to note that the ETC8 document is aimed at engineering landfills to achieve containment of leachate and landfill gas. Furthermore, the recommendations strongly support the use of composite lining systems. In this respect the approach set out in the document is largely consistent with current practice in the UK.

6. DoE WORKING GROUP

Following the inaugural meeting of ETC 8 in Paris in 1989 it was considered appropriate to establish a UK "sister committee" to address the specific needs and problems of this country. Accordingly an informal group was set up in 1990, with membership drawn from all sectors of the waste disposal industry including the DoE, NRA and WRA's ("ie the regulators"), operators (local authority and private contractors), consultants and also site investigation and material testing specialists. It was later expanded to include the vice-chairman of the British Geotechnical Society.

Initially the terms of reference of the group were not clearly defined and it had no official status. However, it was agreed that there was a pressing need for national guidance on landfill design and construction practice, in particular geotechnical aspects. This was subsequently endorsed when in April 1992 the DoE let a contract to the group (under the project management of MRM Partnership) for the drafting of a document entitled "Guidance on Good Practice for Landfill Engineering". It was intended that the document should form a basis for re-writing of part of Waste Management Paper No. 26 dealing with the principles of landfill design.

In order to establish a more efficient and streamlined approach to drafting the various sections of the document a number of small sub-groups were set-up to address the following topics:

- Site investigation
- Site Design and Selection of Liner System
- Mineral Liners
- Flexible Membrane Liners
- Quality Systems
- Monitoring
- Cut-off Walls
- Capping Systems
- Contracts
- Waste Body Stability

The work of the sub-groups is now all but complete and a first draft of the document has been prepared. There is, however, a need to further refine the document to ensure consistency of style and content throughout all sections. This process of editing is now underway, with the aim of producing a final document by the autumn of this year.

7. UK LANDFILL ENGINEERING - THE FUTURE

There is considerable debate in the industry regarding the question of containment, with some holding the view that in certain cases the dilute and attenuate or "controlled release" approach may be acceptable and indeed, preferable. Nevertheless, there is a strong body of opinion, backed up by new UK and European legislation, which supports the containment approach. It seems certain, therefore, that the future of landfill engineering will be based on the need to design sites to provide leachate and landfill gas containment and management. The approach in the UK is based on each site being considered on its own merits related to type of waste, local geological and hydrogeological conditions and the nature and extent of any pollution risks. This approach clearly works in the UK context and is likely to remain in place for the foreseeable future.

An engineered liner and cap are just parts of an overall leachate management system involving minimisation, containment, collection, removal and treatment. Therefore, there is a need to give further consideration to the design of leachate collection and removal systems. In the past, leachate management has often been something of an afterthought. However, for modern containment landfills it must be seen as an essential and integral part of the design process. This is perhaps one area, therefore, where UK experience needs to develop further.

The use of composite mineral/geomembrane liners is likely to become the norm, with additional barriers and intermediate monitoring/leak detection (multiple liners) systems being required in particularly sensitive situations.

Vertical barriers in hard rock quarries and lining over existing waste bodies are also likely to provide challenges for the site designer in the future, as pressure grows to optimise all available air-space.

More rigorous site design and construction quality control will almost certainly be required, with increasing emphasis being placed on independent supervision and certification of the works.

In conclusion, the modern landfill site can clearly be considered to be a major civil engineering project. A typical site may include major earthworks, lining systems, drainage works and heavy engineering. It is imperative, therefore, that such projects are designed in a thorough and professional manner, as any other major civil engineering scheme works. However, it should be recognised that landfill engineering and operation require a multi-disciplinary approach.

REFERENCES

Department of the Environment 1978. Co-operative Programme of Research on the Behaviour of Hazardous Wastes in Landfill Sites - Final Report. HMSO, London.

Department of the Environment 1986. Waste Management Paper No. 26 - Landfilling Wastes HMSO, London.

Harris, R.C. & Lowe, D.R. 1984. Changes in the organic fraction from Domestic Refuse Sites on the Sherwood Sandstone. Quarterly Journal of Engineering Geology, No. 17, pp 57-66.

Anon 1980. Council Directive 80/68/EEC. The Protection of Groundwater caused by Certain Dangerous Substances.

National Rivers Authority 1991. Policy and Practice for the Protection of Groundwater. NRA ,Bristol.

North West Waste Disposal Officers 1986. Guidelines on the Use of Landfill Liners. Lancashire Waste Disposal Authority, Preston.

National Rivers Authority (North West Region) 1989. Earthworks on Landfill Sites. NRA(NW), Warrington.

West Yorkshire Waste Management 1991. Standards for the Lining of Landfill Sites. WYWM, Wakefield.

West Yorkshire Waste Management 1991 Natural Landfill Liners - Guidance Notes for Developers. WYWM, Wakefield.

National Association of Waste Disposal Contractors 1991. Landfill Guidelines No. 11 - Landfill Liners. NAWDAC, London.

North West Waste Disposal Officers 1991. Leachate Management Report. Lancashire Waste Disposal Authority, Preston.

International Socieity of Soil Mechanics & Foundation Engineering, European Technical Committee No. 8, 1991 - Geotechnics of Landfills and Contaminated Land. Ernst, Berlin.

Waste Disposal by Landfill – GREEN'93, Sarsby (ed.) © *1995 Balkema, Rotterdam, ISBN 90 5410 356 6*

Municipal plans for solid waste disposal in Turkey

Yildiz Wasti
Middle East Technical University, Ankara, Turkey

ABSTRACT: Turkey is a country of 58 millions that has seen swift and unprecedented economic growth over the last decade. Rapid urbanization has placed an enormous strain on the waste management resources especially of the larger cities of Turkey, such as Istanbul, Ankara and Izmir. The resulting problems are multi-faceted. A larger amount of solid waste per capita is being generated in consequence of greater purchasing power. Existing refuse dumps that used to be situated at large distances from the built-up areas are now to be found on the periphery of the city. It is the objective of this paper to provide a limited overview of problems related to solid waste management in Turkish cities and to present some reliable data related to waste production, collection and disposal. Various waste management proposals [including sanitary landfills] being considered for implementation by urban metropolitan authorities are presented and case studies are discussed.

1 INTRODUCTION

Turkey is a rapidly developing country with a population of 58 million rising at the rate of about 2.5% per annum. The urban percentage of the population is concomitantly increasing as a result of mass migration from rural areas to the cities. The larger cities of Turkey have each been given the status of a Metropolitan Municipality as a result of the amalgamation of a number of constituent district municipalities. The main function of the metropolitan municipality is the allocation of budget resources among the district municipalities. The metropolitan municipality is also empowered to carry out plans for development requiring large investments for the city as a whole.

The burden of solid waste management is shared by the metropolitan as well as the constituent or district municipalities: storage, collection and transportation of the waste to designated locations are the responsibility of the district municipalities, whereas treatment and disposal are the responsibility of the metropolitan municipality. Apart from a nominal contribution per household called the Sanitation Tax, there is currently no separate charge for municipal waste collection or treatment (Erkip and Kirca, 1991). This poses serious problems as enormous financial resources are required to improve the present methods of collection, storage and disposal of waste and to rehabilitate old dumping sites which now lie close to inhabited areas.

2 CURRENT STATUS OF SOLID WASTE DISPOSAL IN URBAN AREAS

Municipal waste management practice in the past and also at present has consisted of storage in a variety of containers, metal or plastic dust bins, plastic bags, etc., without separation at the source [except by usually unauthorized people in the form of scavenging in uncontrolled conditions at the source or at the final disposal site].

The collection of solid waste is carried out by teams of trucks without any transfer stations, and waste material is disposed of crudely by dumping, largely on land. The enforcement of existing laws and regulations with respect to the storage and

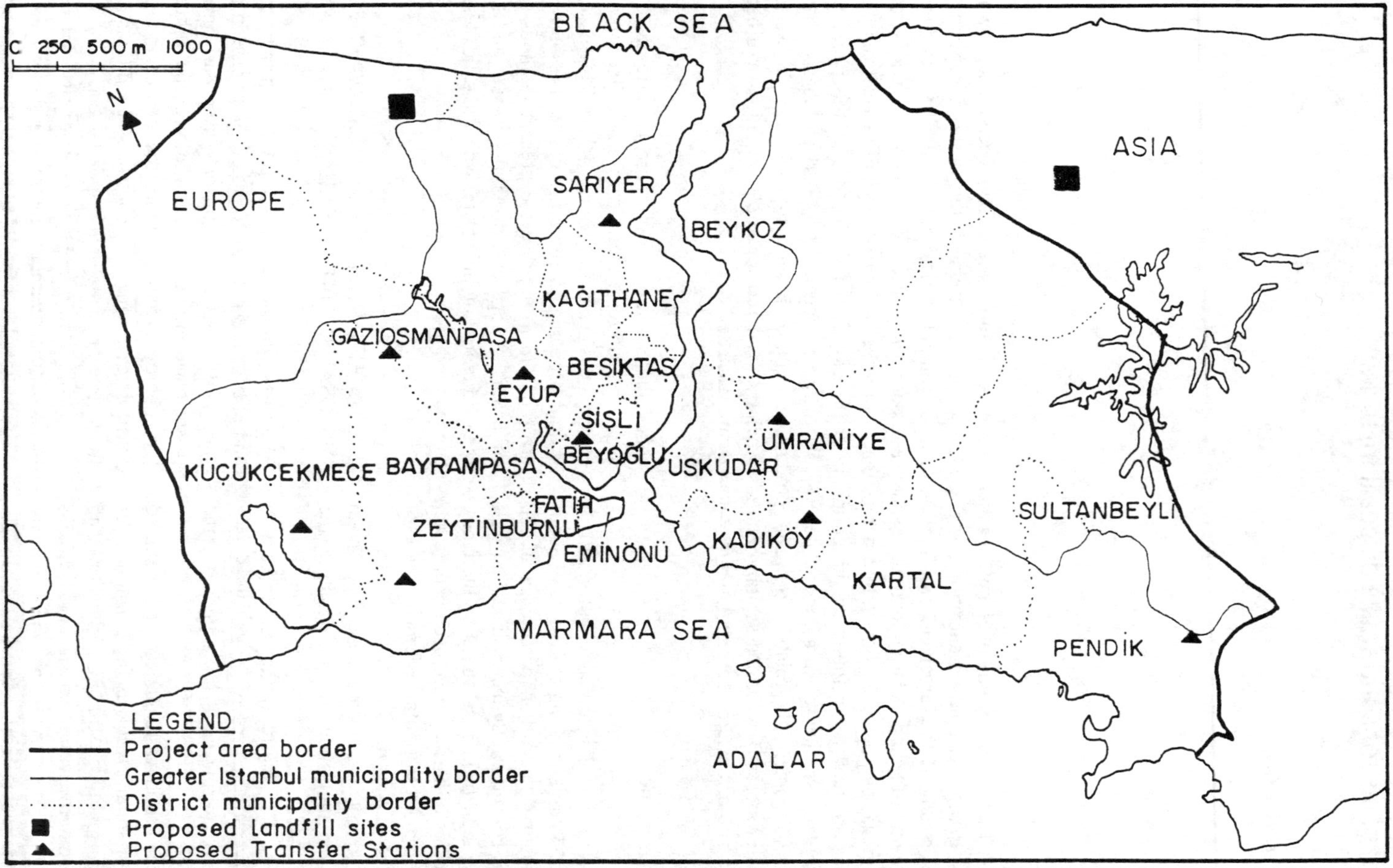

Figure 1. Map of Greater Istanbul

disposal of solid waste is weak. Some waste still finds its way to scattered heaps in the streets.

The continuing settlement of large numbers of people both within and outside the municipal boundaries of large cities in Turkey has put a heavy strain on all services provided by the local authorities, including solid waste management. The resulting difficulties, as well as an increased public awareness of environmental issues, have forced the municipalities to work out plans for the improvement of present solid waste management practices, selection of new landfill sites and rehabilitation of old sites.

The following sections will examine the present state of solid waste storage and disposal in the two main urban areas of Turkey, i.e. the metropolitan municipalities of Istanbul and Ankara, and assess both the progress achieved and envisaged in this highly critical sector of municipal management.

3. THE CASE OF ISTANBUL

The present limits of the Municipality of Greater Istanbul are shown in Fig. 1 along with the boundaries of the 21 individual district municipalities that currently constitute most of the metropolitan area. The 1990 census for the 21 municipalities was 6,885,651. The city population is growing at the rate of 4.7% per year, and estimates for the future population range from 15 to 20 million by the year 2020. The area within the municipal boundaries of Istanbul is 5712 square kilometres including areas of water that divide the city. Istanbul possesses the largest concentration of Turkish light and heavy industry - about 40% in terms of the number of installations, size of the labour force and V.A.T. collected. The unabated influx of capital, labour and associated problems into Istanbul continues.

Geographically, the city is divided into the European and Asian parts with a topography of rolling hills [maximum height 260 metres above sea level] divided by the Bosphorus. The European part is further divided by the Golden Horn. Situated at a latitude of 41 degrees N, Istanbul has warm breezy summers and cool winters, with few snowfalls.

The problem of solid waste management in Istanbul has always posed problems and, even at the present time, the subject is never far from the headlines of the daily newspapers. As mentioned in the report by the Institute of Local Government Studies, University of Birmingham (1978):

Istanbul, one of the great cities of the world, faces management problems of exceptional intensity... We have no doubt that in order to maintain and improve the quality of life of its citizens and to play its critical role in the life of the region and the nation, the Municipality's approach has to change from one in which it is uncritically continuing traditional services to one where it is responding actively to its changing environment.

3.1 Recent studies conducted on solid wastes in Istanbul

With increased demands on the solid waste disposal system, studies were initiated around 1980 by the Turkish authorities. A feasibility study for solid waste management in the Istanbul metropolitan area was undertaken by the World Health Organization [W.H.O.] Regional Office for Europe with the objective of obtaining some international assistance for subsequent implementation (Patrick, 1981). The physical and chemical properties of the solid wastes of Istanbul have been determined to assess their suitability for incineration or composting in a research project sponsored by TUBITAK, the Scientific and Technical Research Council of Turkey (Basturk, 1980). The following paragraphs highlight the main findings of these initial investigations.

Until 1953, waste disposal in Istanbul was carried out by dumping at sea. Later, crude dumping on land was adopted without regard to proper site selection procedures. Sites which became too close to residential areas had to be abandoned without suitable closure procedures and turned into residential areas. On several occasions, accidents caused by the explosion of methane gas were recorded. Attempts by the Directorate of Sanitary Works of the Municipality around 1970 to operate furnace units for incineration were not successful. At times when waste collection has slowed down due to inclement weather or industrial action, fills at the seaside have been resorted to in an attempt to clear the backlog (Basturk, 1980). At the start

Table 1. Typical municipal waste composition percentage by weight

Country/City[a]	1	2	3	4	5	6	7	8	9	10	11	12	13	14	Istanbul		Ankara
Metals	1	1	9	3		3		5	7		2	8	10	5	1.5	1.5	1.2
Paper/board	25	5	45	3	38	10	14[b]	22	32	14[b]	8	30	37	31	12	10.1	8.9
Plastics		1	2		8				3				7			3.1	2.1
Rubber/leather/ wood	7	1		2	12	4		3[c]	7		3	1	6	4	8[f]	6.1[f]	1.8[i]
Textile	3		1						4		4	2	2	2		3.1	1.4
Food and yard waste	44	45	25	60	18	74	56[d]	20	36	50[d]	25	16	26	16	46.5[g]	60.6	50.8[g]
Glass	1	1	11	2		7		6	4		3	8	10	13	3	0.7	1.4
Non-food inorganic	19	46	7	30	24	2	30[e]	43	7	21[e]	55	35	2	29	29[h]	14.6[h]	32.4[h]

[a] 1. Bangkok; 2. Beijing; 3. Berkeley, California; 4. Jakarta, Indonesia; 5. Japan; 6. Korea; 7. Madras, India; 8. New York City; 9. Singapore; 10. Spain; 11. Taiwan; 12. U.K.; 13. U.S.A.; 14. West Germany.

[b] Metal is included under paper and paper board

[c] Includes wood, bones, etc.

[d] Wood only

[e] Includes glass, coconut, shells, fibres, etc.

[f] miscellaneous

[g] organic

[h] ash and cinder

[i] rubble

of 1980 there were six main disposal sites in Istanbul, three each on the European and Asian sides, handling a total of about 3200 tons/day. This would correspond to a figure of 0.9 kg/person/day according to Basturk (1980) and about 0.8 kg/person/day according to the results of the W.H.O. study (Patrick, 1981).

Most of the dumping sites are located in valleys, some of which are very deep. The depth of tipping is not controlled, with the result that at some sites the refuse falls to a depth of more than 20 metres and is left exposed, resulting occasionally in fires. For example, the main disposal site at Habipler on the European side is in a deep valley with a large tipping depth. In seven years, filling to a height of 50 metres over an area of 10 - 15 hectares has taken place. Although the refuse is spread, compacted and periodically covered with soil, numerous fires occur. The site lies in fissured rock, and a serious problem of water pollution due to leachate migrating from the site has resulted in the closure of one of three wells used for the provision of drinking water. The closure of this particular site after covering the top with clay has been proposed in order to prevent further access of rain water to the refuse, failing which it has been suggested that suitable engineering works be carried out to stop the water pollution. It has also been proposed to turn the site into one for the disposal only of inert materials such as construction waste materials and demolition debris (Patrick, 1981).

3.2 Composition of Istanbul municipal solid waste

The information on the composition, physical and chemical characteristics and density of Istanbul solid waste has been given in Basturk (1980) and in the W.H.O. feasibility study by Patrick (1981). The composition of domestic refuse in the Istanbul area varies considerably with season and district. The latter factor is influenced by the standard of living and specially the type of heating fuel [coal or more expensive fuel-oil] used over the 5 - 8 month period when heating is required, and taken into account by dividing the municipal area into four zones (Basturk, 1980) or seven zones (Patrick, 1981). The composition of Istanbul solid waste is given in Table 1 which also includes data from the U.S.A. and several other countries as assembled in the work of Oweis and Kheira (1990). The data on Istanbul is given in two columns; the first was collected over one year by Basturk and the second over 10 months by Patrick, and comprise the average refuse values for all districts. Refuse densities assessed by Basturk range from 220 - 690 kg/m^3 and by Patrick from 203 - 566 kg/m^3 depending on the time of the year and the district. As may be observed, the domestic refuse density is much higher than the typical values of 90 - 100 kg/m^3 in Western European countries, since the solid waste generated in Istanbul consists primarily of wet food waste and coal ash. The suitability of Istanbul refuse for composting has been investigated in the two studies mentioned above considering the moisture content, pH values and the C/N ratio. Basturk (1980) concludes that the refuse is not compostable except for high class areas with fuel-oil heating. Patrick (1981) on the other hand states that the waste is compostable in the poorer coal burning areas outside the winter months, but in fact the composting option is of insignificant importance at present in the Istanbul area.

The average (over the year) calorific values of refuse for various districts in the Istanbul area range between about 450 - 1500 kcal/kg (Basturk, 1980; Patrick, 1981). The analyses using Tanner diagrams indicate that most of the data points lie outside the area of self-sustaining combustion. This is due to the high moisture content during the summer months and the high ash content in the winter months. The W.H.O. study gives some information on industrial and hospital wastes. Around 1980, the total industrial waste was 1,013,678 tons/year. About 4% of this waste was hazardous, which is not atypical for a large industrial area. The proportion of wastes recycled is very high (up to 45%). Some of the hazardous waste is discharged into sewers or surface waters. Hazardous and non-hazardous wastes are not usually separated, but stored in metal containers or open bunkers, or even on open ground. Similarly, hospital wastes are stored in an unhygienic manner.

One of the most recent [1992] studies consists of three technical memoranda for limited circulation prepared by a foreign consultant for the Metropolitan Municipality. This document discusses solid waste generation in Istanbul and presents projections of waste quantities through the year 2020 as well as design recommendations for the solid waste system management of Istanbul. It is pointed out that since the 1981 W.H.O. study:

1. The population of the area has grown from 4.7 million to 7.2 million mainly due to the influx of poor people from the rural areas, whose per capita waste generation rates are expected to be low.

2. The per capita consumption of lignite which results in 15% of ash by weight increased 300% until 1989.

3. Tourism related waste has increased considerably, affecting Eminonu municipality in particular, and this trend is expected to continue.

The report also states that earlier waste generation rate estimates were on the high side, and sets the figures in 1990 at 0.67 and 0.58 kg/person/day for the European and Asian municipalities respectively, excluding the litter and trash collected from major streets and parks, and the 150 tons/day solid waste delivered to the dumps from the military installations. It is expected that the waste generation will rise to about 0.70 kg/person/day by the year 2000 and then decrease as some conversions from coal heating to natural gas take place. It has been determined that approximately 4500 tpd [tons per day] of solid waste was generated in Greater Istanbul in 1990. This amount is expected to be more than 7000 tpd by the year 2000 and nearly 11000 tpd by 2020.

3.3 Implementation plans for Istanbul

Presently in the Istanbul area there are four operating dumps [Halkali, Yakacik, Ümraniye and Aydinli in Pendik] . Plans agreed upon involve two new landfill sites, one on the European side near Çiftalan and the other on the Asian side near Karakiraz, which are each at a distance of about 25 to 30 kilometres from the centre of Greater Istanbul. A system of transfer stations totalling 6 [or 5] on the European side, and 3 on the Asian side, is also envisaged, as shown in Fig. 1. It is estimated that cumulative landfill volumes by the year 2020 are 68 and 43 million cubic metres for the European and Asian landfills. The volume calculations were based on the population estimate of 16.5 million forecast by the year 2020. Other assumptions are that both the landfills will be open by January 1, 1994; the compacted in-place density of the refuse and soil cover will be 800 kg/m^3; approximately 15% of the volume of the landfills will be required for placement of the daily soil cover and 10% of all solid waste will be recycled by scavengers at the sorting/material recovery facility at each landfill. The waste is planned to be compacted into steel, leak-proof cylinders at the transfer stations and transported to the landfills in transfer trucks or trailers.

The European site has been mined for lignite by the open pit method.The Asian site area has been used for clay pits. The surficial geology of both sites is similar: the ridges at a height greater than 120 metres above sea level are covered with Neogene deposits of sand, clay and coal. Cretaceous rock of volcanic origin and consisting of Andesite, Andesitic lava and agglomerate forms the base of the valleys. Neogene soil deposits on the Asian side appear to contain significant amounts of low permeability kaolinitic clays which are suitable for constructing the landfill liner and cap. The surface water divide, separating drainage northwards to the Black Sea from drainage to the south, occurs along the southern boundary of the sites, thus minimizing the amount of surface water run-on that can enter the sites. The topographic features are also favourable.

4. THE CASE OF ANKARA

Ankara is the capital of the Turkish Republic and the second largest city in Turkey, with a population of 2,585,000 in the 1990 census. Ankara saw a steady growth in population since it was chosen in 1923 as the capital of the republic, but there have been large increases both in the area and population of the city, especially between 1950 and 1975. An estimate of the population would be about 3.7 million inhabitants by the year 2000. There are eight district municipalities.

Compared to Istanbul which is astride two continents, the solid waste disposal problems of

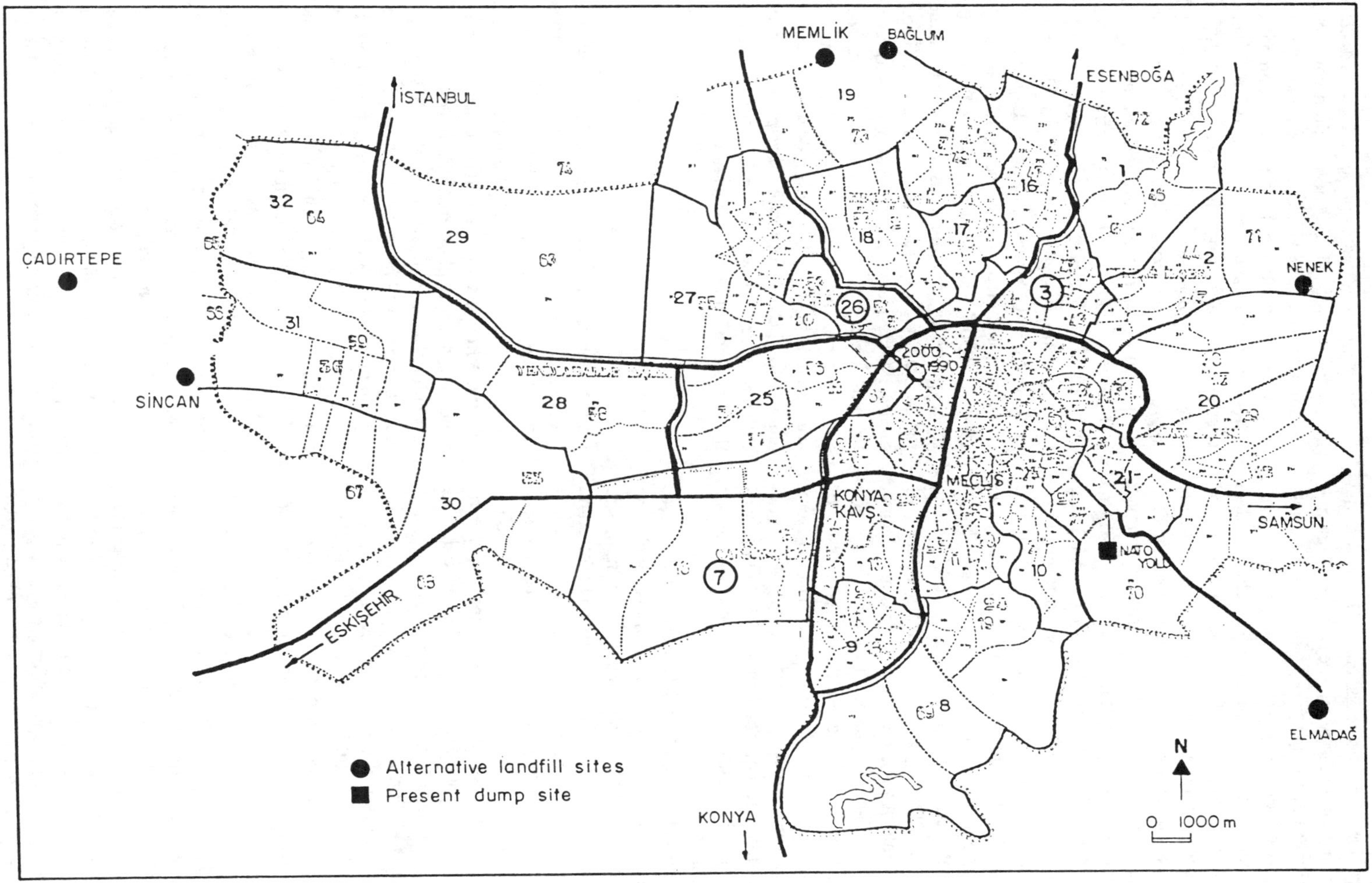

Fig. 2. Map of Greater Ankara

Ankara have not been as acute. However, the percentage of industry in the Ankara region is perceptibly on the increase, particularly in light and service industry.

The daily total amount of waste generated in Ankara varies approximately between 0.56 to 0.62 kg/person depending on the income level of the district, giving an average of about 0.59 kg/person [Kirca et al., 1990]. The results of samples collected over one year starting in April 1981 are reported in Uluatam and Yurteri (1983) and Uluatam et al.(1984). The composition of the waste has been found to vary seasonally and with the district. The amount of ash is very high in the winter months and in poorer areas where cheap solid fuel such as coal is used for heating purposes. The yearly average values of components of solid waste for Ankara are given in Table 1. These figures, however, will need updating because of the compulsory conversion of the heating system to a natural gas based system, which has been progressing for two years and is still continuing. The solid waste of Ankara was found to be unsuitable for composting or incineration.

For the last ten years, and at present, solid waste collected from all Ankara districts, including hospital waste, has been brought to the dump site in Mamak district [Fig. 2]. In spite of some improvements there are serious environmental problems associated with the operation of this site. The material recovery facility here has operated at a loss and was not used to full capacity, due to management problems, and is now closed. The working conditions for the scavengers were not hygienic. In addition, the site lies adjacent to the area marked for the planned development of a large residential satellite town called "Dogukent". An extensive project with an estimated cost of about 16.5 million DM is being contemplated for this site.

The alternative new landfill sites are shown in Fig. 2. The results of the evaluation of all these sites favour the Çadirtepe site near Sincan at close proximity to the waste water treatment plant under construction. However, the Metropolitan Municipality has still to give final approval to the decision on location. In a recent feasibility report (September 1992) prepared for the Municipality, the waste generated is given as 1,767,864 kg/day for the year 1990 and projected as 3,204,608 kg/day for the year 2015. The required volume of landfill between 1993 and 2015 is estimated to be about 40 million cubic metres. The compacted refuse density is estimated to be about 690 kg/m^3 considering the composition given in Table 1 along with typical uncompacted densities and appropriate compaction factors for each waste component. The proposed site contains two valleys with a total volumetric capacity sufficient to last 38 years. The location of a maximum of three transfer stations is given for the four alternative landfill sites [Sincan, Nenek, Baglum and Elmadag] included in the investigation by Kirca et al. (1990). The abovementioned feasibility report proposes the same transfer station locations for the Çadirtepe stations as Sincan given in Kirca et al. (1990). These are area 3 if one station is to be built, areas 3 and 7 for two stations, and areas 3, 7 and 26 for three stations. These area numbers are circled in Fig. 2. Hospital wastes which amount to about 100 tons/day are planned to be collected separately and stored in a special part of the same landfill after disinfection.

5. THE CASE OF IZMIR

Izmir is the third largest city in Turkey with a population of 1,757,414 according to the 1990 census. The first sanitary landfill in Turkey was opened in February 1992 at Harmandali in Izmir. The old dump site in Çigli which is very close to populated areas is planned to be developed as a recreational area, and work has already begun on the drainage, levelling and conversion of the area. The filling at Harmandali is being carried out in four separate locations: Household refuse, commercial waste, etc., is placed in one location, and the other locations are for hospital waste in plastic bags, industrial wastes and waste water treatment plant sludge. The total amount of waste generated is estimated to be 2400 tons per day. Unlike the other two cities discussed, composting has been both possible and feasible in Izmir. About 500 tons per day of waste is being composted at present at the Uzundere composting facility. The abandoned composting facility at Halkapinar is now being used as a transfer station.

Studies have not indicated incineration to be feasible for Izmir.

6. CONCLUSION

The efficient and economic disposal of urban solid wastes in Turkey is still in its initial stages, and the problem is most critical in the major cities of Istanbul, Ankara and Izmir. However, solid waste disposal receives continual media attention, and steps are underway at both the central and local governmental levels to create the urban planning and municipal infrastructure required for the safe disposal of solid waste as well as the development and promotion of a healthier environment.

The case studies discussed in the paper indicate that the sanitary landfill method emerges as the principal and most reasonable method for solid waste disposal when combined with the planned establishment of suitable transfer stations in the metropolitan areas.

ACKNOWLEDGEMENTS

The author acknowledges with gratitude the technical information provided by Mr Ugur Cilasun, Mr Tahir Tümer and Dr Hasan Mani of the Ankara Municipality, as also by Mr Erdal Eren, Managing Director of Nurol Ltd., Ankara. Thanks are also due to Prof.Dr. Ömer Saatçioglu and Prof. Dr. Semra Uluatam of Middle East Technical University, and Prof. Dr. Mehmet Uluçaylı of Dokuz Eylül University.

REFERENCES

Basturk, A. 1980. Ways of solution for mitigating the damage due to Istanbul solid wastes [in Turkish]. TUBITAK Report No. CAG-11, Ankara.

Erkip, N. and Kirca, O. 1991. Solid waste management systems: An approach for analysis and systems design. Journal of Resource Management and Technology, Vol. 19, No. 2, 63-67.

Kirca, O. et al. 1990. Project for the design of the Ankara Metropolitan Municipality solid waste management system [in Turkish]. Sibaren Report, Middle East Technical University, Ankara.

Oweis, I.S. and Khera, R. P. 1990. Geotechnology of Waste Management. London: Butterworths.

Patrick, P.K. 1981. Solid wastes management in the metropolitan area of Istanbul. World Health Organization Feasibility Study.

Uluatam, S. and Yurteri, C. 1983. Investigation on the collection, transportation, evaluation and safe rendering procedures for Ankara solid wastes [in Turkish]. TUBITAK Report No. CAG-40, Ankara.

Uluatam, S., Yurteri, C. and Billur, N. 1984. Effective collection, transportation, evaluation and safe rendering procedures for Ankara solid wastes [in Turkish]. Doga, TUBITAK, Series B, Vol. 8, No. 1, 100-110.

Waste Disposal by Landfill – GREEN'93, Sarsby (ed.) © 1995 Balkema, Rotterdam, ISBN 90 5410 356 6

General report

Y.Wasti
Middle East Technical University, Ankara, Turkey

Y Wasti : The theme of the first session is broad in scope. However the environment and its deterioration must be one of the few topics that can be regarded as truly universal, irrespective of differences of wealth and power amongst countries. But there is currently a large difference in landfill practice in locations which are in relatively close proximity (in a global sense). From the papers submitted to this session we have, on the one hand, an account of an incident at a dump fill in Istanbul resulting in casualties and a description of a waste disposal site in Zagreb which is polluting the groundwater, and on the other hand we have details of the impressive standard of current practice achieved in North America and the high technology lining systems developed in Germany (which allow inspection and repair). The papers in this session generally deal with good landfill practice and associated legislation, regulation and technical guidance.

Seymour and Street describe the improvements in british landfill practice brought about since the 1970's by experience, research and new UK and european legislation policy, e.g. the National Rivers Authority policy for the protection of ground. Furthermore there is the trend towards total containment of landfills rather than relying upon natural containment in low permeability environments or relying on the dilute and disperse principle of leachate becoming safe by attenuation in the unsaturated zone and subsequent dilution in the saturated zone. The latter issue has still yet to be resolved fully, there are still supporters and opponents of the concept. Of special significance is the inauguration, in 1990, of the Environmental Protection Act which requires a landfill operator to acquire a certificate of completion from the relevant Waste Regulation Authority before the waste management licence may be surrendered - currently an operator may surrender a licence at any time and thus end his environmental liability. The new legislation will require operators to achieve higher standards of landfill practice so that a site does not pose a problem even after closure. The drafting of the document Guidance on Good Engineering Practice (Department of the Environment) is very important - this is in line with the work of ETC8 (European Technical Committee 8 of the International Society for Soil Mechanics and Foundation Engineering).

On the other hand **Potter and Yong** estimate that Britain is several years behind North America in terms of environmental protection, particularly with regard to groundwater. The main items discussed by Potter and Yong are; sources of inaccurate determination of hydraulic conductivity of compacted clay liners, the necessity of considering the diffusion mechanism in leachate transport in a compacted clay liner, the effect of leachate type on the hydraulic conductivity of clay liners. The Authors also point to the difficulties encountered in the use of flexible synthetic liners or geomembranes and state that they should not be used as the sole barrier to leachate but in combination with a soil liner or similar. Finally the Authors stress the importance of hydrogeological evaluation of landfill sites even if engineered containment practice is followed.

Forster presents a review of landfill capping which was undertaken on behalf of the Department of the Environment of Great Britain. Waste management policy within the European

Community is also discussed. It appears that current landfill capping technology varies both within a country (because of the delegation of guidance powers within a country) as well as from country to country. But there is an increasing trend towards the development of national minimum standards. Forster points out the importance of providing technical guidance and the implementation of Quality Control and Quality Assurance programmes.

Bishop and Carter deal specifically with german practice for basal seals and landfill covers. A very efficient leak location survey is described.

Belfiore and Magri describe the application of a QC/QA (Quality Control/Quality Assurance) programme to a new municipal waste landfill construction near Milan. It is interesting to note that the cost involved in the development and execution of this programme only amounted to 2.5% of the overall cost. The need for conducting a thorough QC/QA programme during the construction stage of a landfill cannot be over emphasised. Similarly the importance of leak detection in geomembrane systems cannot be overstressed.

Two other papers deal with the evaluation of old landfill sites. **Saarela** presents the methodology followed in assessing the situation of several old sites, of varying capacity, in Finland. He also discusses suitable rehabilitation procedures (mainly the type of final cover for each site).

Darbyshire and D'Arcy describe the methodology, for investigating old landfill sites, adopted by a Waste Regulation Authority (WRA) within North West England. Under the Environmental Protection Act WRA's are required to inspect their areas for sites which might cause pollution of the environment or harm to humans. The emphasis of the paper is on the identification, and risk assessment, of potentially contaminated land. Whilst the Authors state that it is not their intention or desire to formulate a register of contaminated land the General Reporter's feeling is that such a register would be highly useful.

Plainer discusses the preservation of both quality and quantity of water resources. This paper deals with the general principles of water protection and the importance of underground water is emphasised. Examples of the application of some of the tools for resource protection are quoted.

The paper by **Wasti** contains a report on the solid waste management problems of metropolitan areas of Turkey - a rapidly-developing country. The many-sided problems requiring simultaneous solution include; establishment of sanitary landfills (there is only one low-cost site operating currently), operation of existing dump sites so that hazards and threats to human life are minimised, the closure or rehabilitation of some sites, the development of schemes of taxation to finance these projects. Unplanned and illegal growth in the urban areas (including thousands of squatter dwellings on abandoned dump sites, or building developments very close to existing sites) make the size of the problems to be tackled almost unmanageable.

Further issues to be tackled comprise; funding of projects from developed countries with the imposition of technologies which are not sufficiently adapted to local conditions, options for recovery of materials, market surveys for recovered/recycled materials, training and education for both personnel and the public.

The solid waste problem is less acute in rural areas which are less-densely populated and the waste is both smaller in amount and of a less-hazardous nature. Also suitable disposal sites, incorporating natural liners, are usually available in these rural areas.

To highlight the environmental problems of a developing country **Y Wasti** referred to a disaster which had been reported recently on the front page of the most popular newspaper (1 million copies per day) in Turkey. The article concerned an incident at a dump site which was to be abandoned once suitable waste disposal sites had been developed. The incident, which resulted in over twenty casualties with twelve persons being unaccounted for, was a large slide failure which engulfed squatters' houses situated at the edge of the dump. It is thought that the slide was initiated by a methane gas explosion although the cause of the failure has not yet been definitely established.

There is also the usual problem of illegal 'fly-tipping' of waste outside of the municipal dump sites and at Izmir a new low-cost sanitary landfill has been started to try to reduce this problem.

Y Wasti also expounded briefly upon legislation and turkish environmental law. Excessive legislation is often detrimental to the solving of a problem. However without government intervention, usually in the form of regulation or taxation of pollution, the environment is treated in the market-based system as a 'free good', with emissions to the atmosphere costing the polluter only the value of the raw materials wasted and the value of any liabilities that are assessed. However there is also a body of opinion which feels that regulation promotes standardisation of design and construction requirements. As a consequence it severely limits the participation of the technical community in siting and design issues. Hence there is plenty of scope for discussion of environmental legislation.

The turkish environmental law (passed in August 1983) consists of six sections. The law incorporates regulations dealing with air, water, waste disposal and noise pollution. This law is similar to European Community law and in fact it is stated that, in the case of pending legislation, relevant European Community standards and regulations will be enforced.

Turkey issued its first legislative regulations on solid waste in 1991. It includes sections relating to; the recovery of plastic and metal-based packages, collection and transportation of solid waste, the construction and operation of landfills, composting, incineration, and the use ofwater treatment sludge for agricultural purposes.

The current requirements for landfilling of domestic and light industrial wastes are not very stringent. The bottom of the landfill should be at least 1m above the water table and contain a compacted clay liner with a minimum thickness of 600mm and a maximum permeability of 10^{-8} m/sec - unless it is underlain by rock which is only lightly cracked (defined as a layer more than 10 m thick with a permeability less than 10^{-7} m/sec). Leachate collection pipes must be at least 100mm diameter with a fall of not less than 1% and the pipes have to be covered by a granular filter which is at least 300mm thick. The collected leachate is to be treated in accordance with water pollution regulations and the final sealing cap should consist of a suitable thickness of soil shaped to have a minimum slope of 3%. Technical guidance documents on the implementation of this legislation have not yet been prepared although regulations on environmental impact assessment for engineering processes have been issued. However the main problem in Turkey, as in other countries, appears to be the development of efficient mechanisms for implementation of legislation as well as strict enforcement of penalties when the regulations are violated.

Belfiore and Magri
QC/QA program for the construction of a municipal waste landfill.

Bishop and Carter
Waste disposal by landfill - german landfill lining systems.

Darbyshire and D'Arcy
Section 61 - another regulatory nightmare.

Forster
Landfill cover technology within the United Kingdom, the rest of Europe and the United States.

Plainer
Main tools for water resources protection.

Potter and Yong
Waste disposal by landfill in Britain : Problems, solutions and the way forward.

Saarela
Inventory of old sanitary landfills and their rehabilitation.

Seymour and Street
Landfill engineering - UK practice and guidance in a european context.

Wasti
Municipal plans for solid waste disposal in Turkey.

Waste Disposal by Landfill – GREEN'93, Sarsby (ed.) © 1995 Balkema, Rotterdam, ISBN 90 5410 356 6

Discussion

I Peggs stated that geomembranes are not a panacea for landfill lining and from the polymer engineering point of view the emphasis is on ensuring that these materials have long-term durability. Currently there is some rationalisation (or at least less polarisation) regarding use of soil liners and/or geomembranes and it is found generally that the two work best together in composites.

Although there is no one man-made material that is universally applicable for landfill containment currently there is a great tendency to rely very heavily on HDPE (High Density Polyethylene). This may not really be justifiable because, just as there are many different types of steel, there are also many, very different HDPEs and their mechanical durability or resistance to stress cracking can vary by a factor of a thousand. Geomembrane users must become very specific about the type of HDPEs which they wish to use. There can be a lot of difference between the idealised view of a nice integral chemical-resistant liner, and the actual result after installation in the ground. It is very, very difficult with some liner materials to get the desired intimate contact between the liner and the geomembrane. Therefore there has to be a compromise between the theoretically-achievable chemical or mechanical properties of the geomembrane itself and the overall result obtained in the field. It is no good having a membrane with the best chemical resistance in the world if it cannot be installed in the ground to form a leak-tight layer.

Another important item, according to **I Peggs**, is the chemical resistance of the different geomembranes to leachate attack, as mentioned by Potter and Yong. A number of studies have been reported (from Belgium, the United States, France) which deal with both laboratory studies and examination of geomembranes taken from the ground after prolonged (ten plus years) contact with leachate. These are relatively short contact time periods but generally it is being found that leachates arising from municipal waste are nowhere as damaging as was originally suspected. In many studies PVC and HDPE behaved in such similar ways that it is hard to tell the difference between their performance. The Geosynthetics Research Institute at Drexel University has had a PVC membrane in contact with two leachates for 36 months at 85°C and there is no evidence of the plasticizer extraction that is widely talked about.

With reference to the paper by Belfiore and Magri, **I Peggs** believed that there was more to CQA than just being a means "to ensure that the liner is put in according to specification". Specifications are simply numbers/tests/drawings etc. and they say nothing about the practical aspects of installation such as whether a seam has been overheated or not, whether there is a grinding gouge that is just a little bit too deep etc.. It is very necessary to have someone on site who understands what these kind of defects mean to the durability of liners composed of Polyethylene/PVC/Polypropylene, etc.. Polymers are viscoelastic and they are different from the usual construction materials particularly since their properties change with time.

It has been stated that consultants for landfill sites should have earthworks experience - it is suggested that they should also have worked with plastic materials. In the USA (where QA is universally applied) there are now many complaints that the owners are not getting quality or value out of the CQA which is undertaken. CQA is becoming very, very expensive and although the owners receive check lists to confirm that various items have been done there is no active contribution to the quality of the system. Europeans should take note of this prior experience from America.

I Peggs felt that leak detection surveys (described in the paper by Laine and Mosley) should be undertaken much more widely and now that geomembranes are being produced with a conducting layer on the bottom such surveys are much easier to perform. With the inbuilt conducting layer there is no need to worry about the conductivity of the subgrade. The paper mentioned that it is necessary to insulate the soil inside the liner from the soil subgrade so that electrical conduction from the inside to the outside of the cell does not occur. Lane and Mosley achieved this by digging a trench around the outside of the ceils with soil going up to within 2 to 3 metres of the top of the slope. According to I Peggs this is the ideal methodology for optimum sensitivity, but in practice to find the large defects (and even some small ones) it is not essential to have the preceding arrangement because the sensitivity of the equipment can be modified and adjusted to suit. Even though there would be a loss of sensitivity at the edges it is still possible, within a 1½ hectare cell with 300mm of soil on the base, to be able to find small holes (1 or 2mm width) beneath the soil. In one extreme case the technique detected a 600mm slit, which was underneath 1.7m of sand (in a filtration bed in a wastewater treatment plant) only a few metres away from the edge of the cell.

When the soil over a liner is removed (during a leak survey) the fines and the dirtiness of the membrane make it difficult to see small defects. Consequently there is a tendency on site to use a vacuum box to try to find the hole. However this is usually not satisfactory because the underlying clay and the membrane are in intimate contact, having been pressed down by the overburden, so that air cannot be pulled in through the soil and thence through the hole and into the vaccum box.

Referring to the paper by Bishop and Carter, **I Peggs** stated that geomembranes do need protection, especially if HDPE is being used. Furthermore, in addition to the protection of the lining systems currently being installed, efforts should be made to improve current geomembrane types and to develop new materials. In particular it is necessary to diversify in terms of the materials used, rather than simply concentrating on HDPE, and develop suitable geomembranes to fit the range of applications.

L Cowan observed that landfill covering techniques and the current 'keep rainwater out' attitude may actually lead to encapsulation of the waste rather than encouraging the bio-reactor principle. She wondered if delegates had any opinion on the benefits (or otherwise) of allowing rainwater ingress into completed containment landfill sites, in conjunction with leachate treatment, in order to enable degradation of the waste. The question was also raised as to whether such treatment could possibly render the site environmentally benign in a shorter time than otherwise.

With regard to leachate recirculation **M Wallis** opinioned that water was a product of the degradation of waste and so recirculation within an enclosed site would produce an accumulation of water unless leachate was extracted for treatment. In the UK evaporation rates from a landfill site would not exceed rainwater infiltration rates. The inevitable result is that any recirculation system would need an associated leachate treatment plant.

In response **A. Street** said that he hoped that recirculation would be regarded as only a part of the whole leachate management process for a landfill. In the UK the approach is to consider using leachate recirculation in the early stages of a site (probably during the operational stage) as leachate recirculation can be a very valuable aid to reaching a steady state. Even after capping there will be further leachate production due to waste decomposition and some infiltration, therefore in the longer term there will always be a need to extract leachate and treat it prior to disposal.

H Jessberger observed that from old landfills newspapers printed a long time ago have been recovered which have been sufficiently well preserved that the text was still readable. The papers had not been degraded even though these old sites did not have proper, completed final covers. It was suggested that rather than allow uncontrolled ingress of rainwater to aid the decomposition process, perhaps it would be more economical to re-circulate the leachate already in the site thereby avoiding the extra cost of treating the additional leachate produced by the additional rainwater ingress.

A Forster pointed out that D Daniel has questioned the value of the immediate imposition of a final cover over placed waste. Previous research has shown that re-circulation techniques do in fact stabilise leachates and the waste itself so there is a benefit. At the present time there is a movement towards the containment policy but there are situations where the waste is dry so that it will not stabilise and hence it will remain a potential pollution source for an infinite length of time. The natural evolution is to incorporate re-circulation techniques within a covered system so that a bio-reactor is produced and the water content can be controlled. At the present time there are too many instances of severe pollution arising from; leaky caps, caps with no slope to facilitate run-off, poorly engineered caps etc.. Consequently there is a backlash from what has happened in the past. A Forster was of the opinion that re-circulation techniques and moisture content control will become standard in landfilled refuse disposal sites in future.

H Jessberger commented that the german viewpoint would be that in the future there will be no municipal waste and so the problem of degradation will not arise.

D Bishop noted that the 1990 draft of the proposed EC directive only required that landfill should have a 3m thick basal liner (with a coefficient of permeability of about 1×10^{-8}m/sec) and wondered if there was any further information on the likely content of the final EC directive. (No member of the audience was able to give additional information in this matter).

Referring the paper by Laine and Mosley, **H Jessberger** enquired whether or not it was necessary to have a particular type of sub-soil, or a certain electrical conductivity of the ground water, for the leak detection system to work.

In reply **M Lane** stated that the sub-grade soil did not need a particular resistivity for the system to work. It is operable with any type of soil provided that it has some moisture content, ie it is not totally dry. The system has also been used in America in association with brine (which has a very high electrical conductivity) and after modification of the equipment the system had worked perfectly well.

D Bishop queried whether the system would work where there was a coarse granular layer (for leachate drainage) at the base of the landfill and a geotextile between this layer and the underlying geomembrane.

M Lane replied that it would work and cited a recent survey of three landfill sites which contained drainage layers comprised of pea gravel. The method was successful with approximately 300mm of liquid over the pea gravel and also in the drained condition with only moisture in the pea gravel.

A Forster sought information from H Jessberger about the deformation tests which he had conducted in a centrifuge and in particular whether any work had been done on new materials such as bentonite matting and HDPE.

H Jessberger said that it was difficult to conduct such tests in the centrifuge because its primary function is to facilitate testing of a small model to examine modes of behaviour whilst at the same time simulating appropriate body stress levels. Currently work is being started to assess the appropriateness of centrifugal testing for examining landfill lining situations.

T Zimmie commented that he had considered testing liner systems in his centrifuge but the scaling laws would require that the liner be represented by a very thin layer in a centrifugal model - (if models are subjected to enhanced acceleration but the body stress levels and gradients are to be maintained then linear dimensions have to be reduced).

H Jessberger invited A Forster to comment on the proposition that infiltration rate (which is a function of both hydraulic gradient and hydraulic conductivity) is a better parameter for assessing the efficiency of a moisture barrier than the value of the hydraulic conductivity alone.

A Forster replied that, in terms of specifications for sealing layers, the definition of a coefficient of permeability alone can be quite misleading. If a hydraulic gradient is also specified then this is a refinement of the parameters for assessment of the effectiveness sealing layer. Furthermore laboratory-measured permeability values can be affected by the hydraulic gradients applied during the tests. When low permeability soils are being tested hydraulic gradients of the order of 100 may be applied and this situation is not comparable to field conditions.

W Wolski supported the use of infiltration rate as opposed to coefficient of permeability alone and pointed out that unsaturated soils can give very misleading permeability data.

P Pasqualini pointed out that conventional determination of permeability uses water as the permeant whereas the relevant permeability is that relating to the flow of pollutant through a soil. There is also the potential problem of interaction between the soil and the pollutant.

In Finland, according to **J Saarela**, there is a shortage of suitable clay for forming containment liners and investigations are being made into the use of moraine with the bentonite addition.

H Jessberger agreed that suitable sources of lining clay were often not available and that efforts should be made to investigate possible alternatives, eg use of asphaltic concrete layers (Switzerland and Germany), the use of recycled waste to form sealing layers, etc.. In the eastern part of Germany the ground is very sandy and there is no suitable clay available, similarly the Netherlands.

I Martinenghi added that to her knowledge asphaltic concrete was used only in Switzerland and she would be interested to obtain views about this liner material from workers in other countries. I Martinenghi also queried why asphaltic concrete had not been used elsewhere.

A Street replied that from the UK point of view the problem was one of cost and asphaltic concrete is expensive by comparison to other liner systems. So whilst in the UK it has been used in the construction of embankment dams it is not yet viable for landfill lining.

According to **H Jessberger** the technique was currently being used in Southern Germany at two sites where the liner would comprise two layers (each 70mm thick) as a composite lining system beneath a mineral sealing layer.

A Vonk commented that investigations in the Netherlands had indicated that asphaltic concrete was not a good material for lining the bottom and sides of a landfill site. Initially the asphalt has a very high absorption coefficient for contaminants but after a period of 100 years or so the absorption capacity is reached and the contaminants simply flow through the asphalt. Even worse, this asphalt attracts some contaminants so in the long term it will not be an effective barrier.

H Jessberger concluded by adding that there can also be a problem with solvents from the waste and their effect on the stability of the asphalt.

2 Basic principles

Waste Disposal by Landfill – GREEN'93, Sarsby (ed.) © 1995 Balkema, Rotterdam, ISBN 90 5410 356 6

The composition of leachate from waste disposal sites

S.C. Alker
Sunderland University, UK

R.W. Sarsby & R. Howell
Bolton Institute of Higher Education, UK

ABSTRACT: Landfill leachate is a complex cocktail of chemicals subject to processes within a dynamic environment which act to change the composition of the liquid. Little research to date has addressed the characterisation of landfill leachate. This paper contains general results from an ongoing study of the metal ion content of Landfill leachate from sites in the North West of England. An introduction to some of the parameters which affect the composition of the leachate with respect to metal ions is also presented.

INTRODUCTION.

Landfill leachate is produced by the infiltration of fluid into refuse, and comprises materials dissolved in water and other liquids disposed of within the refuse. Within the landfill this liquid collects in areas as it drains through the refuse. In a contained site, the liquid encounters an impermeable barrier and is channelled towards a collection point from where it is usually pumped periodically away for treatment. Older landfill sites in the United Kingdom have operated on the 'dilute and disperse' principle where leachate is allowed to percolate through the soil material beneath the landfill to bedrock. This principle relies on the soil material and bedrock acting as a filter and attenuating the leachate before any ground water is encountered. The principle also assumes that the quantity of leachate leaving the site is relatively small by comparison to the volume of water in the aquifer, so that it is rapidly diluted within the ground water. Hence it is considered that drinking water abstracted from aquifers in the vicinity of landfill sites would have negligible levels of pollution.

Landfill leachate consists of a mixture of chemicals, many of which are toxic to humans and aquatic life. Concern for the purity of ground water reserves in the United Kingdom and the implementation of the Environmental Protection Act 1990, has led to a move to engineer landfill sites by constructing impermeable barriers so that leachate is contained and can be treated to prevent toxic chemicals entering the environment.

Construction of an impermeable barrier for a contained landfill site would appear at first to be straightforward engineering problem. However, when the nature of landfill leachate is considered questions arise as to the efficacy of any barrier to contain landfill leachate. Chemical reaction may occur between the leachate and the barrier material (compacted clay, man-made liners etc.) and facilitate rapid egress of the polluted liquid. The first consideration is the exact nature or characterisation of landfill leachate. A landfill is a dynamic environment with chemicals being dissolved and precipitated by the fluid which moves through the landfill. Not only does the type of refuse, domestic, liquid, industrial waste placed in the landfill affect the composition of the leachate, but leachate composition is also affected by infiltration rates, pH, age and many other factors. (Farquhar 1989, Christensen 1984, Beveridge and Pickering 1983). Therefore composition of the leachate is highly variable.

In attempting to characterise the nature of landfill leachate and to investigate how the above factors affect the composition of leachate, it is necessary to investigate the composition of leachate from a number of sites. Investigations conducted in the United Kingdom, North America and Europe into the composition of leachate, have rarely considered more than a few sites. Each study has looked at different aspects of leachate composition, and has usually concentrated on chemical oxygen demand COD, or biological oxygen demand BOD, which indicate the toxicity to aquatic life, and ammonia leading to methanogenesis and the production of landfill gas. As the studies have been site specific and investigatively specific, it may be difficult to compare the results from them in terms of other parameters and draw meaningful conclusions. In addition to this, differing legislation between countries (Yong 1992) means that some aspects of leachate characterisation in one country may not be relevant to the legal framework in another and so the composition may not be so rigorously investigated.

This paper is presented as a summary of a recent initial investigation into the composition of landfill leachate in the North West of England. The results arise from a study by the lead Author into the composition of leachate from twenty-one sites and are presented in an attempt to characterise landfill leachate in terms of metal ion content. Type of fill, age, and pH are among the factors investigated.

METAL ION CONTENT.

Metal content presents a mixture of toxic ions which tend to accumulate in living organisms and so are not immediately toxic and have often been overlooked in previous studies. Table 1 shows the effects of metal ion build-up in living organisms, which underlines the importance of this aspect.

Table 1. Toxic effects of some metal ions on human health. (After Yong 1992).

Metal	Toxic Effect
Cadmium	Carcinogenic
Chromium	Carcinogenic, Tubular necrosis of kidney
Copper	Carcinogenic, Disorder of copper metabolism - hepatic cirrhosis
Iron	Gastrointestinal irritation
Magnesium	Muscle weakness
Potassium	Diarrhoea, nephrotoxicity
Zinc	Irritability, nausea

PREVIOUS RESEARCH INTO METAL ION CONTENT OF LANDFILL LEACHATE.

As indicated in the introduction, most research to date into leachate characterisation has concentrated on the organic constituents of landfill leachate, with metal ion content being added for completion of data. Legislation specifying tolerable limits of metal contents in leachate waste streams from landfills was introduced in North America in the early 1970's. This was well before effective legislation in the United Kingdom which, although provided with some control in 1974 by the enactment of the Control of Pollution Act, has only really addressed the problem of definable limits of toxic substances in water with the introduction of the Environmental Protection Act(1990), and the Water Act (1989) which defines responsibility for control.

Studies in North America have been conducted since the early 1970's, in an attempt to monitor and control the nature and effect of landfill leachate, whilst the United Kingdom has lagged behind in this respect. A summary of published data from investigations in North America, Europe and the United Kingdom is shown in Table 2, along with ranges of values with respect to metal ion content in leachate samples. The ranges for the United Kingdom and North America are comparable in terms of magnitude of metal ion concentrations, but it should be noted that variations in infiltration rates and climate may lead to the wider ranges of concentrations as seen in the data from North America. Therefore further interpretation may lead to false assumptions.

The behaviour of metal ion content within leachate is not fully understood, as the landfill acts as a dynamic environment in which the number of processes affecting the concentration and mobility of metals in the leachate operate. Attenuation of metal ion content is often related to a predominant process. (Bagchi 1987). These processes are summarised in Table 3.

Therefore the dominant process at a particular site within the landfill will control the relevant proportions of metal ions in the leachate. Whichever process operates depends on a number of factors including pH, temperature, and the presence of other metal ions.

Examination of the data in Table 2, shows that those elements with high mobility generally have the highest concentration whilst those having low mobility generally have the lowest

Table 2. Published data on landfill leachate composition with respect to metal ions. All quantities in mg/l.

Country	Site/Source	Na	Mg	K	Ca	Cr	Cd	Fe	Cu	Pb	Zn
U.K. 1	Compton Bassett	1320	159	858	94	0.06	0.01	9.8	0.17	0.13	0.62
U.K. 2	Bryn Posteg	904	85	195	477	0.1	0.1	242	0.13	0.1	4.9
U.K. 2	Harewood Whin	1150	820	900	2900	0.17	0.01	560	0.05	0.45	0.32
U.K. 2	Chapel Farm 1	1360	420	900	1040	0.05	0.02	340	0.04	0.2	0.09
U.K. 2	Chapel Farm 2	1210	310	840	1030	0.06	0.04	126	0.03	0.13	0.22
U.K. 3	Seamer Carr	2400	420	2050	4100	1	0.003	2050	0.05	0.61	130
U.K. 4	Brofiskin					1	0.2	28	0.04	0.1	0.2
U.K. 4	Maendy					0.2	0.2	300	0.2	0.1	10
U.K. 4	Eastfield Quarry					30		3000	5	3	50
U.K. 4	Rainham					0.3	0.05	1000	0.2	1	10
Switzer -land 5	Uttigen - Berne						2.78	9.0	0.25	0.12	0.71
U.S. 6	EPA	0 - 7700	16.5 - 15600	2.8 - 3770	5 - 4080			0.2 - 5500	0 - 9.9	0 - 5	0 - 1000
U.S. 6	Steiner	100 - 3800						200 - 1700			1 - 135
U.S. 6	Genetelli	450 - 500	160 - 250	295 - 310	900 - 1700		0.4	210 - 235	0.5	1.6	10 - 30
U.S. 6	Ham	106 - 357	22 - 62	107 - 242	111 - 245			20 - 60	0.04 - 0.1		0.04 - 0.5
U.S. 7	Illinois	1360	650	1140	3750	18	0.1	2200	0.48	1.25	104
U.S. 7	Wisconsin	1580	1140	2300	3900	0.5	0.375	1046	0.75	1.65	13
U.S. 7	Sonoma	240	515	74	1030	0.11	0.02	301	1.0	1.1	22
U.S. 7	Cincinnati	750	530	860	2325	0.2	0.36	735	0.71	3.12	52
U.S. 7	Atlanta	75	52.5	620	520	0.03	17	810	0.48	1.16	74
U.S. 7	College Park	37.5	51.5	35	215	0.015	0.01	560	0.15	0.21	6.5
U.S. 7	Winnetha	370	76	450	254	0.04	0.003	1.5	0.10	0.48	0.16
U.S. 8	Snohomish Co.							569	0.07		26.9
U.S. 8	King County							73	0.03		1.12
U.S.	**Range of Values**	37 - 7700	16.5 - 15600	2.8 - 3770	5 - 4080	0.015 - 18	0.01 - 17	0.2 - 5500	0.04 - 9.9	0.21 - 5	0.04 - 1000
U.K.	**Range of Values**	904 - 2400	85 - 820	195 - 2050	94 - 4100	0.05 - 30	0.01 - 0.2	9.8 - 3000	0.03 - 5	0.1 - 3	0.2 - 130
U.K.	**Completion criteria for leachate.**	1500	500	120	1000	0.5	0.05	2	1	0.5	1

1. (Robinson and Gronow. 1992)
2. (Robinson, Barr and Last. 1992)
3. (Barber and Maris. 1984)
4. (DOE. 1978)
5. (Baccini et al. 1987)
6. (Yong et al. 1992)
7. (Chan and DeWalle. 1975)
8. (Mott et al. 1987)

concentration in the leachate.

However, most of these studies have published data on the metal ion content, without reporting information on the factors which affect metal ion content such as age, pH, and type of fill. In addition it has been shown that certain metal ions are associative i.e. the presence of one metal ion indicates the prersence of another. For example Zinc and Lead often occur together, as do Copper and Cadmium. Christensen 1984, showed that the concentration of one of the ions can indicate the presence of the associative

Table 3. Attenuation and Mobility of metal ions in a landfill environment. (After Bagchi 1987).

Metal	Major attenuation mechanism	Mobility in clayey environment
Cadmium	Precipitation, Absorption	Moderate
Calcium	Precipitation, Ion Exchange	High
Chromium	Precipitation, Ion Exchange, Absorption	Low or High
Copper	Absorption, Ion Exchange, Precipitation	Low
Iron	Precipitation, Ion Exchange, Absorption	Moderate
Lead	Adsorption, Ion Exchange, Precipitation	Low
Magnesium	Ion Exchange, Precipitation	Moderate
Potassium	Absorption, Ion Exchange	Moderate
Sodium	Ion Exchange Dilution	High
Zinc	Absorption, Ion Exchange, Precipitation	Low

metal ion. This makes detailed comparison of information between sites problematic.

THE INVESTIGATION

The study reported in this paper includes the parameters of age, type of fill, pH and nature of disposal (in a contained or dilute and disperse site) analysed in conjunction with metal ion content of leachate, in order to obtain a chemical characterisation of landfill leachate. The study is an initial investigation (Alker, 1993), wherein samples of leachate from twenty-one landfill sites were collected and analysed between October 1992 and February 1993. Duplicate samples were taken in High Density Polyethylene 1 litre bottles, at each site. Samples were stabilised with respect to metal ions by the addition of 1 ml of Nitric Acid, immediately after collection. This was added to prevent the sorption of metal ions onto the sides of the container prior to analysis. Samples were taken from leachate chimneys (holding tanks or leachate drain outfalls) by bailer or bucket, at a point before the leachate was treated or pumped to sewer. The leachate was analysed using a Varian AA atomic adsorption spectrometer, in a reducing air acetylene flame. A Jenway flame photometer was used to analyse Sodium, Potassium and Calcium content of the leachate. The organic content of the leachate was not measured. The instruments were calibrated by standard techniques involving standard solutions of the metal ion under test. Results from the apparatus were presented to three decimal places, accuracy only limited by possible ionic interference which has not been accounted for in these results. The results here have usually been presented to a significant number of figures, when compared with completion criteria levels for leachate.

It is possible that variations in flow and infiltration at the collection point may lead to variations in metal content between individual samples. For example if there had been a period of heavy rain prior to collection of the sample in this study, it is possible that the metal ion content as analysed would be diluted relative to the average metal ion content for a particular site. Variations in rate of flow when leachate is being pumped from the site, may also lead to differences between two samples taken at the same time. In order to take account of any variation, two samples were collected as close together as possible, both samples then being analysed and the results presented being the average for the site. In most cases the results for the two samples correlated to at least the second decimal place. The aspect of variation due to flow and sampling technique practised requires further investigation.

Table 4 shows the raw data for concentration of metal ions as analysed from the sites, along with age, pH and type of fill. Type of fill proved difficult to quantify in this initial study, so is restricted to very broad subjective terms of domestic, industrial or mixed fill.

Initial investigations have shown some correlation's between factors, although full interpretation of the data is not yet completed. Most metals show no distinction in ranges of value with regard to type of fill from this study, however, in broad terms Iron appears to occur at higher concentrations at sites taking wholly domestic waste, whilst Chromium and Potassium appear to exhibit relatively low concentrations in ranges of values at sites taking wholly industrial wastes. From Table 4, it can be seen that those

Table 4. Composition of leachate samples from sites in the North West of England, with respect to metal ion content, sampled October 1992 - January 1993. All quantities in mg/l.

Site	C/D	Type	Age YRS	pH	Pb	Cu	Zn	Fe	Cr	Mg	Ca	K	Na	Cd
B F 1	D	D	11	8.3	0.06	0.02	0.01	7.59	0.33	63	60	80	607	0.001
B F 2	D	D	4	8.2	0.04	0.01	0.64	4.8	0.41	19	80	7	712	0.001
B F 3	C	D	0.75	7.6	0.11	0.01	1.15	58.6	0.52	262	58	420	1762	0.014
SWCS	D	D	9	5.5	0.16	0.04	0.04	9.7	0.39	407	490	930	2075	0.01
KLY	D	D	27	5.2	0.09	0.02	0.43	22.3	0.13	18	290	35	800	0.002
A B W	C	D	12	7	0.1	0.03	1.47	39.4	0.29	101	760	870	2088	0.006
A B P	C	D	4	7.6	0.1	0.03	0.34	7.9	0.32	163	625	900	2175	0.009
WIT	C	D	12	7.3	0.03	0.02	0.86	2.9	0.18	72	250	350	1975	0.003
St. H	C	D	17	6.9	0.08	0.02	0.06	2.18	0.26	29	220	55	720	0.005
AL	C	I	16	8.0	0.12	0.04	0.63	8.3	0.17	280	780	180	2020	0.014
H H	D	I	11	7.0	0.05	0.02	0.01	2.17	0.04	19	595	14	787	0.004
WK	D	I	11	8.5	0.08	0.02	0.14	6.16	0.09	22	570	100	2440	0.007
KDL	C	M	3	8	0.08	0.03	1.2	5.8	0.3	133	570	960	1080	0.008
RDCLF	D	M	30	6.8	0.13	0.03	0.1	6.7	0.14	58	390	370	1030	0.004
H P	D	M	25	6	0.12	0.03	0.15	9.4	0.16	171	490	50	1500	0.006
N L	D	M	19	5.6	0.03	0.02	0.09	4.4	0.1	28	450	76	610	0.002
L S V	C	M	19	5.9	0.02	0.02	0.11	9.4	0.34	25	870	39	400	0.003
H C 1	C	M	4	5.7	0.13	0.04	0.35	8.5	0.24	233	960	105	1640	0.009
H C 2	C	M	1	7.5	0.17	0.09	0.25	10.6	0.49	388	760	465	4880	0.025
NWLD	D	M	13	6.5	0.05	0.02	0.2	2.4	0.27	8.2	330	8	1812	0.001
W R C	C	M	10	8.1	0.09	0.17	6.39	7.1	0.35	46	595	10	450	0.02
Range of Values. Domestic		D	0.75 - 27	5.2 - 8.3	0.03 - 0.16	0.01 - 0.04	0.01 - 1.47	2.18 - 58.6	0.13 - 0.52	18 - 407	58 - 760	7 - 900	607 - 2175	0.001 - 0.014
Range of Values. Industrial		I	11 - 16	7.0 - 8.5	0.05 - 0.12	0.02 - 0.04	0.01 - 0.63	2.17 - 8.3	0.04 - 0.17	19 - 280	570 - 780	14 - 180	787 - 2440	0.004 - 0.014
Range of Values. Mixed		M	1 - 30	5.6 - 8.1	0.02 - 0.17	0.02 - 0.17	0.09 - 6.39	2.4 - 10.6	0.1 - 0.49	8.2 - 388	330 - 960	8 - 960	400 - 4880	0.001 - 0.025
Range of Values. Total				5.2 - 8.5	0.02 - 0.17	0.02 - 0.17	0.01 - 6.39	2.17 - 58.6	0.04 - 0.49	8.2 - 407	58 - 960	7 - 960	607 - 4880	0.001 - 0.025

Note : C/D : Refers to Contained site (C) or Dilute and Disperse site (D).
Type - Type of Fill : D = Domestic, I = Industrial, M = Mixed. (These characteristics are based on subjective analysis from information from site on type of fill and as such are not quantifiable).
Metal Deposits in waste at particular sites have been recorded for further analysis.(Alker 1993).

sites taking mixed wastes also have wider ranges of values for metal ion content, reflecting the 'mixed' nature of the waste.

METAL ION CONTENT AND AGE OF LANDFILL.

The mobility of metal ion, along with the type of fill placed affects the concentration of metal ions recorded. It is difficult to correlate the behaviour of metal ions with age of landfill from this study, as only one visit to a particular site was conducted. The picture is complicated by not knowing the 'original' composition or metal content of the refuse placed at each individual site, i.e. there is no point of reference. In addition over the past 30 years there have been significant changes in the composition of both domestic and industrial refuse,

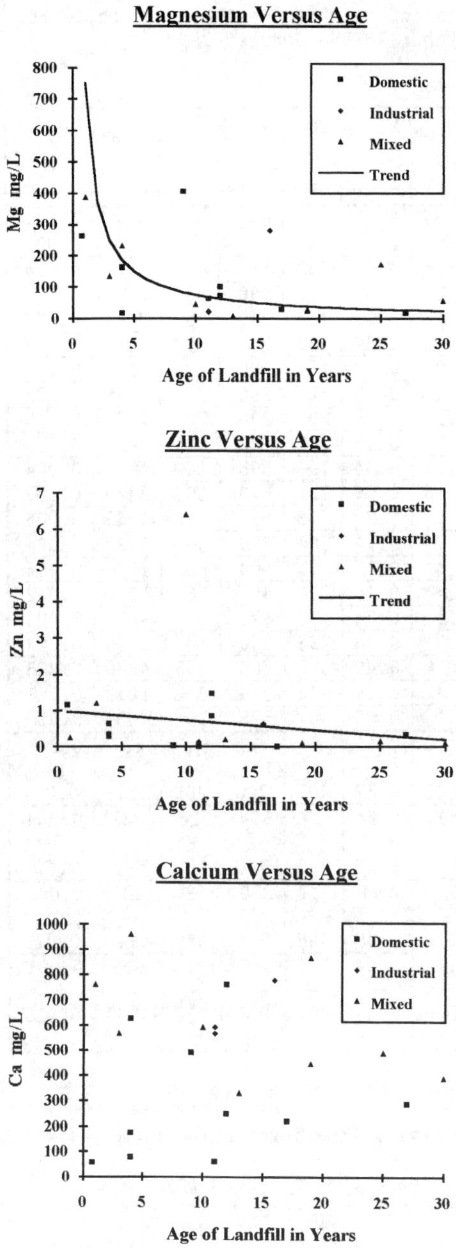

Fig. 1. Metal ion content versus age.

brought about by improvements in living standards, development of the 'throw-away' culture, the decline of traditional industries etc. Thus the data from sites which have different ages by virtue of being constructed at significantly different times in the past may not be compatible.

In Figure 1, which shows the Concentration of Magnesium and Zinc with age, there is a general reduction of metal ion with time. This behaviour agrees with that reported by Farquhar (1989). Some metals as illustrated by Calcium in Figure 1 do not exhibit this trend of decreasing metal ion concentration with time, and this may be linked to the mobility of the metal ion. Further long term investigation is needed to see how metal ion content varies with time at a particular site.

METAL ION AND pH.

Fig. 2 contains data for metal ion content in the leachate with respect to pH, in terms of Magnesium, Cadmium, Chromium and Sodium. All the figures show a general increase in metal ion content with increased acidity or alkalinity, decrease or increase in pH away from neutral. The graphs illustrate a dip in concentrations at pH of 6.5. This corresponds approximately to a neutral pH of 7, when considering that the litre samples were tested after addition of 1 ml Nitric Acid used to stabilise the metal ion content. This would cause a slight shift in results to the acid side of neutral, but cannot be mathematically corrected for in this survey as the 1 litre samples of leachate were not accurately measured in the field. The distribution of metal ions with pH agrees with Chan and Dewalle (1975). For these particular metals depicted in Fig. 2, type of fill does not appear to affect the metal concentration, as points for various types of fill lie both in the low and high pH ranges.

However, on investigation of Sodium versus pH, which is highly mobile, Sodium content is only marginally affected by pH. This graph shows an interesting phenomena possibly related to fill type. The domestic and industrial sites tend to be grouped to the alkaline side of neutral, whilst the mixed sites are generally more acidic. This could reflect the type of waste in a site, but further illustrations using ion content are needed which are not as highly mobile as Sodium, in order to be certain that this could be used as an indicator of type of fill. This aspect needs to be considered further when looking at factors which affect leachate characterisation.

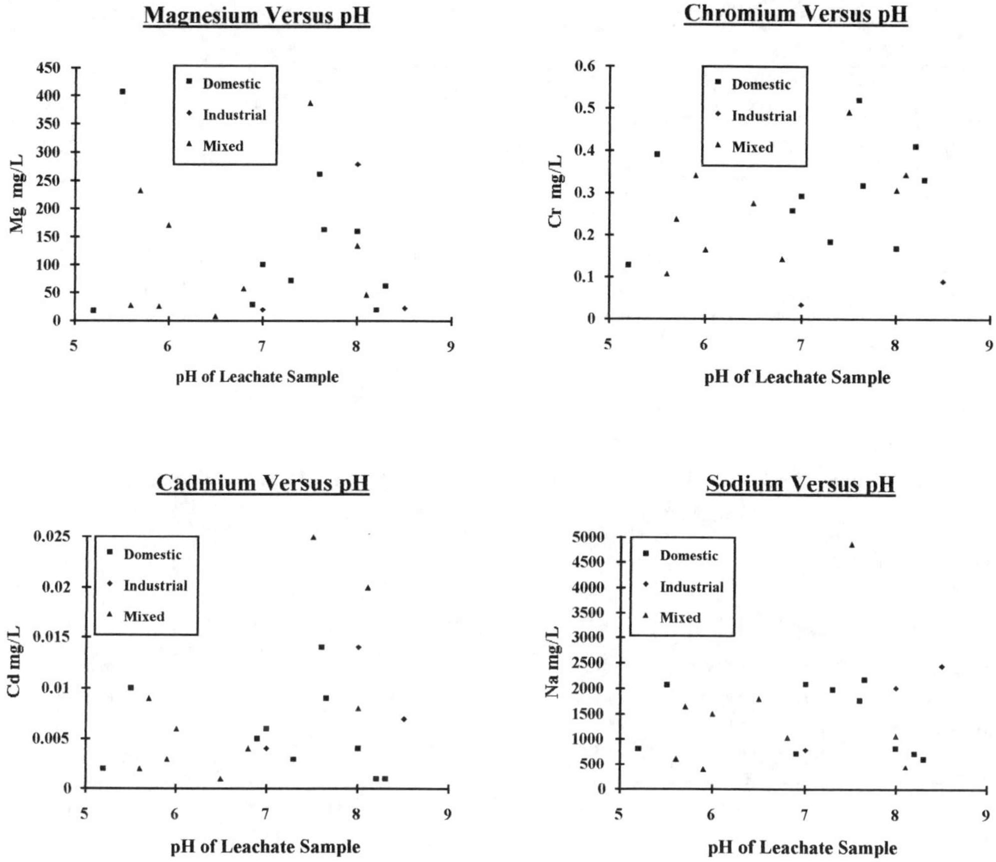

Figure 2 : Metal ion content versus pH.

ASSOCIATION OF METAL IONS.

Figure 3 Shows good correlation between Lead / Cadmium (66.6%), Copper / Cadmium (75.7%), Copper / Lead (37.75%), and Copper / Chromium (26.6%), metal concentrations found in landfill leachate samples collected in this study. This is to be expected from the chemical relationships of metal ions which mean that the presence of one metal indicates the likely presence of another. It has been reported by Christensen (1984), using linear correlation techniques, who also showed significant correlation's between Lead / Cadmium (63%), Copper / Cadmium (77%), Copper / Lead (64%), and Copper / Chromium (62%), metal concentrations found in landfill leachate samples. The percentages have been calculated by correlation of the data for relevant metal ions using a least squares regression technique to produce a line of best fit. The data used in the correlation was not corrected for any error in analysis as previously discussed. Low correlation in data from this study with respect to Copper/Lead and Copper/Cadmium concentrations as compared to Christensen's results, can be seen from the graphs in figure 3, to be caused by data from two individual sites in the survey, skewing the relationship. This aspect requires further investigation as to the cause of this discrepancy.

SUMMARY.

In the United Kingdom there is little published data on the composition and characterisation of landfill leachate. The Environmental Protection Act 1990, puts the onus of completion of a site onto the site owner, with strict limits on the composition of leachate in terms of components which may not have been investigated in depth before i.e. metal ions. Control measures required for

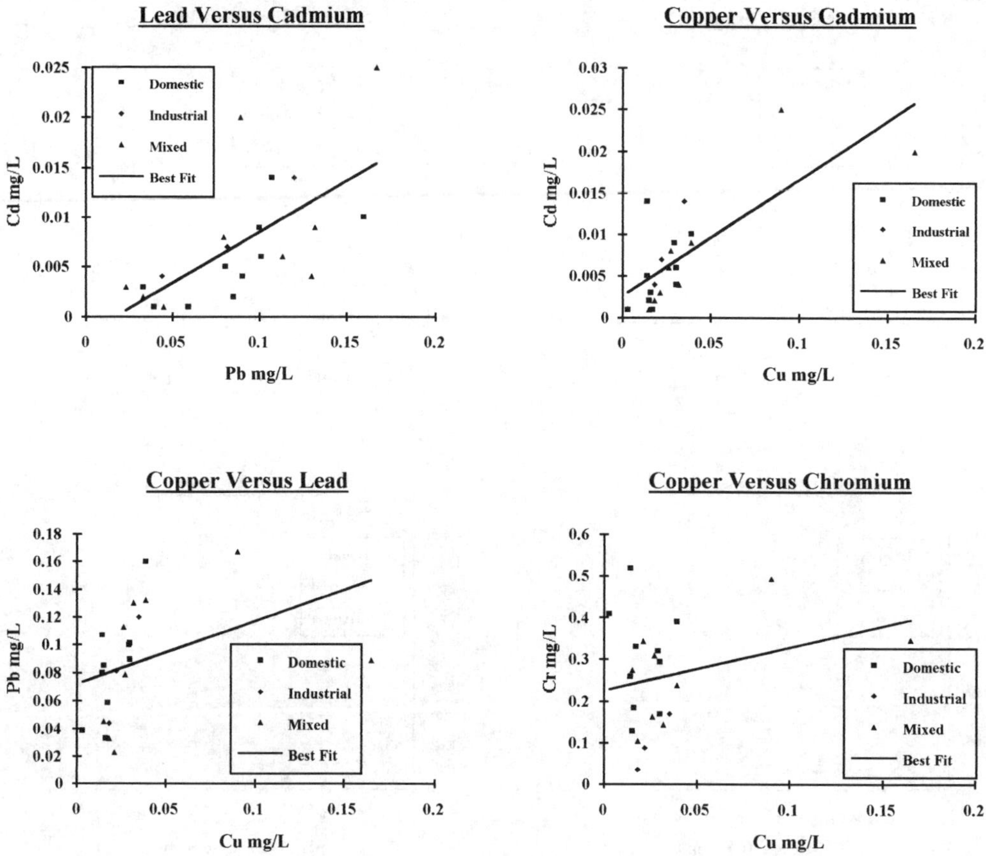

Figure 3 : Metal ion content versus presence of other metal ions.

landfill leachate cannot be fully designed and appraised until the full extent and nature in terms of composition of landfill leachate are fully understood.

Many studies that have been published have tended to concentrate on just a few landfill sites. This is understandable when one considers the many variables associated with landfill sites and generation of leachate. In particular age, pH and the association of metal ions is shown to affect leachate composition. The concentration of these metal ions in the environment has also been shown to produce toxic effects in living organisms, so monitoring and furthur measures may be required to limit the amount of these metal ions generated by leachate, which form a potential threat if released into the environment.

Leachate composition changes with time and continued study of sites selected in this initial survey is required so that leachate composition over time can be more accurately investigated.

REFERENCES.

Alker, S.C. 1993. "An investigation into the characterisation of landfill leachate with respect to metal ion content". M.Sc. Thesis. Bolton Institute.

Baccini, P.; Hensler, G.; Figi, R.; and Belevi, H.; 1987. "Water and element balances of municipal solid waste landfills". Waste Management and Research. Vol. 5 pp. 483-499.

Bagchi, A.; 1987. "Natural attenuation mechanisms of landfill leachate and effects of various factors on the mechanisms". Waste Management and Research. Vol. 5 pp. 453-464.

Beveridge, A.; Pickering, W.F.; 1983. "The influence of sufacants on the adsorption of heavy metals by clays". Water Research Vol. 17 pp. 215-225.

Barber, C.; Maris, P.J.; 1984 "Recirculation of leachate as a landfill management option: benefits and operational problems" Q.J.Eng. Geol. Vol. 117 pp. 19-29.

Chan, E.S.K.; DeWalle, F.B.;1977. "Evaluation of Leachate Treatment. Vol. 1- Characterisation of Leachate". USEPA-600/2-77-186a.

Christensen, T.H.; Tjell, J.C.; 1984. "Leaching form land disposed municipal compost: 4 Heavy Metals". Water Management and Research. Vol. 2 pp. 347-357.

D.o.E. 1978. "Hazardous wastes in landfill sites". Final Report. HMSO.

Farquhar, G.J.; 1989. "Leachate : production and characterisation". Can. J. Civ. Eng. Vol. 16 pp. 317 - 325.

Mott, H.V.; Hartze, K.E.; Yonge, D.R.; 1987. "Metal precipitation in two landfill leachates". ASCE. J. Env. Eng. Vol. 113 pp. 476-485.

Robinson, H.D.; Barr, M.J.; Last, S.D.; 1992. "Leachate collection, treatment and disposal". J.Inst. Waste Env. Management. pp. 321-331.

Robinson, H.; Gronow, J.; 1992. "Groundwater protection in the U.K.: Assessment of landfill leachate source term". J. Inst. Waste Env. Management. pp. 229-236.

Yong, R.N.; Mohammed, A.N.O.; Warkentin, B.P.: 1992. "Principles of contaminant transport is soils". Elsevier Science Publishing BV.

Waste Disposal by Landfill – GREEN'93, Sarsby (ed.) © *1995 Balkema, Rotterdam, ISBN 90 5410 356 6*

Principles of species transport in saturated soils under an electric field

Akram N. Alshawabkeh & Yalcin B. Acar
Louisiana State University, Baton Rouge, La., USA

ABSTRACT: 'Electrokinetic remediation' or 'electrokinetic soil processing' is an emerging technique of extracting chemical species from soils. Based upon the findings of the recent studies in electrokinetic remediation, the chemistry, principles and relative significance of the different transport processes in soils under electric fields are reviewed and discussed.

1 INTRODUCTION

Recent studies in ecosystems have indicated that many areas near urban complexes, metalliferous mines or major road systems contain anomalously high concentrations of heavy metals such as Pb, Cd, Hg, As. Though natural concentrations of such metals in soils are not high, contaminated soils exist due to various reasons including the disposal of urban and industrial wastes, the mining and smelting of non-ferrous metals, and the unauthorized unengineered dumping or disposal of items containing metals, ranging from miniature dry-cell batteries (Ni, Cd and Hg) to abandoned cars and car components, such as Pb-acid batteries. The impact of soil contamination with heavy metals on ground water resources is becoming increasingly significant as the disclosed number of unengineered containment facilities and contaminated sites grow and remediation cost increase.

In the USA, CERCLA (or Superfund) in 1980 has required USEPA to develop criteria for determining priorities among sites that require taking remedial action and to develop and maintain a National Priority List (NPL). Under USEPA regulations, these sites are eligible for remedial action. Though only 149 sites have been cleaned and removed from the NPL since Superfund's inception, the NPL list continues to grow. Recently (1993), USEPA added 33 new sites bringing the NPL total to 1250 sites.

A variety of options may exist in selection of a cleanup remedy at a site, however the efficiency and cost of these options vary widely. Although conventional ground burial and land disposal techniques are often economical, they do not provide the best solution, and in some cases they are not necessarily the most effective solution. There exists an everlasting need to introduce new, innovative, and preferably, in-situ remediation technologies. Several such schemes introduced in the last decade include: incineration, solidification, stabilization, bioremediation dechlorination, flushing, vitrification, washing, thermal desorption, vacuum extraction and chemical treatment (EPA 1991). Each technology exhibits certain advantages and limitations in remediating a myriad of organic/inorganic contaminants encountered in different types of soil deposits. There is not one specific technology that can be considered as a panacea to all types of contaminants and soil deposits.

A major limitation of the most successful remediation technologies is that they are restricted to soils with high hydraulic conductivity and, hence, cannot be used for fine-grained deposits. Furthermore, generally they are not specifically effective in removing contaminants adsorbed on the soil particles. Such adsorption may pose a threat for groundwater and plant contamination. Currently, there is no widely accepted remediation or treatment technology for most hazardous waste sites contaminated with heavy metals. The most common treatment method for such contamination is either containment or disposal.

The challenging demand to develop new, innovative and cost-effective in-situ remediation technologies in waste management stimulated the drive to employ conduction phenomena under electric currents as a soil remediation technology (Acar et al. 1989; Lageman 1989; Acar 1992; Acar and Gale 1992; Acar et al. 1993a; Pamukcu and Wittle 1992; Shapiro and Probstein 1993). The technique uses low-level DC electric potential differences (in the order of few volts per cm) or an electrical current (in the order of mA/cm^2 of cross sectional area) applied through inert electrodes placed within boreholes in an open flow arrangement. The groundwater in the boreholes or an externally supplied fluid is used as the conductive medium. The low-level DC across electrodes placed in a soil mass will cause physico-chemical and hydrological changes in the soil system leading to different electrokinetic processes and species transport to respective electrodes. Extraction and removal is accomplished either at the electrodes or via an external conditioning system. This paper presents the principles of the species transport and removal in soils under an electric field.

2 PRINCIPLES

The applied electrical current (or electric potential difference) leads to electrolysis reactions at the electrodes generating an acidic medium at the anode and an alkaline medium at the cathode. The acid generated at the anode will advance through the soil towards the cathode. This advance is governed by different transport mechanisms including ionic migration due to electrical gradients, pore fluid advection due to prevailing electro-osmotic flow, pore fluid flow due to any externally applied or internally generated hydraulic potential differences, and diffusion due to generated chemical gradients. The alkaline medium developed at the cathode will advance towards the anode by diffusion and ionic migration; however, the counterflow due to electro-osmosis will retard this back diffusion and migration. As a consequence, the acid front will dominate the chemistry across the specimen unless it is significantly buffered and the transport of the front is retarded by the soil in between the electrodes.

This acidification will facilitate desorption of cations from the surfaces of soil particles. Contaminants present in the soil pore fluid or desorped from the soil surface will be transported towards the electrodes depending on their charge. The driving mechanisms for contaminant transport will be identical to those prevailing in the acid-base transport. As a result, cations will be accumulated at the cathode and anions at the anode. Bench-scale studies have shown that heavy metals and other cationic species can be removed from soils with the effluent, they may be electrodeposited or precipitated at the cathode. Treatment of the effluent (such as the use of ion exchange or resin columns) may be used for removal of the different species.

2.1 Electrolysis Reactions

Application of direct electric current through electrodes immersed in water causes electrolysis reactions in the immediate vicinity of electrodes. Oxidation at the anode generates an acid front while reduction at the cathode produces a base front due to the following electrolysis reactions,

$$2H_2O - 4e^- \Rightarrow O_2 \Uparrow + 4H^+ \quad E_o = -1.229 \text{ (anode)} \quad (1)$$

$$2H_2O + 2e^- \Rightarrow H_2 \Uparrow + 2OH^- \quad E_o = -0.828 \text{ (cathode)} \quad (2)$$

where E_o is the standard reduction electrochemical potential which is a measure of the tendency of the reactants in their standard states to proceed to products in their standard states. Secondary reactions may exist depending upon the concentration of available species, e.g.:

$$H^+ + e^- \Rightarrow (1/2)\, H_2 \Uparrow \quad (3)$$

$$M_e{}^{n+} + n\, e^- \Rightarrow Me \quad (4)$$

$$M_e(OH)_n\,(s) + ne^- \Rightarrow M_e + n\, OH^- \quad (5)$$

and M_e refers to metals.

The type of electrolysis reactions ongoing at the electrodes depends upon the availability of chemical species and the electrochemical potential of these reactions. Although some secondary reactions might be favored at the cathode because of their lower electrochemical potential, the water reduction half reaction (H_2O/H_2) is dominant at early stages of the process. Figure 1 presents development of the acid/base profile across the soil in a 10 cm length cylindirical kaolinite specimen.

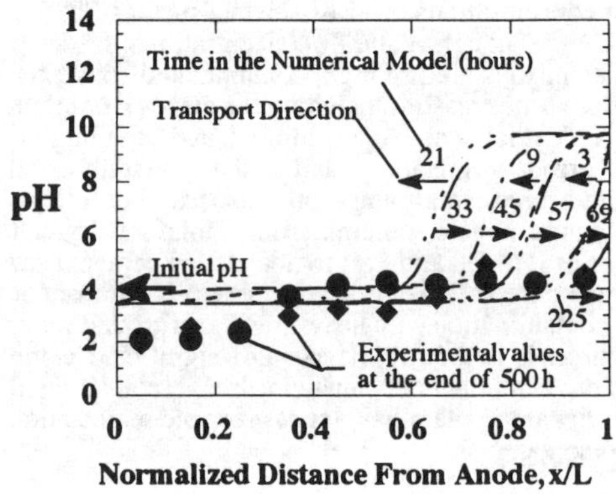

Figure 1. Development of pH profiles across a 10 cm kaolinite specimen (Acar and Alshawabkeh 1994)

Electrolysis reactions at early stages of the process will drop the pH at the anode to below 2 and increase it at the cathode to above 12 depending upon the total current applied (Acar et al. 1989; 1990a and 1990b). At later stages, the acid front advances towards the cathode carrying H^+ and the cationic contaminants and the half cell reactions (H^+/H_2) or [$Me^{+n}/Me(s)$] are expected to dominate.

2.2 Changes in Soil pH

The acid generated at the anode will advance through the soil towards the cathode. This advance is governed by different transport mechanisms including ionic migration due to electrical gradients, pore fluid advection due to the prevailing electroosmotic flow, pore fluid advection due to any externally applied or internally generated hydraulic potential differences, and diffusion due to concentration gradients.

The alkaline medium developed at the cathode due to evolution of OH^- will advance towards the anode by the diffusion and ionic migration; however, the counterflow due to electroosmosis will retard this back-diffusion and migration. Thus, the advance of this front towards the anode will be much slower than the advance of the acid front towards the cathode in case the soil does not have a high buffering capacity. In this case, the acid front will dominate the chemistry across the specimen (Acar et al. 1990a; Alshawabkeh and Acar 1992; Acar and Alshawabkeh 1993, 1994). The decrease in pH value in the soil depends on the amount of acid generated at the anode (Acar et al. 1990a; 1990b) and the buffering capacity of the clay (Harter 1983). Yong et al. (1990) investigates the buffering capacity of various types of clays and demonstrate that the cation exchange capacity and the organic content highly influence the buffering capacity and soil pH. Kaolinitic clays are demonstrated to show much lower buffering capacity due to lower cation exchange capacity (cec) compared to other clay minerals, such as montmorillonite or illite.

2.3 Sorption

Heavy metals and other positively charged species are highly attracted and sorped on the negatively charged clay surfaces. Metals have different sorption characteristics and mechanisms that are also dependent upon the adsorbents. Sorption mechanisms include surface complexation (adsorption) or ion exchange. Although the selectivity sequence is generally a function of size and valence of the cation, it is also affected by the type of clay mineral as depicted in Table 1.

Table 1. Selectivity order for some minerals (Alloway 1990)

Adsorbent	Selectivity Order
Montmorillonite(Na)	Ca>Pb>Cu>Mg>Cd>Zn>Ni
Illite	Pb>Cu>Zn>Ca>Cd>Mg
Kaolinite	Pb>Ca>Cu>Mg>Zn>Cd
Smectite, vermiculite	Zn>Cd>Mn>Hg

Desorption of heavy metals from clay surface is essential in extraction of species from contaminated fine-grained deposits. Acidification of the pore fluid by the electrolysis reaction at the anode and its transport into the soil mass are the fundamental mechanisms that will assist in desorption of these species. The adsorption/desorption mechanism depends upon the surface charge density of the clay mineral, characteristics and concentration of the cationic species, and existence of organic matter and carbonates in the soil. The mechanism is also pH dependent. An increase in H^+ concentration associated with a decrease in soil pH results in desorption of cations by an amount controlled by the soil type (Maguire et al. 1981; Harter 1983; and Yong et al. 1990).

2.4 Precipitation/Dissolution

Dramatic changes in the soil electro-chemistry throughout electrokinetic soil remediation lead to different chemical reactions including precipitation/dissolution of salts and soil minerals. Species transport in soil pore fluid is highly influenced by the formation and dissolution of these precipitates.

The base front generated by electrolysis at the cathode will cause precipitation of most heavy metals and radionuclides at their hydroxide solubility values (Acar and Alshawabkeh 1994). The amount of precipitate differs from one species to another and it is highly dependent on the soil and pore fluid pH. Though the advance of the acid front generated at the anode will cause dissolution of most of the encountered precipitates, each precipitation or dissolution reaction should be treated independently depending upon its solubility product constant.

Acar et al. (1992) and (1993a) recommend the use of different enhancement techniques in order to remove these precipitates from the cathode compartment. The advance of the acid front generated at the anode is expected to cause dissolution of clay minerals. Kaolinite dissolution, which is pH dependent, generates different chemical forms of aluminum and silica. The impact of mineral dissolution on the efficiency of electrokinetic soil remediation is yet to be investigated.

2.5 Contaminant Transport, Capture, and Removal

Free chemical species present in the pore fluid and/or desorped from the soil surface will be transported towards the electrodes depending upon the polarity of their charge. The primary driving mechanisms for species transport are the same as the acid or base transport mechanisms. Ionic migration, advection, together with diffusion will contribute to the movement of species through the soil mass.

At zones of high pH values, both precipitation and sorption will retard species transport. Figure 2 conceptualizes the processes affecting transport of a

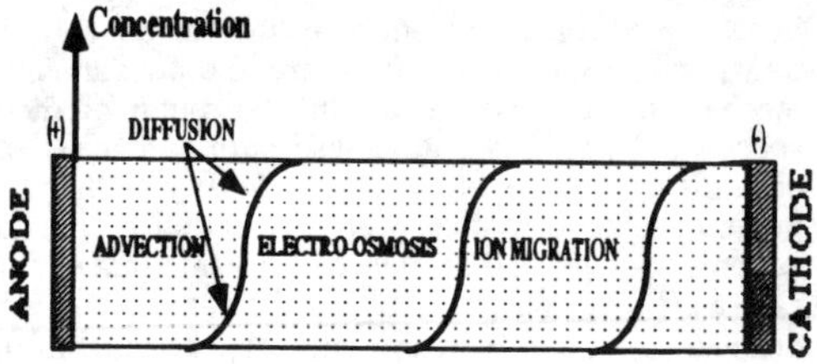

Figure 2. Cation transport processes under an Electric Field

positively charged species.

As a result of the above mechanisms, cations will be captured at the cathode and anions at the anode. Heavy metals and other cationic species will be removed from the soil either with the effluent or they will be deposited at the cathode. The effluent could be treated (such as the use of ion exchange or resin columns) for removal of the excess ions.

3 TRANSPORT PROCESSES

Attempts to study transport processes in fine-grained soils under electric currents demonstrate that ionic migration, electroosmotic advection and diffusion contribute in various degrees to mass transport of heavy metals in the soil pore fluid (Hamed 1990; Hamed et al. 1991; Acar et al. 1992; Pamukcu and Wittle 1992; Acar et al. 1993b). These processes are controlled by several factors including soil mineralogical composition, porosity and tortuousity, electrochemical properties of the chemical species such as ionic mobilities and diffusion coefficients, and pore fluid electrolyte concentration and ionic strength.

3.1 Diffusional Flux

The diffusive mass flux of the j th chemical species, J_j^d, under a chemical concentration gradient is described by Fick's first law,

$$J_j^d = D_j^* \nabla(-c_j) \qquad (6)$$

where D_j^* is the effective diffusion coefficient and c_j is the molar concentration of the jth species, respectively. The effective diffusion coefficient in the porous medium is related to the diffusion coefficient in free solution at infinite dilution, D_j, by (Alshawabkeh and Acar 1992; Shackelford and Daniel, 1991a),

$$D_j^* = D_j \tau n \qquad (7)$$

where n is the soil porosity, varying over a range of 0.1 to 0.7 for fine-grained soils, and τ is a factor representing the effect of tortuosity of the medium. Values for tortuosity factor span over a wide range for different types of soils. The tortuosity values reported by different investigators are as low as 0.01 and as high as 0.84, mostly ranging between 0.20 to 0.50 (Shackelford 1991).

Shackelford and Daniel (1991b) investigate the factors affecting the effective diffusion coefficients, D*, of different inorganic chemicals in compacted fine-grained soils. The molding water content and compaction methods have little effect, if any, on the effective diffusion coefficients. Generally, changing molding water content in compaction results in changes in the soil fabric (Mitchell 1993). According to the results by Shackelford and Daniel (1991b), the change in compacted clay fabric due to different molding water contents have little effect on the effective diffusion coefficient of different chemicals. Consequently, in electrokinetic soil remediation the diffusive mass flux due to a specific chemical gradient is independent of the soil fabric and dependent on soil tortuosity and porosity. The effect of mass diffusion on the overall mass transport is expected to be distinct at the sharply changing pH and concentration fronts prevailing in electrokinetic remediation (Figure 1).

3.2 Migrational Flux

The migrational mass flux of the free ionic species present in the soil pore fluid due to the applied electric field is given by,

$$J_j^m = u_j^* c_j \nabla(-E) \qquad (8)$$

where J_j^m is the migrational mass flux of the jth species, u_j^* is the effective ionic mobility of the jth species. The effective ionic mobility, u_j^*, define the velocity of the ion in soil pores under unit electric field. There is no sound method yet devised to measure the effective ionic mobility (Koryta 1982); however, u_j^* can be theoretically estimated by assuming that Nernst-Townsend-Einstein relation between the molecular diffusion coefficient D_j and u_j, holds true for ions in the soil pore fluid (Holmes 1962);

$$u_j^* = u_j \tau n = (D_j^* z_j F)/RT \qquad (9)$$

where z_j is the charge of the jth species, F is Faraday's constant (96,485 C/mol electrons), R is the universal gas constant (8.3144 J/K mol), and T is the absolute temperature. The effective ionic mobility is a function of the effective diffusion coefficient and species electric charge. Consequently, the effective ionic mobility of a specific ion is dependent on its molecular diffusion coefficient, soil porosity and tortuousity.

A comparison of values of D_j and u_j for some ionic species are presented in Table 2. The relatively high values for H^+ and OH^- are noted. These ions have specifically high values due to their rapid dissociation and association with water. When the value of (F/RT) is substituted in Equation 9, the ratio of the ionic

Table 2. Diffusion coefficients and ionic mobilities for selected ions

Ion	D_j X 10^6 cm^2/s	u_j X 10^6 cm^2/V-s	u_j^*X10^6 cm^2/V-s
H^+	93	3625	760
Na^+	13	519	109
Ca^{+2}	8	617	130
Pb^{+2}	9	736	155
OH^-	53	2058	432
F^-	15	564	118
NO_3^-	19	740	155
SO_4^{-2}	11	413	87

mobility of a specific charged species under a unit electrical gradient to the diffusion coefficient of the same species is about 40 times the charge on the species (40 z_j per volt). Table 2 also presents the corresponding effective ionic mobilities of these species in a soil with a typical porosity of 0.6 and a tortuosity factor of 0.35. The ionic mobility is at least an order of magnitude higher than the diffusion coefficient. This component of the transport process is a significant factor on the efficiency of the electrokinetic remediation process.

3.3 Electroosmotic Advective Flux

Mass flux of the jth chemical species, J_{je}, due to the advective electroosmotic water flow is evaluated by,

$$J_j^e = (c_j/c_w)\, q_e \tag{10}$$

where q_e is the electroosmotic water flux, and c_w is the molar concentration of water (≈ 1).

The theory of Helmholtz-Smoluchowski for electroosmosis has been widely used as a theoretical description of water transport through soils due to electrical gradients (Mitchell 1993). This theory introduces the coefficient of electroosmotic permeability, k_e, as the volume rate of water flowing through a unit cross sectional area due to a unit electrical gradient (cm^2/Vs). Hence, the electroosmotic flux, q_e, is expressed by an empirical relation,

$$q_e = k_e\ \nabla(-E) = k_i\, I \tag{11}$$

where E is the electric potential (V), I is the electric current density (A/cm^2), and k_i is the coefficient of electroosmotic water transport efficiency (cm^3/A s) evaluated by,

$$k_i = k_e/\sigma^* \tag{12}$$

where σ^* is the effective bulk electrical conductivity of the soil (siemens/cm).

The value of k_e is widely accepted to be a function of zeta potential, viscosity of the pore fluid, porosity and electrical permittivity of the soil medium. When the soil pores are treated as capillary tubes, the coefficient of electro-osmotic permeability is given by

$$k_e = \frac{\varepsilon\,\zeta}{\eta}\, n \tag{13}$$

where ε is the permittivity of the medium, ζ is the zeta potential, n is the porosity, and η is the viscosity. Extensive research has been carried out on the zeta potential of the glass-water interface. Hunter (1981) displays the effect of pH and ion concentration in the pore fluid on zeta potential. Figure 3 shows that zeta potential decreases linearly with logarithm of electrolyte concentration (Kruyt 1952; Hunter 1981) and/or the pH of the soil medium.

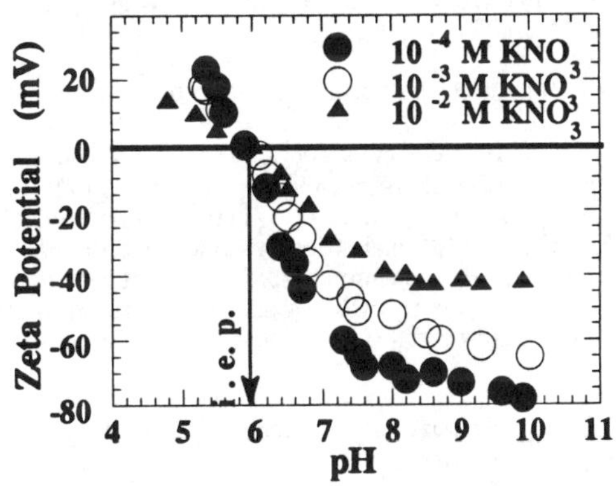

Figure 3. The effect of pH and ion concentration on zeta potential of colloidal TiO_2 (iep-isoelectric point) (Hunter 1981)

The effect of electrolyte chemistry on zeta potential could therefore be represented by an empirical relationship (Kruyt 1952)

$$\zeta = A - B \log C \tag{14}$$

where A, and B are two constants that are evaluated experimentally, and C is the total concentration of the electrolyte presented.

3.4 Migration versus Electroosmosis

While fluid flow under hydraulic gradients is significantly affected by the soil macrostructure (Acar and Olivieri 1989), electroosmotic flow under electric gradients is dependent mainly on porosity and zeta potential and independent on pore size distribution and the presence of macropores. Therefore, electroosmosis is an efficient method to generate a uniform fluid and mass transport in fine grained deposits. Furthermore, presence of electric gradients will result in a uniform charge transport and consequently mass transport in the soil pore fluid. The relative contribution of electroosmosis and ion migration to the total mass transport varies for different soil types and conditions. Similar to Peclet number in solute transport (Acar and Haider 1990), which defines the ratio of advective mass transport to diffusive mass transport could be introduced as a dimensionless number which would define the relative contribution of migrational mass flux to the electroosmotic mass flux,

$$(J_j^m/J_j^e) = (u_j^*/k_e) \qquad (15)$$

Bench-scale studies conducted on electroosmotic soil consolidation demonstrate that values of k_e are in the order of 10^{-5} cm^2/V sec, up to 11.4 x 10^{-5} cm^2/V s for a Na-kaolinite sample with 92% water content (Ballou, 1955).

Recent studies on decontamination of heavy metals from soils using electrokinetics display values in the order of 10^{-6} cm^2/V s for k_e. Table 3 presents a summary of such values reported by Hamed (1990) in investigating removal of Pb(II), Cd(II), and Cr (III). Hamed (1990) also demonstrates that for bench-scale tests, electroosmotic fluid flow would prevail only within the first 100 hours of processing in 10 cm specimens and this flow decreases with time.

Table 3. Relative significance of migration with respect to electroosmosis

Species	Initial Concent. μg/g	Current Density μA/cm^2	Min./Max. k_eX10^6 cm^2/Vs	u_i^*/k_e
Pb^{2+}	125	37	1.0-2.0	77-155
Cd^{2+}	100	37	0.7-8.0	15-165
Cr^{3+}	120	37	0.3-8.0	18-485

Typical values for (u_j^*/k_e) attained from the results of Hamed (1990) are also displayed. These values show that ionic migration is the primary transport process for ionic species. Even when electroosmotic flow is maximum, ionic mobility is at least one order of magnitude higher. It is hypothesized that the drop in pH of the soil due to electrokinetic processing causes the decrease in the coefficient of electro-osmotic permeability associated with the drop in zeta potential; hence, the electroosmotic flow is observed to decrease and eventually stop at later stages of the process. The results of Acar et al. (1989), Hamed et al. (1991), Acar et al. (1993), and Acar et al. (1992) demonstrate this decrease and cessation of electroosmosis at later stages of the testing. In fact, when the species concentration is high, very little electroosmotic transport occurs. In such cases ionic migration is at least two orders of magnitude higher.

3.5 Transference number

Thus far it has been demonstrated that ionic migration is the most significant transport mechanism for removing heavy metals and charged species from soils by electrokinetics. It then becomes necessary to investigate the factors that affect transport of a specific species by ion migration.

When we assume that the current is a result of only ion migration in the free pore fluid (neglecting ion migration in the diffuse double layer, or surface conductance, and assuming the soil particles to be electric isolators; i.e., no conductance through the soil solids), then the total current can be related to the migrational mass flux of species by using Faraday's law for equivalence of mass flux and charge flux,

$$I = \Sigma z_j F J_j \qquad (16)$$

The electric current is related to the applied electric potential gradient by Ohm's law,

$$I = \sigma^* \nabla (-E) \qquad (17)$$

where σ^* is the effective electric conductivity of the medium. Substituting Equation 17 into Equation 16, the effective electric conductivity of the soil is evaluated by,

$$\sigma^* = \Sigma z_j F u_j c_j \qquad (18)$$

Similar to the total current density, the partial current density that represents the contribution of the jth charged species to the overall current density, I_j is evaluated by applying Faraday's law of the equivalence of mass and charge,

$$I_j = z_j F J_j = z_j F u_j^* c_j \nabla (-E) \qquad (19)$$

Constant current is used in most experimental work at

Louisiana State University. Therefore, electric gradients will change depending upon the electrical conductance of the soil,

$$(-\nabla E) = (I / \sigma^*) \qquad (20)$$

Substituting Equation (20) in Equation (19),

$$I_j = \frac{z_j F u_j^* c_j}{\sigma^*} I = t_j I \qquad (21)$$

where t_j is the transport (or transference) number of the ion j, giving the contribution of the jth ion to the total effective electric conductivity, i.e.,

$$t_j = \frac{z_j F u_j^* c_j}{\sigma^*} = \frac{z_j F u_j^* c_j}{\Sigma z_j F u_j c_j} \qquad (22)$$

Consequently, the mass flux of the jth ion due to ion migration is,

$$J_j = \frac{u_j c_j}{\sigma^*} I = t_j \frac{I}{z_j F} \qquad (23)$$

The transport number of the j th charged species depends on its ionic mobility, concentration, and the total electrolyte concentration (or the electrolyte ionic strength) in the pore fluid. The summation of transport numbers of all ions in the soil pore fluid should be equal to one,

$$\Sigma t_j = 1 \qquad (24)$$

The transference number of a species will increase as the ionic concentration of that specific species increases. This implies that when the concentration of a species decreases relative to the total electrolyte concentration in the pore fluid, its transport and removal under electrical currents will be less efficient. Therefore, it is reasonable to assume that the efficiency of removal of a specific species will decrease in time as its concentration in the pore fluid decreases.

The transport of the acid generated at the anode would generally aid desorption of species. However, when this process is considered in conjunction with the transport of a species of interest, the substantial increase in hydrogen ion concentration will result in a complementing increase in the hydrogen transference number decreasing the transference number of other species. If hydrogen ion generation at the anode is not controlled, most of the energy may be consumed by the generation and transport of hydrogen ion across the cell rather than the transport of species of interest.

4 SUMMARY AND CONCLUSIONS

Electric fields applied across a saturated soil mass result in electrolysis reactions, transport of species by electroosmosis, ionic migration and diffusion. These processes are accompanied by sorption between the pore fluid and the soil surface and aqueous phase reactions.

The principles of species transport under an electric field is presented. It is demonstrated that ionic migration is the most significant component of the transport process. The magnitude of mass transport by electroosmosis is at least one order of magnitude less than induced by electrical migration. The efficiency of transport of a species is directly related to its transference number which is related to its ionic mobility and concentration. As the concentration of the species decreases, removal efficiency would also decrease.

While the acid generated at the anode (H^+ ions) would assist in desorption of species from clay surfaces, the increase in its concentration in the pore fluid would decrease the transference numbers of other species decreasing the efficiency in their transport.

Electrokinetic extraction of species from soils requires optimization and careful engineering of the process to meet the preset objectives.

5 ACKNOWLEDGEMENTS

Fundamental aspects of electrokinetic remediation and theoretical modeling of transport of species under an electric field are ongoing studies at Loisiana State University. This project is funded by the Gulf Coast Hazardous Substance Research Center of USEPA at Lamar university in Texas. This support is gratefully acknowledged.

6 REFERENCES

Acar, Y. B. (1992) " Electrokinetic Soil Processing; A Review of the State of the Art" ASCE Grouting Conference, February 1992, Special Publication No. 30, Vol 2.pp. 1420-1432.

Acar, Y. B. and Olivieri, I. (1989) "Pore Fluid Effects on the Fabric and Hydraulic Conductivity of Laboratory Compacted Clay " Transportation Research Record, No. 1219, pp. 144-159.

Acar, Y. B. and Haider, L. (1990) "Transport of Low Concentration Contaminants in Saturated Earthen Barriers " ASCE, Journal of Geotechnical Engineering, Vol. 116, No. 7, pp. 1031-1052.

Acar, Y. B., and Gale, R. J. (1992). *Electrochemical Decontamination of Soils and Slurries*. Patent No. 5,137,608, US Commissioner of Patents and Trademarks, Washington, D.C., USA, August 11, 1992.

Acar, Y.B., Alshawabkeh, A. (1993) " Electrokinetic Remediation Technique; Principals, Implications for Engineering Applications and Current Status," *Journal of Environmental Science and Technology* (in review)

Acar, Y.B. and Alshawabkeh, A. (1994) " Modeling Conduction Phenomena in Soils under an Electrical Current, " 15th International Conference on Soil Mechanics and Foundation Engineering, Oxford and IBH Publishing Co. Pvt. Ltd., New Delhi, India, January 1994 (in press).

Acar, Y. B., Gale, R. J., Putnam, G., and Hamed, J. (1989) " Electrochemical Processing of Soils: Its Potential Use in Environmental Geotechnology and Significance of pH Gradients " 2nd International Symposium on Environmental Geotechnology, Shanghai, China, May 14-17, Envo Publishing, Bethlehem, PA, Vol. 1, pp. 25-38.

Acar, Y. B., Gale, R. J., Putnam, G., Hamed, J., and Wong, R. (1990a) " Electrochemical Processing of Soils: Theory of pH Gradient Development by Diffusion and Linear Convection " Journal of Environmental Science and Health, Part (a); Environmental Science and Engineering, Vol. 25, No. 6, pp. 687-714.

Acar, Y. B., Hamed, J., Gale, R. J., and Putnam, G.(1990b) "Acid/Base Distribution in Electro-Osmosis" Transportation Research Record, No. 1288, Soil Geology and Foundations Geotechnical Engineering 1990, pp.23-34.

Acar, Y. B., Alshawabkeh, A. N., and Gale, R. (1992) " A Review of Fundamentals of Removing Contaminants by Electrokinetic Soil Processing " Proceedings of the Mediterranean Conference on Environmental Geotechnology, edited by Mumtaz Usmen and Yalcin Acar, Cesme, Turkey, May 1992, pp. 321-330.

Acar, Y. B., Alshawabkeh, A., Gale, R.J. (1993a) "Fundamental Aspects of Extraction of Species from Soils by Electrokinetics," Waste Management, Pergamon Press, London, v. 12, No. 3.

Acar, Y.B. , Hamed, J., Alshawabkeh, A., Gale R.J. (1993b) " Cd(II) Removal from Kaolinite by Electrokinetics," Geotechnique, (in press).

Alloway, B. J. (1990) " Heavy Metals in Soils " John Wiley and Sons, New York, 339p.

Alshawabkeh, A., Acar, Y.B. (1992) "Removal of Contaminants from Soils by Electrokinetics: A Theoretical Treatise," Journal of Environmental Science and Health, Part (a), A27(7), pp. 1835-1861.

Ballou, E. V. (1955) " Electro-osmotic Flow in Homoionic Kaolinite " Journal of Colloid Science, Vol. 10, No. 5, pp. 450-460.

EPA (1991) "Innovative Treatment Technologies " Semi-Annual Status Report, EPA/540/2-91/001. No. 2, 50p.

Hamed, J. (1990) "Decontamination of Soil Using Electro-osmosis," A Dissertation submitted to the Graduate School of Louisiana State University in Partial Fulfillment of the Degree of Doctor of Philosophy.

Hamed, J., Acar, Y. B., and Gale, R. J. (1991) "Pb(II) Removal from Kaolinite Using Electrokinetics," Journal of Geotechnical Engineering, ASCE, Vol. 112, February, pp. 241-271.

Harter, P. J. (1983) " Effect of Soil pH on Adsorption of Lead, Copper, Zinc, and Nickel " Soil Science Society of America Journal, No. 47, pp.47-51.

Holmes, P. J. (1962) " The Electrochemistry of Semiconductors " Academic Press, London, 396p.

Hunter, R. J. (1981) " Zeta Potential of Colloid Science " Academic Press, London, 386p.

Koryta, J. (1982) " Ions, Electrodes, and Membranes" John Wiley and Sons, New york.

Kruyt, H. R. (1952) " Colloid Science (I): Irreversible Systems " Elsevier Publishing Company, 389p.

Lageman, R. (1989) "Theory and Practice of Electro-Reclamation," NATO/CCMS Pilot Study, Demonstration of Remedial Action Technologies for Contaminated Land and Ground Water, Copenhagen, Denmark, May 9, 1989, 18 p.

Maguire, M., Slavek, J., Vimpany, I., Higginson, F., and Pickering, W. (1981) " Influence of pH on Copper and Zinc Uptake by Soil Clays " Australian Journal of Soil Research, No. 19, pp.217-229.

Mitchell, J. K. (1993) " Fundamentals of Soil Behavior " John Wiley and Sons, New York,

Pamukcu, S. and Wittle, J. K. (1992) "Electrokinetic Removal of Selected Heavy Metals from Soil" Environmental Progress, Vol. 11, No. 3, pp. 241-250.

Shapiro, A. P. and Probstein, R. F. (1993) "Removal of Contaminants from Saturated Clay by Electroosmosis " Environmental Science and Technology, Vol. 27, No. 2, pp.283-291.

Shackelford, C. D. (1991) " Diffusion of Contaminants Through Waste Containment Barriers" Transportation Research Record, No. 1219, pp. 169-182.

Shackelford, C. D. and Daniel, D. E. (1991a) "Diffusion in Saturated Soil (I): Background " ASCE, Journal of Geotechnical Engineering, Vol. 117, No. 3, March 1991, pp. 467-484.

Shackelford, C. D. and Daniel, D. E. (1991b) "Diffusion in Saturated Soil (II): Results for Compacted Clay " ASCE, Journal of Geotechnical Engineering, Vol. 117, No. 3, March 1991, pp. 485-506.

Yong, R. N., Warkentin, B. P., Phadungchewit, Y., and Galvez, R. (1990) "Buffer Capacity and Lead Retention in Some Clay Materials " Water, Air, and Soil Pollution, No. 53, pp.53-67.

Waste Disposal by Landfill – GREEN'93, Sarsby (ed.) © *1995 Balkema, Rotterdam, ISBN 90 5410 356 6*

Characterization and marketability of bottom ash from municipal incinerators

E. Bahout, J. Hadjinicolaou & R. N. Yong
McGill University, Montreal, Que., Canada

ABSTRACT: Since the bottom ash residues obtained from municipal solid waste incinerators contain considerable amounts of leachable heavy metals and salts, problems have been encountered in deciding upon final disposal scenarios based on concerns about soil and groundwater contamination. Several alternatives, other than landfilling, have been investigated for the ultimate disposal of these residues. The basic factors influencing ash utilization scenarios include concerns of environmental impacts acceptability, technical and economical feasibility, and market availability. A physical treatment process, involving magnetic removal of metal and particle separation into different sizes by sieving, has been suggested as a means for reducing the bottom ash content in heavy metals; namely, Pb, Zn, Cd, Cu, and Cr.

1 INTRODUCTION

Municipal solid waste (MSW) incineration (with energy recovery) has become a widely used alternative, especially in countries where landfill areas are scarce and difficult to site. Fortunately or unfortunately, MSW incineration is considered as a potential viable alternative in any waste management plan (Black and Cunningham, 1991; Gavasci et al., 1990; Sussman, 1989).

The incineration of solid waste produces three main types of solid residues: (1) fly ash (10 to 30% by weight of total residues), which is the fine dust collected from the flue gas by emission control devices; (2) slag (around 1% of total weight); and (3) bottom ash (70 to 90% of total residue), which is the remaining heavy material residue collected at the bottom of the combustion chamber. Although the incineration of the MSW provides a reduction of 70 to 80% by weight, the amount of the remaining residues to be disposed of is still substantial (Hasselriis, 1988; Sawell et al., 1989).

The bottom ash component consists of a metallic and a non-metallic fraction. The non-metallic "inert" material has properties very similar to sand and gravel. This fact leads to considering some useful utilization scenarios of bottom ash other than landfilling, e.g., aggregate material in construction works. Hartlem and Rogbeck (1989) presented the possibility of using "ash-gravel" in road construction and abutment fills. Some implications of such a practice would be the conservation of natural aggregate and the diversion of significant volumes of "waste material" from landfill, thus extending the operating life span of the landfill and saving on the tipping fees.

However, the environmental acceptability of such a practice is still debatable and should be addressed. The overriding concerns as to the best method of solid waste combustion ash management are the content and leachability of heavy metals (e.g., lead, cadmium, copper, zinc, etc.), salts and sulfates contained in the ash and the potential public health effects that may result from improper ash management. In addition, other concerns include the presence of trace organics and inhalation of contaminants from the dust generated by the ash.

Moreover, if ash should be beneficially reused as a raw material, e.g., structural fill for road subbases or building block materials, regulatory requirements are necessary.

The scope of this paper is to present the preliminary results of a bottom ash characterization and the basic considerations for treating MSW incinerator ash residues to improve their suitability in utilization scenarios.

2 INCINERATOR DESCRIPTION

The Montreal Des Carrières incinerator started its operations in 1969. It consists of 4 parallel mass burning incinerator units designed to handle a total of 1200 tons of municipal refuse per day.

The incinerator chamber is fed by gravity via a vibrating feeder, and the refuse moves thereof to the incineration chamber via vibrating grates. To control the formation of acid gases, lime powder is directly injected at the point where the refuse is fed. The flue gases pass through a water walled/boiler economizer heat recovery chamber, enter a multicyclone unit which is followed by a 3-stage electrostatic precipitator.

All three types of ashes are conveyed to a quench tank, then transported by a drag chain through an incline and dumped in a hauling truck.

3 SAMPLING AND TESTING

Since all of the ash streams are combined in the quench tank, it was not possible to obtain a separate bottom ash sample after quenching.

The samples of bottom ash were collected off the finishing grate, just prior to entering the quench tank. Thus, by doing so, it is not known whether or not quenching alters the bottom ash characteristics.

Composite samples of bottom ash were prepared and separated into different particle size fractions; namely, greater than 2 mm (i.e., retained on sieve #10), between 2 and 0.595 mm (i.e., passing sieve #10 and retained on #30), between 0.595 and 0.297 mm (i.e., passing sieve #30 and retained on #50), and less than 0.297 mm (i.e., passing sieve #50). The total amount of lead (Pb), zinc (Zn), cadmium (Cd), chromium (Cr), and copper (Cu) present in these four particle size fractions was determined using the Aqua Regia Digestion method (Walton, 1980).

To determine the metal solubility of bottom ash under non-aggressive leaching conditions, a representative unit sample of bottom ash was crushed to pass sieve #100, before being leached out with distilled water in a batch leach test. The tests were conducted in polyethylene bottles at a liquid to solid ratio (L/S) of 10, by mixing 3g of ash with 30 mL of distilled water. The mixture was then shaken for one day on a horizontal shaker. After the mixing period, the liquid and solid phases were separated by centrifugation and by pressure filtration using 0.45 μm membrane filters. The remaining solids were then reexposed to fresh distilled water at the same L/S ratio and the procedure was repeated for 5 cycles. At the end of the 5th cycle, the L/S was 50. The tests were performed in triplicates and the leachate samples were analyzed for pH, conductivity, redox potential, chloride and sulfate content, and alkalinity. A sub-sample of each leachate was preserved with concentrated nitric acid to be subsequently analyzed for metal content using the Flame Atomic Absorption (FAA).

The total solids content of the bottom ash was determined by sample weight loss after 24 hour period at 104^{0}C according to ASTM method D2216-71. The fixed solids content of each ash was based upon weight loss of a dry sample placed in a 550^{0}C muffle furnace for one hour (APHA standard method 290E).

4 RESULTS AND DISCUSSION

The particle size distribution of the Des Carrières incinerator bottom ash is shown in Figure 1 below. It can be seen from Figure 1 that the bottom ash fraction with size smaller than 2 mm does not constitute more than 20% of the total mass.

The variation of Pb, Zn, Cd, Cr, and Cu content with the particle size of bottom ash is shown in Figure 2.

In general, the bottom ash contained low concentrations (less than 100 mg/kg) of Cd and

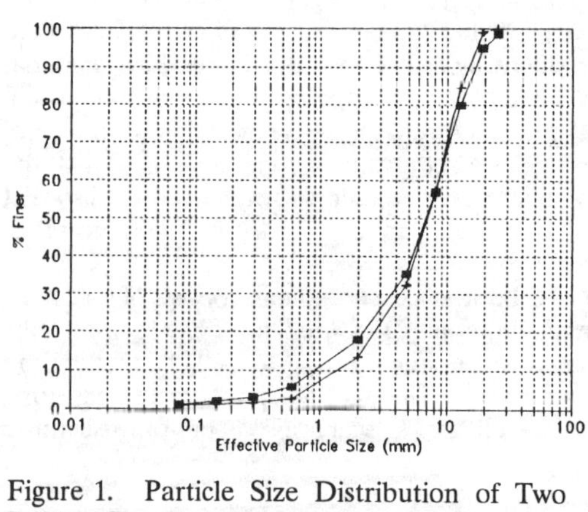

Figure 1. Particle Size Distribution of Two Particle Size Fractions.

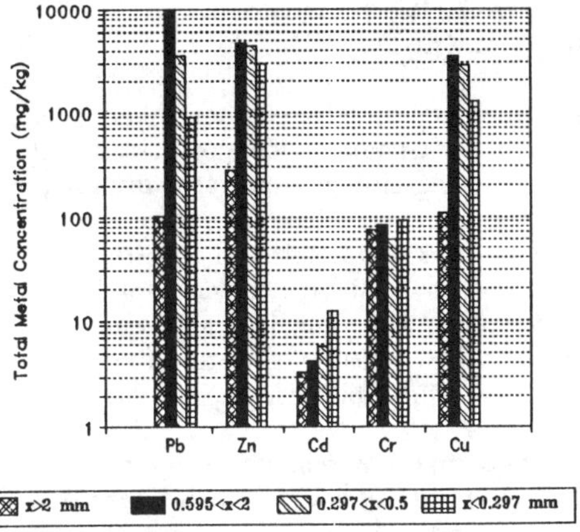

Figure 2. Total Metal Concentration in Four Particle Size Fractions.

Cr, moderate concentrations of Cu, and high concentrations of Pb and Zn.

Referring to Figure 2, it is observed that an abrupt drop in metal concentrations; especially Pb, Zn, Cd, and Cu, occurs when the particle size exceeds 2 mm. Indeed, the fraction of particles with size larger than 2 mm had the lowest metal content, as compared to the smaller size fractions. The highest amounts of Pb, Zn and Cu were found in the particles with sizes between 2 and 0.297 mm.

However, Cr appeared to be equally distributed among all particle sizes, whereas Cd concentration increased with decreasing particle size. Moreover, not only is the total amount of heavy metals (Pb, Zn, Cu) in bottom ash particles smaller than 2 mm greater than that found in particles with larger size, but it is also more leachable (Stegemann and Schneider, 1991; Gavasci et al., 1990).

The percentages of the various heavy metals contained in the bottom ash with particle size greater and smaller than 2 mm are presented in Table 1. The results presented in Table 1 show the interdependence between the metal content of bottom ash and its particle size. It is hence obvious that the removal of the bottom ash with particle size smaller than 2 mm, which constitutes less than 20% of the total weight, would reduce the amount of Pb by as much as 130 times, the amount of Zn by 40 times, the amount of Cd by 6 to 7 times, the amount of Cr by 3 times, and the amount of Cu by 80 times.

Based on these results, it is suggested to discard the fraction of bottom ash with size smaller than

Table 1. Heavy metal content of bottom ash as a function of particle size.

Particle Size	Metal Content (%)				
	Pb	Zn	Cd	Cr	Cu
> 2 mm	0.01	0.03	0.00033	0.0074	0.01
< 2 mm	1.4	1.2	0.0022	0.024	0.8

2 mm, when it is desired to utilize the material in a construction work. By doing so, no more than 20% by weight of the total bottom ash would be discarded, thus leaving at least 80% by weight to be effectively utilized.

The concentrations of cations and anions were again higher in the leachate from the smallest particle size fraction, with the same applying for the pH and buffering capacity.

The total solids content of the bottom ash was greater than 99.6%, indicating that the ashes were essentially dry.

The fixed solid content, on the other hand, was greater than 97%, indicating that the non-combusted carbon and volatile organic content of the bottom ash, reported as loss of weight on ignition (LOI) was less than 3%.

The analyses of the extracted fluids obtained from the batch leaching test at the end of each cycle, were as follows:

- The pH of the leachates remained alkaline throughout the test. It decreased from 12.12, after cycle 1, to 11.38, at the end of cycle 5.
- The chloride dissolution was observed to occur during the first and second cycles. This fact was of course expected since, even under alkaline conditions, chlorides are known to readily dissociate in water.
- The average cumulative fractions of each metal leached out at the end of the fifth cycle were determined by dividing the mass of the contaminant measured in the leachates to the mass of that contaminant in the original sample, expressed as a percentage. Typically, less than 1% of Zn, and less than 2% of Pb were leached out during the test. Referring to Figure 3, it is noted that the amount of Pb and Zn leached out decreased from cycle 1 to cycle 5, i.e., decreased as the pH of the ash decreased from 12.12 to 11.38, indicating therefore the amphoteric nature of the metals. Hence, the pH (and redox potential) environment is an important parameter in that respect. Note that the initial high pH is due to the fact that the ash was not quenched prior to collection.

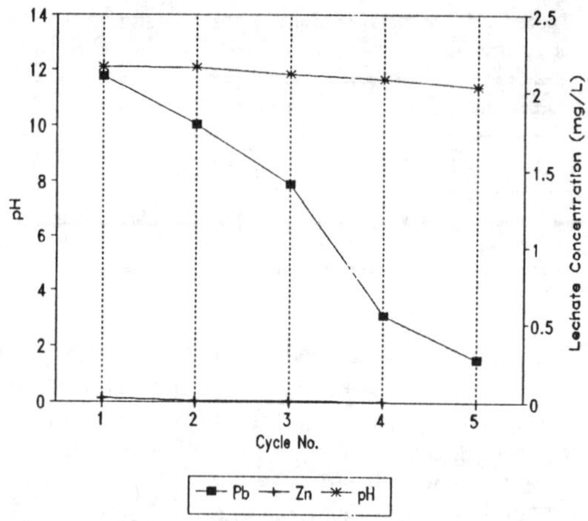

Figure 3. Variation of leachate pH, and Pb and Zn concentrations at the end of each cycle.

What has been presented indicates the advantage of removing the fraction of bottom ash with particle size smaller than 2 mm. Moreover, the fraction of the ash with particle size greater than 2 mm could be subjected to a process where the metallic fraction would be magnetically removed. The cost associated with this "physical treatment technique" should be minimal especially when compared to an electro-chemical treatment technique. Legiec et al. (1988) and Vehlow et al. (1989) have presented the technical possibility of operating a continuous process for heavy metal removal. The extraction studies presented by Legiec et al. (1988) revealed information on the leaching characteristics of the ashes and on the ability of some extraction solutions to separate these metals from the ash matrix. The electrochemical plating techniques was then utilized to recover the metals, in a relatively pure form, from the extraction solution. Although the results were promising, the cost of implementing such a technique on a full-scale basis, without mentioning the operating problems that are likely to occur, has yet to be addressed, and since the amount of metals present in the bottom ash is much less than that found in the fly ash, such a technique would not be warranted for bottom ash conditioning.

Thus, the preferred approach for improving the suitability of MSW incinerator bottom ash in utilization scenarios is to separate the fractions with size smaller than 2 mm.

5 CONCLUSION AND RECOMMENDATIONS

In order to conserve natural aggregates and reduce, or eliminate, the cost associated with MSW incinerator bottom ash disposal, it is strongly recommended to make use of this "aggregate-like" material in construction use. However, the content and leachability of some heavy metals, especially lead and zinc, present in these ashes raised concern about adverse potential effects that accompany any utilization scenario.

Considering the amount of bottom ash, an expensive and laborious treatment technique, such as metal recovery by electrochemical treatment processes will not be advantageous if compared to a process which involves the magnetic removal of the bottom ash metallic fraction and its segregation into two different fractions; namely, the fraction with size greater than 2 mm and the fraction with size smaller than 2 mm.

Preliminary study showed that the fraction of bottom ash with size greater than 2 mm, which consists of around 80% by weight of the total amount; contains a much smaller level of heavy metals which are also less leachable than those contained in the fraction with size smaller than 2 mm.

It is therefore suggested to magnetically remove the metallic fraction from the ash and to discard the fraction with size smaller than 2 mm (less than 20% of total weight), while retaining the remaining part for constructive utilization.

Since the characteristics of solid waste can vary substantially over short periods of time (based upon the nature of the waste burned and the incinerator operating parameters), the remaining residue, i.e., the ash, is very heterogeneous. Consequently, in order to provide statistically valid results which will present a basis for assessing the environmental acceptability of a given ash utilization scenario, it is necessary to take daily samples on an hourly basis over a long time period (e.g., a year) and correlate the results of the ash characterization (which affect the short and long-term behaviour of the ash in the environment) with the quality of the influent waste and the operating parameters of the incinerator.

There is a need to develop some reliable and reproducible laboratory tests to predict leachate characteristics at the site where the ash is utilized in one way or another.

It is essential that the ash residue streams be handled in a well-designed closed-system to prevent the generation of fugitive air borne dust. At this point in time, and before starting to utilize bottom ash on a large scale, it is recommended to wait for additional research and guidelines to regulate bottom ash utilization scenarios, in order not to endanger the public health nor the environment.

REFERENCES

Black, G.A. & J.C. Cunningham 1991. Managing incinerator ash: fly ash separation and stabilization. *Proc. 29th Annual International Solid Waste Exposition (GRCDA/SWANA's).* Cincinnati, Ohio.

Gavasci, R., T. Mangialardi & P. Sirini 1990. Slag and fly ash from MSW incineration plants -characterization and reuse. *Proc. 6th International Conference on Solid Waste Management and Secondary Materials.* Philadelphia, PA, U.S.A.

Glordano, P.M., A.D. Behel, J.E. Lawrence, J.M. Solleau, & B.N. Bradford 1983. Mobility in soil and plant availability of metals derived from incinerated municipal refuse. *Environ. Sci. Technol.* 17: 193-198.

Harlten, J. & J. Rogbeck 1989. Sorted incinerated slag as fill material. *Proc. International Conference on Municipal Waste Combustion.* Hollywood, Florida, U.S.A., Vol. 1.

Hasselriis, F. 1988. How environmental risk from leaching of heavy metals in ash residues from combustion of municipal solid waste can be minimized by control of combustion, emissions and ash residues management. *Proc. 10th Canadian Waste Management Conference,* Winnipeg, Manitoba.

Legiec, I.A., C.A. Hayes & D.S. Kosson 1988. Continuous recovery of heavy metals from MSW incinerator ashes. *Proc. 10th Canadian Waste Management Conference,* Winnipeg, Manitoba.

Poran, C.J. and F. Ahtchi-Ali 1989. Properties of solid waste incinerator fly ash. *Journal of Geotechnical Engineering* 115: 1118-1133.

Sawell, S.E., T.R. Bridle & T.W. Constable 1988. Heavy metal leachability from solid waste incinerator ashes. *Waste Management and Research* 6: 227-238.

Stegemann, J.A. & J. Schneider 1991. Leaching potential of municipal waste incinerator bottom ash as a function of particle size distribution. In *Waste Materials In Construction.* Goumans, J.J.J. R. et al. (editors). Elsevier Science Publishers.

Sussman, D.B. 1989. Municipal waste combustion ash: testing methods, constituents and potential uses. *Proc. International Conference on Municipal Waste Combustion.* Hollywood, Florida, U.S.A., Vol. 2.

Vehlow, J., H. Braun, K. Horch, J. Schneider, & H. Vogg 1989. Semi-industrial testing of the 3R process (Heavy metal extraction, product quality). *Proc. International Conference on Municipal Waste Combustion.* Hollywood, Florida, U.S.A., Vol. 1.

Walton, A. 1980. Méthodes d'échantillonage et d'analyse des sédiments marins et des matières draguées. In *Immersion de Déchets en Mer. Rapport No.1.* Ministère des pêches et des Océans, Ottawa, 1980.

Waste Disposal by Landfill – GREEN'93, Sarsby (ed.) © *1995 Balkema, Rotterdam, ISBN 90 5410 356 6*

Microbial methane oxidation and the effective biological treatment of landfill generated methane

P.A. Barratt
Richards, Moorehead and Laing Limited, Merthyr Tydfil, UK

ABSTRACT: The use of naturally occurring methane oxidising bacteria (methanotrophs) for the metabolism of methane gas has been explored in laboratory experiments, in order to assess the potential for the use of microbiological systems for amelioration of methane gas seeps from active landfills. The work examines the microorganisms responsible for this activity. By using artificially created soil-based and soil-free environments in laboratory microcosms, experiments were undertaken to examine the potential for optimising rates of microbial methane oxidation.
Inoculation of an artificial foam glass medium with methanotrophic bacteria showed oxidation of methane in microcosms during 23 days incubation. However, where a methanotrophic soil was used to transfer a mixed population of methanotrophs to vermiculite particles, methane oxidation rates were significantly faster (4gCH_4/kg/day). Columns of such bacterially colonised vermiculite particles and soil confirmed rapid rates of oxidation using landfill gas as a source of methane.
The paper discusses the feasibility of applying these systems to field situations.

1 INTRODUCTION

The methanotrophs form a specialised group of bacteria which contain the enzyme systems responsible for oxidising methane, in the presence of molecular oxygen, to form the metabolic products carbon dioxide and water:

$$CH_4 + 2O_2 \rightarrow CO_2 + 2H_2O \ (+\ 883\ kJ).$$

It has been realised for some time that these microorganisms are associated with methane in the natural environment. Methane is formed from the anaerobic degradation of organic material, and so may occur in the vicinity of municipal landfill sites. Where methane and oxygen are present simultaneously, methane oxidising populations may therefore flourish (Whalen *et al*, 1990). The presence of these methanotrophic populations has been the subject of much interest in recent years, although the *in situ* activity, and therefore the rates of methane oxidation to be expected in the field, are relatively poorly understood. It is possible however, that such microbial metabolism of methane may contribute to substantial losses of landfill generated methane. Such a significant natural sink for methane could therefore be of applied scientific significance from two points of view:
i) the reduction of potentially hazardous seeps of methane around landfill sites, and other methanogenic environments;
ii) a reduction in the theoretical yield of greenhouse gases from landfills.
A variety of factors are likely to limit the rate of microbial metabolism of methane gas in the field, and these include: rate of methane flux, partial pressure of oxygen, moisture, nutrient concentrations and temperature.
In an attempt to explore the potential for the creation of systems engineered to optimise the rate of methane oxidation in a field application, as an aid to risk reduction at landfill sites, the environmental chemistry of such a system should be understood. This project was conducted to fulfil part of this objective, and the information in this paper relates some of the studies undertaken and the results obtained.

2 METHODS

The study as a whole concerned the following investigative stages:
i) measurement of indigenous methane oxidation rates in landfill cover soil;
ii) isolation and identification of methanotrophic bacteria from soil;
iii) chemical and microbial amendment of soil and its effects on rates of methane oxidation in laboratory microcosms;
iv) physical amendment of soil to improve the edaphic environment for methanotrophy;
v) measurement of methane oxidation in laboratory columns containing an optimised medium for methanotrophs, and utilising landfill gas as a methane source.
Much of the study involving stages (i) to (iii) has been reported previously (Barratt, 1991). This paper concerns stages (iv) and (v).

2.1 Laboratory microcosms

For studies on the metabolism of methane gas in methane-air mixtures, sealed laboratory microcosms were used. These contained a known quantity of soil (or other basal material), plus relevant chemical and/or microbial treatments. Usually, 200g dry weight equivalent of fresh or amended soil was used.
Microcosm containers consisted of a 1 litre glass, screw top Kilner™ jar, sealed with a butyl rubber ring and petroleum jelly. The top of the lid had been adapted to a circular aluminium plate, with a circular sampling port in the centre, which was plugged with a butyl rubber stopper. This allowed the headspace gas mixture (required in the microcosm) to be added to the jar, following evacuation of the microcosm under vacuum, and gas samples to be collected for subsequent chemical analysis. All gas transfers to and from the headspace were undertaken via a hypodermic needle pushed through the rubber stopper. In addition this allowed liquid amendments to be added to the microcosms as necessary during the period of incubation.
All incubations were undertaken at 15°C.

2.2 Soil-free microcosms

To create an active methanotrophic system in a soil-free environment, perlag™, a semi-artificial foam glass medium, was examined. Perlag has a strong, porous capillary structure, giving it enhanced water retentive properties. It is chemically inert and neither leaches nor absorbs mineral ions into or out of the surrounding environment.
Sixteen microcosm jars containing 500g of washed gravel (<10mm particle size), and 200g perlag (<10mm) were prepared. These quantities were approximately equal in terms of volume, due to the low density of the perlag (450 - 550kg/m^3). Gravel was used as a readily available and inexpensive bulking agent to supplement the perlag. These were considered to be advantages if the laboratory tests showed potential for field applications.
A second set of sixteen microcosms was prepared using 350g perlag. No gravel was added to these. To each set of microcosms, four treatments were applied, thereby producing four replicates of each treatment for each basal medium. The treatments were as follows:
A) 45ml sterile distilled water (control);
B) 40ml of basal salts medium, plus 5ml of sterile distilled water;
C) as B, plus a methanotrophic microbial isolate (number 4) at 4.3 x 10^5 cells/g basal medium;
D) as B, plus a methanotrophic microbial isolate (number 7) at 8.7 x 10^5 cells/g basal medium.
The basal salts medium used was adapted from Malashenko and Romanovskaya (1984), and contained KNO_3 (0.1%), KH_2PO_4 (0.1%), NaCl (0.1%), $MgSO_4$ (0.1%), $CaCl_2$ (0.02%), CH_3OH (0.1%), $FeCl_3$ (trace), $CuSO_4$ (trace).
All cells added to the treatments were grown in batch culture, harvested, centrifuged and rinsed in phosphate-buffered saline prior to use in the treatments. The bacterial isolates, 4 and 7, were selected from a methanotrophic landfill cover soil during previous studies (Barratt, 1991), and these were identified to species level during the course of the study.
All microcosms were subsequently filled with a methane:air mixture of approximately 1:3 (v/v). After addition of the methane, time zero samples (0.5ml) were withdrawn for analysis of methane and oxygen by gas chromatography.
Headspace methane was sampled and analysed after 1, 12, 19 and 23 days of incubation at 15°C.

2.3 Soil-based microcosms

The existence of highly active indigenous populations of methanotrophs in field soils gave rise to the question of whether or not it would be

possible to enhance rates of methane oxidation by changing the physical nature of the soil environment to optimise bio-oxidation. In order to test this, vermiculite was chosen as a soil amendment.
In addition to the use of vermiculite as a soil amendment, experiments were designed to examine the possibility of gradually establishing a methanotrophic population on a soil-free medium, by phased dilution of methanotrophic soil with the vermiculite medium over a period of time.
Nine microcosms were established, each containing 200g (dry weight equivalent) of an actively methanotrophic soil collected from a landfill cover soil in the field.
The nine microcosms were then divided into three sets of replicate moisture/nutrient treatments.
The amount of water added as part of these treatments was calculated to produce a final water content of 60% of the water holding capacity (WHC) of the soil, which had been determined previously.
The treatments were as follows:

i) no addition i.e. moisture of fresh soil only;
ii) sterile distilled water;
iii) nutrient broth solution.

All microcosms were sealed and a headspace methane:air ratio of approximately 1:3 (v/v) established. The time zero concentrations of methane, carbon dioxide and oxygen were determined by gas chromatography.
All microcosms were incubated at 15°C, and gas aliquots extracted for subsequent analysis at intervals of 24, 192, 264 and 336 hours.
Original headspace gas concentrations were replenished after the 193 hour sampling.
After 336 hours incubation the contents of the microcosms were altered by diluting the soil with vermiculite in order to achieve a ratio of soil to vermiculite of 3:1 (v/v). Again sampling and analysis of gas was undertaken at various time intervals up to a maximum incubation period of 336 hours.
Similarly, after the previous incubation, further dilution of the soil was undertaken to produce a soil:vermiculite ratio of 1:1 (v/v). Similar total volumes of soil + vermiculite amendment were retained in each microcosm for this next stage of incubation as had been used previously.
Subsequently, a soil:vermiculite ratio of 1:3 (v/v) was created by the addition of further vermiculite, and finally, after 120 hours incubation, all the soil was removed by hand and further vermiculite added to produce microcosms that contained a 100% soil-free medium i.e. vermiculite alone.
Sampling and analysis of headspace gases were undertaken accordingly at each stage.
At each dilution stage moisture amendments were made to achieve approximately 60% WHC of the soil-vermiculite mixture.
The total volume of material in each microcosm was kept approximately equal during the experiment, despite dilution with vermiculite.

2.4 Column systems

In order to simulate a situation that may be found in the field, where landfill gas travels upwards through the ground from a methanogenic environment, laboratory columns were constructed. The design of these columns was as shown in Figure 1.
Four such columns were constructed from 10cm diameter plastic pipe, incorporating vermiculite as the fill medium. Each column had an inlet at the base for the ingress of landfill gas, an outlet for the drainage of liquid, and seven sampling ports at intervals along the height of the column, for the collection of gas samples. The air space beneath the vermiculite support grid and the base of the column allowed for the even distribution of gas as it entered the column, and percolated up through the vermiculite medium.
Landfill gas was generated from two sealed 210 litre plastic bins full of domestic refuse. The gas was collected in butyl rubber bladder bags for use in the experiments.
Each of the four columns was filled with approximately 80cm depth of a vermiculite:soil mixture, produced by mixing 3.3kg fresh, methanotrophic landfill cover soil with 1kg clean vermiculite.
One of the columns represented a sterile control, produced by autoclaving the soil-vermiculite mixture prior to filling the column (30min @ 1.05 x 10^5Pa and 121°C).
Another column represented a non-sterile control, to which was added distilled water only. The two further columns represented the 'treated' columns, where liquid was added in the form of an aqueous solution of nutrients, the composition of which is given in Table 1. One of these treated columns was used purely to monitor the activity of the microbial biomass therein, by means of performing viable counts for bacteria.
All soil-vermiculite was brought to 60% WHC by the addition of the appropriate quantities of sterile liquid, as water or nutrient solution.
All columns were lagged on the outside with

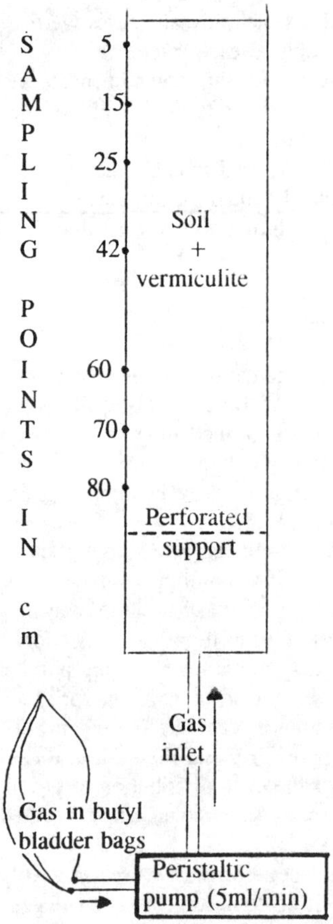

Figure 1. Column design, showing the influx of gas and positions of sampling points (not to scale)

Table 1. Composition of the nutrient solution (pH 7) added to soil-vermiculite columns

Chemical	Concentration
KH_2PO_4	7.5 g/l
NH_4NO_3	7.5 g/l
$MgSO_4.7H_2O$	3.0 g/l
$CaCl_2$	0.15 g/l
$FeSO_4$	0.075 g/l
Yeast extract	0.015 % (w/w)

Table 2. Percentage decline in headspace methane concentration during incubation of soil-free microcosms

	Water	Nutrients	Nutrients + isolate 4
Gravel/perlag			
Day 1	0.8	3.6	4.5
Day 12	11.6	15.8	13.6
Day 19	7.4	5.7	20.4
Day 23	8.5	13.8	37.1
Perlag			
Day 1	0.0	0.0	10.9
Day 12	9.5	13.3	19.9
Day 19	7.5	8.7	22.2
Day 23	9.1	10.0	31.0

sheets of fibreglass matting, in order to exclude light and retain heat, and were then incubated outside at ambient temperatures (approximately 5 - 20°C), during which landfill gas was pumped into the base of the columns at a rate of 5ml/min, via peristaltic pumps. Gases emitted at the top of the columns were vented to atmosphere.

Landfill gas was replenished in the bladder bags as often as was necessary.

After 14 days incubation, to allow the establishment of a methanotrophic biofilm on the vermiculite particles, samples of gas were withdrawn from each sampling port of the three columns (two controls and one test). Sampling was undertaken by pushing a long spinal hypodermic needle through the butyl rubber septum at each sampling port, in order to take a gas sample from the centre of the column at a given depth.

Each sample withdrawn was analysed for methane, carbon dioxide and oxygen.

3 RESULTS AND DISCUSSION

3.1 Perlag microcosms

The results of the changing gas composition seen in the microcosms containing gravel and/or perlag, during 23 days incubation, are summarised in Table 2. Only those results for microbial isolate 4 are shown, as no significant decline in methane concentrations were seen for those microcosms treated with isolate 7.

The results have been presented as the percentage decline in methane concentration, from an initial concentration at time zero of approximately 25% (range 24.0 - 26.4%).

From these results it appears that both the perlag and the perlag-gravel mixture were able to support microbial populations with the ability to oxidise methane. The application of an active methane oxidising microbial inoculum, together with a source of nutrients and methane, appears to have provided an environment conducive to the establishment of a methanotrophic population on the surface of the inert mineral base. A substantial decrease in methane concentration over time was measured in the microbial treatment, as opposed to smaller fluctuations in methane concentration seen in the two controls. The latter, because of the measurement of positive and negative fluctuations in the concentration of methane over time, is likely to have been due to analytical and/or sampling errors. However, despite these sources of error, the results for the microbially treated microcosms indicate a clear downward trend in methane concentration over the 23 days, and are significantly different from either of the two controls ($P > 0.05$).

Although the mean decrease in methane concentration over 23 days was greater for the gravel and perlag mixture, the difference between the use of perlag alone or a mixture of gravel and perlag was not significant.

The average rate of methane oxidation in the microbially treated microcosms was calculated as approximately 0.003gCH_4/kg perlag/day. This is of an order of magnitude observed in some field soils (Pearce, 1989), although it is slow when compared with some rates measured in highly methanotrophic soils (e.g. Whalen *et al*, 1990). It should be noted however that the study represented a period when the methanotrophic population was becoming established on the basal materials. In addition the complete oxidation of methane in the microcosms would have been limited by the availability of oxygen during the course of the incubation. In this respect it is unlikely that the extent of methane depletion was maximal by 23 days.

3.2 Microbial identification

Microbial isolates 4 and 7 were subsequently identified as *Rhodococcus erythropolis* and *Methylobacterium fujisawaense* respectively. Table 2 shows some of the metabolic characteristics of the two bacteria, and highlights the fact that isolate 7, although originally isolated as a methane degrader, subsequently appeared to have lost its ability to utilise methane as a growth substrate.

Table 2. Some metabolic characteristics of the two bacterial isolates used in microcosms

Parameter	Isolate 4	Isolate 7
Gram stain	+	±
Fucose	-	+
Methane	+	-
Methanol	+	+
Citrate	+	+
Catalase	+	+
Oxidase	-	-

The lack of methane oxidising ability in isolate 7 explains why no decline in methane concentration over the two controls was seen in the microcosms using *M. fujisawaense*. This loss of methanotrophic activity during the sub-culturing of such isolates in the laboratory indicates one of the problems associated with the laboratory maintenance of some members of this group of microorganisms. The bacteria had been sub-cultured at 3-5 day intervals, and maintained at 15°C under a 50:50 methane:air mixture.

3.3 Vermiculite microcosms

The results of the analysis of headspace gases present in each of the microcosm systems during the course of incubation are shown in Figures 2, 3 and 4. These represent the methane and oxygen results for the untreated control, the sterile distilled water treatment and the nutrient treated microcosms respectively. Results were converted from percentage values to equivalent volumes of gas in ml.

Each graph illustrates the effects of gradual dilution of the original soil with increasing amounts of vermiculite, such that:

- day 0 - 14 represents 100% soil;
- day 15 - 28 represents 3:1 soil:vermiculite;
- day 29 - 33 represents 1:1 soil:vermiculite;
- day 34 - 37 represents 1:3 soil:vermiculite;
- day 38 onwards represents 100% vermiculite.

The results show that microbial methane oxidising activity, arising from the soil-borne microflora, was successfully transferred to the inert vermiculite medium over a period of approximately 23 days i.e. from the first dilution of the soil at day 14.

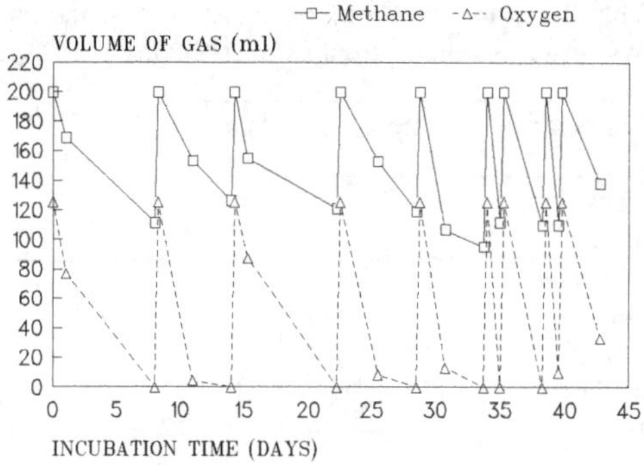

FIGURE 2: MICROCOSM HEADSPACE GAS COMPOSITION CONTROL - NO ADDITIONS

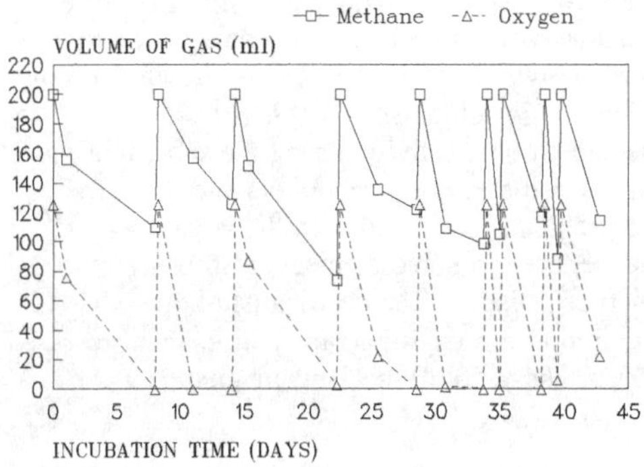

FIGURE 3: MICROCOSM HEADSPACE GAS COMPOSITION STERILE DISTILLED WATER ONLY

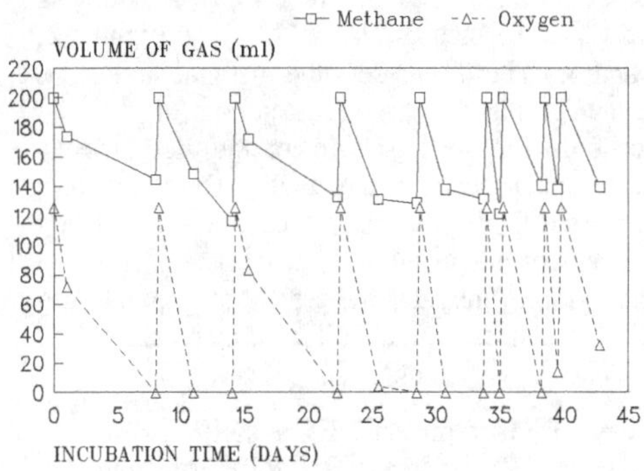

FIGURE 4: MICROCOSM HEADSPACE GAS COMPOSITION NUTRIENT TREATMENT

The rate of microbial methane oxidation seen in the microcosms containing 100% vermiculite was generally greater than that in the original methanotrophic soil. In addition, during the course of the experiment the establishment of microbial biofilm could clearly be seen on the surface of the vermiculite particles. This was typified by a pale brown slime over the particles. This observation was particularly obvious in the microcosms incorporating vermiculite and soil in a ratio of 3:1 respectively.

In the soil-only microcosms (days 0 to 14), assuming that there was approximately 800ml of gas (methane and air) in the headspace, the maximum rate of methane oxidation was calculated to be in the order of 0.15gCH_4/kg.soil/day. This occurred in the soil treated only with distilled water (Figure 3).

The rates measured for the soil:vermiculite dilutions of 3:1 and 1:1 were of a similar order of magnitude to that of the soil. However, at a soil:vermiculite ratio of 1:3 (days 34 - 37), the methane oxidation rate increased in all treatments by approximately two fold, to around 2.7gCH_4/kg/day. This rate is assumed to be lower than the actual rate achieved in the first few hours of incubation following gas replenishment because oxygen became depleted in these microcosms, and almost certainly limited the methane oxidation rate recorded over 24 hours.

The fastest rates of microbial methane oxidation occurred in the microcosms containing vermiculite from which the soil had been removed. Prior to this the vermiculite had been incubated for 23 days in the presence of gradually decreasing proportions of soil. The maximum rate of methane oxidation achieved by this method was approximately 4gCH_4/kg.vermiculite/day.

These results clearly illustrated the suitability of vermiculite as a growth medium for methanotrophs, and led to the development of the vermiculite column studies.

3.4 Column studies

The graphs in Figures 5, 6 and 7 show gas composition with depth in the sterile control, non-sterile control and nutrient treated columns of soil/vermiculite respectively.

The results show that as landfill gas enters the base of the column at a constant flow rate (5ml/min), and travels upwards in the direction of bulk gas flow, so the concentrations of methane and carbon dioxide decline, and that of oxygen increases. This pattern can be seen in the two control columns as well as that treated with nutrients, although the effect is greatly magnified in the nutrient treated column (Figure 7).

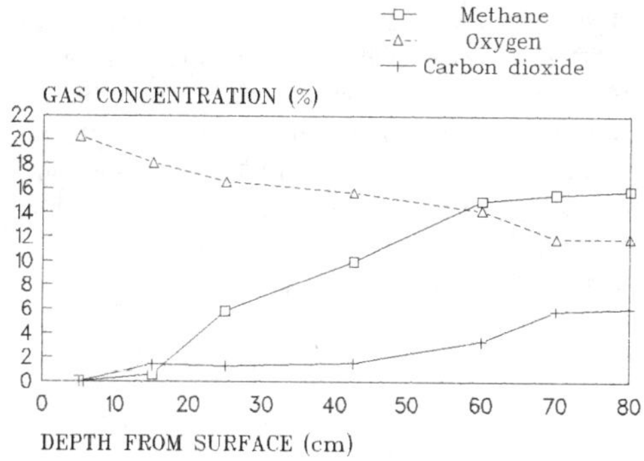

FIGURE 5: GAS COMPOSITION IN COLUMN
- STERILE CONTROL

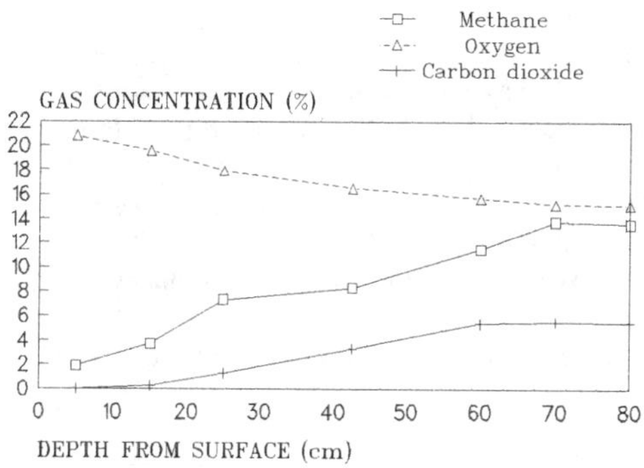

FIGURE 6: GAS COMPOSITION IN COLUMN
- NON-STERILE CONTROL

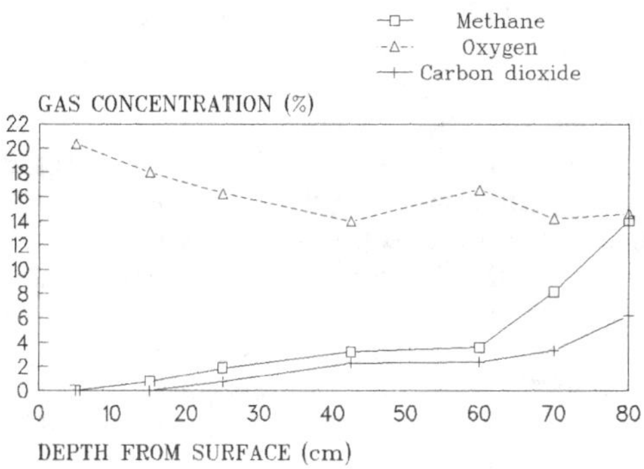

FIGURE 7: GAS COMPOSITION IN COLUMN
- NUTRIENT TREATED

The concentrations of oxygen at depth in all the columns are relatively high (12 - 14% at 80cm), which illustrates that oxygen is unlikely to limit the rate of methane oxidation in this system. This is an important attribute if the system is to work in a field situation, as oxygen in surface soils tends to become depleted at depths much shallower than this, so confining methane oxidation to the surface, where there is an open structure to the soil i.e. a low soil:pore space ratio, and therefore adequate potential for gas diffusion.

The results for the sterilised control show that at 5cm depth methane was not detectable, and at 15cm methane concentrations were <1%. These results, rather than showing oxidation of the methane, may indicate rapid dilution of the methane with air that penetrates the extreme surface layers of the column, although this hypothesis requires further investigation and clarification.

The slope of the graph for methane concentration down the column profile relates closely to the ability of the whole column to act as a biofilter for methane removal. Thus, where the slope rises slowly from a low concentrations of methane at 0 - 15cm depth, methane oxidation can be considered to be significantly reducing methane concentrations in the gas stream. This is clearly the case in Figure 7 where the nutrient treated column maintained methane concentrations below 4% to a depth of 60cm. By contrast the two control columns showed much steeper gradients from 15 - 60cm than Figure 7, indicating that methane was not removed as effectively from the gas stream in these columns.

In general each graph shows that methane and oxygen were inversely related according to depth in the column. This is not a simple relationship however, and is likely to depend on both the microbial consumption of methane and oxygen (in a ratio of $1CH_4:2O_2$), and the atmospheric infiltration of oxygen into the columns.

The decline in methane in both the sterilised and unsterilised controls may be due largely to the physical dilution and absorption of methane as it passes up the column, rather than to microbial oxidation. However, there is a possibility that the sterilised column was not completely sterilised after autoclaving and/or became colonised with airborne methanotrophs during the course of the experiment. The two controls were similar in the gradient of methane depletion, the unsterilised control showing slightly lower methane concentrations throughout, presumably due to microbial methane oxidation. However, there was a marked difference in the nutrient treated column, which contained <50% of the methane present in the control columns at the depths sampled between 15 and 60cm.

Vermiculite appears to have potential for use as a

soil amendment in order to enhance rates of methane oxidation. This principle could be applied in the field, either to landfill cover systems, or to other systems engineered to intercept landfill gas. The use of highly permeable, actively methanotrophic cover systems may be applicable to landfills with low to intermediate rates of methane production, as an aid to the reduction of environmental risk at these sites. Similarly, there is scope for the adaptation of open passive venting trenches around landfill sites, in order to encourage methane oxidation within the trench prior to gas venting to atmosphere.

4 CONCLUSIONS

A methane oxidising microbial population can be established on the surface of inert perlag particles in laboratory microcosms, by inoculating the particles with a methanotrophic bacterium (*Rhodococcus erythropolis*) in the presence of moisture, nutrients and an air-methane gas mixture. The long term stability of such an artificially created population is, however, not known, and the rates of methane oxidation achieved were below those measured in many methanotrophic soils.

Another methane oxidising bacterium (*Methylobacterium fujisawaense*), also isolated from a landfill cover soil, lost the ability to oxidise methane, highlighting the instability of some methanotrophic bacteria and the need to sub-culture and replenish methane-oxygen headspace gas at regular intervals during laboratory maintenance.

An actively methanotrophic field soil can be used as a source of methane oxidising microorganisms, and mixed populations of these microorganisms can be successfully established on vermiculite particles, by gradual dilution of a soil with vermiculite. Rates of methane oxidation achieved by this method reached approximately $4gCH_4$/kg vermiculite/day.

The use of soil and vermiculite as a medium for the growth of methanotrophs in a column system showed that methane concentrations in a continuous gas stream can be reduced significantly.

The use of mineral salts solutions as a source of nutrients for microbial growth has a stimulatory effect on rates of microbial methane oxidation on artificial media.

The advantage of transferring methane oxidisers to a vermiculite medium is that the penetration of atmospheric oxygen into the medium is greater than in soil. Thus methane oxidation may be sustained to greater depths in a particulate profile in a laboratory column. This advantage may be adapted to a field situation in order to establish a passive biofiltration system for oxidising hazardous seeps of methane gas from landfills.

ACKNOWLEDGEMENTS

The work documented in this paper was undertaken with funding from the UK Department of the Environment. The opinions and judgements expressed in this paper are those of the author and do not reflect those of the Department of the Environment. The results of this work may be used in the formulation of government policy, but at this stage they do not necessarily represent government policy.

The work comprising this study was undertaken in the laboratories of Biotal Limited, U.K. The author wishes to acknowledge the company in this respect, and the laboratory assistance of Mrs Angela Langley during the course of the work.

REFERENCES

Barratt, P.A. 1991. Amelioration of landfill gas problems through stimulation of methane oxidation. In Landfill Microbiology: R&D workshop II. November 1991. S.A. Evans and P.S. Lawson (eds.). 67-81. Harwell Laboratories, UK.

Malashenko, J.R and Romanovskaya, V.A. 1984. Ecology and isolation of bacteria growing on natural hydrocarbons (source unknown).

Pearce, F. 1989. Methane: the hidden greenhouse gas. New Scientist. May 6, 1989, p37.

Whalen, S.C, Reeburgh, W.S. and Sandbeck, K.A. 1990. Rapid methane oxidation in a landfill cover soil. Applied and Environmental Microbiology, 56: 3405-3411.

Waste Disposal by Landfill – GREEN'93, Sarsby (ed.) © 1995 Balkema, Rotterdam, ISBN 90 5410 356 6

Automated landfill design and costing utilising computer generated ground modelling

M.S. Bedford
Wardell Armstrong, Newcastle-under-Lyme, UK

ABSTRACT: This paper documents the use of MOSS and UPM (User Programmable Module) in conjunction with LOTUS for optimising landfill design. The design is automated, carried out in 3D and results in a rapid method of determining capital costs per cubic metre of air space.

1 INTRODUCTION

This paper documents the methodology used in developing a series of User Programmable Module Programmes (UPMs) and a LOTUS spreadsheet for the automated design of landfill sites for initial costing purposes. Refinements are required to produce a design for planning purposes and subsequent detailed site design.

The automated design and costing are carried out using two different although integrated computer systems. The initial design work is undertaken on a SUN Spark Station 2, using MOSS software and the costing is carried out on a 486 PC utilising a LOTUS spreadsheet. The computers are on a PCNFS network enabling electronic transfer of data files between the software packages.

2 COMPUTER SOFTWARE

The MOSS Software was initially developed in the 1970s for Highway Design. The company MOSS Systems Limited was formed in 1983 and the software has been developed substantially since then. The User Programmable Module (UPM) was introduced in 1990. The User Programmable Module (UPM), introduced in 1990, allows the functions of the MOSS Systems to be controlled by a user friendly programme written in a basic language. Through the use of UPM, programmes have been written to automate the landfill design process. LOTUS is a standard spreadsheet package which enables data to be manipulated and interrogated graphically. Information produced in the design process, through running the UPMs, is imported to the worksheet created for costing purposes.

3 AUTOMATED DESIGN PROGRAMME APPROACH

The programmes written carry out the following operations:

- designs landfill and restored landform
- designs basal liner, soil and capping models and outputs areas and earthwork volumes
- creates cross sections through all the models for checking purposes
- inserts monitoring wells, venting wells, and leachate drains. Outputs lengths and numbers of the above

- creates a 3D view of the existing, proposed and restored models.

The programmes have taken about 100 hours to develop, producing approximately 2000 lines of programming.

3.1 Site Design

The automated design of the site requires various elements. Initially a ground model is required giving an accurate representation of the existing surface. This can be obtained by surveying the site, or if appropriate purchasing Ordnance Survey digital data. The limit of the landfill can then be decided and added to the system most conveniently by use of a digitiser.

A grade is required across the site, to provide adequate flow in the basal leachate drainage system, and side slope gradients will be dependent on the geotechnical properties of the enclosing strata. A geotechnical engineer should be consulted at an early stage as stability, access and liner integrity are of prime importance. The above parameters provide sufficient data to enable the base of the site to be designed.

A restored surface profile is required. In general gradients of no less than 1 in 20 have been adopted to take account of possible differential settlement and allow free drainings. The programme generates the required profile.

It is essential to visually check the design parameters and basic site design for any obvious errors following which an accurate air space volume can be calculated. It is then possible to repeat the process or to interactively edit the model. When a suitable void design is achieved, ancillary site design detail can be incorporated into the model.

3.2 Gaswells

Waste Management Paper 26 recommends that where waste is placed in thicknesses exceeding 8m an active gas system will be required. The spacing of these wells is normally on a regular grid spacing of about 50 metres. The programme gives the facility to alter this grid interval, and only places wells where the depth of the waste exceeds 8m.

3.3 Leachate Drainage System

The emplacement and phasing of landfill operations is optimised to minimise leachate generation. Water balance calculations can be carried out ostensibly to ensure that all water entering the landfill is absorbed by the waste. Limited credibility is now given to these calculations and more importance is now being placed on the design of an adequate leachate drainage system. The programme is designed to make use of a sump with a riser pipe in the slope of the liner. These leachate drainage pipes are

TABLE 1

ITEM	UNIT
Surface Water Interception Channel	lin m
Lining Surface Area	m^2
Leachate Collection Pipework	lin m
Topsoil Stripping	m^2
Earthworks - cut & fill balance	m^3
Fencing	lin m
Gas Vents - number & meterage	No & lin m
External Monitoring Boreholes	No & lin m

TABLE 2 PRINCIPAL COST ELEMENTS		
WMP 26 REF. NOS	**Items**	**Origin of Data Input**
1	Acquisition of site; purchase, lease or royalty	UNIT COST
2	Surface water interception channel	MOSS
3	Groundwater cut-off drain	UNIT COST
4	Groundwater cut-off up to 6m deep	UNIT COST
5	Groundwater separation (on base)	UNIT COST
6	Culvert (below landfill)	UNIT COST
7	Lining	MOSS
8	Leachate collection	MOSS
9	Leachate removal	MOSS
10	Leachate treatment lagoon	UNIT COST
11	Leachate drains, small diameter sewers	UNIT COST
12	Screen bunds	UNIT COST
13	Topsoil stripping	MOSS
14	Earthworks for bunds and road embankments	MOSS
15	Preparatory excavation for cover material	UNIT COST
16	stockpile for cell 1	UNIT COST
17	Access Road	MOSS
18	Fencing	MOSS
19	Tree planting	UNIT COST
20	Site office and base	UNIT COST
21	Wheel cleaner	UNIT COST
22	Weighbridge	UNIT COST
23	Garages / workshops / fuel stores	MOSS
24	Gas vents	UNIT COST
25	Gas utilisation plant	UNIT COST
26	Site specific requirements eg.	MOSS
27	Detailed landfill design % of total cost	
	Other costs	

MOSS - Information generated from 3D model
Unit Cost - Information generated from cost database.

spaced as required across the base of the landfill.

3.4 Monitoring Boreholes

External monitoring boreholes are required for both groundwater and landfill gas. The programme allows for the borehole spacing to be specified and calculates the depth to 1m below the lowest formation level of the site.

3.5 Elements Quantified From The 3D Design

Table 1 illustrates those elements which are quantified in the MOSS models.

4 THE COSTING

The Lotus spreadsheet calculates engineering costs and capital cost for the design developed within MOSS.

Table 2 shows the principal elements, as detailed in Annex 4 of Waste Management Paper 26 and indicates the source of the input for the costing spreadsheet.

The current design package can generate unit quantities from the MOSS System in approximately an hour of processing time for a site of 2 to 3 million cubic metres. About 30 megabytes of disc space is used during the processing, the majority of which can be retrieved once the programmes are run.

5 CONCLUSIONS

Increased environmental concern and resulting legislation have driven the technology of landfilling away from the philosophy of dilute and disperse to the fully engineered containment system. The programmes enable rapid cost evaluation of engineered containment systems. There are additional facilities in the spreadsheet to enable different types of lining systems to be examined. It also provides graphical representation of costs per cubic metre for different design parameters, and sensitivity analysis by inflating single or multiple cost elements.

Further developments are in progress to enable operational costs to be included in the development of the site. With this additional cost data a financial model can be prepared for the business plan, detailing such items as Start-up Cost, Working Capital, Cashflow Statements and Associated Financial Analysis.

REFERENCES

DoE (1990), Waste Management Paper No. 26 : Landfilling Wastes, London HMSO

DoE (1991), Waste Management Paper No. 27 : Landfill Gas, London HMSO

ACKNOWLEDGEMENTS

Wardell Armstrong for funding the Institute of Waste Management Diploma and Training which made the project possible. MOSS Systems Limited for providing the software to enable this project to be carried out.

Waste Disposal by Landfill – GREEN'93, Sarsby (ed.) © *1995 Balkema, Rotterdam, ISBN 90 5410 356 6*

Micropore-permeability of compacted clay liners

E. Fratalocchi & E. Pasqualini
University of Ancona, Italy

M. Manassero
IG Ingegneria Geotecnica, Turin, Italy

ABSTRACT: The present paper shows that the degree of saturation, S_{ri}, of laboratory and in situ compacted samples can be univocally correlated with their micropore hydraulic conductivity measured in the laboratory. Therefore the evaluation of S_{ri} based on the measurement of water content, dry density and specific gravity of the soil particles (fast and cheap) allows one to limit the number of laboratory hydraulic conductivity tests.

1 INTRODUCTION

A proper design of waste disposals requires an accurate knowledge of the hydraulic conductivity of liners and covers. For compacted earthen liners, the hydraulic conductivity depends on soil structure, which should be as uniform as possible with minimal void space between soil grains (micropores). Nevertheless, since a great number of factors affects the liner construction, earthen liners could contain macropores or secondary features which transmit amounts of flow greater than micropores. The secondary features can result from a lapse during construction quality control or from an adverse combination of soil condition and construction procedures (Trautwein and Williams, 1990). Therefore, compacted clay liners need to be constructed using accurate procedures (Daniel, 1990) designed to eliminate or at least minimize the presence of secondary features; this aspect underlines the importance of construction quality control to ensure proper methods of construction and compaction.

In order to account for the impact of secondary features on hydraulic conductivity, it is necessary to perform large-scale in situ permeability tests. In fact, several case histories (Daniel, 1984; Day and Daniel, 1985; Elsbury et al., 1990) have demonstrated that secondary features can result in actual hydraulic conductivity values greater than those measured by laboratory tests. In situ permeability tests are very reliable both for the final control of the in situ permeability of covers and liners and in preliminary test pads to verify that construction materials and methods will produce the desired in situ hydraulic conductivity (Day and Daniel, 1985; Daniel, 1989). However, since the large-scale in situ permeability tests require a long time for preparation and execution, they are unfit for quality control during construction.

But also the laboratory permeability tests to be performed on undisturbed samples taken from compacted liners often take several days or more to complete, which minimizes their utility as a quality control tools; therefore, the recommended tests for ensuring that the soil is properly compacted are the measurements of the water content and density (Daniel, 1990).

The present paper shows that the measurement of both water content and dry density obtained from the same sample of compacted soil can be used for a rough evaluation of its hydraulic conductivity. In fact, experimental data obtained at the University of Ancona show that the degree of saturation, S_{ri}, of the compacted sample (which reflects both water content and dry density) can be univocally correlated with the hydraulic conductivity of compacted soil measured in the laboratory.

Of course the correlation between S_{ri} and laboratory permeability has the limitation of reflecting only the "micropore permeability" of the compacted earthen liners. Nevertheless it can be considered an important item of evidence for the construction quality control, since the attainment of an assigned value of S_{ri} can be assumed to be a necessary condition (even if not sufficient) to obtain a prefixed permeability value.

Pasqualini and Fratalocchi (1993) have already discussed this topic. The present paper, based on more experimental data, seems to confirm those results.

2 MATERIALS AND METHODS

The present study considers three typical materials used for constructing compacted earthen liners. They have similar plasticity and grading characteristics, as shown in Table 1. For details, see Pasqualini and Fratalocchi (1993).

Table 1. Index properties of the tested materials.

material	Pianfei	Fossano	Orbassano
LL (%)	48	42	37
PL (%)	27	24	22
PI (%)	21	18	15
gravel (%)	2	3	2
sand (%)	12	13	22
silt (%)	61	52	48
clay (%)	25	32	28
Gs (-)	2.766	2.760	2.787

All three materials were compacted in the laboratory using standard Proctor procedure (ASTM D698); Fossano clay was compacted using also modified Proctor (ASTM D1557). The compaction tests were conducted on air-dried soil, which was pulverized to ASTM n°4 sieve in order to obtain small clods; such an accurate preparation and compaction allows to minimize secondary features (Daniel and Benson, 1990).

The permeability tests were performed using flexible wall permeameters, described in detail by Daniel et al. (1984) and Boynton and Daniel (1985). The tests were constant head tests and usually lasted about two or three weeks. Since the hydraulic gradient and the effective stress affect the measured k values, all the permeability tests were performed at the same hydraulic gradient (i = 50) and at the same effective confining pressure ($\sigma'_{average}$ = 100 kPa).

Of course, when the correlation k = f(S_{ri}) is used to estimate the in situ k values, the proper value of σ' should be selected considering the in situ effective pressure.

Some laboratory permeability tests were performed on undisturbed specimens prepared from cubic samples taken from a test pad.

3 RESULTS AND DISCUSSION

Mesri et al. (1994) show that the hydraulic conductivity of natural saturated soft clays of marine and lacustrine origin is univocally correlated to the current void ratio, e, of soils. This univocal correlation between hydraulic conductivity and void ratio does not apply to the unsaturated compacted cohesive soils used for earthen liners. This is illustrated in Figures 1, 2, 3 and 4, which outline the compaction curves with the related values of hydraulic conductivity, k, obtained for the three soils considered: for a given value of dry unit weight, γ_d, (that is, for a given value of void ratio, e) two different permeability values can be obtained, since k depends on the water content, w, of the soil too.

But since S_{ri} = (w G_s / e), where S_{ri} = degree of saturation after compaction (before starting permeability tests) and G_s = specific gravity of the soil particles, S_{ri} is found to reflect both water content and void ratio (or dry density). This suggests a univocal correlation linking k and S_{ri}. In particular the experimental data shown in Figure 4 (de Rosa, 1993) confirm that the increase in permeability of the

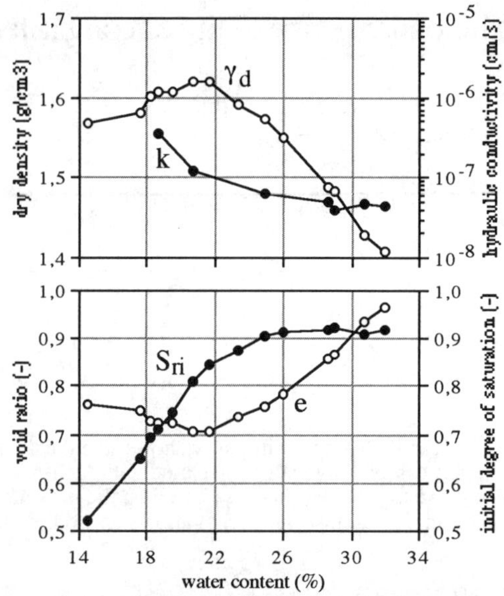

Figure 1 - Pianfei. Standard Proctor compaction. Relationships between hydraulic conductivity and dry density, void ratio, initial degree of saturation.

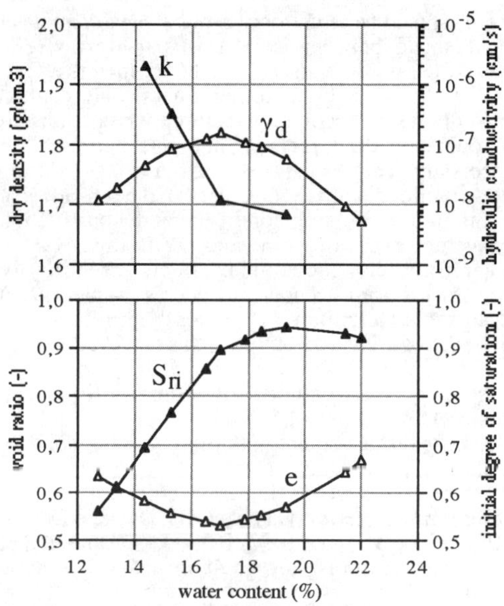

Figure 2 - Orbassano. Standard Proctor compaction. Relationships between hydraulic conductivity and dry density, void ratio, initial degree of saturation.

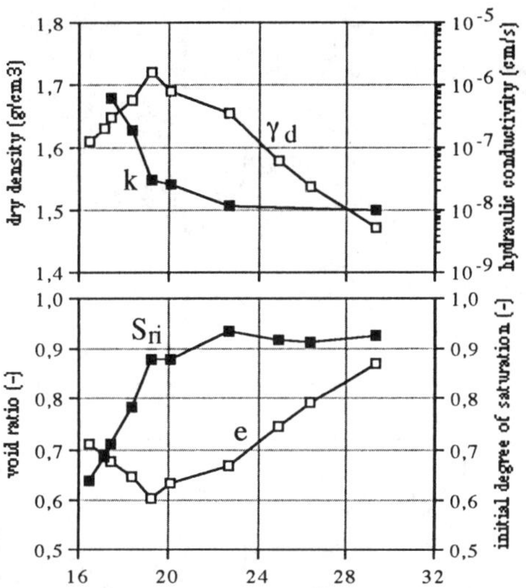

Figure 3 - Fossano. Standard Proctor compaction. Relationships between hydraulic conductivity and dry density, void ratio, initial degree of saturation.

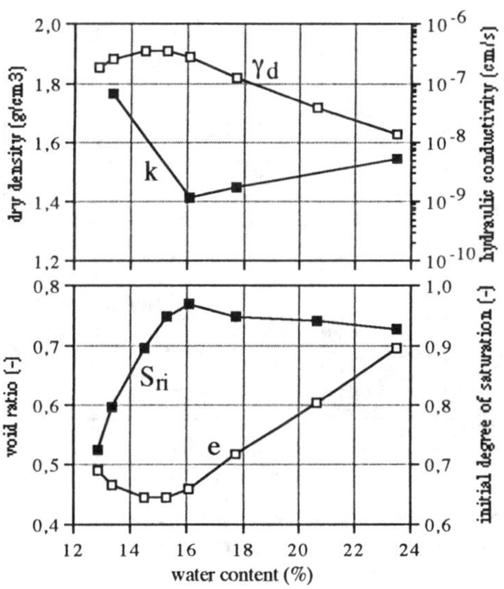

Figure 4 - Fossano. Modified Proctor compaction. Relationships between hydraulic conductivity and dry density, void ratio, initial degree of saturation.

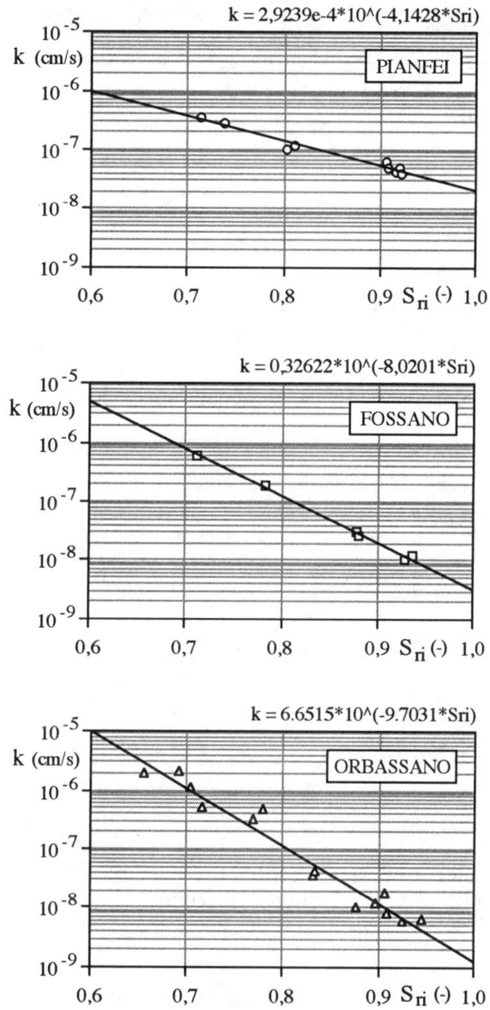

Figure 5 - Standard Proctor compaction. Hydraulic conductivity versus initial degree of saturation for the tested materials.

soil when compacted at high water content (Mitchell et al., 1965; Daniel and Benson, 1990; Leroueil et al., 1992) can be correlated to a decrease in S_{ri}.

Finally, Figure 5 shows that, for each soil, there is a linear relationship between *log*k and S_{ri}, which fits the experimental data with high correlation coefficients (at least for the investigated S_{ri} values). This relationship allows us to estimate both the constant $C_{ks} = (-\Delta log k / \Delta S_{ri})$ and the theoretical value of the minimum hydraulic conductivity, k_{min}, obtainable using the tested soil.

k_{min} is the hydraulic conductivity value evaluated at $S_{ri} = 1$ and it can be considered as an index of the best performance given by the compacted soil examined. Of course the best estimation of k_{min} is possible when the mathematical expression linking k to S_{ri} results from experimental data including specimens with S_{ri} values close to one.

This is evident in Figure 6, which compares the correlation obtained from data on standard and modified Proctor-compacted specimens. In the authors' opinion, the results of Figure 6 do not prove that the relationship between k and S_{ri} depends on the compaction energy, but rather that the mathematical

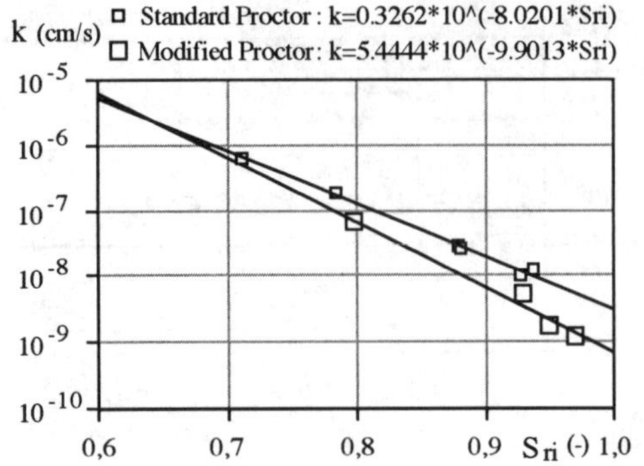

Figure 6 - Fossano. Comparison between specimens compacted with standard and modified Proctor procedure.

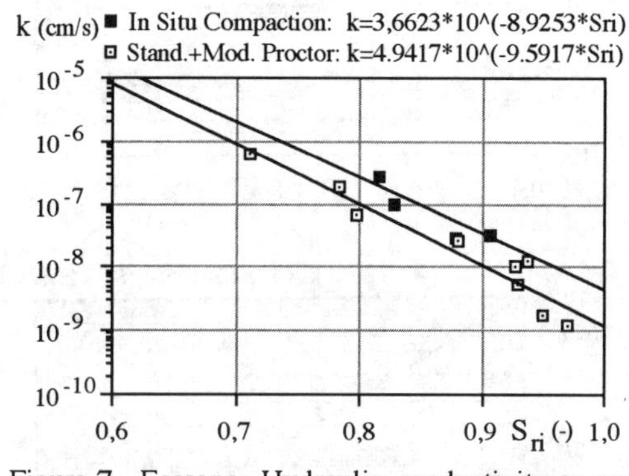

Figure 7 - Fossano. Hydraulic conductivity versus initial degree of saturation: comparison between specimens compacted in the laboratory and in situ.

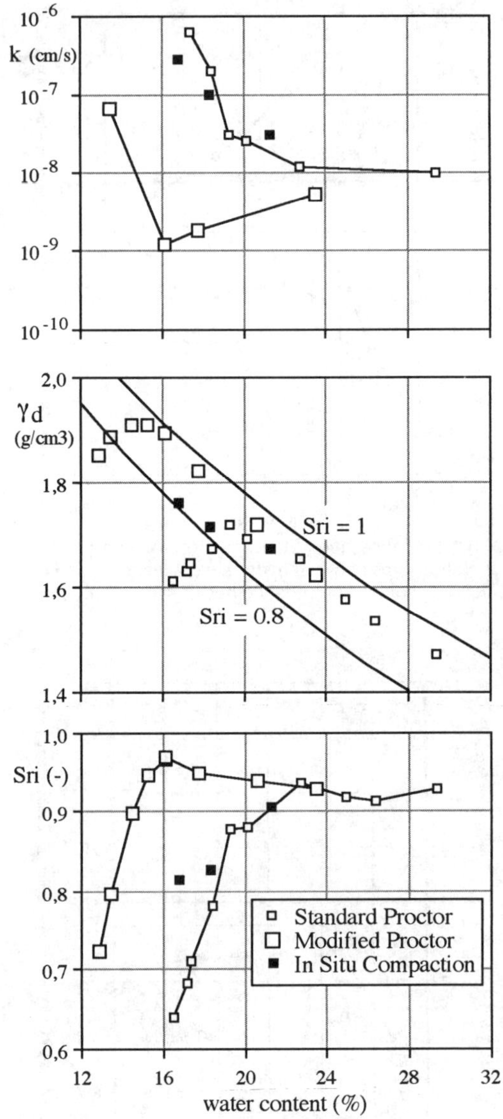

Figure 8 - Fossano. Comparison between specimens compacted in the laboratory and in situ.

expression interpolating the experimental data may depend on the data range.

Finally, the values of dry density and laboratory hydraulic conductivity, measured on undisturbed specimens sampled from a test pad are plotted in Figures 7 and 8 (for details on the test pad see Filipponi, in preparation). The data shown in Figure 8 seem to indicate that the in situ compaction goes between standard and modified Proctor compaction and that there is no apparent relation between the permeability values of specimens compacted in situ and in the laboratory. On the contrary, Figure 7 shows that this relation exists if the permeabilities of both in situ and laboratory-compacted specimens are plotted against S_{ri}.

Table 2 summarises the C_{ks} and k_{min} values for the tested materials. This result seems to support the use of S_{ri} measured on soils compacted in situ as an index of in situ hydraulic conductivity during construction quality control. This is in agreement with the suggestion of a specified degree of saturation, such as that used by Lathi et al. (1987) as a lower bound of the acceptable zone (Figure 8). However, Figure 7 also shows that the three undisturbed specimens taken from the test pad have hydraulic conductivity values above the correlation line obtained from laboratory specimen data; this confirms that the final control should be based on measurement of the in situ hydraulic conductivity by means of laboratory tests on undisturbed specimens sampled from the compacted liner or, preferably, by means of large-scale in situ permeability tests.

Table 2. C_{ks} and k_{min} values for the three materials examined.

material and compaction	C_{ks} (cm/s)	k_{min} (cm/s)
PIANFEI standard Proctor	4.143	2×10^{-8}
ORBASSANO standard Proctor	9.703	1×10^{-9}
FOSSANO standard Proctor	8.020	3×10^{-9}
FOSSANO modified Proctor	9.901	7×10^{-10}
FOSSANO standard and modified Proctor	9.592	1×10^{-9}
FOSSANO in situ compaction	8.925	4×10^{-9}

4 CONCLUDING REMARKS

The results and considerations reported in the present paper must be considered preliminary. Much more information and experimental data are needed to prove the reliability and usefulness of the relationships linking hydraulic conductivity of unsaturated compacted soil to the initial degree of saturation. In particular the authors know that other researchers (Elsbury et al., 1990) found no linear relationship between log k and S_{ri}; however, this apparent contrast could be explained by the following considerations.

The results shown in the present paper concern only specimens obtained by compacting "small clods" (pulverized to pass n°4 sieve = 4.75 mm); on the contrary, Elsbury et al. also prepared specimens by compacting "large clods" (19 mm), in order to take into account the influence of secondary features on the hydraulic conductivity of the compacted specimens and to better represent the soil compacted in the field. As a consequence of these different compaction criteria, Elsbury et al. obtained a wider range of S_{ri}; in fact their S_{ri} values range from almost 30% to 100%, while our S_{ri} values lie between 65% and 95% (see Table 3). It follows that the correlation between k and S_{ri} discussed in the present paper are valid only when the measured k value reflects the micropore permeability of the compacted soil and is not affected by secondary features.

Acknowledgements

The work described in the present paper is a part of the research of national interest, named "Geotecnologie finalizzate alla salvaguardia e al recupero del territorio". The authors are grateful to the M.U.R.S.T. (Ministero della Università e della Ricerca Scientifica e Tecnologica) for partial finantial support.

Table 3. Values of the experimental data analysed in the present paper.

material and compaction	water content (%)	dry density (g/cm^3)	hydraulic conductivity (cm/s)
ORBASSANO	14.40	1.764	2.1×10^{-6}
standard	15.30	1.793	3.2×10^{-7}
Proctor	17.04	1.822	1.2×10^{-8}
	17.30	1.665	5.0×10^{-7}
	19.35	1.775	6.7×10^{-9}
*	15.16	1.806	4.6×10^{-7}
*	17.49	1.813	1.8×10^{-8}
*	19.40	1.758	6.1×10^{-9}
*	14.00	1.793	1.1×10^{-6}
*	17.18	1.802	1.0×10^{-8}
*	19.69	1.738	8.5×10^{-9}
*	14.39	1.728	1.9×10^{-6}
*	18.50	1.720	3.7×10^{-8}
*	18.60	1.718	4.1×10^{-8}
PIANFEI	18.65	1.606	3.6×10^{-7}
standard	20.08	1.579	2.8×10^{-7}
Proctor	20.73	1.621	1.2×10^{-7}
	21.35	1.595	1.0×10^{-7}
	24.84	1.573	6.5×10^{-8}
	28.56	1.488	5.1×10^{-8}
	28.92	1.482	4.0×10^{-8}
	30.77	1.429	4.8×10^{-8}
	31.99	1.408	4.4×10^{-8}
FOSSANO	17.40	1.647	6.2×10^{-7}
standard	18.37	1.675	1.9×10^{-7}
Proctor	19.23	1.720	3.0×10^{-8}
	20.16	1.691	2.6×10^{-8}
	22.67	1.655	1.2×10^{-8}
	29.33	1.474	1.0×10^{-8}
FOSSANO	13.40	1.885	6.5×10^{-8}
modified	16.10	1.892	1.2×10^{-9}
Proctor	17.75	1.820	1.8×10^{-9}
	23.50	1.625	5.2×10^{-9}
FOSSANO	16.79	1.760	2.7×10^{-7}
in situ	18.32	1.714	1.0×10^{-7}
compaction	21.29	1.675	3.1×10^{-8}

* Specimens compacted with energy values slightly different from those provided in the standard Proctor procedure (see Ferranti and Fratalocchi, 1992).

REFERENCES

Boynton, S.S. & D.E. Daniel 1985. Hydraulic conductivity tests on compacted clay. *Journal of Geotechnical Engineering* 111: 465-478.

Daniel, D.E. & C.H. Benson 1990. Water content-density criteria for compacted soil liners. *Journal of Geotechnical Engineering* 116: 1811-1830.

Daniel, D.E. 1984. Predicting hydraulic conductivity of clay liners. *Journal of Geotechnical Engineering* 110: 285-300.

Daniel, D.E. 1989. In situ hydraulic conductivity tests for compacted clay. *Journal of Geotechnical Engineering* 115: 1205-1225.

Daniel, D.E. 1990. Summary review of construction quality control for earthen liners. *Waste Containment Systems: Construction, Regulation and Performance, ASCE* 26: 175-189.

Daniel, D.E., S.J. Trautwein, S.S. Boynton & D.E. Foreman 1984. Permeability testing with flexible wall permeameter. *Geotechnical Testing Journal* 7: 113-122.

Day, S.R. & D.E. Daniel 1985. Hydraulic conductivity of two prototype clay liners. *Journal of Geotechnical Engineering* 111: 957-970.

De Rosa, S. 1993. Comportamento idraulico del rivestimento di fondo in argilla compattata di una discarica controllata. Doctoral thesis, University of Ancona.

Elsbury, B.R., D.E. Daniel, G.A. Sraders & D.C. Anderson 1990. Lessons learned from compacted clay liner. *Journal of Geotechnical Engineering* 116: 1641-1660.

Ferranti A. & E. Fratalocchi 1992. Ricerca sperimentale sul comportamento dei rivestimenti di argilla compattata delle discariche controllate. Doctoral Thesis, University of Ancona.

Filipponi, B. in preparation. Campi prova e prove di permeabilità in sito per l'ottimizzazione della costruzione di una discarica controllata per rifiuti tossici e nocivi". Doctoral thesis, University of Ancona.

Leroueil, S., J.P. Bihan & R. Bouchard 1992. Remarks on the design of clay liners used in lagoons as hydraulic barriers. *Canadian Geotechnical Journal* 29: 512-515.

Mesri, G., T.W. Feng, S. Ali & T.M. Hayat 1994. Permeability characteristics of soft clay. XIII ICSMFE, New Delhi, India .

Mitchell, J.K., D.R. Hooper & R.G. Campanella 1965. Permeability of compacted clay. *Journal of the Soil Mechanics and Foundation Division* 91: 41-65.

Pasqualini E. & E. Fratalocchi 1993. Initial degree of saturation and permeability of compacted clay. IV International Landfill Symposium. S.Margherita di Pula, Italy.

Trautwein, S.J. & C.E. Williams 1990. Performance evaluation of earthen liners. *Waste Containment Systems: Construction, Regulation and Performance, ASCE* 26: 30-51.

Waste Disposal by Landfill – GREEN'93, Sarsby (ed.) © 1995 Balkema, Rotterdam, ISBN 90 5410 356 6

Finite element investigation of the hydraulics of landfill

J.R. McDougall, R.W. Sarsby & N.J. Hill
Bolton Institute of Higher Education, UK

A number of investigators during the 1980's developed mathematical models with the intention of elucidating the internal hydraulic behaviour of municipal solid waste disposal sites, or landfills. The model building is still at an early stage and is drawing upon the expertise of computational hydraulics and groundwater hydrology to define the fundamental equations of motion. Once established, these models can provide valuable information for the design of landfill sites. In particular, bottom liner and drainage layer specification is dependent upon an accurate assessment of soil moisture or leachate transport and accumulation. Moreover, as the most serious threat posed by landfill is that of leakage, either by natural permeation, or via low conductivity pathways, a more accurate representation of the driving force behind any contaminant plume improves pollution modelling.

Hitherto, most models directed specifically at landfill have concentrated separately on flow in the upper unsaturated region or saturated zone on the bottom liner. The relevant equations are well known to hydrologists, e.g., for a leaky, phreatic aquifer with infiltration, the Boussinesq equation (saturated flow) is employed:

$$\nabla[\mathbf{k}\nabla h] + q + I = S_y \frac{\partial h}{\partial t} \qquad [1]$$

where, $\mathbf{\nabla}$ = gradient operator
$\mathbf{k}$ = permeability
h = hydraulic head
q = leakage
I = net infiltration
S_y = specific yield

For unsaturated flow, the combined contributions of soil physics and groundwater hydrology are particularly helpful. Essentially, a two phase mechanism is considered, however if the non liquid phase is assumed to be stationary -(amounting to assuming constant air pressure throughout matrix)- the model reduces to a single equation. Further, if the mechanism of soil moisture transport is considered analogous to molecular diffusion, then a Fickian coefficient of capillary diffusivity can be invoked giving:

$$\nabla[D(\theta)\ \nabla\theta] = \frac{\partial \theta}{\partial t} \qquad [2]$$

where, $D(\theta)$ = capillary diffusion
θ = volumetric moisture content

Note that the dependence of $D(\theta)$ upon the volumetric moisture content means the governing equation is non-linear.

Beyond a steady state analysis, the possibility of varying volumetric moisture contents means the enforcement of diffusion type soil moisture transport may be inappropriate. At low volumetric moisture contents the electrostatic and intermolecular forces are capable of producing large energy potentials and 'diffusion' is an adequate representation.

At higher degrees of saturation soil moisture becomes continuous, matric potential is tempered, and gravitational potential becomes a significant term in the equation of motion. The flow mechanism is thus governed increasingly by hydromechanical forces until eventually at full saturation conventional linear hydraulic head characteristics pertain.

Based on the two phase flow equations but simplified as indicated above, we can write a single general equation for moisture flow in a porous media applicable to both saturated and unsaturated zones:

$$\nabla[\mathbf{k}(\theta)\nabla(p_w+z)] = \frac{\partial}{\partial t}(Sn\theta) \qquad [3]$$

where, S = saturation
n = porosity
p_w = soil water pressure
$\mathbf{k}(\theta)$ = permeability

The right hand side represents the storativity term which can be expanded and manipulated to incorporate more complex behavioural assumptions in the model, e.g. soil matrix deformation and/or variable soil water capacity can be explicitly accounted for. However, the governing equations become increasingly non-linear, for example, Neumann,1973, gives:

$$\nabla[\mathbf{k}\nabla(\Psi+z)] = (C+\frac{\theta}{n}S_s)\frac{\partial\Psi}{\partial t} \qquad [4]$$

where C = $C(\theta)$ specific water capacity
S_s = $S(\theta)$ specific storage
Ψ = $\Psi(\theta)$ soil water potential

It is the aim of the research outlined in this technical note to extend this saturated/unsaturated model of hydrologic phenomena to the particulars of landfill both during construction and progress to some volumetric steady state. At the outset, development of the appropriate mathematical description capable of accounting for the significant volumetric strains and moisture retention characteristics of M.S.W. will be sought. The main difficulties anticipated at this stage and to which this investigation will be addressed arise from:

1] the application of solution methods to highly non-linear governing equations.

2] the difficulty of specifying typical fill properties.

Solution methods - Finite element analysis

The equations cited above arise from the work of computational hydraulics/ groundwater hydrology and solution techniques favour finite difference methods. Where assumptions of matric stability and material homogeneity are tenable then this reflects the 'cheapest' approach. However, in the context of landfill, neither conditions can be realistically upheld, moreover the complex geometry of the flow domain following the incorporation of leachate drainage and pumping facilities suggests that finite element methods may represent a more effective solution method.

Recent research using uncoupled saturated/unsaturated flow models (i.e. saturated flow models treating adjacent unsaturated flow zones as boundary conditions) continue to adopt finite difference methods. Application of finite element methods was however suggested back in the early 1970's (Neumann,1973), in particular to the coupled saturated/unsaturated flow equation(s). Here, real benefits exist due to the avoidance of the need to define the free surface and continual remeshing of the flow region, moreover the effects of seepages in the unsaturated zone are accounted for.

To date only limited evidence of development of these early proposals for the saturated/unsaturated model are apparent. One such case is research by Lam Fredlund & Barbour (1987) who analysed the pattern of seepages through an earth dam during impounding. The governing equations are solved using F.E.M. with a finite difference recurrence scheme in the time dimension. From their model a graphical representation of the progress of the phreatic surface within the dam as it tends to steady state is given, but only the permeability tensor is non linear with the storativity term assumed constant within each time step. This latter term they deduce is analogous to the coefficient of volume compressibility when flow is saturated- a material parameter familiar to geotechnical engineers and routinely obtained as part of site investigation where settlements are of concern- however more sophisticated models are likely to be inhibited by the ability to accurately specify material parameters.

Material properties

It is evident that the universality of any model will be limited by the heterogeneity, of the waste fill, both inter- and intra-site(s). There is no typical cubic meter of M.S.W. to allow precise characterisation, but there is sufficient experimental data to for preliminary estimates of material parameters to be made and for model testing to be performed. In this context soil physics is helpful with a range of empirical techniques which are widely used for the determination of moisture content/suction and moisture content/permeability relationships (Childs/Collis-George,1950). These approaches still require some basic material parameter constants to be supplied but the difficulties of obtaining in situ pore water suctions are generally avoided.

Whether all material properties are so intractable to specification is open to discussion. Improved knowledge of certain material performance is emerging as a result of continuing research, e.g. the change in permeability of clay liners due to cation or acid concentrations. However, the heterogeneity of waste still obfuscates any prediction of the precise mix of leachate chemistry so material parameter inputs may have to be determined on an ad hoc basis.

References

Childs, E.C. & Collis-George,N 1950. Permeability of porous materials. Proc. Royal Society of London. 201.

Lam, L. Fredlund, D.G., & Barbour, S.L. 1987. Transient seepage model for saturated-unsaturated soil sytems: a geotechnical engineering approach. Canadian Geotechnical Jrnl, 24/4.

Neumann,S.P. 1973. Saturated-unsaturated seepage by finite elements. Jrnl Hydraulics Div. A.S.C.E. 99(HY12)

Waste Disposal by Landfill – GREEN'93, Sarsby (ed.) © 1995 Balkema, Rotterdam, ISBN 90 5410 356 6

New lining technology based on Fly-Ash Composite

B.Quant
Polish Academy of Sciences, Institute of Hydroengineering, Gdánsk, Poland

ABSTRACT: The achievements of the Institute of Hydroengineering of Polish Academy of Sciences in Gdańsk in the field of Fly-Ash Composites application in landfill installations are described in the paper. The Fly-Ash Composites are the mixtures of fly-ashes, soluble silicates (water-glass solutions) and/or lime (more often waste lime). The paper presents two principal composites: the Fly-Ash Composite and the Liquid Fly-Ash Composite. The theoretical principle of the composite creation, their basic physico-mechanical properties, their way of formation and mainly the directions of their application are shortly presented. The influence of fly-ashes on the environment is also taken into account.

1 INTRODUCTION

The research on production of strengthening and sealing mass of fly-ashes have been recently carried out in the Institute of Hydroengineering of Polish Academy of Sciences in Gdańsk. As a result of those works several home patent applications were made during last years (Chruściel, Główka 1972, Główka et al. 1986, Kałęcki et al. 1986).

The developed technologies of soil improvement may be divided in two groups, according to two types of applied materials. These materials are called: the Fly-Ash Composite and the Liquid Fly-Ash Composite. Both composites are produced are produced on the base of fly-ash, soluble silicates solution (water-glass solution) and - if needed - lime, most often waste lime. These composites are inexpensive, they base on re-claimed materials (wastes from power industry), in spite of their derivation they are inert for environment and they are widely applicable in building practice.

2 COMPREHENSIVE CHARACTERIZATION OF MATERIALS

From the point of view of the proceeding processes, the presented methods are the typical unisolution silicatization. The fact, that fly-ash itself is a coagulating agent is an essential difference in relation to the conventional silicatization. Strictly speaking, the soluble elements of fly-ash take part of the coagulant. Thus, the soluble silicates solution and soluble parts of fly-ash take part in the process of delayed coagulation, and clearly - delayed gelation.

As it was mentioned above, this paper describes in principle two types of materials. The Fly-Ash Composite is a material, that is produced by mixing, in adequate proportion, the fly-ash and water-glass solution, and next by its compaction. Because of the necessity of compaction, the quantity of solution must correspond to the optimum water content of applied fly-ash.

In the second case - the Liquid Fly-Ash Composite - one deals with similar material. The dif-

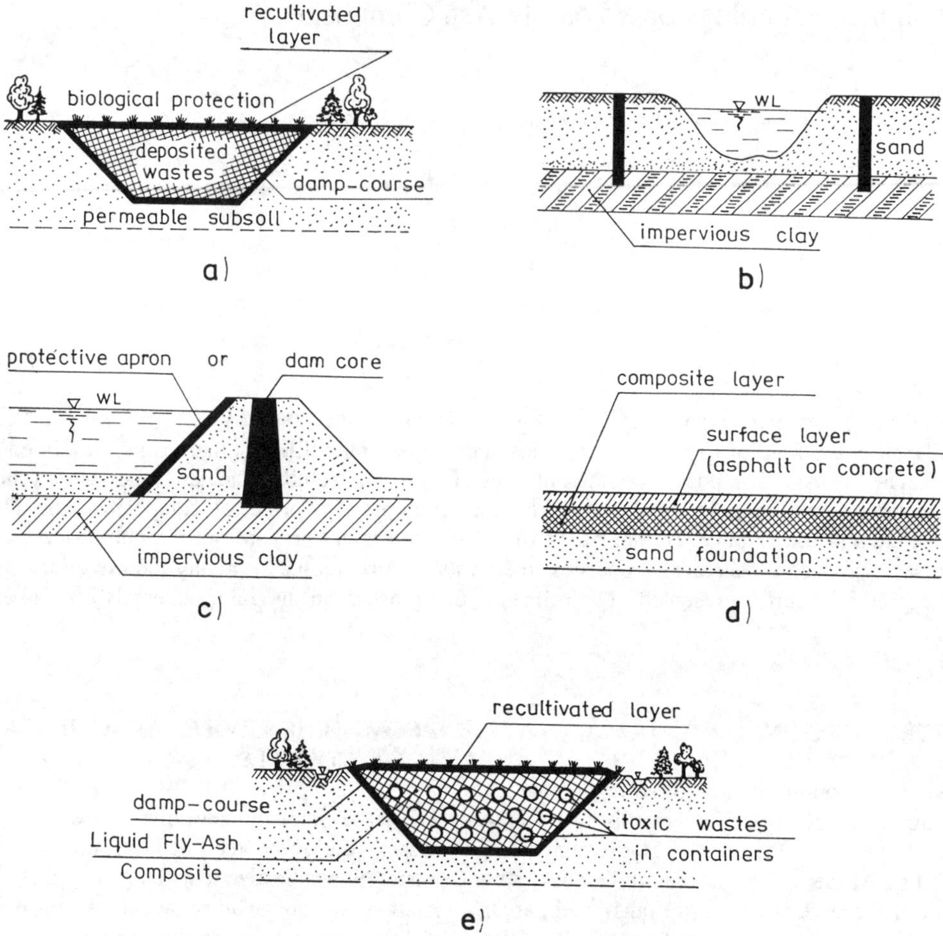

Figure 1. Directions of application of Fly-Ash Composites

ference lies in the quantity of added water-glass solution, which must be much higher that the optimum water content. Quantity ratio of fly-ash to solution can vary in range from 1:0.5 to 1:5.

The mentioned composites differ very much in the technology of formation and in their properties. Those differences will be discussed in further parts of the paper.

3 PHYSICO-MECHANICAL PROPERTIES

3.1 *Fly-Ash Composite*

The unconfined strength (R_s) and the coefficient of permeability (K_{10}) are main physico-mechanical properties of the Fly-Ash Composite. The values of these parameters can vary in wide ranges, depending on many factors. The factors, which mostly influence those two parameters are: the chemical composition (mainly calcium oxide content) and the quantity of colloidal silica added with the water-glass solution.

The value of the unconfined compressive strength varies from a fraction to ca. 5 MPa with the increasing tendency, and the value of coefficient of permeability - in range: 10^{-9}- 10^{-10} m/s with the decreasing tendency in time. The results of investigation of practically applied composites are presented in Tables 1 and 2.

When calcium content in fly-ash is very low, it is necessary to enrich the fly-ash with this

Table 1. Unconfined compressive strength changes in time [MPa].

Age	Object			
[years]	I	II	III	IV
"0"	1.2	1.2	3.2	3.9
1	1.3	1.3	3.6	4.0
2	1.6	2.6	4.5	4.8
3	2.3	3.7	4.7	5.2
4	2.4	3.8	4.9	5.3
5	3.5	3.9	5.0	5.6
6	3.6	3.9	5.2	5.6
7	3.7	4.0	5.2	5.7
8	3.7	---	5.2	---
9	3.8	---	---	---

Table 2. Coefficient of permeability changes in time.

Age	Object			
[years]	I	II	III	IV
"0"	$6.7*10^{-8}$	$9.7*10^{-9}$	$3.3*10^{-8}$	$8.9*10^{-9}$
1	$3.5*10^{-8}$	$4.3*10^{-9}$	$1.2*10^{-8}$	$8.2*10^{-9}$
2	$9.7*10^{-9}$	$9.9*10^{-10}$	$1.3*10^{-8}$	$7.8*10^{-9}$
3	$3.2*10^{-9}$	$8.8*10^{-10}$	$1.1*10^{-8}$	$6.8*10^{-9}$
4	$2.0*10^{-9}$	$2.8*10^{-10}$	$1.0*10^{-8}$	$2.2*10^{-9}$
5	$1.2*10^{-9}$	$2.1*10^{-10}$	$1.0*10^{-8}$	$2.1*10^{-9}$
6	$6.6*10^{-10}$	$2.1*10^{-10}$	$1.1*10^{-8}$	$2.1*10^{-9}$
7	$5.6*10^{-10}$	$2.1*10^{-10}$	$1.3*10^{-8}$	$1.8*10^{-9}$
8	$5.8*10^{-10}$	---	$1.2*10^{-8}$	---
9	$5.6*10^{-10}$	---	---	---

element by adding the material of high calcium concentration. It is recommended to use different kinds of lime, especially the waste lime, such as carbide lime for example.

Two parameters of the highest importance were mentioned above. In many cases more accurate characterization of the Fly-Ash Composite appears to be imperative. It would be useless to describe the methods and the results of investigation. However it might be worth adding, that the investigation, which has been made until now, allows to define the effect of initial composition on such parameters as: frost susceptibility, resistance to static and dynamic water activity, resistance to the activity of many mechanical and chemical agents, modulus of elasticity, rheological properties and many others (Quant 1986, 1989). One should emphasize,that the Fly-Ash Composite formed in accordance with author's instruction, shows the sufficient resistance to the activity of most external agents.

3.2 *Liquid Fly-Ash Composite*

Similarly as in the case of the Fly-Ash Composite, two parameters are of the highest importance in characterization of the Liquid Fly-Ash Composite. These are: the unconfined compressive strength (R_s) and the coefficient of permeability (K_{10}). The last one is almost independent of the material composition, and except for special cases, when technological requirements are not accomplished, it varies most often in range from 10^{-11} to 10^{-12} m/s.

The unconfined compressive strength of the Liquid Fly-Ash Composite depends principally on three factors: calcium oxide content, concentration (density) of water-glass solution and quantity ratio of fly-ash to solution. According to those three factors the strength of composite varies from a fraction to over 10 MPa.

In the instance of Liquid Fly-Ash Composite, the calcium oxide content and the density of water-glass solution must be much higher than in the instance of Fly-Ash Composite. There is no problem with application of higher density solution - the lower dilution of trade product is proper. The condition of higher calcium content is more difficult to execute. Only lignite fly-ashes satisfy that requirement. However it appears, that the difficulties can be overcome by addition of materials with the high calcium content in the quantity, that ensures the reaching of total calcium content not less than 15 per cent.

The Liquid Fly-Ash Composite can be used in the circumstances of high water content or even under water. It is possible to select its composition in such a way, that the dilution with water, inevitable in such circumstances, does not deteriorate the parameters of the composite, and even intensifies its setting.

4 TECHNOLOGY OF FORMATION

4.1 *Fly-Ash Composite*

Low costs and especially simplicity of formation are the most important advantages of the presented method. Each time, the whole practical work consists in several steps turned one after the other consecutively:

- supply of the fly-ash from storage or from Electric Power Station;
- formation of the layer of loose fly-ash on the ground surface;
- addition of lime, if necessary, and accurate mixing with fly-ash;
- preparation of water-glass solution and sprinkling the fly-ash with diluted solution;
- compaction.

The equipment applied in technical formation of the lining made of the Fly-Ash Composite can vary and depends on accessibility, a scale of enterprise and also a performer's inventiveness. The experience, that has been achieved until now, evidences, that the best results are obtained when one applies simple and accessible equipment, such as bulldozer, agricultural fertilizer distributor, soil miller, electric or petrol powered pump (e.g. fire pump) etc. Application of more complicated equipment does not bring better results and increases the costs of enterprise. The authors allow different technologies and equipment, of course after prior consultation.

4.2 *Liquid Fly-Ash Composite*

Technology of the Liquid Fly-Ash Composite formation is almost identical to concrete technology. Components of the composite are blended in a stirrer such as concrete-mixer, and after several minutes, the mixture is poured into a place of destination. The Liquid Fly-Ash Composite does not require compaction.

In case of the Liquid Fly-Ash Composite with short setting time it is necessary to keep the proper order of dosage of components, viz.: water - fly-ash - concentrated water-glass solution. When the composite is applied under water it is recommended to use the water-glass solution with the density not less than 1120 kg/m^3.

5 ENVIRONMENT PROTECTION

The fly-ashes can be a serious threat to the environment. Very often, when the fly-ashes are removed in a wrong way, they really are the source of the environmental pollution, mainly by heavy metals leaching. This problem is often presented in enormous proportions, however it exists and must be included. In Table 3 there are presented the ranges of concentration of some heavy metals in fly-ashes from several Polish Electric Power Stations. Those results are presented in comparison with the medium concentrations of these metals in farming lands in Poland.

The investigation carried out in the Institute of Hydroengineering (Kuziemska et al. 1988, 1993) establishes, that the information about the total contents of heavy metals is not complete. In the most unfavourable circumstances of static leaching, at most (even ?) 1/3 of included metals can be rinsed out. While the filtration of water through the fly-ash is going on, the problem of heavy metals leaching seems to be different. In such appearance heavy metals can be rinsed out in quantity of 70 - 80 per cent of their total contents. Thus, the practical fly-ash threat to the environment depends on the scale of problem.

In the instance of fly-ash utilization as a component of the Liquid Fly-Ash Composite or the Fly-Ash Composite, the above mentioned question can be almost left out of amount. The investigations of composites solubility, repeated many times in the Institute of Hydroengineering have proved, that the leaching of heavy metals from those materials had never exceeded the value of 5 per cent of total content. It must be added however, that in several cases the concentration of determined metals in filtrates exceeded the concentrations permissible for potable water (according to Polish legislation rules). That excesses were small and were found only in very small quantities of filtrates, while the filtration was started. If one includes the value of coefficient of permeability, and so the yield of filtration through the composites layer, one can practically ignore the threat of fly-ash to environment. The exemplary results of such study are presented in Table 4.

Table 3. Heavy metals contents in Polish fly-ashes and soil [ppm].

Metal	Fly-ashes	Soil
Pb	12 - 340	39
Zn	130 - 610	50
Cu	50 - 120	12
Cd	- 5	< 1
Co	12 - 42	6
Mn	300 - 3000	333
Cr	260 - 1950	31
Ni	80 - 180	15

Table 4. Heavy metals leaching in filtration process.

Metal	Degree of leach	Medium conc*	Max. conc.	Potable water
-	%	ppm	ppm	ppm
Zn	3.06	0.962	13.6	5.0
Pb	0.16	0.015	0.21	0.1
Cu	0.08	0.008	0.12	0.5
Cd	0.46	0.009	0.12	0.05

*/ examination were stopped in the moment when the filtrates composition has not differed from water

6 APPLICATION

The formation of horizontal and vertical hydro-isolating cut-off walls is a principal direction of the application of both composites. It refers to sealing of bottoms and slopes of existing and newly built water reservoirs, sealing of hazardous waste dumps, isolation of water reservoirs and waste-dumps with the vertical sheet pile wall, sealing of earth dams and river embankments, and so on.

The use of the Fly-Ash Composite in civil engineering and road construction, as a road and car park foundation and also as a foundation layer under floor in factory buildings, warehouses etc., is a second direction of its application.

Possibility of grass growing on the Fly-Ash Composite predisposes it to application as a cover lining of different landfills, i.e. to the role of recultivated layer. Detaching of deposited wastes from the atmosphere is a principal purpose of such lining.

The Liquid Fly-Ash Composite can be satisfactorily applied in clogging the mine- and bore-holes, stabilization of vertical excavation walls by the gunit method and also in dumps of toxic wastes.

A number of the above mentioned trends of composite applications are shown in Figure 1 enclosed at the end of the paper.

The Fly-Ash Composite has been applied in practice in several places in Poland. The technological projects for next several objects have also been prepared. The costs appeared 3 - 6 times lower in relation to traditional methods (geo-membranes, clay, tar paper on concrete etc.). The assumed physico-mechanical properties of the linings made of the Fly-Ash Composite were achieved each time. It has been stated, that the unconfined compressive strength increased and the coefficient of permeability decreased with time (see Tables 1 and 2).

The Liquid Fly-Ash Composite has not been applied in practice until now.

7 CONCLUSIONS

The paper presents two types of composites based on silicatized fly-ash. One of them was applied in practice and its predicted properties were confirmed. The examination of ground water under and around the waste-dumps proved the efficacy of the applied solution. Apart from mentioned properties, it should be added, that the Fly-Ash Composite is characterized by very good adhesion to soil and to concrete and its shrinkage is almost unnoticeable. Finally, it does not require expansion joints during practical formation.

REFERENCES

Chruściel,J. & W.Główka 1972. The way of the surface soil stabilization. *Polish Patent* No.85921.

Główka,W., H.Kałęcki, J.Nowak & B.Quant 1986. The way of formation of the two-layer damp-course. *Polish Patent* No. 152589.

Kałęcki,H., J.Nowak & B.Quant 1986. The detaching of waste dumps from environment. *Polish Patent* No. 151335.

Kuziemska,I., B.Quant & Z.Sułek 1988. Effect of industrial waters of fly-ash storage from a storage yard of grate wastes on organisms in waters of the Gdańsk coastal region (Northern Poland). *Acta Hydrobiologica* 30: 317-328.

Kuziemska,I., B.Quant & Z.Sułek 1993. Prediction of the influence of grate-waste storage yard in Przegalina on the environment. *Proc. 4th Intern. Symp. on the Reclamation, Treatment and Utilization of coal mining wastes.* Kraków.

Quant,B. 1985. Chemical composition as a base for prediction of parameters of viscoelastic model of Fly-Ash Composite. *Rozprawy Hydrotechniczne* (Hydrotechnical Trans.) 47: 151-191 (in Polish).

Quant,B. 1989. Fly-ashes in hydro- and civil engineering. *Gospodarka Wodna (Water Management)* 5: 105-108 (in Polish).

Waste Disposal by Landfill – GREEN'93, Sarsby (ed.) © *1995 Balkema, Rotterdam, ISBN 90 5410 356 6*

The use of industrial by-products to form landfill caps

R.W. Sarsby
Bolton Institute of Higher Education, UK

S. Finch
Soil Mechanics Ltd, Doncaster, UK

ABSTRACT: This paper relates to a preliminary experimental investigation into the utilisation of two readily obtainable waste materials to make a suitable alternative landfill final capping material for waste disposal landfill sites. The waste materials which were chosen because of their wide spread availability were, Pulverised Fuel Ash and Papermill sludge. By combining these two by-products a material was obtained which came close to satisfying the requirements for a capping material, i.e. low permeability and plasticity. The addition of small quantities of bentonite (not itself a waste product) to achieve complete satisfaction of the capping requirements, and produce a more workable material, was also investigated. The individual materials, and various mixes, were subjected to compaction, classification and permeability tests. The resultant data have been used to identify suitable mix proportions for further research.

ENGINEERED WASTE DISPOSAL

Of the 100 million or so tonnes of controlled wastes being produced each year in the United Kingdom, over 90% goes directly to landfill sites. The remaining wastes usually undergo a pre-treatment process before being deposited in a landfill site. When waste material is deposited into a landfill site it will have the potential to create severe environmental pollution by the accumulation of leachate and landfill gases. The initial approach to landfill leachate control was the 'dilute and disperse' principle whereby leachate was permitted to exit from a landfill site in the belief that it would be attenuated naturally in the ground and become so diluted within the groundwater that no harmful effects would result. Experience with leachate and gas egress has shown this not to be the case and pollutant containment is the currently preferred system.

The infiltration of water into a filled landfill site is kept to a minimum by installing a low permeability cap over the whole site. The cap should be domed in the centre to encourage water run-off and should be covered with a soil as soon as possible after installation. The capping also prevents uncontrolled escape of landfill gas and directs it towards collection points.

In order for a cap to work successfully it must have low permeability and it must be able to undergo large differential settlements without fracture.

In the U.K. the final cap is usually about one metre thick and is customarily constructed of compacted clay. This is an expensive form of construction when a suitable material is not available immediately adjacent to the site. If a suitable natural clay is not readily available man made materials may provide an economic alternative for making the cap. Artificial 'mineral' layer caps have been formed using sand and bentonite mixes.

Synthetic materials commonly used for lining landfill sites are less frequently used as a capping material as they are easily damaged (depending on thickness) and have no self-healing properties unlike clay and they are expensive.

WASTE MATERIALS IN BARRIERS

There are several requirements for alternative construction materials, but most importantly they should be cheap, of low permeability, readily available, durable and inert. Waste material which is widely available should be used, if possible, so there is the advantage of the work being environmentally friendly by re-using waste. In order for a waste material to be considered for use it must be effective in its designated function, and it must not contribute to environmental deterioration. It must also be durable, stable and plastic if it is to be used in the cap where significant bending deformation can occur.

A mineral barrier consists essentially of a skeleton of impermeable solid particles (which give the layer 'thickness' and structure) and a fine-grained low permeability plastic material which infills the voids between the irregularly-shaped solid grains. Both components contribute to the overall low permeability of the layer in addition to their individual contributions. An artificial barrier material can be made from any combination of impermeable solid 'particles' and a fine infiller - example of such combinations are: concrete, bentonite-enriched-sand, grouting of gravel/fissured rock etc.. The actual materials that can be, or are, used depends on the engineering situation being treated and all of the properties required of the artificial material.

MATERIALS USED

Pulverised Fuel Ash (PFA) has been recognised as being a waste material which has potential for use as a low permeability barrier. PFA is the finer portion of the residue from the combustion of pulverised coal in electricity generation and it is produced in vast quantities in most developed countries (UK annual production is about 12M tonnes). This ash is a glassy material composed of spheres which are predominantly silt and fine-sand sized. Fly ash is an excellent bulk filler and it has been used extensively as load-bearing fill in embankments and structures. Permeabilities of less than 1 x 10^{-6} mm/sec have been achieved from compacted PFA although values of 10^{-4} to 10^{-3} mm/sec are more usual. Adding bentonite improves the workability of the mixture without affecting the mechanical properties of the PFA. The concern with fly ash is that it behaves as a rigid material and may be subject to cracking, with the generation of permeable fissures.

In the course of producing various paper products the paper industry generates a waste paper by-product which is commonly mixed with water slurry to form a wet pumpable sludge. Paper sludge has the physical appearance of clay interwoven with cellulose fibres and contains solids ranging in size from colloid and small clay particles to fibres several millimetres long. The fixed portion of the sludge typically consists of kaolin clay with small amounts of aluminium hydrate, titanium oxide, lime and iron. The remainder is hydrated wood dust, fibre debris, starches, resins, proteins, glue, ink and small amounts of organic compounds. Papermill sludges or inert paper solids have traditionally been disposed of in large containment areas or landfill sites. This waste has a high water content (values of 200 to 300% by weight are typical). As-produced the sludge is very weak but if compacted and dried it exhibits considerable strength - plates 1 and 2.

Plate 1 A 1kg easily penetrates fresh sludge

Plate 2 Partially-dried sludge easily supports 1kg

Pulverised Fuel Ashes, papermill sludges and bentonite mixes have the potential for providing an alternative to other types of capping material presently being used. PFA and papermill sludge are particularly attractive because of their low cost. Furthermore these materials have widespread availability and have some favourable geotechnical properties. An important factor is that the artificial material should be able to be spread and worked without difficulty. Papermill sludge could be regarded as a controversial material for use as a final capping material because its long-term behaviour is unknown, however, it has ductility due to its kaolin clay content.

It should be noted that any water or liquid arising from papermill sludge or PFA will not be pure and therefore it could have the potential to contaminate the groundwater systems. However if these materials are used in the cap and not the basal liner as any liquid flowing from the cap will pass into the containment 'bowl' which has been formed by the liner.

Bentonite is an expansive clay mineral formed by weathering of volcanic ash and dust deposits. It is a member of the sodium montmorillonite group which exhibits very high water absorbtion, swelling and shrinkage characteristics. It has a liquid around 400 - 500 with a plastic limit in the region of 60. This material is not a waste product and it is expensively - about £100 per tonne. However its ability to absorb water and its highly plastic nature are invaluable properties.

PREVIOUS WORK

As far back as the late 1970's laboratory and field experiments were conducted on the use of papermill sludge as a basal landfill liner - Stoffel and Ham. They reported that consolidation and biological activity reduced the sludges permeability with time and a municipal leachate had no significant detrimental effect on the sludge permeability during 10 months of contact.

Stoffel and Ham were concerned that if the papermill sludge were to be used as a landfill capping material the differential settlements of municipal refuse might affect the integrity of the sludge cap. Cracks did develop during a 'severe' laboratory settlement test, but the cracks slowly disappeared since the sludge had the capability to flow and remould itself when wet.

According to the Authors laboratory freeze-thaw cycles raised the bulk permeability of the paper sludge but apparently this behaviour was not seen in the field test cells which were subjected to a Wisconsin winter. By placing the sludge layer below the depth of frost penetration the problems of freeze-thaw action could be avoided since the paper would be covered by a large thickness of topsoil and vegetation cover.

Subsequent work has confirmed that papermill sludge has the potential for being a component of landfill sealing caps in terms of its low permeability, ductility, cheapness, availability and freeze-thaw performance - Schultz and Stoffel, Thacker, Pepin, Jedele. Its main drawbacks are high moisture content, when disposed of, and low strength, which makes it difficult to handle and place, and the permeability is not quite low enough. Jedele noted that the sludge strength increased with decreasing moisture content and he added PFA - this lowered the moisture content but did not appreciably affect the permeability.

Pulverised Fuel Ash has been used extensively in the construction industry and in recent times its use as the impermeable grains in barrier mixes has been investigated because of its widespread availability and cheapness.

Bowders et al (1987) proposed that stabilized fly ash could be used as an alternative to clay soils in seepage cut-off applications such as earth dams, impoundment liners and landfill liners. They mixed PFA from various power stations with lime and cement. The strength, durability and permeability characteristics of the compacted mixes depended on the ash quality and amount of stabilizer used in the mixture.

Falling permeability head tests were conducted using rigid wall permeameters with samples being formed using Proctor compaction. At optimum moisture content the permeability of unstabilized PFA ranged from 1.5×10^{-4} mm/sec to 7×10^{-5} mm/sec. The addition of lime or cement resulted in a significant decrease in permeability i.e. by upto an order of magnitude.

The minimum permeabilities obtained with these mixes were all approximately one order of magnitude too high for most liner applications, therefore in an attempt to reduce the permeability further some bentonite was added. Preliminary tests in which bentonite was added to unstabilized PFA produced permeabilities of the order of 1.6 x 10^{-6} mm/sec (5% Bentonite) and 5.0 x 10^{-7} mm/sec (10% Bentonite).

Edil et al (1987) studied the potential for PFA to be a cost effective alternative to soil and synthetic liners. Some permeability values of less than 1 x 10^{-6} mm/sec were found to be achieved from properly compacted PFA. The permeabilities were found to decrease during prolonged permeation over a long time. However, the Authors were concerned that PFA is a relatively rigid material and so is prone to cracking. The addition of 10% bentonite was found to improve the workability of the PFA and did not significantly affect the permeability.

WORK PROGRAMME

The main objective of the practical work was to define mix proportions of PFA, papermill sludge and bentonite which would give a material, suitable for capping landfill sites, with the following properties :-

a) It can be spread and compacted in a controlled manner in the field without difficulty, using conventional plant and methods to form a coherent, uniform layer.

b) The vertical permeability of the placed material should be not greater than 10^{-6} mm/sec at any location.

c) It should be flexible and ductile to withstand both local and global movements due to refuse decomposition and settlement.

The plain waste materials and various mixes of them were subjected to standard light compaction (Proctor) to determine their optimum moisture content, dry density and compactability on the 'wet-of-optimum' side.

Since the creation of the Proctor test increased mechanisation of earthworks has meant that this level of compaction is easily exceeded in the field for conventional earthworks. However a landfill cap is placed and compacted on a highly flexible, elastic surface and so heavy compaction cannot be relied upon. Also the low strength of the mixes being tested means that remoulding is easily achieved and heavy compaction would simply shear and displace the material rather than knead it together to form a coherent mass.

The coefficient of permeability was determined using compaction permeameters and the falling head test. The fixed wall permeameter provides a former for production of specimens and the test has the advantage of being simple and easy to perform. It is recognised that for 'hard' soils compacted dry of optimum within fixed-wall compaction permeameters 'tracking' may occur down the sides of a sample. However because the Authors' mixtures were weak and easily moulded tracking was extremely unlikely to occur. With the high moisture content of the samples saturation effects are not a problem either. Three permeability tests were run on each sample to check for consistency of the results.

MATERIALS PROPERTIES

The compaction curve obtained for the PFA is shown in Figure 1. When carrying out the compaction test a fresh sample of ash was used for each sample to avoid the effects of crushing and degradation of the material. The high optimum moisture content (by comparison to conventional soils) is typical and reflects the nature and mode of production of the material, i.e. from pulverised coal, with the presence of some unburnt carbon etc. The high optimum moisture content of the PFA is also a factor in considering its use in this project since the paper pulp is excessively wet when disposed of and not workable. Addition of dry PFA, which can 'absorb' a lot of water, will produce a workable mix.

Waste papermill sludges exhibit high moisture contents which tend to vary greatly the initial moisture content (by weight) of the sludge used in this research was in the range 270% to 380%.

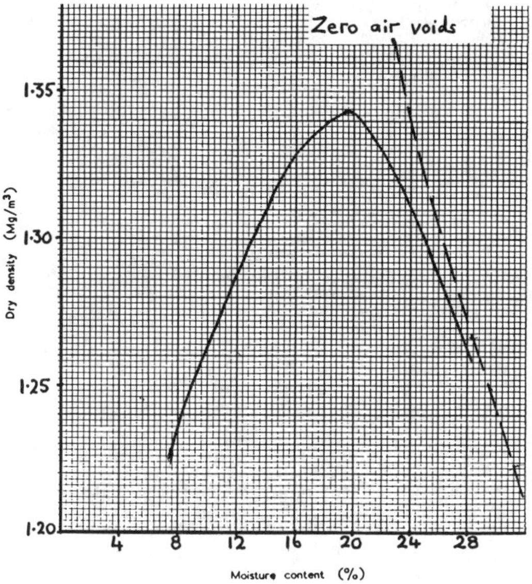

Figure 1 PFA compaction curve

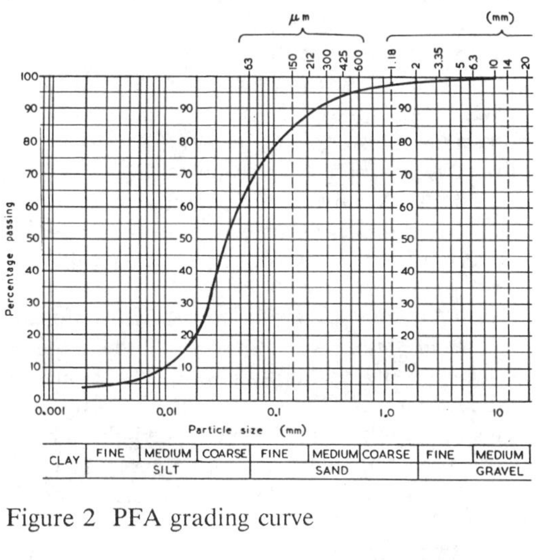

Figure 2 PFA grading curve

The grading curve which is shown in figure 2 confirms the sandy silt nature of fly ash.

These extremely high values are due to the fibrous nature of the sludge which facilitates water retention both between and within the fibres (like peat) and because the fibres themselves have low density (by comparison to soil particles). When a dry piece of the papermill sludge was burned in a crucible it was found to contain 57% fibres and 43% solid grains.

A grading curve (or particle size distribution curve) has little meaning for a paper sludge because of the presence of so many fibres which cannot be sieved or sedimented like conventional soils. Ignition of the sludge does indicate something of its composition since the solid grains were all clay-sized.

It was not feasible to establish a compaction curve for the plain paper sludge. At high moisture contents the material was plastic but too weak for conventional compaction. When the material was dried it acquired significant strength (as indicated in plate 2) but it 'set' into stiff lumps or sheets which would not knead together. At a natural moisture content of 348% the bulk density of a sample formed by hand tamping was 1115 kg/m^3 - with a dry density of 249 kg/m^3.

PERMEABILITY RESULTS

Plain Pulverised Fuel Ash gave a coefficient of permeability of the order of 3 x 10^{-4} mm/sec - this is too high for a landfill cap. The PFA is also non-plastic on its own.

Figure 3 shows that the addition of bentonite to the PFA reduces the permeability and also makes the material flexible. A bentonite addition of about 15% (by weight) would be needed to provide a mix with the requisite permeability for a capping material. This would make this mix too expensive because of the amount of bentonite.

Plain papermill sludge gave a permeability of about 3 x 10^{-6} mm/sec which is only slightly too high for a landfill capping material. However the papermill sludge has a high moisture content which makes it difficult to compact to form a coherent mass.

Figure 4 shows that the addition of small quantities of PFA had negligible effect on the mass permeability to paper sludge. It is only when the PFA content exceeds approximately 50% (by weight) that the permeability begins to be adversely affected.

The initial moisture content of the sludge is about 300% by weight (as in this study) this corresponds to a void ratio of approximately 1.9 - this is the same order as for peat soils. A

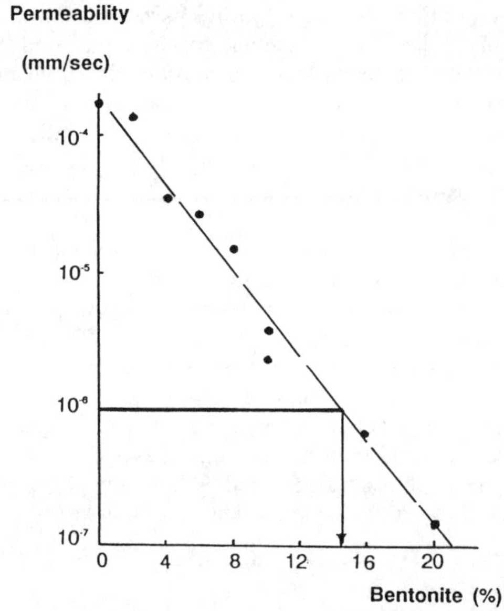

Figure 3 Permeability of PFA - bentonite mixes

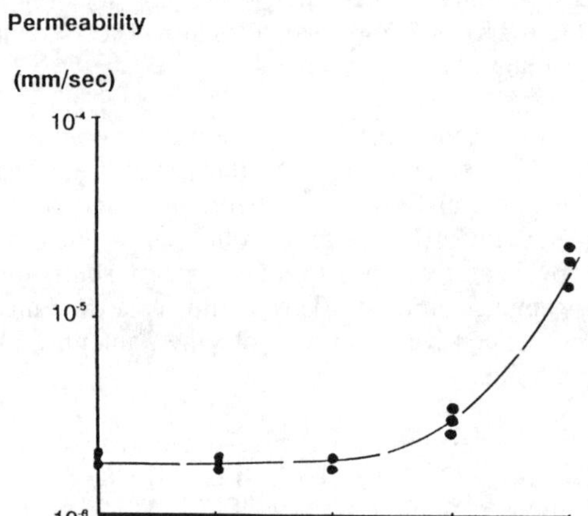

Figure 4 Permeability of PFA - paper sludge mixes

mixture of 50% dry PFA and 50% sludge (by weight) will have a moisture content and voids ratio of about 60% and 0.7 respectively. (Actual values for sludge and cold, dry PFA were; natural moisture content 348% mix moisture content 62% bulk density 1336 kg/m^3, dry density 826 kg/m^3, voids ratio 0.69. If the paper sludge

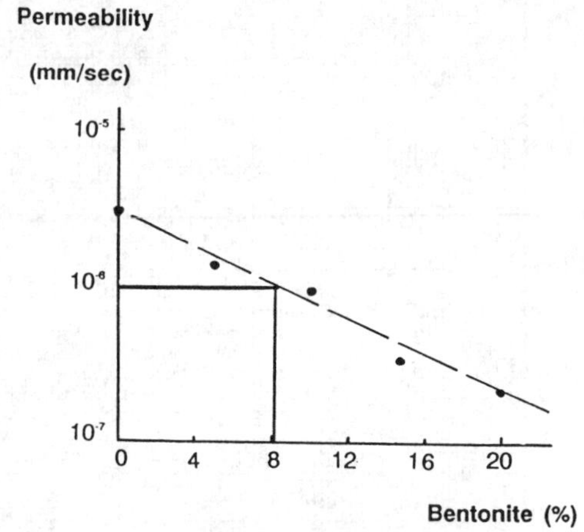

Figure 5 Permeability of paper sludge and bentonite mixes

was mixed with hot, fully-dried PFA the moisture content fell to 60% but the bulk and densities increased significantly, to 1486 kg/m^3 and 927 kg/m^3 respectively, with a reduced voids ratio of 0.51.) These values are much more akin to values associated with soft clay soils and the mixture itself is vastly more workable than the plain PFA. It must also be remembered that the optimum moisture content for PFA is greater than 20% with liquid limits being around 30% so that dry PFA can accept substantial quantities of work and still be workable. The fibrous nature of papermill sludge must not be forgotten either.

As stated previously the permeability of plain papermill sludge was in excess of the required value but the addition of only 8% bentonite produced a mix which was acceptable - figure 5. This addition also made the paper pulp more workable. Although the bentonite may be costly its addition helps to convert the waste materials, i.e. PFA and paper sludge, into a useful and potentially-marketable material. If it is assumed that a waste disposal site would make no charge for the disposal of the modified sludge (as a cap or daily cover) then the owner of the sludge will make a saving of approximately 50% on the normal disposal costs.

At the present time only two tests have been undertaken on mixes containing all three components. These initial tests confirm that mixes in the following region are worthy of in depth investigation, i.e. bentonite content in the range 0 to 10%, PFA content from 0 to 40% and paper sludge from 60 to 100% - all by weight. In addition to permeability tests work will be undertaken to investigate shear strength, plasticity, workability, ductility and cost relationships for mixes in this region.

REFERENCES

Bowders J.J., Usmen M.A. & Gidley (1987) "Stabilized Fly Ash for Use as Low Permeability Barriers", Symposium on Geotechnical Practice for Waste Disposal,, ASCE Geotechnical Special Publication No. 13.

Edil T.B., Berthovex P.M. & Vesperman, K.D. (1987) "Fly Ash as a Potential Waste Liner",Symposium on Geotechnical Practice for Waste Disposal, ASCE Geotechnical Special Publication No. 13.

Jedele L.P. (1987) "Evaluation of Compacted Inert Paper Solids as a Cover Material", Symposium in Geotechnical Practice for Waste Disposal, ASCE Geotechnical Special Publication No. 13.

Pepin R.E. (1985) "The Use of Papermill Sludge as a Landfill Cap", Proceedings of the 1984 NCASI North East Regional Meeting, Special Report No. 85-02.

Stoffel C.M., Ham R.K. (1979) "Testing of High Ash Papermill Sludge for use in Sanitary Landfill Construction", Owen Ayres & Associates, Report for the City of Eau Claire.

Schultz D.T., Stoffel C.M. (1984) "Use of High Ash papermill Sludge in Landfill Caps and Liners" Proceedings of the 1983 NCASI North East Regional Meeting, Special Report No. 84-01.

Thacker W.E. (1985) "Use of Pulp and Papermill Sludges in Liner and Cover Construction", Proceedings of the 1984 NCASI North East Regional Meeting, Special Report No. 85-02.

Waste Disposal by Landfill – GREEN'93, Sarsby (ed.) © 1995 Balkema, Rotterdam, ISBN 90 5410 356 6

The sustainable landfill

M.S.de Silva
Aspinwall & Co., London, UK

R.D.Roberts, C.F.C.Pearson & K.Nicholls
Applied Environmental Research Centre Ltd, Colchester, UK

ABSTRACT: A feasibility study has demonstrated the potential for developing a bioreactor cell rotation landfill which relies on familiar earthwork type construction, but incorporates more elaborate engineering technology to provide a more predictable operating environment. A design has been proposed and fully evaluated in an economic sensitivity and cash flow model which demonstrates the competitiveness of the proposed system with other currently used waste management systems.

1 BACKGROUND AND CONCEPT

Higher environmental operating standards, combined with increasing difficulties in obtaining planning permission, has meant that the economic advantage of local landfills is declining. As this method of disposal accounts for nearly 90% of the UK waste arisings there is increasing pressure to seek alternatives. Combined with public pressure on resource conservation these alternatives have incorporated waste recovery technologies. This concern is reflected in a number of recent EC legislative and policy initiatives designed to promote and stimulate resource recovery. In the UK a number of schemes involving waste separation, recovery and recycling have been implemented on a variety of scales. Such schemes have been primarily concerned with recovery of specific materials such as glass, paper and metals with the residual fraction being disposed to landfill. In a smaller number of cases the residual fraction is processed as refuse derived fuel, anaerobically digested or aerobically composted. The digestion option has a high capital cost and is largely experimental in the UK, whilst the composting option fails to utilize the energy value of this waste fraction.

A need has been identified by the authors to examine the potential for developing a bioreactor cell rotation landfill, which can utilize the residual organics as a substrate for methane generation and subsequently electricity generation. The benefits of such a facility are seen as a smaller landfill-type environment, which can be re-excavated after effective gas generation providing a soil conditioning material as a by-product of the process. The design should allow for the cell facility and infrastructure to be re-used on a rotational basis. The facility should be constructed at low capital cost and incorporate engineering, control and monitoring technology which will lead to predictable operating conditions and electricity generation. The concept should place the facility as a small power generation plant rather than a waste disposal site. The concept should complement existing waste recovery initiatives and add to the short-list of potential waste from energy options. By combining the facility with materials recovery it is anticipated that a significant increase in recovery and recycling of household and commercial wastes can be achieved whilst, at the same time, minimising the continuing requirement for additional landfill space.

2 CONCEPTUAL DESIGN

The research provides a recommendation for a design of sustainable landfill that fundamentally relies on familiar earthwork type construction, but incorporates more elaborate engineering technology to provide a more predictable operating environment. This facility will consist of two banks of 10 cells, each cell being a nominal 70m x 27.5m in surface area, with a smaller base area as a result of the

Table 1 **Cost of Waste Disposal Options and Energy Recovery***

Disposal Option	Cost £/tonne
Incineration	
Mass Burn	15 - 30
Pre-treatment	13 - 25
Modular	-
R D F	
Pellets	10 - 12
Coarse and coal	8
Coarse alone	10 - 12
Thermal Processes	
Gasification	8 - 18
Pyrolysis	-
Liquefaction	~ 15
Biological Processes	
Landfill gas	7 - 25
Digestion	10 - 20
Sustainable landfill	7.80 - 10.20 **

* Net disposal cost, equivalent to 'gate fee' to make operation viable

(after Brown, 1988; Coombs & Coombs, 1989; Wheeler, pers comm).

** this study

side slopes of the dividing walls. In each bank of cells there is a central spine wall with 5 cells back to back on either side. The depth of each cell varies from 10m at the spine wall to 3m at a point furthest from this central wall. This provides a base with a 1 in 5 gradient to facilitate excavation of material on completion of effective gas abstraction. The base is an engineered structure, cast insitu in concrete with a preferential drainage path towards the spine wall. Leachate is channelled at this point into a central duct and collected in a sump to await redistribution. The gas abstraction system consists of 4 vertical wells in each cell placed within a pea gravel column supported by stone gabions. The well is connected at the surface by sacrificial pipework that links into a permanent ringmain constructed within the central spine wall. All cells can be isolated by means of valves.

Process control is maintained by the incorporation of leachate recirculation through the primary gas producing cell. When the next cell in the rotational sequence is to be brought in line, leachate from the primary production cell can be diverted to act as an inoculant to boost the methanogenic process. Waste homogeneity is maintained by ideally using a source separated organic household waste. It is expected that this will be supple-

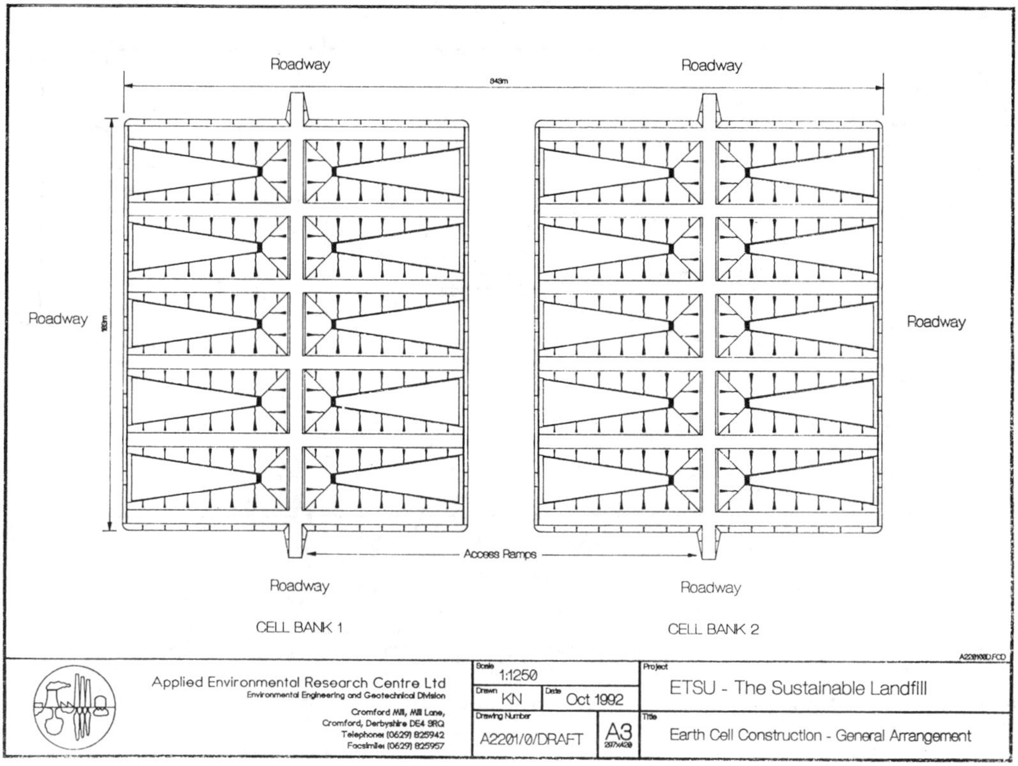

Fig. 1 Basic construction

mented by wastes from the food processing industry and from municipal parks and gardens. The whole is mixed in a Dano drum or similar with minimal liquid addition before transfer to the bioreactor cell.

The control of environmental conditions within the cells is therefore essential to maintain stable methanogenesis. The percolation of water through the refuse can lead to leachates of great organic strength. Leachates with a Chemical Oxygen Demand (COD) as high as 50,000mgl^{-1} can result, largely due to a build-up of VFA's (Harries, 1989). Once a methanogenic population is established, these VFA's are converted into biogas, however, if left unmanaged, the conditions which can develop in the landfill environment may not be conducive to the establishment of a methanogenic bacterial population.

The wastes chosen for the sustainable landfill may exacerbate this problem, resulting in a phenomenon known as 'souring', pH decreasing below the approximately neutral conditions required by methanogens. In many landfills the mixed wastes which are disposed may have an inherent buffering capacity. This is unlikely with wholly organic and putrescible wastes. Techniques must therefore be employed to maintain environmental conditions suitable for microbiological activity. This is best achieved by the use of leachate recirculation which will allow for the monitoring of temperature and pH on-line. A simple dosing system can be incorporated to adjust unsuitable pH.

Filling is achieved by a combination of conventional landfill machinery and a conveyor system. An impermeable membrane cover is placed over the cell on completion of the filling operation, and the unit is sealed.

Alternatives to the basic design are also suggested, the final decision resting on site specific and economic conditions. The basic design and operation remains the same in these alternatives, but options are available for the cheaper construction of the cell base. Care has to be taken in recommending these alternatives in a generic system as detailed in this paper as decisions are site and situation dependent. The basic construction is given in Fig. 1.

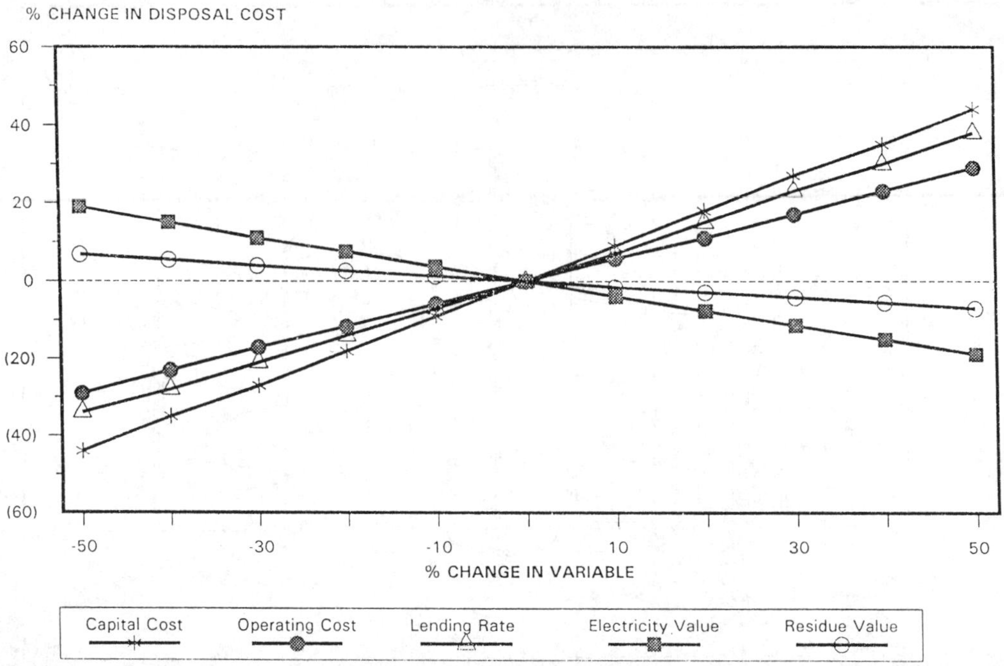

Fig. 2 SENSITIVITY ANALYSIS - BASE CASE

3 AVAILABILITY OF SUITABLE FEEDSTOCK

Feedstock for the sustainable landfill would come from separated MSW, either separated at source, or mechanically. Suitable source separated material is available from the Leeds City Council Scheme. Problems do exist with this method however, caused largely by householders placing organic wastes in plastic bags prior to disposal. This plastics contamination is unsightly and is problematic if a compost afteruse if proposed for this waste fraction. However, the split-bin remains the best method to minimise the level of chemical contamination being transferred into the feedstock. Alternatively, mechanical separation is expected to provide a feedstock which is suitable for methane production but at the expense of the residue quality. The benefit is that standard black bag, dustbin or wheeled bin collections can be maintained. The other primary sources of organic feedstock will be green wastes from civic amenity sites, food processing plant, food retail outlets and municipal parks and gardens. Again, many of these sources may be contaminated by plastics.

Waste generation statistics from Wirral Borough Council have been used to provide the scale for the proposed facility. Wirral MBC provides a good example of a possible source of feedstock that would provide enough methane to fuel a $1MW_e$ generating set. The facility has been scaled on the basis of 40,000 tonnes $annum^{-1}$ organic feedstock (138,000 households). This would indicate compatibility with waste generation information provided by both Milton Keynes Borough Council and Leeds City Council.(Hammerton, Pers Comm; Leeds City Council 1992).

4. COMPATIBILITY WITH CURRENT WASTE DISPOSAL TECHNOLOGY

The sustainable landfill is complimentary to other waste management options; it is wholly compatible with the recovery of dry recyclables and also utilizes low calorific value wastes which would not be suitable for incineration. The main competition may appear to be from anaerobic digestion in fabricated digesters and composting. However, it is the opinion of the authors that the competing options are best considered on a geographical and demographical basis. For example, smaller rural communities may benefit from home composting but at the expense of energy recovery.

A review has been carried out of waste generation statistics, waste handling technology, energy recovery options from waste and cost implications of waste management in order to provide base data for the feasibility of developing the sustainable landfill. This review has proved necessary in order to have a full understanding of the practical implications of the conceptual sustainable landfill on current waste management technology and practice. The study evaluates how the concept may be compatible with current practice by reference to the literature and discussion with the waste management industry. An evaluation of the waste collection schemes throughout the UK has been conducted to determine, firstly, if suitable feedstock can be obtained in the required quantities to operate a sustainable landfill facility. It has also been necessary to determine how the feedstock can be best handled, bearing in mind its organic nature, and the absence of dry bulky materials associated with conventional unsorted MSW.

5. ECONOMIC ANALYSIS

A comprehensive understanding of these aspects of waste handling combined with considerations for microbial enhancement, in order to optimise methanogenesis, have been used to produce a recommended design for the sustainable landfill facility. This recommended design has been used as a base case for an economic sensitivity analysis. Alternatives in terms of construction method and size have also been put forward by way of comparison.

A significant advantage is gained by the sustainable landfill when comparing the economics of waste handling options. The base case considered in the economic analysis indicates a tipping cost of approximately £10/tonne which can be further subsidised by revenue from electricity generation and compost sales to give a gate fee of approximately £7.50/tonne. Even at the base tipping cost (£10/tonne), comparison with the literature indicates that this is considerably less than the cost of both incineration and anaerobic digestion. These figures are exclusive of collection costs. (Table 1).

Changes to the cell floor design reduce the tipping cost further to approximately £8.50/tonne. An economy of scale is also realised on the base design, reducing tipping cost of £7.80 with a facility twice the size of that proposed. Discounting at 8%, 15% and 25% returns a high Net Present Value and a favourable Internal Rate of Return in all cases. Reducing the gate fee below the tipping cost requires a continued reliable income from both electricity generation and residue. A nominal 2p/kWh has been used in the most pessimistic calculations for electricity revenue. The success of the facility would then depend on income from sale of residue.Fig.2 shows basecase sensitivity.

A comparison with current landfilling costs indicates that even using the base case design with high capital cost places the sustainable landfill very competitively at the lower end of private sector charges and in the lower half of public sector charges. The majority of areas producing large volumes of waste are located in regions where landfilling costs are higher, favouring the sustainable landfill option for these areas.

6. CONCLUSIONS

The base case opted for in this study is recommended as a generic design for the sustainable landfill, however, it is evident that cost savings can be made by altering the cell floor construction to decrease capital costs and/or by increasing the size of the facility to benefit from the economy of scale. The former option is dependent on site specific conditions and may prove viable in certain geographic locations. Increasing the size of the facility is a viable option if a greater volume of feedstock can be guaranteed. For example calculations indicate that a doubling of the number of households considered for the base case, to 280,000, will provide enough organic waste to provide methane to fuel a 2 MW_e generator. However, basing the sustainable landfill on 140,000 households is more widely applicable, with only a limited number of UK Boroughs having sufficient households to provide waste for the larger facility. The overall size of the Borough will also influence the volume of feedstock from municipal and commercial sources.

A comparison of the anticipated costs of the sustainable landfill with other currently available waste disposal options are favourable. Capital costs place the sustainable landfill as significantly more favourable technology than either incineration or anaerobic digestion, whilst also undercutting commonly quoted figures for modern landfills. Disposal costs per tonne are also more competitive than incineration and anaerobic digestion whilst falling in the lower range of costs quoted for existing landfills. However, landfill disposal cost is influenced considerably by geographical location, with charges varying throughout the country. The generally excepted view within the waste management industry is that the cost of landfilling will increase as new landfills, with improved environmental protection systems, come on stream. This will further favour the economics of the sustainable landfill.

The sustainable landfill also benefits over composting in urban areas where many households may not have gardens or sufficient space and need for compost. Central collection of organics is also easier in such areas, and the sustainable landfill route will utilize the energy potential of this waste fraction.

The facility can be sited close to urban areas due to its compact nature thus reducing transport costs. Materials recycling would also minimise the volume of residual waste which requires disposal to alternative facilities and result in further transport cost savings.

An assessment of existing waste collection schemes indicates that separation at source is the most compatible with the feedstock requirement of the sustainable landfill. The dual split bin system is the most favourable, but results from the trial schemes still indicate a significant level of plastics contamination by this method. This contamination is caused by householders wrapping organic wastes in plastic bags prior to placement in the split bin. Similar plastic materials entering the aerobic composting pathway are fairly readily removed by hand sorting. Alternatively, organic material provided by the mechanical separation route is seen to be more problematic as a result of a variety of mixed materials depositing with the organic undersize from trommel screening, making the residue after digestion less marketable.

In conclusion, the sustainable landfill is considered to be a viable system for removing a considerable volume of material from the waste disposal stream, and in so doing providing a useful energy source. It is best operated in the presence or in the vicinity of a Materials Recycling Facility (MRF) and is complimentary to current waste disposal options, particularly conventional landfilling and incineration. Removal of the organic fraction from waste destined for a conventional landfill facility will reduce the methane generating potential of that waste. The sustainable landfill utilises the high moisture, low calorific wastes which are not ideally suited to incineration.

The concept, although theoretical, is based on a number of existing waste processing technologies. The authors are currently engaged in obtaining funding for a technical evaluation of the concept using a combination of laboratory and field demonstration scale trials.

BIBLIOGRAPHY

Brown, A. (1988). Refuse: The energy options. In. Landfill gas and anaerobic digestion of solid waste. Eds. Alston, Y.R. and Richards, G.E. UK DEn.

Coombs, J. and Coombs, Y.R. (1989). The prospects for methane recovery from the anaerobic digestion of municipal waste in the United Kingdom. ETSU Report B1234.

Harries, C.R. (1989). Landfill microbiology - work supported at Biotal by the Department of the Environment. In. Landfill microbiology: R & D Workshop. Ed. Lawson, P. and Alston, Y.R. UK.DEn.

Leeds County Council, (1992). A draft integrated waste management strategy for Leeds.

Waste Disposal by Landfill – GREEN'93, Sarsby (ed.) © *1995 Balkema, Rotterdam, ISBN 90 5410 356 6*

The preparation of the cellulose skeleton ion exchangers

S. Sztromajer
Geotechnics Department, Technical University, Łódź, Poland

ABSTRACT: The preparation consists of electrolytic oxidation of the cellulose skeleton materials. The prepared materials acquire cation exchange properties and can be used to remove the heavy metals ions contained in polluted water such as leachate. The expected advantages of the method are its simplicity and low cost.

1 INTRODUCTION

Heavy metal contamination of water is one of the most serious problems in ecology.

It is known that the presence of heavy metals is a special sort of pollution which is particulary burdensome as it interferes with the activity of the enzymes.

The main reason for removing such pollutants is their carcinogen properties; there is no doubt that even a trace of nickel, chromium, lead and cadmium in waters cause an increase in the incidence of cancer.

On the other hand removing such pollution from flowing water is not easy, because of its dilution. There are some, recently introduced, biological methods which depend on the collection of the heavy metals by special plants. These plants have the ability to absorb the unwanted substances in their branches and leaves.

Matsumoto and others have previously presented very interesting methods of removing the cations Cu^{2+}, Ni^{2+}, Cr^{3+} and anions $Cr_2O_7^{2-}$ by means of wool.

The method presented in this paper utilises the properties of such cellulose materials as white cotton textile, waste-paper and wood shavings, to remove metal ions.

These 'attenuation'properties have been known for a long time - neutral solutions of salt can change pH value after filtering though filter paper, or some sorts of the cotton filter cloth. This phenomenon could be explained on the basis of the exchange of the hydrogen ion, previously contained in the hydroxyl group of the cellulose. The ion exchange ability could be increased by means of oxidizing, while the alcohol group $-CH_2OH$ is converted to the carboxyl one $-COOH$. This paper presents in a condensed form an investigation into the efficiency of the electrolytic oxidation of the cellulose materials to produce ion exchangers.

2 EXPERIMENTAL PART

The modification of the cellulose materials was performed in a parallelepiped glass vessel of dimensions 49cm long by 33cm wide by 35 cm high - as illustrated schematically in figure 1.

The vessel was divided into three parts by means of two separators, made of asbestos fabric.

All the electrodes were made of graphite. The electrodes were tubular, with outer diameter 3.8 cm and length 32 cm. To the top part of each electrode, a C-shape silver band was fixed to connect the electrode with the conductor.

The band and the conductor were covered with varnish to prevent oxidation.

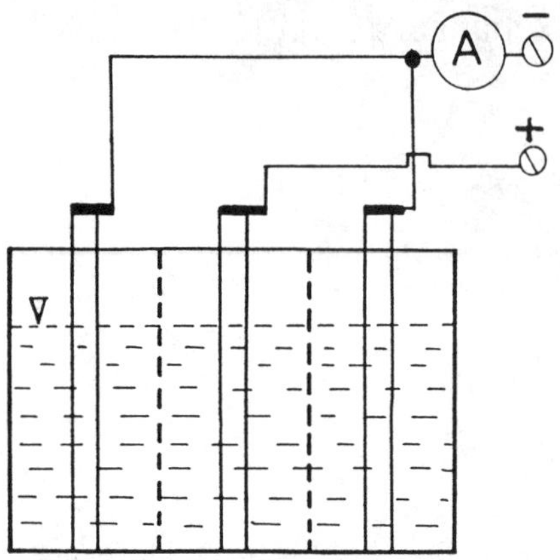

Figure 1 Schematic arrangement of apparatus

The processed cellulose material was located around the anode. The wood shavings were processed in a small cotton bag, which was wrapped around the anode. In this area the intensity of an oxidation seems to be sufficient.

As an electrolyte a 2M water solution of sodium sulphate was used.

The depth of the electrolyte was set at 25 cm.

The current was maintained at the 15 amp, which equals a current density 0,05 amp/cm^2 on the area of the anode. A running time of 120 mins was sufficient when cotton and paper were being modified while wood shavings needed at least 180 mins.

There were some problems caused by the hydrogen and the oxygen emission during process, so the vessel was located under a gas extraction system.

3 RESULTS AND DISCUSSION

The test of the concept was the ability of the modified cellulose to remove heavy metal ions from a water solution.

The test was carried out in the laboratory, however in conditions similar to the field application of the exchangers.

A comparison of the capacity of the 'cellulose exchangers' to that of the standard (carboxyl) ones is given in table 1.

Table 1 The comparison of the standard carboxyl exchangers with the modified cellulose

Producer	Type	Exchange capacity Val/dm^3	Remarks
Permutit Co Ltd (UK)	236	4,1	
Dow Chemical Co (USA)	CCR-2	4,1	
Imacti (Holland)	25	3,0	
Bayer (Germay)	CNP 80	4,5	
Nitrokemia (Hungary)	KCM	3,5	
Experimental "PC"		0,2 - 0,35	Processed cotton textile
Experimental "PP"		0,1 - 0,2	Processed paper
Experimental "PW"	MW	0,05 - 0,12	Processed Wood shavings

The best exchange capacity is obtained from the modified cotton and paper, while the wood shavings are less efficient. Probably this result is caused by fact that cotton and the paper were previously treated while the wood shavings were completely natural.

Maybe a longer processing time would help, more particularly as the diffusion of the oxidising agent is relatively slow.

The anode reactions probably run as follows.

$Na_2SO_4 \rightarrow 2Na^+ + SO_4^{2-}$

$2SO_4^{2-} - 2e \rightarrow S_2O_8^{2-}$

$S_2O_8^{2-} + 2H_2O \rightarrow H_2SO_4 + 2OH^-$

$2OH^- - 2e \rightarrow H_2O_2$

$H_2SO_4 \rightarrow H^+ + HSO_4^-$

$2HSO_4^- - 2e \rightarrow H_2S_2O_8$

$H_2S_2O_8 + H_2O \rightarrow 2H_2SO_4 + O$ (very slow)

$$H_2S_2O_8 + H_2O \rightarrow H_2SO_5 + H_2SO_4$$

$$H_2SO_5 + H_2O \rightarrow H_2SO_4 + H_2O_2$$

$$H_2O_2 \rightarrow H_2O + O$$

Probably the atomic oxygen acts as the main oxiding agent, but the Caro's acid (H_2SO_5) certainly acts stronger if it is in direct contact with the cellulose. It is known that, this acid is a one of the most intense oxidizing agents, in its pure form it reacts explosively with organic compounds. Of course in an aqueous solution such a danger does not exist.

4 CONCLUSION

The present study gives an indication of the possible application of cellulose skeleton carboxyl exchangers to remove heavy metal cations from water. The capacity of such exchangers is low, compared to that of the standard ones, but their cost is also very low.

The proposed method of modification of cellulose materials is easy and the apparatus is simple and cheap, considering it construction and exploitation.

The exhausted exchangers will have been burned, so no regeneration is needed.

The best exchange properties are possessed by cotton textile, then the modified paper with the least efficient being the wood shavings. The difference is probably caused by the slowness of the diffusion of the medium to the exchanger.

REFERENCES

Davidson, G. (1948). Acid Properties of Cellulose. J Text. Inst. vol. 39.

Fedotew, N.P. (1962). Prikładnaja Elektrochimia. Goschimiadat.

Kullgren, (1931). Svensk Kem. Tid. 43 99.

Laws, E.(1981). Aquatic Pollution. John Wiley.

Matsumoto, H.(1979). Nippon Eisegaku. 34,4.

Opieńska-Blauth, J. (1956). Chromatografia bibułowa PWN.

Ott, E. (1950). The chemistry of Cellulose Interscience.

Robek, T. (1960). Teoretyczne podstawy syntezy elektrolitow. PWN.

Rogovin, Z.A. (1953). Chimja cellulozy.

Samuelson, O. (1952). Ion Exchangers in Analytical Chemistry. Uppsala.

Samuelson, O. (1969). Determination of Carboxyl Groups in Cellulose. Svensk Papperstidn Bd58 N19.

Szorigin, P.P. (1937). Zurn. Obszcz. Chim. 7. 188.

Sztromajer, S. (1937). Sposób otrzymywania kationitów przeznaczonych do oczyszczania ścieków. Patent nr. 153563. Warszawa.

Winnicki, N.P. (1978). Polimery czynne.

Waste Disposal by Landfill – GREEN'93, Sarsby (ed.) © 1995 Balkema, Rotterdam, ISBN 90 5410 356 6

The development of lining technology in Germany

Reiner Wackermann
Fachhochschule Darmstadt, Germany

Abstract: The waste area of the future must not be any longer a mixed deposit of waste materials, but only the last link of an extensive process of waste economy.
Deposited is only the remaning waste which has been undergone a primary process.
This deposits are to be constructed and run in a way which ensures that human life and the environment is not endangered. Therefore a multitude of safety measures - multi tombolo system / multi barrier system has to be developped.One of these measures is the tombolo of a base-seal which demonstrates the developement of the deposit technique and it's actuall stage.

1. General

Simultaneously with the economic boom in post-war-aera the volume of waste was encreasing explosively. Only from 1950 to 1961 the volume of domestic waste has doubled in the Federal Republic of Gemany. This waste ended in about 50000 refuse pits without any control and possibility of public influence. It is well-known that many of those early waste areas represent a danger for humans and enviroment. There are called " areas suspected to carry problematic waste" and since the end of the 1970ies they are listed systematically. In the old federal territory up to the present approximately 40500 old waste areas/deposits are registered in the so called "Altlasten-Katastern"(contaminated soil registers)[1].
The-clearing-of-waste-act of june 1972 offered the legal authority for a "well regulated" desposal of waste.The numerous and in their majority disordered small waste areas were closed and the waste was desposed in central, big and systematic deposits or incenerators.In such a manner the number of domestic waste areas was reduced to approximately 330 and in addition to that 47 waste incinerators up to the present.Special measures to protect the subwater were not considered necessary. In August 1986 the new law prepared the foundation for the turning away from the "Wegwerfgesellschaft", i. e. the turning away from the disposal of waste to waste utilization.This law did not only claim that "utilization is of greater importance than disposal" but also that pollution control (soil, air, water, flora, animal world) and human health deverve the same priority (§ 2 AbfG). This means that measures of waste disposal are to be constructed and run according to the "actual standard of the available technique". Concerning this matter the Federal Government promulgates administration prescriptions (§ 4 para. 5 AbfG), in the following designated as TA - waste (technical instructions of waste).

The technical development of security-conceptions for deponies and a development of systems to reduce and recycle the raw material in the waste began with this law.

2. The deponie as a component in a waste econmy-concept

First there must all possibilities of utilization happen like the law of "utilization is a greater importance than disposal. The assumptions are (§3 Abs.2 AbfG):

a) The technical realisation
b) The costs must be real
c) There must been or build a market to sale the new products

Therefore the complete wastespectrum integrated in a waste economy concept must be anali ed and utilization possibilities had to be recognized. Figure 1 shows an example.
The remaining waste has to be treated in a way that emissions of the deposited waste are minimezed, even in the case that all technical methods of protection fail. The deposit itself represents the last link of this whole conception.

3. How to optimize security of a deposit by the "multi-barrier-conception / multible-tombolo conception"

A deposit has to be planned with the purpose not to harm humans and the environment by the disposal of waste today and in future (§2 AbfG). That means that emissions from the deposit are prevented and reduced to the absolute minimum by precautions. This also includes the case of breakdown of one or the other security method. Regarding technical possibilities this can be realized by several (from each other)

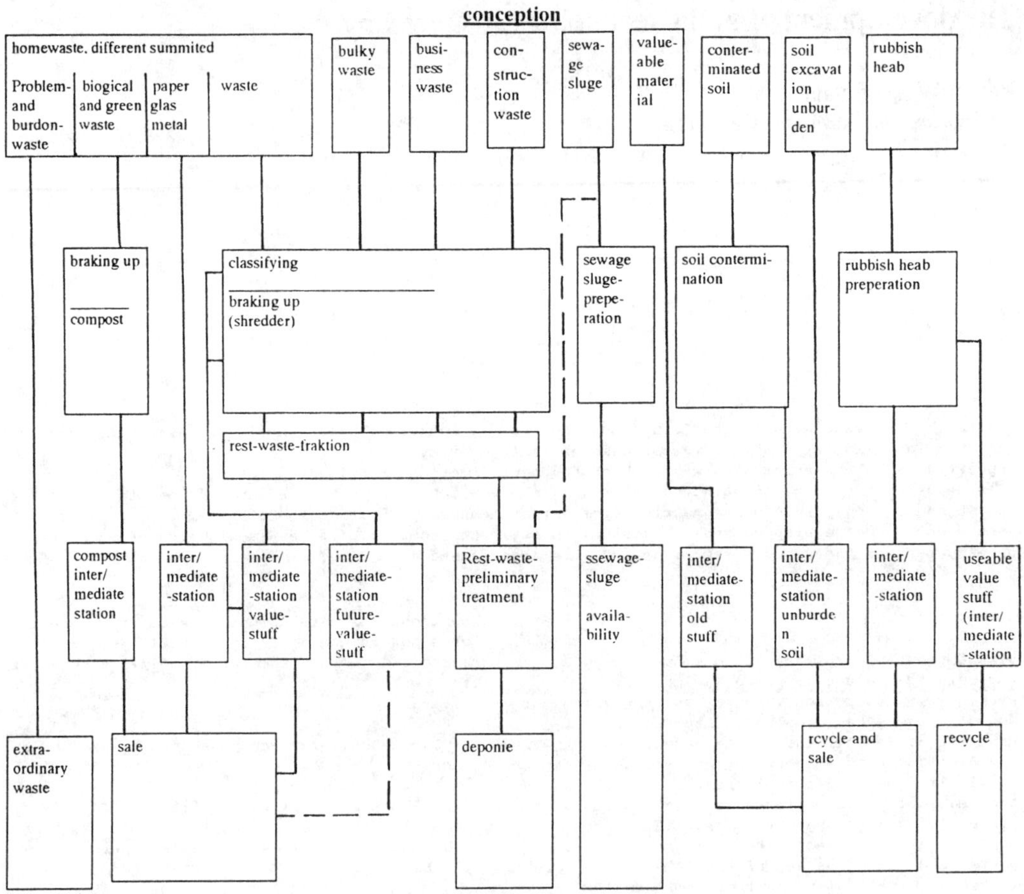

Figure. 1: An example for a waste-deposit-conception

independent security systems, a so-called "multi-barrier-system" (7,8).
Table 1 shows an example.
Depending on the problem an requirement there can be added any number of further barriers.

4. The developement of the deposit technique demonstrated by the barrier / tombolo basis seal(ing up)

Before 1972 (1st waste act) the waste areas generally did not have any technical equipement - no subsoil sealing, no dry-well / seepage pit and no supervision.
In most cases already existent hollows - quarriers, gravel pits, caves and valleys - were used as waste areas / deposits. From 1972 to 1992, the deposit-technique was improved step by step. For large waste areas the consequence was that every new deposit section had to be planned and constructed according to the new technical knowledge.

4.1 Elementary / easy seal(ing up) systems

The first technical methods to discharge the seepage water consisted in split gravel-trenches which were arranged like fish-bones as the condition of the ground was expected / assumed to be so dense that no fugitive water could penetrate the soil of the disposal site.After having recognized that the subsoil-conditions of the majority of the deposits were not as tight as they were supposed to be, the next step was to provide a mineral seal or a plastic seals.
An area drainage with collectors and feeder drains was placed on top of it.

4.2 Combination seal/TA-settlement

Nevertheless, this seal system was improved by combining a plastic seal with a mineral seal (combination sela, figure 2). This basis-seal system was established as technical standard concerning the above mentioned TA-settlement waste.

Table 1 Multi-Barier-Conception
The following table gives an example for a multi-barier-conception

Nr.	Barrier	Function	Realisation
1	geology	ground-water protection	procedure of position search
2	sael	- ground-water protection - minimization of the infiltration rate	- base and if necessary sealing up of the sides - sealing of the surface /seal coat
3	waste / trash	minerulization and immobilization of the residual waste	- picking out the valuable matirials - cold (rot, fermention) hot (combustion) preliminary treatment
4	disposal	- water conservation - purification of the air	registration, derivation and treatment of - seapage water - fill gas
5	running	- minimazation of seepage water and the emissions of dust - immobilization	casting technique - thin-layer-casting - below temporary roof/shelter
6	supervision	- decontamination of waste - water protection - punify of the air	- entry control, e.g. by shallow bunker - building inspectation, e.g. of the production of the base-seal - enterprise supervision, e.g. of the clearing rate of the installation for the prepuration of the infiltration water
7	aftercare and control	- minimization of the infiltration rate - avoidence of incontroled escape of gas - waterprotection - conservation	- sealing-up of the surface, recultivatim and further utilization - supervision of seepage water, surface wateranc gas

It was a special challenge to find out how to protect the plastic seal (usually 2,5 mm HDPE) against damage by the area drainage (usually 30 cm high, graining 16/32 mm). Even heavy fleeces (e.g.1.200 g/m²) were suitable only to a certain degree. This is why mineral protective layers (diameter 10 cm) or combinations composed of fleese and mineral sayer were realized. Moreover, deposits were basically not allowed to be constructed in hollows but as so-called spoil bank deposits. Such a proceeding would gain the advantage that a breakdown of all pumps would not lead to a seepage on the deposit site.

The seepage floots in free fall out of the deposit. The seepage collectors can be constructed in a way that they are flushable and can be controlled by video-recording. But the disadvantage of this seal system is the missing possibility to control (exeptional by Groundwater observation springs from the outside of the deposit) and repair the seal.

4.3 Controlable and repairable seal systems

In the course of a " competition of ideas" by order of the Main-Kinzig Region (3) several deposit-base-seal systems were presented fulfilling the made demand of being controlable and repairable.

a) System DYWIDAG - Steuler

The deposit conception DYWIDAG - Steuler stands from an intermediate depot for e.g. unprocessed waste (especially toxic wastes) to those installations for the mechanical treatment in which the waste is to be prepared for the final stratifiacation.
Certainly these intermediate depots can be change into a final stratification by filling up the control-zone / space with inert material. This presupposed an intensive supervision period of the embedded waste and its casing. The waste

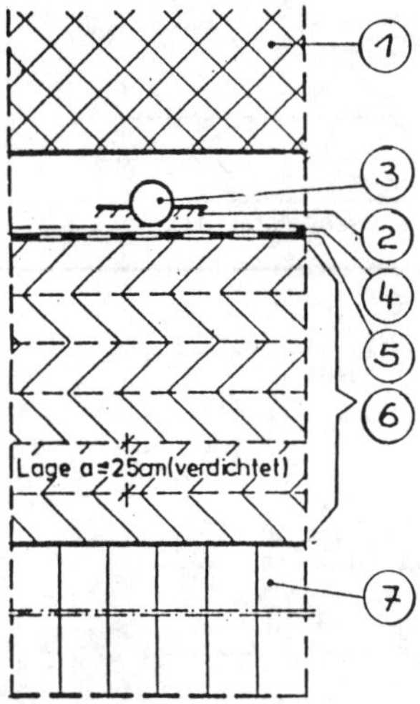

1. Waste
2. Draining stratum
3. Draining pipe
4. Security stratum (geotextil)
5. Sealing element (Synthetic material)
6. Mineral seal
7. Barrier (stones)

Figure 2: Combination-seal

deposition system is a double surfaced, controlable and repairable construction, figure 3. The single elements, directly embedding the waste, consists of a filter layer (respectivly drain-and protection plates) a seal (generally 5 mm thick PEHD) reinforced concrete (prefabricated pieses). At any time they can be controlled and if necessary be repaired. For that prurpose the seal system is surrounded by 2,8 - 3,5 high control - and working space / room. From outside it is enveloped by a reinforced concrete casing.

b) Multi-layered basis seal(ing up) - and control system

The multi-layered basis seal(ing up) systems correspond esentially to a double combination seal constructed one upon the other (see point 4.2)).Thereby the deposit base is devided into segments which are independent from another. The control drainage additionally is equipped with grouting- and bleeding pipes (figure 4). Supposing that the impermeability decreases/ weakens one can inject segmentally the control draining and so far achieve a certain kind of "repair". The mineral seal strata may contain in layers different aggregates to achieve distinct effects like absorption capacity, swelling ability, shrinkage impediment and the conservation of plasticity.

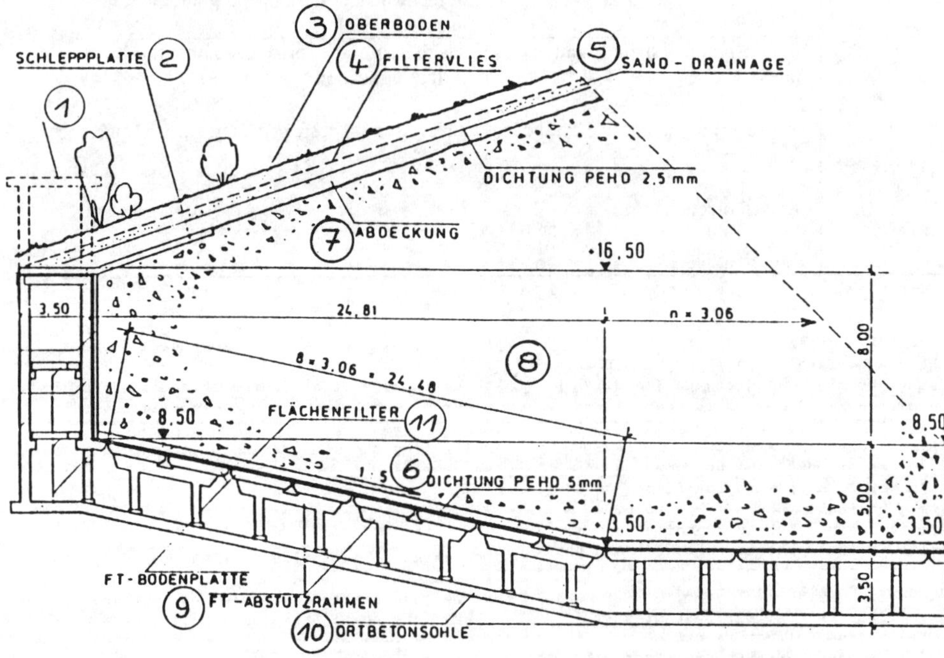

1. Inspection entrance
2. Traction plate
3. Surface
4. Filterfleece
5. Sand - Drain
6. Seal
7. Covering
8. Waste
9. Vertical concrete drop
10. Concrete base
11. Draining stratum

Figure 3: Double surfaced, controlable and repairable deposit system after DYWIDAW - Steuler (3)

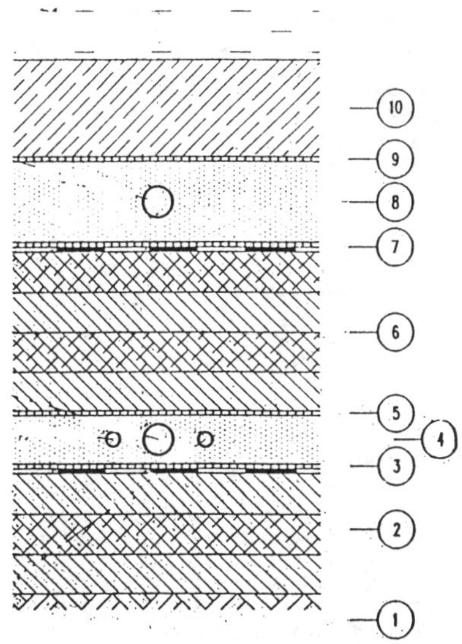

1. Raw planum
2. Mineral sealing
3. Synthetic fleece
4. Controling room
5. Filter fleece
6. Mineral sealing
7. Synthetic fleece
8. Draining stratum
9. Filter fleece
10. Security stratum

Figure 4: Multi-layered basis seal(ing up) system, controlable and only limited repairable

4.4 Realized Controlable Seal Systems

The innovations especially concerning the control of basis seal were accepted without hesitation in the real deposit-planning.

a) Deposit of Wiesbaden

The domestic waste deposit of the capital city of Hessen, Wiesbaden, is one of the first deposits that was equipped with a multi-layered base seal- and control system in accordance with point 4.3 b).
Here, nor grouting neither bleeding pipes were used thus avoiding repairs. In this way, the ability to control the deposit and the second drainage layer could be preserved.
The characteristic of this deposit lays in its spatial exent. This excluded the possibility to flush and control (video recording) the seepage collecting from the outside of the deposit.
Therefor a tunnel which provides the possibility of an inspection on foot and with a length of 700 m below the deposit was constructed.(Figure 5)
The seepage collectors and the control drainage are installed in this inspection tunnel.
From within the tunnel it is possible to carry out control measure and cleaning of the collecting- and control pipes.

b)Residual waste deposit of the Main-Kinzig - Region

For the actually planned residual waste deposit

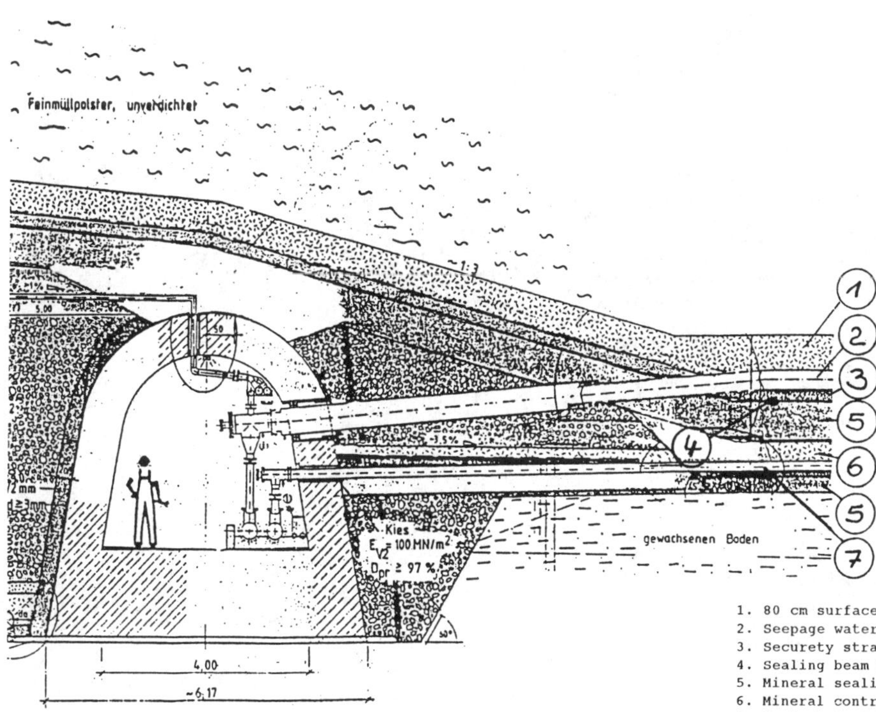

1. 80 cm surface filter
2. Seepage water pipe DN 300
3. Securety stratum
4. Sealing beam HDPE d = 2.5 mm
5. Mineral sealing
6. Mineral controlling stratum
7. Control drain pipe HDPE DN 150

Figure 5: Multi - layered deposit base seal with inspection tunnel (9)

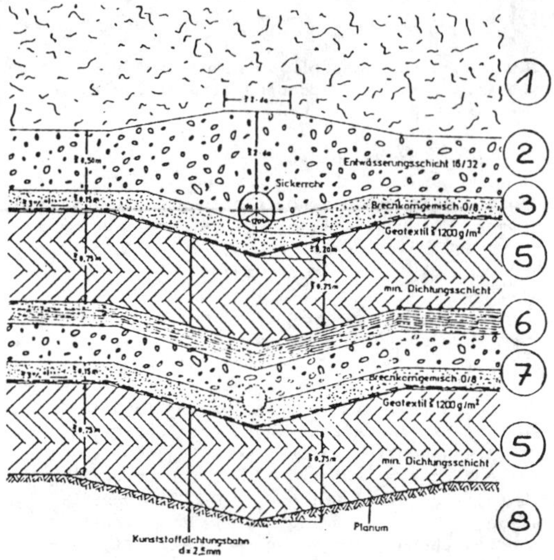

1. Primary processing waste
2. Draining stratum
3. Security stratum geological/synthetic
4. Syntetic seal
5. Mineral seal
6. Asphalt sealing > = 0.20 m
7. Controling draining >.= 0.30 m 16/32
8. Stones

Figure 6: Multi-layered basis seal with an additional asphalt sealing element (5)

of the MK-region it is also intended to install a multi-layered basis seal(ing up) with a control system. The exspected / wanted seal system should be repaireable. In this respect, however, the presented systems, developed at a idea-contest, were not convincing. Especially the quality (impermeability) of a grouting level after having grouted was questioned. For that reason - and instead of a questionable possibility of repair - one decided to use an additional sealing element, an alphalt seal, which is already tried in the hydraulic engineering, figure 6.
Beside its secure and simple production the main advantage of this additional element is that it prevents the draining of the mineral seal to the control drainage.
The last two examples demonstrate that the technical development of constructing deposits makes more progress than the enacting of an administration prescription (TA settlement waste (4)) in order to define / specify and lay the technical stage.

Literature:

[1] Sondergutachten "Altlasten" des Rates von Sachverständigen für Umweltfragen. Bundestagsdrucksache 11/6191 Bonn 1990
[2] Sondergutachten des Rates von Sachverständigen fnr Umweltfragen von Sept.1990 "Abfallwirtschaft". Bundestagsdrucksache 11/8493 Bonn 1990
[3] "Restmülldeponie 2000" Tagungsband zum Hearing am 6. und 7. Juni 1990 Main-Kinzig-Kreis Hanau 1990
[4] TA-Siedlungsabfall Sechste Allgemeine Verwaltungsvorschrift um Abfallgesetz Entwurf Nov. 1991, Bonn.
[5] Bergmann, W. Wackermann, R. Restmülldeponie-Technisches Konzept Main-Kinzig-Kreis Hanau 1992 (unveröffentlicht)
[6] Gesetz nber die Vermeidung und Entsorgung von Abfällen (AbfG) Bundesgesetzblatt, Jahrgang 1986.
[7] Oeltzschuer, H. Die Deponie als technisches Bauwerk, Beitrag in Heft 51, Landesgewerbeanstalt Bayern, Nürnberg 1988.
[8] Stief, K. Das Multibarrierenkonzept als Grundlage von Planung, Bau, Betrieb und Nachsorge von Deponien, Müll und Abfall. 7/1986, Berlin.
[9] Planung des 4. Bauabschnittes der Hausmülldeponie Wiesbaden. Wiesbaden 1990

Waste Disposal by Landfill – GREEN'93, Sarsby (ed.) © *1995 Balkema, Rotterdam, ISBN 90 5410 356 6*

Reassessing methane from UK landfills

Max K.Wallis
School of Mathematics, University of Wales, Cardiff, UK

ABSTRACT: The 1991, 1992 and 1993 statistics for annual methane emissions from UK landfill sites have risen from 0.7 to 1.0 and now 2 Mt per year, due primarily to two changes in methodology. These estimates can be reconciled via more complete assessments with more realistic values of parameters. Increasing deposition to landfill and increasing compaction requires analysis of the dynamic system. It is found that current gassing rates are 20–30% below steady state values based on current landfilling parameters. Reconciliation leads to a much tighter range (1.6 to 2.4 Mt/yr) than ETSU's revision (0.6 to 5.3 Mt/yr). On the 1990 methodology, Government believed that exploitation of landfill gas would curtail UK methane emissions to the atmosphere. However, at the higher generation levels, UK emissions are set to increase substantially, despite the increased exploitation programme facilitated by the NFFO. In view of the importance of methane as a greenhouse gas under the Climate Change convention, the UK policy of landfilling sewage and mixed putrescible waste has to be fundamentally reviewed, in regard to both current controls on landfills and practical limits on efficiency of landfill gas capture and utilisation.

ESTIMATION METHODS

Methane emissions from UK landfills have been reported annually in 'DEPWS' (DoE 1991, 1993a); figures are normally revised retrospectively with changes in calculation procedures to ensure consistency in each set, but the 1993 edition gives incomplete landfill figures [1]. In the last 3 years, details of methodology and parametric values have become available.

1.1 Empirical Degradation

Prior to 1991, UK estimates were based on total annual MSW landfill with assumed C (carbon) content and degradability (Richards 1989). This calculated the total mass of methane per yr (in tonnes) from

$$M = W \times p \times d \times f \times 16/12 \qquad (1)$$

with
W in tonnes/yr landfill (20.8 Mt adopted);
p – fraction of degradable C (0.21);
d – fraction of matter that actually decays (0.25)
f – fraction of C into CH_4 (0.5; other into CO_2)
16/12 converts C into CH_4 in mass terms.

W = 20.8 Mt/yr was clearly an underestimate; UK domestic waste was commonly taken as 20 Mt/yr (DoE 1992), but records are poor and incomplete (Aumonier et al. 1993). Domestic and commercial landfill together exceed 30 Mt/yr and industrial (non–construction but including sewage and food processing waste) landfill is about 15 Mt/yr or more (Table 1). The methane generation rate adopted earlier amounting to 35kg CH_4/t was too low; an ERL study had over double this (76kg/t; Manley et al. 1990) and the recent EC study (Bouscaren et al. 1992) recommends 55kg/t. Thus the previous 0.7Mt CH_4/yr would more than triple. However, current emissions relate partly to past lower rates of landfilling, implying dynamical correction may be needed.

1.2 Landfill gas exploitation study

The ERL report for ETSU (Manley et al. 1990) studied 75 of the 453 largest landfills in England & Wales (those with over 200,000 tonnes in place) and provided the basis for the 1992 DEPWS estimate of 1.0 Mt methane (DoE 1991). Based on new pumping measurements at selected sites, ERL calculated methane production from the 453 sites of 3.5Mt/yr, with potential capture efficiency of 55%. The 3.5 somehow scaled down (DoE 1990a) to 1.4Mt CH_4/yr (Aitchison pers. comm. 1992), the original figure being considered too high and subject to considerable uncertainty, partly because it implied total gas production of 222m^3/t higher than commonly inferred (~135m^3/t); the generation efficiency related to an uncertain model [2].

TABLE 1 Waste tonnages (UK, 1990)

Source	Waste Arisings (Mt) WMP1[a]	ETSU–93[b]	landfill (Mt)	methane (kg/t)[d]
Domestic (incl.Civ.Amenity)	20	28	27.5±2	53
Commercial	15	14.6	7.5±1	108
Industrial	32)	56[e]	15±2	45
Demolition & Construction	32)		–	–

Notes: [a] Waste Management Paper No.1 (DoE 1992); 96% of domestic and 52% of commercial & industrial wastes go to landfill; [b] grossed up from figures given for England & Wales alone, by a factor 1.12; [c] present estimates in Mt [2]; [d] "best estimate" of ETSU–93; [e] sum of classifications 'industrial' and 'inert', but evidently excluding mine and quarry wastes that exceed 100 Mt/yr.

The ERL study was motivated towards estimating the exploitable resource of landfill gas; the adaptation for estimating total atmospheric emissions (Aitchison 1993) was obviously incomplete. It was based on the lower limit of 1.4 Mt/yr (above). Numerous smaller landfill sites were ignored and a 12.5% reduction was assumed to result from oxidation in surface soils; capture and combustion of gas from sites that collect it (20% of the 453) was assumed 100% effective.

1.3 Stoichiometric Approach

The 1993 ETSU estimate (Aumonier et al. 1993, hereafter called ETSU–93) started from stoichiometry, assuming degradation of a chemical C,H,O, N,S combination, with 53% of the C converting to methane. The models give 110–250m³ methane per tonne, 2 or 3 times higher than data from real landfills. Aumonier et al. therefore sought reductions for aerobic decomposition (<1%), loss in leachate (~2%), microbial biomass (~5%) and oxidation in surface soils (~20%) to give the methane figures of Table 1, for average refuse compositions. The figure for domestic waste is close to the EC standard of 55kg/t (Bouscaren et al. 1992; this however without surface soil oxidation), but the commercial figure is twice as high, probably because paper decays by less than the theoretical d=55% of ETSU–93.

A wide range of 0.6–5.3 Mt/yr methane emissions was calculated, taking p=23–32% degradable carbon, but ignoring industrial waste as uncertain (though 30% extra is guessed) and adopting high values for gas utilised and flared. A major uncertainty is the oxidation in surface soils.

2 CAPTURE & UTILISATION

Records are kept of methane combustion in on–site heat or power generation, given as totalling 0.11 Mt/yr by ETSU–93 or 0.15 Mt/yr at Jan.'93 by Maunder (1993). Flaring devices are in place at only 20% of the ERL sites, that produce some 0.28 Mt/yr methane (Aitchison 1993). However, only a proportion of this is captured and flared. While the ERL calculations give average 55% capture at existing UK sites, the individual values range from ~20% to over 100%, so are unreliable. Entrapment and collection technology is at an early stage and relatively inefficient; pipework and valves to flare systems leak; flares themselves allow as much as 15% of gas through unburned (burning at less than optimum levels with fluctuating flow; perhaps 5% leaks on average); gas from older sites that lack impervious lining migrates into adjacent soils and strata. A UK government study (AFRC 1988) gave actual capture rates as under 50%. Experience in the USA (EMCON 1990) finds the efficiency of methane capture ranges from 40–90%. The upper level is feasible only at best and if the gas is flared rather than used for fuel. Thus at least 20% of the methane escapes, inclusive of leakage in the collection system and slippage through the flare. Let's suppose therefore that the efficiency of capture plus flaring at the 90 ERL sites with flare devices is 50% (0.14 Mt/yr), so the utilised and flared total is 0.25Mt/yr (1990 data).

3 RECONCILING ESTIMATION METHODS

The lower stoichiometric ETSU–93 limit would have an implausibly high fraction of methane utilised or flared (0.47 Mt/yr or 44%), while the upper figure is much higher than the 4.0 Mt/yr inferred by simply scaling up the ERL results to the full 64 Mt/yr landfilling total. This implies that the stoichiometric values are not useful, though the approach may give information on the start–up and final decay phases of landfills. Using updated emission figures to reconcile the other two estimation methods should allow derivation of tighter and more useful methane totals. Peer et al. (1993) put emphasis on a 'regression model' rather than reconciliating the other two, but their approach is in effect quite similar to the present one.

3.1 Use of site emission measurements

The data from pumping trials was grossed up by ERL (Manley et al. 1990) to relate to the 453 large sites in England & Wales (those with >0.2 Mt in place in 1989, and dating pre–1987). Allowance for smaller sites, for the rest of the UK and for oxidation in surface soils has to be made. Information on oxidation is scanty; Aitchison (1993) quotes up to 25%, presumably including older sites, while ETSU–93 has a 'best estimate' of 20% with values 10–50%. Peer et al. (1993) cite 10%, while Whalen et al. (1990) found high levels in an old landfill (7 yrs post completion, 1.5–3 m soil cover). They found the oxidation rate to be sensitive to soil moisture and to temperature, as Nozhevnikova et al. (1993) confirm. As oxidation varies with season and with type and age of site, it is unclear what single values represent. The 50% estimate of Whalen et al. relates to near optimal conditions. Pending proper characterisation, let's divide sites [3] into large and active ones (ERL's 453), the remaining gassing landfills (about 1000 medium–sized and/or completed) and others (some 10,000), and assume up to 12.5% oxidation for the ERL sites, 15–30% for the other gassing ones and 100% for the remainder.

The ERL sites potentially emit ≥1.4 Mt/yr adjusted for oxidation to give ≥1.225–1.4 Mt/yr. The average age of waste in place in the trial sample was 5.5–6 yr, the domestic waste fraction of 47% was a little above average and the tonnage in place (370 Mt) amounted to nearly 6 yrs landfill. The intermediate sites therefore account for up to half the landfilling over the last 12 yrs. Several of the ERL tests were on landfills over 12 yrs old; some methane comes from landfilling up to 30 yrs ago, though decreasing amounts if a 15–yr half–life applies to the slowest degrading component (ETSU–93). Some 45% of landfill waste is in this long–lived category (data of Tables 1,2) and if past landfilling rates were lower the contribution from >12yrs is relatively small. Potential emissions of the intermediate sites are thus estimated at 1.3–1.8 Mt/yr, reduced via oxidation to the range 0.43–1.53 Mt/yr.

The total has to be increased by 12% to allow for Scotland and N.Ireland (14% extra population; 6% allowed in D.En. (1990a), 20% in ETSU–93) giving ≥1.85–3.28 Mt/yr. Deduction of that flared and utilised (sect. 2) gives ≥1.6–3.0 Mt/yr (based on the 1.4 Mt/yr minimum of DoE (1990).

3.2 Emission factor method

The product p x d as chosen by Richards (1989) equated to 100 m^3/t landfill gas, whereas higher figures 130–150 m^3/t are common. ERL give p = 0.245 in domestic and 0.335 in commercial waste. Let's adopt d = 0·30, 0·22 ± 0·02 (in the 23, 15% putrescibles and 34, 60% paper of UK domestic, commercial garbage resp.) to be compatible with 130–150 m^3/t. Note that recent trends to increased waste compaction, to larger landfill sites and to oxygen–excluding cover, all suggest higher methane production.

The CH_4:CO_2 ratio is higher than 50:50 and varies with age. The Pollution Inspectorate gives 65:35 (Rae 1989), but values in the range f=0.5–0.7 are 'common' (AFRL 1988); ERL adopted 0.55±0.1, though their data average 62:38 over 14 sites; stoichiometric theory has supported values near 0.5, like ETSU–93's f=0.53, but are on the low side of field data; let's adopt f = 0.55–0.65.

The figure for total annual domestic landfill has to be augmented by commercial and industrial waste (incl. food processing and sewage). Re–estimate domestic and commercial landfill as 35±3 Mt/yr [4] and take industrial putrescible landfill as 15±2 Mt/yr (but with lower p=0.185 as ERL) giving an extra 26–40% [4]. With these factors, Equ.(1) results in 2.11–3.39 Mt/yr (before subtracting combustion). However, a correction is needed for a dynamic system if landfilling was lower in past years.

4 DYNAMIC CORRECTION

The ERL study takes gassing rates to decline exponentially with time after emplacement (after an initial build–up) with three categories of waste–decay constants k_i = .05–.69 /yr. Suppose landfills are increasing at rate r (≃ 3% /yr): the reduction due to lower landfilling in the past is then

$$\int_{-\infty}^{0} M e^{rt} e^{kt} \, dt \Big/ M \int_{-\infty}^{0} e^{kt} \, dt = \frac{k}{r+k} \quad (2)$$

The mean dynamic reduction factors are calculated from

$$F(r) = \Sigma f_i/(1 + r/k_i) = \Sigma f_i/\,(1 + rT_i/\ell n2) \quad (3)$$

for f_i the fraction in each waste category with decay half–life T_i. ERL's f_i, p and updated T_i (ETSU–93) then give the results of Table 2.

The average F for r=3%/yr weighted according to waste fractions (Table 1) is thus about 0.70, while the average for r=2%/yr comes to 0.77. It's not clear that landfill tonnages have increased at 3%/yr; however, the trend to increased compac–

TABLE 2. Dynamic reduction factor F

T =	–biodegradables– 5	9	15 yrs	inert T = ∞	p	Dynamic 'F' $r = 2\%$	3%
domestic	15%	10%	24%	26%	0.245	0.77	0.70
commercial	11	13	43	23	0.335	0.75	0.64
industrial	15	7	15	38	0.185	0.79	0.72
spoil	4	0	4	72	0.040	0.79	0.71
civic amenity	18	12	23	27	0.265	0.78	0.71

Note: the calculation uses the 3–component decay model of ERL, with compositions f_i in %. The fractions of moisture when added to the biodegradable and inert fractions make up 100%; moisture and the inert fraction are ignored in calculating F.

tion might give higher decay fraction (d) in newer deposits, and the trend to larger sites may mean lower oxidation. These trends suggest the use of a higher value of r than simply the growth rate of landfilling tonnages.

5 DISCUSSION

The correction for a dynamic system is significant in the reconciliation of site measurements with emission factor methodology. The value F = 0.7 gives 1.48 to 2.37 Mt/yr from the latter, compared with ≥1.85 to 3.28 Mt/yr from the former (sect. 3.1). There is uncertainty in extrapolating for past tonnages and in guessing oxidation rates (allowed for implicitly in the emission factor approach) and it is reassuring that the two ranges overlap. The ERL measurements give higher numbers, suggesting that the theoretical dynamic correction may be overdone. Taking F = 0.77 (of r=2%/yr), the range common to the two methods is 1.85 to 2.61 Mt/yr, or 1.6 to 2.4 Mt/yr after deducting that flared and utilised (referred to the year 1990).

The Environment White Paper (HMG 1990) stated that energy recovery from landfill would reduce Britain's emissions substantially in the next 30 years. That was when landfill emissions were put at 0.7 Mt/yr and at 22% of the UK total, rather than 2 Mt/yr and 40% of the total (DoE 1993a). The high prices paid for electricity from landfill gas under NFFO (the 'Non–Fossil Fuel Obligation') has encouraged use for generation, with 90% of the 0.09 Mt/yr recorded at Dec.1992 having NFFO support (Maunder 1993). A further 32 projects have been approved under NFFO, totalling 0.1 Mt/yr. Energy Paper 58 (D.En. 1990b) forsaw landfill gas exploitation reaching *up to* 0.75 Mt/yr by the year 2005, an increase of 0.65 Mt/yr over 15 yrs. Evidently, such small amounts may not even counter increasing landfilling (>2%/yr of 2 Mt/yr surmised). Moreover, electricity consumers' continuing subsidy via NFFO of landfill gas management is surely politically insecure; stronger regulation and landfill levies are on the current government agenda.

The UK government is currently considering whether the environmental externalities of landfilling could be corrected via landfill levies (DoE 1993b). Leachate has recognised if unquantified pollution costs, but what of methane that is a relatively inert pollutant as greenhouse gas? If methane pollution is costed relative to CO_2 [5], having 25–50 times its potency, I have estimated £13–27 per tonne landfill (Wallis 1992) higher than current direct landfilling costs (£7.50–22.50/t) and higher than the £10/t notional levy considered to cover all externalities (DoE 1993b). With 50% collection and combustion efficiency on a typical managed site, the externality cost is halved. This £7–13/t landfill is an appreciable extra on top of the dumping cost. On a more modern site, designed from the beginning for gas extraction and use, 75% capture is realistic. If an equivalent methane tax were imposed, the £10–20 differential with–or–without extraction appears sufficient incentive to stimulate investment in landfill gas control measures.

The US EPA (1992) reports an optimistic "up to 90%" with properly designed and contained sites. However, EMCON (1990) say 25–35% escapes if the gas is utilized in gas reciprocating engines, when less vigorous pumping to maintain higher methane content is required. Leakage through the control–flare is additional. It is doubtful that 90% capture and utilisation can be demonstrated in practice, over the full lifetime and under subsidence due to degradation, including leakage in the collection equipment and slippage through combustion devices. AFRC (1988) pointed out the losses over the initial and final stages of the landfill, when exploitation is not economic.

Once leakage rates over a site's lifetime at 30 or 50% are recognized, this must be a significant problem (ENDS 1990). It calls into question the whole policy of permitting mixed waste in landfills, unless several times lower leakage is demonstrated to be feasible without excessive expenditure and control procedures. Limiting the co–disposal of putrescibles and paper, in favour of composting and recycling or incineration is an alternative. Aerobic treatment of landfills can be more effective in limiting methane emissions (US–EPA 1992) though it is a cost rather than benefit in energy terms. Maintenance of aerated, non–waterlogged surface soils (Nozhevnikova et al. 1993) and cultivating methane oxidising bacteria in them can also be effective in cutting emissions (DoE–sponsored research underway at Essex Univ.).

NOTES

[1] The 1993 edition of DEPWS (DoE 1993a) tabulates 1.000 Mt for each of the four years 1988–91, but admits in a footnote that the figure relates to 450 landfills considered viable for gas utilisation. A full estimate of 1.4–4.0 Mt with mid–point of about 2 Mt is attributed to the Watt Committee but this is misleading. Figures from the ETSU study as reported to the Watt Committee in Jan.'93 were 1.27–3.13 Mt with preferred value of 2.07 Mt (1990 base). The range widened to 0.6–5.3 Mt with 'best estimate' 2.0 Mt in the published version ETSU–93, but none of the figures have the endorsement of the Watt Committee's Methane Working Group.

[2] ERL used a time–dependent model, assuming differential degradation coefficients. This is analogous to the 'Scholl Canyon' method advocated by the U.S. EPA (Peer et al. 1990), which obtains methane emission rates depending on environmental variables like temperature and moisture content, which would be site–dependent. ETSU–93 followed this, assuming half–lives 5, 9 and 15 yrs for readily, moderately and slowly degradable material, dependent on site characterisation. However, environmental variables are supposed not to affect the total emission over a 20–30 years gassing lifetime; since we would be interested in integrated greenhouse impacts (GWPs) evaluated over longer times than this, calculating the instantaneous emissions adds unnecessary complexity and uncertainty. It is only for the purpose of checking sites for their actual emissions that the Scholl–Canyon procedure can be useful.

[3] The total of Gassing Landfills exceeds 1000 (defined as those needing remedial treatment). D.En. (1990b) said 1300 sites might require gas management, while the Environment White Paper (HMG 1990) stated there are 4000 controlled landfills of which 1000 need remedial measures. HMIP in 1989 said 1006 sites in England & Wales, of which 60% are closed (but gassing continues for 10 or even 20 years). That is known to be an underestimate, as Councils had not completed surveys; DoE's Pollution Inspectorate give 1500 (Rae 1989), while FoE have published a revised estimate of 1480 gassing landfills (ENDS 1992). A Water Research Council study for the DoE has almost 12,000 sites (ETSU–93), of which 3321 were 'operational' (England & Wales 1990), though most are small.

[4] In the DEPWS statistics (DoE 1993a) it appears that landfilling has grown at nearly 2% over 1980–88 and been constant ever since; this is implausible when recycling and waste reduction in the UK has had little penetration. New Zealand, for comparison, records growth of 5% /yr. Since landfilling became a sensitive issue and the EC started compiling waste statistics, there has been political incentive to play down the UK's waste record. Much trade (commercial) waste is collected by local authorities and landfilled – domestic and trade waste together were given as 28 Mt in the 1989 House of Commons Environment Committee, 2nd Report. An independent report records 80% of waste disposal authorities in England & Wales as disposing of 26 Mt in 1990 (FoE 1991), which implies a UK total of 35 Mt/yr, allowing 12% for Scotland and N.Ireland and 96% landfill fraction.

The ERL study quotes 50 Mt/yr industrial landfill with a substantial C–fraction (18.5%), but figures of 14 Mt/yr (ENDS 1992) or 16.6 Mt/yr (Table 1: 52% of the WMP1 industrial category) seem more plausible. The ratio industrial:domestic landfill in western Germany is 31±9% (Bingemer & Crutzen 1987). The 15±2 Mt/yr with ERL's data on composition would amount to an extra 33±7%. ETSU–93 gives a guess of 30% for commercial plus industrial compared with domestic waste methane, which is implausibly low on their own figures that have commercial waste alone at 25% of the domestic waste landfilled.

[5] Global costs of greenhouse gases. The costs of global climate change induced by greenhouse gases may be extremely high – £100's per tonne C – but extremely uncertain. One can instead consider costs of removing greenhouse gases, eg. via planting forests or via extracting and burying CO_2 from appropriate types of power stations. The former has been costed at £5–10/t and the latter at £11/t (current exchange rates). The EC's proposed energy/Carbon tax at the rate of $10 /b.o.e., that was estimated to take us half way to stabilizing CO_2 emissions, gives a similar cost.

REFERENCES

AFRC (Agriculture & Food Research Council) 1988. Basic study of landfill microbiology and biochemistry, ETSU B1159

Aitchison E 1993. Methane emissions from UK landfill sites, in "*Methane Emissions*" 92–98. The Watt Committee on Energy, London.

Aumonier, S, P. Lawson & A. Shepherd 1993 [ETSU–93]. An assessment of methane emissions

from UK landfills, Energy Technology Support Unit report to DoE, *CWM* 063/93
Bingemer, H.G. & P.J. Crutzen 1987. The production of methane from solid wastes, *J.Geophys.Res.* 92: 2181–2187
Bouscaren, R., Ch. Houllier & B. Crozet 1992. Possibilities for control of greenhouse gases other than CO_2 in the European Community, Ch.1, Methane, *CITEPA Report*, Paris
D.En. (Dept.of Energy) 1990a. National assessment of landfill gas production, ETSU *Renewable Energy Technology Summary* 198
D.En. (Dept.of Energy) 1990b. *Energy Paper 58*, Dept. Trade & Industry, London.
DoE (Department of the Environment) 1991. *Digest Environmental Protection and Water Statistics* (DEPWS), no.14, HMSO London
DoE (Department of the Environment) 1992. *Waste Management Paper No.1*, HMSO, London.
DoE (Department of the Environment) 1993a *Digest Environmental Protection and Water Stastistics* (DEPWS), no.15, HMSO London.
DoE (Department of the Environment) 1993b. Landfill costs and prices. Correcting possible market distortions (Coopers & Lybrand study) HMSO London
ENDS 1990. *ENDS Report* 189, 18–21, Environmental Data Services, London
ENDS 1992. *ENDS Report* 206, Environmental Data Services, London
EMCON 1990. EMCON Associates to 'Energy and Environment 90', *3rd Int.Landfill Gas Conf.*, ed. G E Richards & Y R Alston
FoE 1992. *A Survey of Local Authority Recycling Schemes in England & Wales*, Friends of the Earth, London
HMG 1990. *This Common Inheritance*, para. 5.70 [Environment White Paper] Cmd. 1200, HMSO London
Manley. B.J., D.G. Wilson & H.S. Tillotson 1990. *National assessment of landfill gas production*, [ERL study] Environmental Resources Ltd., ETSU B1192
Maunder D.H. 1993. Landfill gas to energy, the current UK position, paper to ETSU Seminar on Landfill Gas, Solihull; also pers.comm. to Dr T Morris, DTI, Dec.1992
Nozhevnikova A.N., A.B. Lifshitz, V.S. Lebedev, & G.A. Zavarain, 1993. Emission of methane into the atmosphere from landfills in the former USSR, *Chemosphere* 26: 401–417.
Peer, R.L., S.A. Thorneloe & D.L. Epperson, 1993. A comparison of methods for estimating global methane emissions from landfills, *Chemosphere* 26: 387–400.
Rae G 1989. The control of landfill gas, Dept.of Energy's *Review*, no.7, 8–9.
Richards, K.M. 1989. Landfill Gas: Working with Gaia, *Biodeterioration Abstr.* 3: 317–331.
US–EPA (US Environment Protection Agency) 1992. *Options for reducing methane emissions internationally* 1, Global Change Division EPA, Report to Congress
Wallis, M.K. 1992. Externality costs of landfills, *Warmer Bulletin* 35, p.19.
Whalen, S.C., W.S. Reeburgh, & K.A. Sandbeck 1990. Rapid methane oxidation in a landfill cover soil, *Appl.Environ.Biol.* 56: 405–411

Waste Disposal by Landfill – GREEN'93, Sarsby (ed.) © 1995 Balkema, Rotterdam, *ISBN 90 5410 356 6*

General report

M.S.de Silva
Aspinwall and Co., London, UK

M de Silva: My aspiration is to summarise the papers and to bring together the thoughts on basic principles.

McKendry took us through the benefits of a risk assessment approach for landfill design and brought out the value of computer-aided design for landfills to assist in making decisions rapidly when necessary. An important point was that a large leap has been made in design philosophy (particularly in the UK) and the lack of good data on field performance is tending to restrain developments in design. The comment (from the Department of the Environment) that despite the fact that there have been several leaks from landfill sites none of them has had any significant impact on potable water sources is particularly encouraging.

Wackermann illustrates some rather elaborate german basal lining systems. Under some circumstances such elaborate systems would be appropriate but each case must be carefully considered.

The paper by **Bedford** brings across the benefits to be derived from computer-aided support for landfill design such as their use in design and costing, water balance calculations long-term monitoring and control.

Barratt covers the application of bio-technology in landfill. This is a component which has to be appreciated, landfilling of refuse is not just an engineering situation. The chemistry and biology of the waste is complementary to the engineering solution in devising the best way of dealing with this landfill. No longer is a landfill operation regarded as an exercise in entombment of refuse. The modern landfill has to be thought of as a living system.

Wallis offers a challenge to conventional calculation of gas production from refuse - perhaps some of the basic assumptions made in evaluating gas yields need to be re-thought.

Gas control also requires serious attention and it is necessary to address the question of why we design landfills - to maximise gas production or to reduce gas production. Obviously there is a fundamental difference in philosophy of each of these approaches. Both can be appropriate, according to individual circumstances, as can the basic principle of opting for a dilute-and-disperse site or a containment site. Each site has to be treated as an individual case and an appropriate solution found.

Other contributions on the future of landfilling, its design and the use of various materials include those of **Bahout Hadjinicolaou and Yong** and **Quant** who discuss the use of Pulverised Fuel Ash (PFA) materials in basal and lining systems. Again this fits in very well with the concept of waste minimisation which is a major area for the future. Here we have a novel way of dealing with the PFA to reduce its potential environmental toxicity and to enable it to be used as a construction material.

Sztromajer's paper also relates to the use of biotechnology, i.e. cellulose skeleton-ion exchange mechanisms, for contaminated water clean-up. There is obviously potential application here for leachate and waste water treatment. The system is claimed to be low-cost and can utilise materials which are themselves waste products.

Alshawabkeh and Acar discuss species transport in saturated soils under electric fields. This is a very sophisticated option for treating wastes and waste liquids. It employs physio-chemical techniques instead of biotechnology. Whilst the principles of the process have been known for some time there is little experience of the field performance of this process. The greatest interest in the system to date has been for the remediation of contaminated land.

The paper by **de Silva, Roberson, Pearson and Nicholls** concerns the sustainable landfill and extension of the principles of controlled engineered landfill in combination with waste minimisation, maximisation of gas production so as to use it to its full potential as a bio-reactor or energy generator/liberator.

As a summary there are several points to be made:

a) Many of the basic principles have been established but it is necessary to evaluate how complementary they are with current legislation (bearing in mind that the legislation must allow some flexibility in the chosen path). The driving force for development should be environmental protection and providing that a particular solution can achieve this then legislation must allow for this flexibility.

b) The compatibility of the principles of engineering and the biological and chemical demands of the landfill requires further consideration. A better integration of engineering and biotechnology is needed to either maximise of minimise gas production.

c) Each site must be considered as unique.

d) There is scope to operate in an environmentally-sympathetic way, when dealing with landfill, and there are various low-cost alternatives that can be used, in addition to the more high-tech methods.

e) The question of why the landfill is being designed needs to be addressed - is it to influence gas production or is it to provide a containment repository for the waste.

f) It is frequently asked as to whether landfill will continue, in the long term, to play such an important role in waste management. It is believed that it will for the foreseeable future but that there will be a complementarity with the need for waste recycling and minimisation. Furthermore it is considered that the two schools of thought about landfill practice, i.e. dilate and disperse or containment, will continue to exist and that given specific circumstances both solutions may be acceptable or correct.

Alshawabkeh and Acar
Principles of species transport in saturated soils under an electronic field.

Bahout Hadjinicolaou and Yong
Characterization and marketability of bottom ash from municipal incinerators.

Barratt
Microbial methane oxidation and the effective biological treatment of landfill-generated methane.

Bedford
Automated landfill design using computer-generated ground modelling.

de Silva, Robertson, Pearson and Nicholls
The sustainable landfill : The potential for developing a bioreactor cell rotation landfill.

McKendry
Risk assessment of engineered containment landfill designs.

Quant
New lining technology based on fly ash composites.

Sztromajer
The preparation of cellulose skeleton ion exchangers.

Wackerman
The development of lining technology in Germany

Wallis
Reassessing methane from UK landfills.

Waste Disposal by Landfill – GREEN'93, Sarsby (ed.) © 1995 Balkema, Rotterdam, ISBN 90 5410 356 6

Discussion

K Gothheil outlined a special aspect of composite liners ie the behaviour under thermal load. This was currently being researched. In municipal waste landfills temperatures may rise to 80°C and this poses a question about the water content of the mineral liner at the base of the landfill.

Five years ago three test cells were built (at the University of Karlsruhe) consisting of a mineral lining layer overlain by HDPE. The surface of mineral liner is heated upto 40°C and with the base being maintained at 15°C, ie a gradient of 25°C across a thickness of 75mm. Three materials (for the mineral layer) were tested, a fat clay, a silty clay, a silty clay with 3% bentonite.

Two years after the start of the test the cells were opened and desiccation was observed, together with significant shrinkage cracking. Neutron measuring probe tubes were then installed and the cells were re-closed. Five years from the start of the testing the shrinkage cracks are estimated to have progressed almost to the base of the mineral liner.

To study this phenomenon under natural conditions, ie with impervious subsoil, under a landfill lining system, a test site has been built with a plan area of 250m^2. This site incorporates an electric heating system, to maintain a temperature of about 40°C, and 2½m of static load cover to simulate the waste. The water content of the mineral liner is measured using tensiometers and neutron probe measuring tubes. Five months after starting the mineral layer is desiccating but at a much slower rate than in the laboratory. Gas pressure in the liner increases with temperature and this may encourage desiccation. Conclusions are; never build a composite liner on pervious subsoil, drainage layer beneath a mineral liner of a composite system speeds up desiccation of the mineral liner, the ideal arrangement would be a mineral liner between an upper HDPE cover and a lower asphalt liner so that water content of the clay cannot change.

With regard to K Gottheil's comments on cracks and fissures in composite liner systems, **H Jessberger** wondered if there was any indication that these gaps may re-close. Whilst there may be a possibility of desiccation cracking of mineral layers in composite liners, if the geomembrane subsequently ceases to act as a barrier then there will be re-wetting of the mineral layer which may lead to sealing of the fissures.

K Gottheil stated that for shrinkage cracks it depends on the width of the fissures as to whether or not they will close upon re-wetting of the liner. The HDPE liner is likely to function 5 - 15 years. Shrinkage cracks in the mineral layer will be about 0.5 - 1.0mm wide. Subsequent closure will depend on this width, whether there is any infilling, the static load applied to the liner (if there is 20 - 30m of waste deposited then this imposes a vertical stress on the clay layer), and the nature of the material composing the mineral layer. The field material in the case reported by Gottheil was a natural silty clay.

S Clamp commented that for a multi-layer liner system a saturated drainage layer could be incorporated between the barrier element and this would avoid desiccation cracking. This layer would also provide a useful monitoring system for leakage. However, the head in this drainage layer should not be so high as to cause hydraulic fracture of the liner. The whole system would consist of a composite liner (eg HDPE and clay layer) over the saturated drainage layer with a further lining layer (eg HDPE or clay) beneath.

S Clamp was very much in support of the risk assessment approach to design. The design should be on the basis of a maximum appropriate seepage rate with due consideration being given to the sensitivity of each specific site.

According to S Clamp more field data of liner performance is needed, particularly for the long term behaviour. It was stressed that more field monitoring is urgently needed to assist future development of design, not just to check for leakage. At the same time there is a requirement for the development of better monitoring equipment.

The need for performance monitoring was supported wholeheartedly by **M McKendry**.

S Jefferis proposed that designers should try to prevent cracks in the liner from being localised and large, ie it is better to have a lot of small cracks spread throughout the mineral layer. If granular bentonite is used in mats then a few large joints are obtained between the particles (giving a high permeability in the swelled material) and manufacturers are now moving towards the use of powdered bentonite to give many small joints between particles. S Jefferis proposed that fibre reinforcement should be introduced into clay to prevent major localised cracking so that upon re-wetting the amount of swelling needed at any location in the clay is quite small.

Any soil swelling investigations must pay attention to the water chemistry of the material. Desiccation and re-wetting can cause local high concentrations of chemicals and the re-wetting agent (water or leachate) may be significantly different from the water removed during desiccation. There is the possibility of osmotic flow through liners (in addition to Darcian, diffusion and thermal flows). The direction of the flow will depend on chemical concentrations and may lead to desiccation of layers within a composite liner. It is vital to remember the long time scale of processes within landfill.

P McKendry wondered about the effect that degree of compaction has on the shrinkage /swelling propensity of a clay layer. He suggested that by compacting the clay highly to achieve low initial permeability the tendency to crack may be enhanced.

S Jefferis added that 'fat clays' may not be the best material for lining (because of their crack potential) and that it might be advantageous to add granular material to a clay to act as crack inducers, thereby producing a lot of small cracks rather than a few large ones.

H Jessberger supported the risk assessment approach to design as he believed that the effects and aims of engineering works should be considered before becoming involved in design details. In his opinion risk was a function of the probability of system failure and the severity of the resulting effects. However, the methodology of risk assessment (as applied to chemical plants for instance) was not directly applicable to landfill sites because in the latter only a few not very important, active safety elements exist. The important barrier elements are passive, and we do not have sufficient data to characterize the behaviour of these elements, particularly in the long term. Also it is very difficult to define failure of the system, eg if the liner permeability increases at what value correspond to failure. Therefore it is not possible to calculate definitely the risk associated with a landfill site. However, we can use a numerical risk assessment value to compare different systems. To conduct the risk assessment numerical values are needed for the effects of particular 'loading cases' and at the present time we have insufficient data to provide consistent values.

P McKendry commented that Risk Assessment is a methodology and it suffers from a lack of plausible data. It is a useful test for comparing possibilities and for applying sensitivity analyses.

J Dunn reported that risk assessment is widely used in the USA for hazardous waste sites but unfortunately, even though risk is a normal part of life, the public want landfill sites to have zero risk and this is an impossible situation.

P McKendry concurred that there is no such thing as zero risk, everything leaks and even the United States Environmental Protection Agency now accepts that if an emission rate is below a certain value that it equates to zero leakage.

M Wallis commented that complete containment was not feasible with regard to gaseous emissions and the options available were capping, together with collection of the landfill gas, or controlled aeration of the surface of the landfill with substantial oxidation. The oxidation rates are quite sensitive to temperature (so landfill behaviour in winter is different from that in summer) and the water content of the surface layers. Control of the preceding parameters causes a major problem. In the USA they are conducting experiments on artificial aeration by feeding air in via sub-surface pipes. Such developments require quite different technology from that currently associated with landfill sites.

P McKendry opinioned that at the present time we have no realistic control over the sensitive biological process which takes place inside the landfill bioreactor. A decision on whether or not to cap a site depends on what is the goal with respect to the landfilled refuse, ie is it to provide a waste repository or to facilitate waste stabilization. If a landfill is capped to keep out water then the degradation process is slowed down which in turn leads to the site having a pollution potential for a very long time. Promotion of degradation leads to a more rapid stabilisation but the higher production rate of leachate may present problems with treatment capacity. Currently there is little control over the tipping within a landfill site and attitudes towards landfilling and filling methods will have to change if we are to move towards the bioreactor approach to refuse disposal.

With regard to landfilling practice **L Cowan** pointed out that current statutory regulations such as the requirement for waste to be covered on a regular basis as it is deposited, prevent a controlled fluid movement through the waste thereby inhibiting the creation of a managed bioreactor.

K Seymour said that within the National Rivers Authority (NRA) there was no consensus view on the best form of landfill cap. The Authority's overriding concern was minimisation of the possibility of pollution escaping from a landfill site, whether by ensuring that the underlying liner was fully effective or by minimising the amount of leachate within the site.

According to **S Hoekstra** the philosophy in the Netherlands is to concentrate on the formation of an effective cap rather than on a basal liner, because of the unreliability of the latter item. Moreover the cap is readily accessible for remedial work if necessary. Furthermore rainwater which enters the waste does not flow through it uniformly, there are preferential paths and so refuse decomposition is not uniform and it will not occur throughout the entire waste mass.

M de Silva suggested that the best way to reduce the time needed for landfilled refuse to stabilize was to cap the site and control the inflow of water into the waste. The moisture content of the waste is quite critical to microbial activity and it is not easy to control moisture movements in practice as operational factors such as daily cover will affect moisture flow. Furthermore the waste materials decompose in different ways, for instance with paper (which is heavily lignified) the aerobic stage is important for degradation and there is little or no evidence that the lignin is broken down in an anaerobic environment.

S Clamp felt that complete landfill containment or the development of a refuse bioreactor were often viewed as mutually exclusive whereas the sensible approach was to have the bioreactor completely contained so that inputs and emissions could be fully controlled.

A Forster felt that the lack of proven technology with regard to many parts of an engineered landfill system, eg long-term behaviour of liners, clogging of leachate collection systems, was the reason for the regulatory authorities' caution towards development of the bioreactor approach.

S Jefferis stressed the urgent need for drainage systems which will continue to work for a long time without clogging to enable the leachate to be controlled.

With regard to the cost of landfilling **R Wackermann** pointed out that lining systems may appear to be expensive but they should be considered in terms of their cost per tonne of waste contained. For example, for a 50m deep waste deposit with a lining system which costs £100 per square metre the expenditure is only £2/m^3 (approximately £2/tonne) of waste. This cost is small in comparison to the costs associated with other means of waste disposal, eg in Damstadt it costs approximately £150/tonne to incinerate waste.

P McKendry commented on a comparative risk assessment he had made of four liner systems, ie single, double, composite, double composite. By considering the inherent problems of each system the probability of exceeding a significant leakage rate was established with a 'normal leakage rate' being of the order of 2000 kg/hectare/day. Furthermore the cost implication (between £120k to £480k per hectare) of each option could be assessed. These data had been used to derive a cost/benefit ratio which indicated the double composite to be the best system - however the difference between the four systems was not dramatic.

3 Engineering behaviour

Waste Disposal by Landfill – GREEN'93, Sarsby (ed.) © 1995 Balkema, Rotterdam, ISBN 90 5410 356 6

Water balance evaluation of different multi-layered landfill final cover systems

A. Forster
MRM Partnership, Bristol, UK

ABSTRACT: The consequences of excessive infiltration into a landfill can be severe in terms of leachate remediation and/or treatment costs and the polluting potential of a site. However, little published information exists on the performance of different final cover systems. A three year research project is being carried out to provide an evaluation of infiltration rates through various cover designs.

A series of small scale lysimeter cells have been constructed at the BFI owned Little Packington Landfill site, near Birmingham. A 1m clay seal was placed at varying slopes and using different compaction techniques. Each clay cap is overlain by a 5mm thick synthetic lateral drainage layer and 0.5m of topsoil which has been grassed. Monitoring comprises an automatic weather station, soil moisture probes, output from each lysimeter and runoff from the drainage and topsoil layers (not differentiated).

The study enables an accurate water balance assessment for specific cap designs to be made, which may assist in identification of an effective final cover for landfills.

1 INTRODUCTION

Total containment is now becoming the dominant strategy for waste management within the United Kingdom. Final covers form an integral part of this approach.

The design and quality control during installation of a landfill cover will determine how much, if any, precipitation will percolate into the underlying waste. It is now common for site licenses to specify surface sealing materials with a hydraulic conductivity of not greater than 1.0×10^{-9}m/s; this is equivalent to 31.5mm/annum at unit hydraulic gradient. However, there is very little information yet published on whether caps of such materials, particularly those laid in an engineered fashion, in practice allow more, or less, than 31.5mm/annum of percolation to occur.

As part of a programme of waste management research, funded by the Department of the Environment, MRM Partnership were commissioned to conduct a field trial which assesses the nature and rate of infiltration through different clay cover design systems. At present, a three year timescale has been set for completion of all stages of the project.

2 DESIGN

Efficient moisture collection and measurement was essential to the success of the design.

Utilising experience from previous experimental work (Ref. 1) an initial design was drafted. Further modification to include drainage and topsoil profiles was thought to allow a more accurate simulation of full scale

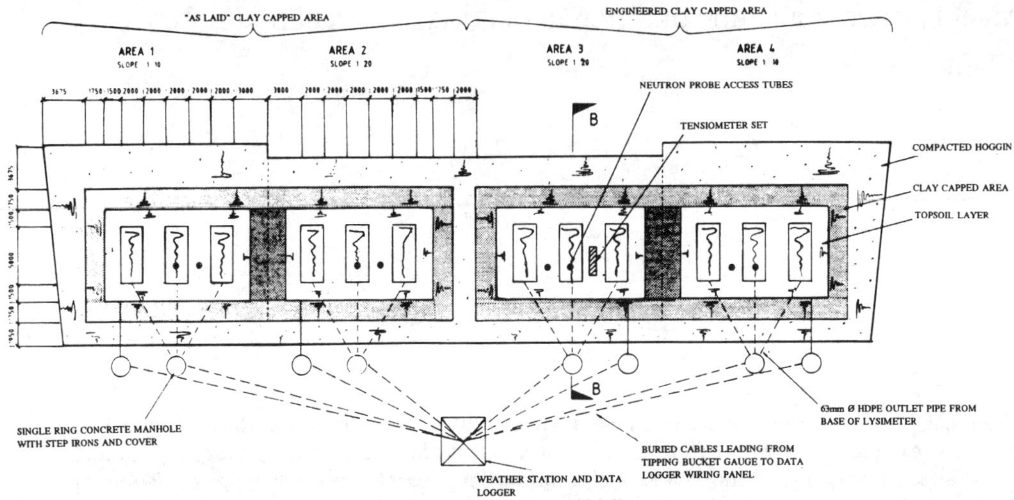

FIGURE 1 : Plan showing test field area indicating dimensions of collection and monitoring system. (All dimensions are in mm)

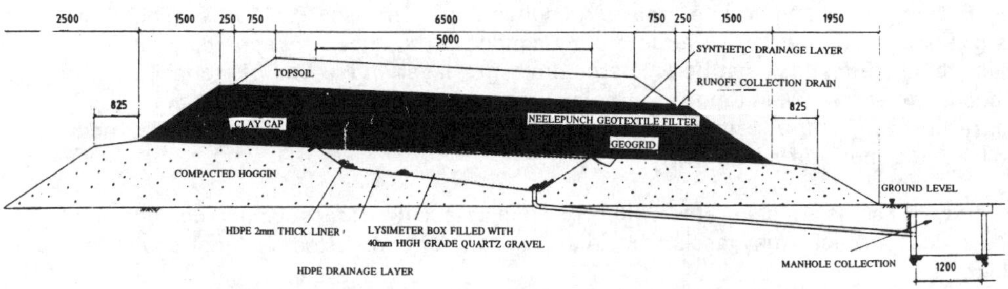

FIGURE 2 : Section through lysimeter cell and cover profile. (All dimensions are in mm)

landfill conditions. The overall field layout and cover profile is summarised in Figures 1 and 2.

Twelve lysimeter cells were constructed beneath four experimental capping systems, these include:

At slopes of 1:10

- Three engineered to a typical landfill specification (ie., $<1.0 \times 10^{-9}$ m/s).
- Three "as laid" non-engineered sealing layers.

At slopes of 1:20

- Three engineered to a typical landfill specification (ie., $<1.0 \times 10^{-9}$ m/s).
- Three "as laid" non-engineered sealing layers.

Each clay seal is overlain by a 5mm thick synthetic lateral drainage layer and 0.6m of topsoil which has been grassed.

On-site weather conditions are recorded by an automatic weather station. Flows from each lysimeter cell, drainage above the clay seal and climatic parameters are monitored hourly and stored on a Campbell Scientific CR10 data logger (Figure 1). Weekly site visits allow downloading of the stored data onto a laptop PC for subsequent analysis.

Lysimeter cells and connecting pipework were fabricated on-site using a high density polyethylene (HDPE) material (Plate 2). To ensure an accurate real time response to percolation events a non-absorbent, coarse grained quartz aggregate underlain by a HDPE drainage net was placed in each lysimeter cell

(Plate 1). Drainage from each cell is measured via a tipping bucket gauge (0.1 litre capacity).

To avoid damage to the lysimeter cells and pipework during placement of the clay seal a HDPE geogrid was placed over each lysimeter cell to dissipate the load from earthmoving plant. A non-woven geosynthetic fabric was placed over the geogrid to prevent "fines" from the overlying clay washing into the collection and monitoring system.

In Areas 1 and 2 (Figure 1) clay was placed in three ~ 33cm layers and tracked over twice between each layer by a PC 300 Excavator and represents the "as placed", non-engineered capped area. In contrast, for Areas 3 and 4 compaction of each ~15cm layer was carried out using a static smooth wheeled roller (4 passes) and a tamping roller (4 passes) in accordance with Department of Transport earthworks specification (Ref. 2). Moisture control and compactive effort was monitored using a calibrated nuclear density meter.

A synthetic lateral drainage layer was placed immediately above the clay sealing layer (Plate 3). A central core drainage fabric sandwiched between a trammel filter fabric provides a preferential flow pathway for infiltrating water above the capping layer. Flow from each drainage area (approximately 8.5m x 12m) is collected by a downslope guttering system and monitored individually using four tipping bucket gauges (1.0 litre capacity).

Topsoil was loose tipped avoiding soil compaction and damage to the drainage layer below. A 0.7m soil layer was laid, using Ministry of Agriculture, Fisheries and Food specifications, which has subsequently settled to a final thickness of approximately 0.6m. Seeding of the test area with a grass/clover/ley mix was timetabled to coincide with the start of the growing season (ie., late September).

Soil moisture flux over time throughout the full cover profile is monitored weekly using neutron probe and tensiometer recordings. Over the field study area, eight neutron probe profiles at 10cm depth intervals and one tensiometer profile at 20cm increments will allow identification of any variations in moisture that may occur (Figure 1).

3 CONSTRUCTION

Construction work commenced in July 1991. Where possible, standard earthmoving plant was used in an attempt to replicate full scale operational conditions seen elsewhere at the Little Packington site.

Extrusion welding and electrofusion coupling provided a water tight HDPE collection system. Leak testing for all lysimeter cells and connecting pipework was conducted before and after the placement of the clay seal layer. No leaks were detected.

Preliminary laboratory testing of on site clays identified the control parameters to be used during placement. A Mercia Mudstone source provided the material for the sealing layer. The mudstone has the following typical properties: 56% clay, 34% silt, 10% sand, 45% liquid limit, 21% plastic limit, 25% plasticity index, 31% natural moisture content, 20% optimum moisture content and a maximum dry density of 1.65 Mg/m^3.

Field tests using a nuclear density measuring device revealed sufficient compaction (ie, ≥ 95% of laboratory Proctor density) was achieved on Areas 3 and 4 only. Significant (ie, > 25%) non conformance was recorded on Areas 1 and 2. Moisture content in all four areas was fairly uniform. a summary of test results is given in Table 1.

Spraying of the clay surface with water at regular intervals prevented excessive desiccation during the construction phase. However, modification of the final clay profiles allowed one week to elapse before placement of the drainage and subsequent layers, during which some minor desiccation occurred (ie, <5cm in depth).
Topsoiling and seeding was completed by late September 1991, however due to unsuitable weather conditions full scale monitoring of climatic and flow parameters did not commence until late March 1992. Installation of soil moisture instrumentation occurred in late April 1992.

PLATE 1
On site fabrication of lysimeter cells using 2mm HDPE geomembrane.

PLATE 2
Infilled lysimeter cells prior to placement of the clay sealing layer.

PLATE 3
Lateral drainage layer above the clay sealing layer.

TABLE 1 : Summary of test results taken during placement of the clay sealing layer

	"As Placed" Clay Seal		Engineering Clay Seal	
	Area 1	Area 2	Area 3	Area 4
No. of Tests	23	17	41	34
Av. Moist Content (%)	29.3	29.1	29	28.9
Moist Content Range	27.1-32.8	28.2-30.1	27.3-31.3	27.4-30.6
Av. Proctor Density (Mg/m^3)	1.39	1.31	1.41	1.42
Av. Proctor Density (%)	95.88	89.72	96.50	96.56
Proctor Density Range	1.25-1.45	1.17-1.36	1.36-1.44	1.36-1.43
No. of non-conformances	6	17	2	1
% Failure	26	100	<5	<3

Surveying on completion of each profile layer maintained a high degree of accuracy throughout the construction phase.

4 TEST FIELD CONSTRAINTS

The field study area is located on an area of old (ie, > 10 years) inert fill. Thermal and subsidence effects on the cover design caused by waste is therefore not considered. Although some studies have attempted to incorporate these effects (Ref. 3, Ref. 4) it was felt that an accurate assessment of cap performance under optimum conditions would identify essential design principles for use in full scale landfill applications.

Hydrological studies on such a small scale have to contend with problems of the representativeness of construction methods. For example, it is probable that a greater degree of compaction was achieved in Areas 1 and 2 (non-engineered areas), due to space constraints than would normally be expected using similar placement techniques at a working landfill site.

Using a single clay source materials for the sealing layer variation in hydraulic performance of each test area can be attributed solely to slope and compaction differences. Caution however, must be exercised when interpreting the results for use with other natural clay-like materials.

In an attempt to minimise measurement error, where possible, on site calibration and fielding testing, using destructive and non destructive techniques, were carried out.

5 PRELIMINARY RESULTS AND COMMENTS

A full scale monitoring programme started in April 1992 and a detailed assessment of cover performance for the period April 1992-93 is currently being drafted for review by the Department of the Environment.

It is anticipated that seasonal variations will occur in percolate rates, the magnitude of which may be highly dependant on the effectiveness of the drainage layer above the clay seal. Moisture profile monitoring will also identify periods of saturated and unsaturated flow conditions prevail.

Variation in flow rate above and below the clay sealing layer as a result of differences in slope and/or placement techniques has not yet been fully assessed.

ACKNOWLEDGEMENTS

I would like to thank BFI Packington Ltd for their co-operation with this project and also my fellow MRM colleagues, Keith Knox and Peter Dumble for their advice and support.

REFERENCES

1. KNOX, K. (1991). Water Balance Methods and their Application to Landfill in the UK. (Draft version) Dept. of Environment December 1991.

2. DEPT. OF TRANSPORT (1986). Specifications for Highway Works. (Part 2) August 1986.

3. MELCHIOR, S.G AND MEIHLICH, G. (1989). Hydrological studies on the effectiveness of different multi- layered landfill caps. Proc. Int. Landfill Symp. (Vol 1) 9-13 Oct. 1989 Sardinia.

4. HOEKS, J. et al (1987). Bentonite liners for isolation of waste disposal sites. Waste Man. Res. 5, 1987.1

Waste Disposal by Landfill – GREEN'93, Sarsby (ed.) © 1995 Balkema, Rotterdam, ISBN 90 5410 356 6

Determination and assessment of the mechanical properties of waste materials

H. L. Jessberger & R. Kockel
Ruhr-Universität Bochum, Germany

ABSTRACT: In the geotechnical design of landfills attention must be given to the mechanical poperties of the deposited waste materials as these define landfill stability and deformations which are usually assessed applying Soil Mechanics principles. For non soil-like wastes there are existing no generally accepted testing methods for determination of appropriate parameters of the waste for stability analysis.

Results from different element and model tests on municipal solid waste show that no shear failure but large deformation occurs. Failure Parameters φ and c cannot be clearly defined. From the results of first centrifugal model tests with waste the conclusion can be drawn that such tests are a suitable instrument for assessing a waste structure phenomenology. Pre-loading of the waste has a significant effect on the strength and deformation properties of the material.

1. INTRODUCTION

A waste landfill or waste pile has to be regarded as an engineered structure for which long-term safety has to be proved on the basis of an appropriate geotechnical design. This means the waste body has to be stable against slope failure. Furthermore, the deformation behaviour and stresses imposed on subsoil, liner systems and internal structures have to be assessed. Figure 1 gives a schematic illustration of possible failure mechanisms and deformation problems.

Aside the design of new landfills stress and deformation analyses of waste bodies are also required at the remediation or extension of abandoned landfills or landfills in operation. Typical examples are the repair of old basal drainage systems by the help of open cuts (Jessberger/Kockel, 1991) or the substitution and design of failed leachate collection/monitoring shafts.

Such analyses are done by applicating Soil Mechanics principles. Input data are appropriate geotechnical parameters of the waste. By reference to recommendation R 2-6 and R 3-6 of the "Recommendations on the Geotechnics of Landfills and Contaminated Land" (ISSMFE/ETC 8, 1991) waste can be subdivided into soil-like waste and other, non soil-like waste. In the case of non soil-like waste, which may include municipal solid or mixed waste, Soil Mechanics are only limited applicable. Therefore an appropriate assessment of waste body stability, stresses and deformations requires extensive knowledge of the mechanical properties of these wastes ("Waste Mechanics").

In this paper the results of laboratory and first centrifugal model tests on waste from municipal landfills are presented (municipal solid waste = MSW). Significant mechanical mechanisms of this material are discussed and conclusions on the stability and deformation behaviour of waste bodies are drawn.

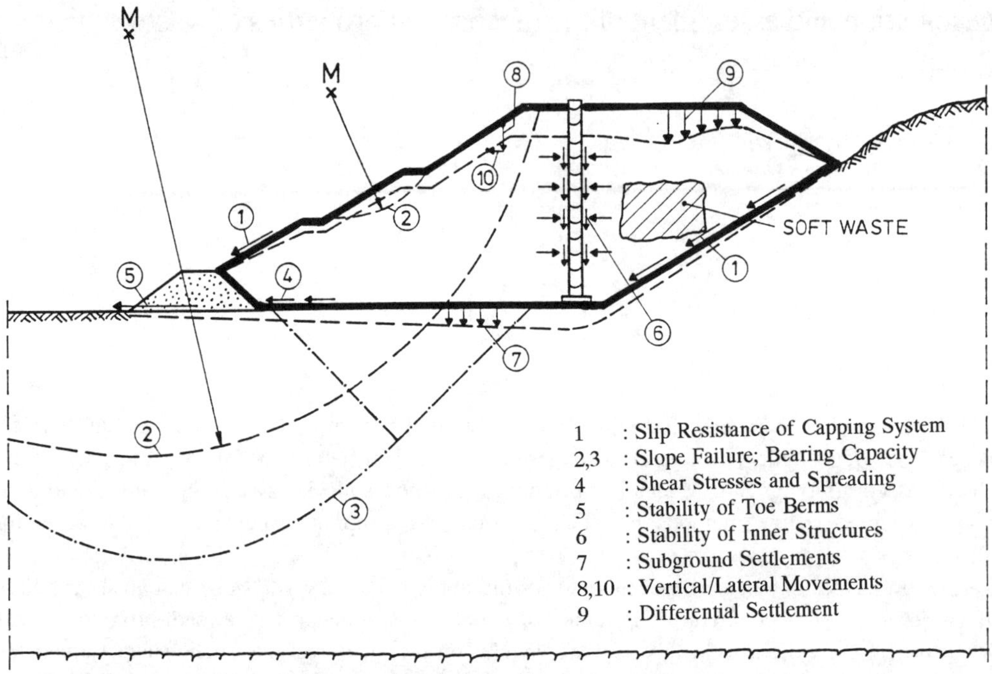

Figure 1 Stability and deformation of a landfill

2. GEOTECHNICAL WASTE PARAMETERS

Geotechnical parameters of primary concern are those describing strength and deformation properties of waste. Equivalent to Soil Mechanics the shear parameters - angle of internal friction φ and cohesion c - and the modulus of deformation E_s are used in design calculations. Both may be derived from different test methods, e.g.:

- Laboratory tests (element tests)
  - Triaxial compression test
  - Direct shear test
  - Confined compression test
  - Unconfined compression test
- In situ tests
  - Slope failure test
  - Cone penetration test
  - Plate load test
  - Pressuremeter test
- Model tests
- Back analysis from
  - Waste body deformations
  - Waste body failures

In testing MSW the chosen test method has to take care of the specific material characteristics, e.g. by the use of large test equipment or by in situ testing. MSW is very heterogeneous in comparison to soils as it consists of a large scatter of materials like food and garden waste, plastics, soils, textiles, wood, paper, metal, glass, etc. These materials extremely vary in their mechanical properties and geometry. As a result MSW shows high compressibility and strength. Latter can be explained by a "Reinforced Earth" - like effect as there are fibrous elements which are interlocked and therefore act like reinforcement. For this reason failures with shear planes through the waste material have only been reported when there are weak zones or stability reducing hydraulic situations.

Furthermore, the mechanical properties of MSW are time-dependent. Due to biochemical degradation processes of organic material the composition of the deposited waste changes with time. The resulting pore space may cause additional settlements and a weakening of the waste structure. This may lead to reduced strength properties of the waste body.

Different authors have reported on geotechnical waste parameters and on the basis of laboratory and in situ tests, back analysis or personal experiences shear parameters illustrated in Fig. 2 have been recommended. Some authors prefer the specification of parameters of fresh and old waste taking degradation processes into consideration. However, the shear parameters of MSW show broad variations and the responsible geotechnical engineer has to decide which value of φ and c has to be introduced into the required geotechnical proofs at landfill design.

A better understanding of the relevant mechanical properties of MSW can be provided by conducting triaxial and confined compression tests. Additional results from first centrifugal model tests on vertical slopes of MSW yield further information on the behaviour of a waste structure.

3. TRIAXIAL COMPRESSION TESTS

At the Soil Mechanics and Foundation Engineering Division of the Ruhr-Universität Bochum several triaxial test series on MSW samples have been conducted. The available test equipment comprises a large (∅ 30 cm, h = 60 cm) and a small (∅ 10 cm, h = 20 cm) triaxial compression cell.

In the tests a 1 to 3 year old waste sampled at a municipal landfill has been used. Before testing the material has to be processed. In a first step large constituents which are expected to falsify testing results are seperated. In the following this waste without large constituents is defined as untreated waste (uW). For assessing influences from waste treatment tests were conducted either using untreated waste (uW) or

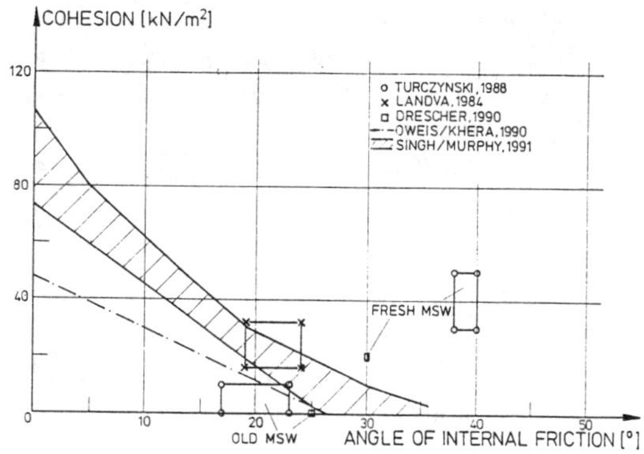

Figure 2 Shear parameters of municipal solid waste according to different authors

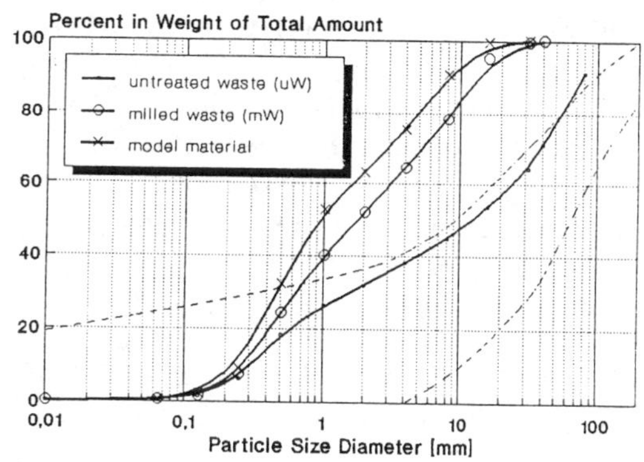

Figure 3 Grain size distribution of untreated, milled and model waste

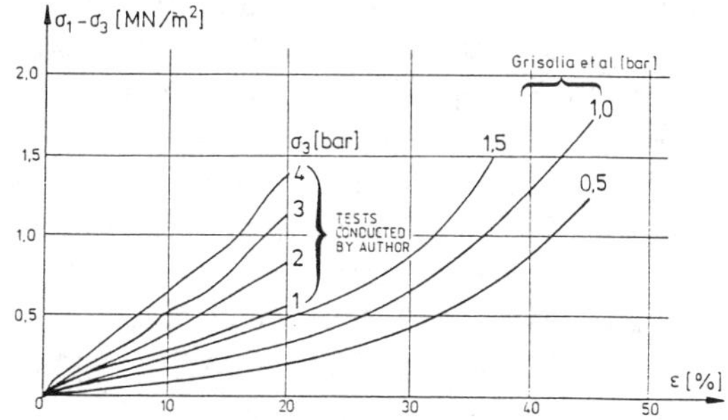

Figure 4 Results of triaxial compression tests on reconstructed and milled waste

milled waste (mW). The grain size distribution of both materials is shown in Fig. 3, that also includes the cuve of the so called model material (see section 5).

Test conditions for triaxial tests are as follows:

- "CD-test"
- isotropic consolidation ($\sigma_1 = \sigma_2 = \sigma_3$)
- constant confining pressure ($\sigma_2 = \sigma_3$ = const.)
- constant strain rate: 1 %/min
- measurement of σ_1, σ_3, ε_1, ε_v

The comparison of test results with uW and mW shows up minor differences. Therfore the following explanations refer to milled, 1 to 3 year old waste. In Figure 4 typical stress-strain-curves for mW are presented. Tests by Grisolia et al. (1991) conducted on specimen of "reconstructed waste" confirm these results:

- Failure cannot be observed even at high sample compression
- The waste hardens with deformation

An attempt to explain these effects is the following model of thought:

Obviously the high compressibility of MSW is responsible for the observed material properties. At isotropic consolidation a relatively large pore space of the sample can be observed. These pores may be differentiated into macropores between the waste particles and into micropores representing the pore space within the particle. The effective "particle to particle" contact surface can be assumed as relatively small. With increasing deviatoric stress $\sigma_{1\text{-}3}$ a compression takes place which is described by the vertical compression ε_1 and volumetric compression ε_v (Fig. 5). The decreasing gradient of these curves with specimen compression indicates an increase of horizontal strain ε_3 with increasing ε_1.

At early test stages the horizontal strain is about zero. At the same time a high vertical compression - that means reduction of pore space - takes place. This might result in an increase of the effective contact surface so that the matrix stresses acting on these surfaces probably do not change significantly. Therefore a failure at early stages of the test cannot be expected. At states of additional vertical compression the horizontal strain ε_3 increases probably indicating an increase of matrix stresses at the same time. But even at ε_1 = 20% a limit value is not reached.

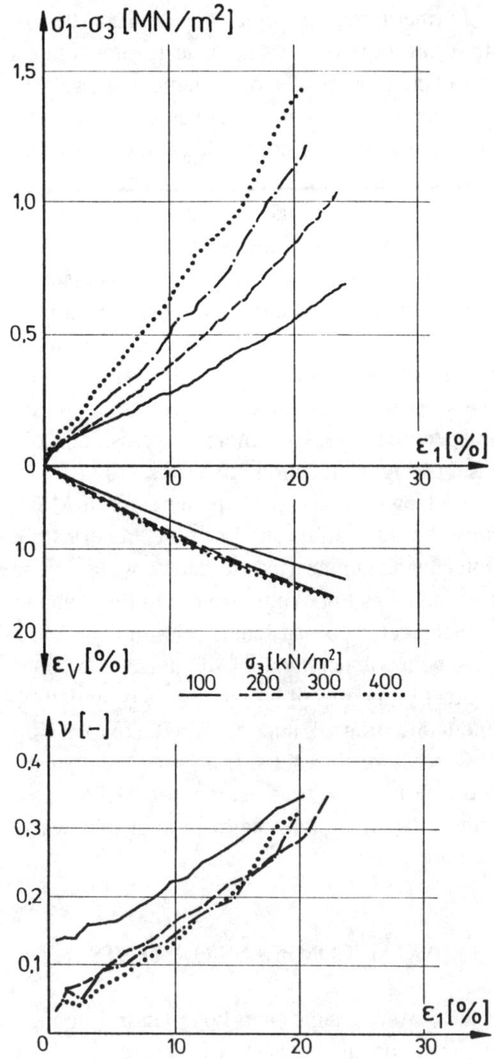

Figure 5 Triaxial compression test on milled, 1 to 3 years old MSW

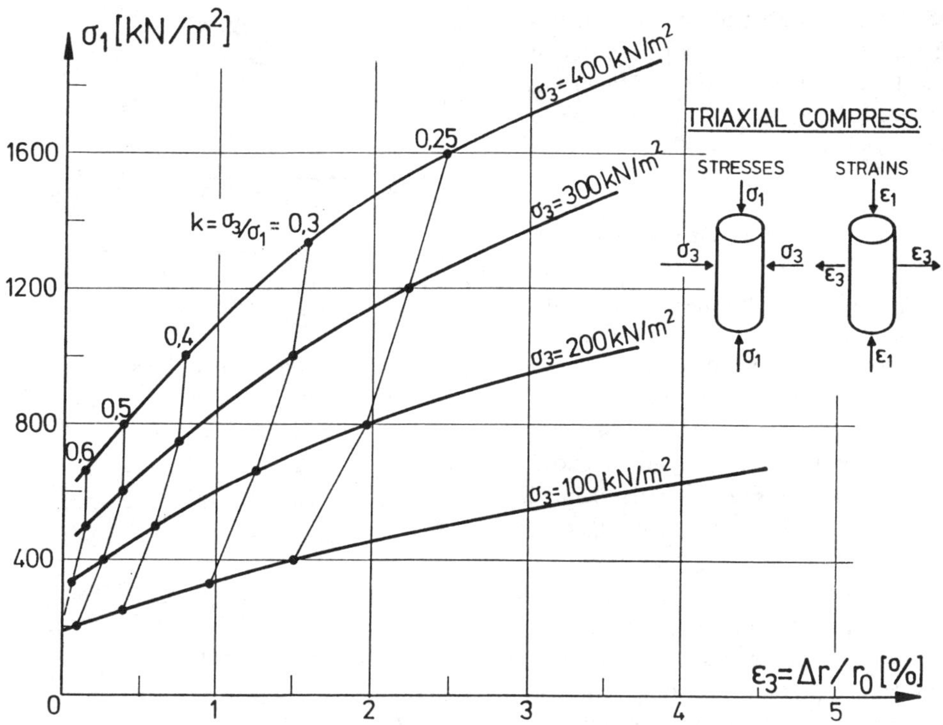

Figure 6 Stress-strain-relationships for MSW derived from triaxial compression tests

The lower diagram in Fig. 5 shows the Poisson ratio versus ε_1 defined as follows for triaxial compression tests with constant confining pressure:

$$\nu = \frac{\Delta \varepsilon_3}{\Delta \varepsilon_1}$$

The curves for different confining pressures confirm an increasing trend of ε_3 with ε_1. A tendency of the almost linear relation towards reaching a limit value cannot be observed. Furthermore, the Poisson ratio does not vary with the confining pressure; the deviation of the curve σ_3 = 100 kN/m² is probably due to technical problems at sample preparation.

On the basis of these test results the type of failure significant for MSW cannot be clearly defined. It is obvious that failure parameters φ and c according to the Mohr-Coulomb theory do not describe satisfactorly the failure condition. A possible way of approaching suitable shear parameters is the application of a limiting strain criterion (Jessberger/Kockel, 1991). The effectiveness of such a criterion has to be proved in further research.

Probably the stress-deformation behaviour of a waste body is of more importance for the geotechnical design of landfills than stability considerations. Fig. 6 gives an idea of the stress-strain situation of waste bodies at states of high stress. The diagram is derived from the results of several triaxial compression tests on milled MSW and may be used for a first approach of the ratio K of horizontal to vertical principal stress dependent on the possible horizontal strain ε_3. It can be realised that for K = const. an increase in horizontal strain leads to reduced minor principle stress whereas K increases with the major principal stress at ε_3 = constant.

It must be noticed that the diagram presented in Fig. 6 is only valid for "normally consolidated" material. However, creep effects of the tested MSW might significantly influence triaxial testing results. By further tests with different stress paths these results shall be confirmed and a more detailed description of the stress-strain behaviour is intended.

4. CONFINED COMPRESSION TESTS

For assessing landfill settlements confined compression tests in a large-scale compression cell (Fig. 7) are conducted. Landfill surface settlements can reach 10 to 30% and even more of the original landfill height. One should distinguish into mechanical and degradation induced settlements according to Fig. 8. By the help of confined compression tests on waste materials load induced, mechanical settlements of a landfill can be assessed. In combination with a suitable approach for settlements caused by biochemical degradation processes, e.g. a correlation between gas production and settlements, a total settlement prognosis may be possible.

For the confined compression tests up to 15 years old, untreated MSW (uW) has been used. The waste is unsaturated and the sample is compacted up to dry densities of about 7 to 10 kN/m². After pre-loading at 25 kN/m² constant loads (50/ 100/ 200/ 400/ 660 kN/m²) are applied for several days and sample compression is measured with time.

A typical time - settlement diagram is shown in Fig. 9. The maximum sample compression reaches a considerable degree of more than 30 % of the original sample height. Using the semi-logarithm scale an initial settlement can be identified during the first 10 to 30 seconds after load application followed by a more slowly increase of sample compression. This part may be defined as primary compression. Although these settlement curves look quite similar to the settlements of cohesive soils this material behaviour can better be explained by creep effects rather than by consolidation effects.

From different tests on MSW the stiffness modulus has been derived for the chracterisation of the load dependent settlement of MSW. For

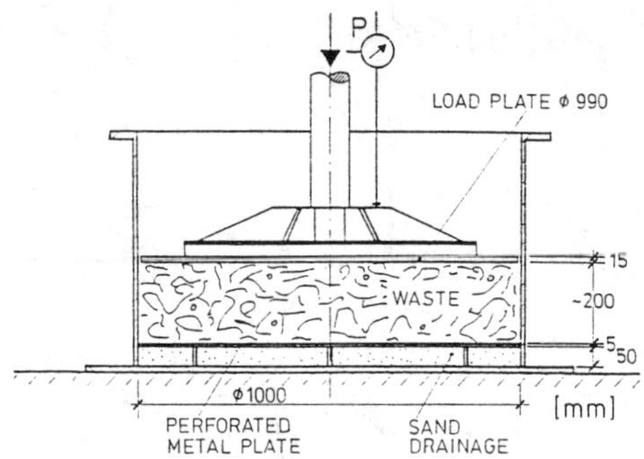

Figure 7 Compression cell ∅ 1000 mm

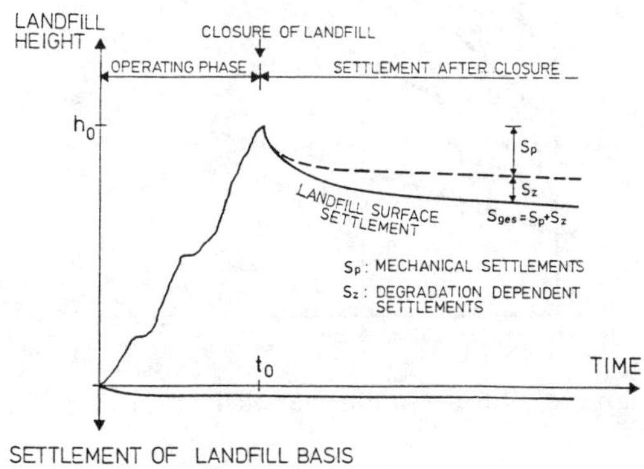

Figure 8 Scheme of landfill settlements

each single load step the stiffness modulus E_s is defined as:

$$E_s = \frac{\Delta\sigma}{\Delta\varepsilon}$$

It should be noticed that this stiffness modulus is derived from the virgin compression line and therefore only yields information on settlements of waste that never before has been subjected to loads higher than those applied in the confined compression tests. Further tests on MSW are planned for determining the modulus of recompression, too.

Fig. 10 shows the load dependent range for the stiffness modulus of MSW. The results are confirmed by data from other authors. But it should be emphasized that the presented stiffness modulus derived from confined compres-

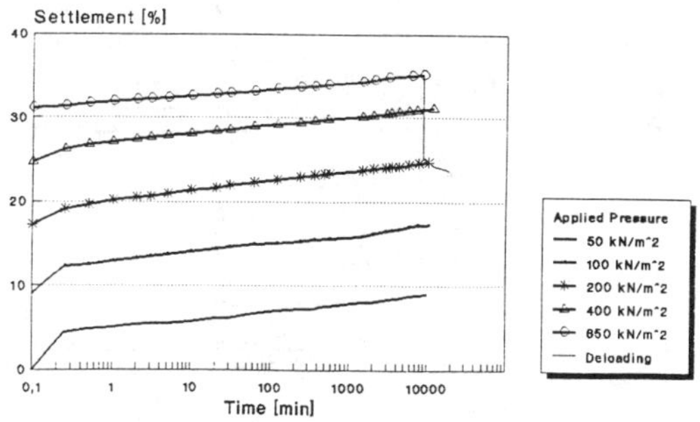

Figure 9 Time - settlement curves of MSW (uW)

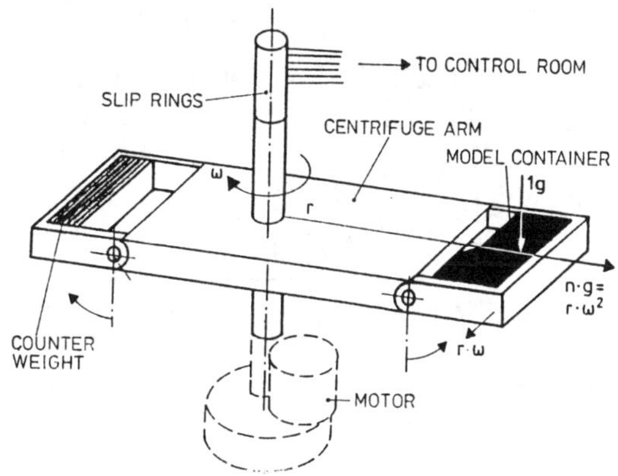

Figure 11 Geotechnical centrifuge with model (Jessberger/Güttler, 1988)

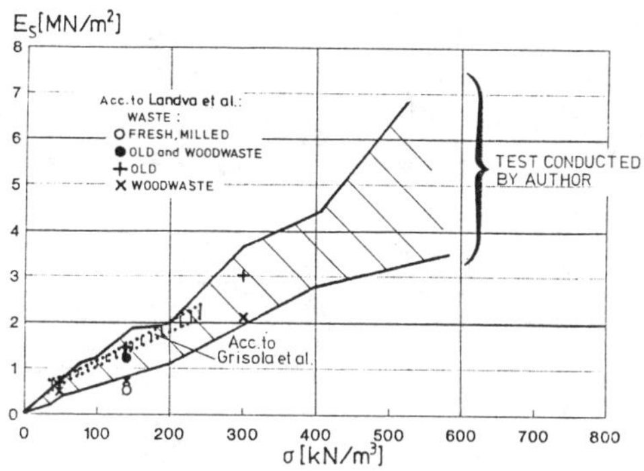

Figure 10 Load dependent range of stiffness modulus of MSW

sion tests only represents the mechanical settlements. Therefore further research should aim at the determination of degradation dependent deformations. Continuous field measurements of landfill deformations are necessary to assess load- and time-dependent settlements and to provide realistic forecasts of the waste body deformation.

5. CENTRIFUGE MODEL TESTS

For the assessment of the system behaviour of waste structures and for understanding realistic failure mechanisms centrifuge model tests on vertical slopes of MSW have been conducted at the Soil Mechanics and Foundation Engineering Division of the Ruhr-Universität Bochum (Jessberger/Kockel, 1992). In particular, these model tests are aiming at the following:

- Gaining informations about realistic failure mechanisms of waste bodies
- Assessment of strength properties of MSW
- Testing the applicability of centrifuge model tests for the analysis of waste structure behaviour

In a centrifugal model test a 1:n - scale model of a prototype is brought into a high acceleration field according to Fig. 11. Stress - states of the prototype in the 1g-acceleration field are identical to those of the model in the ng-acceleration field (Jessberger/Güttler, 1988).

In the tests the prototype of a two-dimensional vertical slope has been modelled. As model-material a fine-milled, 1 to 3 year old waste from a municipal landfill is used. According to scale requirements most of the grains > 10 mm have been manually seperated. The grain size distribution of this modelmaterial is shown in Fig. 3. The visual assessment shows a peat-like, fibrous structure.

The material is placed and compacted in layers of about 5 cm comparable to the waste placement at waste disposal sites. In order to reduce friction effects from the side walls of the rectangular model-container panes of glass have been inserted. After that the model is swung up in the acceleration field for "consolidation" where large irreversible settlements occur (Fig. 12a and b). The g-level 50g is chosen according

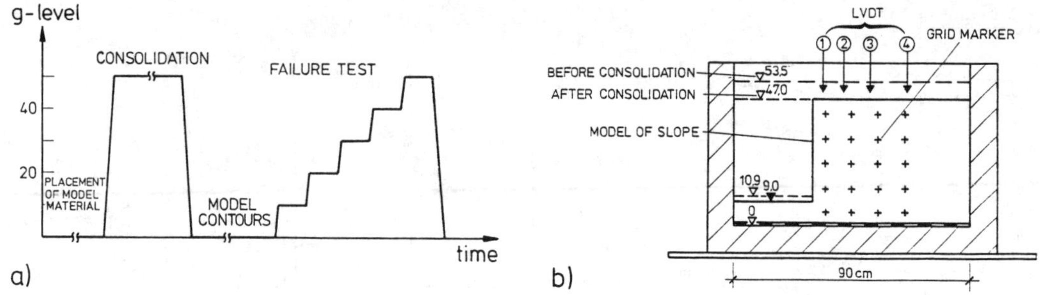

Figure 12 Scheme of a test design and model preparation

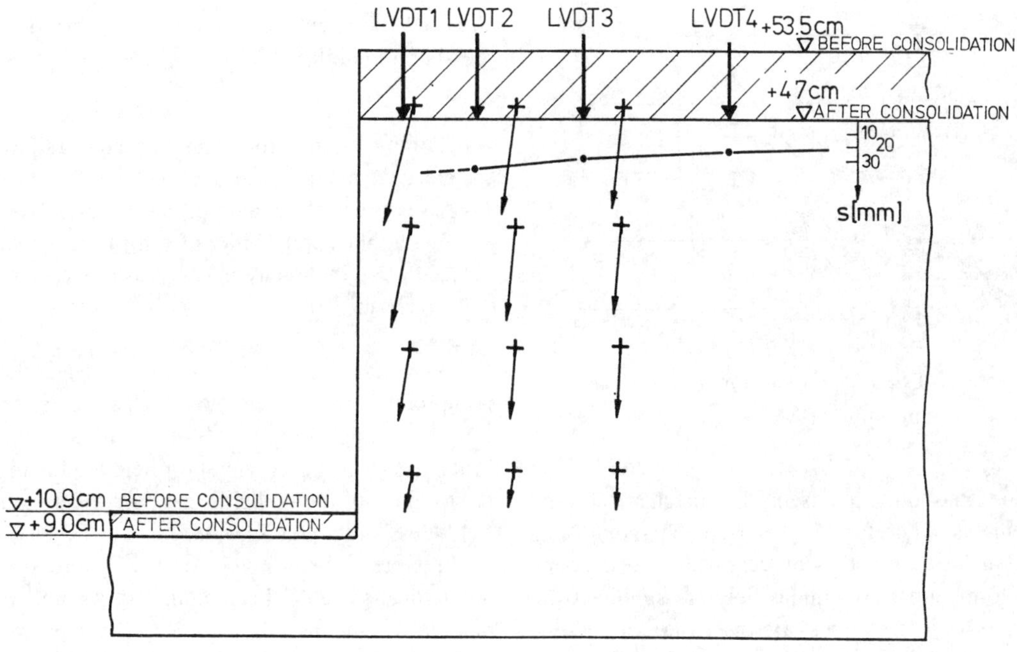

Figure 13 VMB 3 - Slope deformations at 50g related to the slope-geometry "before consolidation"

to the level which is expected at slope failure. The last step of model preparation is the modelling of the vertical slope (Fig. 12b). Grid markers are used for measuring slope deformations.

By gradually increasing the acceleration field (Fig. 12a, failure test) it is intended to reach a slope failure induced by self-weight of the waste material. The increase of acceleration means at the same time the increase of slope height according to scale requirements. E.g. at 25g a model slope of 40 cm height would correspond to a prototype slope of 25·0.4 = 10 m. During the test surface settlements are measured and the slope deformations are observed by camera.

During the tests large vertical and horizontal slope deformations have been observed. With increasing g-level the slope tends to incline forward (Fig. 13) and in the upper part of the slope horizontal tension stresses develop. Due to the inclination a first failure of the upper slope can be observed for prototype heights of about 15 to 17 m. The failure contours, fissures and

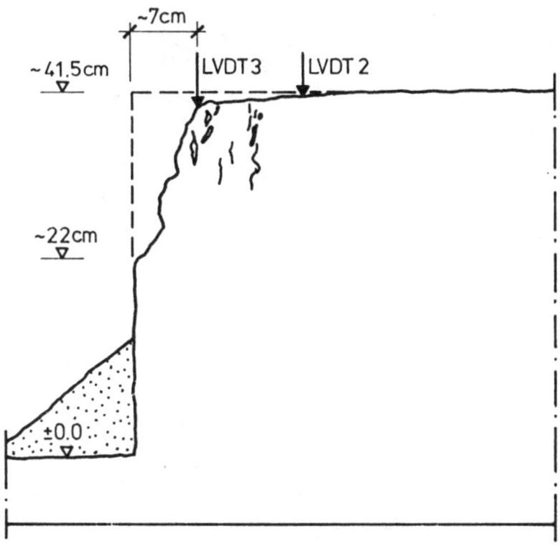

Figure 14 VMB 2 - Fissures and cracks after first failure of the upper slope part

cracks of such a model after stopping the centrifuge are illustrated in Fig. 14. The main direction of cracks is vertical indicating horizontal tension stresses; obviously the maximum tensile stress of the waste material has been exceeded.

A further slope failure can be observed at g-levels of about 60 or more equivalent to a prototype slope height > 20m. This failure extends over the slope region that has formerly been weakened by open tension induced cracks.

It has to be noticed that the lower part of the slope is stable and shows up only minor deformations. This effect is probably due to the high overburden pressure during consolidation as this pressure might lead to a better interlocking of waste particles and therefore to higher material strength.

6. SUMMARY AND CONCLUSION

A suitable geotechnical design of a waste landfill requires extensive knowledge on the mechanical properties of municipal solid waste. In this paper results of triaxial compression, confined compression and model tests are presented.

Laboratory and model tests show that significant mechanical material properties of MSW are the high compressibility and high strength due to a "Reinforced Earth" - like behaviour. Shear planes across the waste material have not been observed neither in the triaxial nor in model tests. Model test results show that the material is capable of significant tension stresses and that the failure mechanism is different to the one to be expected for soils. Consolidation or preloading of the material has significant influence on its strength and deformation properties.

In assessing landfill stability special consideration should be given to the state of stress and deformation of the waste body. A first approach for assessing the stability on the basis of Soil Mechanics principles is the determination of shear parameters from triaxial tests using a limiting strain criterion. An evaluation of effects from degradation processes is not possible on the basis of the presented test results. Further research on waste materials with a high state of degradation is necessary.

There is need of describing the mechanical properties of a complete waste structure. Centrifugal model tests are a suitable method for the assessment of system behaviour and may also be used in Waste Mechanics. Moreover, in situ tests and extensive landfill monitoring programs are necessary for a better evaluation of the mechanical waste properties and the waste body behaviour.

ACKNOWLEDGEMENT

The research fonds by the German Research Board (DFG) and the fruitful cooperation with the partners (RWTH Aachen, TU Braunschweig, TH Darmstadt) of the joint research project on "Waste Mechanics" are gratefully acknowledged.

LITERATUR

Die Bundesregierung: Technische Anleitung zur Vermeidung, Verwertung, Behandlung und sonstigen Entsorgung von Siedlungsabfällen (TA Siedlungsabfall). Entwurf vom 22.11.92.

Jessberger, H.L., Kockel, R: Mechanical properties of waste materials. XV Ciclo di conferenze di Geotecnica di Torino, Torino, 1991.

ISSMFE/ ETC 8: Geotechnics of Landfills and Contaminated Land - Technical Recommendations. Verlag Ernst & Sohn, Berlin, 1991. (2nd Edition 1993 in preparation)

Turczynski, U.: Geotechnische Aspekte beim Aufbau von Mehrkomponentendeponien. Dissertation, TU Dresden, 1988.

Landva, A.O., Clark J.I., Weisner, W.R., Burwash, W.J.: Geotechnical Engineering and Refuse Landfills. 6th National Conference on Waste Management in Canada, Vancouver, British Columbia, 1984.

Drescher, J.: Standsicherheit von Deponiekörpern, Festigkeit des Deponiegutes. Vortrag beim 6. Nürnberger Deponieseminar, 1990.

Oweis,I.S.,Khera,R.P.: Geotechnology of Waste Management. Butterworth & Co, 1990.

Singh,S.,Murphy, B.: Evaluation of the Stability of Sanitary Landfills. In: Landva, A., Knowles, G.D. (editors): Geotechnics of Waste Fills - Theory and Practice. ASTM STP 1070, Philadelphia, Okt. 1990, pp. 240 - 258.

Grisolia,M., Napoleoni,Q., Sirini,P., Tancredi,G.: Geotechnical Behaviour of Sanitary Landfill based on Laboratory and In Situ Tests. XV Conference di Geotecnica di Torino, Italien, 1991.

Jessberger, H.L., Kockel, R.: Stoffverhalten von Abfallstoffen im Hinblick auf die geotechnische Bearbeitung von Abfalldeponien. In: 1. Zwischenbericht zum Forschungspaket "Geotechnik der Abfallstoffe", Deutsche Forschungsgemeinschaft, September 1992.

Jessberger, H.L., Güttler, U.: Geotechnische Großzentrifuge Bochum - Modellversuche im erhöhten Schwerefeld, Geotechnik, Heft 2 (1988).

Waste Disposal by Landfill – GREEN'93, Sarsby (ed.) © 1995 Balkema, Rotterdam, ISBN 90 5410 356 6

Leachate permeation of compacted clay

N. Kalteziotis
Greek Public Works Research Centre, Athens, Greece

R.W. Sarsby
Bolton Institute of Higher Education, UK

K. Lam
Universal Dockland Ltd, Hong Kong

ABSTRACT: Samples of compacted clay have been subjected to sustained permeation by leachates from landfilled refuse. Control tests have been conducted using water permeation. As yet there is no indication that leachate has had any adverse effect on the permeability of this lining clay.

1 INTRODUCTION

The effectiveness of a clay lining in preventing leachate egress depends on its ability to maintain a very low permeability while in contact with these permeants. The main cause for concern is that leachates, by virtue of their chemical composition, may have an adverse effect on the structure of the clay and hence its permeability.

Changes in permeability of some soils are known to occur when they are permeated with certain chemical substances - these changes have been mainly attributed to ion exchange and mineral dissolution. Ion exchange may cause either an increase or decrease in the thickness of the adsorbed double layer surrounding the clay particles. If the double layer thickness is reduced the effective void space is increased and the overall permeability will probably increase.

2 LEACHATE COMPOSITION

A landfill is a constantly changing environment with chemicals being dissolved and precipitated by the fluid which moves through the landfill. The composition of the leachate is affected by type of refuse, water infiltration rate, pH, age and many other factors (Farquhar 1989). Most of the organic components are free volatile fatty acids and their concentration decreases with increasing age of the landfill as does BOD/TOC/COD. The leachate pH tends to increase with time, i.e. from an initial acid state towards neutral and

Table 1 Typical composition of leachates (Waste Management Paper No 26)

Determinand	Recent wastes *	Aged wastes *
pH	6.2	7.5
+ COD	23800	1160
+ BOD	11900	260
+ TOC	8000	465
Ammoniacal-N	790	370
Chloride	1315	2080
Sodium	960	1300
Magnesium	252	185
Potassium	780	590
Calcium	1820	250
Manganese	27	2
Iron	540	23
Nickel	0.6	0.1
Copper	0.12	0.3
Zinc	21.5	0.4
Lead	8.4	0.14

* All units of mg/ℓ except for pH.

\+ COD = Chemical oxygen demand
BOD = Biological oxygen demand
TOC = Total organic content

many samples show high concentrations of iron and zinc metal ions. In the United Kingdom the head of leachate in a landfill is typically maintained at 1m or less, by periodic pumping out of the liquid. This has the effect of decreasing the concentration of the leachate over the life of a landfill cell. Typical composition data are presented in Table 1.

3 EFFECT OF PERMEANT COMPOSITION ON PERMEABILITY

As early as 1954, Michaels and Lin (1954) studied the sensitivity of the permeability of kaolinite to various permeants. Although effects were difficult to quantify because of the mixing of different fluids permeating the same sample, the data suggested an increase in permeability following permeation with the organic solvents acetone and dioxane.

Since that time tests have been conducted using a wide range of permeants including methanol, acetic acid, hydrochloric acid, sodium hydroxide, heptane, landfill leachate. Permeation tests on both natural clays and kaolinite mixtures have been undertaken using various forms of permeameter. Some investigations, particularly those with high chemical concentrations, have observed significant changes in permeability as a result of sustained permeability. On the other hand the passage of landfill leachate through a compactedclay soil appears to have little effect on its permeability - Daniel and Liljestrand (1984), Brunelle et al (1987).

4 TEST PROGRAMME

Samples of compacted clay have been subjected to long-term leachate permeation. The two clay soils tested have both been used to form lining layers for engineered waste disposal sites. The test specimens were formed using heavy compaction with moulding moisture contents of optimum +3%, i.e. a typical condition for a clay liner.

The leachate was obtained from a nearby Municipal Solid Waste landfill site which has been operating for 6 years. The composition of the leachate is indicated in Table 2.

Table 2 Test leachate composition

Parameter	Value *
pH value	7.2
Conductivity	5010
Ammoniacal nitrogen	110
Nitrous nitrogen	-
Nitric nitrogen	-
Oxygen absorbed	-
COD	800
TOC	312
Chloride	510
Sulphate	72
Phosphate	-
Metals :	
Lead	0.01
Cadmium	<0.01
Chromium	0.02
Copper	0.01
Zinc	0.25
Manganese	0.45
Iron	1.80

* All mg/L except for pH and conductivity (μs/cm)

After forming the soil specimens they were saturated and their water permeability was measured using the falling head test. Constant leachate heads were then applied to some samples whilst control specimens had continuous water permeation applied to them.

5 TEST RESULTS

The permeability values measured so far are presented in tables 3 and 4 - permeation of the two soils is still continuing.

Prior to the commencement of sustained permeation, i.e. upto week zero, each soil was saturated and an essentially 'constant' value of permeability obtained. This value was 'constant' in the sense that repeat permeability tests were undertaken over a period of only a few days.

The values in table 3 suggest a distinct increase in permeability, for both soils, with the length of time of sustained water permeation. However the data suggest that a steady state value is obtained fairly rapidly and in some cases it is virtually identical to that at the start of testing.

Table 3 Values of permeability (x 10^{-6}mm/sec) (Water permeation)

Soil	Sample	Duration of permeation (weeks) 0	5	10	15	20
A	1	0.50	0.56	0.59	0.65	0.64
	2	0.47	0.55	0.60	0.66	0.66
	3	0.42	0.54	0.54	0.53	0.53
B	1	0.42	0.44	0.42	0.40	0.41
	2	0.35	0.38	0.38	0.36	0.36
	3	0.32	0.38	0.37	0.35	0.36

Table 4 Values of permeability (x 10^{-6} mm/sec) (Leachate permeation)

Soil	Sample	Duration of permeation (weeks) 0	5	10	15	20
A	1	0.31	0.21	0.18	0.16	0.15
	2	0.32	0.21	0.18	0.13	0.13
	3	0.26	0.21	0.20	0.17	0.15
	4	0.32	0.28	0.22	0.18	0.16
B	1	0.20	0.19	0.17	0.14	0.12
	2	0.23	0.21	0.18	0.14	0.14
	3	0.26	0.23	0.18	0.15	0.13
	4	0.36	0.27	0.19	0.13	0.12

The ratio of the 20-week permeability to the initial value ranges between 1.0 and 1.3, - for practical purposes this corresponds to no change in permeability with time.

Table 4 contains the leachate permeation data. There is a slight difference between the initial permeability of these samples and those quoted in Table 3 because of discrepancies in moulding moisture contents.

All samples show a clear trend of decreasing permeability with duration of permeation. The reductions are quite significant with the 20-week permeability values ranging between 33% to 61% of the initial value. Whilst it may not be practical to take account of the permeability decrease caused by the passage of leachate for actual liners at least it gives engineers confidence that these conventional water permeation tests are not totally inappropriate or unsafe.

6 REFERENCES

Anon (1986), Waste Management Paper No. 26 - Landfilling Wastes, Department of the Environment.

Daniel, D.E. and Liljestrand, H.M. (1984). Effects of landfill leachates on natural liner systems, Geotechnical Engineering Report GR83-6, University of Austin, Texas.

Farquhar, G.J. (1989), Leachate : production and characterisation, Can.Journ.of Civ.Eng., vol 16.

Michaels, A.S. and Lin, C.S. (1954), The permeability of kaolinite, Ind. and Eng. Chem., vol 56.

Waste Disposal by Landfill – GREEN'93, Sarsby (ed.) © 1995 Balkema, Rotterdam, ISBN 90 5410 356 6

Recovery of scrap rubber tires from landfill for construction uses

J.C. Lyons & D.G. Hitchens
Geotechnical Engineering Division, Lehigh University, Bethlehem, Pa., USA

D.A. Monticello
Dunn Corporation, Parsippany, N.J., USA

ABSTRACT: Currently enormous quantities of tires are thrown away each year. The majority of these tires end up in landfills causing numerous environmental and health concerns. This paper investigates the feasibility of using scrap rubber tires as a light weight aggregate in soil. Presented is a preliminary laboratory investigation of the compressive strength of soil/tire, soil/tire/oil composites, and the potential of using tire pieces as a light weight aggregate in bricks and other construction uses.

1 INTRODUCTION

Scrap rubber tires are a world-wide problem with a majority of the waste tires being dumped into landfill areas. The size and severity of the problem is increasing each year not only by occupying valuable land but also by threatening the public's health as (1) Fire-Hazard - once set ablaze, it is almost impossible to extinguish such a fire which causes serious air pollution problems: and (2) Health Hazards - the debris/water trapped in the empty spaces of the tires provide an ideal breeding ground for insects, rodents, as well as bacteria.

Based on an U.S. EPA survey, production of rubber tires for automobiles increased 42 percent from 1960 to 1988. Currently, an estimated amount of 280 million waste tires are discarded annually in the United States alone and from this amount, only a small portion of tires are recovered for recycling purposes (5.6 percent recovered in 1988). Current methods for reuse of scrap/waste tires may be grouped into three general categories:

(1)fuel source or a raw material in the production of other polymeric materials.

(2)additive as a part of asphalt-rubber mixtures for use as a substitute in asphalt paving materials mixtures.

(3)lightweight aggregate for concrete mixtures or as a soil reinforcement element, either cut into small pieces, shredded, or as whole tires.

If the waste tires are used as described in categories (1) and (2), additional refining processes are needed, consequently requiring investment and creating more environmental problems. To utilize large amount of waste scrap tires within a short duration, category (3) has the obvious economic and environmental benefits. Currently, a research program on the use of scrap rubber tires for construction is being carried out at Lehigh University which this paper outlines:

(1) Examines the bonding mechanism between soil/tire and other admixtures such as lime/flyash/cement in the environment based on environmental geotechnical principles.

(2) Presents a series of experimental laboratory data including:

a)Using scrap tires as lightweight aggregate with various combinations of cement/petroleum oil and tire sizes.

b)Stress strain characteristics of such materials evaluated in comparison with conventional construction material.

All test results are summarized in graphical or tabulated form.

2.0 EXPERIMENTAL STUDY

In 1992 two research projects on the use of scrap rubber tire as a construction material were under taken at Lehigh University. One experiment evaluated the compressive strength of soil/tire mixtures and soil/tire/petroleum mixtures, and observed the tire soil interactions. The other experiment investigated the use of scrap rubber tires as aggregate in adobe bricks.

2.1 Materials

(1) Soil: The soil used in the preparation of samples was obtained from the vicinity of Bethlehem Pennsylvania. These are classified as silty clay with Liquid Limit = 38, specific gravity = 2.79 and is classified as ML by the Unified Soil Classification (ASTM D2488-84). The optimum moisture content, as determined by ASTM D698-70 to be 15.0%. All of the soil samples passing #4 US sieve.

(2) Rubber Tire: Materials used in the studies were obtained from tire sidewalls, such that no metal was included. The tire specimen sizes consisted of circular pieces of tire 1 inch and .5 inch in diameter and .25 to 1.25 inches in thickness. These pieces were obtained by punching the sidewalls with a metal punch. Additionally random sizes were used in the compressive strength experiment. The random sizes varied from .125 inch to 1.5 inch in diameter and approximately .25 inch to 2 inch in thickness.

(3) Flyash: The flyash used was obtained from Pennsylvania Power and Light's Martin Creek Power generation plant. The flyash originated from the combustion of bituminous coal. It possessed a water content of 27%, an in-situ void ratio of .84 and is classified as non plastic according to the Unified Soil Classification System.

(4) Petroleum: The petroleum contaminated samples were produced by adding clean 10W40 motor oil.

(5) Lime: Anhydrous lime (Ca0) was added to reduce cracking on the surface by adsorbing the moisture from the sample as it dried. The anhydrous lime expanded to fill the voids between the soil and tire on the surface.

2.2 Laboratory Study

(1) To investigate the compressive strength of soil/tire mixtures and soil/tire/petroleum mixtures, samples were made and compacted in a standard proctor mold using ASTM Method (698-64T). The mixtures comprised of 1 inch, .5 inch and random tire piece sizes mixed in with soil in the amount of 0, 5, 10, 15, and 20 percent by weight of dry soil. This was done to show the relationship between tire shape and the resulting compressive strength. Optimum moisture content was used for compaction.

The soil/tire/petroleum mixtures consists of 0, 5 10 and 20 percent oil content (to dry soil weight). These soil/petroleum mixtures were all combined with 10% of .5 inch diameter tire pieces. These samples were also tested for the compressive strength using the methods stated above.

(2) Adobe bricks: The earth bricks were made by mixing 0, 2, 4, 6, 8 percent of 1 inch and .5 inch tire pieces by weight with local soil. The water content used was approximately 20%. This made it easier to form the bricks in the mold. The mixtures were compacted into a 8.1875 x 4 x 2.25 inch steel molds. The bricks were then oven dried at 150 degrees Fahrenheit for a period of four days, this was done with the mold removed to facilitate even drying. Additional bricks were made with a 1:4 ratio of flyash to soil with 6% of tire by weight in the form of 1 inch and .5 inch tire pieces. Also some bricks were coated with lime to reduce cracking. Lime coating was applied to the bricks after 2 hours, This enabled the bricks to be handled without deformation yet they were wet enough for the lime to hydrate and react with the soil to fill the cracks. The bricks were subjected to a three point loading test. Additionally they were subjected to freeze thaw tests.

3 RESULTS AND DISCUSSION

(1) Seven experiments were performed on two groups of the soil/tire and soil/tire/oil samples. The tires sizes were .5 inch 1 inch and random sizes. Tire contents varied from 0% to 30% by weight and the oil content varied from 0% to 20% by weight. The corresponding stress strain results are in graphical form in Figures 1 through 5.

Experimental data indicates that compressive strength decreases when tire pieces are added to the soil. Additionally, when comparing tire contents of 5%, the 1 inch tire piece sample has a compressive strength that was much lower than .5 inch tire piece sample (Figure 1). If 10% oil is added the strength increases for the 10% and 20% rubber tire amounts (Figure 2).

(2) The results from the investigation on the use of

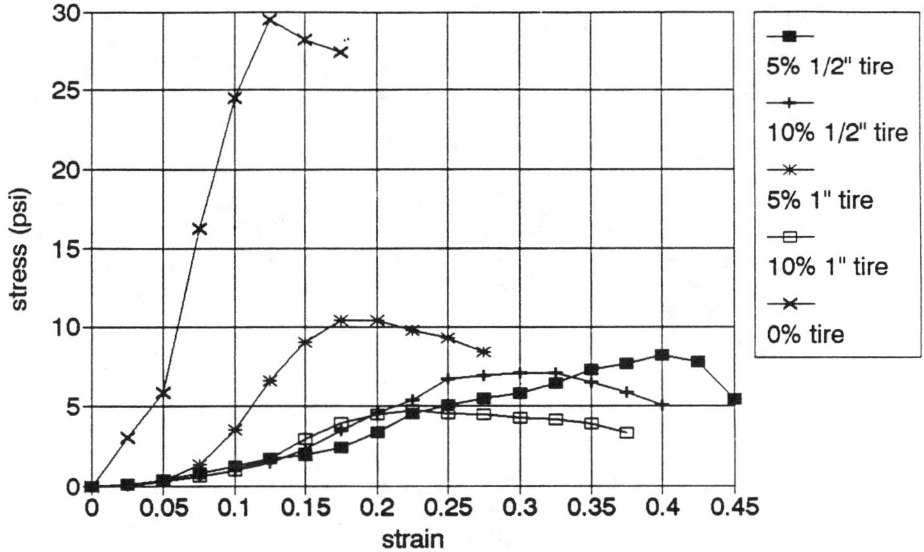

Figure 1. Plot of the stress versus strain for unconfined compression of different tire percentages and sizes.

scrap rubber tires as an aggregate in adobe bricks. The tire sizes used were .5 inch and 1 inch diameter pieces. The amount of scrap rubber tire in each sample ranged between 2% to 8% by weight, with 8% tire content being the maximum working amount of tire. This produced bricks with dry densities ranging from 105 to 113 lb/ft^3. The load deflection results are presented in Figures 6 through 10. Note that the initial values of load should be ignored, as they are the seating load on the sample.

The load deflection tests on the adobe bricks demonstrated that the strength of the bricks were significantly decreased with the addition of rubber tire. The .5 inch rubber tire pieces produced bricks of higher strength (approximately 50% higher) than the bricks made of 1 inch pieces. Additionally the .5 inch diameter aggregate of rubber tire produced bricks which were approximately 30% more elastic than the bricks consisting of no aggregate. The bricks made with 1 inch pieces of rubber tire exhibited even greater elasticity.

The use of 25% flyash by weight made the mixture harder to work with and decreased the ultimate strength by 20% to 50% (Figure 9), when compared to similar bricks made without flyash. The elastic properties seemed to remain constant with the other bricks of similar tire content.

Dehydrated lime (CaO) was used in both investigations to seal and reduce the cracks present in the samples. The lime did significantly reduce the amount and size of cracks, but did not change the ultimate strength of the samples (Figure 10).

Also, significant cracking was noticed around the tire pieces near the surface of the samples. This resulted in spalling of the surrounding soil when the bricks were subjected to freeze thaw tests. The spalling is due to the different coefficients of thermal expansion between the materials.

4. SUMMARY AND CONCLUSIONS

The experimental data indicates that the desirable strength of the soil/tire and soil/tire/oil mixtures are obtained. In general the smaller the amount of tire added the higher the strength. Additionally, the higher the tire content the more elastic the mixture becomes. If 10% oil is added the strength increases for all tire contents. The addition of flyash reduces the ultimate strength of the samples by 25% to 50%. Dehydrated lime reduced the amount surface cracking between soil and tire, but did not significantly effect the overall strength.

Failure occurred at the soil tire interface for both tests. Better binding between the two materials will drastically increase the strength of the composite. This may be possible by partially melting the tires causing the nearby soil grains to bond to the tire

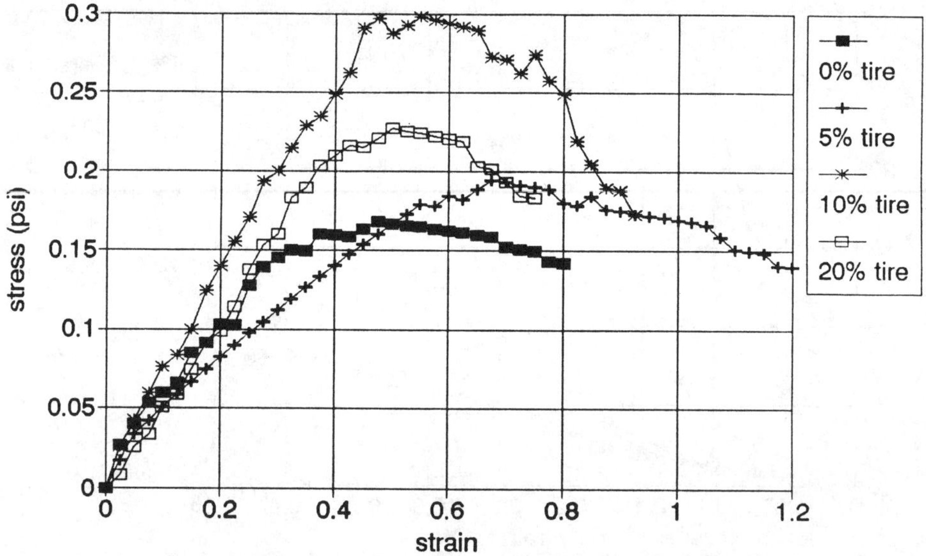

Figure 2. Plot of the stress versus strain for unconfined compression of soil containing 10 percent oil and various amounts of .5 inch tire pieces.

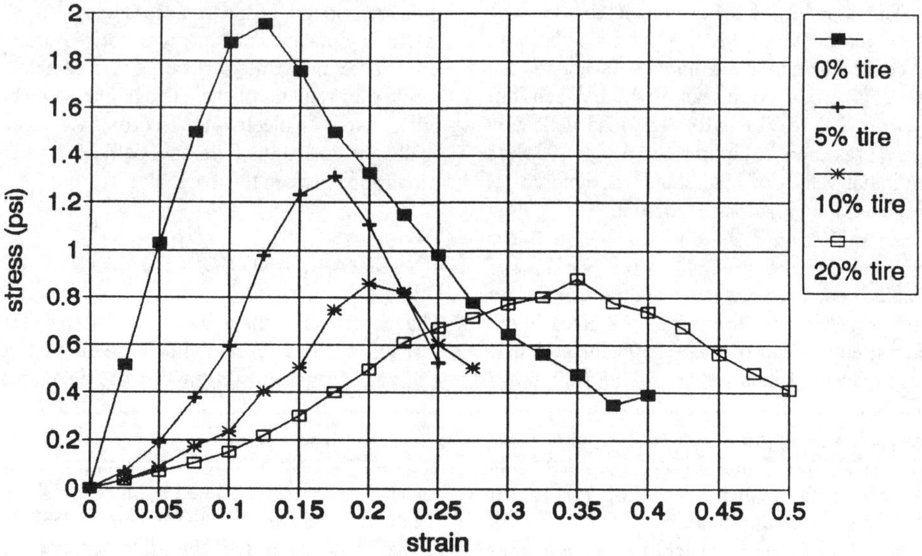

Figure 3. Plot of the stress versus strain for unconfined compression of soil containing various amounts of .5 inch tire pieces.

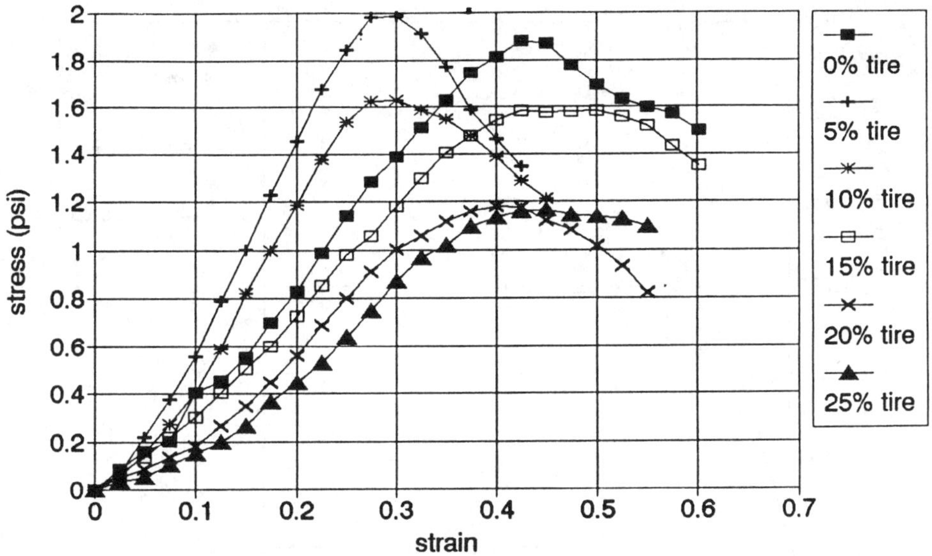

Figure 4. Plot of the stress versus strain for unconfined compression of soil containing various amounts of 1 inch tire pieces.

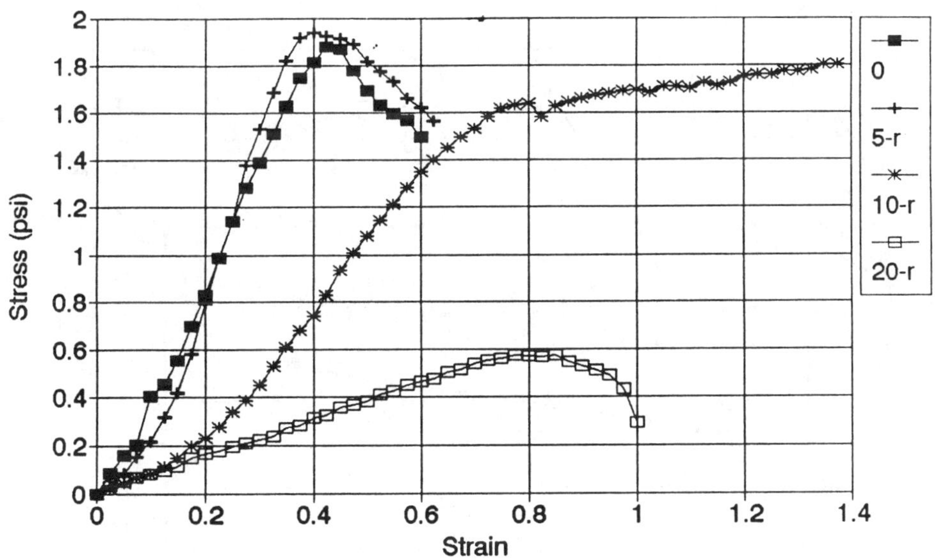

Figure 5. Plot of the stress versus strain for unconfined compression of soil containing various amounts of random sized tire pieces.

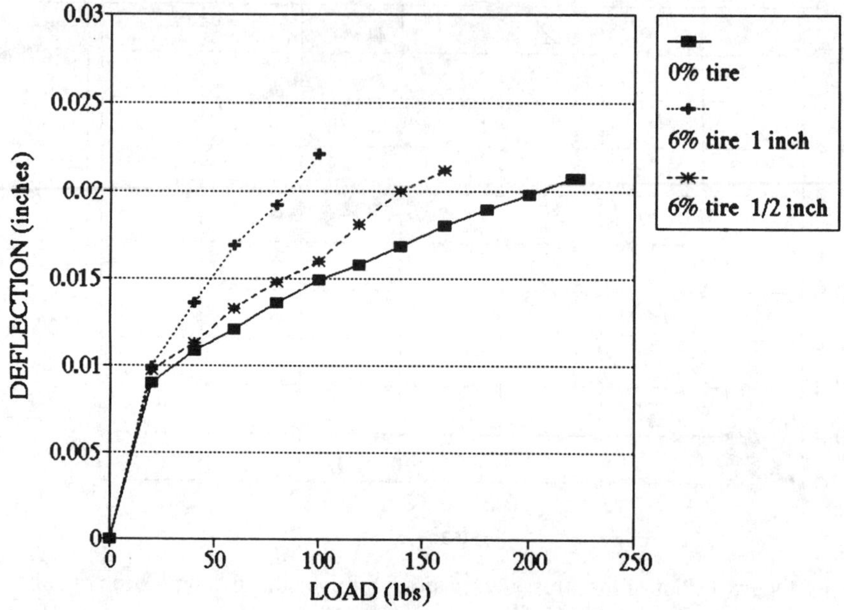

Figure 6. Plot of deflection versus load of adobe bricks containing 6 percent tire pieces of .5 inch and 1 inch in comparison to no tire.

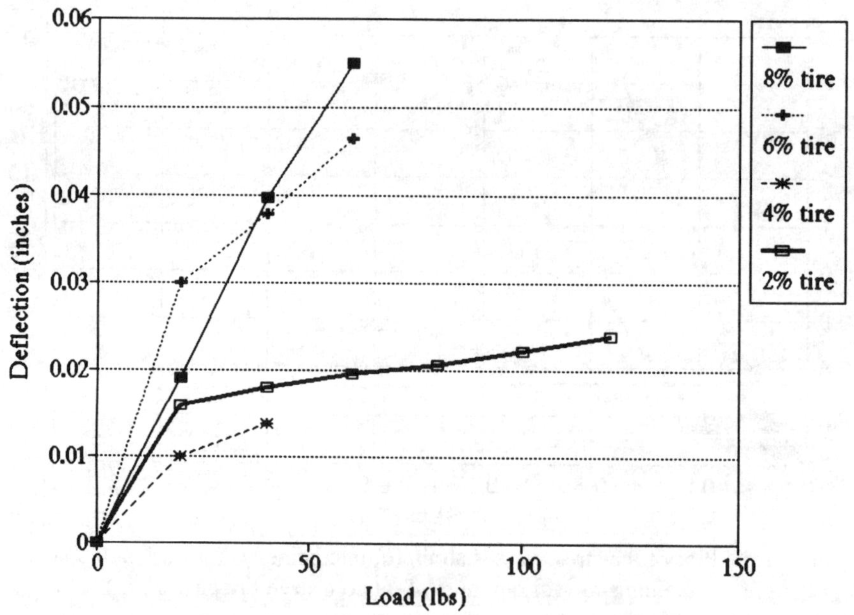

Figure 7. Plot of deflection versus load of adobe bricks containing various amounts of 1 inch tire pieces.

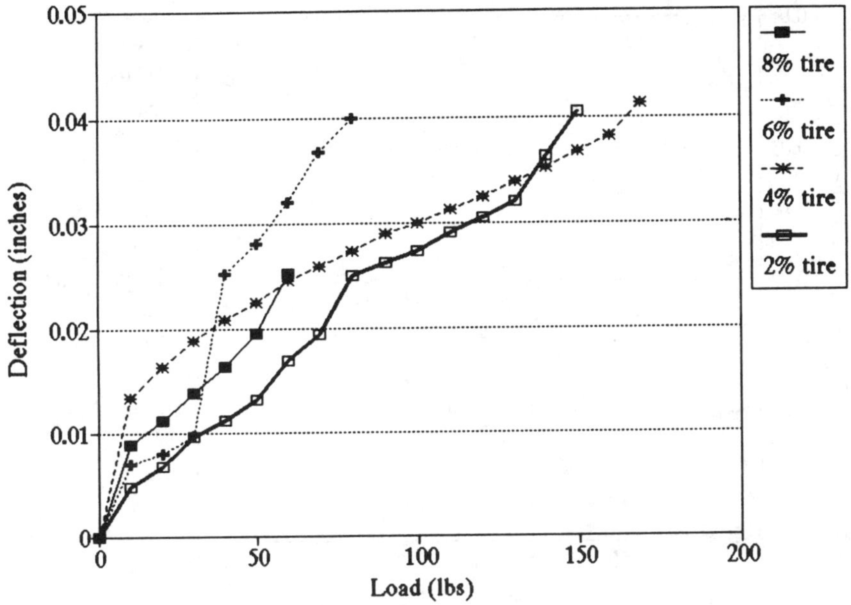

Figure 8. Plot of deflection versus load of adobe bricks containing various amounts of .5 inch tire pieces.

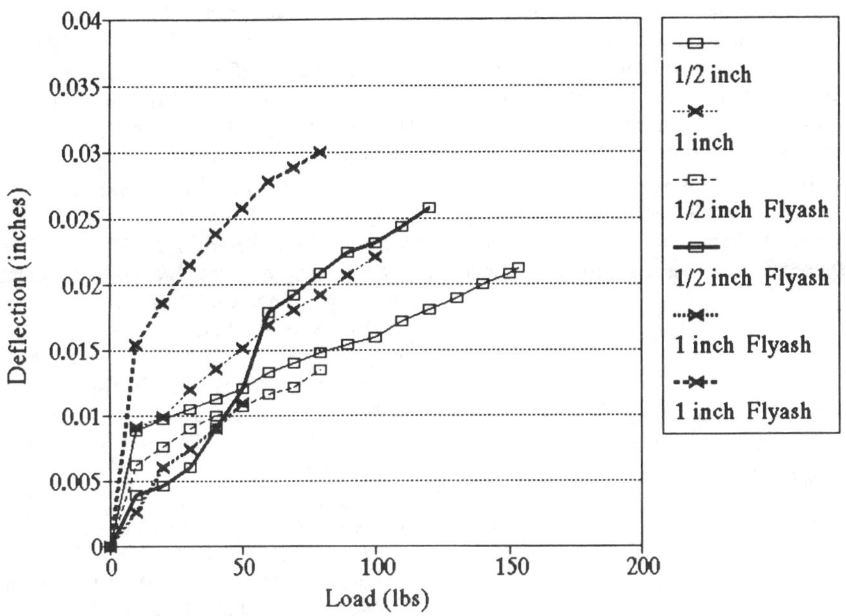

Figure 9. Plot of deflection versus load of adobe bricks comparing the strength of bricks with 25% flyash to 0% flyash with 6% tire of 1 inch and .5 inch sizes.

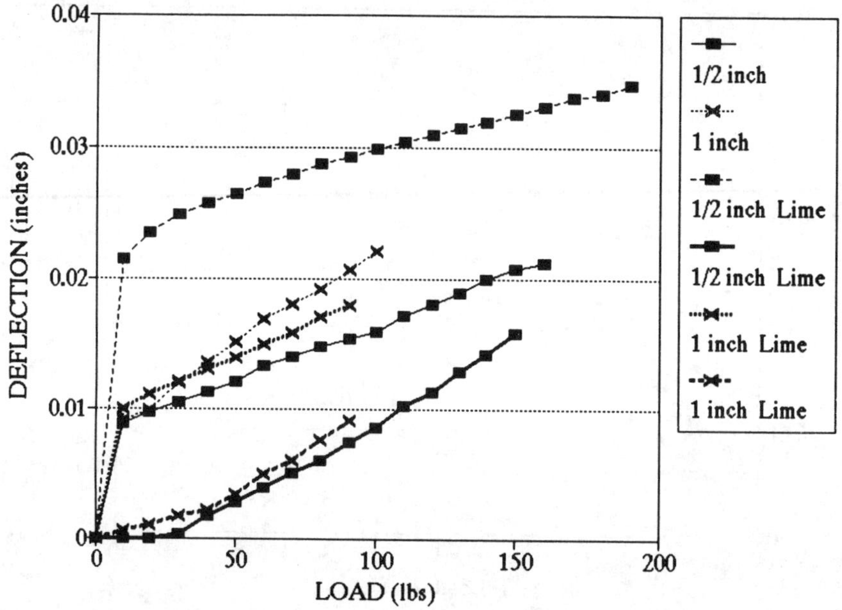

Figure 10. Plot of deflection versus load of adobe bricks comparing bricks with 6% tire of .5 inch and 1 inch tire pieces to bricks of similar tire content with no lime treatment.

pieces. Also, the addition of cement may increase the bonding between the soil and tire.

Overall, the use of scrap rubber tire as a light weight aggregate in construction materials shows promise. Further research needs to be conducted to better quantify the characteristics of soil/tire and soil/tire/oil mixtures. The advantage of using scrap rubber tires in this manner are that it reduces disposal of the rubber tires and provides a lightweight aggregate material, thus providing a two fold savings.

Acknowledgments

The authors would like to state their appreciation to Dr. H.Y. Fang and Mesut Pervizpour for their review of this paper. Additionally they express thanks to Janice Ruhl for her earlier contributions.

References

Ahmed, I. & Lovell, C.W. 1992. *Use of Waste Products in Highway Construction.* In Usmen, M.A. & Acar, Y.B. (eds.) Environmental Geotechnology, p.409-418. Rotterdam: Balkema.

Chae, Y.S. & Gurdziel, T.J. 1976. *New Jersey Fly Ash as Structural Fill.* In H.Y. Fang (ed.), New Horizons in Construction Materials, p. 111-126. Lehigh Valley PA: Envo Publishing.

Chaney, R.C., Dengler, L.A. & Fang, H.Y. 1983 Pulverized Coal Ash as Structural Fill: Discussion. *Journal of Geotechnical Engineering* 109:1356-1359.

Davidson, J.I. & Sereda, P.J. 1978. Measurement of Linear Expansion in Bricks Due to Freezing. *ASTM Journal of Testing and Evaluation* 6:144-147.

Fang, H.Y., Hitchens, D. & Hontz, D. 1992. Use of Scrap Rubber Tires for Construction Materials. *Proceedings, 24th Mid-Atlantic Industrial Waste Conference:* 263-275. West Virginia University.

Koerner, R.M. & Schoenberger, R.J. 1976. *Synthesis of Construction Materials from Waste Products.* In H.Y. Fang (ed.), New Horizons in Construction Materials, p. 111-126. Lehigh Valley PA: Envo Publishing.

Pamukcu, S., Hijazi, H.M. & Fang, H.Y. working paper. Study of Possible Re-use of Stabilized Petroleum Contaminated Soils as Construction Material.

Waste Disposal by Landfill – GREEN'93, Sarsby (ed.) © 1995 Balkema, Rotterdam, ISBN 90 5410 356 6

Dynamic analysis of the behaviour of hydraulically deposited uranium mill tailings under seismic action

J.M. Martínez Santamaría & A. Santos Moreno
Laboratorio de Geotecnia del Centro de Estudios y Experimentación de Obras Públicas (MOPT), Madrid, Spain

J.L. Santiago
Empresa Nacional de Residuous Radiactivos (ENRESA), Madrid, Spain

ABSTRACT: The inactive uranium mill tailings originated by a plant, formely in operation in Andujar (Spain), were disposed of by building at the site, through hydraulic filling, five adjacent tailing piles. A dynamic analysis of the behaviour of the tailings, in the final configuration of the disposal, under seimic action, has been carried out by CEDEX. A summary is given of the result of triaxial dynamic tests, under anisotropic consolidation, performed on tailings with various degrees of plasticity. The dynamic analysis of the behaviour of the final tailing pile is shown, along with the conclusions of the study.

1 INTRODUCTION

The Empresa Nacional de Residuos Radiactivos, S.A. (ENRESA) is conducting the reclamation of an inactive uranium mill tailings disposal site at the town of Andujar in the south of Spain. The Andujar milling plant was in operation from 1959 to 1981 and all the tailings generated were disposed by hydraulic filling in five adjacent piles which were constructed at different periods between 1959 and 1981. Each individual tailings pile is confined by a peripheral dike, constructed with cycloned sand by the upstream method. The tailings slurry was deposited using multiple discharge perimeter spigotting.

The reclamation of uranium mill tailings piles requires the insitu characterization of their material properties in order to prepare the stabilization and remedial action designs. The geotechnical characterization of the Andujar tailings was conducted by the Centro de Estudios y Experimentaicón de Obras Públicas (CEDEX) of the Spanish Ministry of Public Works and Transportation (MOPT).

As a basic requirement of the regulations for inactive uranium mill tailings final disposal, the behavior of the embankment under seismic action, with return period not less than 1000 years, must be analyzed. This has been done through use of QUAD-4 Computer Program, applied to a previously zoned embankment, with due consideration of the nature and dynamic behavior of the different fractions of tailings which result from the construction procedure of the disposal waste.

2 INDEX PROPERTIES AND GENERAL CHARACTERISTICS OF THE TAILINGS

Four material types were identified in the tailings pile:

1) "Sand" < 50% passing sieve no. 200 (ASTM) and PI (plasticity index) = 0.

2) "Silt" - 0<PI<10% and generally, liquid limit W_L < 30%)

3) "Clayey silt" - 10%<PI<20% and generally, 30% < W_L < 50%)

4) "Clay" - PI > 20% and generally, W_L > 50%.

Figure 1 shows a typical cross section of the tailings pile. It can be observed that former peripheral dikes of individual piles become partially embedded when a new tailings pile is built adjacent to a formely existing pile.

Despite the fact that the construction process has produced a net deposit of sand at the peripheral dike, both the perimeter spigotting and the limited dimensions of the individual piles have permitted the deposit of fine beds of sand even at the central portions of the piles. This creates a banded texture of the tailings where the different categories of materials may appear finely interbedded in samples obtained at the central portion of an individual pile. This banded texture may be dependent upon the flow rate and composition of the slurry being hydraulically placed in the pile. Table 1 shows, in number of samples and percentage, the variation of interbedded materials ("Sand" to "clay") as obtained in borings placed at the central axis or at the edges of the existing piles.

3 DYNAMIC PARAMETER OF TAILINGS

3.1. Field characterization

"Cross-hole" and "down-hole" techniques were used to evaluate the dynamic moduli (deformation modulus and shear modulus) of the tailings at very low strains. Shear wave velocities within the tailings pile vary between 140 to 190 m/sec. Corresponding shear moduli were determined between 336 and 616.8 Kg/cm^2, which confirm the low strength and open structure characteristics encountered at the tailings pile.

Laboratory testing

Figure 2 shows the diagrams of variation of shear modulus ratio and damping ratio with shear strain, as obtained in clayey tailing materials by the resonant-column method.

Dynamic triaxial tests on tailing samples of "clay", "silt" (which includes and clayey clay silt) and "sand" were performed under anisotropic consolidation. The failure criterion under uniform cyclic deviatoric stress was a permanent strain of 5%. Figure 3 shows the dynamic strength cur-

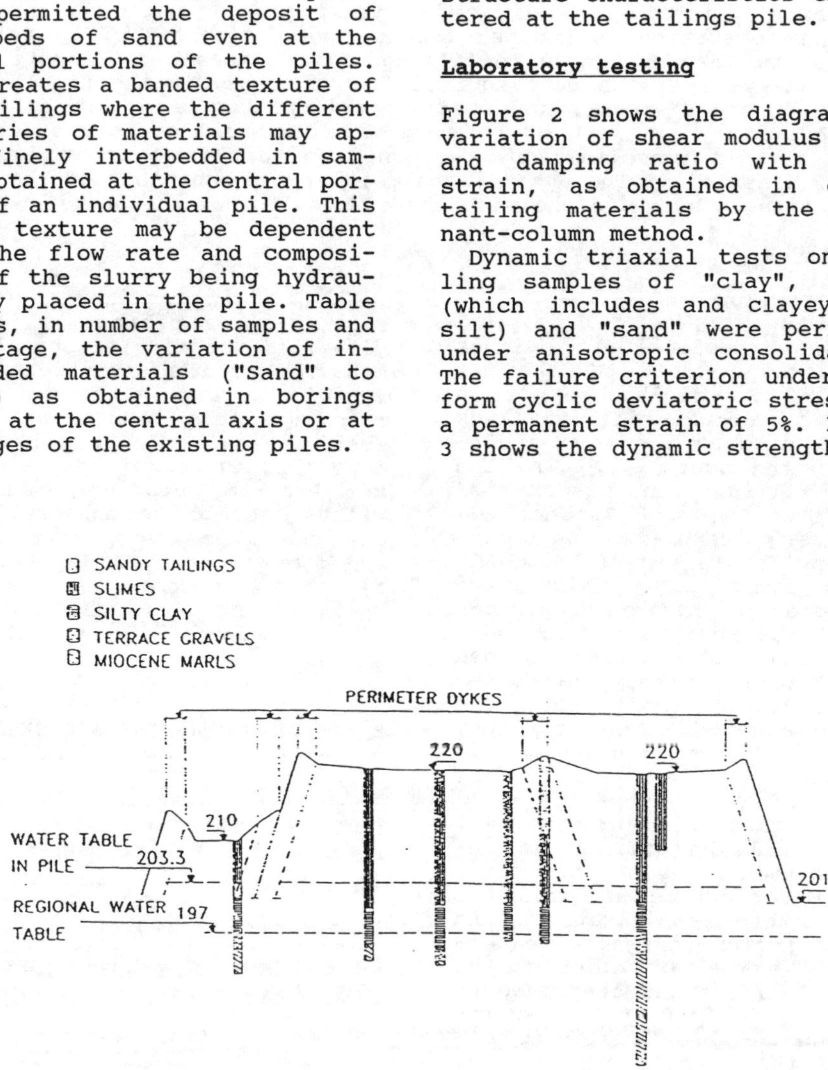

Figure 1. Tailings pile-cross section

Table 1. Distribution of materials in borings

	Borings at pile edge No. of samples	%	Borings at pile center No. of samples	%
Total no. samples	108	100	74	100
No. of samples sand	44	41	16	22
No. of samples silts	20	18	5	7
No. samples clayey silt	27	25	13	17
No. of samples clay	17	16	40	54

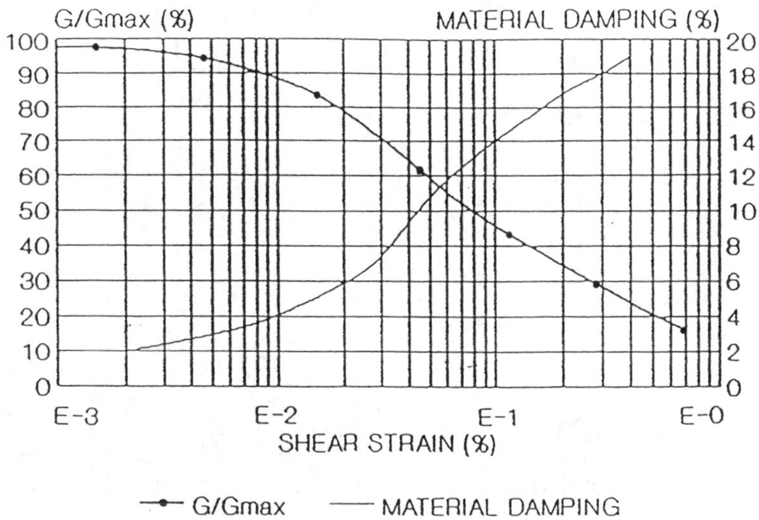

Figure 2. Shear modulus and damping vs. Shear strain for tailing

ves (stress ratio vs number of cycles) for the three types of tailing materials. A significant difference in their behaviour under dynamic loads is found, the "sand" being the most susceptible to liquefaction.

4 DYNAMIC ANALYSIS OF THE TAILING EMBANKMENT

Figure 4 shows the discretization of a half section (because of symmetry) of the waste embankment. The locations of predominant "sand" and "silt" material or "clay" material are shown in this figure.

Quad-4 computer program has been used, assuming for all embankment materials, whether clayey or silty and sandy, a value for the initial dynamic shear modulus of 260 Kg/cm_2 (520 Ksf). For the natural soil, underlying the embankment, a constant value for the dynamic shear modulus, of

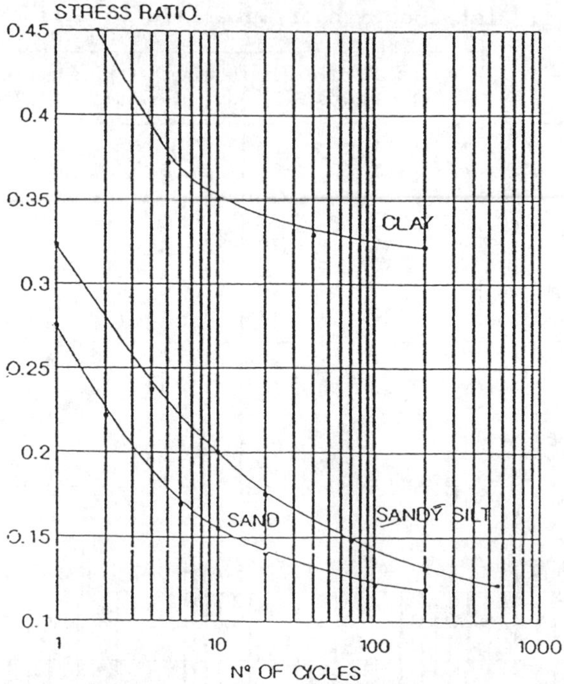

Figure 3. Dynamic strength of tailings

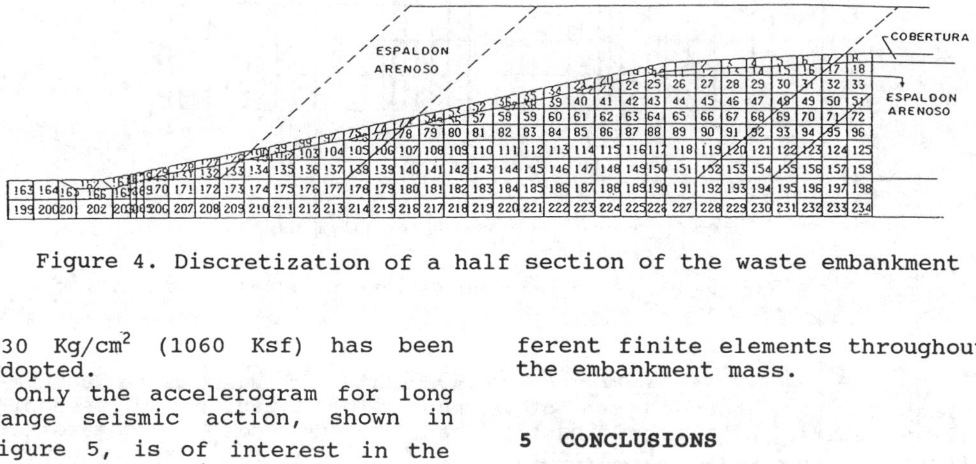

Figure 4. Discretization of a half section of the waste embankment

530 Kg/cm^2 (1060 Ksf) has been adopted.

Only the accelerogram for long range seismic action, shown in Figure 5, is of interest in the response analysis of the tailing embankment.

Table 2 gives the comparison between dynamic strength for the embankment materials for uniform cycles of seimic action, as obtained in laboratory, and dynamic response in equivalent uniform cycles (K.L. Lee method), of different finite elements throughout the embankment mass.

5 CONCLUSIONS

By application of the well known QUAD-4 computer program, the dynamic response under seismic action of an embankment made of uranium mill tailings, deposited by a hydraulic fill method, has been performed. The two-dimensional approach is justified by the embankment configuration.

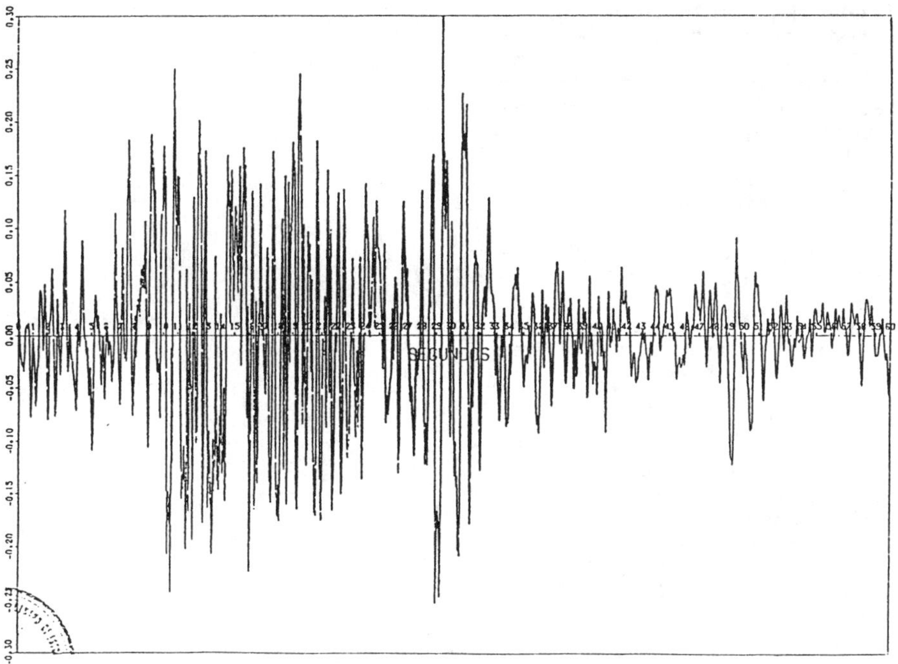

Figure 5. Seimic action, A_{max} = 0.36

Table 2. Estudio dinamico del terraplen de Andujar (ENRESA)

RESUMEN DE VALORES DEL ANALISIS DINAMICO PARA EL TERREMOTO LEJANO

NUM DE ELEMENTO DIN.	EST.	TAU MAX. ANAL. DINAMICO Tn/m2	PSF	σy ANALISIS ESTATICO Tn/m2	0.65*TAU MAX/σy OBTENIDA DEL ANALISIS DINAM.	RELACION TAUlab./σy ENSAYOS (Kc=2)			NUM. CICLOS EQUIVALENTES DE 0.65*TAUMAX/σy
						ARENA	LIMO ARENOSO	ARCILLA	
29	188	2.060	412	7.120	0.188	-	0.142		18.1
30	189	2,210	442	7.301	0.197	0.160	0.208		8.3
41	170	2.245	449	7.107	0.205	-	0.118	0.517	1.0-(1000)
42	171	2.810	563	8.384	0.218	-	0.100	0.517	1.0-(1600)
59	152	2.850	570	7.212	0.257	-	0.193		12.2
60	153	3.220	644	8.764	0.242	-	0.203		9.2
65	156	3.420	684	14.030	0.158	-	0.100	0.517	1.0-(109)
67	160	3.800	760	15.050	0.164	-	0.193		12.8
69	162	2.990	598	20.869	0.093	0.275			1.0
90	141	4.040	808	10.310	0.255	0.275			1.0
92	143	4,280	856	18.903	0.147	0.275			1.0
108	109	3,330	667	10.256	0.211	-	0.163		36.7
109	110	3.510	702	11.740	0.194	-	0.153		55.3
137	80	2.480	496	11.038	0.146	0.114	0.142		126.6
138	81	3.570	715	10.915	0.213	-	0.162	0.334	33.5
142	85	4.390	878	16.578	0.172	-	0.150		54.1
144	87	5.025	1005	19.318	0.169	0.120	0.140	0.517	1.0-(125)
146	89	6.190	1238	20.010	0.201	-	0.139	0.517	1.0-(125)
148	91	4.810	962	25.360	0.123	0.275	0.350		1.0
151	94	5.365	1073	27.407	0.127	0.275	0.350		1.0

The results obtained can be considered on the safe side, because of the fact that, despite the predominance of non plastic materials in certain zones of the waste mass, the method of deposition has fostered a banded structure of the tailings, in which the plastic layers are confining and reinforcing the non-plastic layers, potentially unstable in certain areas of the embankment.

The possibility of introducing the real configuration of the embankment in its closure form, as well as the possibility of taking care, although in a simplified form, of the predominant textures found in the different zones of the embankment section, are the basic advantages of a method of seismic response analysis based on finite element discretization and nonlinear shear stress-shear strain relationship.

REFERENCES

LEE, K.L. et al. (1972). "Number of equivalent significant cycles in strong motion earthquakes". Proceedings of "The international Conference On Microzonation" for Safer Construction Research and Application", Seatle, Washington, October.

IDRISS, I.M. et al. (1973). "QUAD-4A computer program for evaluating the seismic reponse od soil structures by variable damping finite element procedures", Report No. EERC 73-16, College of Engineering, University of California, Berkely, California.

Waste Disposal by Landfill – GREEN'93, Sarsby (ed.) © 1995 Balkema, Rotterdam, ISBN 90 5410 356 6

Sorption capacity of shredded waste tires

J.K. Park, J.Y. Kim & T.B. Edil
University of Wisconsin-Madison, Wis., USA

ABSTRACT: This paper reports the sorption isotherm experimental results for shredded scrap vehicle tires with various organic compounds. Tire chips were found to have 2.5 to 10% of the sorption capacity of granular activated carbon (GAC) on the volume basis. The sorption equilibrium occurred within two days for 0.6, 1.3 and 2.5 cm size tire chips, although the sorption rates for 0.6 and 1.3 cm size tire chips were slightly faster than 2.5 cm size tire chips. Scrap vehicle tires could replace granular materials in a primary leachate collection system and remove potentially toxic organic compounds in landfills significantly.

1 INTRODUCTION

The management of scrap tires has become a growing problem in recent years. Of the scrap tires, 70 to 85% are currently landfilled or stockpiled. Whole tires are difficult to landfill because they tend to float to the surface and also take up large volume of valuable landfill space. Stockpiles of scrap tires are located in many communities, resulting in public health, environmental, and aesthetic problems. Each year over 240 million scrap tires are generated in the United States. Additionally, about 2 billion waste tires have accumulated in stockpiles or uncontrolled tire dumps across the country (U.S. EPA 1991).

Tires are principally composed of vulcanized rubber, rubberized fabric containing reinforcing textile cords, steel or fabric belts, and steel-wire-reinforced rubber beads. The most commonly used tire rubber is styrene-butadiene copolymer (SBR) containing about 25% styrene by weight. Other elastomers such as natural rubber (cis-polyisoprene), synthetic cis-polyisoprene, and cis-polybutadiene are also used in tires in varying amounts. A typical composition of tire rubber is shown in Table 1.

Table 1. Rubber compounding composition (from Dodds *et al.* 1983).

Component	Weight Percent
SBR	62.1
Carbon black	31.0
Extender oil	1.9
Zinc oxide	1.9
Stearic acid	1.2
Sulfur	1.1
Accelerator	0.7

Park et al. (1991) investigated the permeation of organic compounds through SBR gaskets used for potable water distribution systems and found high organic compound sorption capacity for SBR. Many hazardous organic compounds have recently been detected in leachate from industrial hazardous waste and municipal solid waste landfills (Gibbons et al. 1992). However, earthen liners and geomembranes (flexible membrane liners) were found to have limited abilities for containing organic compounds (Edil et al. 1992a; Sakti et al. 1992). It is postulated that hazardous organic compounds found in landfills can be removed significantly by installing a layer of shredded tire chips in a proper location in landfills.

The principle objectives of this research were (1) to determine the sorption capacity of shredded tires for hazardous organic compounds, (2) to estimate the partition and diffusion coefficients of organic compounds in tire chips, (3) to assess the retardation of organic compound movement by installation of shredded tire layers, and (4) to develop design guidelines for use of shredded tires in landfill. In this paper, only the results of the first two objectives are reported.

2 MATERIALS AND METHODS

2.1 *Tires*

Shredded and ground tires were obtained from a local tire processor. No attempt was made to separate tires based on different manufacturers or sizes. The densities of tires with and without metal material were measured to be 1.22 and 1.15 g/cm^3, respectively. Before tests, tires were washed with deionized water, dried at a 40°C oven for a day, and stored in a desiccator.

2.2 *Organic Compounds*

Methylene chloride(MC), trichloroethylene (TCE), toluene, and *m*-xylene were selected for testing. These organic compounds are the most frequently detected compounds at the waste disposal sites (Plumb and Pitchford 1985). The organic compounds selected covers wide ranges of the aqueous solubility (S) and the molecular weight (M.W.). The physical/chemical properties of these organic compounds are summarized in Table 2.

Table 2. Properties of organic compounds tested.

Organic compounds	M.W. (g/mol)	Density (g/cm^3)	S (mg/L)	log K_{ow}
MC	84.93	1.327	16,700	1.25
TCE	131.39	1.467	1,100	2.53
Toluene	92.14	0.867	515	2.69
m -Xylene	106.17	0.866	200	3.20

2.3 *Sample Analysis*

Gas chromatographic analysis was performed on a Varian 3600 gas chromatograph with 60 m long 0.25 mm I.D. Supelcowax-10 megabore column and a flame ionization detector (FID). An aliquot of 2 mL sample was directly injected into the column using a gas-tight micro-syringe. Helium was used as a carrier gas.

The column temperature was programmed to initially hold at 60°C for five minutes, then to climb to 230°C at a rate of 15°C/min and then to hold for another one minute. A Varian 1093 Septum Equipped Programmable Injector was programmed to start at 50°C, then to immediately climb to 230°C at a rate of 150°C/min, and then hold for one minute.

Calibration curves were produced using standards at 0, 10, 20, 30, 40, and 50 mg/L. The FID detection limits were approximately 0.5 mg/L for methylene chloride and TCE, and 0.3 mg/L for toluene and *m* -xylene, respectively.

2.4 *Sorption Experiments*

A series of batch sorption tests were conducted using tire chips. The batch sorption test consisted of submerging tire chips in a 350 mL glass reactor spiked with target organic compounds and measuring the concentration periodically. Four glass reactors were used and rotated by a rotary agitator. All the experiments were conducted at a 20°C constant temperature room.

2.5 *Data Analysis*

The partition coefficient, K, expresses the ability of the tire chip to sorb organic compounds from the solution. K is the ratio between the organic compound concentration in the medium and the organic compound concentration in sorbate at equilibrium as follows:

$$C_s = KC_l \tag{1}$$

where C_s is the organic compound concentration in tire chips (mg/L) and C_l is the organic compound concentration in the liquid (mg/L). K can be determined from the experimental data using the following relationship:

$$K = \frac{M_{eq} \times \rho}{M_s \times C_{eq}} \times 10^6 \tag{2}$$

where M_{eq} is the weight of organic compound sorbed at equilibrium (g), M_s is the initial weight of tires (g), ρ is the density of tires (g/cm^3), and C_{eq} is the concentration in the medium at equilibrium (mg/L).

The rate of diffusion is expressed in terms of diffusion coefficient. The mass balance for pure Fickian diffusion through a flat plate is expressed as follows:

$$F = -D\frac{dc}{dx} \tag{3}$$

where F is the mass flux ($mg/cm^2\cdot sec$); D is the diffusion coefficient (cm^2/sec), c is the concentration of organic compound in tires (mg/cm^3), and x is the distance along the direction of diffusion (cm).

Change in the concentration of diffusant at any point with time in the material diffused can be expressed by Fick's second law. The experimental system is represented by the condition that a plain sheet of tire chip is suspended in a well-stirred solution of limited volume. The mass transport of organic compounds in a tire chip and the relationship between the organic compound concentration in the tire chip, c(x,t) and the organic compound concentration in the solution, C(t) can be expressed as follows:

$$\frac{\partial c}{\partial t} = -\frac{\partial F}{\partial x} = -\frac{\partial}{\partial x}\left(-D\frac{\partial c}{\partial x}\right) = D\frac{\partial^2 c}{\partial x^2} \tag{4}$$

$$-A\frac{\partial c}{\partial t} = -\frac{A}{K}\frac{\partial C}{\partial t} = \pm D\frac{\partial c}{\partial x} \tag{5}$$

$$\text{at } x = \pm l \text{ and } t > 0$$

where 2A is the thickness of the solution in contact with both sides of tires (cm), C is the concentration of the organic compound in the solution (mg/L), t is the time (sec), and $2l$ is the thickness of tire chip (cm).

The thickness of the solution in contact with both sides of the tire chip, 2A, is the ratio of solution volume to the surface area of the tire chip. The boundary and initial conditions are:

$$C = C_0,\ c(x,0) = 0 \quad \text{at } -l < x < +l \tag{6}$$

The analytical solution of Eq. 4 subject to Eqs. 5 and 6 is approximately given as follows (Crank 1975):

$$\frac{C_t}{C_0} = \exp\left(\frac{T}{\alpha^2}\right) \times \text{erfc}\left(\frac{T}{\alpha^2}\right)^{0.5} \tag{7}$$

where $T = \frac{D \times t}{l^2}$ and $\alpha = \frac{A}{K \times l}$.

The sorption half-time, $t_{1/2}$, is defined as the time when C_t/C_0 is 0.5. Eq. 7 can be converted to the following equation by using the sorption half-time, $t_{1/2}$ (Reynolds et al. 1990).

$$t_{1/2} = 0.585 \times \frac{A^2}{K^2 \times D} \tag{8}$$

The recorded weight gain data can be evaluated based on % weight gain of the organic compound by the tire chip, where % weight gain was calculated by the following expression (Berens 1985; Park et al. 1991):

$$\%\ \text{weight gain} = \frac{100\, V_1 \times \rho_1}{(1 - V_1) \times \rho_2} \tag{9}$$

where V_1 is the volumetric fraction of organic compound sorbed per unit volume of tire chip, ρ_1 is the density of organic compound (g/cm^3), and ρ_2 is the density of tire chip (g/cm^3). V_1 is a function of the organic compound (weight gain) sorbed and is related as:

$$V_1 = \frac{M_w/\rho_1}{(M_i/\rho_2) + (M_w/\rho_1)} \tag{10}$$

where M_w is the weight gain calculated from the concentration changes in the liquid phase during the experiment (g), and M_i is the initial weight of the tire chip (g).

3 RESULTS AND DISCUSSION

A series of batch sorption tests were conducted using only a single compound to assess the sorption capacity of tire chips. The K and D values were estimated using Eqs. 2 and 8, respectively. The concentration changes over time was predicted using Eq. 7 with these estimated K and D values and compared with the measured concentrations to assess the accuracy of this data analysis method. The predicted curve and the observed data are plotted in Figure 1. It can be seen that the predicted curve using the estimated K and D values is in good agreement with the observed data.

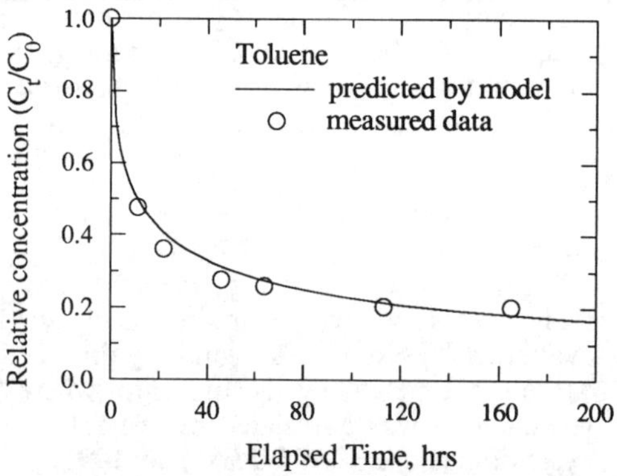

Figure 1. Comparison between the calculated curve and observed data.

Table 3 summarizes the single-solute sorption test results for tire chips along with X/M values for granular activated carbon (GAC), which are calculated based on the parameters reported by Dobbs and Cohen (1980). Because of its higher octanol-water partition coefficient and lower solubility, *m*-xylene had 2 to 3.5 times greater partition coefficients than toluene and TCE.

Another series of batch sorption tests was conducted to investigate the effect of organic chemical mixtures on sorption capacity. MC, TCE and toluene were added simultaneously to a reactor. The experimental conditions and results for the batch sorption test are summarized in Table 4. Tire chip size varied from 0.6 to 2.5 cm and tire/liquid ratios from 0.75 to 1.12% by volume. MC had a very low partition coefficient due to its high polarity and aqueous solubility. Toluene had the greatest partition coefficient, closely followed by TCE. There seemed to be no significant difference between the sorption capacities for the multi- and single-solute systems. It can be seen that tire chips had 1 to 4% of the GAC sorption capacities. GAC has approximately 2.5 times lower apparent density than tire chips. Therefore, the sorption capacity of tire chips per unit volume could be increased to approximately 2.5 to 10% of that of GAC.

Figure 2 shows the changes in the liquid phase concentration over time for three different tire chip sizes. The half sorption times for TCE and toluene were approximately three times less for 1.3 cm size tire chips than with 2.5 cm ones. However, the half sorption times for 0.6 cm size tire chips were practically the same as the 1.3

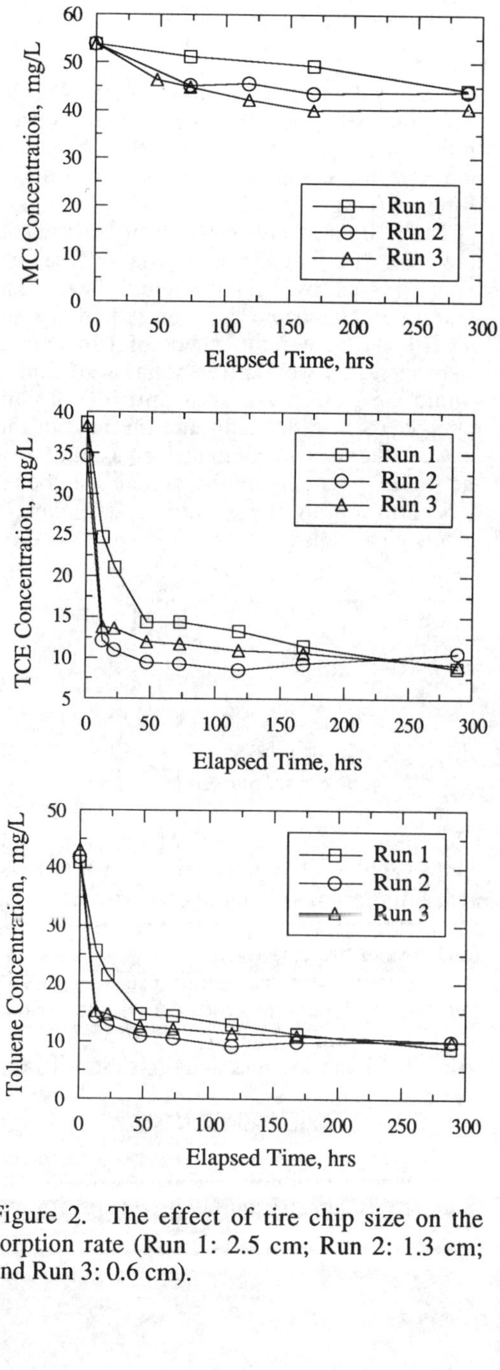

Figure 2. The effect of tire chip size on the sorption rate (Run 1: 2.5 cm; Run 2: 1.3 cm; and Run 3: 0.6 cm).

Table 3. Experimental conditions, sorption capacities, and partition and diffusion coefficients from single-solute tests on tire chips

Organic compounds	Tire/liquid ratio (% by volume)	C_{eq}/C_o	$t_{1/2}$ (hr)	TIRE X/M (g/kg)	GAC X/M (g/kg)	K	D (cm^2/sec)
MC	1.22	0.476	120	0.9	23	89.0	1.23×10^{-7}
	2.83	0.431	100	0.4	21	45.4	1.37×10^{-7}
	4.65	0.421	95	0.3	20	28.2	1.59×10^{-7}
	5.96	0.418	80	0.2	20	22.0	1.71×10^{-7}
TCE	0.72	0.373	30	3.0	154	230.4	0.74×10^{-7}
	2.09	0.167	8	1.3	103	234.4	0.65×10^{-7}
	3.89	0.106	2.7	0.8	77	208.5	1.01×10^{-7}
	5.26	0.080	1.7	0.6	65	206.3	0.92×10^{-7}
Toluene	1.23	0.202	11	2.6	72	316.1	1.06×10^{-7}
	2.66	0.096	3	1.3	49	345.0	0.79×10^{-7}
	3.77	0.083	1.7	0.9	46	280.0	0.92×10^{-7}
	4.21	0.074	1.9	0.8	44	285.9	0.79×10^{-7}
m -Xylene	0.74	0.135	6.8	5.1	122	856.1	0.24×10^{-7}
	2.45	0.062	2.1	1.7	105	601.8	0.37×10^{-7}
	3.52	0.046	1.1	1.2	100	569.1	0.35×10^{-7}
	4.34	0.038	1.2	1.0	96	555.4	0.33×10^{-7}

Table 4. Experimental conditions and estimated dimensionless partition coefficients and diffusion coefficients from multi-solute tests.

Run no. % tire/liquid ratio Tire chip size	Organic compounds	C_{eq}/C_o	$t_{1/2}$ (hr)	X/M[1] (g/kg)	Weight gain (%)	K	D (cm^2/sec)
Run 1	MC	0.866	NA[2]	0.56	0.06	13.69	NA[2]
1.12%	TCE	0.296	27	1.99	0.20	209.90	1.37×10^{-7}
0.6 cm	Toluene	0.275	24	2.17	0.22	233.10	1.59×10^{-7}
Run 2	MC	0.807	NA[2]	1.03	0.10	28.64	NA[2,3]
0.82%	TCE	0.267	8	2.54	0.25	332.29	NA[3]
1.3 cm	Toluene	0.237	8	3.19	0.32	389.27	NA[3]
Run 3	MC	0.789	NA[2]	1.16	0.12	35.13	NA[2,3]
0.75%	TCE	0.277	8	3.01	0.30	344.49	NA[3]
2.5 cm	Toluene	0.254	8	3.50	0.35	387.63	NA[3]

X/M[1] : X = Organic compound sorbed in the tire chips (g); and M = weight of tire chips used (kg).
NA[2]: Since the concentration decrease of methylene chloride was too small to observe the $t_{1/2}$, it was not possible to estimate the diffusion coefficient.
NA[3]: The same method for the diffusion coefficient could not be used because the shape of tire chips changed. The diffusion through sides of chips could not be ignored.

cm size tire chips. The equilibrium concentrations for different tire chip sizes were almost the same. The equilibrium reached within two days for TCE and toluene; thus, the reaction rate may not be a limiting factor. This finding indicates that the size of tire chips need not to be smaller than 1.3 cm. Practical tire chip sizes would range from 1.3 to 2.5 cm.

Batch sorption tests were conducted at 4, 35, and 55°C and the results are shown in Figure 3. There was no statistical difference between the sorption capacities measured at different temperatures. This response suggests that the sorption mechanism is mainly by physical bonding. The desorption was found to be irreversible which implies that the organic compound molecules are trapped in the tire chip polymer matrix.

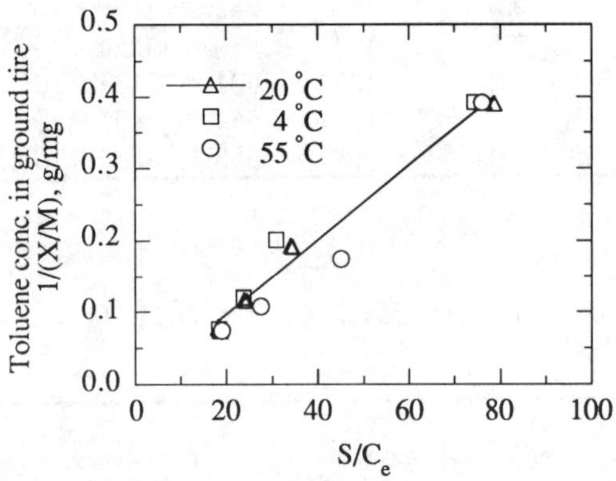

Figure 3. Sorption capacity of toluene at three different temperatures.

The measured partition coefficient exhibited a logarithmic relationship with the octanol-water partition coefficient (K_{ow}) as shown in Figure 4. The diffusion coefficient, on the other hand, was found to vary with the molecular diameter (D_m) as shown in Figure 5. The following empirical relationships are obtained for these parameters:

$$\log K = 0.824 + 0.614 \log K_{ow} \quad (11)$$

$$\log D = -4.426 - 5.001\ D_m \quad (12)$$

These relationships allow the prediction of the partition and diffusion coefficients for various organic compounds, which, in turn, allows the

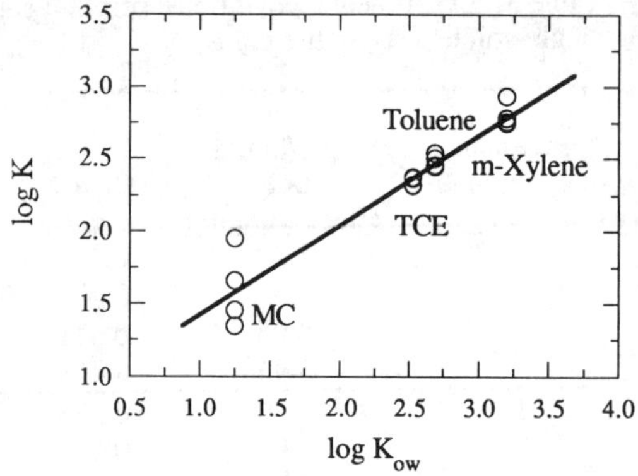

Figure 4. Relationship between K_{ow} and K.

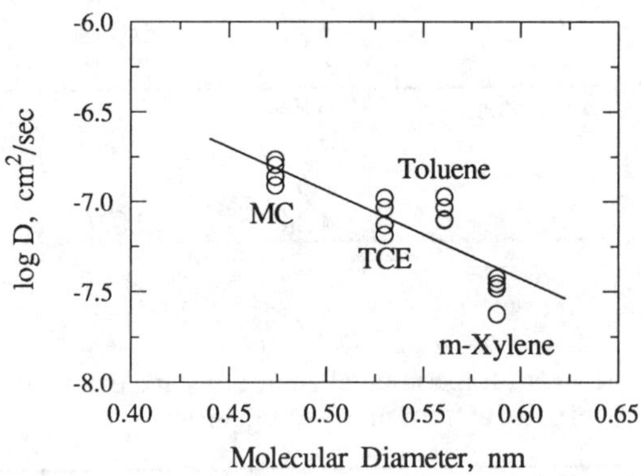

Figure 5. Relationship between molecular diameter and D.

estimation of the sorption capacity and rate. The thickness, *d* (cm), of the tire chip layer needed to remove a given amount of organic chemicals may then be determined in a waste management system using the following relationship:

$$d = 3.83 \times 10^{-6} \cdot \frac{M_a}{(1 - n) \cdot \rho_t} \quad (13)$$

where n is the porosity of tire chip layer, ρ_t is the density of a tire chip (g/cm^3), and M_a is the required mass of tire chips per unit area (kg/ha). If 90% of these compounds need to be removed from the leachate, then the mass of tire chips required can be estimated as follows:

$$M_a = \frac{f \bullet Q_r \bullet t_d \bullet \rho_t \bullet 10^{-3}}{(1 - f) \bullet 10^{0.824 + 0.614 \log K_{ow}}} \quad (14)$$

where f is the fraction of organic compound to be removed, Q_r is the leachate generation rate (m^3/ha/yr), and t_d is the design landfill life time (yr).

For a typical landfill in North America, this procedure yielded the following tire chip thicknesses for benzene, TCE, *m*-xylene, and PCP removal: 32.8, 18.4, 7.3, and 0.5 cm, respectively. The thickness of 32.8 cm required for benzene removal should then be used in this example since it is thickest for the compounds assessed. If a 30-cm thick tire layer is used, then the removal efficiencies of benzene, TCE, m-xylene, and PCP are expected to be 89.2, 93.5, 97.4, and 99.8%.

As shown in the aqueous phase batch isotherm tests, tire chips could sorb significant amounts of organic compounds. The installation of a 10 to 30 cm thick tire chip layer over the landfill liner could result in three major advantages: (1) provide an effective avenue of disposal for scrap vehicle tires, (2) substitution of sand and gravel typically used in the drainage layer of the leachate collection system by tire chips (tire chips behave as excellent drainage material as shown by Edil et al. 1992b) , (3) retardation of potentially hazardous organic compound movement from landfills by sorption. The following additional advantages are also likely to be achieved: (1) decreasing persistent organic compound concentration in liquid phase, (2) furnishing a longer detention time for acclimation and biodegradation (3) providing ideal sites for microbial consortia to grow, and (4) maintaining constant liquid phase concentration by sorption/desorption.

Another application of tire chips in landfills is its use as a vapor filter layer to remove volatile organic compounds emitted from the landfill to the atmosphere. It is estimated that approximately a 10-cm thick tire layer in a landfill would result in an effective and economical vapor filter and provide another reuse/disposal option for scrap vehicle tires. As a rule of thumb 10 automobile tires are needed to generate a 0.01 m^3 of tire chips (10-cm thick layer over 1 m^2 area).

4 CONCLUSIONS

From the batch isotherm experiments using shredded tire chips, the following observations are made:

(1) Tire chips are found to have relatively high organic compound sorption capacities. Tire chips have 2.5 to 10% of the sorption capacity of GAC on the volume basis.

(2) The sorption equilibrium occurs within two days for 0.6, 1.3 and 2.5 cm size tire chips, although the sorption rates for 0.6 and 1.3 cm size tire chips are slightly faster than 2.5 cm size tire chips. Recommended tire chip sizes range from 1.3 and 2.5 cm.

(3) The partition coefficient increases logarithmically with octanol-water partition coefficient and diffusion coefficient decreases with increasing molecular diameter.

A primary leachate collection system could employ tire chips to substitute granular materials and to remove potentially toxic organic compounds in landfills. Furthermore, a tire chip layer could be installed as a filter layer in a landfill cover system to prevent volatile organic compounds such as trichloroethylene and vinyl chloride from escaping into the atmosphere.

ACKNOWLEDGEMENTS

This investigation is based on work supported by the University of Wisconsin System Groundwater Research Advisory Council and the Wisconsin Department of Natural Resources Waste Tire Management or Recovery Grant Program.

REFERENCES

Berens, A.R. 1985. Prediction of organic chemical permeation through PVC pipe, *Jour. AWWA*, 77(11):57-64.

Crank, J. 1975. *The Mathematics of diffusion* , 2nd edition, Clarendon Press, Oxford, U.K.

Dobbs, R.A. and J.M. Cohen 1980. *Carbon Adsorption isotherms for toxic organics*, EPA-600/8-80-023. Cincinnati, Ohio: Environmental Protection Agency

Dodds, J., W.F., Domenico, D.R. Evans, et al. 1983. *Scrap tyres: a resource and technolgy ealuation of tyre pyrolysis and other selected alternative technologies*, U.S. Department of Energy Report EGG-2241.

Edil, T.B., J.K. Park, & P.M. Berthouex 1992a. Attenuation and transport of volatile organic compounds in clay liners, *Proc. of the Mediterranean Conference on Environmental Geotechnology*:: 77-86. Edited by Usmen & Acar, Cesme, Turkey.

Edil, T.B., P.J. Fox & S.W. Ahl 1992b. Hydraulic conductivity and compressibility of waste tire chips, *Proc. 15th Annual Madison Waste Conference:* 49-61.

Gibbons, R.D., D. Dolan, H. Keough, K. O'Leary & R.O'Hara 1992. A comparison of chemical constituents in leachate from industrial hazardous waste & municipal solid waste landfills, *Proc. 15th Annual Madison Waste Conference:* 251-276.

Park, J.K., L. Bontoux, T.M. Holsen, D. Jenkins, & R.E. Selleck 1991. Permeation of polybutylene pipe and gasket material by organic chemicals, *Jour. American Water Works Association*, 83(10):71-78.

Plumb, R.H. and A.M. Pitchford 1985. Volatile organic scans: implications for ground water monitoring, *Proc. NWWA/API Conf. on Petroleum Hydrocarbons and Organic Chemicals in Ground Water - Prevention, Detection, and Restoration.* Dublin, Ohio: Natural Water Well Association.

Reynolds, G.W., J.T. Hoff, & W.G. Robert 1990. Sampling bias caused by materials used to monitor halocarbons in groundwater. *Environ. Sci. & Technol.*, 24(1):135-142.

Sakti, J. P., Park, J. K., and Hoopes, J. A. 1992. Transport of aqueous organic chemicals through HDPE Geomembranes, to be presented in *Water Forum '92.*: Baltimore, MA.

United States Environmental Protection Agency 1991. *Market for Scrap Tires*, EPA/530-SW-90-074B.

Waste Disposal by Landfill – GREEN'93, Sarsby (ed.) © *1995 Balkema, Rotterdam, ISBN 90 5410 356 6*

Factors influencing geomembrane interface friction

E. Pasqualini & D. Sani
University of Ancona, Italy

M. Roccato
Società Italiana per il Gas p.a., Italy

ABSTRACT: The evaluation of shear resistance interfaces parameters between geomembranes and compacted clays and/or geomembranes and other geosynthetics is essential for assessing the stability of liner and cover systems. Therefore, a strong effort has been recently done to evaluate this resistance by means of lab-tests. This paper has the aim to examine some factors which could influence the results obtained in the laboratory.

1 INTRODUCTION

The use of composite liner and cover systems is generally recommended in the design of waste landfills design, since they proved to give both very low hydraulic conductivity and great effectiveness against chemical diffusion (Daniel, 1991; Manassero & Pasqualini, 1991). However, the use of composite liner and cover systems requires that greater attention be paid to the stability analysis of landfill, for the presence of the interfaces whose resistance against in-plane shear can be very low and thus act as potential planes of failure (Mitchell & Mitchell, 1991).

As the most critical part of any stability analysis is the selection of proper values for the shearing resistance along the interfaces, a large number of experimental devices has been recently built. Table 1 shows an outline of some of these equipments.

At present an increasing number of experimental data is available. On the other hand, relatively little is know about the factors which could influence the interface shearing resistance, whose understanding could explain the differences observed where some experimental data are compared. These differences can be due to the variety and properties of used materials (roughness, polymers, textures, deformability, tickness, ...), or to the particular environmental conditions (dry, wet and submerged condition, relative specimen position, temperature, abrasion, dust, handling, ...).

In this paper some of these factors are considered. This preliminary study is focused on the interfaces, where a geomembrane is present, since the available interface friction parameters for sliding resistance between smooth-surfaced geomembranes and a variety of soils and geosynthetic materials are significantly less than the friction parameters of the interface without geomembrane (Martin et al., 1984; Negussey at al., 1989; Mitchell et al., 1990; O'Rourke et al. 1990; Yegian & Lahlaf,1992; Pasqualini et al. , 1993).

The shear boxes used at the Geotechnical Laboratory of the University of Ancona were obtained using modified Wikeam-Farrance 25000 direct shear testing apparatus. The modified spilt boxes were designed taking into account the following:

-the area of the contact surface between the tested materials was kept constant during the shear test to avoid possible sources of inaccuracy in the evaluation of normal and shear forces applied to test specimens;

-the tests were performed up to high relative displacement to determine the residual shear resistance;

-the tests were done on dry, wet and submerged specimens (in the presence of water or leachate) to reproduce the possible in-situ conditions;

- the tests were performed both on geosynthetic-geosynthetic and geosynthetic-soil interfaces.

For details on the equipment see Sani (1992).

The test program of this preliminary stage of investigation is outlined in Table 2.

Table 2 summarizes only the tests performed on interfaces with geomembrane. In particular the interfaces indicated with the letters A,B,C and D refer to one kind of smooth geomembrane and four different geotextiles; interface E refers to another kind of smooth geomembrane and one kind of geotextile. The letter F is a geonet-geomembrane interface; G represent a combination between a smooth geomembrane and a compacted natural clay.

Table 1. Outline of some equipments used to investigate the shear resistance interfaces.

References	Equipment	Dimension
Martin et al.,1984	Shear box	10cm x 10cm
Saxena & Wong, 1984	Shear box	28cm x 28cm
Koerner et al.,1986	Shear box	10cm x 10cm
Williams & Houliahan, 1987	Shear test	30cm x 30cm
Mitchell et al.,1990	Shear box	7cm x 7cm
Mitchell et al.,1990	Shear box	5cm x 5cm 7cm x 7cm
Mitchell et al.,1990	Pullout box	43cm x 25cm
O'Rourche et al.,1990	Shear box	6cm x 6cm
Matichard et al.,1991	Shear box	10cm x 10cm
Matichard et al.,1991	Half box rigid plane	25cm x 25cm
Matichard et al.,1991	Inclined plane box	25cm x 25cm
Matichard et al.,1991	Shear box	30cm x 30cm
Matichard et al.1991	Inclined plane box	100cm x 100cm
Yegian & Lahlaf ,1992	Static shear box	20cm x 30cm
Yegian & Lahlaf ,1992	Shaking table	91cm x 81cm
Pasqualini et al.,1993	Shear box	14cm x 14cm
Pasqualini et al.,1993	Shear box	7cm x 7cm
Pasqualini et al.,1993	Shear box	6cm x 6cm 10cm x 10cm

Most interfaces were tested under both dry and submerged conditions. In addition the wet condition was reproduced to simulate the situation which might occur on the slopes of the landfill where the liquids (and/or leachate) flow and do not stagnate. To reproduce wet interfaces the materials were immersed in water, for a pre-established time, immediately before starting the tests (one hour was considered convenient and adequate).

In this first stage of the research the wet and submerged tests were always performed using tap-water (leachate has not been used yet).

Interface shear tests were executed by shearing the interface samples under strain-controlled loading at the displacement rates of about 0.12 mm/min.

The normal stress applied to the sample interfaces during testing was varied from a minimum of 25 KPa to a maximum of 294 KPa.

Table 2. Examined interfaces.

	Interface	
A	Smooth LDPE (1mm thickness)	HDPE 450 nonwoven needle-punched
B	Smooth LDPE (1mm thickness)	PET 300 nonwoven needle-punched
C	Smooth LDPE (1mm thickness)	PP 500 nonwoven needle-punched
D	Smooth LDPE (1mm thickness)	PP 470 woven slit film
E	Smooth HDPE (2mm thickness)	PET 400 nonwoven needle-punched
F	Smooth LDPE (1mm thickness)	HDPE 500 (3.5mm thickness)
G	Smooth LDPE (1mm thickness)	Compacted Clay

2 RESULTS AND DISCUSSION

This represents only a preliminary contribution to the study of the factors which influence the shear resistance parameters.

The following factors are considered: number of shear cycles; relative position of interface samples; temperature; difference between dry and wet interface shear resistance; relative displacement; normal stresses; wetting of compacted clay materials.

2.1 *Influence of shear cycles*

In order to evaluate the typical behaviour of interface shear strength versus shear displacement, most of the interface samples were subjected to more than one cycle of shear at the same normal stress.

A typical result is shown in Figure 1 which refers to a geomembrane-geotextile interface (interface E in Table 2).

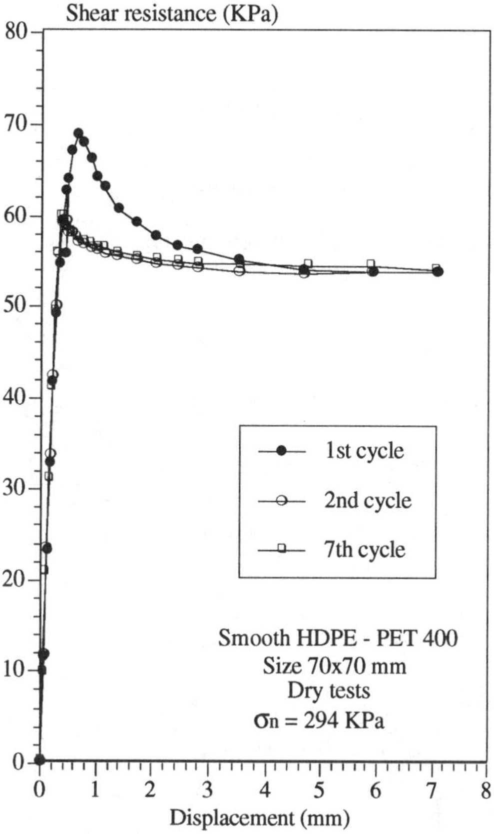

Figure 1. Influence of shear cycles on geotextile-geomembrane interface (after Pasqualini et al., 1993).

Usually, during the first shear cycle a well defined peak resistance is obtained at very small deformation; after that, the resistance tends towards a residual value. In the subsequent cycles of direct shear, the difference between the peak and residual resistance is less marked; moreover, the value of the residual resistance is not significantly affected by the number of shear cycles. Therefore, as the peak resistance is well defined (and not always) only in the first cycle and mobilised at very small displacement, the results of the tests should be analysed referring to the residual shear strength which is more useful from a practical point of view.

Seed et al. (1988) attribute these results to a possible "polish-effect" of the geotextile on the geomembrane; moreover, it is also possible that after the first shear cycle the fibres of the geotextile are smoothed and oriented.

A different trend is observed when the influence of shear cycle is investigated referring to the geomembrane-geonet interface (interface F in Table 2). A characteristic of this kind of interface is the significant increase in shear resistance after the first cycle. This is shown in Figure 2, where the residual friction angles of three different shear cycles are compared. This increase in shear resistance is ascribed to an increase in roughness of the low-density polyethylene liner surface due to the previous passage of the high-density polyethylene geonet.

2.2 *Influence of relative position of interface samples*

To ensure that the area of the interface does not change during the test, the bottom sample is designed to be greater than the top sample (see Figure 3).

Now, if the geosynthetic material of the bottom sample is more deformable than that of top sample, the latter tends to penetrate the former as a consequence of the vertical load applied. When this happens the result of the test can be misleading and the evaluation of the residual friction angle questionable (see, for example, the experimental results shown in Figure 4).

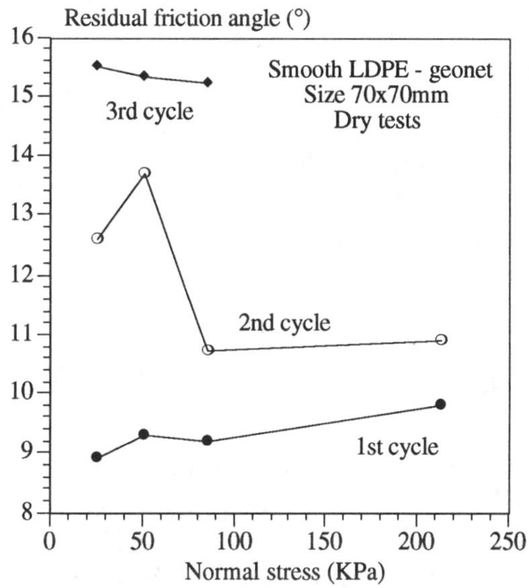

Figure 2. Influence of shear cycles on geomembrane-geonet interface.

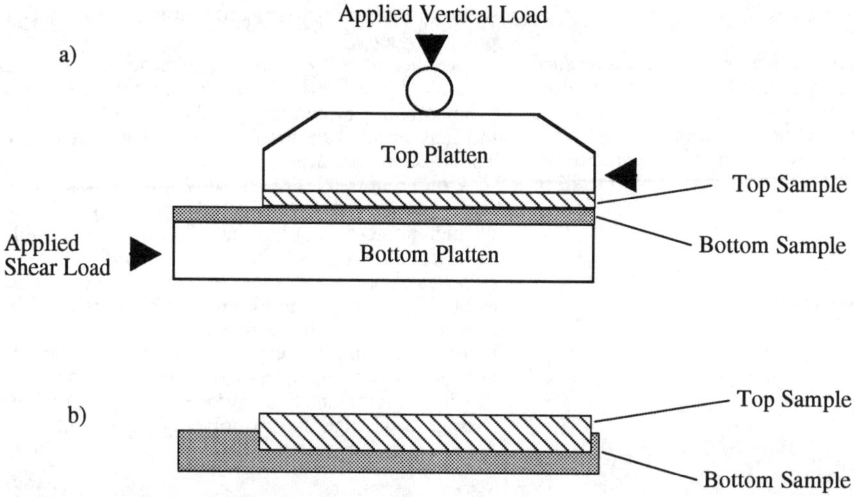

Figure 3. Schematic illustration (not to scale) of direct shear interface. a)Sample configuration. b) Amplified effect of t he penetration of the top sample into a more deformable bottom sample. (After Pasqualini et al., 1993).

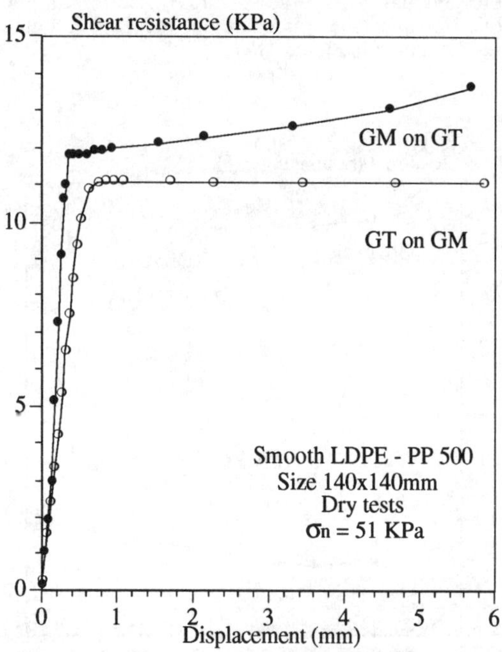

Figure 4. Influence of the relative position at geomembrane (GM) - geotextile (GT) interface (after Pasqualini et al., 1993).

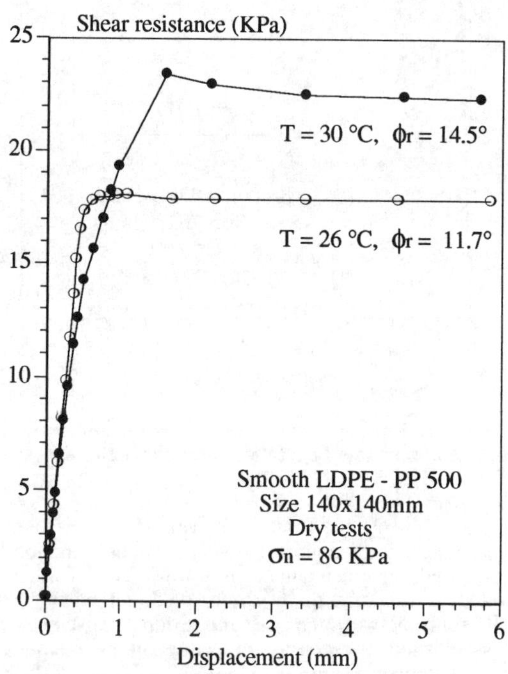

Figure 5. Influence of the temperature (after Pasqualini et al., 1993).

2.3 *Influence of the temperature*

The parameters of temperature is usually very important for its great influence on all polymeric materials.

This aspect was not investigated methodically. The room temperature of the laboratory was found by chance to influence the shear resistances of the examined interfaces. As these series of tests were performed in the very hot summer of 1992, the range of temperature "investigated" varied from 25°C to 30°C.

The results reported in Figure 5 and in Table 3 show that the influence of the temperature on the shear resistance is not negligible, at least for the geomembrane-geotextile interfaces considered and in the range of temperature examined.

Table 3. Residual friction angles versus temperature for three different geomembrane (GM) - geotextile (GT) interfaces.

Int.	Normal stress (KPa) σ_n	Residual friction angle ϕ_r Temperature 26°C-27°C	Residual friction angle ϕ_r Temperature 29°C-30°C
B	25	13.8°	15.9°
B	86	12.6°	16.0°
C	25	12.4°	14.7°
C	86	11.7°	14.5°
D	25	12.2°	17.4°
D	86	13.3°	17.7°

Int. : kind of the interfaces.
Notes: Size 140x140mm; dry tests; GT on GM.

These preliminary results suggest that this aspect should be investigated systematically over a wider and more significant range of temperature (see also Mitchell et al., 1992). They also stresses the importance of taking into account the temperature when experimental data from different laboratories are compared.

At present, a temperature-conditioned large box is being built where the direct shear apparatus, operating at the University of Ancona, will be housed. In this way it will be possible to investigate systematically the effect of the temperature, which will be varied from about 0°C to about 45°C (for details see Stella, in preparation).

2.4 *Dry versus wet condition*

The interface A,B,C and D (see Table 2) were tested both dry and wet, but not submerged, at normal stress ranging from 25 KPa to 86 KPa.

The tests carried out in the wet condition gave shear resistances systematically lower than those obtained in the dry condition. Table 4 shows an example of the results obtained.

Table 4. Residual friction angles in dry and wet conditions at the geomembrane (GM) - geotextile (GT) interface.

Int.	Normal stress (KPa) σ_n	Residual friction angle ϕ_r Dry	Residual friction angle ϕ_r Wet
D	25	12.2°	11.1°
	51	12.8°	11.1°
	69	12.9°	10.7°
	86	13.3°	10.7°

Int. : kind of the interfaces.
Notes: Size140x140mm; temperature 27°C; GTonGM

2.5 *Influence of normal stress on shear resistance*

The geomembrane-geotextile and geomembrane-geonet interfaces were sheared at normal stress ranging from 49 KPa to 294 KPa.

Interface E (see Table 2) was tested both dry and submerged. The results shown in Figure 6 indicate that the residual friction angles, both dry and

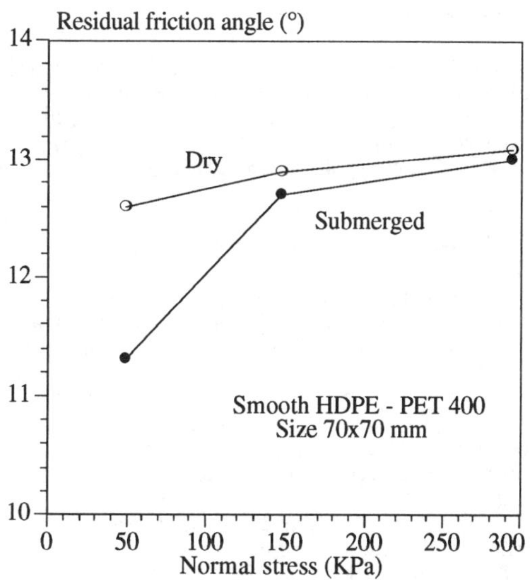

Figure 6. Influence of normal stress on dry and submerged shear resistance at the geotextile-geomembrane interface (after Pasqualini et al., 1993).

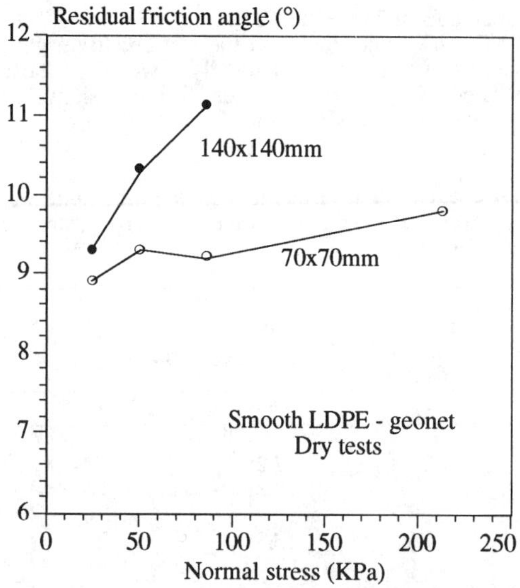

Figure 7. Influence of normal stress and dimension interface on shear resistance at the geonet-geomembrane interface (after Pasqualini et al, 1993).

submerged, depend on normal stress, but the difference between dry and submerged angles tends to decrease with increasing normal stress. Moreover, the residual friction angles obtained seem to increase with the normal stress applied.

The tests on interface F (see Table 2) were carried out also to investigate the influence of interface dimensions on the results.

The comparison of the results in Figure 7 shows that the smaller interface gave lower friction angles. Moreover, the results shown in Figure 7 confirm that the geonet-geomembrane coupling produces interfaces with very low shear resistance, which seems to increase with increasing normal stress.

2.6 *Influence of relative displacement*

As the sample dimension was reduced, it was possible increase the applied normal stress and displacement to highlight the importance of performing the tests up to large relative displacement.

In Figure 8 are presented three typical experimental shear-displacement curves obtained at three different normal stresses, showing that the residual shear resistance is reached only after a large relative displacement.

This is more evident in Table 5a and 5b which the values of the residual friction angles are evaluated at different deformation.

The results obtained seem to indicate that a relative deformation greater than 5-6 % should be reached for a realistic evaluation of theresidual friction angle of the geomembrane-geotextile interface.

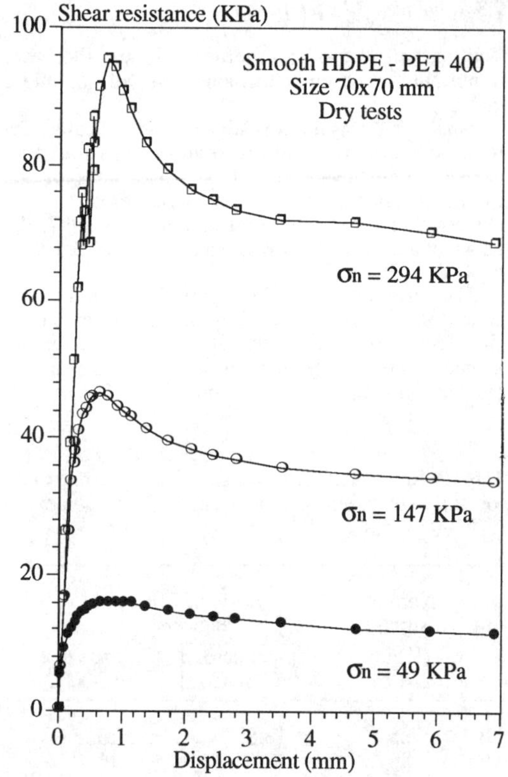

Figure 8. Examples of shear-displacement relationship (after Pasqualini et al., 1993).

2.7 *Influence of wetting clay material*

Both dry and submerged direct shear tests were performed on this interface (interface G in Table 2).

Initially, the tests were performed in dry conditions. The clay was compacted in the Standard Proctor mold and then three specimens were trimmed and placed into the upper split box. Normal stresses of σ_n = 66 KPa, σ_n = 120 KPa, σ_n = 214 KPa were applied and then the samples were sheared at a strain rate of about 0.12 mm/min.

A second series of tests was performed on specimens prepared in the same way. However, before shearing they were submerged and soaked for 24 hours under a light surcharge of 25 KPa. After this period of soaking, the final vertical loads were applied, and the specimens were immediately sheared at a strain rate of about 0.12 mm/min. Total normal stresses applied to these samples were σ_n= 66 KPa, σ_n = 120 KPa, σ_n = 214 KPa.

Both dry and submerged tests were repeated for two different dry densities and water contents.

Table 5a. Values of the residual friction angles evaluated at different deformations.

Def. (%)	Residual friction angle ϕ_r Dry		
	σ_n=49KPa	σ_n=147KPa	σ_n=294KPa
2	13.5°	14.1°	14.9°
3	13.0°	13.8°	14.3°
4	13.0°	13.6°	14.0°
5	12.9°	13.5°	13.8°
6	12.9°	13.3°	13.7°
8	12.6°	13.1°	13.5°
10	12.6°	12.9°	13.1°

Def. : deformation.

Table 5b. Values of the residual friction angles evaluated at different deformations.

Def. (%)	Residual friction angles ϕ_r Submerged		
	σ_n=49KPa	σ_n=147KPa	σ_n=294KPa
2	11.7°	13.2°	13.4°
3	11.4°	13.0°	13.2°
4	11.4°	12.9°	13.2°
5	11.2°	12.9°	13.1°
6	11.2°	12.7°	13.0°
8	11.2°	12.7°	13.0°
10	11.2°	12.7°	12.9°

Def. : deformation.

Table 6. Values of shear resistance, τ_f, for two different dry density, γ_d, and water content, w, of clay material.

Int.	Normal stress	Test dry		Test submerged
	σ_n (KPa)	ϕ_r (°)	τ_f (KPa)	τ_f (KPa)
	66	13.8	16.3	14.5
H1	120	15.6	35.2	25.7
	214	13.8	53.1	46.4
	66	13.0	15.3	14.0
H2	120	11.1	24.4	24.4
	214	10.3	39.5	39.8

Notes: Dry and submerged tests were sheared at the same strain rate; Int. : kind of the interfaces.
Clay sample before shear test :
H1 : γ_d =1.62 g/cm^3 w=w_{opt} = 21%
H2 : γ_d =1.55 g/cm^3 w=w_{opt} = 26%
Clay sample after shear test - closet to the interface :
H1 : w = 28%
H2 : w = 27%

The results, summarized in Figure 9, show that when the clay is compacted at high water content, the dry and submerged interface resistances are almost equal; moreover, the submerged interface resistance seems to be independent of the water content " as-compacted" of the clay specimens.

This behaviour can be explained by considering the increase in the water content values after soaking measured at the interface of the clay specimens at the end of the submerged direct shear tests, which is shown in Table 6. This confirms the results obtained by Seed and Boulanger (1991), who pointed out the potentially important effect of wetting these interfaces during field construction.

These tests, whose results are shown in Figure 9, were carried out with the aim to investigate the influence of a possible increase in water content at the geomembrane-clay interface. For this reason the same strain rate was imposed during both dry and submerged tests. Of course, when the tests are performed to obtain the actual undrained and/or drained shear interface resistance, the strain rates must be appropriately selected.

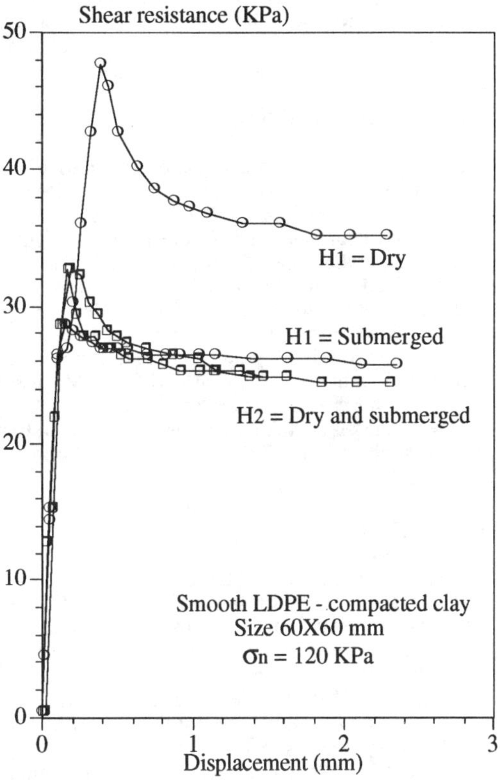

Notes: H1 : γ_d =1.62 g/cm^3 w=w_{opt} = 21%
H2 : γ_d =1.55 g/cm^3 w=w_{opt} = 26%

Figure 9. Influence of wetting clay material (after Pasqualini et al., 1993).

3 FINAL REMARKS

This paper analyses some factors which might influence the results of direct shear tests performed on different interfaces when geomembrane liner are present.

In particularly, paper shows:

-the strong influence of temperature on the shear resistance of the geomembrane-geotextile interface;

-the important effect of wetting compacted clay materials on smooth geomembrane-clay interface shear strengths.

On the contrary, the influence of normal stress on dry and submerged friction angles seems to be less important. These results must be considered very preliminary and will be verified by means of a larger number of experimental data.

Overall, it was observed that the experimental data presented do not always agree with those obtained by other researchers. This can be explained by differences in materials, laboratory conditions and testing apparatus. Therefore, data available in literature must be considered only as reference values for a first-step design, which should then be checked by tests carried out on the specific materials while trying to reproduce the most probable in situ conditions.

ACKNOWLEDGEMENTS

The work described in this paper is a part of the research project named " Aspetti Geotecnici Relativi agli Impianti di Smaltimento dei Rifiuti" performed in cooperation with Ingegneria Geotecnica S.r.l. (Turin) and Società Italiana per il Gas p.a. (Turin). The latter also provided financial support.

Finally, this research forms part of the research project of national interest named "Geotecnologie Finalizzate alla Salvaguardia e al Recupero del Territorio" and funded by M.U.R.S.T. (Ministero dell'Università e della Ricerca Scientifica).

The writers wish to express their gratitude to both for this support.

REFERENCES

Daniel, D.E. 1991. "Compacted Clay and Geosynthetic Clay Linings." *Proc.XV CGT.* Italy.

Grisolia,M.; Napoleoni,Q.; Sirini,P.; Tancredi,G. 1991. "Geotechnical Behaviour of Sanitary Landfill Based on Laboratory and in Situ Test." *Proc. VII* Int. Conference on Solid Waste Management and Secondary Materials, Philadelphia, PA U.S.A.

Jessberger,H.I. & Kockel,R. 1991. "Mechanical Properties of Waste Materials." *Proc.XV CGT.* Italy.

Koerner,R.M; Martin,J.P.; Koerner,G.R. 1896. "Shear Strength Parameters between Geomembranes and Cohesive Soils." Geotextiles and Geomembranes, 4:21-30.

Manassero, M. & Pasqualini,E. 1991. "Aspetti Geotecnici nella Progettazione delle Discariche Controllate." *Proc. XV CGT.* Italy.

Manassero, M. & Pasqualini, E. 1991. "La Funzione di Impermeabilizzazione negli Impianti di Smaltimento dei Rifiuti e negli Interventi di Risanamento Ambientale." L'Ingegnere anno LXVI. Italy

Martin,J.P.; Koerner,R.M.; Whitty,J.E. 1984. "Experimental Friction Evaluation of Slippage between Geomembranes, Geotextiles, and Soils." *Proc.* Int. Conf. on Geomembranes, Industrial Fabrics Association International, Denver, Colo., pp.191-196.

Matichard,Y; Delmas,P.; Soyez,B.; Mathieu,M. 1991. "Stability of Lining Systems on Slopes." *Proc. III* International Symposium, Sardinia '91. Italy.

Mitchell,J.K. & Mitchell,R.A. 1991. "Stability of Landfills." *Proc. XV CGT.* Italy.

Mitchell,J.K.; Seed,R.B.; Seed,H.B. 1990. "Kettleman Hill Waste Landfill Slope Failure. I: Liner-System Properties." *Geotechnique* Vol 116, 4:647-668.

Mitchell,J.K.; Seed,R.B.; Seed,H.B. 1992. Closure to the "Discussion of Kettleman Hills Waste Landfill Slope Failure. I: Liner-System Properties." *Geotechnique* Vol.118, 4:645.

Negussey,D.; Wijewickreme,W.K.D.; Vaid,Y.P. 1989. "Geomembrane Interface Friction." Canadian *Geotechnique* 26:165-169.

O'Rourke,T.D.; Druschel,S.J.;Netravali,A.N. 1990. "Shear Strength Characteristics of Sand-Polymer Interfaces." *Geotechnique* Vol.116 3:451-469.

Pasqualini,E.; Roccato,M.; Sani,D. 1993. "Shear Resistance at the Interface of Composite Liners". *Proc. IV* International Symposium, Sardinia '93. Italy.

Sani,D. 1992. "Caratteristiche di Resistenza all'Interfaccia dei Sistemi di Rivestimento Compositi." Tesi di Laurea, Università degli Studi di Ancona, Facoltà di Ingegneria. Italy.

Seed,B.R. & Boulanger, R.W. 1991. "Smooth HDPE-Clay Liner Interface Shear Strengths: Compaction Effects." *Geotechnique* A.S.C.E. Vol.117 4:686-693.

Seed,B.R.; Mitchell,J.K.; Seed,H.B. 1988. "Slope Stability Failure Investigation: Landfill Unit B-19, Phase I-A, Kettleman Hills, California." Report UCB/GT/8801. U.S.A.

Seed,B.R.; Mitchell,J.K.; Seed,H.B. 1990. "Kettleman Hills Waste Landfill Slope Failure. II: Stability Analysis." *Geotechnique* A.S.C.E. Vol.116 4:669-690.

Stella,M. (in preparation) "Indagine Sperimentale sul Comportamento delle Resistenza al Taglio di Interfacce Geosintetico-Geosintetico e Geosintetico-Terreno." Tesi di Laurea, Università degli Studi di Ancona, Facoltà di Ingegneria. Italy

Yegian,M.K. & Lahlaf,A.M. 1992. "Dynamic Interface Shear Strength Properties of Geomembrane and Geotextiles." *Geotechnique* A.S.C.E. Vol.118 5:760-779.

Waste Disposal by Landfill – GREEN'93, Sarsby (ed.) © 1995 Balkema, Rotterdam, ISBN 90 5410 356 6

Mechanical properties of wastes in a sanitary landfill

J.M. Sánchez-Alciturri, J. Palma, C. Sagaseta & J. Cañizal
Division of Ground Engineering, University of Cantabria, Santander, Spain

ABSTRACT: A series of in situ tests have been carried out on a sanitary landfill. Results of plate loading tests and static and dynamic penetration are presented. Interpretation of these results is difficult because of the lack of experience with this kind of materials. Some ideas of general application are presented and mechanical parameters are obtained for the particular landfill analyzed and compared with other cases reported in the literature.

1 INTRODUCTION

Design and operation of sanitary landfills involve a variety of geotechnical problems, for which a knowledge of the stress-strain-strength-time behaviour of the material (waste and earth cover) is required. Typical examples of these problems are the predictions of: (i) amount and rates of settlement, which are important to evaluate the actual capacity of the landfill and its potential use after closure, and (ii) stability of the landfill lateral slopes, which in some cases can reach a height over 50 m.

The adequacy of laboratory tests for this purpose is limited. Undisturbed samples are difficult to take, and in laboratory compacted samples it is not possible to reproduce adequately in situ conditions (age, temperature, drainage, gas, etc.).

For this reason, in situ tests and performance monitoring of actual landfills are specially useful for the evaluation of the mechanical behaviour of solid wastes. This paper describes their use in a particular case, and some comments on the interpretation of results of general applicability.

2 SITE DESCRIPTION

Meruelo sanitary landfill is the main urban waste disposal centre in the region of Cantabria, in Northern Spain. The disposal rate is about 115,000 tons per year, corresponding to a population of 300,000 people. Table 1 shows typical waste composition. The initial bulk density is about 0.2 Mp/m^3, and it is increased to 0.8-1.0 Mp/m^3 after compaction. The water content is 48%.

Table 1. Meruelo landfill waste composition.

Components	% by weight
Organic matter	52.0
Paper	15.7
Paper board	6.2
Plastics	8.3
Wood	2.4
Metals	3.2
Glass	4.1
Earth and ashes	1.4
Others	6.7

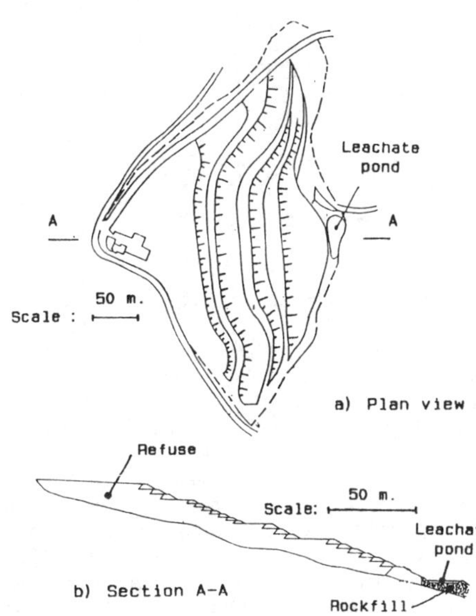

Fig. 1. Meruelo landfill

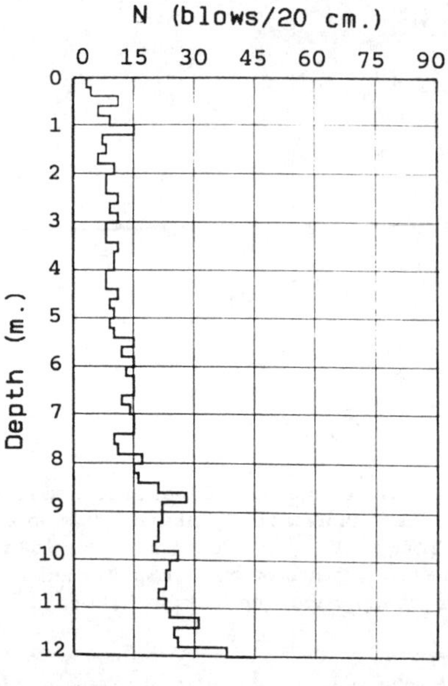

Fig. 2. Dynamic Borros penetration test

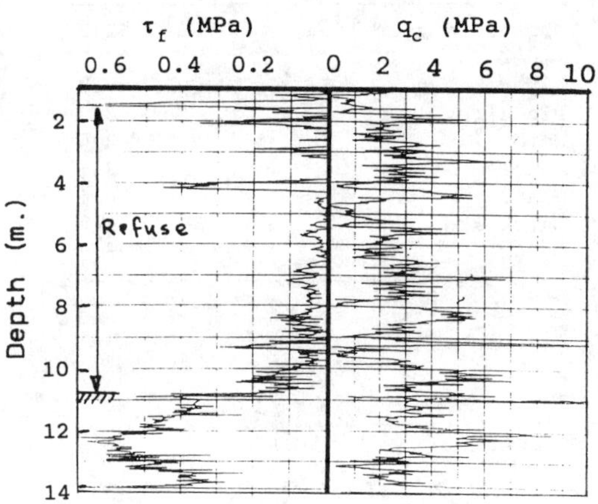

Fig. 3. Cone penetration test

The Meruelo landfill is located in a valley, the front slope being raised from an initial embankment by the 'upstream' method, with dykes 2 m. high. The final difference in elevation between the toe and the crest is about 50 m. Mean slope is 3(H):1(V), with three flat levels (fig. 1).

Landfilling began in December 1988, and finished in January 1992. Operation involved heavy compaction of the wastes with sheepfoot rollers, and covering each layer of waste with 20 cm. of silty clay. At the beginning of operation, the thickness of the waste layers was 0.5 m., but it was increased to 2.0 m. at the final stages.

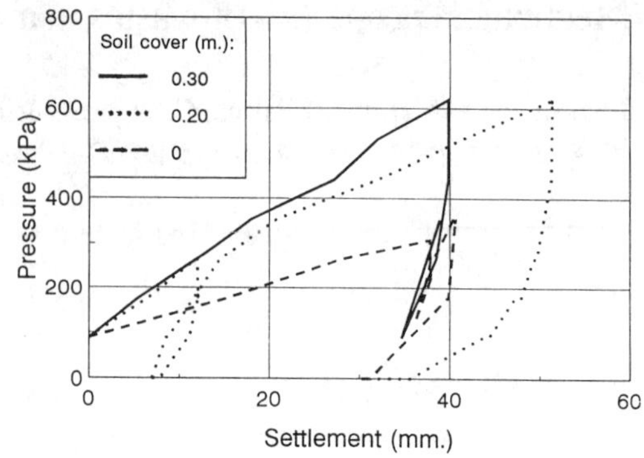

a) Plate diameter: 0.30 m.

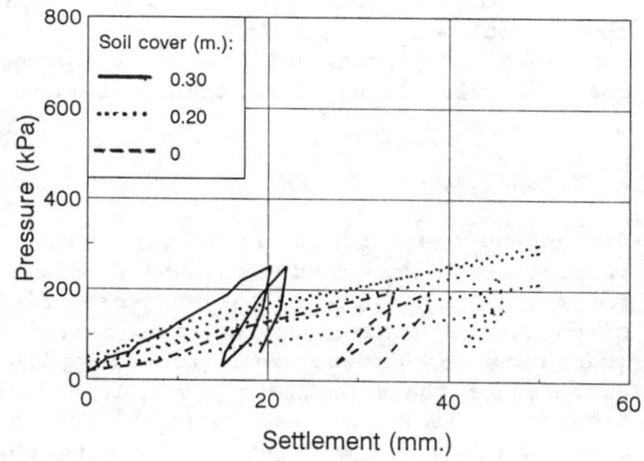

b) Plate diameter: 0.45 m.

Fig. 4. Plate loading tests

The landfill was monitored along its construction period and after closure. The aims of the study were to ascertain the design bases for future landfills, and to assess the stability of the front slope, which in this particular case can be a critical point, due to the configuration of the landfill (fig. 1). The analysis of the movements is presented in detail in a companion paper, and here they are only used as an aid for the evaluation of the waste properties.

A series of in situ tests was carried out, consisting on static and dynamic penetration tests, and surface plate loading tests.

3 TEST RESULTS

In Borros dynamic penetration tests, the resistance increases roughly linearly with depth. Typical values are 10 blows/20 cm. at a depth of 5-10 m. (fig. 2).

Static cone penetration tests, CPT, were also run. Fig. 3 shows a typical example.

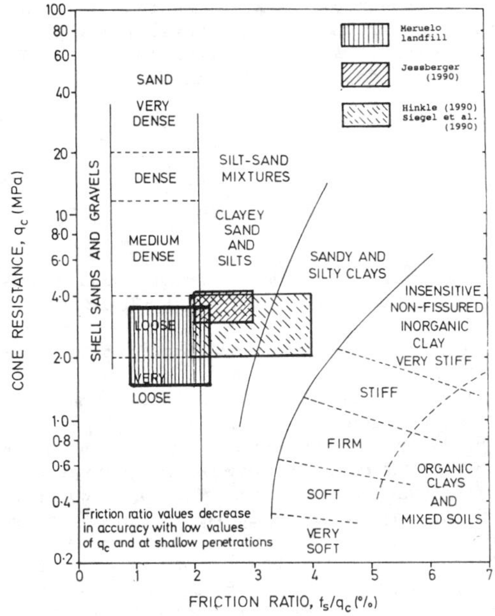

Fig. 5. Schmertmann's soil type chart

Resistance also increases with depth, and at 10 m. the tip pressure is about 3 MPa.

With respect to the plate loading tests, they have been performed on plates of 30, 45 and 75 cm. of diameter. The surface of the waste was covered by a layer of compacted soil, and the tests were run in places with a cover with thickness of 30, 20 and 10 cm., and without cover (Sánchez-Alciturri et al. 1993). In fig. 4 the resulting pressure-settlement curves have been plotted.

Finally, monitoring of landfill movements has two parts: surface settlements after completion of each flat level (fig. 1), and horizontal movements of the front slope.

Surface settlements are analyzed in detail in a companion paper. They reach values of 400-500 mm. (3% of the landfill height) after three years.

Horizontal movements of the slope are relatively small (less than 100 mm.) and addressed inwards, thus indicating that slopes are far from instability.

4 INTERPRETATION

Interpretation of the results of mechanical tests in wastes is subjected to many uncertainties, due to the lack of a conceptual reference model of behaviour for this material. The analysis is usually made on the basis of models and methods established for soils. In fact, wastes are composed of individual particles with a certain interlocking, and this supports this approach to some extent. However, there are significant differences with soils: the void ratio is very high, which implies a large volumetric compressibility; the 'particles' are of very different natures, and some of them are weak and very deformable; there is a process of decomposition with time, which causes self-induced settlements and a variation of properties with time.

In spite of these differences, interpretation of the tests on wastes by means of concepts derived from soil behaviour can be useful, at least at the present state of knowledge. On this basis, the concepts of 'friction angle' and 'cohesion intercept' are commonly used. On the other hand, wastes are usually non saturated. Therefore, interpretation of test results in terms of undrained situation, with no volume change (ϕ=0) may be an unrealistic approach, and an analysis in terms of equivalent c-ϕ can be more adequate, as it will be shown later.

4.1 Penetration tests

The use of penetrometers provides an easy and inexpensive way to investigate the behaviour of waste in situ and to evaluate the relative variation of mechanical properties with time.

In the static penetration tests performed in Meruelo landfill, the tip resistance is of about 1-3 MPa, and the friction ratio (shaft friction/tip resistance) of 1-2%. These values agree with other cases reported in the literature (Hinkle, 1990; Jessberger, 1990; Siegel et al., 1990). In the Schmertmann (1978) interpretation chart, the above results fall in the range of loose sands to soft clayey sands and silts (fig. 5).

Interpreted as an undrained resistance (ϕ=0), the measured tip resistance would correspond to an undrained shear strength, s_u, of 50-150 kPa. Under the more realistic assumption of non-cohesive material, the equivalent friction angle would be of about 28°-35°.

4.2 Plate loading tests

Plate loading tests are usually performed on top of landfills after closure, once the final sealing cover has been installed. The usual aim is to evaluate the bearing capacity for potential use of the landfill area. In the present case, the tests have been run during landfill operation, in order to analyze the mechanical behaviour of the waste material.

In any case, the interpretation of the test results must take into account the layered nature of the ground. For a closed

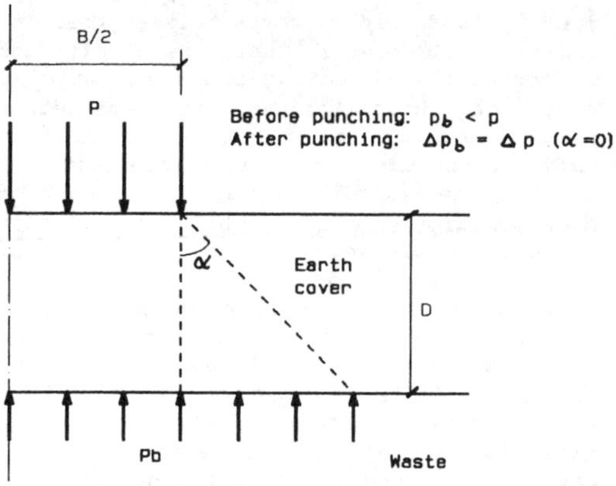

Fig. 6. Cover punching

landfill, the thickness of the final cover is about 0.5-1.0 m., and hence the use of a large plate is necessary to get useful results. In a landfill in operation, the earth cover on each layer is of about 10-40 cm., and so a smaller plate can be used.

Two modes of failure were postulated by Sowers (1968): punching of the earth cover, or rotational failure involving earth cover and waste. However, little evidence is found of the second mechanism, and the punching mode seems to be the most frequent case. Some results of laboratory triaxial tests (Jessberger, 1990; Landva and Clark, 1990; Siegel et al., 1990), indicate that stress-strain response of compacted waste under shear loading is a monotonically increasing curve, not reaching any limit load. The deformation modulus is very low, of the order of 0.5 to 2.0 MPa, and at shear strains over 20% no signs of failure are observed. Moreover, at large strains, a slight upward inflection is produced, indicating a material stiffening.

Under these conditions, the load-settlement response of a plate load test on a landfill can be represented as follows: a first stage of elastic response, until punching of the earth cover is produced. After punching, a new linear range follows, in which the increment of applied load is entirely transmitted to the waste, with no further cooperation of the cover.

The punching pressure is governed by the strength and thickness of the earth cover. For a circular plate, it is given by:

$$p_p = p_b + 4.s_u.\frac{D}{B} \qquad (1)$$

where s_u is the shear strength of the cover and p_b is the unit pressure transmitted to the waste, which is a fraction of the applied pressure, p_p. Assuming that the pressure p_b results from a lateral spreading of the vertical stress at an angle α (fig. 6), the punching pressure results:

$$p_p = s_u.\frac{(1 + 2.\tan\alpha.\frac{D}{B})^2}{\tan\alpha.(1 + \tan\alpha.\frac{D}{B})} \qquad (2)$$

The pressure p_p has an upper bound corresponding to the rotational failure of the cover itself. This limit load is about $6.s_u$, and can be reached before punching for high D/B ratios.

The slope of the initial part of the pressure-settlement curve of fig. 6 depends on the deformability and strength of the waste and of the cover. However, the response after punching can be more easily interpreted. If the shear strength of the cover has been reached, any further applied load is transmitted entirely to the waste, with no cooperation of the cover. Under these conditions, the incremental load-settlement law is given by:

$$\Delta p = \frac{E_w}{1-\nu_w^2}.\frac{\Delta s}{B} \qquad (3)$$

where E_w and ν_w are the Young's modulus and Poisson's ratio of the waste.

In order to check this response, a numerical simulation of the plate loading test has been carried out, varying the involved parameters (geometry, strength and stiffness) within wide ranges. A finite element model has been applied, using the program CRISP (Britto and Gunn, 1987). Soil cover has been modelled as elastic-perfect plastic, and waste as linear elastic material.

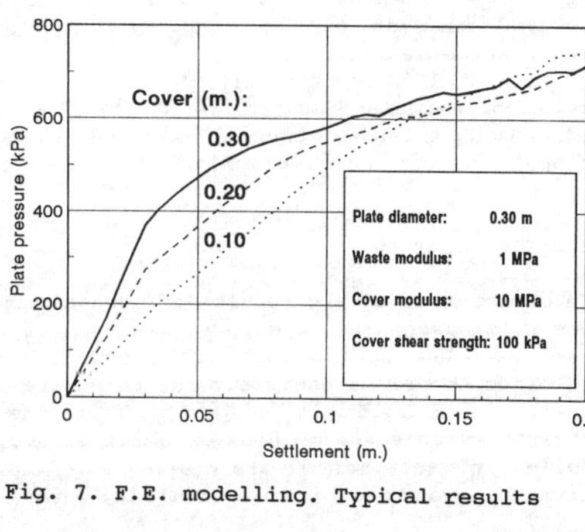

Fig. 7. F.E. modelling. Typical results

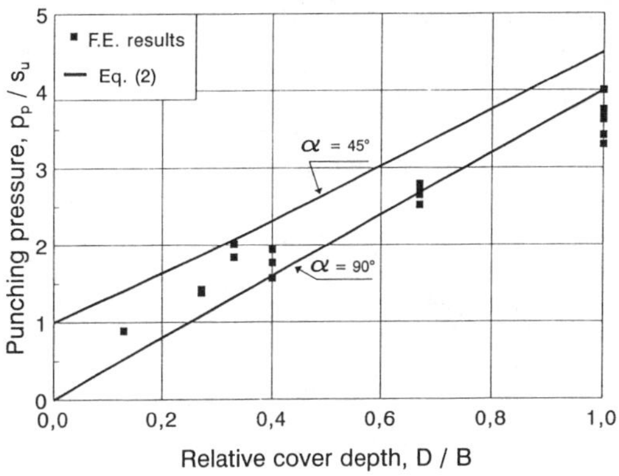

Fig. 8. F.E. modelling. Punching pressure

Fig. 7 shows a typical result. The above described features (initial elastic range, punching failure and post-punching linear response) are evident. In fig. 8 the obtained values of the punching pressure are compared with the predictions of eq. 2. As can be seen, for low D/B ratios, lateral spreading is important, at angles higher than 45°. For D/B greater than 0.5, the punching load is even lower than the value for infinite spreading ($4.s_u.D/B$). This may be due to the vicinity of the conditions of rotational failure of the cover.

The results of the actual tests performed in Meruelo landfill follow the above described behaviour, although a wide scatter is observed (fig. 4). Shear strength of the cover material, determined from laboratory and in situ tests is in the range 100-150 kPa. This implies a punching pressure of 200-600 kPa, and in some of the tests it may not have been reached. The Young's modulus of the waste deduced from the tests without cover is in the range 1.0-2.0 MPa. As predicted (eq. 3), the final slope of all the curves seems to tend to this value. After complete unloading, irrecoverable settlements are very large. This indicates that the virgin compression curve is of elastoplastic, strain-hardening nature, and the above determined moduli are equivalent elastic parameters.

5 SHEAR STRENGTH

In the preceding sections, it has been stated that the observed behaviour of the wastes does not show a definite shear failure. However, in a number of publications relevant shear strength parameters (cohesion and friction angle) are proposed for wastes in landfills. Several reasons can be invoked to explain this apparent contradiction.

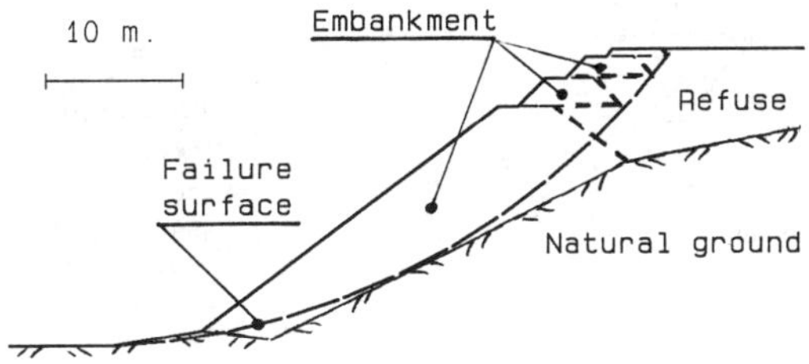

Fig. 9. Slope failure in the Cabezón de la Sal landfill, in Cantabria (Spain)

First, the above described behaviour has been observed in small or medium scale tests (laboratory or in situ), and it is possible that at larger scales a shear failure is better defined.

On the other hand, wastes are placed in situ and covered with earth. This results in a layered material, in which the earth layers can play the role of making the shear response more 'brittle'.

Finally, in many cases, a shear strength is defined on the basis of unacceptable strains, rather than a true failure.

In most of the few reported cases of slope instability in landfills, failure has not taken place in the waste material itself, but instead it has been initiated at the base foundation ground (Dvirnoff and Munion, 1986; Tonteri and Lindroos, 1987), or along the sealing layers, which in the case of synthetic materials can have a very low friction angle (Mitchell et al., 1990). Fig. 9 shows the failure registered in the landfill of Cabezón de la Sal, in Cantabria (Spain). The failure involved the natural foundation ground, and only 10% of the failure surface was developed through the wastes.

Several authors have presented values of shear strength of landfills, obtained from laboratory or in situ tests, or from back-analysis of observed failures. It has become usual to present the results in a plot of cohesion vs. friction angle (c-ϕ diagrams), such as fig. 10.

These diagrams must be looked at with care. The results of laboratory drained triaxial or shear tests provide an envelope in Mohr's plane, and hence the two values of 'c' and 'ϕ' can be determined. However, when in-situ tests or field performance are back-analyzed, there is only one equation (failure condition or value of ultimate load), and hence the result is a relationship linking the two unknown parameters.

In some cases, this line is presented as the final result, and it must be born in mind that only one particular (unknown) point of the line corresponds to the true parameters.

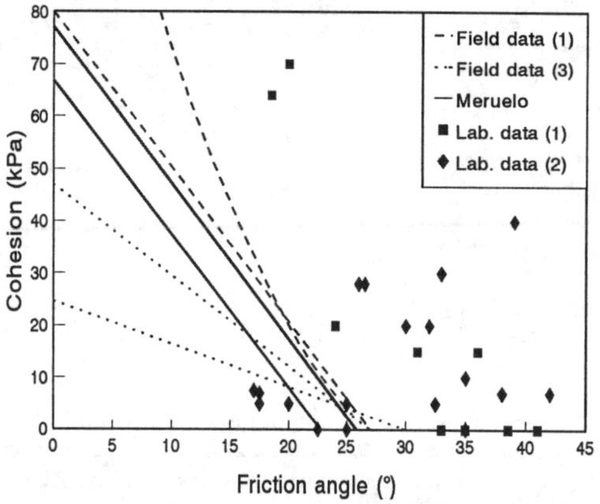

Sources: (1) Singh and Murphy (1990) (2) Jessberger (1990) (3) Oweiss and Khera (1990)

Fig. 10. Shear strength parameters. Summary of field and laboratory data

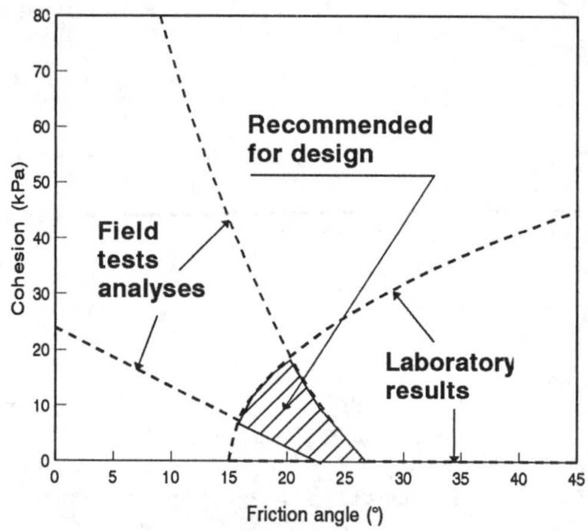

Fig. 11. Interpretation of fig. 10. Recommended design parameters.

In other cases, an additional assumption is made, and individual values of 'c' and 'ϕ' are reported. The most usual additional assumptions are either that of purely cohesive (ϕ=0) or purely frictional (c=0) behaviour. This procedure is valid in normal soils, where there is a basis for assuming one of these two types of behaviour. However, in wastes, there is no firm basis for this decision. As a consequence, for field tests it is better not to do any assumption about c or ϕ being zero. As it was stated in section 4, from the two alternatives, c=0 condition seems more realistic than ϕ=0.

Fig. 10 shows strength parameters measured in wastes. The dots represent laboratory tests (triaxial or shear tests), and they have been taken from the compilations presented by Jessberger (1990) and Singh and Murphy (1990). The lines correspond to back-analysis of field performance, included in the compilation of Singh and Murphy (1990) or presented by Oweiss and Khera (1990).

The lines of Meruelo landfill correspond to a global safety factor between 1.3 and 1.5, which seems a reasonable assumption, given the horizontal movements recorded at the front slope.

As can be seen, there is a wide scatter in the results, but some trends are observed:

(i) The results of laboratory tests indicate that a significant friction is mobilized. Most of the values of the friction angle are in the range 25°-35°. No value lower than 17° has been measured (in the original compilation by Singh and Murphy (1990) a value is reported of ϕ=0, but it corresponds to an unconsolidated-undrained test). The cohesion intercept is very variable, but usually below 30 kPa.

(ii) The lines corresponding to field data diverge in the zone of low friction angle, giving values of s_u from 20 kPa to over 100 kPa. However, all the lines converge for ϕ greater than 15°.

(iii) In this zone (ϕ>15°), the lines of field data fall in the lower bound of the results from laboratory tests.

These trends are shown in fig. 11, where a zone has been marked as recommended for design. This zone corresponds to the intersection between the bands of field and laboratory tests. It roughly coincides with the proposal of Singh and Murphy (1990), taking a lower bound of 15° for the friction angle.

CONCLUSIONS

Design and operation of sanitary landfills require a knowledge of the mechanical behaviour of the material, formed by interbedded wastes and earth cover layers. For this purpose, in situ testing and back analysis of field performance data can be used.

Although the wastes are not soils, some useful results can be obtained if the tests are interpreted in terms of methods and models well established for soils.

Wastes are a very deformable material, which can sustain very large shear strains without reaching a definite failure. So, the strength parameters are usually defined for a limiting strain condition.

The available evidence from laboratory and in situ tests shows that the shear strength of the wastes is better defined as cohesive-frictional (c-ϕ), and interpretation as undrained behaviour (ϕ=0) gives unacceptable high values of the shear strength, s_u.

In the laboratory triaxial or shear tests reported in the literature, a significant friction angle is usually measured, always higher than 15°. For this range, a reasonable agreement is observed between laboratory and field data. This enables the definition of proposed ranges for the strength parameters, which can be taken as preliminary design values.

REFERENCES

Britto, A.M. and Gunn, M.J. 1987. "Critical State Soil Mechanics via Finite Elements". E. Horwood Ltd. U.K.

Dvirnoff, A.H. and Munion, D.W. 1986. "Stability failure of a sanitary landfill". Proc. Int. Symp. Env. Geotechn. (H.Y. Fang, ed.), Bethlehem, Pa, 25-35.

Hinkle, R.D. 1990. "Landfill site reclaimed for commercial use as container storage facility". Geotechnics of Waste Fills-Theory and Practice. (A. Landva and G.D. Knowles, eds.) ASTM, STP 1070,331-344.

Jessberger, H.L. 1990. "Stoffeigenschaften von Abfall im Hinblick auf Standsicherheitsuntersuchungen an Abfalldeponien". Neuzeitliche Deponietechnik. (H.L. Jessberger ed.) Bochum. 171-193.

Landva, A.O. and Clark, J.I. 1990. "Geotechnics of waste fill". Geotechnics of Waste Fills- Theory and Practice. (A. Landva and G.D. Knowles, eds.) ASTM, STP 1070,86-103.

Mitchell, J.K., Seed, R.B. and Seed, H.B., 1990. "Kettleman Hills waste landfill slope failure. Parts I and II."J. Geot. Engg. ASCE 116:4,647-691.

Oweiss, I.S. and Khera, K. 1990. "Geotechnology of waste management". Butterworths. London.

Sánchez-Alciturri, J.M., Palma, J., Cañizal, J. and Sagaseta, C. 1993. "Post-closure bearing capacity of a sanitary landfill". Coll. Intern. Environnement et Géotechnique. Paris. 601-608.

Schmertmann, J.H., 1969. "Dutch friction cone penetrometer exploration of research area at Field 5, Elgin Air Force Base, Florida". Contr. Rep. S-69-4. U.S. Army Engrs. Wat. Exp. Stn. Vicksburg, Miss.

Siegel, R.A., Robertson, R.J. and Anderson, D.G. 1990. "Slope stability ijnvestigations at a landfill in Southern California". Geotechnics of Waste Fills- Theory and Practice. (A. Landva and G.D. Knowles, eds.) ASTM, STP 1070,259-284.

Singh, S. and Murphy, B. 1990. "Evaluation of the stability of sanitary landfills". Geotechnics of Waste Fills- Theory and Practice. (A. Landva and G.D. Knowles, eds.) ASTM, STP 1070,240-258.

Sowers, G.F. 1968. "Foundation problems in sanitary landfills". J. San. Engg. Div. ASCE 94:SA1,103-116.

Tonteri, K.J. and Lindroos, P.T. 1987. "Stability problems of landfill site founded on peat bog". Proc. Xth. ECSMFE. Dublin, Ireland, 1,337-342.

Waste Disposal by Landfill – GREEN'93, Sarsby (ed.) © 1995 Balkema, Rotterdam, ISBN 90 5410 356 6

Three years of deformation monitoring at Meruelo landfill

J.M. Sánchez-Alciturri, J. Palma, C. Sagaseta & J. Cañizal
Division of Ground Engineering, University of Cantabria, Santander, Spain

ABSTRACT: As a part of a research program on geotechnical problems involved in waste disposal, a sanitary landfill has been instrumented in Meruelo, in the region of Cantabria, Spain. It is located in a valley, with an initial retaining dyke whose height is increased by the 'upstream' method up to a final value of 50 m. At present, the records of three years of monitoring are available, and they are analyzed in the paper. They include surface settlements in points with different thickness of waste (up to 16 m.) and horizontal and vertical movements of the front slope for stability assessment.

1 INTRODUCTION

Sanitary landfills are the most widely spread method for waste disposal. Their relatively low cost and their capability of providing sanitary control make them attractive for developing countries. Furthermore, the improvement of placement methods with increasing environmental protection contribute to maintain their role in more developed countries. The new techniques tend to decrease the volume occupied by the wastes, and to use sites (valleys or slopes) which in the past were considered unsuitable for this purpose.

Decomposition of the wastes causes the development of significant deformations of landfills along time. The resulting settlements have been analyzed and modelled by several authors, the work of Sowers (1973) being a pioneer on this topic. These decomposition processes are governed by factors such as the nature of the wastes, placement method, compaction tools and climate.

An accurate prediction of the amount of settlement of a landfill in operation is important to evaluate its true capacity, and hence its active life. Furthermore, after the landfill has been closed and sealed, its bearing capacity as foundation depends on the remaining settlement rate. For these purpose, observation of performance of actual landfills, with measurement of movements, is the only sound basis for a prediction. In the case of landfills with important slopes, measurement of the slope movements can be used as a means to control the stability during and after construction.

The results of a monitoring of this kind are presented and analyzed in this paper. The main aim is to present detailed results of the measurements, with enough data to enable interpretation.

2 SITE DESCRIPTION

The Meruelo sanitary landfill is the main urban waste disposal centre in the region of Cantabria, in Northern Spain. The disposal rate is about 115,000 tons per year, corresponding to a population of 300,000 people. Table 1 shows typical waste composition. Some index properties of the wastes are: initial bulk density 0.2 Mp/m^3; water content, 48%; pH 6; heating value 1400 Kcal/kg.

Table 1. Meruelo landfill waste composition

Components	% by weight
Organic matter	52.0
Paper	15.7
Paper board	6.2
Plastics	8.3
Wood	2.4
Metals	3.2
Glass	4.1
Earth and ashes	1.4
Others	6.7

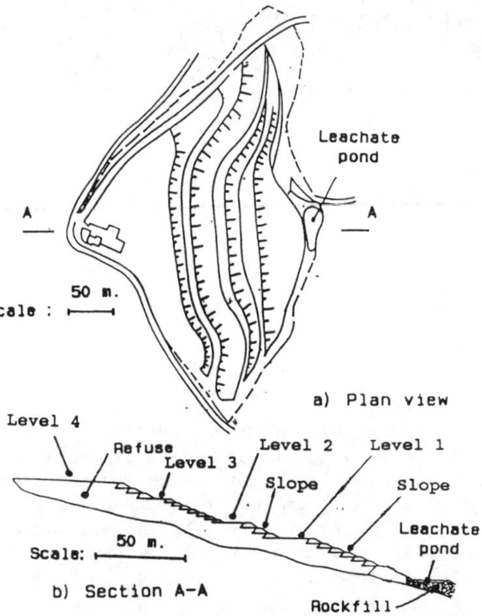

Fig. 1. Meruelo landfill

The landfill is located in a valley. The wastes are contained by an initial embankment, whose height is increased from the initial 15 meters by means of smaller dykes, 2.5 to 1.5 m. high. These dikes are superimposed by the 'upstream' method, partially founded on the previous dike and partly on the wastes. The final height of the landfill (difference in elevation between the toe and the crest) is about 50 meters. Three berms, 20 m. wide, were created at intermediate levels. The mean slope (including berms) is 3(H):1(V) (fig. 1). So, the final configuration includes four slopes and four flat levels, numbered 1 to 4, level 4 being the final surface. The natural ground profile has a relatively steep slope (fig. 1), so the thickness of waste under each level is fairly constant, in the range 14-18 meters for the central part of the landfill.

Table 2. Construction and measurement dates

Level No.	Date of completion	First settlement record (days from completion)
1	March 1989	Jan 1990 (306 days) (*)
2	April 1990	Aug 1990 (125 days)
3	July 1991	Jan 1992 (214 days)
4	Jan 1992	Sep 1992 (261 days)

(*) Measurements were discontinued from May 1990 to September 1992

Landfilling began in December 1988, and it finished in January 1992, and the landfill was sealed. In Table 2 the precise construction periods for each level are given. According to the compaction degree achieved, Meruelo is a medium to high density landfill. The wastes were extended and compacted using a sheep-foot roller specially designed for this purpose. Then, the daily earth cover 20 cm. thick was extended and compacted with a vibratory roller. At the beginning of operation, the thickness of the waste layers was 0.5 m., but it was increased to 2.0 m. from level 2. Based on in situ density measures using a nuclear probe, the following average values have been found:

- landfill mass (wastes+cover): 1.2 T/m^3
- closing dykes: 1.8 T/m^3
- rockfill toe: 2.0 T/m^3
- natural foundation ground: 2.0 T/m^3

The value of 1.2 T/m^3 for the interbedded wastes and daily cover means a unit weight for the compacted wastes of 0.8 to 1.0 T/m^3.

A central rockfill drain was installed to collect the leachate, which is evacuated through the closing embankment to a pond located at the toe.

3 MONITORING SCHEDULE

Landfill deformations were monitored along its construction period and after closure. The aims of the study were to ascertain the design bases for future landfills, and to assess the stability of the front slope, which in this particular case can be a critical point, due to the configuration of the landfill (fig. 1).

A series of in situ tests was also carried out, consisting on static and dynamic penetration tests, and surface plate loading tests. They are interpreted in a companion paper. Here they will be referred to as a guide for interpretation of the measurement of landfill movements.

Two magnitudes are being measured: horizontal movements of the front slope, and settlements of the landfill surface, at the four existing flat levels (fig. 1).

The displacements have been measured by surveying. Closed levelling is used for the settlements and angular intersection for the horizontal displacements. The reference points were installed in the landfill surface and slope, and so they were subjected to many construction hazards. Measurements were taken initially twice per week, and with decreasing frequency after the first three months. At present, a monthly schedule is followed for all the readings.

Fig. 2 shows the location of all the points installed for control of vertical and horizontal displacements.

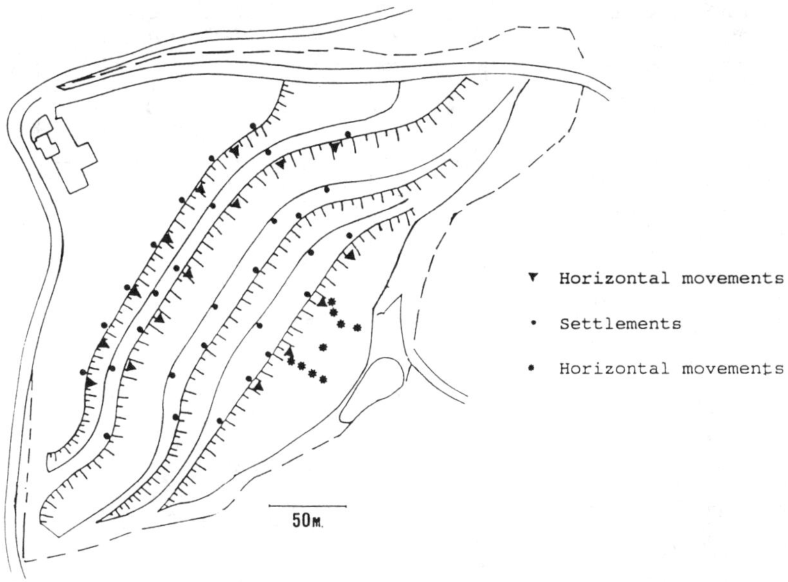

Fig. 2. Monitoring points

3.1 Horizontal movements

Horizontal displacements are measured in magnitude and orientation. Therefore, the components paralel and normal to the slope can be calculated. These latter are the most important ones for the control of stability. They are considered positive when addressed inwards, and negative when they are out of the slope.

Horizontal displacements of the front slope were initiated in the lower slope (below level 1) in January 1990, about ten months after completion of level 1. The measured values represent relative displacements with respect to the first reading (17-01-90). Nine points have been measured until November 1992, although five of them were affected by construction hazards. Then, after closure of the landfill, 16 new points have been installed, 4 of them at the same position of the existing points in the lower slope (below level 1), and the other 12 points in the slope between levels 2 and 3. At the same time, new fixed reference bases were installed for long-term surveying. In this paper, only the first series of readings (lower slope) are presented.

3.2 Settlements

Settlements have been measured in the four horizontal levels. As for the horizontal displacements, the values measured at each point represent relative movement with respect to the first reading,as detailed below (see Table 2).

In level 1, completed in March 1989, 11 points were installed and controlled since January 17, 1990, 306 days after completion of filling. In May 1990, level 1 was used for storage of containers. All the points were destroyed, and no further measurement could be taken until September 1992, when 7 new points were installed. So the results for level 1 are not considered in this paper.

In level 2, 8 points were controlled from August 3, 1990, 125 days after completion of this level. Some of the points were also damaged by the landfill operation, and replaced by new ones. Only three points (26, 27, 28) have continuous readings through all the time.

In level 3, measurements started in January 30, 1992, 214 days after completion. Eight points are being controlled, with no incident.

In the final surface (level 4), 7 points have been installed after top sealing. Measurements have started in September 1992, and they are not presented in this paper.

4 MEASURED MOVEMENTS

4.1 Horizontal displacements

The measured displacements normal to the slope are summarized in Table 3 for the 4 undamaged points. The displacements of two

Table 3. Horizontal displacements normal to the slope below level 1

		Horiz. movements (mm)			
Age (days)	Points	5	6	7	8
323		2	-3	1	5
357		11	10	7	9
385		13	16	9	11
419		16	14	8	11
447		16	19	13	12
476		22	23	14	14
501		23	23	10	18
531		37	24	19	16
561		28	21	17	14
594		35	24	17	16
662		28	31	12	21
693		38	44	22	21
718		36	39	21	28
750		35	39	23	27
849		47	48	27	28
879		47	51	23	26
915		51	53	25	32
947		59	57	28	25
977		56	62	32	30
992		57	66	35	34
1053		60	69	41	41
1083		61	69	41	44
1113		59	69	40	45
1143		61	66	37	41
1179		61	63	38	42
1214		66	65	39	41
1240		68	67	41	48
1273		62	67	42	51
1304		64	64	41	47
1333		69	75	47	54
1353		72	77	54	58

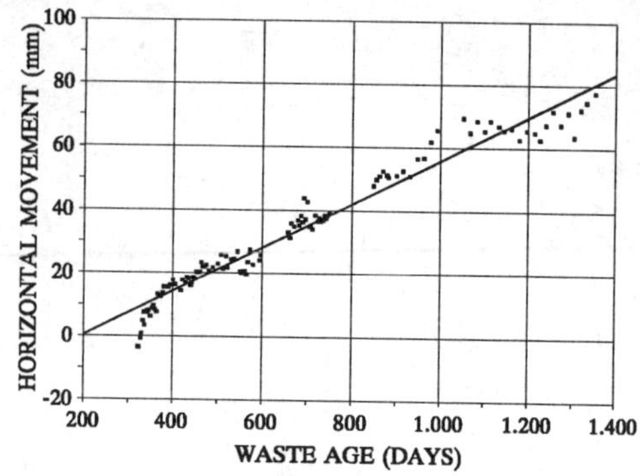

a) Point Nº 6

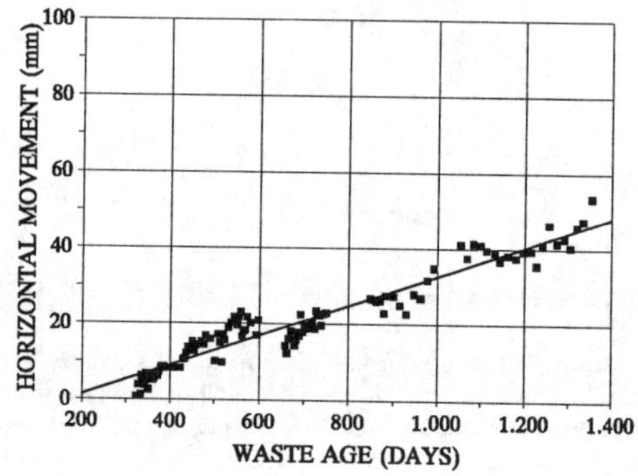

b) Point Nº 7

Fig. 3. Horizontal movements normal to slope

are positive (addressed inwards), indicating that the slope is far from instability, because they are located in the zone of emergency of the potential failure surfaces.

The magnitude of the displacements is small, and they reach values between 50 mm and 100 mm after 1000 days. It must be pointed out that 10 months elapsed between the completion of fill level 1 and the start of the measurements. Hence, the registered displacements are not absolute values, but only relative to the first record.

4.2 Settlements

As described in section 3.2, only in levels 2 and 3 there are continuous readings of settlements for the period 1990-92. A summary is given in Tables 4 and 5. In level 2 there are measurements along three years in three points. In level 3, the period for records is one year, and there are 8 points of control.

In fig. 4, the settlements of level 2 have been plotted. The settlement after three years is of about 450 mm. (range 379-489 mm.), corresponding to 3% of the thickness of the waste (15 m.). The settlement rate decreases along the measuring period from about 50 mm. per month at the first stages to 5 mm. per month at the final part.

In level 3, the measured settlements during the measuring period (one year) range from 111 mm. (point 14) to 413 mm. (point 12). In fig. 5, the settlements of three representative points have been plotted.

5 INTERPRETATION

As stated above, it is not the aim of this paper to give a complete interpretation of the measured movements, which are still in progress. However, some comments can be made.

Table 4. Settlements at level 2

		Settlement* (mm)		
Age (days)	Points	26	27	28
125		0	0	0
158		51	59	39
181		76	104	79
207		153	144	114
227		173	165	121
262		214	207	132
293		244	232	152
323		275	252	168
350		288	259	175
429		352	306	219
461		373	321	233
489		391	337	249
517		399	343	254
550		403	346	261
579		410	353	268
597		417	359	273
655		435	373	288
688		442	377	295
716			380	288
746			394	307
776			396	309
811			400	326
837			407	333
870			419	340
901		466	433	345
933		468	431	353
964		473	443	360
992		488	473	380
1006		484	489	379

* Waste thickness, H:15 m for all points

Table 5. Settlements at level 3

	Points Settlements* (mm)							
Age(days)	7	8	9	10	11	12	13	14
214	0	0	0	0	0	0	0	0
232	23	24	16	21	20	32	17	8
246	39	42	28	37	39	63	34	18
259	51	54	37	49	46	85	47	26
278	70	71	50	64	62	111	55	30
289	80	80	55	72	71	130	64	33
307	103	100	71	90	91	161	84	44
319	111	105	77	98	100	176	91	50
336	103	94	63	87	90	177	80	34
354	124	112	80	106	110	210	103	50
366	132	115	83	112	119	227	110	54
380	145	120	92	122	131	248	123	64
395	156	124	97	132	140	266	131	68
413	170	135	106	139	153	289	142	77
430	170	130	96	130	145	290	131	62
444	186	143	106	141	160	312	144	72
458	197	152	113	148	167	326	150	77
476	209	151	118	152	175	342	155	79
488	211	139	116	150	177	348	154	81
507	222	145	123	158	189	369	165	91
520	227	147	127	163	198	385	173	100
535	235	152	130	166	204	400	181	106
549	241	155	134	171	212	413	187	111

* Waste thickness:
H=14 m for points 8,9,10,11
H=16 m for points 7,12,13,14

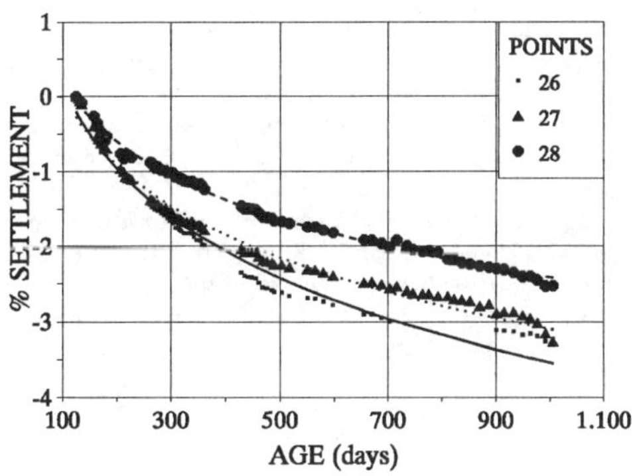

Fig. 4. Settlements at level 2

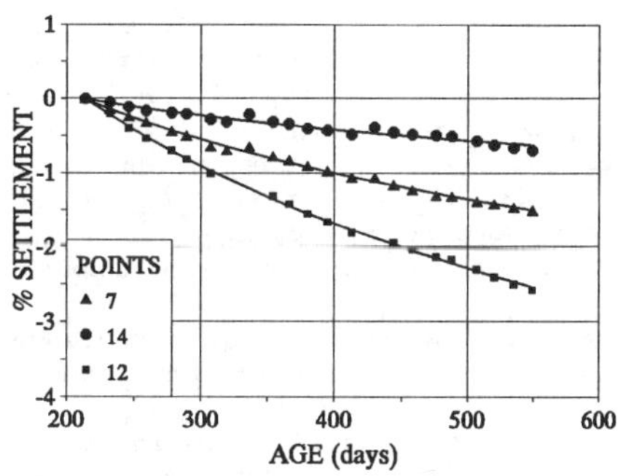

Fig. 5. Settlements at level 3

5.1 Horizontal displacements

As commented above (fig. 3), the horizontal displacements at the toe of the slope are small and addressed inwards, and increase at a almost linear rate with time, of 1.5-2.5 mm/month. However, there is a periodic oscillation about the average trend. At some given moments, the displacements decrease, i.e., the incremental movements are addressed out of slope. These changes seem to be periodical, with a period of about one year, and this could mean that they have a seanonal basis. In any case, the outward incremental movements are small (2-4 mm.) and are recovered after two or three months.

5.2 Settlements

The recorded settlements in level 2 are about 400 mm. after three years. In level 3, after one year, the settlements are 100-250 mm., with the exception of one point, which settles 400 mm.

In figs. 4 and 5 the logarithmic law proposed by Sowers (1973) is fitted to the measured settlement values in both levels, with a good agreement.

It must be remembered that these values do not represent total settlements, but only relative values with respect to the first reading. For this reason, they are more easily interpreted in terms of settlement rates.

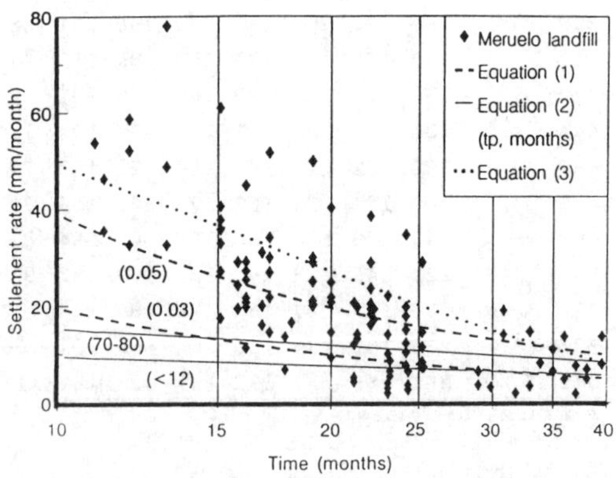

Fig. 6. Rate of settlements versus time

Fig. 6 shows the measured values of the settlement rate (difference of settlement between consecutive readings divided by the time elapsed between them). The origin of time for each level has been taken as the mid time between start and end of the corresponding construction period (the 'medium fill age', defined by Yen and Scanlon, 1975).

As expected, the settlement rate decreases with time. Although there is some scatter, a definite trend is observed.

Several authors have tried to interprete measured settlements or settlement rates with some fitting formulae. These 'models' are not based on theoretical considerations, but they are simple fitting procedures. From settlement formulae, settlement rates can be obtained by differentiation, and conversely, settlement predictions can be obtained by integration of settlement rate formulae. All of the proposed or derived models give laws of settlement rate, "v", decreasing with time.

Sowers (1973) adopted a behaviour similar to secondary consolidation of soils, in which settlement varies linearly with the logarithm of time. This implies an inverse law for the settlement rate:

$$v = 0.434 \cdot H \cdot C'_{\alpha} \cdot \frac{1}{t} \qquad (1)$$

where C'_{α} is a coefficient. Fitting the data gathered by Sowers results in a range between 0.03 and 0.09 for C'_{α} with most data in the lower part of the range.

Equation (1) gives an infinite initial settlement rate, with endless settlement process, resulting in a "final settlement" tending to infinite.

Yen and Scanlon (1975) proposed a logarithmic decrease of the rate:

$$v = v_1 - b.\log t \qquad (2)$$

where v_1 has the meaning of the settlement rate at t=1. This law has the advantage of having two parameters, so in principle it can give a better fit than Sowers' law. In the paper by Yen and Scanlon (1975) a discrepancy is found between both approaches, but it is due to a different definition of the settlement rate (secant or tangent). When the same definition is used in both models a fair agreement is obtained.

Equation (2) gives also an infinite initial rate, but the rate becomes zero for a given time, and hence there is a final settlement value corresponding to that time.

Yen and Scanlon (1975) give values for v_1 and b for different construction times (t_c) and landfill heights (H). For the present case, with height in meters, time in months and settlement rate in mm/month, the following values can be appropriate: for $t_c \leq 12$ and $H < 12$, $v_1 = 15.6$ and $b = 6.3$; for $70 \leq t_c \leq 80$ and $12 < H < 24$, $v_1 = 26.4$ and $b = 11.4$. The final settlement is about 5-6% of the landfill height, and the time required is between 100 and 200 months.

Gandolla et al. (1992) used an exponential law like in a typical decay process:

$$v = r_o \cdot H \cdot e^{-k.t} \qquad (3)$$

where r_o has the meaning of an initial unit rate. The authors give values for r_o and k, obtained from fitting settlements observed in laboratory large scale tests. The measured settlements are several times larger than the foregoing values, probably due to the lack of compaction of the wastes in the tests. However, the formal structure of equation (3) can be maintained, and it will be used in the following together with the other approaches.

Eq. (3) gives finite values for the initial rate and for the final settlement, which is reached in an infinite time.

In figure 6, the results of equations (1), (2) and (3) have been plotted and compared with measured settlement rates. For equations (1) and (2), the parameters given by the respective authors (Sowers, 1973; Yen and Scanlon, 1975) have been used. Equation (3) has been fitted to the measured values, which is for $r_o = 6$ and $k = 0.06$ (H in meters, t in months and v in mm/month).

As can be seen all the equations agree reasonably well with measured settlement rates for times over two years (25-40 months). For earlier stages the measured settlement rates are larger than that predicted by equations (1) and (2).

The best fit given by the negative exponential law (3), is reasonable because eq. (3) has been fitted, instead of using parameter values proposed in the original papers. In order to test the quality of this fitting, the predicted final settlement can be computed. Integration of equation (3) gives:

$$s = \frac{r_o \cdot H}{k} \cdot (1 - e^{-k.t}) \quad (4)$$

The final settlement is, for $t = \infty$:

$$s_f = \frac{r_o \cdot H}{k} \quad (5)$$

For the above values of r_o and k, it results a settlement of 1.5 m, 10% of the landfill height. This is a reasonable figure for an engineered landfill.

6 CONCLUSIONS

Results of three years of movement monitoring in a municipal sanitary landfill are presented, as well as the relevant characteristics of the landfill.

Horizontal movements at the toe of the landfill slope are small and addressed inwards, increasing at an almost linear rate with time, with a periodic oscillation about the average trend, with a period of about one year.

Measured vertical movements reach 3-4% of the landfill height and can be well fitted to a semilogarithmic law, with settlement rates decreasing with time. Three settlement rate prediction laws, two derived from empirical settlement-time formulae and one of empirical nature have been compared. Checking the different laws for the settlement rate, a good fitting is found after two years of construction, with measured rates higher than predictions for earlier stages.

REFERENCES

Gandolla, M., L. Dugnani, G. Bressi & C. Acaia 1992. The determination of subsidence effects at municipal solid waste disposal sites. *Proc. 6th Int. Solid Wastes Congress*. Madrid.

Sowers, G.F. 1973. Settlement of waste disposal fills. *Proc. 8th ICSMFE:* 207-210. Moscow.

Yen, B.C. and B. Scanlon 1975. Sanitary landfill settlement rates. *ASCE Journal of the Geotechnical Engineering Division* 101, GT5: 475-487.

Waste Disposal by Landfill – GREEN'93, Sarsby (ed.) © 1995 Balkema, Rotterdam, ISBN 90 5410 356 6

Resuspension of fine-grained dredged material discharged in open water

B. Schuppener
Federal Waterways Engineering and Research Institute, Berlin Branch, Germany

F. Schwieger
Federal Waterways Engineering and Research Institute, Coastal Branch, Hamburg, Germany

ABSTRACT: When discharged from barges soft fine grained material is eroded by the flow forces and a turpidity cloud of resuspended material develops causing considerable ecological danger. Two types of tests are described by which the resuspending process and the parameters were investigated. In fall tests specimen are dropped in a waterfilled glass-cylinder. In towing tests soil specimen are towed in a tank and the loss of material is recorded. Both types of test show that the undrained shear strength und the flow velocity are the most important parameters. They define 3 separate regions of different erosion properties: a region where erosion is negligible, a second region where erosion is continuous, and a third region where samples are torn apart. An analytical model based on the test results is presented by which the mass can be determined which is resuspended when dredged material is charged in open water.

1 INTRODUCTION

When fine-grained dredged material is discharged from barges at open-water disposal sites, it is eroded by the flow forces and a turbidity cloud of suspended dredged material develops causing considerable ecological dangers. On the one hand, the turbidity cloud has a very great capacity for absorbing oxygen dissolved in water. This oxygen depletion can be a hazard to the flora and fauna of a given stretch of water. On the other hand, it is especially fine-grained dredged material that is frequently loaded with pollutants which can be released in the turbidity cloud and carried away in an uncontrolled manner. Estimating the hazardous potential of a turbidity cloud requires a soil-mechanical and hydraulical model describing the development of the turbidity cloud not only in qualitative but also in quantitative terms.

The present paper will show the methods of testing used to investigate the resuspending process and the findings obtained. Based on these, a soil-mechanical and hydraulical model will be presented which is useful for a quantitative estimation of the amount of dredged material resuspended during dumping at open-water disposal sites.

2 TESTS

The resuspending process is mainly governed by two parameters: the undrained shear strength c_u of the dredged material and the velocity of fall v (Pointdexter-Rollings, 1990 and Raudkivi, 1981). To determine the numerical value of the resuspension rate and the influence of these parameters the following types of tests were performed:

- falling tests using ball shaped samples in glass cylinders filled with water, and

- towing tests in which mud bodies were horizonatlly towed in a test tank.

2.1 *Falling tests*

A glass cylinder consisting of 4 individual segments of a total height of 4 m was set up for carrying out the falling tests. Around the cylinder, a scaffold was erected to form a working platform at about 3.5 m height.

To avoid excessively frequent replacements of water in the glass cylinder, which is very rapidly contaminated during mud tests, initial pre-tests using balls of different densities and different diameters were carried out. Only afterwards did the main tests commence. During these tests, a placement hopper had to be used because the majority of mud samples was too soft for preparing ball-shaped falling bodies. Only samples of greater strength were an exception. However, using the placement hopper produced falling bodies which, by good approximation, were also ball-shaped.

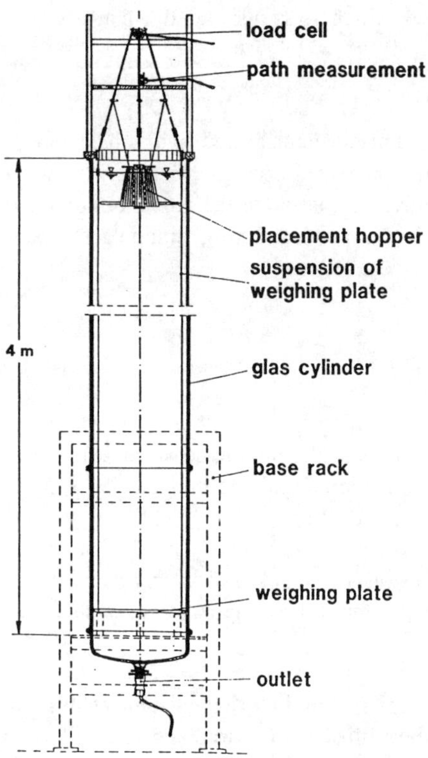

Figure 1: Schematic view of test set-up with sedimentation balance

Figure 1 shows a schematic view of the test set-up. The placement hopper is arranged at the top of the glass cylinder below the water surface. It is suspended on a load cell. The load cell is connected through ropes to the weighing plate located at the cylinder bottom. Therefore, this arrangement is useful not only to measure the starting weight in the placement hopper but also the residual weight of the falling body when hitting the bottom.

The falling process itself is observed by means of a nylon thread provided with an anchor which is embedded in the mud when the placement hopper is filled. This thread is wound around a reel so that a turning angle pick-up can be used to record the falling path as a function of time.

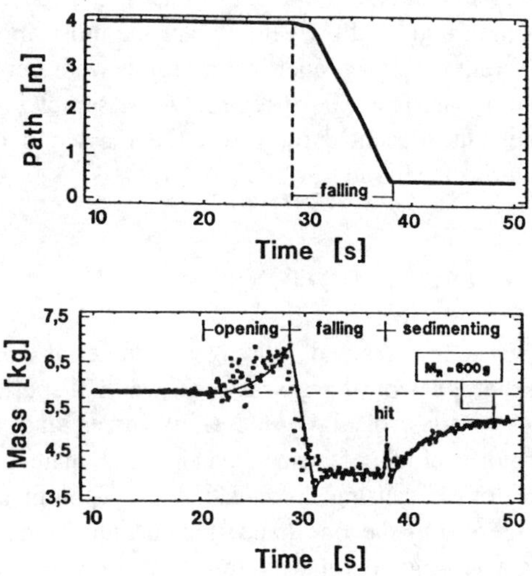

Figure 2: Curves showing examples of the results of a falling test

By way of example, Figure 2 shows the data recorded during such a falling test. The upper graph contains the path-time curve. The falling process starts with a very short acceleration phase. Then the sample sinks at a constant velocity to hit the base plate after about 29 seconds. The velocity of fall is obtained from the slope of this path-time curve.

The lower graph illustrates the weight-time curve. During the first 11 seconds, the data recorder shows the constant total weight of the sedimentation balance with the mud still kept in the placement hopper. This is followed by an unsteady weight increase attributable to disturbances when the placement hopper is manually opened. Then the falling process starts, i.e. the weight is suddenly decreased because only the placement hopper and the base plate are suspended from the load cell now. The residual sample hits the plate during the 38th second. Afterwards, the weight increase during settling of the material eroded in the course of falling can be clearly observed. The weight difference between the starting weight and the weight measured on the base plate after a sedimentation time of 10 sec was defined as the resuspended amount of mud M_R still kept in suspension. In the test shown in Figure 2, this amount weighs about 600 g.

The tests included mud balls of different sizes. As the ball size changes the velocity of fall, it was thus possible to investigate the effect of the velocity of fall.

In summary, the falling tests in the glass cylinders produced the following results:

⇒ The resuspension rate R increases with the velocity of fall v.

⇒ There is a lower limiting velocity v_u below which the erosion forces of the water are so small that the resuspending process is negligible.

⇒ In addition, there is an upper limiting velocity v_o. As soon as this velocity is exceeded, the continuous resuspending process stops and the mud sample is torn into partial lumps.

2.2 *Towing tank tests*

While the velocity in the falling tests was limited by the size of falling bodies, the tests in the towing tank permitted the velocity to be directly selected and adjusted. The tank has a length of 120 m, a width of 2.70 m and a water depth of 1.35 m. Above the tank, a rope-driven towing carriage is mounted which can be continuously accelerated up to a velocity of 4 m/sec.

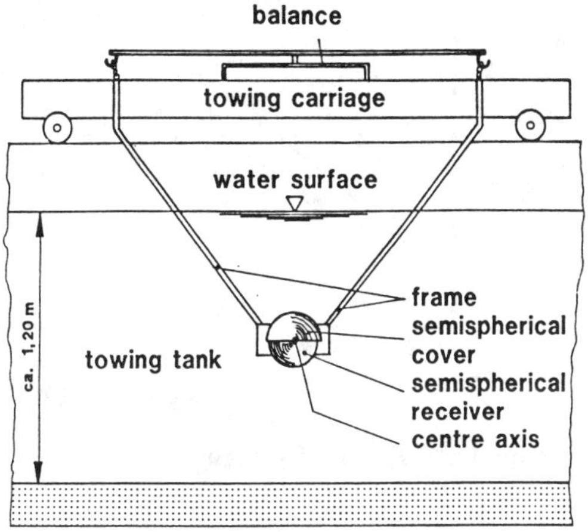

Figure 3: Towing carriage with soil sample

For carrying out the tests, a frame to be immersed in the tank is mounted under the towing carriage (see Figure3). The lower end of this frame holds the container for the ball-shaped mud sample. It consists of two semispherical parts, with the mud being placed on the lower receiver and covered with the upper cover.

The test starts by weighing the water-immersed frame holding the soil sample and recording the starting weight M_A. Then the test run is performed. During the acceleration phase the mud sample is protected against erosion by the semispherical cover. As soon as the measuring run is entered, the cover is turned back exposing the mud sample to the flow. When leaving the measuring run the cover is again turned over the mud sample to prevent further erosion during deceleration. This ensures that the mud ball surface is exposed to erosion only as long as it is moved at a constant velocity in the area of the measuring run. For test evaluation, the sample is weighed again. Then the resuspension rate can be determined from the weight difference to the starting weight, the sample surface and the duration of erosion.

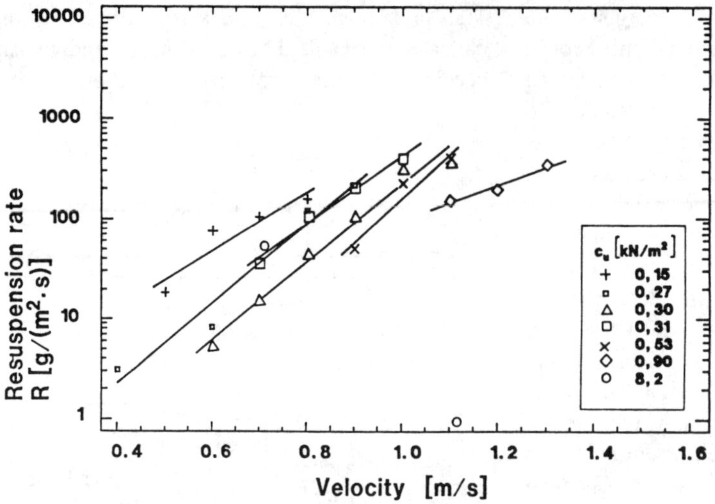

Figure 4: Effect of velocity on resuspension rate R

Figure 4 contains a summary of the results of all series of tests carried out on mud of different undrained shear strengths c_u and at different towing velocities. For samples having the same strength, an additional regression analysis was performed. The following conclusions can be drawn from the graph:

⇒ The towing tests, too, show that there is a lower limiting velocity v_u below which no erosion occurs. The straight lines start when test velocities are reached at which the measured resuspending process is significant for the first time.

⇒ As soon as the lower limiting velocity v_u is exceeded, continuous erosion starts which increases exponentially when the velocity v rises.

⇒ When the upper limiting velocity v_o is exceeded, discontinuous, very strong "erosion" begins similar to the falling tests when the sample is torn apart to form partial bodies. Due to the discontinuity in the erosion behaviour the straight lines in Figure 4 stop at the upper limiting velocity v_o.

In Figure 5, the upper and lower limiting velocities v_o and v_u are shown as a function of the undrained shear strength c_u. The regression curves of the two velocities separate 3 regions of differing erosion properties:

- the lower region where erosion is negligible,
- the middle region where erosion is continuous, and
- the upper region where samples are torn apart.

It will be appreciated that the velocity range where the resuspending process is continuous, which was initially considered the essential range, is relatively small. From the march of the two curves it may also be supposed that above a certain undrained shear strength the two limiting velocities v_u and v_o will have the same value. When a specific strength is exceeded, continuous erosion will cease to occur, and the cohesive soil will be discontinuously torn apart to form lumps. This is also confirmed by observations during pre-tests in the towing tank where at higher velocities and higher shear strengths the mud sample suddenly broke into partial lumps without showing any erosion plume. However, it was impossible to determine the exact point by measurements.

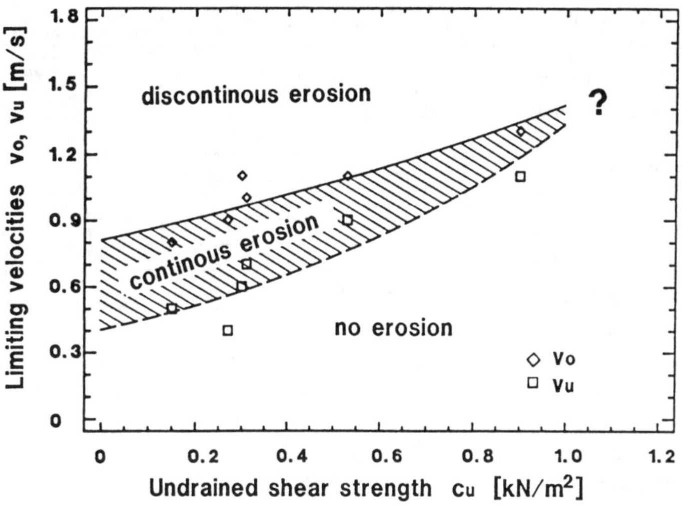

Figure 5: Effect of undrained shear strength c_u on limiting velocities v_u and v_o

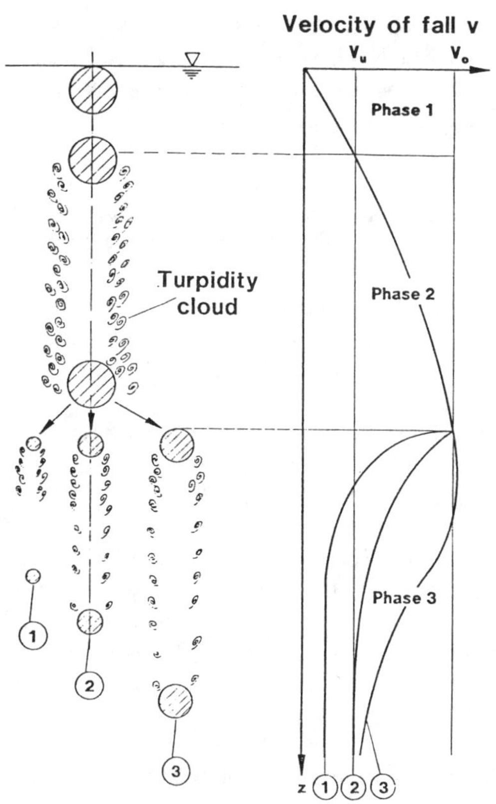

Figure 6: Schematic falling process of mud when discharged at open-water disposal sites

3 MODEL FOR ESTIMATING THE AMOUNTS OF RESUSPENDED FINE GRAINED DREDGED MATERIAL

Following the results of the falling and towing tests, the following phases must be distinguished when fine grained dregded material is discharged at open-water disposal sites (see Figure 6):

⇒ After leaving the barge, the mud mass is accelerated during the first phase. Erosion does not occur before the lower limiting velocity v_u is reached.

⇒ Continuous erosion and, accordingly, the resuspending process starts only when the lower limiting velocity v_u is exceeded in the second phase. During that phase, the mud mass is further accelerated up to the upper limiting velocity v_o.

⇒ As soon as the upper limiting velocity v_o is exceeded, the mud breaks apart to form partial bodies at the beginning of the third phase.

As regards the erosion behaviour in the third phase, the partial bodies can be divided into 3 size classes:

⇒ The small partial bodies are strongly decelerated and fall to the bottom at a velocity which is smaller than the lower limiting velocity v_u. Therefore, they are not eroded, and their velocities of fall remain constant because of their constant weight.

⇒ The medium-sized partial bodies are also decelerated. Their velocities are between the two limiting velocities v_u and v_o so that they are eroded. Due to the weight loss their velocities decrease. When the water is sufficiently deep, continuing erosion will make them so small that their velocities approach the lower limiting velocity v_u at which their weight is no longer reduced so that their velocities of fall remain constant.

⇒ During the falling process, the mud body progressively disintegrates from its surface to the core of the mass. As the core of the mud mass is initially protected against erosion, there may also be partial bodies whose velocities of fall still exceed the upper limiting velocity v_o for a short time. As soon as they are exposed to the free flow, however, they will also disintegrate into partial bodies having a velocity below v_o. Subsequently, they will behave like the medium-sized partial bodies .

Thus, the upper limiting velocity v_o results in an upper limit for the size of partial bodies into which a mud mass will disintegrate. Larger partial bodies will disintegrate immediately because, as a result of their higher weight, they are accelerated to reach velocities exceeding the upper limiting velocity v_o.

The upper limiting velocity v_o and the associated maximum size of partial bodies into which a mud mass disintegrates form the main components of the soil-mechanical and hydraulical model for estimating the amount of resuspended dredged material. Using this model, it will be safe to assume the following simplified and idealised conditions:

⇒ The mud discharged from a hopper barge or dropped from a dredging shovel onto the water surface is homogeneous as regards its undrained shear strength c_u.

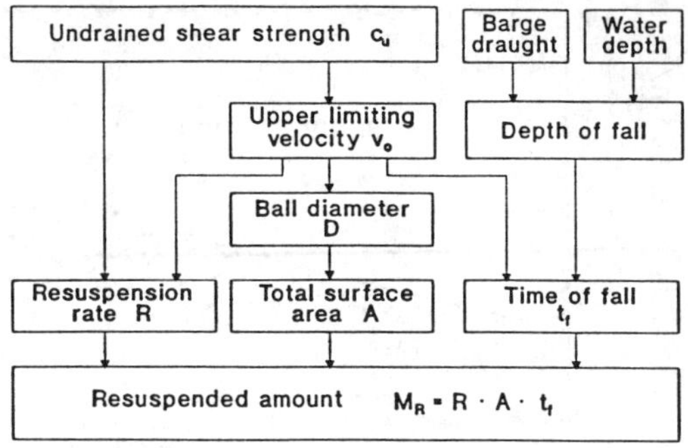

Figure 7: Flow chart of the model for estimating the resuspended amount of mud discharged at open-water disposal sites

⇒ As soon as the falling process begins, the mud body disintegrates to form ball-shaped partial bodies of a size corresponding to the upper limiting velocity v_o.

⇒ Therefore, as early as at the time $t=0$, the partial bodies have assumed a velocity v_o at which they fall to the bottom of the water.

⇒ Other known parameters are the weight and the specific gravity of the mud, the depth of the water, and the draught of the barge.

Based on these conditions, the resuspended amount of mud can be determined in the following steps (see Figure 7).

The starting point of the model of calculation is the undrained shear strength c_u. This value is used in a first step to determine the upper limiting velocity v_o at which the partial bodies formed from the disintegrating mud sink to the bottom of the water. Using the upper limiting velocity v_o and the undrained shear strength c_u, the resuspension rate R can then be determined from the test results.

For determining the total surface A of the partial bodies formed from the mud mass, first the diameter D of the partial body must be determined which sinks to the bottom of the water at a constant rate having the value of the upper limiting velocity v_o. The diameter results from the equilibrium condition that the buoyancy weight of the mud ball equals the flow resistance (Kozeny, 1953, and Bokuniewicz, 1978). Using the diameter D, the weight and the surface area of the ball-shaped mud bodies can now be calculated. Thus, the weight of the barge load under buoyancy and the weight of the partial bodies leads to the number of partial bodies and their total surface area A into which the barge load disintegrates.

This provides the numerical values for the three quantities of resuspension rate R, total surface area of partial bodies A, and time of fall t_f which are required for determining the amount M_R of the resuspended dredged material and, accordingly, for estimating the hazardous potential.

One objection to this model could be that the mud in the barge is by no means homogeneous and, therefore, is torn apart to form much smaller partial bodies having a larger total surface area A compared to the area resulting from the upper limiting velocity v_o which determines the weight. Given a larger surface area, the resuspended mass will also increase. On the other hand, the smaller partial bodies lead to a lower velocity of fall and, therefore, a lower resuspension rate R. When the velocities of fall are lower, however, the time of fall t_f rises. As it is not obvious which of the three influences is dominant, a comparative calculation was carried out. In this calculation, it was assumed that the barge load disintegrates to form partial bodies of half the size obtained at the upper limiting velocity v_o. The comparison shows that when partial bodies are smaller, lower amounts of mud are resuspended. Thus, the assumption that the size of partial bodies is determined by the upper limiting velocity v_o is on the safe side.

When the mud has a relatively high undrained shear strength, however, the model may be too much on the safe side. For such a type of mud, the upper limiting velocity is higher and, therefore, the acceleration phase is longer during which the resuspension rate R is considerably lower than the one assumed in the model.

To account for cases like this, the model can be refined without any problems by approximating the acceleration phase using several velocity stages.

4 CONCLUDING REMARK

The model for estimating the amount of fine-grained dredged material resuspended in a turbidity cloud during placement at open-water disposal sites is essentially based on element tests and their quantitative evaluation. Although the model of the placement process was subjected to a number of simplifications, it still covers the main hydraulic and soil-mechanical parameters. Whether or not they are included with sufficient accuracy can be assessed only when there is also a satisfactory match between the prediction according to the model and tests under natural conditions. The model is still to pass these tests.

5 REFERENCES

Bokuniewicz, H. J.; Gebert, J.; Gordon, R. B.; Higgins, J. L.; Kaminsky, P.; Pilbeam, C. C.; Reed, M.; Tuttle, C. 1978. Field study of the mechanics of the placement of dredged material at open-water disposal sites. *Technical Report D-78-7, WES,* Vicksburg, Mississippi

Kozeny, J. 1953. Hydraulik. *Springer-Verlag,* Vienna

Pointdexter-Rollings, M. E. 1990. Methodology for analysis of subaqueous sediment mounds. *Technical Report D-90-2, WES,* Vicksburg, Mississippi

Raudkivi, A. J. 1981. Grundlagen des Sedimenttransports [Foundations of sediment transport]. *Sonderforschungsbereich 79, "Wasserforschung im Küstenbereich",* Universität Hannover

Waste Disposal by Landfill – GREEN'93, Sarsby (ed.) © 1995 Balkema, Rotterdam, ISBN 90 5410 356 6

Consolidation of fine-grained dredged material in an open-water disposal site

F. Schwieger
Federal Waterways Engineering and Research Institute, Coastal Branch, Hamburg, Germany

B. Schuppener
Federal Waterways Engineering and Research Institute, Berlin Branch, Germany

ABSTRACT: The stress dependent consolidation has been determined for suction dredged and mud pumped sediments which represent the spectrum of typical materials deposited in open-water disposal sites. Laboratory compression tests were performed using a modified test cylinder which allowed for oxygen-free sample testing and extraction of relatively large amounts of pore water.

The results of the consolidation test were applied to an one-dimensional mathematical model to calculate the long term consolidation behavior of an open-water disposal site.

1 INTRODUCTION

The formation of an open-water disposal site for fine-grained dredged material affords a number of steps including dredging, transportation, and deposition of the material (see Knöpp, 1989). The dredging method used is especially important since it influences the pore water content of the material that is to be deposited at the site. At the extreme ends of the spectrum of possible methods we have on the one hand, suction and reclamation dredging which greatly increases the water content of the river or canal sediment, and on the other hand, use of grabs or mud pumps which leave the water content of the sediment almost unchanged. The suction and mud pump methods produce mixtures containing about 20% and 40% solid material respectively. In the course of time, pore water will be pressed out as the material consolidates thereby releasing pollutants. It is for this reason that the consolidation process is in our view an essential criterium for the evaluation of dredged material deposition in an aquatic milieu.

In this paper we present laboratory compression test results with which sediment consolidation as a function of stress could be determined. This stress dependent consolidation function was then applied to a model to estimate the long term settlement and the emergence of pore water.

2 LABORATORY EXPERIMENTS

Laboratory experiments on the consolidation behavior of cohesive dredged material including the extraction of the pore water were carried out at the Federal Waterways Engineering and Research Institute. Dredged material samples were taken from the lock canals of Drakenburg, Dörverden und Langwedel on the Middle-Weser. The dredged material had a regionally typical average grain size distribution and consisted of 10% clay, 75% silt, and 15% sand. With a total of 10% organic substances the sediment had an average grain density of $\rho_s = 2.55$ g/cm³.

Dredged material samples were taken using a piston type Beeker-Sampler with which it was possible to obtain samples not exposed to air. This was important because pore water samples obtained from the laboratory compression tests were to be analysed for pollutants. Oxidation of the dredged material must be avoided so that unstable chemical bonds (e.g., heavy metal bonds) do not form and falsify the subsequent analysis of the extracted pore water. In the laboratory, the samples were handled in a glove box which had an oxygen-free atmosphere after being flushed with argon. This involved removing the sample from the sampler cylinder, homogenizing it, and inserting it in a specially designed compression test cylinder. This cylinder was developed in close cooperation with the Institute for Soil Science at the University of Hamburg and is shown in Fig 1.

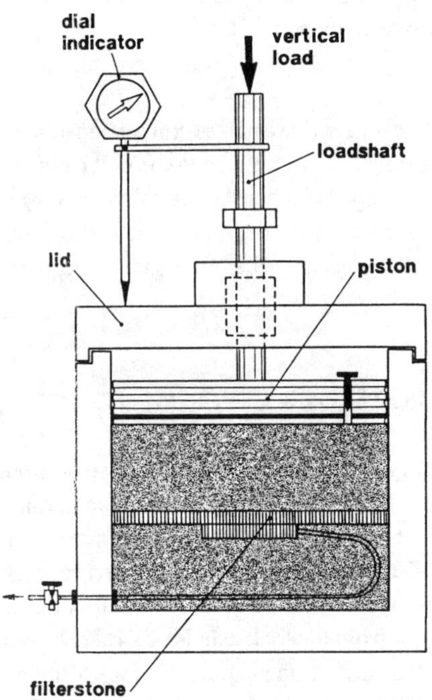

Figure 1: Inert compression test cylinder

It is airtight, constructed of chemically neutral materials, and has a diameter of 200 mm so that during multiple loading steps at least 100 ml pore water are obtained for each step. The middle position of the filter stone for pore water drainage is unique and only a single extraction is necessary. We refer to this cylinder with the name "inert" because it could be loaded under oxygen-free, i.e., inert gas conditions.

A second test cylinder of conventional design but with the same large diameter of the inert cylinder was constructed. A comparison test between this cylinder and a conventional cylinder with a diameter of 70 mm was made to check that the soil mechanical boundary conditions are similar for the two cylinders. Loading of the cylinders was similar to that in an underwater disposal site. The tests showed that the coefficient of compressibility measured for both test cylinders agreed within the range of experimental error.

Results of the chemical analysis of the pressed out pore water performed at the Institute for Soil Science showed that it is absolutely necessary that the pore water be extracted in an oxygen-free manner (use of the inert test cylinder) if it is to be considered to be representative of pore water in an underwater disposal site. This has been verified with experiments using partially oxidized sediment. Further evaluations with respect to ecological aspects are extensively described in the report of the ARGE Schlicktechnik (1992).

2.1 Compression tests for determining the consolidation

The consolidation process for cohesive sediment material may be clarified with the help of a spring-piston model as shown in Fig. 2 (Schultze & Muhs,1967). Two pistons in a cylinder are separated a certain distance and connected by springs. The springs correspond to the grain skeleton and the small openings in the piston model to the sediment layer's porosity. The water between the pistons constitutes the water filled pore volume. If the upper piston experiences a load, the springs will deform, but only to a degree depending on how much water escapes through the holes in the pistons. Initially, the applied load is carried by the water and as the water gradually escapes, the springs take up the load until equilibrium is again established.

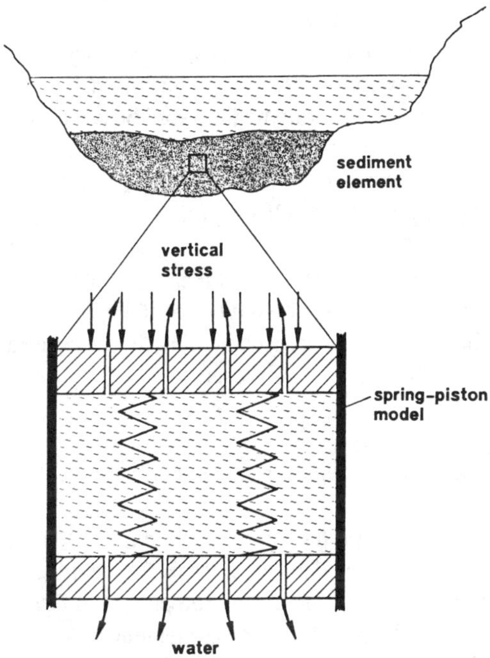

Figure 2. Spring-piston model for consolidation

The compression tests give us stress dependent functions for the coefficients of compressibility and permeability from which the coefficient of consolidation may be calculated.

2.2 Compression test results

The results of all experiments are summarized in Fig. 3. The stress dependent coefficient of compressibility E_s was determined from the stress vs. settlement experiments and corresponds to the deformation modulus for an uni-axial deformation with constricted lateral elongation. The permeability coefficient k was determined after each loading step and its dependency on stress is shown on the log-log plot as a straight regression line. The coefficient of consolidation c_v, which is defined as $c_v = k \cdot E_s/\gamma_w$ (where γ_w = specific gravity of the water), may be derived from the graphs for E_s and k.

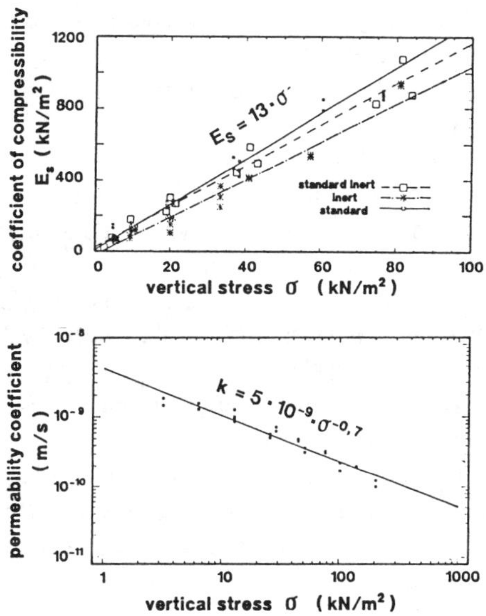

Figure 3. Compression test results.

3 MODEL EXPERIMENTS

There are a number of theories to mathematically describe consolidation in cohesive soil, e.g., the one dimensional theory of Terzaghi (1925)

$$\frac{\delta u}{\delta t} = c_v \cdot \frac{\delta^2 u}{\delta z^2}$$

where u = pore water pressure, t = time, z = depth coordinate, and c_v = coefficient of consolidation. The solution of the partial differential equation allows to calculate the time dependent settlement. This may be done with a mathematical model due to Johnson (1976) provided by the Technical University of Hannover. The model uses the method of finite differences (FDM) and allows an increase in dead weight as occurs when several layers are installed.

3.1 Model conditions

The following assumptions and simplifications were made:

- One-dimensional consolidation theory was assumed.
- The dredged material was considered to be completely water saturated.
- The underground was taken to be incompressible.
- For a hydraulically permeable underground, the hydraulic permeability was assumed to be so high that the pore water could drain unhindered.
- The increase in dead weight resulting from the step-wise introduction of the layers was immediately noticable.
- An increase in the moist specific weight was not taken into account.
- The model only considered the primary settlement, i.e., that due to the pressing out of the pore water, and not secondary settlement due to plastic creep.

The model was run in several versions which used different gravel pit geometries, deposition heights, layer thicknesses, as well as various dredged material consistencies:

- The deposition heights were chosen according to the water depth of the Middle-Weser gravel pits and were here assumed to be 4, 8, and 12 m.
- The layer thicknesses were deduced from the barge sizes and their annual workload. It was assumed that the annually deposited layer was 1, 2, or 4 m thick.
- The calculations were made for both a permeable and an impermeable underground. The underground of the Middle-Weser consists primarily of margalite clay stone layers, so that for the case when the gravel has been removed, the gravel pit bottom was considered to be impermeable with drainage occurring only in the upward direction. For the case of a pit partially filled with gravel, the underground was assumed to be permeable. However, the drainage capability of the lower gravel layer will gradually decrease as the pore volume fills with fine-grained material and the sediment is compressed at the mud/gravel boundary where the stress has a maximum to form a filter skin (self-compaction effect, see Albiker 1986).
- The calculations were performed for two types of sediment:

1. Suction dredged resuspended Material A of high water content that must first resettle before consolidation begins (Kiekbusch 1984).
2. Grab or mud pump dredged Material B with a natural water content that is pre-consolidated through its self weight.

The time dependent settlement for 300 years was calculated for both types of sediment using the same stress dependent functions deduced from the compression tests.

3.2 Model results for settlement

Fig. 4 shows the time dependent settlement results for the 32 cases tested bythe model. Sample results taken from the curves in Fig. 4 for the 30 and 300 year settlement times are shown in Table 1. The settlement ofpermeable and non-permeable underground is more or less identically. The settlement increases with the number of layers and reaches between 15 % and 27 % of the fill height. Cases with several layers show a higher but more slowly attained settlement than cases with fewer layers which show a less but more rapidly attained settlement. The settlement for the same fill heights but different number of layers is after 300 years still not the same. This is due to the incomplete reduction of the pore water overpressure. The final settlement should be the same for cases with the same fill heights but different number of layers. The model does not achieve this status in part because an one-dimensional consolidation theory is used which does not take into account variable densities and layer heights. This effect is stronger for thick layers than for thinner ones. The model cases run with the thinner layers thus give better settlement estimates.

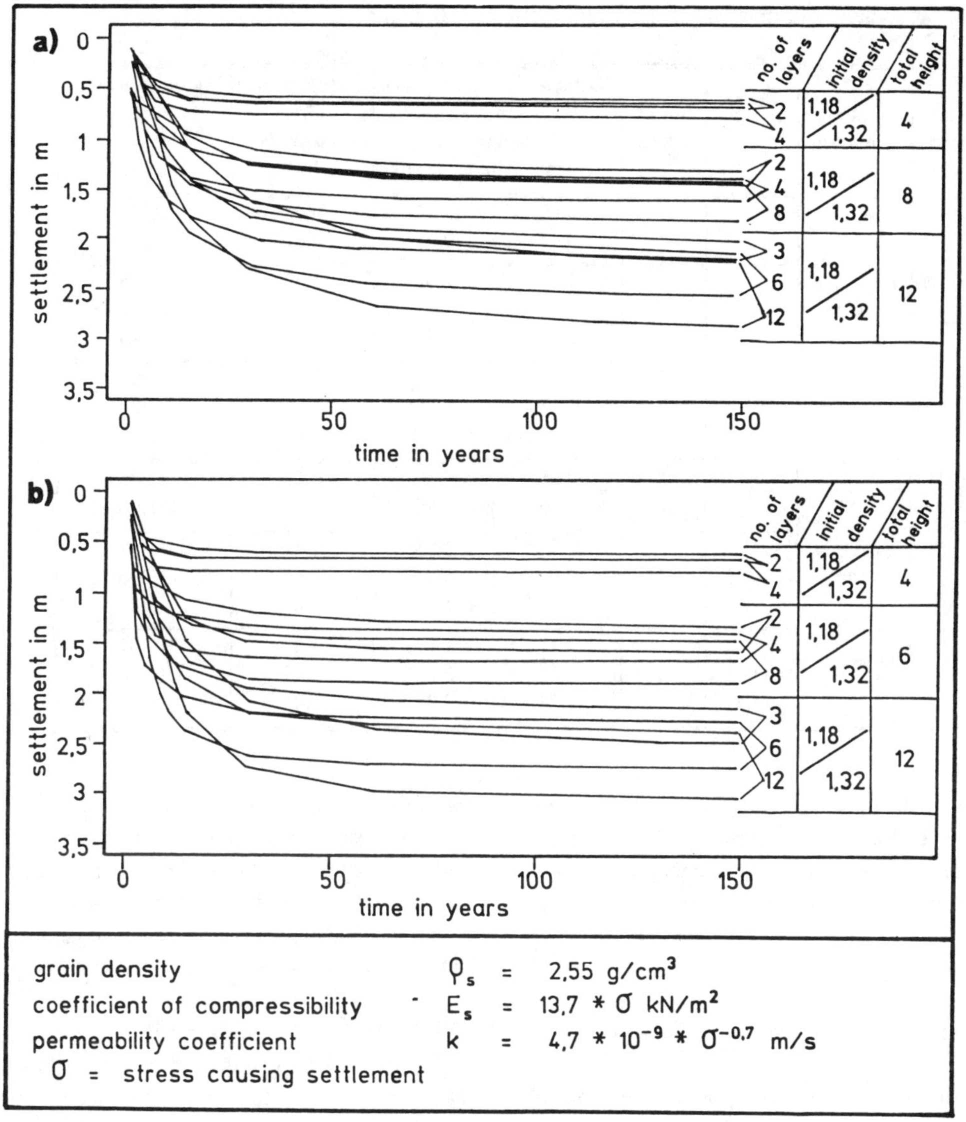

Figure 4. Settlement for a) impermeable and b) permeable underground.

The average material density only increases slightly for all cases after 300 years. The density of the suction dredged Material A remains below that of a natural sediment found in the lock canals. This is caused by the different methods of formation of the sediments. The river or canal sediment is formed grain by grain over a long period, whereas the gravel pits are rapidly filled with compact material.

The settlement of the re-sedimentated Material A is less than that of the pre-sedimentated Material B after 300 years. This is because Material B experiences a higher self weight consolidation stress. The rate of settlement is higher for the cases with a permeable underground, i.e., the seepage length is shorter, than for the impermeable cases. As a result of the self-compaction effect mentioned in Section 3.1

Table 1. Settlement and material density after 30 and 300 years.

Material type	Initial conditions	Fill height [m]	no. of layers	Settlement after 30 years [m] Underground permeable	impermeable	Settlement after 300 years [m]	Average density after 300 years [g/cm³]
A "suction"	ρ = 1,18 [g/cm³] w = 3,0	4	2	0,60	0,58	0,61	1,212
			4	0,66	0,63	0,68	1,217
		8	2	1,19	1,08	1,32	1,215
			4	1,41	1,25	1,46	1,222
			8	1,47	1,24	1,55	1,228
		12	3	1,94	1,71	2,12	1,221
			6	2,17	1,78	2,35	1,229
			12	2,06	1,61	2,46	1,236
B "mud pumped"	ρ = 1,32 [g/cm³] w = 1,5	4	2	0,65	0,65	0,66	1,385
			4	0,76	0,76	0,77	1,398
		8	2	1,31	1,23	1,39	1,389
			4	1,64	1,50	1,66	1,406
			8	1,85	1,64	1,89	1,421
		12	3	2,16	1,97	2,27	1,401
			6	2,63	2,28	2,74	1,420
			12	2,74	2,31	3,04	1,437

and the hydrostatic pressure at the bed acting against the pore water pressure, the drainage conditions at a permeable underground will become progressively worse and eventually be similar to those of an impermeable underground.

3.3 Model results for pressed out pore water

In order to account for possible ecological consequences resulting from drainage of polluted pore water during the consolidation in underwater disposal sites, the settlement model was also used to estimate the time dependent drainage of pore water volume. The results are shown for an impermeable underground in Table 2 as volume per year per disposal site area. The values in the table are averages for the time intervals shown.

Analogous to the settlement results, the table shows that:

- The pressing out of pore water volume is highest at the beginning of a model run and has ceased after about 30 years,
- Pore water production is highest when several thick layers quickly fill the site and more gradual when the site is slowly filled with thinner layers. This is because the maximum consolidation stress is reached earlier the former case.

We have not shown the results for the permeable case. The results indicate high pore water volumes for this case since the self-compaction effect and the external hydrostatic water pressure were not considered by the model. This is unrealistic for a subaquatic disposal site where only small amounts of pore water would flow into the gravel layer.

4 CONCLUSION

The results of the laboratory and model experiments on the settlement and pressed out pore water for an open water disposal site may be summarized as follows:

Table 2. Pore water volumes for an impermeable underground.

Material type	Initial conditions	Fill height [m]	no. of layers	Average upward water flow per time interval in years [l/(yrs · m²)] 0 - 8	9 - 15	16 - 30	31 - 300
A "suction"	ρ = 1,18 [g/cm³] w = 3,0	4	2	56	11	4	0,1
			4	61	14	3	0,1
		8	2	97	20	11	1
			4	105	34	11	0,5
			8	74	70	18	0,8
		12	3	146	39	18	1,3
			6	120	62	26	1,5
			12	74	70	35	2,6
B "mud pumped"	ρ = 1,32 [g/cm³] w = 1,5	4	2	69	9	2,6	--
			4	82	11	2,2	--
		8	2	118	20	9	0,6
			4	145	30	9	0,3
			8	126	56	16	0,7
		12	3	189	37	14	1
			6	183	68	23	1
			8	126	109	36	2

- The experiments were performed for materials obtained using two extreme types of dredging methods. We therefore assume that our results are pertinent for actual dredging and deposition measures that lie within the bandwith of these extremes.

- After underwater deposition, the initial volume reduction resulting from settlement is very small so that it is unfeasible to exploit this volume for other purposes. Major settlement through self-weight and buoyancy has occurred after 30 years. Additional loading with gravel to accelerate the settlement is not reasonable since this would produce an almost impermeable boundary between the load and the upper surface of the sediment. If a capping layer is necessary for ecological reasons, it must be remembered that the firmness at the sediment boundary remains very weak and is approximately that of the sediment itself. The increase with depth of the firmness is negligible.

- The pore water will primarily be emitted into the gravel pit. Even when the underground is permeable, most of the pore water will flow upwards from the underground because of the self-compaction effect and reduction of permeability due to the higher stress. Thus only a short term flow of pore water into the groundwater would occur.

Through the use of the presented model, we are able to make reasonable estimates of the consolidation behavior of open-water disposal sites. Nevertheless, it would be very desirable to do in situ measurements during the dredging and deposition phases in order to verify the theoretical boundary conditions incorporated in the model. Establishing a control open-water disposal site for long term measurements would be very useful and is planned for this year.

REFERENCES

Albiker, B. 1986. Die Selbstdichtung von Schlammlagerfeldern. In Wasser + Boden 11, p. 558-563.

ARGE Schlicktechnik, 1992. Ökologische Auswirkungen bei der Verbringung schadstoffbelasteten Baggergutes aus der Mittelweser in aufgelassenen Kiesgruben, Hamburg.

Johnson, L.D. 1976. Mathematical model for predicting the consolidation of dredged material in confined disposal areas, WES Vicksburg, Mississippi, USA.

Kiekbusch, M. 1984. Laborversuche zur Simulation der Schlickentwässerung im Spülfeld. Fachseminar Baggergut, Strom- und Hafenbau, Hamburg.

Knöpp, H. 1989. Flußsedimente und Hafenbaggerschlämme sowie Beseitigung und/oder Verwertung von Fluß- und Hafenschlämmen, Müllhandbuch 3009 und 3475, Lfg. 4/89, Berlin, E. Schmidt Verlag.

Schultze, E. & H.Muhs 1967. Bodenuntersuchungen für Ingenieurbauten, 2. Auflage, Berlin/Heidelberg/New York, Springer Verlag.

Terzaghi, K. 1925. Erdbaumechanik auf bodenphysikalischer Grundlage, Leipzig/Wien, Franz Deuticke.

Waste Disposal by Landfill – GREEN'93, Sarsby (ed.) © 1995 Balkema, Rotterdam, ISBN 90 5410 356 6

General report

W.Wolski
Agricultural University of Warsaw, Poland

W Wolski : The topic of this session is the "Engineering behaviour of waste". Only three papers (the one by Jessberger and Kochel and the two written by Sanchez-Alciturri, Palma, Sagaseta and Canizal) deal exactly with the session topic. The other papers deal with areas related to the engineering behaviour of waste. The articles by Park, Kim and Edil, and Lyons, Monticello and Hitchens are concerned with wastes used for special purposes, whilst construction aspects of waste disposal is the theme of submissions from Pasqualini, Sani and Roccato, and Forster. In general the papers submitted to this session give the impression that among geotechnical engineers there is no clear definition of waste. Therefore the first (and very important) question to be considered is whether it is possible, or not, to define waste from the geotechnical point of view.

Forster describes a project involving comprehensive field tests on the effectiveness of clay cover used in different ways. Four clay caps were tested; two layers were recompacted to achieve a landfill specification in terms of an appropriate coefficient of permeability, two others were placed as laid without any compaction. Each layer of the covers had a different inclination, i.e. 1 to 10 and 1 to 20. Underneath were located lysimeters and very careful measurements were made of water percolation.

The project is ongoing and the monitoring programme has only been in operation for one year. Hence the variations in flow rates above and below the sealing layer resulting from differences in slope and placement technique have not yet been fully assessed. However it is anticipated that there will be a high dependence of cover effectiveness on the drainage layer above the clay seal. This is very important and to be expected, but it would also appear important to ensure a sufficient thickness of topsoil to prevent desiccation or freezing (depending upon the climatic conditions applied to the sealing layer).

The Paper by **Schwieger and Schuppener** contains a one-dimensional model for dredged material deposited in open water. Based on laboratory experiments the model enabled calculation of the long-term consolidation behaviour of newly-sedimented materials. The laboratory tests were conducted in a specially designed compression cylinder.

Settlements were predicted for periods of 150 years and longer. Settlement analyses indicated negligible influence of substrate permeability. Settlements increased with the number of layers in which the sediment was deposited - and the consolidation process commenced later and became slower as the number of layers increased.

Even though non-linear parameters were used in the model there are simplifying assumptions which strongly affect the predicted consolidation behaviour. It is therefore important to conduct in-situ field monitoring for back analysis and confirmation of the proposed model.

It is essential to underline the environmental aspects of this paper /study even though it is not intimately concerned with refuse placed in landfill. The proposed model used to predict the time dependent pore volume of water escaping from the sediment is important for ecological consequences when the dredged sediment is polluted.

Santamaria, Moreno and Santiagos present an approach to stability evaluation of a material susceptible to liquefaction in seismic conditions, i.e. uranium tailings deposits completed 10 years ago. The trailings do not represent a danger because they originate from uranium but because of their liquefaction potential.

There are different types of materials in the deposits, i.e. sandy materials, slimes, silty clay, etc.. The paper shows the application of a computer program published by Idris (1973) adjusted to suit this multi-layer situation. The properties of the four tailings materials, i.e. sands, silts, clayey silts and clay were determined by the resonant column method in the laboratory and also using dynamic triaxial apparatus. Very importantly in-situ tests were also performed - both down-hole and cross-hole tests. Hence the Authors obtained the shear wave velocity in the range 140 - 190 m/sec - this seems rather low for a 10-year old deposit, according to recent measurements which I have made on newly deposited slimes in a polish tailings dam.

There was good correspondence between in-situ and laboratory tests and the Authors concluded that the clay, of course, is the least sensitivity to liquefaction and the sands are the most. It would be helpful if the Authors had used the additional classification of liquefiable soils introduced by Carrier, i.e. the SFR index (sands-to-fines ratio).

The contribution by **Schuppener and Schwieger** is essentially a continuation of the preceding article by Schwieger and Schuppener and describes experiments which result in the model which is used to estimate the amount of resuspended mud discharged from barges in an open-water disposal site. Clouds of such fines in the water are very dangerous ecologically if one considers absorption of oxygen in salty water and also if one appreciates that there are pollutants in this discharged material.

The Authors' approach is based on the assumption that the resuspension process is mainly governed by two parameters, i.e. the undrained shear strength (Cu) and the falling velocity of the particles. Large scale laboratory tests proved these assumptions. Unfortunately the Authors do not define the material used - it is important to take into account the expansive and dispersive properties of the discharged material. It is considered that the dispersion potential will be critical to the evaluation of this process rather than Cu or falling velocity.

Lyons, Monticello and Hitchens give a good example of how geotechnics can help environmental protection and at the same time provide benefits in construction problems, i.e. the use of scrap tyres (of which there is a large surplus) as a lightweight aggregate in soils.

The paper by **Park, Kim and Edil** provides a good theme for discussion, as to whether scrap tyres could replace granular material in a primary leachate collection system for the removal of potentially toxic organic compounds from landfills. In a similar way to the work of Lyons et al this process reduces the volume of scrap tyres (and is thus environmentally beneficial) and at the same time uses them as a construction element in landfill engineering.

The article by **Sanchez-Alciturri, Palma, Sagaseta and Canizal** on the monitoring of the Meruelo landfill is a model example of how in-situ tests of landfill have to be interpreted to evaluate the engineering behaviour of waste. This paper is a very good example of how to approach engineering problems and hence derive appropriate solutions. Similarly the contributions of **Pasqualini, Sani and Roccato, Jessberger and Kochel, and Sanchez-Alcituri, Palma, Sagaseta and Canizal** (on mechanical properties of waste) are such excellent papers that they deserve to be read and they will not be paraphrased here.

A review of the session papers immediately throws up the following points :

i) As stated previously, how can waste be defined (and its properties quantified) in a geotechnical sense.

ii) It is feasible to analyse the behaviour of "soil-like-waste" (in the terminology of H Jessberger) using routine geotechnical procedures/methods such as would be used for peat soils. However for the "non-soil-like-waste" this process cannot be applied and we need to use observational methods. The major problem is to define the boundary between the two states, i.e. when we can use geotechnical methods and when we cannot. Moreover there is a kind of transition zone between "soil-like-waste" and "non-soil-like-waste". The question is whether it is possible to define conditions which enable us to distinguish between these two types of waste and minimise the transition state which we cannot analyse. If this can be done we can then select appropriate methods and tests.

iii) Definition of the overall mechanical behaviour of waste material is a major problem - in a similar way to the difficulty of defining the mass properties of rockfill in large dams. It is significant that none of the papers of the session contained information on the evaluation of the mass permeability of refuse.

Forster
Water balance evaluation of different multi-layered landfill final cover systems.

Jessberger and Kochel
Determination and assessment of the mechanical properties of waste materials.

Lyons, Monticello and Hitchens
Recovery of scrap rubber tyres from landfill for construction uses.

Park, Kim and Edil
Sorption capacity of shredded waste tyres.

Pasqualini, Sani and Roccato
Factors influencing geomembrane interface friction.

Sanchez-Alciturri, Palma, Sagaseta and Canizal
Mechanical properties of wastes in a sanitary landfill.

Sanchez-Alciturri, Palma, Sagaseta and Canizal
Three years of deformation monitoring at Meruelo landfill.

Santamaria, Moreno and Santiagos
Dynamic analysis of the behaviour of hydraulically-deposited uranium mill tailings under seismic action.

Schuppener and Schwieger
Resuspension of fine-grained dredged material discharged in open-water.

Schwieger and Schuppener
Consolidation of fine-grained dredged material in an open-water disposal system.

Discussion

Commenting on the data presented by Pasqualini and Roccato, **T Zimmie** pointed out that for seismic design dynamic friction angles should be used as opposed to static ones. Even though the friction angles are very similar there are a variety of different interfaces (for deformed membranes the normal force will probably affect the friction angle). For regions of seismic activity (California, Greece etc) the value of dynamic interface friction is very important for landfill design.

B Vonk commended Schwieger and Schuppener on their paper concerning consolidation of soft clays and contributed his experiences with the consolidation of polluted clay soil dredged from the bottom of a lake. The dredging and transport of the soil is performed hydraulically so that the fluid has a density of about 1.2 Mg/m^3. The fluid is pumped into a landfill site and left so that the solids can settle out and consolidate. Because of the low fluid density consolidation occurs under very low effective stresses. For this stress region B Vonk's experience was that the relation between coefficient of compressibility and vertical stress appears to be non-linear as opposed to the linear model used by Schwieger and Schuppener. This non-linearity at low stress levels cannot be detected using oedometer tests because these have inherent inaccuracies at stress levels less than 50 kpa approximately.

C Sagaseta offered a solution to a discussion (relating to predictions of landfill settlement with time) that started about 18 years ago. In 1973 Sowers proposed a predictive model similar to that for secondary consolidations of soils, ie settlement being dependent upon logarithm of time (Proc 8th ICSMFE, "Settlement of Waste Disposal Fills"). Two years later Yen and Scanlon (ASCE Journal of Geotechnical Engineering Division, "Sanitary Landfill Settlement Rates") presented data in terms of rate of settlement, ie the increase in settlement in one month, as a function of the age of the landfill. When they compared their results with the predictions from Sower's model that found that the latter predictions were 3 or 4 times greater than their measurements. Attempts have been made to explain the discrepancy in terms of difference of climatology, humidity etc but C Sagaseta proposed that the apparent discrepancies results from a simple algebraic error. Yen and Scanlon defined a settlement rate which was the derivative of settlement with respect to time, whereas from the Sowers model the settlement rate was equal to settlement divided by time, ie the mean settlement rate during the life of the landfill. Thus two different definitions of settlement rate, ie either tangent or secant, were used and for the ages of the landfills analysed the ratio of these two settlement rates are between 3 and 5. So if a consistent definition of settlement rate is applied the agreement between the data of Sowers and those of Yen and Scanlon is as good as can be expected in environmental engineering.

R Jones asked E Pasqualini for further information on the optimum size of direct shear box for determining soil geomembrane shearing characteristics, also the selection of a suitable shearing rate to ensure drained or undrained conditions at the soil membrane interface.

E Pasqualini replied that the selection of shearing conditions is made on the usual soil mechanics principles for stability, ie immediate stage (undrained), long-term behaviour (drained). The most important thing is not to evaluate the contact shearing resistance for the soil at the compaction water content because this will change at the interface. For shear testing the sample should be as large as possible using either laboratory equipment manufactured specially for the purpose or field trails - normal laboratory equipment is usually inadequate.

A Obinwa queried whether it was safe to use clay as a lining material bearing in mind that some laboratory studies had demonstrated that the clay properties could be modified by leachate penetration.

T Zimmie replied that the idea behind composite liners (composed of clay and geomembrane layers) was to overcome this potential problem by having two barrier materials which are not susceptible by attack by the same substances. Furthermore the permeability of compacted clays is not significantly affected by leachate permeation because this liquid does not contain high concentrations of particular chemicals. However desiccation cracking may occur.

D Daniel confirmed that in his experience leachate from municipal solid waste did not increase the permeability of compacted clay - in fact it tended to decrease it, probably due to biological activity plugging the clay pores. The situation may be different with Bentonite layers particularly where they are used in specialised situations, eg petroleum hydrocarbons are present.

P McKendry enquired whether it was possible to make reliable estimates of the landfill settlement that will occur over a selected time period due to either surcharging of the waste or decomposition of the refuse.

D Daniel stated that he expected municipal solid waste to settle by about 10 - 15% of its original thickness over a period of several years. Settlements of upto 30% have been recorded for older sites but now the refuse is subjected to heavy compaction during placement (to increase the intake of the site). However more important than the amount of settlement is the question of how much the capping layer will distort since this distortion and bending leads to poor performance of the cap.

J Dunn supported the general comments by D Daniel but emphasised that every site is different in terms of waste composition, wetness of the waste, site construction etc. Furthermore the nature of refuse has changed significantly with time and wastes are becoming more heavy. Significant differential settlement may occur at the edge of the site where the depth of waste beneath the cap changes rapidly.

In reply to a question by **L Martineghi, H Y Fang** stated his belief that the basic behaviour of waste could be represented by a modified form of Terzaghi consolidation theory.

D Grossman cautioned that slope stability analyses associated with landfill, eg failure on an inclined refuse liner interface, may be appropriate for the first few years of life of the interface but in the long-term decomposition will change the shear strength characteristics of the interface significantly. The shearing resistance may well decrease so that the factor of safety against sliding is reduced. A further major problem is that the basal drainage system is likely to stop working after a time and there will be an increased fluid level within the landfill, particularly in areas of high precipitation, and this will also reduce the factor of safety.

H Jessberger concurred that the effect of waste degradation on its shear strength properties was very important and illustrated the point by referring to the behaviour of a sheet pile wall driven through old waste.

The waste was in a temporary valley and the sheet piling was installed to give access to the leachate drain in the bottom of the valley. For the design of the wall the waste was assumed to have a friction angle of 20° and a small cohesion. During excavation of the waste the deformation of the sheet-pile wall was monitored and was found to be much higher than excavated. The excessive displacement of the wall was considered to be due to movement occurring along a deep slip surface (downhill along the valley sides) within the highly decomposed waste at depth. This lower waste was also very wet due to fracture of a leachate drainage pipe.

C Sagaseta pointed out that many geotechnical engineering works rely on effective drainage and that it is common practice to monitor the performance of these systems. Hence it should be normal practice to monitor the performance of all components of a landfill site if the site is to function properly in the long term.

M Manassero believed that it was important to be able to estimate the final settlement of a waste deposit and how long this would take to occur. He outlined a simple curve fitting method for doing this with the relationship between settlement (s) and time (t) being taken as $s = C - be^{-at}$ (this form of equation also describes biological degradation and some chemical reactions). By taking field measurements of settlement at different times and fitting these to the preceding equation the data may be extrapolated to predict the final settlement.

4 Pollutant control

Waste Disposal by Landfill – GREEN'93, Sarsby (ed.) © *1995 Balkema, Rotterdam, ISBN 90 5410 356 6*

Rubber-kaolinite and rubber-bentonite liners

Gokhan Baykal, Aydin Kavak & Mehmet Alpatli
Bogazici University, Istanbul, Turkey

ABSTRACT: Waste tire chips supplied from tire retread industry were mixed with kaolinite and bentonite, and free swell, swelling pressure and rigid wall, falling head hydraulic conductivity tests were conducted using gasoline as the permeant. The basic idea behind this application is the utilization of the physical property of rubber; expansion upon contact with gasoline, which compensates for the decreased diffused double layer thickness of clay and keeps the hydraulic conductivity of the composite within the permissible limits. The lower hydraulic conductivity values obtained with the rubber addition is due to the decrease in the size of the interclod pore space with the pressure exerted by the swelling of rubber. Workability and ease of compaction of this composite liner material makes it a good candidate for field application. The composite liner can retain large volumes of permeant.

1 A NEW COMPOSITE LINER SYSTEM

Environmental pollution due to petroleum based contaminants is a major hazard. Petroleum refineries produce huge amounts of wastes which have to be disposed safely. Also underground petroleum storage tanks may leak and pollute the ground water. Considering the fact that there are millions of underground storage tanks in the world, the potential for pollution is enormous. Thousands of kilometers of petroleum pipelines have been constructed which might leak frequently. There is a need for a safe and economic lining system which will capture the leaking or waste petroleum and will not permit petroleum to pollute the environment. The use of conventional clay liners is not a safe practice against petroleum based permeants. Due to the low dielectric constant of the petroleum products, the thickness of the diffused double layer of clays decreases causing an increase in hydraulic conductivity (Mitchell 1987). The thicker the diffused double layer thickness of clay, the higher will be the effect of petroleum based permeants on the hydraulic conductivity. At low concentrations and low hydraulic gradients the effect of permeants may not be very critical, however in the case of gasoline, the permeant is concentrated and its effect on hydraulic conductivity becomes critical. Permeation with pure hydrocarbons can increase the hydraulic conductivity upto three orders of magnitude (Anderson et al 1985) In practice, bentonite or other highly plastic clays are used as liner materials. This makes them more vulnerable when they are used against petroleum contamination.

To overcome this problem, waste rubber tire chips were mixed with kaolinite and bentonite and were compacted. When gasoline is permeated through this composite liner, rubber expands upon contact with gasoline. This physical property of rubber is the main idea behind this new composite liner system. The expansion of rubber upon contact with gasoline exerts a pressure on the interclod pore walls causing a decrease in their size causing a reduction in hydraulic conductivity. When the optimum amount of rubber is added into the soil, the hydraulic conductivity increase, due to shrinking diffused double layer can be compensated. The free swell and swelling pressure of the composite upon gasoline permeation is an important design consideration. If very small magnitudes of swelling pressure is obtained with low volume percentages of rubber, the system will not work effectively. On the other hand if the swelling pressure is too high, then the integrity of the liner will be a problem. The compressive stress acting on the liner due to the overburden and storage facilities should not be exceeded. This study was initiated to give an idea about considerations to be taken during design. The actual performance in the field will be further studied using double ring infiltrometers.

2 METHODOLOGY:

The objective of the testing scheme was to observe the swelling of rubber upon contact with gasoline. Free swell tests and swelling pressure tests were conducted using both consolidometer and CBR apparatus. Free swell of only rubber was also determined by placing rubber and gasoline in a burette and measuring the volume change. Swelling pressure measurements will give the magnitude of the pressures applied by the expansion of rubber. If swelling pressure exceeds the compressive stress acting on the liner, the liner will expand and its stability will be lost. To observe all aspects of the swelling effects 4, 7 and 10 percent by weight of rubber kaolinite and rubber bentonite specimens were compacted and swelling tests were conducted. As a control mixture rubber sand mixtures with the same weight percentages of rubber were tested.

It was extremely hard to conduct rigid wall falling head hydraulic conductivity tests on rubber added bentonite due to low hydraulic conductivity of this composite and due to low degree of saturation. Compacted specimens were tested upto one month using water as the permeant. On the same specimen gasoline was introduced as the permeant and the test was continued. Although no major increase was observed from the test data, when the sample was physically examined, it was seen that the top three centimeter section became completely nonplastic. It was obvious that if more time was spent for testing, the nonplastic section of the specimen would have progressed and finally all sample would have become nonplastic causing a major increase in hydraulic conductivity. Considering the immiscible behaviour of the gasoline, it was decided to compact the specimens with gasoline and then conduct the hydraulic conductivity tests. This considerably shortened the testing time and would represent the worst case where the whole thickness of the liner became nonplastic. At the same time it will show whether the swelling of rubber will work or not. However it must be kept in mind that when gasoline is mixed with bentonite or kaolinite, it makes the clays nonplastic; this fact should be considered in the evaluation of hydraulic conductivity test results.

The physical and engineering properties of Kaolinite and Bentonite used in this study are given in Table 1.

Table 1 The properties of kaolinite and bentonite

	Kaolinite	Bentonite
Specific Gravity	2.615	2.503
Liquid Limit (%)	30	298
Plastic Limit (%)	20	44
Plasticity Index	10	254
Optimum Moisture Content (%)	21.5	36
Dry Density kN/m^3	15.55	12.00

Rubbers are special kinds of polymers possesing long, flexible molecules which can be crosslinked to form a three dimensional molecular network. Rubbers can be crosslinkable high polymers (Allen 1972). Sixty to seventy percent of worlds rubber production is for tire manifacturing. Styrene-Butadine Rubber (SBR) constitutes most of the rubber used for tire production. All synthetic rubbers are petroleum by-products, and they are increasing their share in the tire industry (Hahn 1970). The rubber used in this study is the byproduct of tire retread industry and was obtained from the outermost part of the tire. Before retreading the tire, the outermost part of the tire is chipped off using a drum with protruding steel nails. When the tire is pressed on the rotating drum, the tire chips are obtained. The outermost part of the tire is a polyisoprene which is a synthetic rubber having a similar chemical structure to that of natural rubber. The rubber particles have lathlike shape of varying lengths. To be able to control the test variables, rubber particles were sieved and a standard gradation curve was used throughout this study.

Rubber particles were immersed in gasoline and their swelling behaviour was observed. A major portion of the swelling of rubber occurred in the first half hour following the immersion after which the swelling rate has decreased (Figure 1). The swelling of rubber is a physical process. The final swell of rubber after one week of test were found to be approximately 2.5 times of its original volume.

The workability of rubber soil composite is very good. The compaction curves for rubber bentonite shown in Figure 2 states that the mixture can be compacted efficiently. The optimum water content increases and the dry density decreases with the addition of higher percentages of rubber.

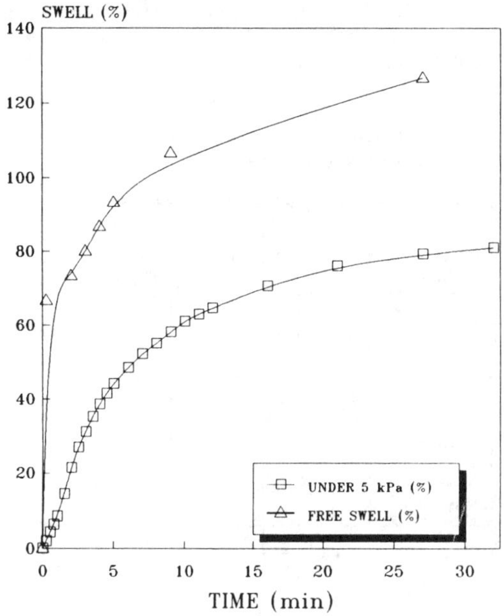

Figure 1 Free swell of rubber immersed in gasoline

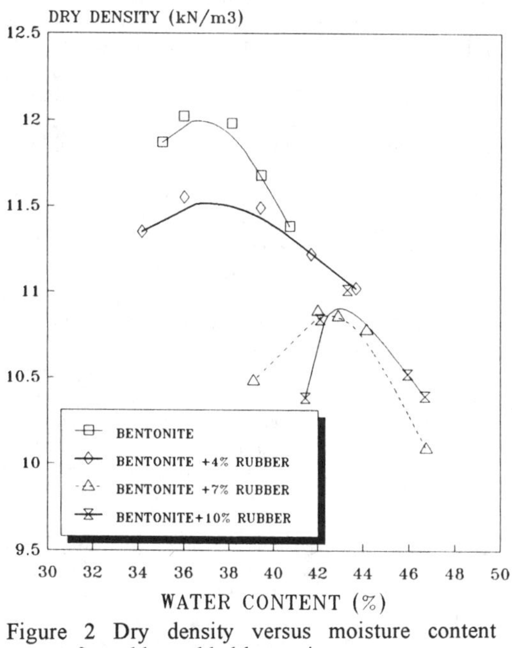

Figure 2 Dry density versus moisture content curves for rubber added bentonite

3 RESULTS AND DISCUSSIONS

The hydraulic conductivity tests were conducted on two sets of specimens; rubber added kaolinite and bentonite compacted with water and permeated with water and gasoline, and rubber added kaolinite and bentonite compacted with gasoline and permeated with gasoline. The hydraulic conductivity values of samples compacted with water are presented in Table 2.

Table 2. Hydraulic conductivity of rubber added kaolinite and bentonite compacted with water.

Rubber Percent by Weight	Permeant Type	Hydraulic Conductivity 10^{-7}cm/sec
Kaolinite		
0	water	1.080
10	water	2.790
0	gasoline	104.0
10	gasoline	0.990
Bentonite		
0	water	0.363
0	gasoline	0.524
4	gasoline	1.220
7	gasoline	0.535
10	gasoline	0.474

The hydraulic conductivity of kaolinite increased approximately 100 times when gasoline was permeated through the sample. Addition of ten percent by weight of rubber decrreased the hydraulic conductivity value back to 10^{-7} cm/sec. The tests with bentonite was very hard to conduct due to the low permeability of bentonite. With the low hydraulic gradients and no backpressure it was impossible to achieve saturation. On one sample which was tested for a month and a half, a nonplastiic zone of approximately three centimeters was observed. If the samples were tested even longer, it is likely that the nonplastic zone will progress and make the whole sample nonplastic. The results seen in this table do not reflect the potential hazard in the long term. Gasoline changes bentonite into silt by shrinking fully the diffused double layer and making it nonplastic. Expansion of rubber chips would help in reducing the hydraulic conductivity in case of gasoline permeation. Considering the time limitations for testing, it was decided to compact the specimens with gasoline and measure the hydraulic conductivity. This would simulate the worst scenario considering that the whole liner is saturated with gasoline. Although the samples compacted with gasoline are not directly comparable to the samples compacted with water, because gasoline made kaolinite and bentonite nonplastic, it might still give an idea about the expected behavior of rubber. The hydraulic conductivity values are presented in Table 3.

Table 3. Hydraulic conductivity of rubber added kaolinite and bentonite compacted with gasoline.

Rubber Percent by Weight	Permeant Type	Hydraulic Conductivity 10^{-7} cm/sec
Kaolinite compacted with gasoline		
0	gasoline	134.4
4	gasoline	573.6
7	gasoline	1074.0
10	gasoline	328.2
Bentonite compacted with gasoline		
0	gasoline	199.0
7	gasoline	132.0
10	gasoline	7.43

For the kaolinite compacted with gasoline the hydraulic conductivity value was 1.3 times, the permeabilty of the sample compacted with water. With bentonite the increase in hydraulic conductivity was 563 times when the sample was compacted with gasoline. A decrease in hydraulic conductivity is only observed with 10 percent rubber added bentonite. Lower weight percentages of rubber were not efficient when gasoline was used for the compaction of samples. As mentioned earlier, the test results obtained by using gasoline for compaction are not reliable for the simulation of the actual case, however an idea can be obtained about the high hydraulic conductivity values, which should be expected in the long term, when the whole thickness of liner is saturated with gasoline, and the whole liner becomes nonplastic. For this case only high amounts of rubber will be sufficient.

Considering the working mechanism of this proposed liner system, prediction of swelling pressures when gasoline is permeated becomes critical. The percent swell and swelling pressures of compacted rubber kaolinite and bentonite are summarized in Table 4.

Table 4 Percent swell and swelling pressure of compacted rubber added kaolinite and bentonite

Rubber percent by weight	Immer-sed in	Swell %	Swelling Pressure (kPa)
Kaolinite			
10	water	0	54
10	gasoline	0.19	142
Bento-nite			
0	water		385
0	gasoline		103
10	water	4.02	225
10	gasoline	3.13	208

The swelling behavior is demonstrated better in Figures 3 and 4. In Figure 3, the swelling pressure of 10 percent rubber added kaolinite and bentonite immersed in water and gasoline is presented. In kaolinite the swelling pressure is 54 kPa when the sample is immersed in water. An increase in swelling pressure to 142 kPa is measured when the sample is immersed in gasoline. For bentonite with 10 percent rubber the swelling pressure is measured to be 208 kPa. for immersion in water. When the same sample is immersed in gasoline, the swelling pressure is 225 kPa. These two values are close to each other but different mechanisms are involved. Immersion in water affects bentonite and the swelling is due to high plasticty of bentonite. There is no contribution of rubber in swelling. In the case of immersing in gasoline, bentonite becomes nonplastic and it does not contribute in swelliing process; this time rubber expands and causes an increaase in swelling pressure. As expected kaolinite sample showed a lower swelling pressure with water. However the swelling pressure with gasoline was considerably higher. For all type of clays, with rubber addition high swelling pressures should be expected. This is the key mechanism that will make rubber addition in liners feasible. The confinement is extremely important. If the liner can not expand under the swelling pressure due to the expansion of rubber, all the pressure will be utilized in contracting the interclod pores, thus causing a decrease in hydraulic

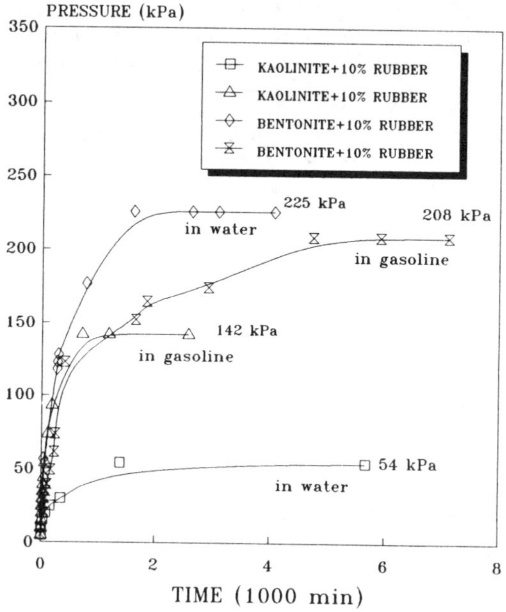

Figure 3 Swelling pressures of rubber kaolinite and rubber bentonite upon immersion in gasoline

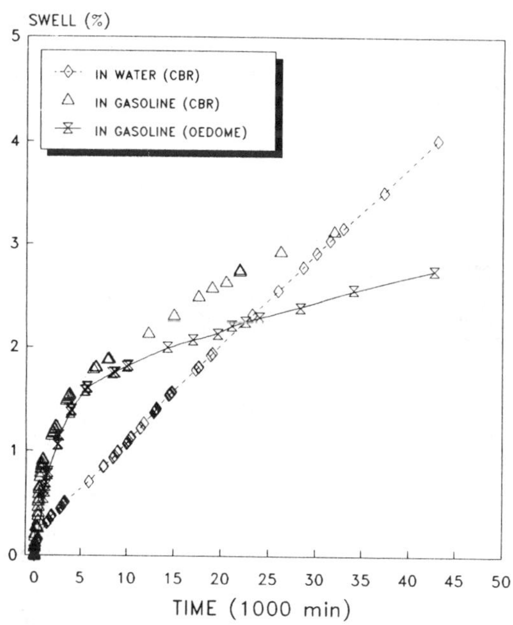

Figure 4 Free Swell of rubber bentonite immersed in gasoline

conductivity. The swelling percentages of 10 percent rubber added bentonite are presented in Figure 4. CBR test and oedometer were used in conducting the swell tests. The percent swell values measured in CBR and the oedometer have similar trends. The tests were conducted upto 30 days. Three and four percent swell values were measured with oedometer and CBR respectively when the ten percent rubber added samples were immersed in gasoline.

The results of the hydraulic conductivity tests are not conclusive. The tests should be conducted for longer periods of time. Field tests will give the real feasibilty of using rubber in clay liners. However this study demonstrates the potential for a new composite lining system for a special application. Besides hydrocarbon based permeants, any other organic permeant that will cause expansion of rubber can be blocked with this system. As long as adequate confinement pressure is present, even high percentage of rubber added sand can be used as a liner for special applications. This fact is demonstrated with bentonite compacted with gasoline. Gasoline makes bentonite completely nonplastic. With the addition of ten percent by weight of rubber, this silt like material has a hydraulic conductivity of 7.43×10^{-7}cm/sec . With higher percentages of rubber this value may even be lower. No confinement was provided for the above sample during testing. In the field there will be a confinement on the liner which will further decrease the measured hydraulic conductivity.

In addition to low hydraulic conductivity the composite liner has a very interesting feature; rubber takes in a considerable amount of gasoline and does not release it. This way hundreds of cubic meters of gasoline will be captured by rubber chips in a typical liner system and these will not be released back into the liner. Rubber swells upto two and a half times its volume when immersed in gasoline. This gives an idea about rubbers gasoline retaining capacity.

Another major advantage of this lining system is high volume utilization of waste rubber. For every one million cubic meter of rubber added liner, approximately two hundred thousand cubic meters of waste rubber will be utilized. Two hundred thousand cubic meters of tire chips will make possible the disposal of at least two million waste tires. Thus the disposal of waste rubber will be achieved in large volumes. Availability of rubber in large quantities, and ease of compaction of the material makes this composite liner very attractive. Crack propagation in the liner will be minimized with the reinforcing effect of lath shaped rubber chips. Its potential as a liner will be promoted especially in areas with a possibility of petroleum contamination. The lining system will be efficent in both lining of underground petroleum storage facilities as well as lining of surface petroleum tanks. The tests will be continued in the field to obtain more data about the performance of these liners.

4 CONCLUSIONS

A new liner system by adding rubber chips into clay is developed which is very suitable especially against hydrocarbon based permeants. The physical property of rubber; expansion upon contact with gasoline, is the key idea behind this liner system. High swelling pressures are obtained when gasoline contacts with rubber which will cause contraction of interclod pores causing a decrease in hydraulic conductivity.

A new dimension in lining technology is presented; permeant retaining capacity of rubber added liner. The composite liner will hold large volumes of permeant which will not be released back into the liner. This way until the retaining capacity of the rubber in the liner is reached no leak through the liner will be expected. The amount of permeant that can be retained in the liner is at least equal to the volume of rubber added into the liner.

Its potential to minimize cracking , its availability and ease of handling and compaction makes this liner system attractive. At the same time large volumes of waste tires will be utilized in these liners. The studies will be continued in the field by constructing a test section and measuring the hydraulic conductivity using double ring infiltrometers.

REFERENCES

Allen, P.W. 1972. *Natural Rubber and Synthetics*. C.Lockwood ed., Granada Publishing Ltd.

Anderson, D.C., K.W. Brown and J.C. Thomas 1986. Conductivity of compacted clay soils to water and organic liquids. *Waste Management and Research*. Vol. 3, No. 1: 339-349.

ASTM D4546-87 Method A. Free Swell Test.

ASTM D4546-87 Method C. Swelling Pressure Test.

Baykal, G., N. Yesiller and K. Koprulu 1992. Rubber clay liners against petroleum based contaminants.*Environmental Geotechnology*. Rotterdam: Balkema.

Hahn, W.G. 1970. *The petroleum industry*. Mc.Graw Hill, N.Y.

Mitchell, J.K. and F.T. Madsen 1987. Chemical effects on clay hydraulic conductivity. *Geotechnical practice for waste disposal 87. ASCE, N.Y.:87-116.*

Yesiller, N. 1991. Hydraulic conductivity of compacted waste rubber-clay mixtures as earthen liners for petroleum based contaminants. *MSc Thesis. Bogazici University, Turkiye.*

Waste Disposal by Landfill – GREEN'93, Sarsby (ed.) © *1995 Balkema, Rotterdam, ISBN 90 5410 356 6*

Compacted clayey soils as cover material in landfills

A. Bouazza
USTHB, Institut de Genie Civil, Bab Ezzouar, Algeria

ABSTRACT: The urban and industrial wastes disposal is becoming an acute problem because it is requiring more free spaces in peripherals areas which much often are indispensable for urban development . The need to expand is very often linked to environmental protection, in this case the groundwater protection and reduction of gaz migration which puts the surrounding constructions at risks. In this paper, the results of a preliminary investigation on the suitability of some algerian clayey soils as cover material for landfills are presented.

1 INTRODUCTION

The urban and industrial wastes is a crucial problem confronting many municipalities. Safe disposal sites are at premium in many areas, and indiscriminate disposal can constitutes a potential source of soil and groundwater contamination. It is well known that the major problem associated with landfills is the generation of leachate.

The main problem which prompted the present preliminary study was linked to the state of the refuse landfill situated at Oued-Smar 15 kms east of Algiers. This landfill receives approximately 580000 t/year of wastes of all sorts. The typical refuse in the Oued-Smar landfill consists of 72 % food (including fibres, shells, etc...), 9 % papers, 8 % plastics, 4 % textiles, 1 % glass, and 5 % of other materials (the composition percentage is per weight). 15 years after its opening, it shows the signs of a very ill designed landfill (Bouazza, 1993). The main signs that can be reported are as follow:

-Unpleasant and unhealthy odours are emerging from the site.

-Repulsive aspects which play a negative role on the neatness and the protection of the site.

-Presence of a permanent smoke which when it reaches the eastern highway, makes the traffic very dangerous.

The visit to the site showed that the cover material used is composed of demolition wastes, sands and soil which is brought to the site by local builders.

The main focus of the present study is to determine the suitability of two compacted soils, in terms of hydraulic conductivity, as cover material landfills. The soils were compacted with standard proctor effort at several water contents. The hydraulic conductivity (k) was measured in a rigid wall permeametre.

2 COVER MATERIAL

The main purpose of the cover material is to minimize the amount of precipitation infiltrating into the waste mass, and hence reduce the quantity of leachate generated at the bottom of the landfill. The relative suitability of soils for various functions are illustrated in table 2 (Lutton et al, 1979).

Two clayey soils were used for the present study. There properties are presented in table 1.

3 TEST PROCEDURE

Compaction tests were performed by the standard proctor. Upon completion of compaction, the top surface of the specimen was trimmed and then it was placed between the end plates of a compaction mold permeametre . The compaction mold permeability tests were performed by flowing the liquid upward through the 100 mm diametre compacted specimen (116 mm high). The hydraulic gradient used in the present study ranged from 20 to 100. The permeant liquid was de-aired 0.005 N Ca So_4. All tests continued until the hydraulic conductivity (k) was steady.

Table 1: Soil properties

Properties	Soil 1	Soil 2
Spec. gravity of solids	2.64	2.70
Optimum water content (%)	28	19
Maximum dry density (kN/m^3)	13.65	15.94
Liquid limit (%)	56	55
Plasticity Index (%)	22	41
Particle size distribution		
silt & clay (<0.075mm)	96	79
sand (0.075-4.75mm)	4	21
uscs symbols	MH	CH

4 RESULTS AND DISCUSSIONS

The results of the permeability tests are shown in fig. 1. Each data point in the figure represents the final hydraulic conductivity (steady state). It can be seen from this figure that the hydraulic conductivity of soil 1 is pratically independant from the hydraulic gradient whereas the hydraulic conductivity of soil 2 decreased as the hydraulic gradient increased. It is thought that this decrease is probably due to the migration of clay particles at higher gradients.

In terms of suitability as cover material and referring to the Lutton et al table, one could notice that the hydraulic conductivity criteria is verified for both soils. However, it should be noticed that soil 2 is more resistant to cracking and more impedant to gaz migration than soil 1.

Table 2: Ranking of soil types according to performance of some cover function (Lutton et al, 1979).

Soil type uscs symbol	Impedance of water percolation	K (cm/s)	Support vegetation	Impedance of gaz migration	Resistance to water erosion	crack resistance	frost resistance
GW	10	10^{-2}	10	10	1	1	1
GP	12	10^{-1}	10	9	1	1	1
GM	7	5 10^{-4}	6	7	4	4	3
GC	5	10^{-4}	5	4	3	7	5
SW	9	10^{-3}	9	8	2	2	1
SP	11	5 10^{-2}	9	7	2	2	1
SM	8	10^{-3}	2	6	4	5	2
SC	6	2 10^{-4}	1	5	6	6	4
ML	4	10^{-5}	3	3	7	10	6
CL	2	3 10^{-8}	7	2	8	8	8
OL			4		7	8	7
MH	3	10^{-7}	4		9	9	9
CH	1	10^{-9}	8	1	10	3	10
OH			8				9

Ranking: 1 (best) to 12 (poorest)

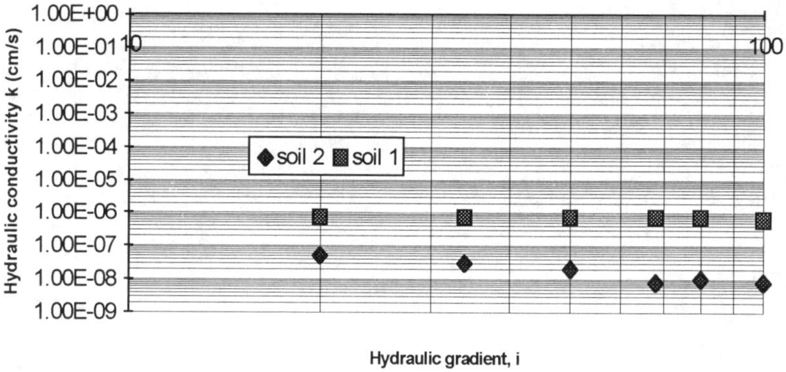

Fig. 1 Hydraulic conductivity vs hydraulic gradient

5 CONCLUSIONS

Landfill technology is still in its infancy in a less developed country like Algeria, especially when it comes to the practicalities of construction. However, local authorities are getting aware of the problems that can be encountered will ill designed landfills. The present preliminary work has shown that some clayey soils around Algiers can be suitable as cover material. However, the present "recipe" needs to be in contact with much more aggressive permeants in order to know its behaviour fully.

REFERENCES

Bouazza, A. (1993). The geotechnical engineer and the environment protection in Algeria. Proc. Int. Conf. on the Environment & Geotechnics : From decontamination to protection of the sub soil, Paris, France. (In french).

Lutton, L.R., King, K.S. & Reades, D.W. (1979). Design & construction of covers for solid waste landfills. Report to the US EPA, N° EP 600/2-79-165, Washington D.C., USA.

Waste Disposal by Landfill – GREEN'93, Sarsby (ed.) © 1995 Balkema, Rotterdam, ISBN 90 5410 356 6

Geochemical interaction processes between landfill clay liner materials and organo-metallic landfill leachate

J.P.L. Dearlove
University of Sunderland, UK

ABSTRACT: This paper describes the results of a preliminary investigation into the geochemical processes relevant to the interaction of landfill leachate with boulder clay. Two leachate sample pretreatment methods have been evaluated. The incomplete destruction of the organic component in landfill leachate may cause significant inaccuracies in the determination of some metals by atomic absorption spectrophotometry.

The chemical composition of three landfill leachates were determined. Samples of boulder clay were immersed in two of the leachate samples. In general, Na, K and Mg concentrations for the clay increase while in the leachate they decrease indicating metal uptake by the clay.

Leachate ultrafiltration experiments indicate that the concentration of particles >1000 nm is 50 mg/l while the concentration of colloidal particles (1000nm to 1.5 nm) is between 160 and 200 mg/l. Fe and Mn are associated with these colloidal particles. Al, Cu and Zn are not associated with the colloids in this particular leachate sample.

1 INTRODUCTION

Containment landfill sites in the UK may be developed through utilisation of a natural clay or other low permeability base or by installation of an engineered liner. These liners are required to produce a physical barrier to leachate flow with a permeability of at least 10^{-9} m/sec (DoE, 1992). Despite the low permeability of natural and engineered clay liners some leachate migration will occur in response to a hydraulic head of leachate. Landfill leachate may vary in composition from a relatively innocuous to a highly toxic liquid.

On passing slowly through the clay base of a landfill site contaminants dissolved in the leachate will interact chemically with the clay by such processes as ion exchange, physical sorption and precipitation (Dearlove, 1992). In particular metal contaminants are removed from the leachate. These natural attenuation processes are based on predictions which assume that the landfill leachate behaves geochemically as an aqueous solution. Geochemical processes observed from water:rock interaction are thought to be directly transferable to leachate:rock interaction processes. This assumption fails to recognise a highly significant geochemical component of leachate from household waste not commonly found in natural groundwaters. This component is the presence of organic carbon at concentrations in excess of 465 mg/l for leachate from aged waste and 8,000 mg/l for leachate from recent waste (Davies and Hammond, 1991).

Organic carbon in the form of colloidal particles (particles in the size range 1000 to 1 nanometres (nm)) has the potential to sorb high concentrations of metals onto surface sites. This organic component may thus compete with the clay matrix in the sorption of metal ions from solution. In many cases the cation exchange capacity (CEC) of the organic material far exceeds that of many natural clays. Under certain chemical conditions inorganic components within the leachate may also form colloidal particles. Colloidal particles also have the property of remaining completely in suspension in an aqueous medium. Thus colloidal particles and in particular organic colloidal particles may act as carriers for trace metals in landfill leachate. Metal ions in solution may be removed from an aqueous phase by ion exchange, sorption or precipitation onto the geomatrix while metal ions sorbed preferentially onto the surface of colloidal particles will by-pass these processes. However metals transported on colloidal particles may be removed by physical processes such as ion exclusion and molecular filtration because of their size (Dearlove, 1992). Metal migration from a site may therefore be enhanced or reduced depending on the filtration capacity of a given clay liner.

The aim of this work is to evaluate appropriate analytical techniques to determine the concentration of particulates over a range of sizes from 1000 nm to 1.5 nm and to measure the concentrations of metals which occur purely in solution and those which are bound to the colloidal particles. These data will then be evaluated in light of various laboratory experiments aimed at measuring the physical and chemical properties of a variety of landfill site clay liner materials.

2 EXPERIMENTAL PROGRAMME

A number of experiments are in progress to evaluate the interaction between landfill leachate and clay materials. Samples of boulder clay were collected from the base of an active landfill site in Northumberland, England. A number of landfill leachate samples from another landfill site in the NW of England were collected from household waste which has been undergoing anaerobic degradation for more than 8 years. These leachate and clay samples are thus representative of many established landfill sites in the UK developed on a containment principle using natural glacial clay deposits.

Relatively short-term experiments include the simple immersion of clay samples in landfill leachate. The concentration of metals in the leachate and the concentration of metals sorbed onto the boulder clay can provide a crude indication of which metals undergo exchange reactions between the clay and the leachate.

Long-term laboratory column experiments for a constant head of leachate are in progress although results from these experiments were not available at the time of writing. Transparent plastic tubes were driven into the boulder clay to a depth of approximately 250 mm. The tubes were then dug out to leave a tightly sealed plug of clay at the base on the tube. The base of the tube was then cut to leave a 200 mm plug of clay. The tube was then filled with 250 ml of leachate. The head of leachate (130 mm) and clay thickness (200 mm) give a hydraulic gradient of approximately 1.5. Initial estimates indicate a permeability for the clay of approximately 10^{-9} m/s. The leachate is monitored periodically for variations in pH, Eh and dissolved oxygen (DOX) although no significant variation with leachate depth or time has been observed.

Further measurements of the physical properties of the clays including porosity, shear strength and plasticity are planned. The effects of variations in these properties will be incorporated into the results of the laboratory chemical experiments.

3 LEACHATE SAMPLE COLLECTION

The samples of landfill leachate were collected on two separate occasions. Samples A1 and A2 were collected in November 1992 from a leachate monitoring well installed in an area of waste in excess of 8 years old. Sampling procedures aimed at minimising potential changes in the physical and chemical properties of the leachate through interaction with air were used. Approximately 10 times the final sample volume were passed through a narrow-necked sample bottle before being filled to capacity and stored at $4^{o}C$ prior to analysis. Sample L1 was collected in April 1993. It represents a bulked average leachate sample from a large area of waste which is stored in a holding tank prior to aerobic treatment. Although this sample is not truly representative of leachate still in contact with the waste mass and has undergone some interaction with atmosphere it does represent a chemically stable, "average" leachate.

4 ANALYTICAL METHODS

4.1 Leachate sample preparation

In order to accurately measure the concentration of metals in organo-metallic landfill leachate it was first necessary to evaluate the suitability of various leachate sample preparation techniques. Metal concentrations in leachate samples were measured using a Varian SpectrAA10 flame spectrophotometer. Trace metals determined using a Varian SpectrAA300 furnace spectrophotometer.

Leachate samples with no sample pretreatment, partial sample oxidation and total sample oxidation were prepared. The purpose of this work was to determine the effect of the organic component in the leachate on absorbance signal suppression. Untreated leachate was simply aspirated directly into the spectrophotometer. Partial oxidation of leachate samples involved gently heating the sample on a hot plate in a teflon beaker, adding 5% by volume conc. HNO_3 (PrimaR grade), further gentle heating followed by a further addition of 5% by volume conc. HNO_3. The sample was then allowed to cool before being aspirated. The complete oxidation of leachate samples involved the following procedure; Gently heat sample, add 10% by volume conc. HNO_3 (PrimaR grade), continue to heat then add 10% by volume conc. $HClO_4$ (PrimaR grade). Cover beaker and allow to reflux until straw coloured liquid becomes colourless. If sample fails to clear a further 10% by volume conc. $HClO_4$ was added. Reduce sample volume to approx. 1 ml, add 10 mls of 6M HCl (PrimaR grade) and repeat volume reduction and acid addition. Transfer quantitatively to a suitable volumetric flask and make up to original sample volume using distilled/deionised water.

Chemical oxygen demand (COD) and chloride were determined by titration after the procedure of Rump and Krist (1985).

4.2 Sample filtration

Leachate samples were filtered in the laboratory using an Amicon 8010 flat bed ultrafiltration cell. A series of 10ml samples were first filtered using a Whatmans PTFE 1000 nm filter. A 10ml sample of 1000 nm filtrate was then passed through each of the following filters; Whatmans PTFE 500nm, Whatmans PTFE 200nm and three Diaflo YM series ultrafilters (YM100, YM30 and YM10) with filtration cut-off values of 100,000 molecular weight (MW - approx. 10nm), 30,000 MW (approx 3nm) and 10,000 MW (approx. 1.5nm). This procedure produces 6 filtrates with upper filtration cut-off limits of 1000, 500, 200, 10, 3 and 1.5 nm. All filters were weighed before and after filtration (and air drying) to determine the weight of particulates removed from a known volume of leachate. The organic component of the filtrate samples was then fully oxidised using the procedure described earlier.

5 RESULTS AND DISCUSSION

The results for untreated, partially oxidised and fully oxidised leachate samples are given in table 1. No significant difference occurs between the three sample treatment methods for K. The high error for the partially oxidised sample indicates the level

Table 1. Metal concentrations in untreated, partially oxidised and fully oxidised leachate.

Process	K	Fe	Ni	Al	Cu
			(mg/l)		
Untreated	454	3.5	2.0	0.5	0.1
leachate +/-	56	0.1	0.3	0.1	0.1
Partially	260	3.3	2.1	0.5	0.1
oxidised +/-	180	0.1	0.2	0.1	0.1
Fully	435	5.3	0.5	2.4	0.3
oxidised +/-	35	0.6	0.1	0.6	0.1

Quoted errors are standard deviations of triplicate analyses.

of difficulty in achieving reproducible results. Mn and Mg also follow this trend with values in mg/l for untreated, partially oxidised and fully oxidised of 0.2+/-0.1, 0.3+/-0.2 and 0.2+/-0.1 for Mn and 85+/-7, 77+/-3 and 72+/-3 for Mg respectively.

The results for Fe, Al and Cu show that analysis of untreated and partially oxidised leachate samples may underestimate true metal concentrations from around 35% to 80%. Ni concentrations in untreated and partially oxidised samples overestimate concentrations by a factor of 4.

Table 2 shows the results for the three leachate sample A1, A2 and L1. These results are typical of landfill leachate and show the great variability in chemistry, even from the same monitoring well such as samples A1 and A2.

Table 3 shows the results of partial analyses for boulder clay samples.

Table 2. Chemistry of landfill leachate samples A1, A2 and L1.

Element	A1	A2	L1
		(mg/l except pH)	
pH	8.0	7.9	8.1
DOX	0.4	1.0	1.3
COD	4876(50)	1283(45)	1360(30)
Cl	2800(55)	1350(30)	1500(50)
Na	2200(210)	1050(65)	1150(50)
K	670(60)	460(40)	435(35)
Ca	40(7)	50(5)	nd
Mg	50(3)	110(10)	72(3)
Fe	0.3(0.1)	0.4(0.1)	5.3(0.6)
Mn	0.1(0.1)	0.1(0.1)	0.22(0.01)
Ni	0.4(0.1)	0.2(0.1)	0.5(0.1)
Co	0.2(0.1)	0.1(0.1)	nd
Cu	0.1(0.1)	0.2(0.1)	0.30(0.01)

nd = Not determined. DOX = Dissolved oxygen. Data in brackets are standard deviation errors of triplicate analyses.

Table 3. Chemical analyses of boulder clay samples BC1, BC2 and BC3.

Element	BC1	BC2	BC3
		(wt%)	
Mg	10.0	30.0	20.0
K	7.2	7.7	7.2
Na	0.6	1.2	0.6
Fe	0.3	0.2	0.2
Mn	0.05	0.02	0.04
Ni	0.02	0.02	0.02

Sample BC1 represents an untreated blank sample. Sample BC2 and BC3 were immersed in leachate samples A1 and A2 respectively for a period of 2 weeks. Sample BC2 shows an increase in Mg, K and Na over the control sample BC1. All other elements analysed occur at similar concentrations to the control sample. The resultant leachate from the clay:leachate interaction, sample A1/E (Table 4), has lower concentrations of the elements K and Mg than the original leachate (Sample A1 - Table 2).

Table 4. Chemical analyses of leachate samples A1/E and A2/E .

	A1/E	A2/E
	(mg/l)	
K	448+/-30	364+/-30
Mg	40+/-4	70+/-7
Fe	0.3+/-0.1	0.2+/-0.1
Mn	0.1+/-0.1	0.1+/-0.1
Ni	0.5+/-0.1	0.2+/-0.1
Co	0.3+/-0.1	0.1+/-0.1

These data suggest that the elements K, Mg and, by inference,

Na (due to its similar chemical behaviour) have been sorbed onto the surface of the clay, probably by cation exchange reactions. Boulder clay sample BC3 (Table 3) and the leachate A2/E (Table 4) show similar results for Mg but not Na and K. The concentration of Na and K in sample A2 are lower than for sample A1. It is possible that the concentration dependence of cation exchange reactions may be responsible for these differences. Further research is needed to investigate the effects of varying metal concentrations on exchange and precipitation reactions. The results of the long-term column experiments may provide some additional information.

Table 5 shows the results of filtration experiments to determine the concentration of particulates of varying size ranges. These data indicate that there are 3 to 4 times the concentration of organic and inorganic particulates in the colloid size range 1000 to 1 nm (160 to 200 mg/l) than in the > 1000 nm size range (50 mg/l). The greatest concentration of colloidal particles occurs in the smallest size range 200 to 1.5 nm.

Table 5. Concentrations of particulates in landfill leachate sample L1.

Particle size range	Concentration (mg/l)
>1000 nm	50
1000 to 500 nm	10
500 to 200 nm	10
200 to 1.5 nm	140 to 180

Table 6a and 6b show the results for metal concentrations in filtrates of the landfill leachate sample L1. Analyses for Fe and Mn show a decrease in the concentration of these elements with decreasing filter pore size. This suggests that these metals are either bound to organic colloidal particles or they form inorganic colloids which are being progressively removed by filtration. The Fe data indicate that approximately 2 mg/l of iron, approximately 38%, is associated with a colloid phase. Particles > 1000 nm have Fe concentrations of

Table 6a. Metal concentrations in filtrates of landfill leachate.

Filtrate	Zn	Fe (mg/l)	Mn	Ni
Unfiltered	0.3	5.3	0.22	0.5
+/-	0.1	0.6	0.01	0.1
<1000 nm	0.3	4.5	0.23	0.5
+/-	0.1	0.1	0.01	0.1
<500 nm	0.4	3.8	0.21	0.4
+/-	0.1	0.1	0.02	0.1
<200 nm	0.3	3.7	0.21	0.5
+/-	0.1	0.2	0.01	0.1
<10 nm	0.3	3.1	0.18	0.4
+/-	0.1	0.1	0.01	0.1
<3 nm	0.4	3.0	0.18	0.5
+/-	0.1	0.1	0.01	0.1
<1.5 nm	0.3	2.5	0.16	0.4
+/-	0.1	0.1	0.01	0.1

Table 6b. Metal concentrations in filtrates of landfill leachate.

Filtrate	Na	K	Al (mg/l)	Mg	Cu
Unfiltered	1150	435	2.4	72	0.30
+/-	240	35	0.6	3	0.01
<1000 nm	1200	430	1.8	71	0.30
+/-	20	40	0.1	3	0.01
<500 nm	1675	310	1.4	69	0.20
+/-	15	90	0.4	3	0.10
<200 nm	1540	300	1.6	71	0.28
+/-	10	100	0.1	3	0.03
<10 nm	705	255	2.9	45	0.35
+/-	75	15	0.1	5	0.20
<3 nm	565	255	3.1	50	0.37
+/-	45	15	0.4	5	0.01
<1.5 nm	455	455	2.9	42	0.37
+/-	75	15	0.2	5	0.01

Quoted errors in tables 6A and 6B are standard deviations for triplicate sample analyses.

0.8 mg/l. The 1.5 nm filtrate thought to represent the concentrations of Fe in solution is 2.5 mg/l.This concentration of iron occurs on 160 to 200mg/l of particulate material (Table 5), a loading factor of 10 to 12.5 micrograms Fe per milligram of colloidal matter. Mn has a lower loading factor of 3 to 4 micrograms of Mn per milligram of colloidal matter.

Zn, Ni, Al and Cu show no significant variation in concentration with filtration. This suggests that these elements are not bound to colloidal particles and are thus present as complexed or free ions in the leachate.

Na, K and Mg show a decrease in concentration for the 10 nm, 3 nm and 1.5 nm Diaflo membrane filters in comparison to the concentrations for the Whatmans filters (1000, 500 and 200 nm). K also shows an inconsistently high concentration for the 1.5 nm filtrate. These data are thought to be the result of interferences between these elements and either the Diaflo filter membranes or the sodium azide preservative applied to these filter membranes. This preservative was removed by washing and soaking the membranes in distilled/ deionised water for 24 hours prior to use in accordance with the manufacturers instructions. It is uncertain at this time whether these elements are colloid bound or occur in solution. The results for exchange reactions between leachate and clay (Tables 3 and 4) suggest that these elements form free ions in solution which are available for ion exchange with the clay matrix.

6 CONCLUSIONS

The preliminary results of leachate:clay interaction experiments appear to show that readily exchangeable cations in landfill leachate such as K, Na and Mg appear to be sorbed onto the surface of the boulder clay. Fe and Mn show significant concentrations associated with the colloidal phase in the leachate, either bound to organic colloids or forming separate inorganic colloids. Zn, Ni, Al and Cu are not bound to colloidal particles and Ni does not appear to be involved in exchange reactions with boulder clay.

ACKNOWLEDGEMENTS

The author would like to thank Mr A. Mistery for trace element analyses of leachate.

REFERENCES

DoE 1992. Waste Managment Paper No. 26: Landfilling wastes. HMSO.

Davis, J. and Hammonds, J. 1991. The control of leachate. *Proc. of Planning & Engineering of landfills*. University of Birmingham, 10-11th July, 1991.

Dearlove, J.P.L. 1992. The effect of organic colloids on water:rock interaction geochemistry - a discussion document. *Proc. 8th International Conference on solid waste management & secondary materials*. Philadelphia, USA. Nov 15-18th, 1992.

Rump, H.H and Krist, H. 1985. *Laboratory manual for the examination of water, wastewater and Soil*. VCH Publishing.

Waste Disposal by Landfill – GREEN'93, Sarsby (ed.) © *1995 Balkema, Rotterdam, ISBN 90 5410 356 6*

Utilization of expansive soils for lining of waste disposal landfills

M.A. El-Sohby & M.T. El-Bakey
Department of Geotechnical Engineering, Al-Azhar University, Cairo, Egypt

A. M. Elleboudy
Department of Geotechnical Engineering, Banha University, Cairo, Egypt

ABSTRACT: Expansive soils are usually removed by huge amounts from construction sites and replaced by compacted cohesionless materials to minimize the effect of the swelling pressure on the foundations. A trial was made to utilize these discarded swelling clays as a lining material for the waste disposal sites. This investigation was focused on their hydraulic conductivity, swelling pressures, and shear strength. The aim of this study is to consume these undesirable geomaterials and provide costless effective substance for lining of waste disposal landfills.

INTRODUCTION

The production of wastes is constantly increasing and presents a tremendous problem to the society and nature today. Several types of dumping in the open areas, waterways, and seas have been outlawed in most countries because it is hazardous to the environment. A much cleaner alternative is deposition in landfills sealed by economical lining material such as compacted clay.

In Egypt, foundation engineers used to excavate a considerable depth of expansive clays and shales, and refill the construction sites with compacted cohesionless soils to form a satisfactory bearing stratum and minimize the detrimental effect of high swelling pressures on the structures. Accordingly, dumping of piles of excavated unwanted swelling soils has formed a major problem, and these soils were considered as undesired waste materials.

Since overpopulated cities, such as Cairo, produce huge amounts of domestic waste materials and heaps of industrial wastes, and has large areas of waste dumps, engineers were faced with the challenge arising from the need to save their cities from environmental polution. These circumstances inspired the researchers with the idea of using the discarded swelling clays for lining of waste disposal landfills.

Investigations were carried out to characterize a swelling clay formation existed in a wide area north east of Cairo to evaluate its suitability as a liner for the waste disposal sites. All the physical and mechanical properties were determined in the laboratory. A special attention was given to its hydraulic properties and its effectiveness as a hydraulic barrier. The swelling pressures and axial free swell, and shearing strength were determined for compacted specimens. Compaction and remoulding with various techniques and moisture contents were employed to achieve the desired performance.

SOIL CHARACTERISTICS

The expansive soil formation used in this research is located in an arid desert environment in a new suburban area north east of Cairo. The average values obtained from several representative samples indicated that it is mainly composed of 53% clay, 45% silt, and 2% sand. Its natural dry unit weight was about 18.2 kN/m^3. The average liquid limit was 80% , plastic limit was 28 %, and the shrinkage limit was 15.8 %. The clay fraction was mainly composed of mixed layers of montmorillonite and vermiculite. The particles above 2 micron were mostly silicate minerals. The dominant adsorbed cation was calcium. The cation exchange capacity was 29.23 meg/100g, and the external specific surface was 39.76 cm^2/gm. The pH value was 7.55. The activity of soil was 0.98. More details about this soil was documented in previous work of the first and third authors (El-Sohby and El-Bakey, 1987).

COMPACTION AND SAMPLE PREPARATION

Fabric of soil exerts a dominant influence on its hydraulic conductivity, so it would be desirable to match fabrics in laboratory compacted samples with those developed in the field. Three compaction methods were evaluated by R.J.Dunn (1984): static; impact using ASTM D1557 type compactor; and kneading using a Harvard Compactor over a range of densities and water contents typical to many compacted clay liners, 90 to 95 % of maximum dry density at water content 2 - 3 % points above optimum moisture content determined by ASTM test method D 1557. Static compaction would develop a more flocculated structure than impact or kneading methods and thus higher hydraulic conductivities are expected. This agrees with the results of tests on the effect of compaction on swelling, shrinkage, and stress - strain characteristics of clays done by Seed and Chan (1959) and hydraulic conductivity of clays measured by Mitchell et al (1965). Static compaction also produced replicate samples of best quality and seems better suited to use than kneading or impact methods commonly used in the laboratory for clay soil. However, kneading by sheep - foot rollers is the best compaction technique for clays in the field, since it produces more dispersed structure due to the introduction of large shear strains during compaction which results in a lower hydraulic conductivity than static compaction (Mitchell,1990). Thus, in the laboratory testing program, the modified AASHTO compaction test was used to determine the maximum dry density and the optimum moisture content. The average maximum dry density of the compacted samples was 17.2 kN/m^3, and the optimum moisture content was 17.00 %. Then, the samples were prepared for oedometer and triaxial testing by static compaction. The soil was first oven - dried and then pulverized to destroy the clay clods which may extremely affect the values of the measured hydraulic conductivity. For the oedometer test, the required weight of the oven - dry soil was calculated to get the required dry density which was the maximum dry density determined by the AASHTO test. The sample was mixed with 5 % moisture content and poured into the mould, and statically compressed untill it reached the height corresponding to the required density. The triaxial test samples were prepared from the same pulverized dry soil after mixing it with different amounts of water to achieve 5, 10, 15, 20, and 25 % moisture contents. The required density was achieved by static compaction. An extractor was used to get the specimen out of the mould to be enclosed in the rubber membrane and mounted in the triaxial cell.

SHEARING STRENGTH AND SWELLING PROPERTIES

A quick undrained triaxial test was performed on the prepared samples to determine the shearing strength parameters of the soil at different moisture contents. The results of these tests are summarized in Table 1.

Table 1. Results of triaxial tests

Moulding Moisture Content	Shearing Strength Parameters	
	c_u (kN/m^2)	Φ^0
5 %	80	45
10 %	200	40
15 %	230	30
20 %	250	14
25 %	150	5

The swelling characteristics were determined using the different pressure method, the preswelled sample method and the double oedometer test. For a ring diameter of 6.30 cm, initial sample height of 1.575 cm, and initial moisture content of 5 %, the swelling pressure measured by the first method was 1130 kN/m^2. The second method gave an axial free swell 24.17 % when the sample was inundated under a light load of 10 kN/m^2. The swelling pressure determined by the preswelled sample method was 900 kN/m^2. The double oedometer test gave a value of 1000 kN/m^2 for the swelling pressure,

SOLID WASTE PROBLEMS

The investigated swelling soil was intended to be used for municipal soil waste landfills. These wastes represent a major cause for environmental pollution in overpopulated cities when are dumped into open areas. When the solid wastes are confined into landfills, they produce leachates which contain extracted, disolved, or suspended inorganic and organic materials. The most common municipal landfill pollutants are total dissolved solids, chlorides, sulphates, methane, oils and greases, plastic and other organics, in addition to bacteria and algae (U.S.EPA, 1980). Leachates are generated in solid waste disposal sites when water is allowed to infiltrate the site. Inorganic and organic materials are then extracted and dissolved from the waste as the water percolates through the decomposing refuse. The leachate mounds on the bottom of the landfills and starts seeping through the bottom as soon as enough hydrostatic head has built up. The composition of the leachate depends on the type and quality of waste. The amount of contamination of ground water that takes place depends on the hydrogeology of the area, and on the attenuation capacity of the soil. Dispersion of the leachate plume is three - dimensional in that a plume not only extends laterally, but often sinks. Given enough time the plume can disperse over many square kilometers, contaminating the ground water supply of the entire community (Alther, 1987). Thus, the solid waste disposal sites should be constructed in a geologically favorable areas where the ground water table is deep enough to avoid its pollution. It should be at least 1.50 m below the bottom of the facility. Moreover, landfills must be covered with rather impermeable layer to prevent surface water from infiltrating into the landfill site. The lining soil and the cover, if initially wetted, should be protected from drying out to avoid desiccation of clay and the resulting cracking which increases the in - situ hydraulic conductivity up to 1000 times larger than values measured in the laboratory (Day and Daniel, 1985). Alther (1987) stated that the local clays of favorable character can be used when possible to line the landfill disposal sites, and if its permeability is more than 10^{-7} cm/sec, bentonite liners can be used. Bentonite is an excellent material for lining but it is rather expensive and not always available. Thus, the authors have tried to find a more economical alternative that also has expansive properties and satisfies the requirement of having low hydraulic conductivity to function as a good sealer.

HYDRAULIC CONDUCTIVITY

Low hydraulic conductivity is a prime reason for choosing clay as a major component of waste containment barrier systems. Therefore, hydraulic conductivity was measured in the laboratory for the clay samples during the performance of the consolidometer tests. These conductivities are the saturated hydraulic conductivities of the tested specimens which where initially compacted by static compression to dry density of 17.2 kN/m^3 and moulding water content of 5 %, and became fully saturated during the test by inundation to determine the swelling pressure and the axial free swell. Mitchell and Jaber (1990) confirmed that the consolidometer permeameter system may offer the best means for quantitative assessment of chemical interaction, induced volume changes, and for testing clays under confining stress states that are most representative of those in the field. Moreover, the compatibility of a clay liner material with leachate can be assessed in the laboratory, using consolidometer tests under the overburden stresses expected in the field, and preferably using a permeant that is as close in composition as possible to the leachate or liquid waste expected in a specific waste containment facility. Table 2 gives the values of the average hydraulic conductivity of the tested samples as measured by the consolidometer permeameter during the performance of the swelling tests.

Table 2. Hydraulic conductivity of testes samples

Pressure (kN/m^2)	m_v x10^{-3} (m^2/kN)	c_v x10^{-5} (cm^2/sec)	k x10^{-9} (cm/sec)
50	0.82	8.69	7.13
100	0.71	8.11	5.76
200	0.47	7.46	3.51
400	0.27	6.73	1.82
800	0.12	6.05	0.80
1600	0.07	5.44	0.38

The values shown in the table ranged between 0.38 x10^{-9} and 7.13 x10^{-9} cm/sec. Figure 1 illustrates the effect of the applied pressure on the hydraulic conduvtivity.

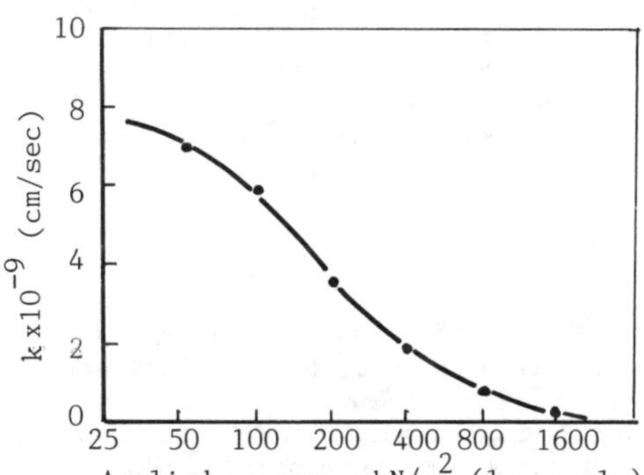

Fig. 1 Effect of applied pressure on hydraulic conductivity

It is logical that it increases with the increase of the applied pressure. The values in Table 2 indicate that the tested soil has a very low hydraulic conductivity and can serve as a good liner since the U.S. environmental Protection Agency guidelines require that the compacted clay component of a double liner system be at least 1.00 m thick with a maximum saturated hydraulic conductivity of 1 $x10^{-7}$ cm/sec (U.S.EPA, 1980). The swelling properties of the soil enhances its ability as a sealer since the swelling soils are usually compacted in the field on the wet side of the optimum and any further increase in moisture content after placement will cause expansion of soil which clog the pores and increase its retention capability.

CONCLUSIONS

Utilizing the expansive geomaterials which are usually removed from the building construction sites will hit two targets at the same time; the consumption of these undesirable materials, and providing a costless substance with the required low hydraulic conductivity and sufficient shearing strength and durability to serve as a lining material for domestic solid waste disposal sites. Testing the swelling soil in the laboratory to determine its hydraulic conductivity and swelling and strength properties is only a promising begining for making a full use of this soil in the field. Placement conditions and details of construction of the clay liner determine the fabric and structure of the compacted clay, and consequently its actual properties, specially the hydraulic conductivity and strength. Hence, a careful design and implementation can produce a liner that satisfies the required specifications. The execution of a successfull liner using the swelling soils will be the next challenging step in this optimistic research to be certain that this material can retain its integrity and retention capability over a long time inspite of any post - construction changes.

REFERENCES

Alther, G.R. 1987. The qualifications of bentonite as a soil sealant. Engineering Geology, 23: 177-191.

Day, S.R. and Daniel, D.E. 1985. Hydraulic conductivity of two prototype clay liners. Journal of Geotechnical Engineering Division, ASCE, Vol. 111, No. 8: 957-970.

Dun, R.J. 1984. Laboratory measurement of fine - grained soil fluid conductivity. Engineering Geology, 21: 215-223.

El-Sohby, M.A., Mazen, S.O., and El-Bakey, M.T. 1987. Evaluation of shear strenth parameters of expansive soils. Proceedings of the Eight Asian Regional Conference on Soil Mechanics and Foundation Engineering, Vol. 1: 163-166, Kyoto, Japan.

Environmental Protection Agency, 1980. Lining of waste impoundment and disposal facilities, SW-870, U.S. EPA, Washington D.C.

Mitchell, J.K., Hooper, D.R. and Campamella, R.G. 1965. Permeability of compacted clay. Journal of Soil Mechanics and Foundations Division, ASCE, 91 (SM4): 41-65.

Mitchell, J.K. and Jaber, K. 1990. Factors controlling the long - term properties of clay liners. Proceedings of Symposium on Waste Containment Systems : Construction, Regulation, and Performance. ASCE Geotechnical Special Publications No. 26: 84-105.

Seed, H.B. and Chan, C.K. 1959. Structure and strength characteristics of compacted clays. Journal of Soil Mechanics and Foundations Division, ASCE, 92 (SM6): 105-134.

Waste Disposal by Landfill – GREEN'93, Sarsby (ed.) © 1995 Balkema, Rotterdam, ISBN 90 5410 356 6

Bacteria and tree root attack on landfill liners

H.Y. Fang
Lehigh University, Bethlehem, Pa., USA

ABSTRACT: Bacteria in landfill can cause changes in the characteristics of leachate, and consequently, affect the liner performance. Also landfill sites are generally planted with trees to enhance evapotranspiration and for the purpose of beautification of unattractive landfill sites, but these trees also have a detrimental side effect. The roots seeking nutrition and water are capable of and will penetrate liners, consequently, the leachate in the landfill can migrate through or along the roots and root hairs and thereby leak into the surrounding areas or aquifers.

This paper presents the mechanisms of soil-bacteria interaction and soil-root interaction in landfill areas. All explanations are illustrated by the recently developed particle energy field theory which is based on physico-chemical and biological principles. Control of or minimization of these attacks on landfill liners is also discussed in the paper.

1. INTRODUCTION

Liners are the major structural elements for controlling landfill sites' liquid waste leaking into surrounding areas or aquifers. In general, engineers have taken all necessary measures to make sure that these controlling facilities are safe and durable. However, some additional factors that the author believes are important, but about which there is very little available published data, and are serious enough to be brought to our attention, namely, bacteria and tree root effects on landfill liners.

There is no proven direct relationship between bacterial activities and the performance of liners; however, indirectly bacteria can cause change in the characteristics of pore-fluid (leachate), and consequently, will affect the liner performance.

Landfill sites are generally planted with trees to enhance evapotranspiration and for the purpose of beautification of unattractive landfill sites, but these trees also have a detrimental side effect. Tree roots seeking nutrition (food) and water are capable of and can penetrate clay liners or synthetic membranes which line landfill areas. Also, the pore fluid (leachate) in the landfill can migrate through or along the roots and root hairs and thereby leak into the surrounding areas.

The purpose of this paper is to present the mechanism of soil-bacteria interaction and soil-root interaction in landfill areas. All explanation will be illustrated by the recently developed particle energy field theory which is based on physico-chemical and biological principles (Fang, 1991a). Control or minimization of these attacks on landfill liners are also discussed in the paper.

2. BACTERIA ATTACK LINERS

It has been observed that bacteria can attack the liner and cause the instability of controlling facilities. However, the mechanics of how bacteria attack the liner is not clearly understood. It is important for us to know the mechanisms between bacteria and clay liner interaction, especially for the long-term stability of liners.

2.1 Nature of Bacteria

Bacteria are single cell life forms. Four major elements, carbon, hydrogen, oxygen and nitrogen compose about 98.5 percent by weight of the atoms of the cells. Phosphorus and sulfur together make up the additional percent. The most abundant molecule in all living cells is water, which usually makes up about 70 percent of the cell.

In general, there are two basic types of bacteria: heterotrophic and autotrophic. The heterotrophic type of bacteria requires energy and carbon directly from soil's organic matter, however, the autotrophic type requires energy from oxidation of mineral constituents. Table 1, shows the classification of geomicrobiology for engineering purposes. As indicated in Table 1, the autotrophic type can change the soil/rock constituents which leads to a change of soil/rock properties.

2.2 Soil-Bacteria Interaction

Most bacteria exist in an aerobic condition above the groundwater table, in general, existing at a shallow depth from ground surface, especially in the A-horizon, as shown in Figure 1. In examining Figure 1, indications are that in one gram of soil can exist four million bacteria. Disturbed soil has more bacteria than undisturbed because the air (oxygen) creates more bacteria. The rate and degree of bacterial activity depends on the availability of nutrients, microorganisms, substrate composition and temperature. In cold weather, bacterial activities are less, as flooding, fluctuating water table or construction activities disturbing the ground soil will accelerate the aerobic process. In the anaerobic condition, bacteria exist primarily below the groundwater table; in comparison to the aerobic condition, the bacteria activities are a slower process.

As indicated in Figure 2, there are three basic possible routes that may cause a soil to change behavior by bacterial activities. Bacteria can transfer in an ion-exchange changing the soil-water structure, or changing the characteristics of pore

TABLE 1 CLASSIFICATION OF GEOMICROBIOLOGY FOR ENGINEERING PURPOSES (After Fang, 1986)

		Types & Descriptions	Influenced Depth			Effects on Geotechnology
FLORA		Tree or Plant Roots	Horizons A	B	C	
FLORA		Tree-Root System Soil-Root System Root-Water Mechanics	x	x		Ground settlement; Broken drainage pipes; Damage highway pavement; Damage landfill liners.
FAUNA	Animals	Macro-Animalss Rodents, Millipedes, Centipedes, Spiders, Mites, Ticks, Slugs, Snails, Insects, Ants and Earthworms.	x	x		Damage top seals of clay liners; Top portion of slurry wall; Surface erosion.
FAUNA	Animals	Micro-Animals Protozoa Amoeba, Flagellates, Ciliates, Infusoria.	x	x		Bacteria corrosion; Landfill liners.
FAUNA	Animals	Nematoda Feed on decaying organic matter; Feed on earthworms, protozoa & bacteria.	x	x		Bacteria corrosion;
FAUNA	Bacteria	Heterotrophic Energy & carbon directly from the soil organic matter.	x	x	x	Bacteria corrosion; Clay liners; Slurry walls.
FAUNA	Bacteria	Autotrophic Energy from oxidation of mineral constituents.	x	x		Bacteria corrosion

* Soil profile and horizons (see Figure 1)

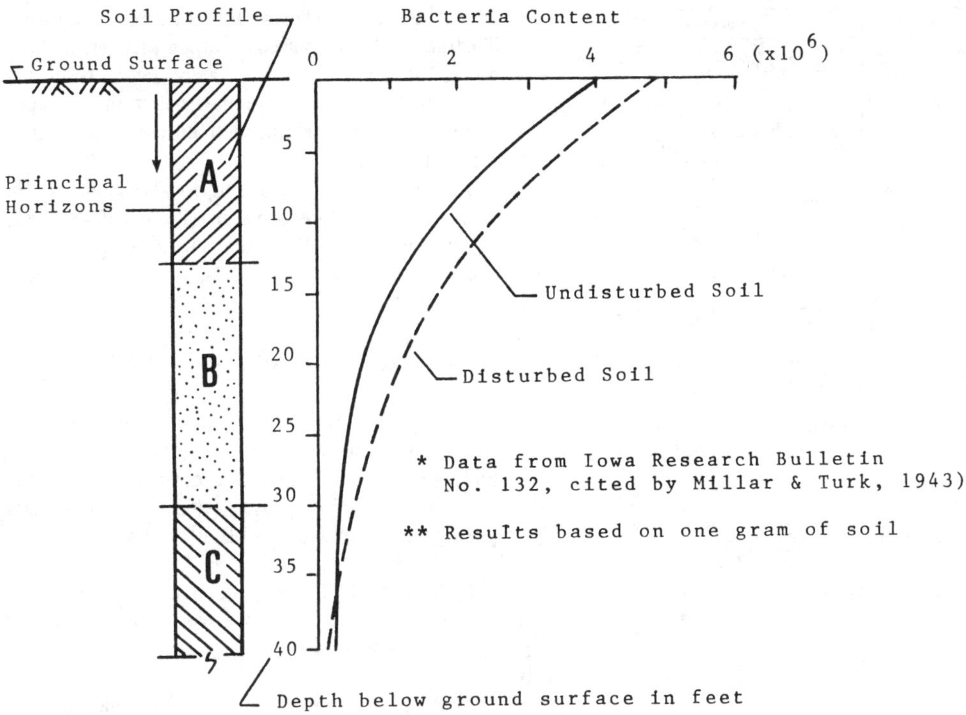

Figure 1 Distribution of bacteria in soil layers

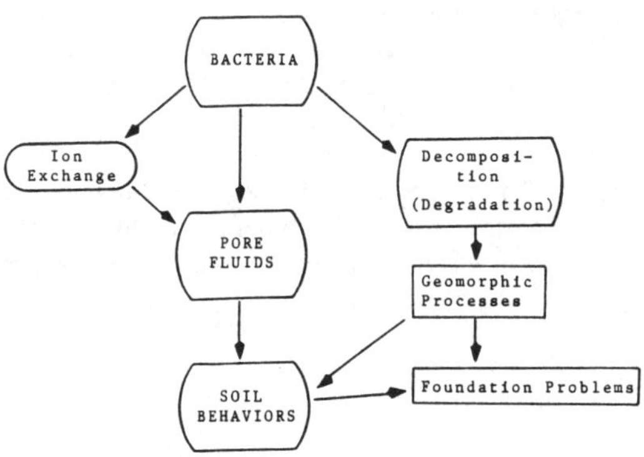

. Soil Parameters
. Volume Change (Shrinking & Swelling)
. Compaction (Maximum Density & Optimum Moisture Content)
. Hydraulic Conductivity

Figure 2 Bacteria-soil-water interaction (After Fang, 1987a)

fluid, or through decomposition (geomorphic process) of soil constituents.

2.3 Bacteria Decomposition Process

Three major decomposition processes in landfill areas are illustrated in Figure 3, namely mechanical, chemical and physico-chemical and bacterial decomposition processes. However, the bacteria process is not as visual in comparison with the other process.

During the decomposition process in the landfill areas, the ground temperature increases. When temperatures are high, detention times are long and sulfate (S) concentrations are appreciable. The difficulty is always associated with reduction of sulfates to hydrogen sulfide (H_2S). Hydrogen sulfide is often blamed for the corrosion of various geostructural members, however, in sanitary landfill areas, the refuse is nonhomogeneous and contains organic matter, biological changes caused by decomposition constantly occur. Figure 4 shows the typical soil-water-bacteria interaction. Bacteria can change hydrogen sulfide to sulfuric acid in a landfill

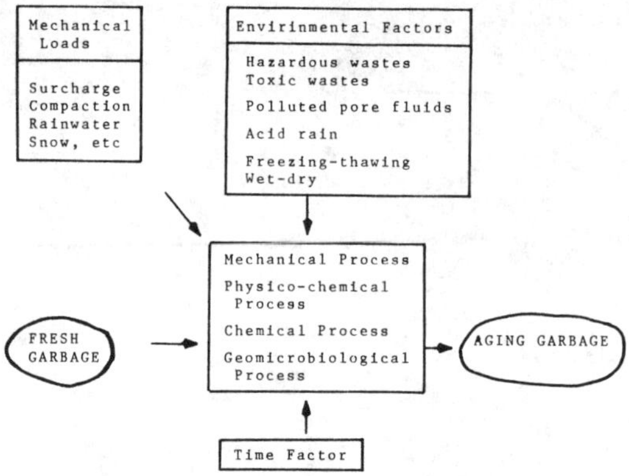

Figure 3 Aging process (geomorphic) stages between fresh and aging garbage

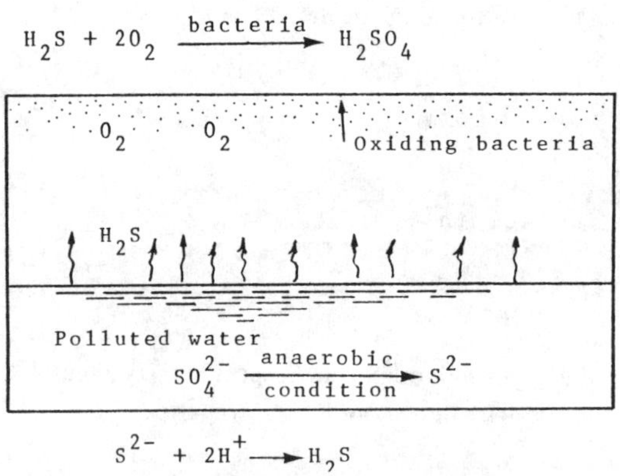

Figure 4 Formation of hydrogen sulfide in landfill area resulting from oxidation of hydrogen sulfide to sulfuric acid

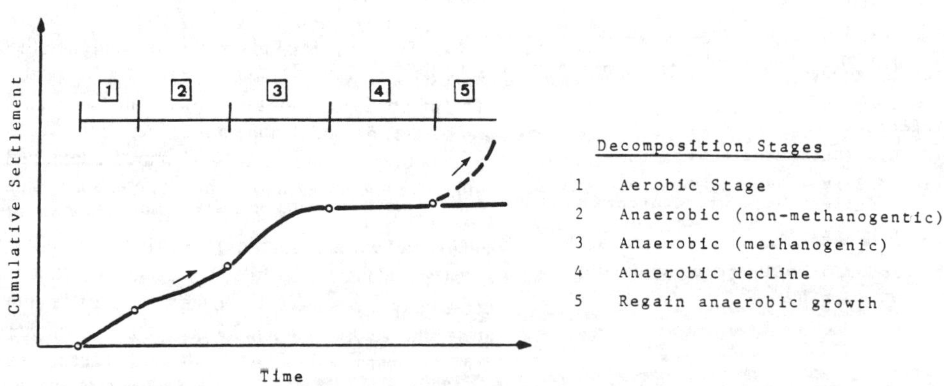

Figure 5 Schematic diagram illustrating the cumulative settlement versus time due to decomposition of organic matters (Based on Wardwell, Charlie and Doxtader, 1982)

area. Figure 5 shows the schematic diagram illustrating the cumulative settlement versus time due to decomposition of organic matter. Basically, there are five decomposition stages with all stages related to the bacteria activities in the refuse material.

Bacteria can transfer in an ion-exchange which changes the soil/water structure (Kelley, 1948; Fang, 1987b), or changes the characteristics of pore water as shown in Table 2.

3. TREE ROOT ATTACK ON LINERS

The landfill areas, in general, are planted with trees/vegetations for the purpose of beautification of undesirable landfill sites. To plant trees or vegetation also has a detrimental side effect. Two important phenomena on how tree roots attack liners:

(1) Tree roots looking for nutrition (food), the landfill area is an ideal place. Roots are capable of and can penetrate liners.

(2) Pore fluid (leachate) in landfill will seep out through root hairs by suction force.

3.1 Soil-Root Interaction

In a study reported by Fang (1991b), tree root interaction indicates that the shape of the tree branches is proportional to the shape of the tree roots. This means, that tall trees or vegetation have a deeper or larger distribution of roots. In order for the tree to be stable above ground, the roots serve

TABLE 2 EFFECT OF BACTERIA ON PORE FLUIDS

(a) $H_2S + 2O_2 \xrightarrow{\text{Bacteria}} H_2SO_4$

(b) $CH + O_2 \xrightarrow{\text{Bacteria}} CO_2 + H_2O$

(c) $CH_2O + O_2 \xrightarrow{\text{Bacteria}} CO_2 + H_2O$

(d) $C_6H_{12}O_6 + 6O_2 \xrightarrow{\text{Bacteria}} 6CO_2 + 6H_2O$

(e) $SO_4^{2-} + \text{Organic Matter} \xrightarrow[\text{Bacteria}]{\text{Anaerobic}} S^{2-} + H_2O + CO_2$

(f) $\text{Organic Sulfur Compounds} + O_2 \xrightarrow{\text{Bacteria}} CO_2 + H_2O + SO_4^{2-}$

(g) $\text{Organic Nitrogen Compounds} + O_2 \xrightarrow{\text{Bacteria}} CO_2 + H_2O + NO_3^-$

(h) $\text{Organic Phosphorus Compounds} + O_2 \xrightarrow{\text{Bacteria}} CO_2 + H_2O + PO_4^{3-}$

i) $\text{Organic Sulfur/Nitrogen Compounds} + H_2O \xrightarrow[\text{Decay}]{\text{Anaerobic}} CO_2 + H_2S + CH_4 + NH_4^+$

Data from: Dickson (1974); Sawyer and McCarty (1978); Stevenson (1982); Bitton and Gerba (1984)

just as foundations do for building structures, the taller or bigger the tree, the deeper or wider the root distribution.

Roots are always looking for food and water and landfill sites provide all these necessities for growing tree/vegetations (Figure 6). The force produced by tree roots is strong enough to penetrate the clay liners or synthetic membrane. For example, bamboo roots can penetrate into a 6-inch thick concrete wall. Many roadways, pavements and drainage facilities have been damaged by live bamboo roots, other common trees planted around landfill areas are silver maples, willows and poplar. These tree roots will also damage liners, but less seriously in comparison with bamboo roots.

3.2 Soil-Root-Water Interaction

All higher plants require immense quantities of water, practically all of which is absorbed from the soil through their root systems. Special organs are provided for absorption which include the tip of roots and innumerable root hairs (Figure 7). The root tips are always seeking new 'feeding ground'. Garbage dump site is one such rich food storage area. Usually, only a comparatively small quantity of water is taken in through the leaves. Water is one of the essentials for all life processes. Some of it is decomposed into its elements in plants and these elements become fixed in the tissues. Furthermore, all chemical processes within plant cells take place in water mediums.

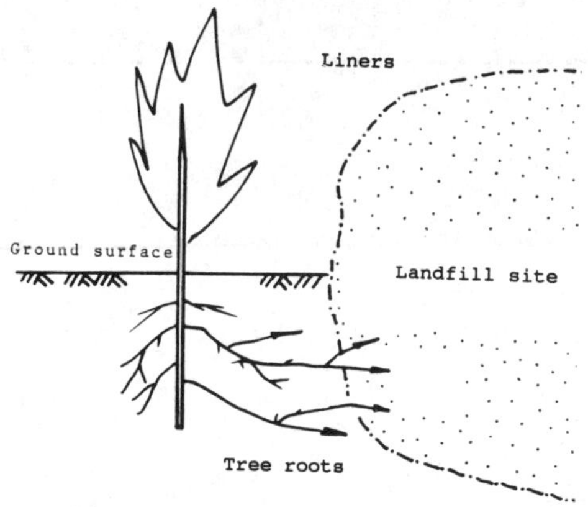

Figure 6 Tree roots looking for nutrition in the landfill, consequently roots will penetrate liners

Moisture enters roots by the process of osmosis. Because the plant cell within the root has greater concentration of soluble material than the soil moisture, hence water passes in to equalize the concentration. How leachate seeps out from landfill through root hairs is illustrated in Figure 7. This phenomenon of moisture migration between root hairs and voids of soil is part of capillary action.

4. CONTROL OR MINIMIZATION OF BACTERIA AND TREE ROOT ATTACK ON LANDFILL LINERS

4.1 Control or Minimization of Bacteria Attack

As discussed in Section 2.2 and Figure 4, indications are that bacteria will cause changes in pore fluid behavior. In general, they change from weaker acid to stronger acid as indicated in Table 2. Soil acidity is important to a geotechnical engineer because of its corrosive effect on metals and other construction materials; on the other hand, acidity is sometimes desirable because of its catalytic effect on certain reactions employed in soil stabilization (Winterkorn and Pamukcu, 1991).

Recent research results indicate that when pH values decrease, the static shear strength on laboratory prepared liner specimens (sand-bentonite mixtures) increases (Niak, 1986). However, contradicting results are found when the pH value decreases, the dynamic shear modulus also decreased as reported by Du, et al. (1986) which is explained as when the bonding energy created by acid between soil particles is destroyed by dynamic loading. In other words, the dynamic loading can cause more damage than static loading alone in the

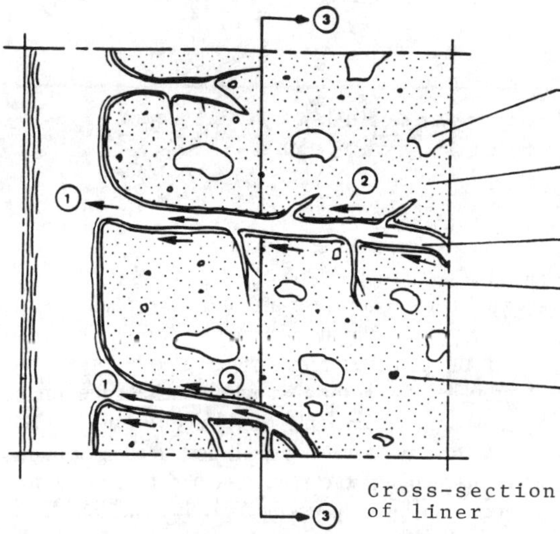

(1) Pore fluid (leachate) in landfill seep out through root hairs by suction force

(2) Pore fluid (leachate) in landfill seep out between root and soil by capillarity action

Figure 7 Mechanism of soil-root-leachate interaction at landfill area landfill liners, therefore, precautions must be taken when landfill sites are in an earthquake zone or any other dynamic loading can be applied.

4.2 Control or Minimization of Tree Root Attack

Some vegetation such as squash, spread their roots laterally and widely within the surface 1-foot zone. Some root systems are coarse and open, whereas others consist of masses of fine fibrous roots. The ability of roots to obtain nutrient substances also varies greatly. Even though, we cannot explain wholly the quantity of nutrients produced by a landfill that is excepted by roots, the forces within the plants and around their feeding roots may also have a controlling influence.

Plant species differ not only in their water requirements, but also in their adaptation to water conditions. Some plants are so constituted that they can tolerate an abundance of soil water, whereas others can endure limited moisture. Therefore, selecting types of tree/vegetation must be carefully planned. Aggressive types such as bamboo, silver maples, willows, poplars should be avoided, especially the bamboo family.

5. SUMMARY AND CONCLUSION

(1) Liners are the major structural elements for controlling landfill liquid waste leaking into surrounding areas or aquifers. Most liners are made of clay and synthetic membrane. In general, engineers have taken all necessary measures to make sure that these controlling facilities are safe. However, some additional factors such as the bacteria and tree root attack on landfill liners are believed to be an important factor requiring further investigation.

(2) Bacteria can cause a change in the characteristics of leachate from landfill as shown in Table 2 and consequently will change the performance of liners.

(3) Dynamic loading can cause more damage than static loading alone in the landfill liners, therefore, precaution must be taken when landfill sites are in an earthquake zone or any other dynamic loading can be applied.

(4) Unnoticed hairs of tree/vegetation roots play an important role for pore fluid (leachate) seeping out from landfill or damaged liners as these roots search for nutrition (Figure 6). Tree/vegetation planted around the landfill must be considered (Figure 7).

(5) For selecting types of tree/vegetation, aggressive types such as bamboo, silver maples, willows, poplars should be avoided, especially the bamboo family.

REFERENCES

Allinger, N. L., Bigelow, M. J. and McAllister, H. C. (1976), *An Introduction to General, Organic, and Biological Chemistry,* Wadsworth Publishing Co. Belmomt, CA, 582p.

Bitton, G. and Gerba, C. P. (1984), editors, *Groundwater Pollution Microbiology,* John Wiley & Sons, NY, 377p.

Dickson, T. R. (1974), F*rom Atoms to Attitudes,* John Wiley & Sons, NY, 361p.

Du, B. L. , Mikroudis, G. K. and Fang, H. Y. (1986), Effect of pore fluid pH on the dynamic shear modulus of clay, Hazardous and Industrial Solid Waste Testing and Disposal, ed. by Lorenzen, et al. *ASTM STP* 933, pp. 226-239.

Evans, J. C. (1991), Geotechnics of hazardous waste control systems, Chapter 20, *Founation Engineering Handbook*, 2nd ed. Van Nostrand Reinhold Co. NY, pp. 750-777.

Fang, H. Y. (1986), Introductory remarks on environmental geotechnology, *Proc. lst Intern. Sym. on Environmental Geotechnology*, v. 1, pp. 1-14.

Fang, H. Y. (1987a), Effect of bacteria on behavior of soil-water systems, *Proc. lst Intern. Sym. on Environmental Geotechnology*, v. 2, pp. 290-294.

Fang, H. Y. (1987b), Effect of ion-exchange reaction on soil-water behavior, *Proc. lst Intern. Sym. on Environmental Geotechnology*, v. 2, pp. 297-300.

Fang, H. Y. (1991a), Theory of particle energy field and its applications to environmental geotechnology, *Proc. 2nd Intern. Sym. on Environmental Geotechnology*, v. 2, pp. 277-307.

Fang, H. Y. (1991b) The use of bamboo inclusions for earth reinforcement, *Proc. Biotechnical Stabilization Workshop*, The University of Michigan, D. H. Gray, ed. pp. A-33 to A-62.

Kelley, W. P. (1948), *Cation Exchange in Soils*, Reinhold Publishing Corp, NY, 144p.

Millar, C. E. and Turk, L. M.(1943), *Fundamentals of Soil Science*, John Wiley & Sons, NY, 462p.

Naik, D. (1986), Effect of temperature and pore fluid on shear characteristic of clay, *Proc. lst. Intern. Symposium on Environmental Geotechnology*, v. 1, pp. 382-390.

Sawyer, C. N. and McCarty, P. L. (1978), *Chemistry for Environmental Engineering*, McGraw-Hill Book Co. NY, 532p.

Stevenson, F. J. (1982), *Humus Chemistry*, John Wiley & Sons, NY, 443p.

Wardwell, R. E., Charlie, W. A. and Doxtader, K. A. (1982), Test method for determining the potential for decomposition in organic soils, *ASTM STP* 820, pp. 218-229.

Weir, W. W. (1936), *Soil Science, Its Principles and Practice*, J. B. Lippincott Co. Chicago, 615p.

Winterkorn, H. F. and Pamukcu, S. (1991), Soil stabilization and grouting, Chapter 9, *Foundation Engineering Handbook*, 2nd ed. Van Nostrand Reinhold Co. NY, pp. 317-378.

Waste Disposal by Landfill – GREEN'93, Sarsby (ed.) © 1995 Balkema, Rotterdam, ISBN 90 5410 356 6

Steeply sloping lining systems – Stability considerations using reinforced soil veneers

C.D.Hall & A.J.T.Gilchrist
Netlon Limited, Blackburn, UK

ABSTRACT: A soil cover layer to a lining system is often required to prevent construction and operational damage. In landfill capping, a soil cover layer is required in returning the site to agricultural or amenity use.

The stability of the soil cover layer must be considered at the design stage especially as 'steep' side slopes may have been incorporated to maximise the landfill capacity. Structural geogrids provide a tension sustaining component which allows the designer to achieve the required factor of safety against sliding failure of the soil veneer on steep slopes.

This paper summarises the practice of reinforced soil veneers.

INTRODUCTION

The practice of veneer stabilisation in landfill engineering can be traced to the United States of America (Chouery-Curtis and Butchko 1991). Where failures of soil cover layers on geomembranes began to be experienced as lining of containment facilities became mandatory. One of the factors common to most of the sliding failures was the low interface friction angle between various synthetic components, or synthetic components and soil layers. Side slopes of 12° - 18° were commonly being designed and constructed yet the critical interface friction angle between some components can be as low as 8°.

In the UK, side slopes of the lining system are often required to be in excess of 20°. As the 'doming' of capping systems becomes more prevalent, these too can present sliding stability concerns. The resulting increase in the sliding forces can lead to excessive strain and eventual rupture of critical components of the lining or capping. It is evident that, in these circumstances, a tension sustaining component can provide a beneficial restoring force.

THE GEOGRID COMPONENT

Geogrids are planar structures formed by a regular network of tensile elements having apertures, of appropriate geometry, which set up an interlock with adjacent soil. The attendant transfer of soil loads to the geogrid thereby performs the function of soil reinforcement and soil veneer stabilisation.

As a structural component in a soil veneer, there are four essential geogrid design properties:

1. Long-term stress strain characteristics which are used to identify the safe design strength.
2. The junction strength where soil load transfers from the transverse bars of geogrid to the longitudinal ribs which accumulate load along the slope.
3. Robustness during installation.
4. Polymer durability throughout the design life.

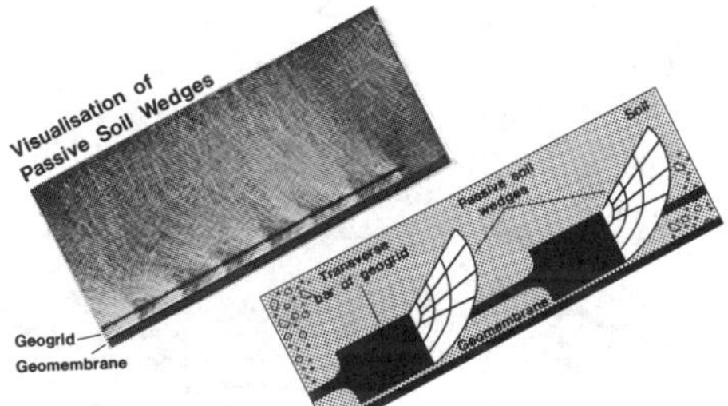

Figure 1 The local effects of a geogrid

Figure 1 illustrates how a geogrid functions within the soil veneer. The upper portion of the figure shows a visualisation of the bearing of the soil on transverse ribs and the lower portion conceptualises the localised passive soil wedges. These abutment forces need then to be wholly and fully transferred to the longitudinal ribs and ultimately to an anchorage arrangement at the top of the slope.

STABILITY ANALYSIS

A commonly used design method is based on a rigid block analysis using simplifying, but conservative, assumptions. Figure 2 shows the forces involved in the sliding stability of the soil layer on a geomembrane. This simple diagram illustrates the essence of the problem. Examination of the term for Factor of Safety shows that stability problems arise for a soil cover layer on a geomembrane at gradients of 1:4 (vertical:horizontal). It is evident that an extra upslope force has the potential to preserve a desired factor of safety at steeper angles.

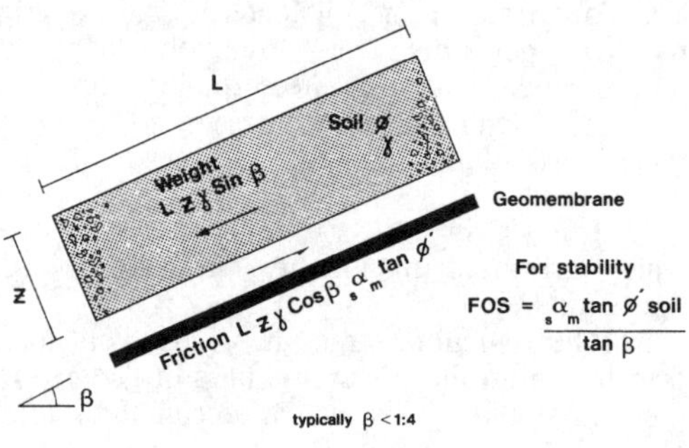

Figure 2 The stability of a soil cover layer

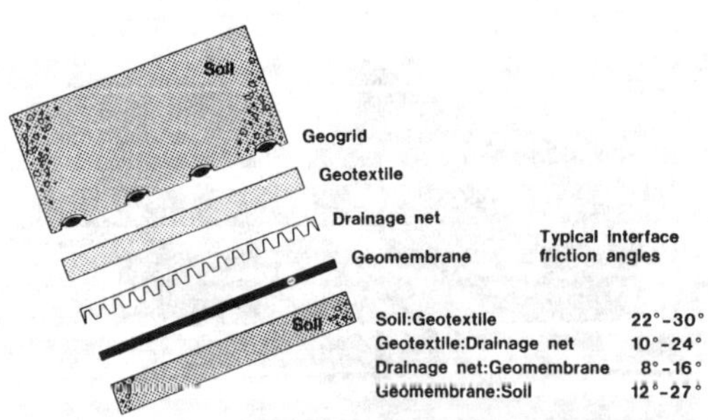

Figure 3 Reinforced soil veneer in landfill lining and capping

Real lining systems are more complex, as shown in Figure 3. A variety of protection and drainage components are often incorporated above and below what is the essential component - the geomembrane. Each interface needs to be examined separately to identify which has the least frictional value. Many of the values overlap, as shown in the table, and a design cannot be finalised until the most critical interfaces have been examined. There is no substitute for large shear box testing of the proposed components to verify the final design.

A feature of the design method which has not received critical scrutiny until recently concerns the strain condition of the geomembrane. One of the aims of landfill lining design is to ensure that the geomembrane can perform its containment function without the need to sustain excessive strain. Veneer stabilisation therefore must perform the duty of controlling geomembrane strain.

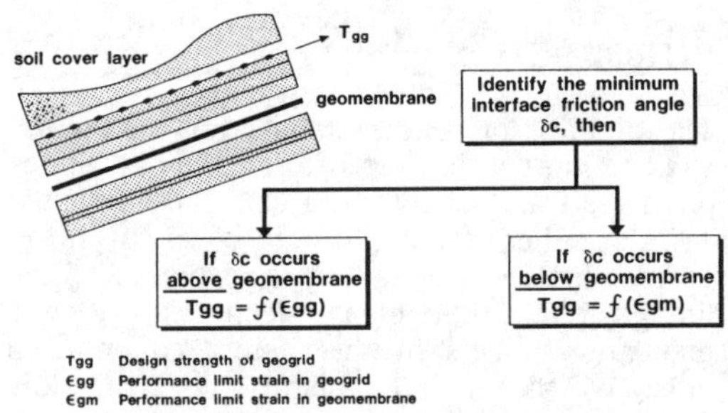

Figure 4 Reinforced soil veneers strain conditions

As Figure 4 shows, if the critical interface occurs above the geomembrane then the tensile force required in the geogrid Tgg is a function of the strain ε_{gg}, in the long-term, that the geogrid can sustain. The geomembrane is thus 'protected' by the geogrid in the strain sense. If the critical interface occurs below the geomembrane - such that the geomembrane is within the sliding block - then Tgg is a function of the allowable long-term strain in geomembrane ε_{gm}.

CASE STUDY

INTRODUCTION

Lochaber District Council, within the Highland

Region, owns and operates a landfill site at Brackletter to accept domestic waste from Fort William, Spean Bridge and the surrounding areas.

DESCRIPTION OF SITE

The site of the landfill is part of a sand and gravel quarry which is now partially worked out. The area in use has been divided into four cells numbered 1 to 4 as shown in Figure 5.

Although four cells are available, filling has been carried out in two distinct phases with cells 3 and 4 being filled first. These two cells were overfilled to form a dome approximately 8m high above the surrounding ground level. The dome was finished off with the daily soil cover.

The area is renowned for its outstanding natural beauty and the landfill has been designed to be returned to leisure use once filling is complete.

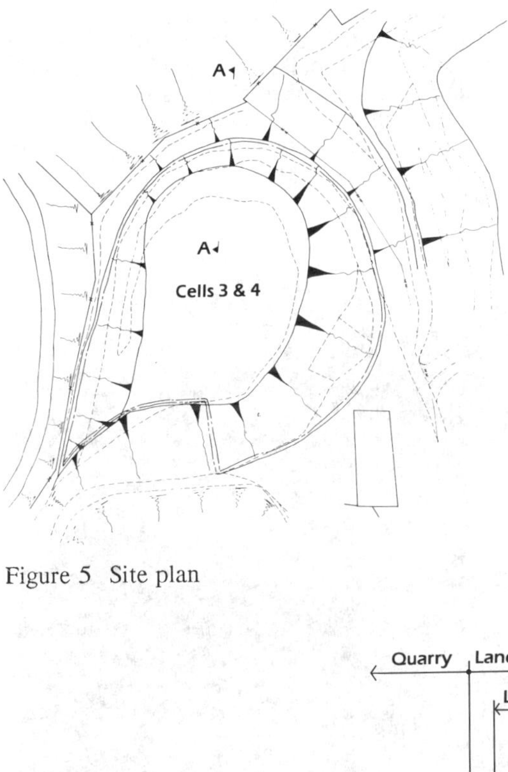

Figure 5 Site plan

DEFINITION OF PROBLEM

In order to maximise the capacity of the cells the overfilling was profiled to a slope of 1 vertical to 2 horizontal. The required capping detail was as shown in Figure 6.

However, with a slope angle of some 27° from the horizontal, the sand and topsoil layer were prone to sliding on the geomembrane. Although the internal angle of friction of the sand was higher than the slope angle, the critical interface angle was that of between the sand and the geomembrane. This angle was in the order of 8° to 12°. Downslope instability therefore had to be addressed and an engineered solution obtained.

DESIGN

Based on the theory described above, a reinforced soil veneer system was devised to achieve the required factor of safety. The slope length (L) was 19.0m and the total soil cover thickness (z) was 1.0m.

The anchorage of the geogrid at the top of the cap was designed as a horizontal anchorage length within the level top section of the cap. This eliminated the need for a fixing detail such as a trench and though it

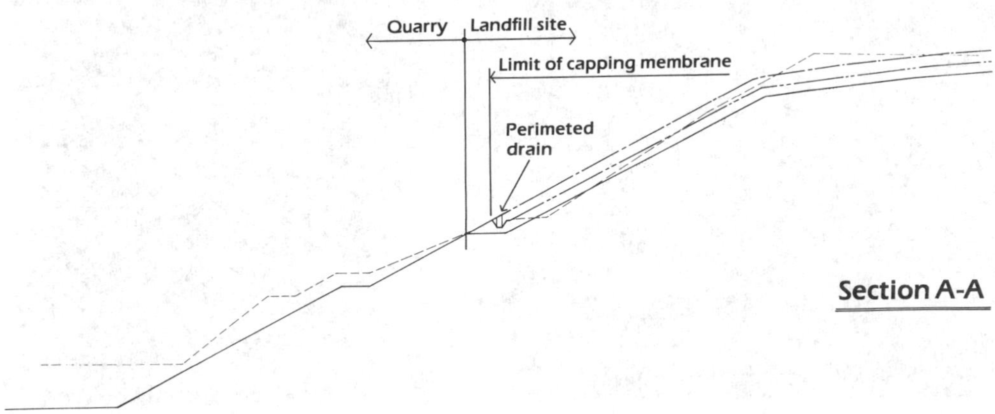

Figure 6 Capping detail

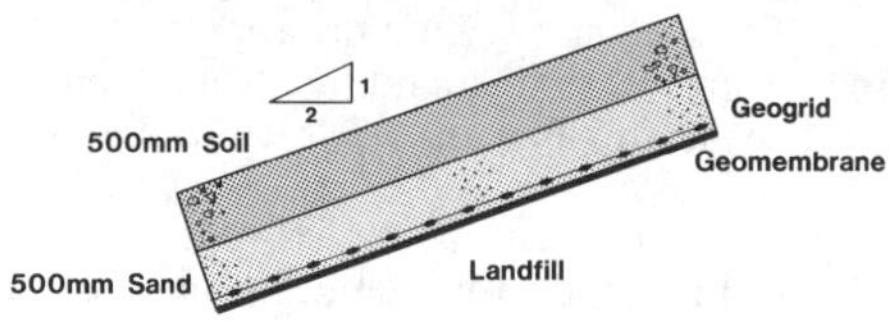

Figure 7

Figure 8

used slightly more geogrid it simplified the construction.

The final design section is shown in Figure 7.

CONSTRUCTION

The refuse cap was initially profiled to the required angles and the access berms were constructed. The methane drainage and venting system were then installed.

A separate subcontract was let to install the capping liner (and the base liner for the adjacent cell).

Once the lining subcontract was completed, the main contractor installed the geogrid veneer system. Tensar SR110 geogrid was rolled down the slope as shown in Figure 8 having provided sufficient length at the level top section to install the anchorage detail. Adjacent lengths were simply abutted, with no formal connections. Joints along the length of a geogrid strip were made with a proprietary bodkin which provides a full strength connection.

Figure 9

To aid the contractor's access to the site the cover layer was dozed over the geogrid from the top using a small bladed machine: the normal process is to fill from the bottom towards the top of the slope. No problems arise using this procedure, providing the anchorage design allows for the machine induced forces.

The contractor then placed the topsoil and created the final slope angles before seeding.

SUMMARY

The veneer has now sustained its first winter in an area renowned for heavy rainfall at approximately 2 metres per annum. In these circumstances, pore water pressure considerations should also enter the stability evaluation. The vegetation has established and there is no evidence on any of the slopes of soil slippage or creep.

The cap has taken on a natural appearance and blends in with the local environs, as shown in Figure 9.

REFERENCE

Chouery-Curtis, V.E. and Butchko, S.T. 1991. Structural Geogrids used to Stabilise Soil Veneer Covers. Geosynthetics '91 Conference, Atlanta.

Waste Disposal by Landfill – GREEN'93, Sarsby (ed.) © *1995 Balkema, Rotterdam, ISBN 90 5410 356 6*

Selection of clays for use as landfill liners

R. M. Jones, E. J. Murray & D. W. Rix
Murray Rix Partnership, Hinckley, UK

R. D. Humphrey
Lincs Lab., Lincoln, UK

ABSTRACT: The need to design adequate landfill lining systems has been highlighted by recent publications and legislation. The careful selection of materials to form low permeability barriers is essential in engineered waste disposal. The term 'suitable' (for use as a natural clay landfill lining material) is introduced. Individual geological deposits within Lincolnshire have been classified as unsuitable, marginal or suitable based primarily on plasticity criteria. The term 'acceptable' (for use as an earthworks material) is also introduced. Use of the Moisture Condition Value test as a means of acceptability testing, earthworks control and a guide to the degree of achievable compaction is discussed. The results presented were obtained during a study of possible sources of natural material for landfill liners within Lincolnshire.

1 INTRODUCTION

In landfill operations the safe containment of waste and control of leachate is of paramount importance in the mitigation of ground and groundwater pollution and the protection of public health. Recent publications and legislation including the National Rivers Authority Groundwater Protection Policy 1992, Control of Pollution Act 1974, Environmental Protection Act 1990, and various EC Directives call for more stringent control on the containment of waste. Careful selection of materials coupled with rigorous design of containment cells are a vital part of engineered waste disposal systems.

The linings to these cells generally consist of low permeability clays possibly in conjunction with a synthetic liner or liners. This paper explores the use of a classification scheme to identify possible sources of natural materials suitable for use as lining material to landfill sites. It is based primarily on a study carried out within Lincolnshire. The classification is applied to individual strata identified from geological maps and use is made of available published and ground investigation information. Suitability criteria laid down by the National Rivers Authority (NRA) are appraised and the classification of material is based on these criteria and recent work on the relationship between plasticity and permeability.

Material suitability relates to material type and its potential to form a barrier of sufficiently low permeability. Within the study suitability is based primarily on the collation of plasticity data, and is closely linked to the definition of 'material characteristics' as given in BS 5930:1981

For individual landfill sites it is essential to examine in more detail the properties of the clay proposed for use as a lining. Thus in addition to the definition of suitability of material the term acceptability in earthworks is introduced and is closely linked to the 'mass characteristics' of a soil as defined in BS5930:1981. It relates to the ability of a material to be satisfactorily excavated, transported, placed and compacted. The definition of acceptability is comparable with that adopted by the Department of Transport in the Specification for Highway Works (1991) which is often used as a guide to compaction requirements for clay linings.

Given the above it is apparent that if a material is unsuitable it is also unacceptable however a material which is suitable will not necessarily be acceptable.

2 MATERIAL SUITABILITY

It is generally taken that the overriding requirement of a material for use as a landfill lining is its capability of achieving a permeability of 1 x 10^{-9} m/s or less following compaction. This degree of permeability is sometimes defined as 'practically impervious' and often taken as distinguishing clays from higher permeability silts (Somerville 1986). In terms of plasticity the division between clays and silts is known as the A-Line as shown on Figure 1. In the study, data was collected from a large number of sources, but as might be expected, relatively few permeability results were obtained for specific deposits. It thus became necessary to identify potentially suitable deposits using other more readily available soil parameters such as plasticity. In practice permeability is related to a large number of factors and materials classed as silt based on plasticity data could achieve the required low permeability if adequately compacted within an acceptable moisture content range. Nevertheless because of the lack of test data and in accordance with the NRA requirements materials plotting below the A-Line were defined as unsuitable as were those materials with greater permeability e.g. sands and gravels. Conversely clays which plot above the A-Line were deemed as suitable or marginal in terms of permeability.

The NRA also define suitable materials as those clays with a Liquid Limit (LL) of less than 90%, a Plasticity Index (PI) of less than 65% and a clay content of greater than 10%. The upper limits to LL and PI specified by the NRA are based on criteria defined by the Department of Transport (1991) for modern earthworks plant. Materials in excess of these limits can give rise to problems in stability, deformation and compaction in earthworks. However these limits preclude the use of extremely plastic clays which would exhibit very low permeability characteristics and if used with care may be adopted as a lining material. Sources of such materials are scarce and have not been identified within Lincolnshire. Further study of the use of such deposits would be warranted. However for the purpose of the study materials above these limits were defined tentatively as marginal as shown on Figure 1.

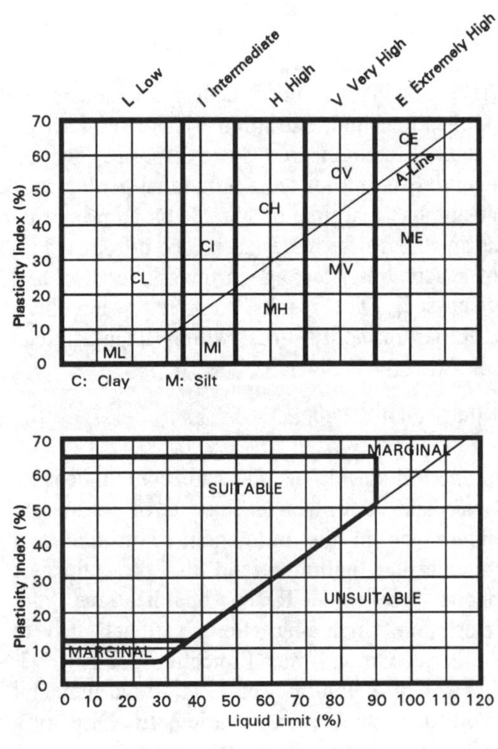

FIGURE 1: PLASTICITY AND SUITABILITY OF MATERIALS

From a standard Plasticity Chart the lower limit of PI for suitable clays would be 6% in accordance with the NRA requirements. However the permeability of clays varies with PI and Murray et al (1992) suggest there is a notable increase in permeability where the materials have a PI of 12% or less. A PI of 12% is also considered to correspond closely with a clay content of 10% (Skempton 1953) which forms one of the NRA suitability criteria. Results of percentage clay in deposits were sparse and the adoption of a lower limit to PI of 12% was taken as an alternative to clay content in collating the data and classification in terms of suitability. Examination of the relationship between permeability, clay content and plasticity suggests that clays with a PI in the range 6% to 12% should be classed as marginal.

Figure 1 illustrates plasticity with respect to the suitability of materials. The availability of plasticity results from a wide variety of sources means initial classification of many deposits can easily be achieved in a study similar to the one undertaken.

For any deposit which includes both discrete cohesive and granular components the use of plasticity results as a method of classification will present a biased view since only the cohesive materials would have been selected for testing. Highly variable materials should be classed as marginal or even unsuitable since the end product, the landfill liner, must be homogeneous and conform to a specified standard.

The NRA preclude the use of organic material within landfill liners. However Postlethwaite (1991) suggests that under managed conditions peat could be used successfully within a composite lining system. There must obviously be reservations about the use of such highly organic materials and detailed investigation of these deposits would be necessary to justify their use.

Consideration of the degree of variability, as highlighted by the pattern of plasticity results, and effects of likely constituents of a deposit are important in classification of suitability.

3 MAPPING OF MATERIAL SUITABILITY

In the study initial identification and location of sources of potentially suitable material for landfill linings were based on published geological maps and memoirs. Identification of unsuitable materials i.e. rocks, sands and gravels was straightforward from such information however, for the cohesive deposits a more detailed assessment was required.

Figure 2 shows plasticity results for Alluvium and Fen Deposits. The wide spread of results is

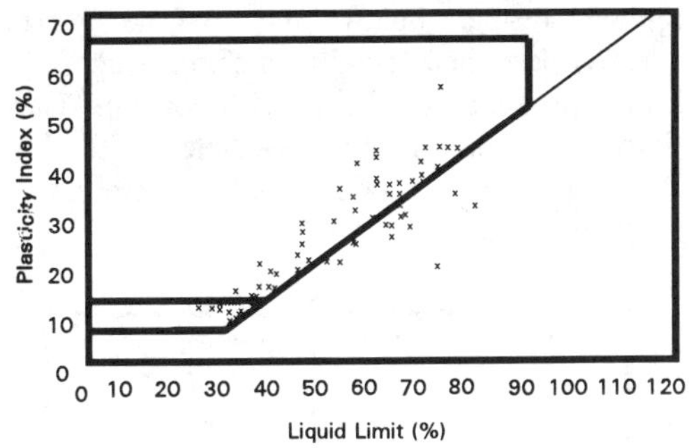

FIGURE 2: PLASTICITY CHART: ALLUVIUM AND FEN DEPOSITS

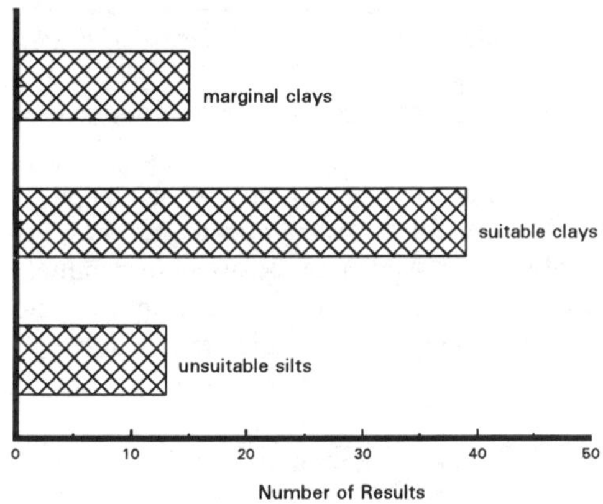

FIGURE 2A: COMPARISON OF PLASTICITY RESULTS

to be expected from an inherently variable material, and highlights the presence of clays of low to high plasticity along with silts. Figure 2A compares the number of results of marginal and suitable clays and unsuitable silts. The majority of results fall in the suitable clay category. However granular materials are also present within this deposit but cannot be represented on Figure 2 or 2A. An overall assessment of this material would suggest that due to the wide variation in results a marginal classification would be appropriate.

By contrast the results for the Lower Lias Clay, Figure 3, show a much less variable pattern with mainly intermediate and high plasticity clays. Although a proportion of results fall within the silt category, none fall within the marginal clay range. Those samples yielding results below the A-Line were generally described visually as comprising silty clays suggesting a significant clay content. The relatively uniform results associated with the Lower Lias Clay are considered to represent a suitable material.

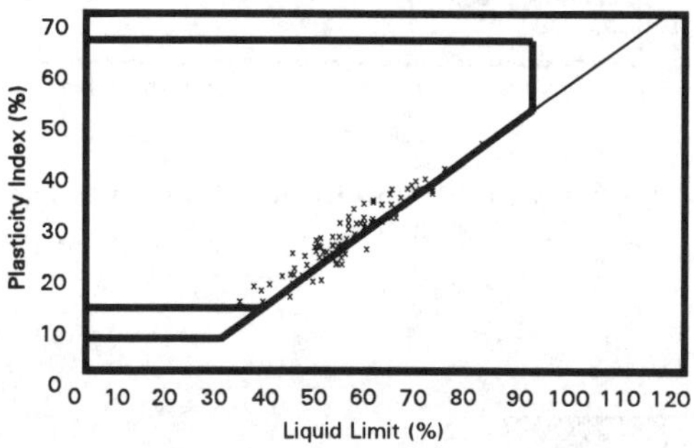

FIGURE 3: PLASTICITY CHART: LOWER LIAS CLAY

Following similar appraisals a map of material suitability was drawn up for the County. The map may be used to facilitate planning and as a basis for geographical and economic appraisal of potential landfill sites.

4 ACCEPTABILITY IN EARTHWORKS

Acceptability in earthworks relates to the excavation, handling, trafficability and compaction characteristics of the material in achieving a low permeability barrier. Problems may be experienced in general earthworks with materials which are too soft or too wet, whilst with very stiff materials the degree of compaction necessary to obtain the required permeability may not be achievable. Adequate compaction of suitable materials is essential to form a low permeability barrier. Structural features such as fissures and laminations in the emplaced lining may provide preferential seepage paths and hence increased permeability.

Interstratified or mixed deposits comprising both suitable and unsuitable materials will affect the acceptability of the strata as a whole. The ease of identification and segregation of different classes of material from a single deposit needs to be considered where such cases arise.

5 ACCEPTABILITY CRITERIA

In most earthworks operations it is strength which governs material acceptability. However in linings to landfill sites the overriding requirement is a low permeability. In achieving the desired permeability and in placement of the materials it is necessary to place restrictions on the moisture content, undrained shear strength and density of the emplaced material. These parameters usually form the basis of control measures in place of permeability because of the protracted time associated with permeability measurements. The use of alternative, less time consuming tests to establish limiting values of acceptability is advocated by the Department of Transport (1991). However the definitive parameter, in this case permeability, of placed material must be verified.

As an alternative control criteria the Moisture Condition Value (MCV) (Parsons and Boden 1979) may be used. The MCV is a compaction test which produces a degree of compaction which generally lies between those achievable using the BS 2.5kg and 4.5kg rammer tests. The use of the MCV test in general earthworks is described by Cobbe and Threadgold (1988). The MCV test provides rapid and reproducible results in determining the acceptability of materials, which combined with a maximum air voids requirement to control the degree of compaction may be used to assess the overall adequacy of the placed lining.

In adopting any form of control criteria the relationship of permeability to the control parameters needs to be established. Figure 4 illustrates the relationship between permeability and moisture content. The general pattern of decreasing permeability with increasing moisture content over a wide range of results is consistent with work by Murray et al (1992), Needham (1991), Seymour (1992) and Parkinson (1991). Close to and dry of the optimum moisture content the increase in permeability is likely to be attributable to difficulties in remoulding clays which are likely to be around or dry of their plastic limit. Wet of the optimum the general decrease in permeability is consistent with remoulding of the material and removal of discontinuities such

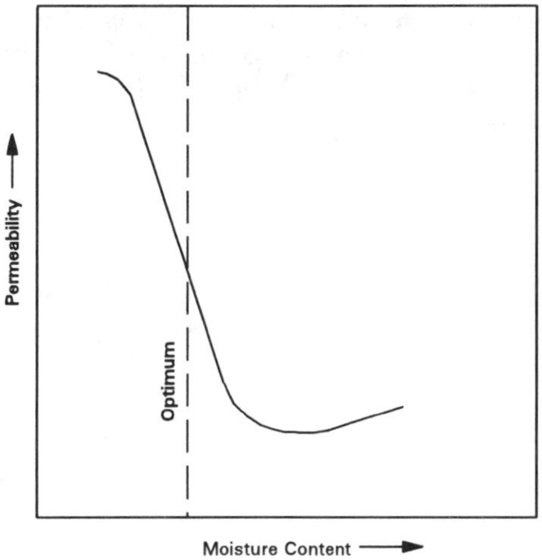

FIGURE 4: PERMEABILITY AGAINST MOISTURE CONTENT

as fissures. Remoulded wet clay was used as a seal in canal construction, where it was puddled against the sides and bottom of the canals to produce an almost watertight seal. However modern plant and equipment are unlikely to be able to traffic or compact such very wet and soft materials. Thus appropriate limits need to be placed on acceptability.

On the basis of the foregoing discussion the lower limit of moisture content or upper limit of MCV would be governed by the permeability requirement, whereas the upper limit to moisture content or lower limit to MCV would be likely to be controlled by shear strength (Murray et al 1992).

In order to ensure adequate compaction, limits should be applied to the density and air voids content of the emplaced material. Compaction trials are essential to establish the control parameters to be adopted and total reliance on laboratory testing is unwise.

6 SUMMARY

1. Based on a study within Lincolnshire a classification of natural deposits for use as landfill liners, using the concept of material suitability, is discussed. In the study materials were defined as unsuitable, marginal or suitable primarily based on plasticity data. Update and reassessment of the classification is appropriate as further information becomes available.

2. The definition of suitability takes into account the requirements of the NRA and more recent work on permeability of recompacted natural clay.

3. Following classification of clays as either suitable or marginal, consideration is given to material acceptability in earthworks. The proposed lining material must have acceptable mass characteristics of strength and structure to facilitate handling, trafficking and recompaction to form a low permeability barrier.

4. The overriding requirement of any landfill liner is low permeability but placement of material cannot be readily controlled by permeability tests. Establishment of relationships between permeability and other soil properties enables control of earthworks operations by more rapid acceptability tests. The use of the MCV test is discussed.

5. In control of earthworks the lower limit of moisture content or upper limit of MCV is governed by the maximum permeability requirement of 10^{-9}m/s.

6. The upper limit of moisture content or lower limit of MCV would generally be controlled by the strength of the material and handling, trafficking and compaction requirements.

ACKNOWLEDGEMENTS

The authors would like to thank Linc Waste for their co-operation and permission in allowing this paper to be published.

REFERENCES

British Standards Institution 1981. *Code of practice for site investigations BS 5930*

Cobbe M.I. and Threadgold L. 1988. Compaction control by the moisture condition test - a new approach. *Journal of the Institution of Highways and Transportation,* Dec 13 - 18.

Department of Transport 1991. *Specification for highway works.* HMSO

Murray, E.J., Rix, D.W., and Humphrey, R.D. 1992. Clay lining to landfill sites. *Quarterly Journal of Engineering Geology* Vol 25 No 4 p371 - 376.

National Rivers Authority North West Region 1989. *Earthworks to landfill sites.*

Needham, A. 1991. Clay liners - the answer lies in the clay. *Surveyor Special Supplement to Wastexpo International '91,* 18 + 19 Sept, Cannock, Staffs 11 - 13

Parkinson, C.D. 1991. The permeability of landfill liners to leachate. *The Planning and Engineering of Landfills,* Midlands Geotechnical Society

Parsons, A.W. and Boden, J.B. 1979. The moisture condition test and its potential application in earthworks. *Transport and Road Research Laboratory Supplementary Report 552.*

Postlethwaite, N. 1991. Dynamic permeability - peat as a landfill liner. *The Planning and Engineering of Landfills,* Midlands Geotechnical Society.

Seymour, K.J. 1992. Landfill linings for leachate containment. *The Institution of Water Management* Vol 6, No 4 August, 389 - 396.

Skempton, A. W. 1953. The colloidal activity of clays. *Proceedings 3rd International Conference on Soil Mechanics and. Foundation Engineering.* (Switzerland) Vol I, p 57.

Somerville, S.H. 1986 .Control of groundwater for temporary works. *Construction Industry Research and Information Association. Report 113*

Waste Disposal by Landfill – GREEN'93, Sarsby (ed.) © 1995 Balkema, Rotterdam, ISBN 90 5410 356 6

The use of lake sediment in sealing up building ground work as well as the bottom and the buttresses of dumping ground

J. Kokowski
Department of Geotechnics, Academy of Agriculture, Poznań, Poland

J. Kokowski
H. Kokowska Geo-Ecol-Bud, Poznań, Poland

ABSTRACT: The paper presents a new method for sealing up subsoil for construction and, particularly, for dumping ground with a soil composite and gythia suspension. The composite is made using in-situ ground with 5 to 25% addition of lake sediment (gythia). A particular role of chemically active part of gythia - $Ca(HCO_3)_2$ was indicated. Most findings were documented using original methodology. Positive effects of subsoil sealing were obtained even with the composite of gythia with gravel strongly permeable for water, which yielded the water permeability index K below 1.0×10^{-10} m/s. Such good results facilitated suggesting technologies for applying these composites to construction practice (dumping grounds, cross-cuts, roads, reservoirs, pipelines and wells).

1 INTRODUCTION

Waste dumping grounds require ecological locations and, first of all, appropriate sealing of their bottoms and walls to accomodate leachates dangerous for soil and ground water.

The subsoil constituting the bottom of a dumping ground outside the saturation zone in the tract of land with the thickness of 3 m (EWG) should have permeability index K-1.0×10^{-9} m/s achieved by natural or artificial means. Such prepared subsoil should be additionally sealed up with foil, the best with HDPE laid on geotextiles. Such a construction will prevent infiltration of leachates into subsoil, diffusion, disruption caused by local subsoil dislocations, and, to a large extent, flood scouring.

Recent studies indicate that each foil, independently on its manufacturing technology, density and environment where it is used is subjected to creep and ageing-crumbling processes expressed by the loss of elasticity. More and more frequent failures found in foil seals indicate that disruptions may also be caused by high temperatures appearing in dumped deposits which bring about strong contraction of clayley subsoil. The subsoil delaminated due to the contraction sucks foil into cracks resulting eventually in its disruption and failure.

In Poland the dumps are made in used gravel or sand excavations as well as in clay or loam enclaves. Polish soils including cohesive ones such as Wurm, Riss and Mindel clays and natural loam deposits practically never fulfill the requirements for limiting water permeability to the range of K-$1.0 \times 10^{-9(-10)}$ m/s. To obtain the required "K" this kind of clay or loam sediments must be loosened, homogenized and finally densified with the energy of at least 2744 KNm/m^3. Sporadically there are attempts at making foil backing with ash mixtures, hydrophobized, stabilized (with cement, lime, ash) or bituminizated mixtures. More and more often various composites made on basis of cohesive soils or ashes additionally activated with different chemicals such as waterglass, calcium chloride, lime, etc., are applied.

A good supplement for the above mentioned methods for sealing the dumps, slopes and other constructions with sub-foil backing may be the new technology suggested by the authors. In this method the main sealing material is gythia, so called, lake chalk which commonly occurrs in great quantities in Poland. There are districts whose documented geologically gythia deposits exceed 1 billion tons.

2 CHARACTERISTICS OF THE LAKE SEDIMENT

Lake chalk (gythia, lake sediment) is calcareous

Table 1 Occurence of snails and molluscs in lake sediment from the Zielona Góra District deposit.
B. Brzezie, Sz. Szumiąca, Zb. Zabór

Ecol* Group	Taxons	Location LS* B	Sz	Zb
1 L	Acanthinula aculeata (Müller)			I
5 O	Vallonia pulchella (Müller)		II	I
5 O	Pupilla muscorum (Linné)	I	I	I
10 P Pp	Pisidium obtusale (Lamarck)		II	
10 P	Valvata cristata Müller	IV	V	
10 P	Lymnea palustris (Müller)	II		
10 SP	Bathyompalus contortus (Linné)		III	
10 SP	Gyraulus acronicus (Ferussac)	VI	V	IV
10 SF	Valvata piscinalis (Müller)	III	III	I
10 SF	Bithynia tentaculata (Linné)	IV	IV	V
10	Bithynia - Operculum		I	
10 SF	Pisidium milium Held		II	
10 SF	Sphaerium corneum (Linné)	II	III	
10 FS	Pisidium nitidum Jenyns		III	
10 FS	Lymnea peregra (Müller)	II		
10 S	Lymnea stagnalis (Linné)	II		
10	Lymnea sp. juv.	III	II	I
10 S	Planorbarius corneus (Linné)	II		
10 S	Planorbis carinatus Müller	I		
10 S	Armiger crista cristatus (Linné)		II	
10 S	Armiger crista nautileus (Linné)		III	
10 S	Acroluxas lacustris (Linné)	I		I

Ecological group(EC*) , 1 L-Forest species
5 O Open area species , Lake sediment(LS)
10 Water species from different habitals periodically drying reservoirs(10 Pp) small overgrowing reservoirs (10 P) lakes and ponds(10 S) and running water(10 F)

sediment-sludge deposited in tunnel-valley and melting lakes as well as in old river-beds.
Gythia is easily soluble in water and during stirring it resembles txotrophic susponcion. In has three main components: calcium carbonate ($CaCO_3$), organic substance and terrigenic material with pelite and aleurite (silt, mud-dust) fractions. Moreover, it contains small (weight) fossil communities (mollusks, snails, ostracoda). The age of malacofauna sediments was determined in a broad stratigraphic range including Allerod, earlier dryas and preboreal and boreal phase of Holocen proven by the presence of Pupilla muscorum (Table 1). Analogous age range was

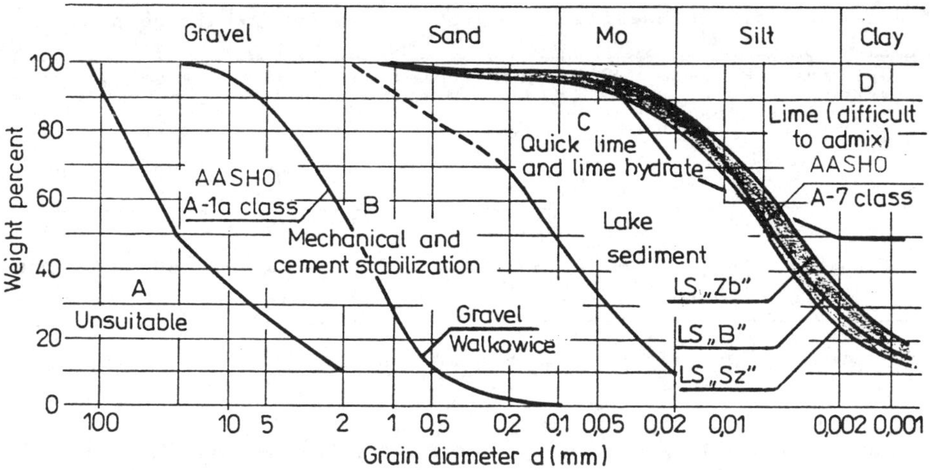

Figure 1 Granulation of gravel and lake sediment used for composites and suspensions

assumed for carbonate lake sediments in the northern Poland (Stasik 1972) and in the Carpathians (Gerlach 1972).

In the experiment discussed in this paper lake sediment from the Zielona Góra District was used sampled at three sites: Szumiąca (LS"Sz"), Zabór (LS"Zb") and Brzezie (LS"B"). In this distric the amount of the documented class C2 gythia is over 70 million tons.

3 MALACOFAUNA COMMUNITIES IN THE LAKE SEDIMENT

In the investigated lake sediment 22 taxons of snails and molluscs were found. The frequence of their appearance was given in seven groups out of 10 suggested by V. Loziek (1964) while the quantity of moluscs was classified into 6 nummerical groups Table 1): I - individual 1-5 specimens; II - 6-9; III - 10-29; IV - 30-99; V - 100-300; VI - 301-1000. It is worthwhile noticing that in the Lubuski Lake District sediments as many as 40 taxons of snails and molluscs were recognized (Alexandrowicz 1980).

4 PHISICO CHEMICAL PROPERTIES OF THE LAKE SEDIMENT AND GRAVEL

Table 2 presents parameters of the gythia used in tests as sealing-setting material in composite with gravel (Fig. 1). It should be stressed out that

Table 2 Physical end chemical parameters of lake sediments from the deposits Brzezie, Szumiąca, Zabór from which experimental material nas taken

Parameter		Values mean	extremal
Water content	(%)	81.16	52.60 - 95.10
Specific grafity	(KN/m^3)	15.04	14.23 - 15.86
Volume grafity	(KN/m^3)	23.37	24.21 - 27.50
Plasticity limit	(%)	38.60	37.12 - 40.32
Liquid limit with	(%)	58.30	50.98 - 59.36
Consistency index		2.11	1.02 - 3.04
$CaCO_3$ content	(%)	73.00	55.20 - 98.60
$Ca(HCO_3)_2$ *	(%)	10.30	8.10 - 11.60
Ignition 773K	(%)	62.00	62.40 - 64.60
Loss 1073K	(%)	95.50	92.50 - 97.30

*Determined with Dronineau method in 0.2 normal ammonium oxalate with titration with 0.1 normal $KMnO_4$

with respect to the content of the main component CaO and other macro- and microcomponents, lake sediments are similar to carbonates and chalk of older geological eras (Mesosoic, Paleosoic). As seen from the test results gythia is free from any harmful components and is an "ecologically pure" deposit (Table 4).

In selecting soil and gravel for sealing it was assumed that with other non-cohesive and, especially, cohesive soils the sealing effect of gythia will be higher and considerably cheaper. Water permeability of the gravel chosen for sealing was presented in Table 3.

Table 3 Water and air permeability of gravel used for gythia composite

Ec*	Optimum water content Wopt	Maximum bulk density	Porosity	Water permea-bility	Air permeability of		
			n	K	wet	dry sample	dry
KNm/m^3	(%)	(KN/m^3)		(m/s)	PPm	PPs_I **	PPs_{II}
201	10.0	15.68	0.408	$2.09*10^{-3}$	2500	2600	3350
603	9.4	16.12	0.392	$1.16*10^{-3}$	1100	1500	2000
1206	9.0	17.96	0.321	$1.11*10^{-3}$	780	1180	1420

*Ec compaction energy British Standard 603 KNm/m^3 ** Air permeability

Table 4 Chemical analysis and water permeability index of lake sediment used in this study

Additive	Deposit Szumiąca	Lake Sediment Zabór	Brzezie
$CaCO_3$ (%)	96.50	81.93	86.20
$Ca(HCO_3)_2$	11.80	8.91	10.32
$MgCO_3$	1.80	1.96	2.04
CaO	53.03	44.78	47.12
MgO	1.40	1.35	1.20
K_2O	0.24	0.22	0.18
Na_2O_3	0.70	0.90	0.82
R_2O_3	0.30	0.25	0.35
Fe_2O_3	0.05	0.05	0.10
Al_2O_3	0.25	0.20	0.25
SiO_2	0.25	0.18	0.21
K (m/s) "in situ"	$<1.0*10^{-10}$	$<1.0*10^{-10}$	$<1.0*10^{-10}$
P_H	8.6	8.0	8.1

Sulphur (SO_3), manganium, lead and zinc accurred in trace guantities. No arsen, titanium or copper was found

5 PREPARATION AND TESTING OF SUSPENSIONS AND GYTHIA COMPOSITES

The process of preparing mixtures, composites, hydrofobrome composites and the composites stabilized with 5, 10 and 25% of gythia in relation to dry mass of gravel was carried out in a laboratory mixer controlled electronically and supplied with a two band pulverizer. The activators, such as Portland cement (cp 35), "Calgon" - mixture of sodium 6 - metaphosphate and sodium carbonate (Rząsa, Kokowski, 1983), Base-Seal (Soil Compactor and Stabilizer), plastmal 2 were added in water solution according required dilutions. The water content of the composites at compacting was on the level of W_{opt} 1.1% (maximum water content) determined with the test No 9 British Standard B.S. 1377, ASTM-Des. 698-58T and Des. D 1557-58T.

The gythia suspensions were made with rotary mixers (3600 rot/min) working for 5 minutes after sieving gythia through field vibratory sieve. The water content of the suspensions was from 70 to 200% depending on the kind of used gythia. The necessary components, activators and anti-contraction substances were dosed into gythia together with make-up water (wood-dust, a little sand, chlorides). In laboratory tests the gythia suspensions were tested with equipment used for evaluation of thixotropic solutions (Marsch's funnel, filtration press, shearmeter, water content with the method of upturned test tube, density with Baroide scales). In the water permeability tests the CBR cylinders were used loading the composite with 5 m high water column for 30 days at 283 K. The air permeability was determined using the apparatus according to the uathors' idea (Kokowski 1933) with the following formula:

$$PP = \frac{V \times h}{A \times t \times p}\left[10^{-8}\frac{m^2}{Pa \times s}\right], \left[\frac{cm^4}{G \times min}\right]$$

where:

A - sample area (m^2, cm^2)

V - volume of the air (m^3, cm^3) passing through composite sample till stabilization of pressure,

h - sample height (m, cm)

t - measurement time (s, minute)

p - assumed constant pressure of feeding air - 0.98 KPa (100 mm Hg)

The measurement range at the 0.98 KPa pressure in the SI system is:

- for 0.50 mm nozzle ... 2×10^{8} to 80×10^{-0} m^2/PaxS
- for 1.50 mm nozzle ... 70×10^{-8} to 4000×10^{-8} m^2/Pa x s

Table 5 Results of tightness tests of gravel composite basing on Szumiąca gythia

Conc.* comp.	Dry density	Porosity	Gythia	composite	permeability			
			A I R			W A T E R		
		n	wet	dry	dry	no cp 35	with cp 35	
			no cement Sample				2(%)	5(%)
(%)	(KN/m^3)		PPm	PPs$_I$	PPs$_{II}$	K(m/s)		
			Compaction energy E_c 201 KNm/m^3					
5	17.8	0.32	385	350	440	$2.4 \cdot 10^{-4}$	$1.0 \cdot 10^{-6}$	$1{\cdot}3 \cdot 10^{-7}$
10	18.7	0.28	310	283	355	$1.3 \cdot 10^{-4}$	$1.6 \cdot 10^{-7}$	
25	16.9	0.34	100	48	54	$1.6 \cdot 10^{-6}$	$1.0 \cdot 10^{-7}$	$1.0 \cdot 10^{-9}$
			E_c 603 KNm/m^3					
5	18.8	0.28	72	65	73	$1.9 \cdot 10^{-5}$	$1.0 \cdot 10^{-8}$	$1.0 \cdot 10^{-9}$
10	19.3	0.26	70	65	82	$5.0 \cdot 10^{-6}$	$1.0 \cdot 10^{-9}$	
25	20.5	0.30	25	13	14	$9.0 \cdot 10^{-7}$		
			E_c 1206 KNm/m^3					
5	19.3	0.26	3	18	19	$1.4 \cdot 10^{-8}$	$1.0 \cdot 10^{-9}$	
10	20.5	0.21	3	3	3	$1.0 \cdot 10^{-9}$		
25	18.4	0.28	2	2	2			

(Conc. comp.*) Concentration composite — Water content at compacting: 9.0 - 10.0 %

The Szumiąca deposit amounts to > 3 mln tons

Table 6 Results tightness tests of gravee composite basing on Zabór gythia

Conc.* comp.	Dry density	Porosity	Gythia	composite	permeability			
			A I R			W A T E R		
		n	wet	dry	dry	no cp 35	with cp 35	
			no cement Sample				2(%)	5(%)
(%)	(KN/m^3)		PPm	PPs$_I$	PPs$_{II}$	K(m/s)		
			Compaction energy E_c 201 KNm/m^3					
5	16.0	0.39	250	400	430	$5.8 \cdot 10^{-5}$	$1.0 \cdot 10^{-6}$	$1{\cdot}0 \cdot 10^{-7}$
10	15.0	0.43	30	340	352	$3.6 \cdot 10^{-8}$	$1.2 \cdot 10^{-7}$	$1.0 \cdot 10^{-8}$
25	14.5	0.44	12	85	90	$1.4 \cdot 10^{-8}$	$1.0 \cdot 10^{-8}$	$1.0 \cdot 10^{-9}$
			E_c 603 KNm/m^3					
5	16.9	0.36	107	175	185	$1.0 \cdot 10^{-5}$		
10	15.9	0.39	13	90	95	$1.5 \cdot 10^{-8}$	$<1.0 \cdot 10^{-9}$	
25	15.4	0.42	7	45	45	$4.9 \cdot 10^{-7}$		
			E_c 1206 KNm/m^3					
5	17.2	0.35	84	140	147	$5.3 \cdot 10^{-6}$		
10	16.2	0.38	8	50	63	$8.0 \cdot 10^{-7}$	$<1.0 \cdot 10^{-9}$	
25	15.9	0.39	1	2	3	$<1.0 \cdot 10^{-9}$		

(Cnnc. comp.*)Concentration composite — Water content at compacting: 9.0 - 10.0 %

The Zabór deposit amounts to >2 mln ton

Moreover the following parameters were determined in the composite actvated with cement: compressive (R_c), shearing (R_t), bending (R_g), tearing (R_r), tearing in split sleeve (R_rT) strength, deformation moduli (M_{VI}, M_{VII}), California Bearing Ratio (CBR) in dry and wet system after 7, 28, 180, and 360 days.

6 RESULT ANALYSIS

6.1 Description of the modification processed occurring in gythia composites

The studies on modification of soil with lake sediment itself and gythia activated with cement Calgon, Base-Seal, etc. are under way. A certain index in evaluation of this process was the content of chemically active lime fraction in gythia ($Ca(HCO_3)_2$ determined with Drouineau method (1979).However, even now it is possible to observe that the dissociation phase starts at mixing components together and from that moment it decreases continually terminating after 3 - 6 days. The gel phase starts at mixing and is still observed after 180 days (further studies have not been carried out) but it is supposed to last up to 500-600 days. The noelithic phase is slightly observed, runs paralelly (narrow band) and continues non stop. The carbonization phase reaches its maximum after 360 days and does not stop. The action of these phases is similar to that described by A. Kezdi (1979).

6.2 Water permeability

Analysing the collected material it should be stated that the required water permeability K-1.0x10^{-9} is obtained by the mixtures of gravel with 10 and 25% of gythia densified with the energy of 1206 KNm/m^3 (Tables 5 and 6). The composites stabilized with an addition of 2 and 5% of cp 35 reached the required water permeability at 10 and 25% addition of gythia densified with the energy of 603 KNm/m^3. Increasing the densifying energy to 1206 kNm/m^3 leads to total sealing of the composite with "K" haveing the values lower than 1.0x10^{-10} m/s. The permeability results for the other non-cohesive soils (all-in aggregates, sands) and clay met the water permeability criteria already at addition of 10 and 5% of gythia and 2 and 5% of cp. 35.

The additions of plastmal 2 (Germany) and Base-Seal in the proportion of 0.43 and 1% (in relation to soil dry mass) were revealed in the composites with low gythia content (5 and 10%). The role of the a.m. preparations, first of all, was important in facilitating obtaining higher density particularly at vibratory densifying.

The third gythia used in the research analysed within the following work (LS "B" - Brzezie) also yielded very favourable results. The already mentioned criterion 1.0x10^{-9} m/s was obtained for the mixtures of 10% gythia with gravel and the composites stabilized with cement in the proportions of 2 and 4% and densified with the energy of 603 KNm/m^3 (material in the Department archieves). In the composites with fine, medium and coarse sand the secessary proportion of gythia was 10% with an addition of 3% of cement or 0.48% of Base-Seal which secured water permeability values on the level of 1.0x10^{-9} m/s. For each mixture or composite an addition of 2-5% cement facilitated improvement of mechanical parameters necessary in the geological zones subjected to local subsoil dislocations or expected flood scourge. In clayley sands and clays a sufficient addition turned out to be the dose of from 7 to 10% of gythia at 2% addition of cement.

6.3 Evaluation of the activation effect

The effect of such preparations as Base-Seal, plastmal 2, RRP - 235, Dinoran, etc./ is controversial at least concerning gythia composites. At present, cement, maybe high-grade cement seems to be the best activator. Its action consists in creating strong mechanical bonds, and particularly, in activating pozzolan potential in gythia. Also the cost of cement additive is competitive with respect to other preparations.

6.4 Mechanical properties

Some of the determined mechanical properties are given in Table 7. California Bearing Ratio and the moduli were examined after 7 days of dry and wet treatment (CBR cylinder with the composite immersed in water). After 28 days of wet treatment CBR exceeded 200 and after 180 days only the depth of 2.5. m/m could be reached where CBR exceeded 450 (maximal range of ring dynamometer). Removal of the composite from the ring required application of a chisel.

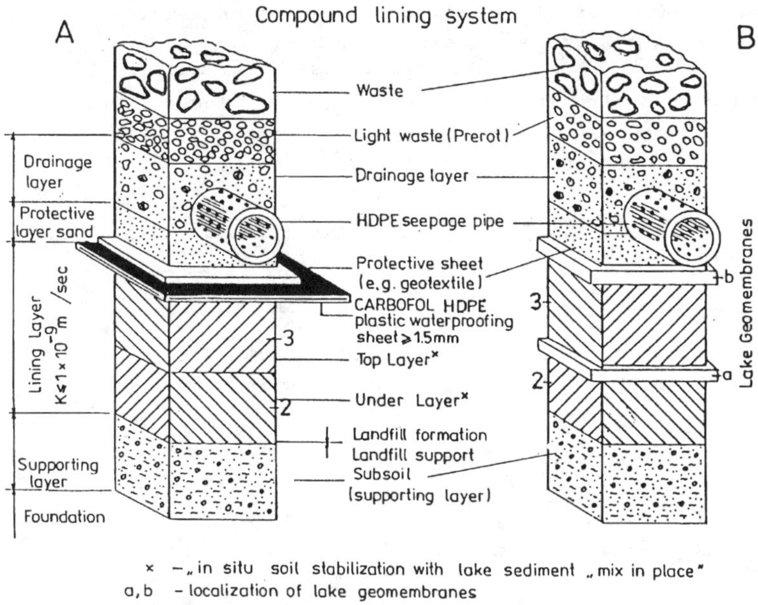

Figure 2.1 Landfill bottom with composite sealing layer under fail and hydromechanically laid lake sediment geomembrane

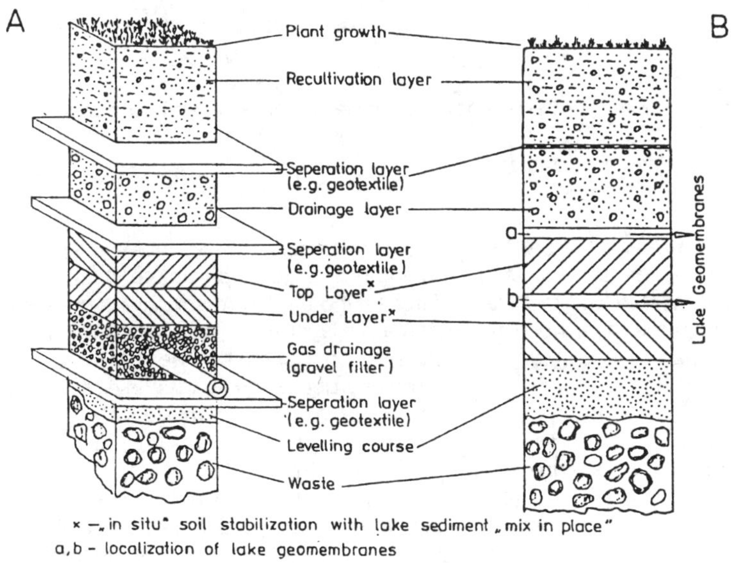

Figure 2.2 Sealing landfill capping system

7 APPLICATION OF THE COMPOSITES

The gythia mixtures and activated composites can be used for sealing up bottom, walls and for capping dumping rounds (Figs 2.1 and 2.2). In case of overground dumps the whole surface of the bottom and the embankement base can be covered with 2 or 6 layers of gythia composite with "in situ" soil using the "mix in place" method. The under layers are modified with gythia with an

Table 7 Some mechanical properties of the composite basing on Brzezie lake sediment activated with 5 % addion of cement cp 35

Conc.* comp.	Unconfined strength (MN/m^2)								CBR	Mv_7
	R_c Compression			R_t Shearing		R_r Tearing		R^T_r **	7	7
(%)	7	28	180	7	28	7	28	7	days	MN/m^2
	after the period days air-damp-curing									
Samples										
5	0.22	0.65	1.68	0.02	0.08	0.02	0.07	0.06	50	40
10	0.83	1.54	3.62	0.10	0.20	0.08	0.15	0.08	79	60
25	1.70	2.56	6.04	0.22	0.30	0.02	0.26	0.19	140	90

(Conc.comp.*) Concentration composite
** Tearing strength-in bipartte tube

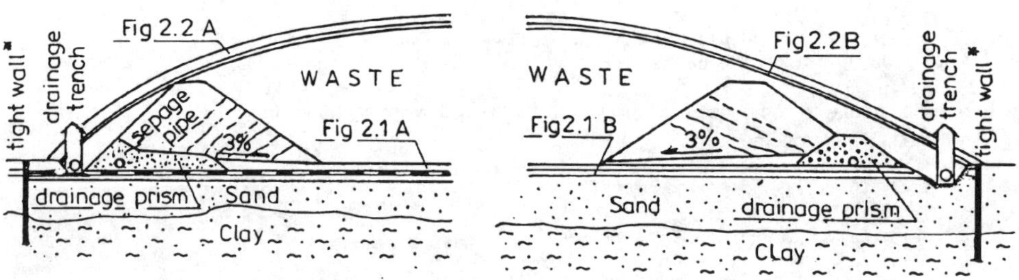

* **Figure 3** Overground dumps with internal leachant drainage. Trench turned into drainage durning capping sealing
* tight wall - gytia thixtropic clayley suspension

addition of plastmal or Base-Seal while the top layers, laid directly under the foil are activated with an addition of 2-5% cement or Calgon. Depending on the layer design the layers can be interlaced with gythia geomembranes (Figs 2.1 a, b).
The geomembrane suspension is pumped or sieved through field vibratory sieve into earth reservoir and supplemented with water and components determined in laboratory tests (cements, Base-Seal, plastmal, anti-shrink additions). Such prepared gythia hydromass in form of thixotropic suspension is transported by mud pupps onto the earlier made composite layer with distance fittings facilitating an even spread of the lake sediment. The water content of the hydromass, depending on the kind of gythia, ranges from 70 to 200%. In favourable topographic conditions gythia can be pumped directly out of deposit onto prepared surface using hydromechanic technology (if gythia is pure and has required parameters).
After setting the geomembrane is covered with a geotextile layer and with sand or the material for the next layer of the composite.
On many dumping sites the layers lying directly under foil should be made of the composite stabilized with cement and reinforced with "AKZO" or "HATELIT" net.
The mixtures for this kind of layers are made in stationary mixers putting the composite on earlier laid "AKZO" or "HATELIT" net densifying it with vibrating roll and eventually with pneumatic one. Fig. 3 illustrates a concept of an overground dump with the bottom sealed with gythia composite with drainage system for leachates. Sandy embankements of the dump are not sealed and leachates flow along sloping impermeable floor of the composit into the dump bowl. At higher levels of the leachates in the bowl they permeate

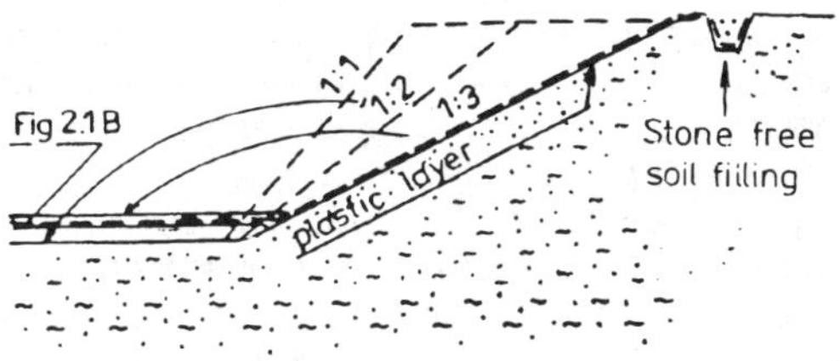

Figure 4 Taking soil from landfill embakements for lake composites on previonsly compacted subsoil

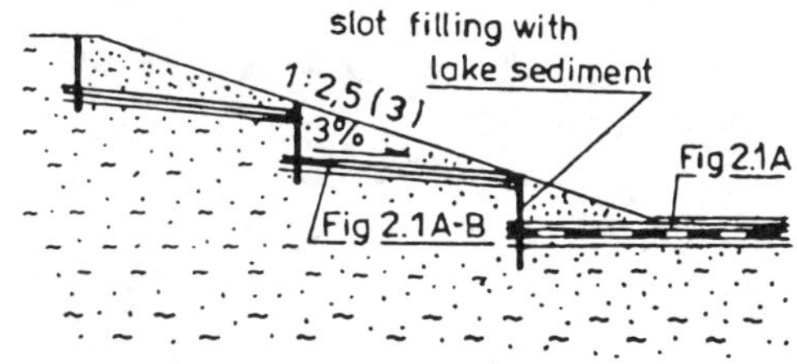

Figure 5 Sealing landfill embankements with terrace method. Verical partitions filled with lake sediment suspension

through slope body according to earlier calculated depression curve and, in places of expected pipings, are contained into the drainage laid in the outside slope. The drainage is laid in breakstone prism additionally areated. Its geommetry depends on the assumed scheme or on expected quantity of leachates (Figs 3a, b). In this desing the embankements are subject to recultivation while the bleeding ditch is transformed into a pipeline with control wells.

Fig. 4 illustrates obtaining soil from excavation slopes with an initial slope of 1:1. The bowl subsoil is compacted with rolls till the parameters according to the energy of 2744 KNm/m^3 are achieved. Later the material is taken from the slopes for making each composite or stabilized composite layers. Usually 3 to 6 layers are made with the thickness of 0.20 to 0.25 m with the "mix in place" method. When the slope of 1:2.5 is reached the slopes are compacted by a special set of rolls, e.g. "Bomag" and subsequently sealed up according to the assumed rules. There is also the possibility of sealing the slope with the composite and small milling machine cutting narrow vertical slots without using specialized equipment (Fig. 5).

Presently, small modern milling machines cutting slots 1.5 to 2.5 m deep have the efficiency of 150 to 300 m/h.

The vertical slots 0.10 to 0.30 m wide will be filled with gythia hydromass with volume weight from 12.0 to 15.0 KN/m^3 which depends considerably on the kind or type of the gythia. In many cases it is enough to mix gythia intensly in the deposit adding small amount of activators.

The remaining figures (Figs 6, 7 and 8) present securing of constructions cut deeper into the ground such as landfills, water reservoirs and trenches. Their subsoils are sealed with the composites activated with cement in which, at earlier descussed proportions, the CBR index after 28 days reaches the values within 200-450. Side surfaces around the construction shell or the slot for tight wall are filled with suspension or plastic

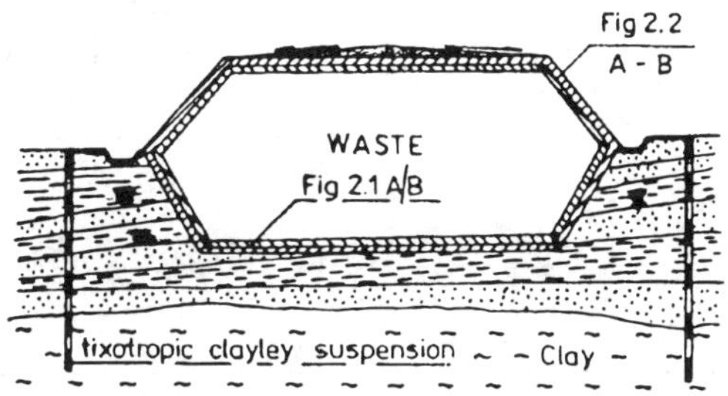

Figure 6 Cutting of water flow into landfill with trenches filled with thixotropic suspension

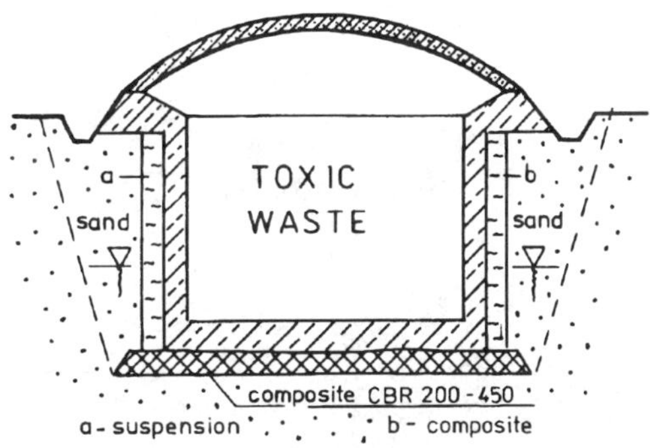

Figure 7 Sealing up storage reservoir for waste. Bottom - the composite stabilized with cement Rc>4,0 MPa Suspension poured around casing

material haveing the ability to control placement of each layer. The composite adheres ideally to concrete or steel constructions and to remove it a chisel is needed. In the situations where high temperatures can be expected the gythia composite, and particularly the suspension, should be additionally reinforced with, e.g. wood dust obtained from grinding shop or filters. The amount of this component ranges from 3 to 10% of the addition of gythia. Its action is unbelievably effective and eliminates shrinking even at drying

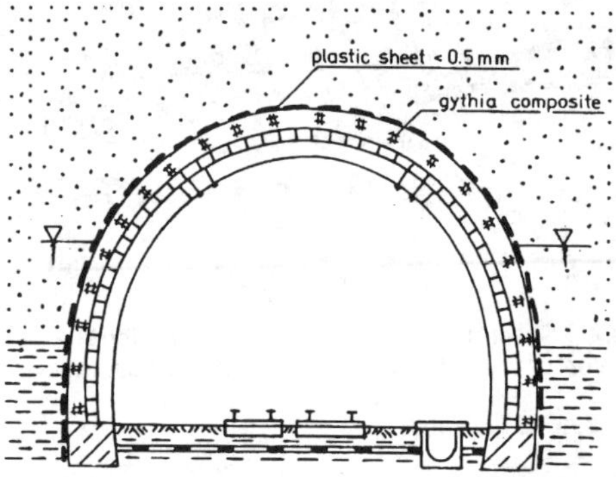

Figure 8 Cutting of water flow into trench with composite or plastic lake sediment suspension

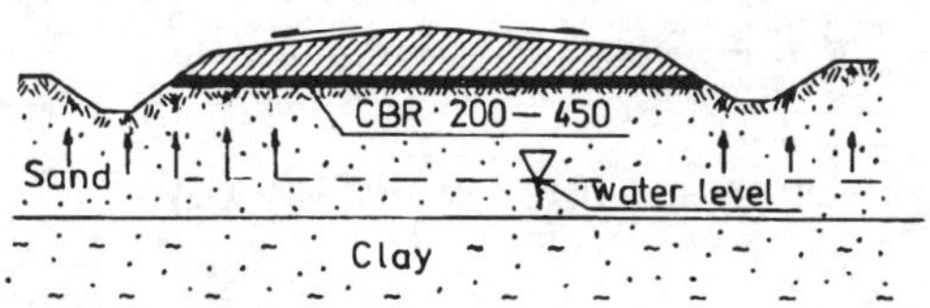

Figure 9 Cutting of layer in cohesive subsoil made of lake sediment composite under road surface

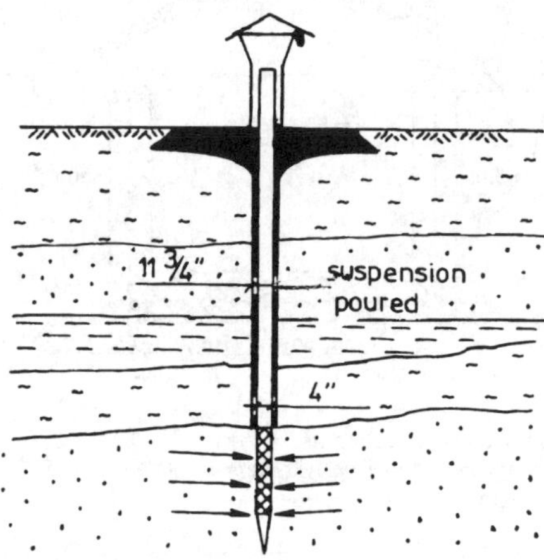

Figure 10 Sealing up piezometer and well with thixotropic lake sediment suspension

in the temperature of 373 K.

Fig. 9 illustrates enormous possibilities for application of gythia composites as sealing layers of the roads founded on cohesive soils. Absolute tightness is obtained insensitive to any frost damage. California Bearing Ratio can be modelled at will up to the value of 450. Depending on the load to be applied to the road soil or soil mixture (skeleton, dust, loam) should be selected and then modified with gythia activated with cement or other preparations.

Fig. 10 documents already applied by the authors protection of pipe columns (plastic lead free pipes) monitoring investigated ground water tables. The authors install over 600 piezometers of this kind all over the country. The composite or the suspension after filtering are absolutely unwashable and, first of all, as chemically clean as almos none other soil origin material.

8 CONCLUSIONS

The deposit of mineral lake deposit (gythia) containd various amounts of calcium carboante with granulation included in dust fraction and of colloidal loam. The lake deposit deposits are extremally easily wahed out and their loosening is comparable to working the dense plastic slime. The lake sediments are easily turned into thixotropic state and the amount of necessary activator or activators is determined in laboratory. Gnerally, most of the gythias can be pumped directly from the deposit following earlier loosening with, e.g. vibroflote set producing a crater from which gythia sludge is aspirated by pumps and transported to the destinied place (the suspension thickness is within the range from 12-15 KN/m^3. In this layout the missing part of water is provided into the crater from the nearby running or stagnant water source. However, is seems that the most important issue on this stage is to find appropriate activators which would be stronger than common and cheap cement and which would liberate the enormous binding force present in the lake sediment ($CaCO_3$ reaches 100%, $CaCO_3$ reaches 100%, CaO - 56% and $Ca(HCO_3)_2$ up to 13%). At present the binding force of this material is liberated only in slow pozzolana, reactions strengthened with cement, calogen, BASE-SEAl, Plastmal 2, RRP-235, Dinoren. A good illustration of this can be the fact that even now the author obtains, CBR for gravel-sand composites of the order of 200 or 350. Discovery of the most favourable activators will limit the neccesity of using so high doses of lake sediment (according to present trials, eg. 25%) in the composite while retaining the required elasticity (tear In partitionad sleeve, tear till exceeding critical strain). Also efforts should be started to apply gythia for covering subsequent layers of waste deposits with thin covers made with hydromechanic technique. The lake sediment could be, transformed (drying, grinding) into dry mixes with the activators and then it could be used at langer distances from its deposits (transportation costes of wet gythia) and manufacture of varions sanitary mats saturated to a different degree with calcium carbonate (gythia). These mats covering subsequent layers of waste would decompose gradually.

In the end the authors wish to stress out the fact that in Poland there are enormous deposits of lake sediment deposits whole thickness reaches up to 20 m. Their expolitation is subsidized by the government (soil fertilization) and it is usually carried out with the open pit technology (down to water table or 2 m below) while the remaining part is wasted. Hence an appropriate and planned expolitation would lead to obtaining huge amounts of this valuable natural resource.Then we shall have enough gythia to correct soil pH where, as it is commonly known, it significantly inhibits plant sorbtion of heavy metals (lead, copper, mercury) and to carry out modern ecological waste economy including sealing up bottom and embankements of dumping grounds.

9 REFERENCES

Alexandrowicz, S.W. 1979. Micropalenotological surwey of lake sediments in northern Poland. Archiwum Instytutu Geologii Surowców Mineralnych. AGH, Kraków.

Gerlach, T. Koszarski, I. Koporowa, W. 1972. Sediments lacustres postglaciares dans le depresion de Jasło-Sanok. Studia Geomporphol. Carpat.-Balkan. Vol. VI. Kraków.

Kezdi, A. 1979. Stabilized earth roads Akademia Kiado, Budapest.

Kokowski, J., Rząsa, St. 1983. A pipette method for analyzing the mechanical compositon of the soil. Zeszyty problemowe Postępów nauk Rolniczych z. 220.

Kokowski, J. 1992. Portable field TAPP apparatus for air permeability mesurement. Roczniki AR Poznań, seria melioracje 93.

Lozek, V. 1964. Quartarmollusken der Tschechoslowakei Rozpr. Ustr. Ust. Geol. Sv. 31, Praha.

Stasik, J. 1972. Genesis of sedimentation basins on sandrie azeas. Zeszyty Problemowe Podstaw Nauk Rolniczych z. 107, PWN, Warszawa.

Waste Disposal by Landfill – GREEN'93, Sarsby (ed.) © 1995 Balkema, Rotterdam, ISBN 90 5410 356 6

Dissolved silica in landfill leachates and its possible significance for clay barrier stabilities

D.A.C. Manning, J. Owen & C.R. Hughes
Department of Geology, University of Manchester, UK

ABSTRACT: Determination of the silica contents of landfill leachates coupled with detailed geochemical modelling of compositional data permits the use of theoretical mineral phase stability diagrams to predict the stability of clay minerals which are in contact with leachates, with subsequent implications for the integrity of clay barriers. Leachate data available so far show a range in silica contents from 0 to approximately 23 mg/L Si, in most cases in excess of values appropriate for saturation with respect to quartz. Modelling of these data suggests that, like many natural waters, leachates commonly may be compositionally buffered by smectite-illite assemblages, reflecting the presence of both minerals in clays with which the leachates come into contact. The approach discussed here may offer a means of using chemical data for leachates to monitor the the integrity of clay barriers remotely.

The purpose of this paper is to present preliminary data obtained in a continuing study of the geochemistry of landfill leachates. The principle aim of the study is to use leachates as analogues for naturally occuring but inaccessible organic-rich sedimentary waters, and so to contribute to our understanding of sediment diagenesis. However, it has become clear that the study also has implications for landfill practice, particularly with respect to prediction or monitoring of the stability of clay barriers to leachate migration.

1 MINERALOGY OF CLAY LINERS

The use of clays as a liner or barrier material designed to prevent the loss of migrating leachates from landfill waste disposal sites has a long history, developing from the use of sites within naturally occuring clay deposits, through the use of imported and engineered clays to form a barrier in sites with non-clay bearing or variable bedrock characteristics to modern practise of using a combination of imported clay and artificial (such as HDPE) lining materials (Bath 1993). The properties of clays which are exploited in their use as a barrier material include: (1) their potential to form a hydraulic seal as a consequence of their low hydraulic conductivities, (2) their ability to adsorb organic solvents, and (3) their cation exchange capacity and large specific surface areas, which enhance the removal of metals from solution by both ion exchange and sorption. The clays within a barrier thus help isolate leachate from neighbouring hydrogeological systems by acting as a seal which is both chemically and physically reactive.

Selected properties of representatives of major clay mineral groups, which represent those which occur most commonly in surface environments, are summarised in Table 1. In this table, it can be seen that the adsorptive properties of individual species depends on the structure of the clay and the nature of the interlayer site, which influences the ion exchange and shrink-swell behaviour. Thus kaolinite, which lacks interlayer cations and has no net charge on the faces of sheets which make up its structure, has only limited ability to exchange cations and to adsorb water or organic species. Illite contains interlayer potassium which is stably held within the structure, and so again there is limited cation exchange capacity at neutral pH and limited water adsorption. The smectite family of clays excels in its ability to exchange cations, and this arises as a consequence of chemical variation within the sheets which make up the structure, inducing charge imbalances within the interlayer site which must be satisfied by interlayer substitution of cations. It is believed that the adsorption of water and polar organic species is facilitated by reaction with the interlayer cation and/or the clay sheets to produce stable coordination complexes (Skipper et al., 1993), the net result of which is that layers of water or organic molecules lie between the interlayer cations and the aluminosilicate sheets.

The smectite family of clay minerals is one of the most reactive groups, which is one reason for its suitability as a barrier material. However, smectites not only show desirable reactions within the interlayer site (which may also influence the mechanical properties of the clay, such as its ability to shrink or swell) but they also react to form other clay minerals. Of particular importance from the geological point of view is the transformation of smectite to illite, which is well known to occur as a consequence of sediment burial (Hower et al, 1976). Evidence from oil wells suggests that this transformation takes place at temperatures of the order of 80°C. However, it has recently been shown that under certain conditions the smectite-to-illite transformation takes place at much lower temperatures (down to 35°C) under conditions of high solution pH (Eberl et al, 1993). The major consequence of this reaction is that many of the desirable properties of a smectite clay are lost as it is transformed to an illite, resulting in deterioration of the sealing and adsorbtive properties of a clay barrier.

In addition to the smectite-to-illite transformation, other reactions may take place to produce kaolinite, again reducing the ability of a clay liner to perform as expected.

Table 1. Composition and properties of representatives of the major clay mineral groups.

mineral/formula	[1]structural type	[2]basal spacing	interlayer water	glycol adsorption
kaolinite $Al_2Si_2O_5(OH)_4$	1:1	7.1Å	none	none
illite $K_{1-1.5}Al_6Si_6O_{20}(OH)_4$	2:1	10Å	none	none
"Na bentonite" $Na_{0.7}Al_{3.3}Mg_{0.7}Si_8O_{20}(OH)_4.nH_2O$	2:1	~12.5Å	1 layer	2 layers [3]17Å
"Ca bentonite" $(1/2Ca)_{0.7}Al_{3.3}Mg_{0.7}Si_8O_{20}(OH)_4.nH_2O$	2:1	~15Å	2 layers	2 layers [3]17Å

Bentonites are bulk clays composed predominantly of smectites, usually with minor cristobalite and other accessory minerals such as zircon; the properties cited are for Na or Ca montmorillonite, which is the dominant clay mineral constituent of bentonite.
[1]Structural type refers to the proportion of tetrahedral:octahedral layers in the aluminosilicate sheets of the clay structure (Velde, 1993).
[2]Approximate basal spacings are given for the bentonites; they are very variable, depending on the degree of hydration of the interlayer site and other factors which are sample specific.
[3]Interlayer spacing when fully glycolated.

Table 2. Chemical reactions involving kaolinite, illite and smectite, with dissolved silica and ionic species.

$$3Al_2Si_2O_5(OH)_4 + 2K^+ = 2KAl_3Si_3O_{10}(OH)_2 + 2H^+ + 3H_2O \quad (1)$$

kaolinite — illite

$$K_1 = (aK^+)^2/(aH^+)^2$$

....

$$xNa_{0.7}Al_{3.3}Mg_{0.7}Si_8O_{20}(OH)_4.nH_2O + K^+ + (2.1x-1)H^+ =$$

Na-bentonite

$$KAl_3Si_3O_{10}(OH)_2 + 0.7x\,Na^+ + 0.7xMg^{2+} + (8x-3)SiO_2 + (nx+(6.1x-3)/2)H_2O \quad (2)$$

illite

(x = 3/3.3)

$$K_2 = (aNa^+)^{0.7x}.(aMg^{2+})^{0.7x}.(aSiO_2)^{(8x-3)}/(aK^+).(aH^+)^{(2.1x-1)}$$

....

$$xNa_{0.7}Al_{3.3}Mg_{0.7}Si_8O_{20}(OH)_4.nH_2O + (2.1x)H^+ + (8-6.1x)/2H_2O =$$

Na-bentonite

$$Al_4Si_4O_{10}(OH)_8 + 0.7x\,Na^+ + 0.7xMg^{2+} + (8x-4)SiO_2 + nxH_2O \quad (3)$$

kaolinite

(x = 4/3.3)

$$K_3 = (aNa^+)^{0.7x}.(aMg^{2+})^{0.7x}.(aSiO_2)^{(8x-4)}/(aH^+)^{2.1x}$$

1.1 Chemical controls on the stability of clay minerals

A number of chemical reactions can be written which describe the transformations of clay minerals during reaction in the presence of aqueous solutions (Table 2). The variable stoichiometry of the smectites in particular complicates the formulation of the reaction coefficients, leading to unwieldy expressions for the equilibrium constants. However, these reactions clearly demonstrate: (1) that the kaolinite-illite reaction is independent of fluid silica content, (2) that the smectite-illite reaction involves the release of silica to the fluid, and so is not independent of fluid silica content, and (3) that the smectite-kaolinite reaction also involves loss of silica to the fluid. Thus it is essential to determine the silica contents of any fluid which might be reacting with the clay assemblage, in order to assess the significance of smectite breakdown. These reactions also demonstrate that smectite stability can be enhanced by increasing the dissolved silica content of a coexisting fluid, which will drive reactions (2) and (3) to the left. The smectite-illite reaction involves a very small change in free energy, and need not require any change in the aluminosilicate sheet structure (Garrels, 1984).

2 PREDICTION OF CLAY STABILITY

As shown in Table 2, reactions involving clay minerals are dependent on the composition of the reacting fluid, and so it is possible to construct theoretical phase stability diagrams which can be used to predict the mineral assemblage stable in the presence of a fluid of a particular composition. The construction of such diagrams is discussed in detail in texts such as Garrels and Christ (1965) and Stumm and Morgan (1970) and numerous examples are given by Bowers et al (1984). The complex stoichiometry of the clay minerals of interest complicates the construction of stability diagrams, and so it is conventional to simplify compositions to particular species. In this paper, model compositions for potassium smectite ($K_{0.3}Al_{1.9}Si_4O_{10}(OH)_2$) and illite ($K_{0.8}Al_{1.9}(Al_{0.5})Si_{3.5})O_{10}(OH)_2)$) will be used, following the procedure of Garrels (1984). Using this simplification Figure 1 has be drawn, to illustrate the relative stability of illite, kaolinite, smectite and other minerals as a function of fluid composition.

3 APPLICATION TO LANDFILL LEACHATES

The use of stability diagrams is widespread in the consideration of geological problems, and previously has been applied to landfill leachates by Chen and Bowerman (1974). However, it is unusual for the silica contents of leachates to be determined routinely (cf Robinson, 1989), preventing the routine prediction of clay mineral stability. We have undertaken a number of analyses of leachates derived from several sites, for which dissolved silica has been determined in addition to major cations and both inorganic and organic anions. Samples were filtered to 0.2μm within 24 hours of collection, diluted by a factor of 10 and stored in a refrigerator prior to analysis. Cations (including Al and Si) were determined by ICP and AAS; anions by ion chromatography. Propionate values are not given because of interference with bicarbonate. Al was consistently below a detection limit of 0.1 mg/L in the sample, indicating that mineral particulate matter had been removed by filtration. In all samples of leachate so far examined, Si varies between 0 and 23 mg/L Si, with some indication that elevated values occur in leachates with low organic acid or COD contents and which are methanogenic (see later). COD, BOD and ammoniacal nitrogen have not been determined for these samples, but have been obtained independently by the site operators as part of the routine monitoring programme.

Table 3. Compositional data for selected leachate samples

sample	1	2	3	4
pH	7.4	6.7	6.8	6.8
Ca	155	612	358	202
Mg	287	125	138	315
Na	6330	890	46	4146
K	613	219	16	499
Fe	8	11	0	0.2
Si	12	0.9	3.7	17
Cl	1470	860	50	5250
acetate	3050	1550	2.0	4.0
butyrate	650	350	<0.5	<0.5

Examples of the leachate compositions are give in Table 3.

The data given in Table 3 represent those required for solution modelling, in which charge balance is achieved by adjustment of bicarbonate.

The calculated phase stability diagrams can only be used once the fluid compositional data have been recalculated as species activities. This approach is essential as a means of taking into account the fact that some elements will be present as more than one chemical species, depending on the formation of complexes in solution. Thus chloride, for example, will be present as free Cl^- and as chloride complexes of metals, such as $CaCl^+$, $MgCl^+$ and others. Recalculation of the chemical analysis requires knowledge of the complex species which are present and their thermodynamic stability constants. These are not fully known, especially for complexes involving metals and organic acid anions in solution, although the relevant database is growing. The approach taken here has been to use the well-established geochemical solution modelling package EQ3NR (Wolery, 1992) together with the SUPCRT 92 database (Johnson et al, 1992) to recalculate leachate compositions using the available thermodynamic data for complexes which are assumed to be present. Such an approach is necessarily only a model when dealing with fluids as complex as leachates (especially as determination of composition is incomplete), but it allows predictions to be made which can then be tested by experiment or observation.

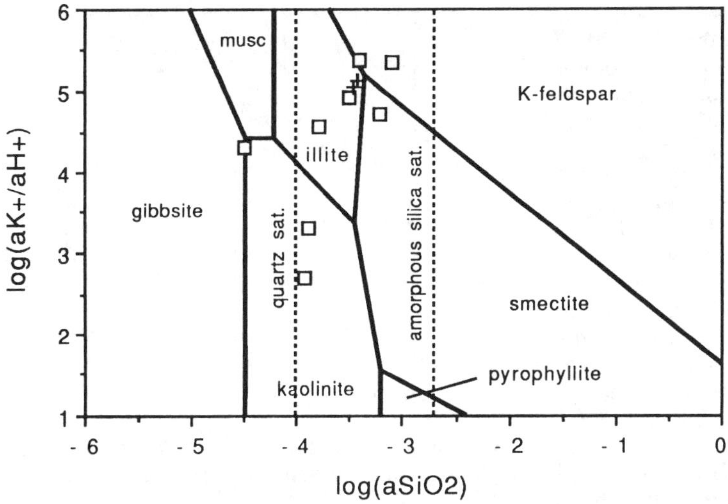

Figure 1. Leachate compositions plotted on the log(aK^+/aH^+) vs. log($aSiO_2$) stability diagram (recalculated from Garrels, 1984; 25°C). Open squares relate to samples analysed as part of this study; crosses represent analyses from Chen and Bowerman (1974), recalculated using EQ3.

Fluid compositions for the leachates considered here are plotted in Figure 1 on the log(aK^+/aH^+) vs. log($aSiO_2$) phase stability diagram. Data for samples analysed in this study have been augmented with data from Robinson (personal communication) and with those published by Chen and Bowerman (1974). The compositions plotted here have been selected from the larger database which is being accumulated as part of this study, and represent extremes in variation shown by the silica and organic acid anion contents. Even though the number of samples is limited, the overall pattern is consistent with the pattern observed for natural waters in contact with clays (Garrels, 1984), in that many of the points plot close to the illite-smectite field boundary. All but one lie between the limits on dissolved silica activities imposed by either quartz saturation or amorphous silica saturation, which suggests that the silica contents of the leachates are controlled by reactions which do not necessarily involve quartz, such as clay transformation reactions. One sample plots close to the gibbsite-muscovite-kaolinite triple point; this sample comes from a site where codisposal of chemical waste is undertaken, and so may represent a special set of circumstances. None of the points lie within the predicted stability field for smectite, which suggests that in every case the leachate composition is controlled not by smectite alone but by a mixture of minerals. The coincidence of many of the plots with the smectite-illite field boundary suggests that both minerals may be present and able to influence the composition of the leachate.

In sites where a bentonite barrier is present and where samples have been taken from leachate known to be in contact with the clay, these observations could be interpreted to suggest that the barrier might contain illite as well as smectite, either as an original component or as a consequence of reaction after installation. Much more detailed work, with complete knowledge of the history of the site being studied, is needed to substantiate such conclusions, but the leachate compositional data plotted in Figure 1 clearly suggest that this approach has potential merit as a means of remotely monitoring the mineralogical integrity of clay barriers to leachate migration.

4 SILICA IN LANDFILL LEACHATES

A number of chemical changes are known to take place as landfill leachate evolves, particularly from the acetogenic to the methanogenic stages (Robinson, 1989). As part of this study,we are sampling individual leachate wells repeatedly in order to monitor changes in silica content as a function of time, but we have yet to obtain a suite of samples which show the transition from acetogenic to methanogenic conditions. However, some indication of the behaviour of silica can be obtained by investigating whether or not there is a correlation between silica content and parameters such as TOC, total volatile fatty acid (VFA) content or acetate content, all of which are high during acetogenesis and low during methanogenesis. Included in this are data from a larger separate study carried out by Robinson (personal communication) for which we have determined leachate silica contents. In addition, a number of studies of the behaviour of silica in natural waters have drawn attention to a possible association between high silica levels and high levels of organic acid anions, which has been used as evidence to suggest that silica solubilities are enhanced by the formation of complexes with organic species (Bennett and Siegel, 1987). If complexing between silica and organic species does occur, it might be expected that the silica contents of leachates would correlate with dissolved organic acid anion contents, or total organic carbon.

The silica contents of the leachates analysed in this study vary from 0-23 mg/L Si, and according to plots against TOC, VFA and acetate contents (Figures 2 and 3) there is no correlation between the silica content and the content of the organic constituents. These plots suggest that high silica contents are found in leachates with low TOC or low fatty acid contents. As the samples considered here are taken from mature cells which are anaerobic, it appears that the high silica levels are associated with methanogenic rather than acetogenic leachates. Much more work is required to determine the stage at which high leachate silica levels are achieved.

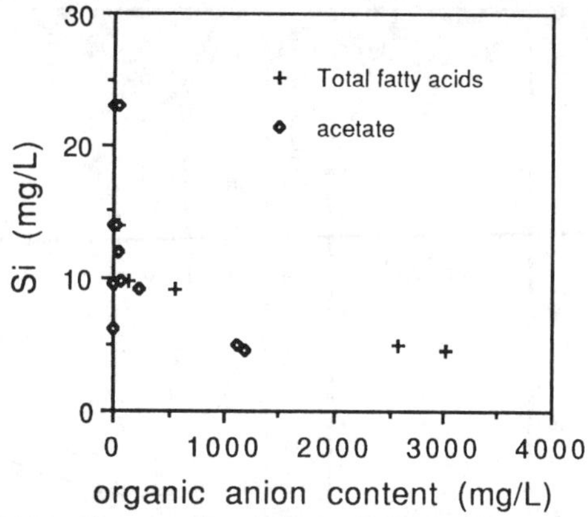

Figure 2. Silica contents of leachates (as Si) plotted against organic anion contents.

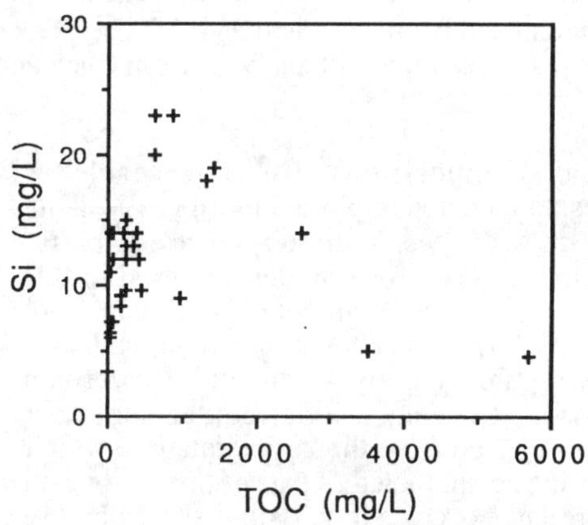

Figure 3. Silica contents of leachates (as Si) plotted against total organic carbon (TOC) contents.

5 CONCLUSIONS

For the leachates studied here, it has been shown that their compositions are consistent with contact between leachate and a mixed clay assemblage, consisting of illite and smectite. None of the samples show evidence of contact with smectite alone, which suggests that either clays introduced with refuse or impurities within the clay barrier have reacted with the leachates. The extent to which the reaction takes place will depend on the rate at which the leachate is generated, as the solubility of individual clay minerals inwater is so low that a small amount of mineral will effectively buffer the fluid composition. Similarly, an assemblage of more than one mineral in contact with leachate will tend to buffer the composition of the leachate, providing the opportunity to use leachate compositions as a means of monitoring the state of the mineral assemblage present within a barrier. Clearly, care has to be taken in the evaluation of leachate data, to ensure that other controls on fluid composition can be taken into account (such as might arise from the codisposal of chemical waste). However, the potential value of this approach seems to justify the routine determination of leachate silica contents as they offer the potential to provide a relatively inexpensive way of monitoring the mineralogical integrity of a clay barrier. This study is continuing, and we would welcome access to material derived from degraded clay barriers in order to test some of the predictions which arise from the chemical modelling.

ACKNOWLEDGMENTS

Jennifer Owen acknowledges support from the Natural Environment Research Council in the form of a post-graduate studentship. We are grateful to a number of companies who have supplied leachate samples for analysis, and to Howard Robinson for providing both samples and data. We would like to thank Jan Gronow for her comments on a draft of this paper, but the authors alone remain responsible for its shortcomings.

REFERENCES

Bath, A. H. 1993. Clays as chemical and hydraulic barriers in waste disposal: evidence from pore waters. In D. A. C. Manning, P. L. Hall and C. R. Hughes (eds.), Geochemistry of Clay-Pore Fluid Interactions, p. 316-330. London, Chapman and Hall.

Bennett, P. C. & Siegel, D. I. 1987. Increased solubility of quartz in water due to complexing by organic compounds. Nature 326:684-686.

Bowers. T. S., Jackson, K. J. & Helgeson, H. C. 1984. Equilibrium activity diagrams. Berlin, Springer-Verlag.

Chen, K. Y. & Bowerman, F. R. 1974. Mechanisms of leachate formation in sanitary landfills. In T. F. Yen (ed.) Recycling and disposal of solid wastes: industrial, agricultural and domestic, p.349-367. Ann Arbor, Ann Arbor Science Publishers Inc.

Eberl, D. D., Velde, B. & McCormick, T. 1993. Synthesis of illite-smectite from smectite at Earth surface temperatures and high pH. Clay Minerals 28: 49-60.

Garrels, R. M. & Christ, C. L. 1965. Solutions, minerals and equilibria. New York, Harper.

Garrels, R. M. 1984. Montmorillonite/illite stability diagrams. Clays and Clay Minerals 32: 161-166.

Hower, J., Eslinger, E. V., Hower, M. E. & Perry, E. A. 1976. Mechanism of burial metamorphism of argillaceous sediment: I. Mineralogical and chemical evidence. Geological Society of America Bulletin 87: 725-737.

Johnson, J. W., Oelkers E. H. & Helgeson, H. C. 1992. SUPCRT 92: software package for calculating the standard molar thermodynamic properties of minerals, gases, aqueous species, and reactions from 1 to 5000 bars and 0° to 1000°C. Computers and Geoscience 18: 899-947.

Robinson, H. D. 1989. Development of methanogenic conditions within landfills. Proceedings of the Second International Landfill Symposium, Porto Conte (Sardinia), paper XXIX.

Skipper, N. T., Refson, K. and McConnell, J. D. C. 1993. Monte Carlo simulations of Mg- and Na-smectites.In D. A. C. Manning, P. L. Hall and C. R. Hughes (eds.), Geochemistry of Clay-Pore Fluid Interactions, p. 40-61. London, Chapman and Hall.

Stumm, W. & Morgan, J. J. 1970 Aquatic chemistry. New York, Wiley.

Velde, B. 1992. Introduction to clay minerals. London, Chapman and Hall.

Wolery. T. J. 1992. EQ3NR, a computer program for geochemical aqueous speciation calculations: Users guide and documentation (version 7). Berkeley: UCRL-MA-110662 part III, Lawrence Livermore Laboratory, University of California.

Waste Disposal by Landfill – GREEN'93, Sarsby (ed.) © 1995 Balkema, Rotterdam, ISBN 90 5410 356 6

Use of electrokinetics to control the migration of heavy metals from landfill sites

Sibel Pamukcu & Gary Newhart
Lehigh University, Bethlehem, Pa., USA

J. Kenneth Wittle
Electro-Petroleum, Inc., USA

ABSTRACT: The electrokinetic process is an emerging technology for in-situ soil decontamination, in which chemical species, both ionic and nonionic are transported to an electrode site in soil. These products are subsequently removed from the soil via collection systems engineered for each specific application. Electrokinetics refer to the movement of water, ions and charged particles relative to one another under the action of an applied direct current electric field. In a porous, compact matrix of surface charged particles such as soil, the ion containing pore fluid can be made to flow to a collection sites under the applied field. The process has been demonstrated in the laboratory to show the viability of the electrokinetic treatment to move heavy metals through soil, with hydraulic conductivities as low as 10^{-8} cm/s, to a directed location

1 INTRODUCTION

The remediation of contaminated soils and ground waters is considered to be the most costly and time consuming activity in a site cleanup. Often 'sites' are contaminated by various materials ranging from inorganic to organic species, and the clean up plan consists of terminating the source of the contamination, followed by removal of the contaminated matrix and placement of the contaminated matrix in a secure landfill. This technique, though effective in the initial site cleanup, is just a transferal of the problem to another location. In-situ treatment of the contaminated soils and ground waters has become a preferred option to the 'dig it up and bury it elsewhere' mentality. Long term pump and treat methods are often very expensive and meet with limited success due to the difficulties encountered while hydraulically controlling contaminant movement towards a pre-engineered collection point. Other individual in-situ methods such as bioremediation and chemical treatments meet limited success due to the processes inherent difficulty to place and transport the necessary nutrients and chemicals. In light of the mentioned methods of containment and decontamination, electrokinetics is a promising technique to decontaminate soils and ground waters affected by inorganic and organic contaminants (Shapiro and Probstein, 1993; Wittle and Pamukcu, 1993).

Electrokinetic containment and subsequent decontamination of a contaminated matrix involves three electrokinetic processes: electroosmosis, electrophoresis and electrolytic migration of ionic and polar species. Electroosmosis produces the rapid flow of fluid (water, leachate) during the decontamination process, and may be the dominant containment/decontamination process in low permeability soils. Electrophoresis is the migration of charged colloids, with respect to a electrical potential gradient. In matrices with low hydraulic conductivity, electrophoresis is not the dominant process due to matrix hindrance of colloid movement, whereas in more permeable soils, this component may be very influential. Electrolytic migration is the migration of ionic species which are present in the pore water fluid (ground water, leachate). Electrolytic migration also refers to the dissociation of the water molecule into H^+ and OH^-, which migrates to the cathode and anode respectively. This process is responsible for the current conduction in a soil-water system, as well as pulling layers of water molecules and the chemical species harbored within the fluid. In considering the electrokinetic migration of inorganic contaminants within a soil fluid matrix, akin to that encountered in a landfill, the generation of an acid front originating from the anode, migrating towards the cathode would determine the metal speciation and solubility of many of the metals commonly detected. When a water molecule dissociates, it is reduced to H^+ at the anode and oxidized to OH^- at the cathode. The acid front at the anode propagates towards the cathode, whereas the basic front generated at the cathode propagates towards the anode. The migration of the acid front is enhanced by two factors: H^+ ions moves more quickly through a soil-fluid matrix than OH^- ions, and the electroosmotic flow from the anode toward the cathode enhances the H^+ ion migration. Acidic conditions are generally beneficial in the stripping of inorganics from the soil matrix, due to cation exchange with the soil matrix surface, as well as keeping the inorganic complexes in solution.

The containment and extraction of inorganic contaminants by the electrokinetic method proceeds according to the presented theory when the contaminant is in solution, or at least of an insufficient size as to allow the colloid (suspension, emulsion) not to be excluded from transport through the soil matrix.

2 BACKGROUND

The components of electrokinetics have been defined as electroosmosis, electrolytic migration and electrophoresis. Many investigators have shown that the electroosmosis component is the dominant electrokinetic process in low ion concentration systems, whereas in high ion concentration systems, electroosmosis and electrolytic migration are the primary components, with electrophoresis contributing primarily in permeable soil-fluid systems (Pamukcu and Wittle, 1992; Pamukcu, et. al, 1991; Mitchell 1986; Probstein, et. al, 1986). The host fluid (often ground water) transport of inorganic contaminants depends on the properties of the soil matrix in which the fluid flows, and the electrochemical properties of the contaminant which the fluid is transporting. Ground water transport of an inorganic species is related to the fluid's velocity through the soil matrix, the diffusion/dispersion coefficient that is occurring in the soil-fluid system, and the physical adsorption of the contaminant to the soil matrix and chemical interactions which occur at the soil-fluid interface.

The electroosmotic phenomenon was presented to the scientific community by Reuss in 1808. He reported that when a direct electric current (DC) was applied across a porous media, the fluid in the system moved towards the cathode of the electric field. This phenomenon was mathematically explained by Helmholtz in 1879, and revised by Smoluchowski in 1921. The current Helmholtz-Smoluchowski theory describes electroosmotic velocity of a fluid of a certain viscosity and dielectric constant, through a surface charged porous medium with a known zeta potential, with respect to an electrical gradient. An approach by Spiegler (1958) considers the interaction of the mobile phase components (fluid and soil) to each other and the resultant frictional forces of these components with the matrix's pore wall. Gray and Mitchell (1967) reported that the electroosmotic flow with in a soil matrix increases with increasing fluid content. The investigators also showed that electroosmotic flow decreases with an increase in the electrolytic concentration of the pore fluid; the electrolyte retardation effect is evident in soils with high anion retention capacities (Pamukcu and Wittle, 1992; Wittle and Pamukcu, 1993). Kahn et al. (1993) proposed a modified theory of electroosmotic flow through soil matrices. The investigators theorized that the true electroosmotic flow is directly proportional to the current carried by the surface of the charged soil matrix.

Electrokinetics has been used to dewater clays , silts and fine sands by the construction industry since as early as the 1940's. Applications have stabilized embankments, dewatered foundation excavations, and increased the loading capacity of construction piles (Casagrande, 1949). Segall et al. (1980) demonstrated that the technique could be used to decontaminate soils affected by inorganic and organic materials. Soils contaminated by the surplus of chemical compounds used presently and in the past had been remediated by removal to land fills, secure or otherwise. The acknowledgment that this is simply a transfer of the contaminant to another site, has led to the interest in in-situ remedial techniques. Acar et al. (1991) and Hamed et. al. (1991) reported the migration of an acidic front from the anode towards the cathode in an electrokinetic system is significant in the removal of inorganic metals from clayey soils. More recently, Kahn (1991) developed a method to predict contaminant removal from a soil matrix, based on electroosmotic flow predicted by the soil matrix surface charge. Experiments using O-nitrophenol and zinc were conducted by the researchers, in which good agreement between the measured removal and the predicted removal occurred.

3 INVESTIGATION

An extensive laboratory program was undertaken to investigate the removal of heavy metals from clay matrices, using the electrokinetic technique. The clay soil matrices were prepared in the laboratory with either synthetic ground water (inorganic salt concentrations representing a selected ground water type), distilled water, or a humic substance solution, then contaminated with a predetermined concentration of a heavy metal, and compressed to a compact saturated soil sample. Each specimen type (clay, pore solution and metal type) was prepared in triplicate sets and exposed to identical electrokinetic test conditions, with varied treatment times with in the identical sets. The initial contaminant metal concentrations used in the sample preparations, were based on field values reported in Environmental Protection Agency (EPA) documents. This paper reports the results of the efforts to remediate the test soils contaminated with As, Cd, Pb, Sr, U and Zn.

3.1 Materials

The clays used in the soil matrix preparation were a commercial grade of both Georgia kaolinite and sodium montmorillonite (bentonite). The sandy soils were prepared with washed and sieved New Jersey beach sand; sand retained by the no. 100 sieve was used. The variation in soil mineral matrix was used so as to investigate the remedial possibilities in a clayey soil matrix (Na-montmorillonite) with many available bonding sites, and in a sandy (10% Na-montmorillonite and sand) mixture which simulates a matrix with reduced available surface bonding sites. The pore water solutions used in this investigation were distilled water, a synthetic ground water solution and a humic substance solution. The synthetic ground water was prepared following a formula of the ground water salt concentrations typically found in a mid-western United States geographic region. The pH of the synthetic ground water was measured to be 7.9. The synthetic The humic substance solution was prepared by dissolving a commercially available humus stock starting material in distilled water. The measured concentration was 900 mg/L, with a starting TOC level of the solution 257 mg Carbon/l. This compares favorably to an aquifer rich in organic carbon. The pH of the soil slurries were neutral to slightly acidic.

3.2 Procedures

The slurries were mixed at 100% water content for the kaolinite clay, at 50% water content for the sand-Na-montmorillonite clay mixture, and at 200-300% water content for the Na-montmorillonite clay (by dry weight). The mixing was achieved using a hand held mixer in approximately 3 gallon portions. The slurry was consolidated in Lehigh University's geotechnical laboratory consolidometers. The consolidometers are capable of applying incremental sustained stresses whilst allowing the excess pore fluid to drain out of the consolidation units. The procedure creates a homogeneous, near saturated soil matrix, compacted by a constant pressure to a constant density. The soil samples were consolidated to 200 kPa pressure. At the end of the consolidation period (approximately 24 hours for the kaolinite clay, and 7 days for the Na-montmorillonite slurries, and somewhat less than 7 days for the Sand-Na-montmorillonite clay samples), the consolidated soil tube was removed. The extruding top portion of the soil sample was trimmed and collected for water content and initial quantitative chemical analysis. Samples of the pore waters were also collected for quantitative chemical analysis to assess the concentration of metal retention by the soil matrix. Upon completion of initial soil collection and sample weighing, the sample tubes were mounted in the electrokinetic apparatus and prepared for testing. The electrokinetic apparatus is shown in Figure 1.

The tests were conducted under a constant potential of 30 volts DC across the electrodes. The voltage in the soil and the current generated were measured concurrently with the fluid flow measurements. The tests were conducted for a minimum of 24 hours. A number of randomly selected specimens were tested for times ranging at minimum 24 hours to a maximum of 50 hours. At the completion of each individual experiment, the electrode chamber fluid was collected. The collected cathode fluid's pH consistently ranged from 10 to 12, whereas the anode chamber fluid's pH were consistently in the 2 to 3 range. The soil sample was extruded from the sample tube, and sectioned into anode, center and cathode portions. The sections were extracted and analyzed for the sample's target metal concentration and compared to that sample's starting metal concentration. While dissecting the soil sample, 5 pH measurements at equal intervals between the anode and the cathode sides of the soil were collected and recorded.

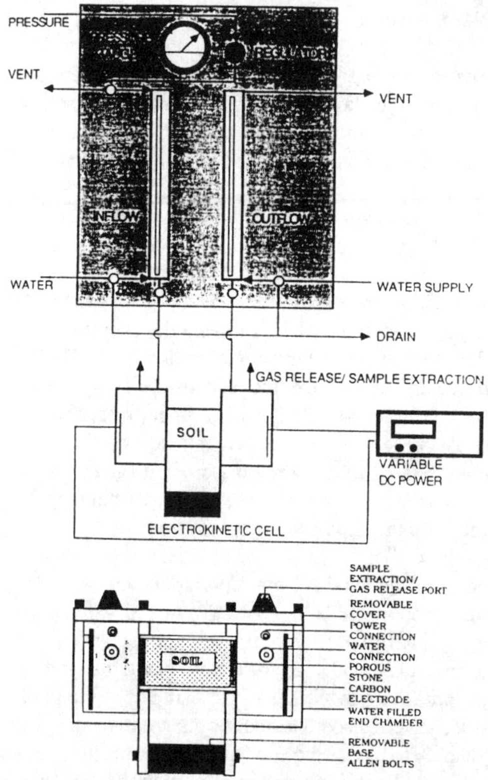

Figure 1. Diagram of the Electrokinetic Apparatus and Control Panel

3.3 Results

Ion transport in compact soils by electrokinetic migration was observed for all metals in all soil materials. Therefore clean-up of soils may be achieved either by clean water displacement as in pumping well extraction, or simply by the effective movement of ionic species in solution, even when the liquid phase may be stationary. Electroosmotic flow was achieved in all soil types. The current efficiency of the electroosmotic transport (volume of liquid transported toward cathode per quantity (amp. hr) of electricity) is high in soil-contaminant combinations for which electroosmotic flux is the dominant mechanism of contaminant transport. This appears to be the case when cation concentration is low, or the contaminant remains undissociated and poorly adsorbed onto clay surfaces. This agrees well with the Donnan principle and earlier observations by Gray and Mitchell (1967) and Probstein and Renaud (1986). Current efficiency may be indirectly related to removal efficiency for some ionic species. An increase in current efficiency may signal the removal of current carriers from within the pore space by precipitation into insoluble salts or by adsorption onto clay surfaces. Anion retention capacity of clay, as well as it's cation retention, appears to play an important role in current efficiency determination of soil-contaminant systems.

Long term tests, with strontium in kaolinite soil exhibited migration patterns which were dependent on time and spatial variation of the voltage gradients in the soil. The transport mechanism of strontium deviated significantly from the plug flow prediction of the existing convection, diffusion and adsorption models of contaminant transport. The experimental and theoretical results coincided well after 4 pore volumes of through flow with the modified approach of time and spatial variation of voltage gradients. The transient and spatial variation of the oxidation and reduction state of the soil appeared to be one of the controlling factors in the transport mechanism.

3.3.1 Electrokinetic Flow of Water

The flow from the anode chamber into the soil is designated as inflow, whereas the flow from the soil into the cathode chamber is designated as outflow. In saturated soils, these two quantities measured over time should be equal to each other under steady state flow conditions, that is with no change in the volume and saturation state of the soil medium. An example of the inflow, outflow and current variation with duration of electrokinetic treatment with constant 30 volts applied across the electrodes for kaolinite clay with strontium is shown in Figure 2. The majority of the measured flows exhibited steady state behavior with close matching of the inflow and outflow curves. The current densities, measured in current per unit cross sectional area of the specimen ($A=9.89\ cm^2$), varied considerably from one soil type to another as well as from one metal to another, largest being for the montmorillonite soils. As current densities increased, the deviation from steady state flow increased, due to the volume change and gas generation in the soil pores. Soil types with low current efficiencies were those with high current densities and non-steady state flows. In a few cases for which the ionic concentrations were fairly large, such as zinc, the net flow to the cathode site ceased and at times reversed toward the anode. This is consistent with

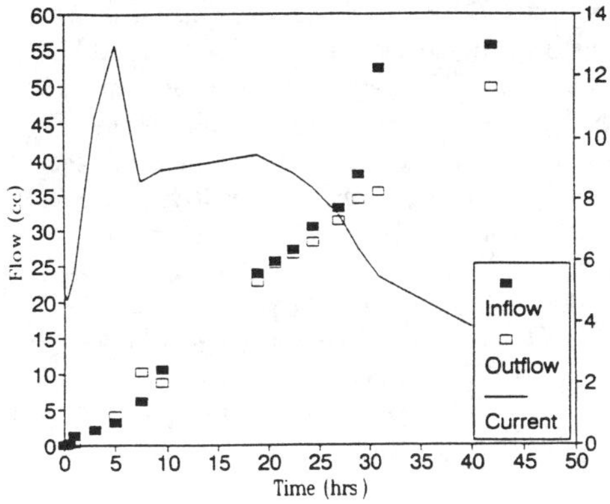

Figure 2. Electroosmotic Flow and Current Variation With Time in Kaolinite Clay With Distilled Water and Strontium

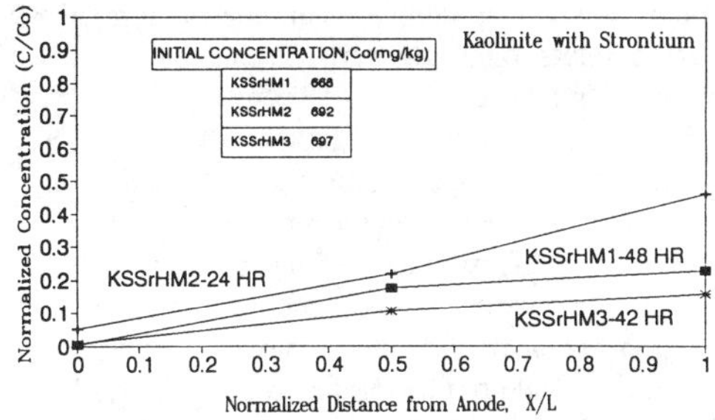

Figure 3. Concentration Profiles of Strontium in Three Replicate Kaolinite/Distilled Water Soil Specimens After Electrokinetic Treatment

Donnan principle that the electroosmotic flux toward cathode is directly proportional to the ratio of concentration of water to concentration of cations, $[H_2O] / [+]$. Therefore at high concentration of the cation, the ratio becomes small and so does the electroosmotic flow toward cathode.

Kaolinite soils showed constant current efficiencies with little dependence on the pore fluid composition (distilled water, simulated ground water, humic solution). The volume of inflow (normalized by the soil water volume) per moles of electrons transferred varied from 4 to 250 moles^{-1} for all of the metals and all the pore fluid types. The current efficiencies for montmorillonite soils appeared to be more dependent on the metal ion type. For all of the soil specimens, the current efficiency remained constant with moles of electrons transferred, in general. This observation indicated a steady state transport of liquid with no major changes in the hydraulic and electrical resistivity of the soil system during the period of electrokinetic treatment. The consistency of data also confirmed the intended functioning of the equipment with minimum volume change and electrolytic reactions in the soil. Therefore, in a field situation if water is supplied at the anode end, the transport is expected to continue at a steady state for extended periods of time with no major changes in the physical state of the soil. This was demonstrated by long term tests with strontium, up to 18 pore volumes of water transport through kaolinite soil for which the long term current density remained around 0.14 mA/cm^2

3.3.2 Electrokinetic Migration of Metal Ions

All the metal ions migrated in the soil under the influence of the applied electric field. The rate and direction of migration, and the actual percent removal of the metal from the soil varied significantly with the metal type. An example of the strontium metal concentration measured at three points in three replicate kaolinite soil samples, normalized by the original measured concentrations are presented in Figure 3. In some of the soil-contaminant pairs treatment time was staggered to observe the transient behavior of the transport. Since only three points were sampled along the soil length, the actual concentration profile may not be well represented by the linear distribution assumed between the consecutive points.

Observing the concentration profiles, it becomes clear that none of the metals are transported via a plug flow mechanism. In this case, the plug flow mechanism would best be characterized by a "step function" shaped profile of contaminant concentration that marches in time from the anode toward the cathode end. In the majority of the profiles there is significant reduction in concentration of the metals in the first half (anode to center) of the soil specimen and an increase in concentration at the cathode end of the soil. This concentration is probably due to the formation of hydroxide salts and also the increased cation retention capacity of the soil resulting from an increased pH at the cathode end of the soil at the soil-water interface. The pH profile of soil after an electrokinetic (E-K) test is typically given in Figure 4, (kaolinite with strontium). High pH favors formation of hydroxide salts, however, when the metal tends to remain in solution for a wide range of pH, such as strontium, the increase in concentration at the cathode end diminishes in time as more of the metal is released into the cathode chamber, as observed in Figure 3.

The rate of migration and ultimate decontamination appear to be highly dependent on the type of soil-metal combination. Judging from the speciation distribution of metals with pH and the oxidation/reduction state, the metals that have complex chemistry exhibited removal and migration trends that could not be predicted analytically. The following trends were observed for each of the metals tested:

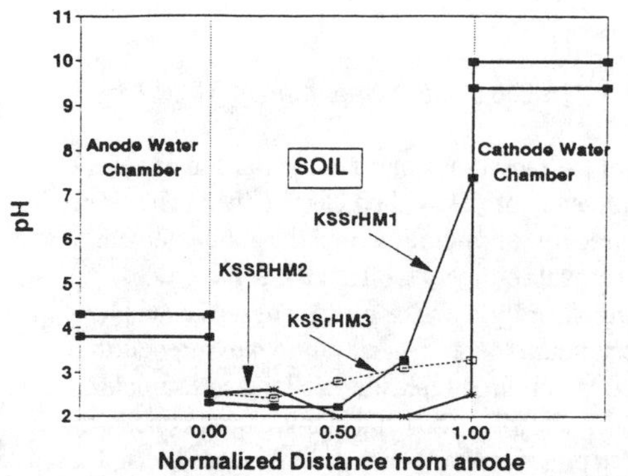

Figure 4. pH Distribution From Anode to Cathode Chambers in 3 Duplicate Specimens of Kaolinite Clay With Distilled Water and Strontium After E-K Treatment

A. Arsenic: Arsenic was in anionic ($HAsO_4^{=}$) form in the initial mixing state into the slurry. Figure 5 shows the fraction of arsenic found at three different locations in the soil and the electrode chamber waters after the 24 to 48 hours of electrokinetic treatment of high As concentration soil specimens. Arsenic showed substantial reduction from its initial concentration at both the cathode and anode regions of the soil specimens. Very little removal was achieved into the electrode water chambers. Most of the product appeared to have accumulated at the center of the soil specimens by the end of the short term treatment. Arsenic speciation and solubility is greatly affected by the pH and redox potential of the soil. The availability of arsenic , in the form of arsenite (As(III)) increase at low redox potential and alkaline conditions. The average pH of the center section was 2.9 ranging from 1.8 to 3.0. Subsequent analysis of redox potential variation of kaolinite clay during electrokinetics showed that, in general, the anode end of the soil remains in oxidizing state, while the cathode end is in reducing state. At high redox levels (oxidizing state) the majority of As is found in arsenate form (As(IV)) which is not as soluble as As(III), and tend to be retained on the oxide surfaces of the clay minerals (Masscheleyn et al., 1991). Accumulation of arsenic at the center of the soil specimens can be explained by the low pH and the oxidizing state of the soil at that location. Accumulation was evident at the center of the soil samples for most of the specimens except for the low concentration case in which the metal appeared not to migrate substantially.

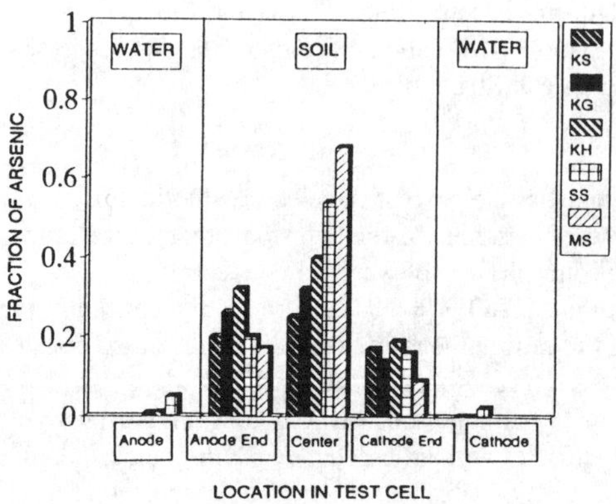

Figure 5. Average Distribution of Arsenic in Soil and Electrode Chambers Waters After E-K

B. Cadmium: Figure 6 shows the fraction of cadmium found at three different locations in the soil and the electrode chamber waters after the 24 to 48 hours of electrokinetic treatment of high Cd concentration soil specimens. Cadmium is a divalent cation which showed a migration path toward the cathode. Cadmium exhibited a maximum concentration at the cathode end of the soil sample. This is attributed to the increase in the hydrolysis with an increase in pH at the cathode end. The average pHs measured at the cathode ends were 5.3, 8.1 and 9.0 for the distilled water, ground water and the humic solution specimens of kaolinite clay, respectively. Until around pH 8, Cd remains in its divalent cationic form. Beyond this value it starts forming complex species which are either charged positively or negatively or neutral. The tendency and the abundance of these products control the removal rate until the acid front reaches the cathode region of the soil. However, with the high pH prevailing at the soil-water interface of the cathode end of the soil, a thin layer of precipitate would form at the interface making it difficult for cadmium to be removed into the water chamber. There is also an increase in soil adsorption capacity with increasing pH which would contribute to the accumulation of the metal at this region (Sposito, 1984; Basta and Tabatabai, 1992).

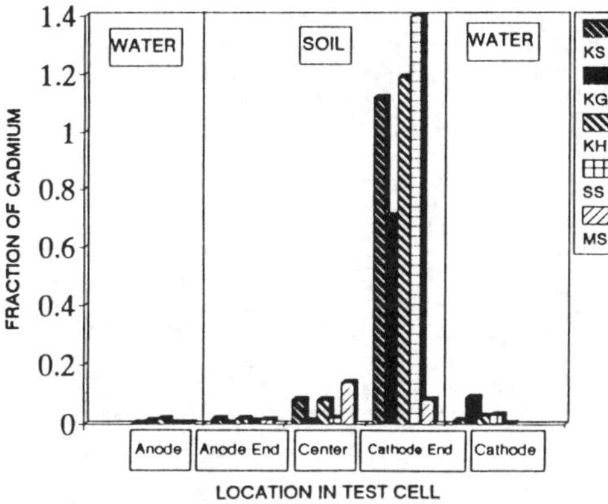

Figure 6. Average Distribution of Fraction of Cadmium in Soil and Electrode Chamber Waters After E-K

C. Chromium: Chromium was introduced to the soil in anionic form, ($Cr_2O_7^{=}$) chromate. Chromate (chromium in the hexavalent oxidation state) is an anion which carries a negative two charge. Figure 7 shows that approximately 30% of the Cr was extracted into the anode chamber by the end of 24 to 48 hours of treatments. In this case, the electromigration of the ion took place in the opposite direction of the electroosmotic water flow. There was accumulation of the metal at the discharge end (anode end of the soil) owing to the increased anion retention of the soil at low pH and also to the deceleration of ionic migration due to electrophoretic and relaxation effects (Kortum and Bockris, 1951).

D. Lead: Lead is one of the preferentially adsorbed metals by clay minerals (Basta and Tabatabai, 1992). It is mostly in the form of a divalent cation below pH of 9. Above pH 9 the PbO is the stable species (Dragun, 1988). At sufficiently high pH (>11) it forms an anionic species of hydrolysis product, $Pb(OH)_6^{-2}$, which is expected to migrate in the opposite direction of electroosmotic flow. Other hydrolysis products of lead which occur at pH levels greater than 6 are: $Pb_2(OH)^{+3}$, $Pb_4(OH)_4$ and $Pb_6(OH)_8^{+4}$. These species would exhibit increased ionic velocities due to their higher valances. The effect of high pH on the clay adsorption of metal becomes more significant at high concentrations of the metal, as shown by Basta and Tabatabai (1992). The accumulation of lead at the cathode end of the soil specimens is shown in Figure 8. This accumulation is attributed to: (i) precipitation and anion species formation, (ii) increased concentration of lead at the discharge end (cathode region of soil), (iii) increased adsorption and retention of lead on clay due to high pH and increased concentration at the cathode region of the soil.

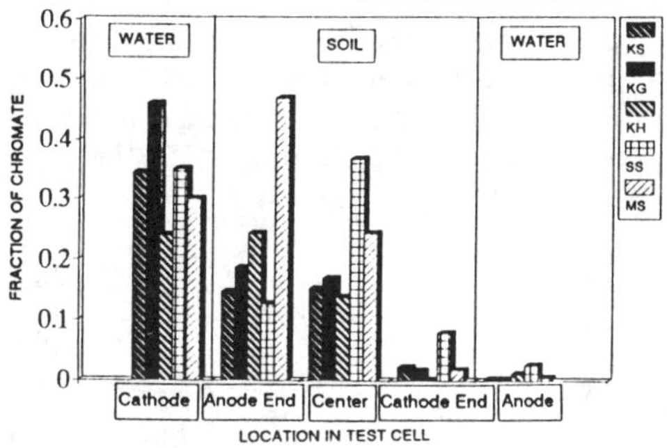

Figure 7. Average Distribution of Fraction of Chromium (anionic form) in Soil and Electrode Chamber Waters After E-K

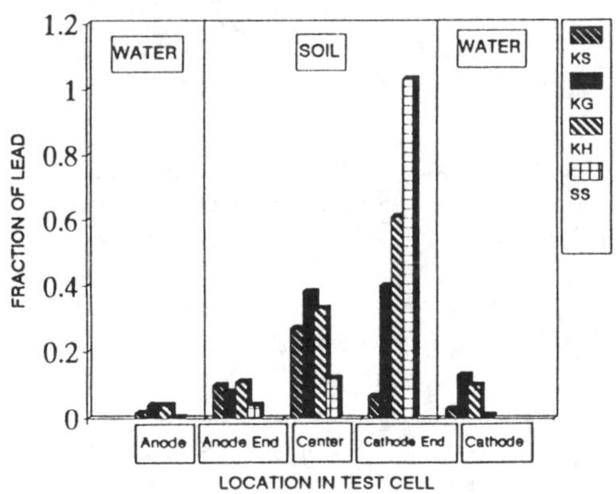

Figure 8. Average Distribution of Fraction of Lead in Soil and Electrode Chamber Waters After E-K

E. Strontium: Strontium remains a divalent ion for a large range of pH values and is stable to hydrolysis throughout the pH range (2 to 10) of the electrokinetic tests. Figure 9 shows fraction of the metal in the soil and electrode chamber waters after the short duration of electrokinetic treatments. The accumulation of the metal at the cathode end is due to retardation effects of electrophoresis and relaxation as the concentration of the cation increases at the cathode region (Kortum and Bockris, 1951) The mechanism which triggers the accumulation is probably the increased cation retention capacity of the clay at high pH levels. As the pore volume fraction of water transported through the soil increases, the concentration profiles shift down as illustrated in Figure 2. In that particular Figure, the pore volume fraction of water transported through soil for the specimen labeled as KSSrHM2 was 0.48, whereas it was 0.92 and 1.26 for the specimens labeled KSSrHM1 and KSSrHM3, respectively. Subsequent long term tests on strontium contaminated kaolinite clay with distilled water showed nearly complete removal of the metal after about 4 pore volumes of water transport (Wilkowe, 1992).

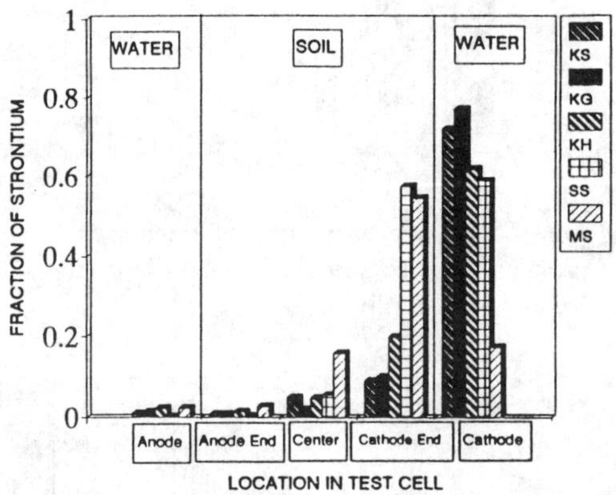

Figure 9. Average Distribution of Fraction of Strontium in Soil and Electrode Chamber Waters After E-K

Figure 10 shows the variation of strontium concentration remaining at the three locations of soil specimen with pore volume of water transported for 4 specimens of kaolinite clay with distilled water and similar initial concentration of strontium (average 718± 50 mg/kg). As observed, at about constant rate of electroosmotic water flow (1.25 ± 0.5 cc/hr), which corresponded to an average electroosmotic hydraulic conductivity of 3.5 x 10^{-5} cm/sec, the variation of strontium concentration in soil with pore volume of water flow is nonlinear. The data indicated that after about one pore volume of water flow, the concentration reduction of the metal at each location became steady with pore volume of water flow. Analysis of this data shows the relative contributions of electromigratory transport (at the beginning of the treatment) and the electroosmotic transport of strontium (after 1 pore volume of water flow) to the overall decontamination process when most of the metal constituent is expected to be present in the pore water.

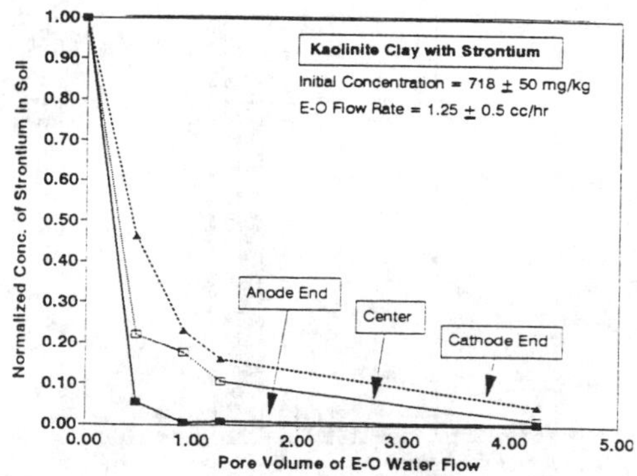

Figure 10. Variation of Strontium Concentration With Pore Volume of Flow in Kaolinite Clay Specimen

F. Uranium: Uranium was introduced in the soil as the uranyl salt from a series of hydroxide complexes. Below the pH of 6, they are cationic in nature. At approximately pH 6, the $UO_2(OH)_2.H_2O$ precipitate (Borovec, 1981). At higher pHs, the anion hydroxide species such as $UO_2(OH)_3^-$ and $UO_2(OH)_4^=$ occur which would then migrate in the opposite direction of flow. These species may never be able to travel back to the anode chamber since they would encounter low pH environment on the way and tend to change form and sign. However, as long as the high pH gradient prevail at the soil-water interface at the cathode end, uranium would be difficult to transport into the cathode water chamber. As observed in Figure 11, little uranium was transported to the cathode chamber, while most of it accumulated at the cathode region of the soil specimens. There was as much uranium measured in the anode water chamber as in the cathode chamber, probably due to the diffusion of the metal.

G. Zinc: Zinc is amphoteric for which the divalent cation is stable below pH 7.7. The neutral species ($Zn(OH)_2$) predominates above pH 9.1. The anionic species, $Zn(OH)_3^-$, $Zn(OH)_4^=$ become significant after about pH 11. Due to the amphoteric nature of zinc and its tendency to form polynuclear hydrolysis species, similar removal fractions were achieved at the anode and cathode waters. Accumulation was mostly observed at the center of the specimens except for Na-montmorillonite soil as shown in Figure 12.

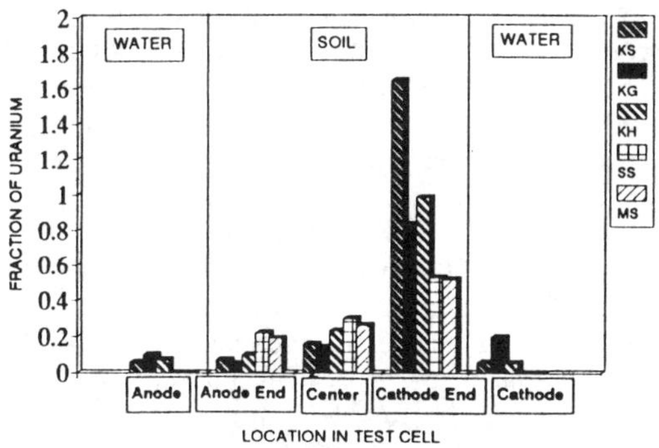

Figure 11. Average Distribution of Fraction of Uranium in Soil and Electrode Chamber Waters After E-K

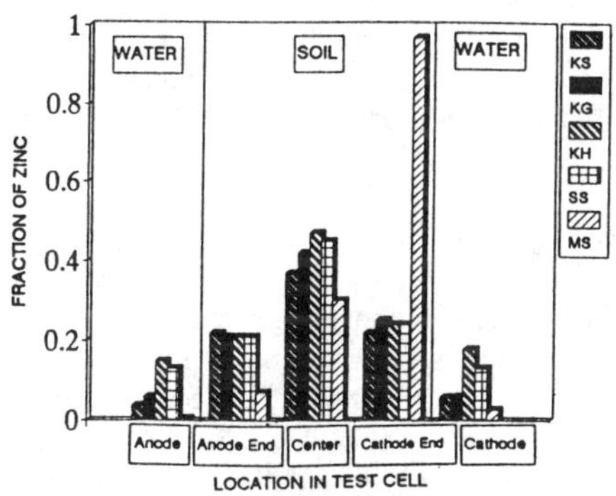

Figure 12. Average Distribution of Fraction of Zinc in Soil and Electrode Chamber Waters After E-K

In an overall analysis of the data, the best removals from about 85 to 95% was achieved over 24 to 48 hour period for Cd, Cr, and Sr in all the soil types. During this time period, the average potential gradient in the soil remained constant at 3 volts/cm . In these experiments, the soil types ranged from pure clays (kaolinite, montmorillonite) mixed with either distilled water, simulated ground water or humic substance solution; to a sand-clay mixture. Among the three soil types (kaolinite, montmorillonite and sand-montmorillonite) tested, kaolinite showed the highest efficiency of electroosmotic water flow, followed by the clayey sand. Kaolinite exhibited constant flow efficiency regardless of the nature of the metal ion in the mixture, whereas, flow through montmorillonite and clayey sand was influenced by the type of metal ion present.

The water flow toward the cathode was lower in mixtures containing higher concentrations of the metal for the same quantity of electricity. As an example, although more water per electron was transmitted towards the cathode for the low concentration Cd case, it didn't necessarily result in higher removal of the metal. Assuming a linear distribution of the metal concentration between the anode and the center of the soil, the percent removals were computed as 84% for the high concentration case, and 16.5% for the low concentration case. At lower concentrations, the current carried by the clay surface may constitute a larger portion of the total current and result in a larger net flow of water in the cathode direction, or electroosmosis, as observed by Napier in 1846. At higher concentrations, the current is easier to pass through the soil, because the bulk of it would be carried through the pore space. The quantity of true electroosmotic flow to the cathode may remain the same, however the flow would probably be counteracted by the movement of a larger concentration of anions that drag water molecules along. The anion retention of the clay then plays an important role in the development of a net water flow to the cathode. That is, the higher the anion retention capacity of the soil the higher the net flow toward the cathode. This is best illustrated by the pore volume of water transport to cathode and percent removal or concentration reduction of metal calculations for 4 different metals presented in Table 1.

As observed, the variations in concentration reduction percentages for Cr, Cs, Sr and U appear to depend on the soil matrix, and the metal and pore fluid type, with little or no correlation to the volume of water flow during the short term treatment. For the cationic species (Sr^{++}, Cs^{+}, UO_2^{++}) the volume of water, represented as the fraction of the pore volume, ranged between 0.25 and near 1.00. The flow for the anionic form of Cr, ($Cr_2O_7^{=}$), was consistently lower than that of the cationic species. This demonstrated the effect of water dragging action of the migrating ions on the net flow towards the cathode. In the case of the anionic species $Cr_2O_7^{=}$, as the anion migrates towards the anode, it drags along sufficient amount of water to measurably reduce the net flow towards the cathode. Obviously, both the anion and cation retention capacity of the soil, which subsequently effect the ratio of the cations to anions that are free to migrate, influence the quantity of water movement towards the cathode.

Table 1. Average Percent Reduction of Metal Concentration (at the lowest concentration location) and Pore Volume Fraction of Water Transported Toward Cathode Chamber

Metal	KS✧ % Rm ①	KS✧ PV ②	KH % Rm	KH PV	KG % Rm	KG PV	MS % Rm	MS PV	SS % Rm	SS PV
Cr	93	0.2	97	0.2	95	0.2	95	0.1	97	0.1
Cs	72	0.7	74	0.3	77	0.6	55	0.9	89	0.3
Sr	98	0.4	96	0.9	99	0.4	92	0.5	99	0.6
U	79	0.4	70	0.7	85	0.3	44	0.8	33	0.6

✧KS: Kaolinite/Distilled Water;
MS: Na-montmorillonite/Distilled Water;
SS: Sand/10% Na-mont.\Distilled Water;
KG: Kaolinite/Ground Water;
KH: Kaolinite/Humic Soln.

①% Rm : percent removal at the location of lowest concentration achieved

②PV: Pore volume of water transported to the cathode chamber during treatment

4 CONCLUSIONS

This study describes some of the past and present efforts to develop the electrokinetic technology into a well engineered method to use in pollution migration control and subsequent soil decontamination. In laboratory investigations, up to 99% heavy metals removal is reported. The metals included surrogate radio nuclides such as cobalt, strontium, and uranium; anionic species of chromium and arsenic, and other metals such as cadmium, lead, nickel and zinc. The degree of removal of these species from the soil, by the electrokinetic treatment process was assessed through the individual species' profile generated across the soil between the electrodes. Removals of some of metal species up to 99% were achieved at the anode or the cathode end of the soil upon completion of a 24 to 48 hours of treatment, or a maximum of 1 pore volume of water displacement toward the cathode chamber. Transient pH change through the soil had an dramatic effect on the metal removal, as evidenced by accumulation of the metals at the discharge ends of the soil specimens. This accumulation was attributed to the precipitation of the metal and increased cation retention capacity of the soil material found in the high (basic) pH environment of the cathode end. In general, the reduced mobility and dissociation of the ionic species as they encounter areas of higher ionic concentration in their path of migration results in the accumulation of the metal species at the discharge ends of the soil specimens. The presented results indicate that the control of heavy metal migration from landfill sites is feasible by the electrokinetic enhancement of metal species transport in soils. The development of this in-situ technology for a wide range of applications hinges on a better understanding of the transient chemical and physical processes during the application of direct current through contaminated soil systems.

ACKNOWLEDGMENTS

This research was funded by the Office of Technology Development, Office of Environmental Restoration and Waste Management, U.S. Department of Energy, managed by Argonne National Laboratory, Contract No. 2112406.

REFERENCES

Acar, Y.B., Gale, R.J., Putnam, G., and Hamed, J., 1989. Electrochemical Processing of Soils: Its Potential Use. In Environmental Geotechnology and Significance of pH Gradients, Proc. of the 2nd Int. Sym. on Envir. Geotech., ol. 1, pp. 25-38. Envo Publishing, Bethlehem, Pa.

Acar, Y.B., Gale, R.J, Putnam, G., Hamed, J., and Wong, R., 1990. Electro Chemical Processing of Soils: Theory of pH Gradient Development by Diffusion and Linear Convection. J. Envir. Sci. and Health, Part (A); Envir. Sci. and Eng, 25 (6), pp. 687-714..

Basta, N.T., Tabatabai, 1992. M.A., Effect of Cropping Systems on Adsorption of Metals by Soils: III Competitive Adsorption. Soil Science, Vol. 153, No.4, pp. 331-336.

Borovec, Z, 1981. Chemical Geology, Vol. 32, pp. 45-59.

Casagrande, L.1949. Electro-Osmosis in Soils. Geotechnique, London, England, Vol. 1, No. 3, p. 159.

Dragun, J.,1988. The Soil Chemistry of Hazardous Materials, The Hazardous Material Control Research Institute. Silver Spring, Maryland.

Gray, D.H., and Mitchell, J.K., 1967. Fundamental Aspects of Electro-Osmosis in Soil. J. of the Soil Mech. and Found. Div., ASCE, SM 6, pp. 209-236.

Hamed, J., Acar, Y.B. and Gale, 1991. R.J., Pb(II) Removal From Kaolinite by Electrokinetics. J. of Geotech. Eng. Div., ASCE, Vol. 117, No. 2, pp. 241-271.

Khan, L. I., Study of Electroosmosis in Soil: A Modified Theory and Its Application in Soil Decontamination, 1991. Dissertation, in partial fulfillment of requirements for the degree of Doctor of Philosophy, Lehigh University, Bethlehem, Pa.

Khan, L.I., Pamukcu, S., and Fang, H.Y., 1993. A Modified Theory for Electroosmosis in Soil. J. of the Geotechnical Eng. Div., ASCE, New York, (revised , working paper).

Kortum, G. and Bockris, J.O'M., 1951. Textbook of Electrochemistry. Vol I &II, Elsevier Publishing Co., New York, Amsterdam.

Masscheleyn, P.H., Delaune, R.D., Patrick, W.H. Jr., 1991. Effect of Redox Potential and pH on Arsenic Speciation and Solubility in a Contaminated Soil. Env. Sci. and Technol., Vol. 25, No. 8, pp. 1414-1419.

Mitchell, J.K, 1986. Potential Uses of Electro-kinetics for Hazardous Waste Site Remediation Presented at USEPA-University of Washington Workshop on Electrokinetic Treatment of Hazardous Waste Treatment. Seattle, WA.

Napier, J.,1846. Phil. Mag., 29(10).

Pamukcu, S., Khan, L.I. and Fang, H.Y., 1991. Zinc Detoxification of Soils by Electro-Osmosis. Transportation Research Record No. 1288, pp. 41-46, TRB, Washington, D.C.

Probstein, R.F. and Renaud, P.O., 1986. Quantification of Fluid and Chemical Flow In Electro-Kinetics. Proc. of the Workshop on Electro-Kinetic Treatment and Its Application In Environmental Geotechnical Engineering For Hazardous Waste Site Remediation, US EPA, University of Washington, Seattle, Wa.

Segall, B.A., O'Bannon, C.E. and Matthias, J.A.,1980., Electro-Osmosis Chemistry and Water Quality, J. of the Geotech. Eng. Div., ASCE, Vol. 106, No. GT10, pp. 1143-1147.

Shapiro, A.P. and Probstein, R.F., 1993. Removal of Contaminants from Saturated Clay by Electroosmosis. Environ. Sci. Technol., Vol. 27, pp. 283-291.

Spiegler, K.S., 1958. Transport Process In Ionic Membranes. Transactions of the Faraday Society, vol. 54 Part 9, pp. 1408-1428

Sposito, G.,1984. The Surface Chemistry of Soils., Oxford University Press, New York.

Wilkowe, A.,1992. Modeling of Electrokinetic Decontamination of Soils, Thesis in partial fulfillment of requirements for the degree of Master of Science, Department of Civil Engineering, Lehigh University.

Wittle, J.K. and Pamukcu S., 1993. Electrokinetic Treatment of Contaminated Soils Sludges and Lagoon., DOE/CH-9206, Argonne National Laboratory, Chicago, Il.

Waste Disposal by Landfill – GREEN'93, Sarsby (ed.) © *1995 Balkema, Rotterdam, ISBN 90 5410 356 6*

Selection of soils for compacted clay lining

R.W. Sarsby
Bolton Institute of Higher Education, UK
M. Williams
Stanger Consultants Ltd, Manchester, UK

ABSTRACT: The most important parameter for a compacted soil which is being used to provide a barrier within a waste containment system is its permeability. Using a soil properties database an attempt has been made to correlate the influence of various soil parameters with permeability for soils compacted at optimum moisture content. A preliminary predictive equation has been generated which takes account of variations in particle size distribution and plasticity index. The equation was found generally to predict permeabilities within a range of 0.5 to 2 times the actual values, though in some instances it overestimated or underestimated by an order of magnitude. Generally the equation was found to provide a fairly good indication of the acceptability of a material as a compacted clay liner, for the soils studied.

INTRODUCTION

The three methods currently used within the United Kingdom for the disposal of domestic and industrial solid waste are recycling, incineration and containment within landfill sites. The UK disposes of the majority of its annual domestic and industrial solid waste production, including the ash residue from incinerated waste, by landfilling within containment sites.

Waste disposal sites are most commonly constructed within large topographical depressions, whether natural or man made. Migration of leachate or any other contaminated fluid, from the waste containment site is prevented by a barrier beneath and around the waste. The barrier may be low permeability soil (either natural or placed and compacted so as to be relatively impermeable), or a very low permeability man-made, membrane (such as high density polyethylene) used in conjunction with an underlying 'fail-safe' layer of low permeability soil. Whatever system is used part of the liner comprises a mineral layer and assurance is needed that the permeability of the soil layer is no higher than the assumed value. If the permeability is lower than anticipated then this is a bonus although we must avoid excessive over-estimation of permeability as this will lead to rejection of suitable materials and excessive cost.

One of earliest stages in the development of a landfill site is to identify and approve a suitable source of clay liner material.

Generally a maximum permeability of 10^{-6}mm/s is specified for a compacted soil liner. Low permeability is generally associated with clay soils and hence the use of the term 'clay liner' but other soil types may exhibit suitable hydraulic conductivities. The term 'compacted clay' basically refers to any suitable soil that displays the required low permeability, and in some soils the clay content is minimal. Both clays and silts and various admixtures of these with sand have been used in their compacted state as barriers in waste disposal sites. The identification of sources of suitable soils and the checking of compacted liners relies on laboratory permeability tests, which take many days and which can be subject to large experimental errors. In addition there is often a discrepancy between laboratory and in-situ values of hydraulic conductivity. Consequently there is a role for a simple predictive empirical relationship, between permeability and other soil properties, which can be used to strengthen the selection criteria.

As an extension to the basic equation presented by Darcy numerous workers have attempted to calculate permeability using soil properties that are amenable to mathematical treatment such as porosity, void ratio and particle size.

Various formulae have been derived for cohesionless granular media; Hazen (1892), Terzaghi (1925), Korzeny (1927), Loudon (1952), with varying degrees of success. Such formulations are unsatisfactory for fine-grained cohesive soils and recourse has been made to empirical equations derived from particular laboratory measurements of permeability; Fraser (1935), Carrier and Beckman (1984). The limited number of such relationships reflects the number of factors influencing the permeability of a compacted clay which is particularly sensitive to subtle changes in soil structure.

METHODOLOGY

The way in which particular factors affect hydraulic conductivity was investigated empirically from an examination of available laboratory data for various soils. These data generally comprise; particle size distribution, index tests, compaction and permeability values. The data were analyzed using a 'database approach', in that a store of soil data was established and information was then sorted and examined for correlations. Most of the data was from previous studies undertaken at Bolton Institute. Data were also obtained from existing literature, where possible, to form as large a data set as possible.

The data were normalised and sorted into data matches using a program written by Williams (1992). These matches were examined visually for any correlations between permeability and other soil variables. Statistical numerical correlation techniques were then used to more quantitatively describe the data correlations and in addition the soil data were analyzed using empirical formulae produced by previous workers.

Since Lambe (1954) many workers have studied experimentally the permeability of fine-grained soils. In general terms the permeability has been found to be controlled by the properties of the permeating fluid, the soil structure and the interaction between the permeating fluid and the porous medium. The structure of compacted clays is mainly influenced by particle size distribution, compactive effort, moulding water content, mineralogy. Generally the higher the clay content the lower the permeability and compacted soils exhibit their minimum permeability when compacted slightly wet of optimum (typically + 2% to + 4%). Dry of this value the permeability changes rapidly with moisture content whereas on the wet side permeability increases only slightly with increasing water content.

As permeability can be influenced significantly by moisture content it is important to establish a common reference point to allow comparisons between different soils. Consequently the Authors decided to study the effect of various variables upon the value of permeability at optimum moisture content (k_{opt}) and not the minimum permeability. This reference point was chosen because most of the available permeability data related to soils at, or slightly wet of, optimum. Also the point of optimum compaction provides a good point of commonality between soils, because at optimum compaction the soils have reached their lowest attainable void ratio. The minimum attainable permeability could only be established for eight of the soils in the database. Permeability at optimum moisture content (k_{opt}) is greater than the minimum value (typically by a factor of between 1.2 and 6.0 approximately) and other permeabilities resulting from compaction wet of optimum. Therefore, the (k_{opt}) value could safely be regarded as an upper limit or 'worst case' value of permeability for use in the assessment of clay for lining purposes.

ANALYSIS

The grading curves for all the soils stored in the database are illustrated in figure 1. It can be seen that the data relate to a large variety of soils. The general and numerical analyses of the data gave very similar results, and suggested relationships do exist between some of the soil variables and k_{opt}. The strongest and most consistent correlations were found between (k_{opt} and soil variables which were related to clay mineral content.

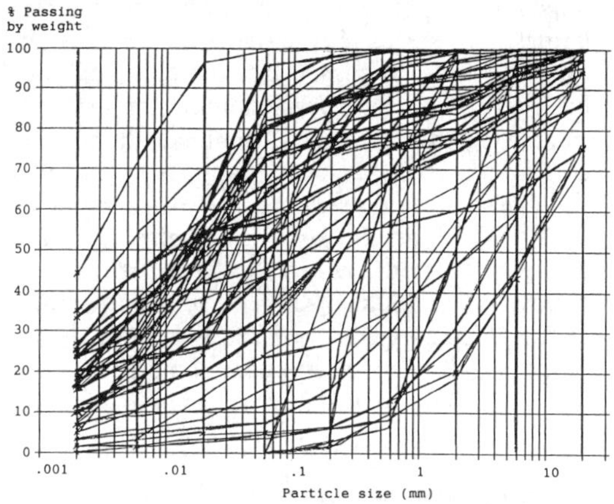

Figure 1. Grading curves in the database

There were no strong, consistent correlations with simple grading parameters such as d_{10}, d_{30}, clay content, sand content. This is because permeability is determined not by any particular particle size but by the whole grading curve because of the effects of infilling. However in general terms k_{opt} did show some correlation with the amount of fine material present particularly when this was expressed by d_{10} (the particle size for 10% by weight passing).

In general soils have typical permeabilities (in terms of orders of magnitude) as follows:

Clean gravel	$k \approx 10^3$ mm/s
Clean sand	$k \approx 10^0$ mm/s
Uniform silt	$k \approx 10^{-3}$ mm/s
Fat clay	$k \approx 10^{-6}$ mm/s

The following best-fit equation was chosen to express the foregoing general relationship between permeability and particle size in terms of d_{10} (expressed in mm):

$$k_{opt} = \frac{100^{\log d_{10}}}{4} \quad \text{mm/s} \qquad (1)$$

This formula generally predicted permeabilities of the expected order of magnitude for the samples in the database, but it also produced some significant overestimations and underestimations. As said before the whole of the grading curve affects the permeability and no factors were introduced to try to account for particle size distribution.

Examination of the database suggested that the most critical or 'effective' range of the grading curve for reducing permeability was the sub-d_{10} sizes. The more this section of the grading curve becomes non-uniform, the greater is the likelihood that permeability will be reduced, since an increased particle size variation suggests a greater infilling of voids by finer particles. An increase in the value of (d_{10}/d_o) was seen to reduce the permeability. Similarly non-uniformity in the plus d_{10} sizes may also reduce permeability due to infilling, but probably not to the same extent as the sub-d_{10} sizes.

Several parameters were tried for representation of the effects of the larger particles but efforts were quickly concentrated on only two, ie coefficient of uniformity, CU (equal to d_{60}/d_{10}) and coefficient of curvature CZ (equal to $d_{30}^2/d_{10}.d_{60}$). Individually neither of these parameters showed any true correlation with k_{opt}, since any value of CU or CZ can cover a variety of grading curves (differing in either relative particle sizes or shapes). However when the permeability data were divided into groups corresponding to the general soil types used to derive equation (1), then there was a clear correlation between cu and permeability, i.e. as cu increased then k_{opt} decreased. There was no such clear correlation with cz. Hence equation (1) was modified to

$$k_{opt} = \frac{100^{\log d_{10}}}{\left(\dfrac{d_{10}}{d_o}\right)^2 \times CU} \quad \text{mm/s} \qquad (2)$$

This formula reduced the high permeabilities predicted for soils with relatively large d_{10} values and generally resulted in slightly underestimated values of k_{opt}.

In agreement with the findings of Carrier and Beckman there was a good inverse correlation between k and plasticity index (PI). Intuitively one would expect that high plasticity is a reflection of a significant clay mineral content which in turn promotes low permeability. Similarly some correlation between k_{opt} and voids ratios/specific volume was to be expected.

Hence the final equation was,

$$k_{opt} = \frac{100^{\log d_{10}}}{\left(\frac{d_{10}}{d_o}\right)^2 \times CU} \times \left\{\frac{25\ (1+e)}{(1+PI)}\right\} \text{ mm/s} \quad (3)$$

The minimum permeability is then taken as 0.5 k_{opt} and occurs when the compaction moisture content is 2% wet of optimum.

Figure 2 show the relationship between measured and predicted permeabilities for several predictive equations.

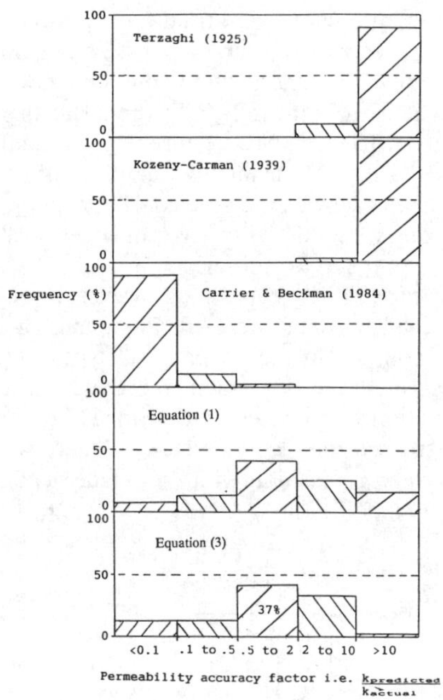

Figure 2 Relative accuracy of permeability predictions

Although equation (3) produces both overestimation and underestimation of measured permeability (k_{meas}) approximately 40% of the predictions were in the range 0.5 (k_{meas}) to 2.0 (k_{meas}) and 70% were in the range 0.5 (k_{meas}) to 10.0 (k_{meas}). This compares very favourably with the accuracy of previous empirical equations.

Furthermore it must be remembered that the equation is intended as a tool for rapid initial selection of appropriate soils for compacted clay linings. For such soils the permeability must not exceed 1×10^{-6} mm/sec but there is no lower limit. If the actual permeabilities of the soils from the database are examined then using equation (3) to make an initial selection of a suitable soil would result in erroneous acceptance of 1 soil out of 68 (for this soil the measured coefficient of permeability was 1.6×10^{-6} mm/s as opposed to a predicted value of 0.13×10^{-6} mm/s and erroneous rejection of 5 soils out of 68 (four of these soils were tested by the same person and the accuracy of their data is questionable).

CONCLUDING REMARKS

Using a database of results from permeability tests an empirical equation has been derived, using both visual and analytical correlations (together with anticipated trends). This equation can be used to aid selection of suitable lining soils (for waste containment) on the basis of likely permeability. Acceptance based on the permeability predicted by the equation correlated very well with acceptance based on measured permeability and at the same time the equation was not unduly conservative with regard to rejection of suitable soils.

REFERENCES

Carrier, W.D. and Beckman, J.F. (1984). "Correlations between index tests and properties of remoulded clays", Geotechnique 34 (2).

Fraser, H.J. (1983). "Experimental study of the porosity and permeability of elastic sediments", J. of Geol.

Hazen, A. (1892). "Some physical properties of sand and gravel with special reference to their use in filtration", 24th Annual Report, Mass. State Board of Health.

Korzeny, J. (1927). "Uber Kapillare Leitung das Wassers in Boden", Akad. Wiss. Vol 136, Pt. 2a., Vienna.

Lambe, T.W. (1954). "The permeability of fine-grained soils", ASTM STP 163.

Loudon, A.G. (1952). "The computation of permeability from simple soil tests", Geotechnique 3 (1).

Terzaghi, K. (1925). "Erdbaumechanik auf Bodenphysikalischer Grundlage", Publ. Deuticke, Vienna.

Williams, M.R. (1992). "Prediction of the permeability of compacted soils", MSc thesis, Bolton Institute, November.

Waste Disposal by Landfill – GREEN'93, Sarsby (ed.) © 1995 Balkema, Rotterdam, ISBN 90 5410 356 6

Assessment of deformed mineral sealing layers

Rainer Scherbeck
Prof. Dr.-Ing. Jessberger + Partner Ltd, Bochum, Germany

Hans L. Jessberger
Institute of Soil Mechanics and Foundation Engineering, Ruhr-University Bochum, Germany

ABSTRACT: Deformability of mineral sealing layers is of fundamental interest when analyzing the maintainance of the barrier function of lining systems. Since detailed geotechnical investigations on this subject have not been carried out yet, a research program was initiated using centrifuge model technique and laboratory testing. From the centrifuge model tests, cracking mechanism was identified to be the most critical liner´s response on differential deformation whereas shearing mechanism could be accepted when assessing the barrier function. Laboratory tests on the tensile behaviour of compacted cohesive soils gave information about the ultimate tensile strain mineral linings could withstand without cracking. The results from the reported investigations were condensed into recommendations for the ETC 8 concerning assessment of deformed mineral sealing layers.

1 PROBLEM AND PREVIOUS WORKS

The mineral sealing layer - the main component of a liner system in waste landfills - has to withstand different influences without loosing it´s technical function which is described by the sealing effect of the low permeable mineral layer. The influence of deformations seemed to be one of the most important since differential deformations could lead to a critical loading of the liner and to a loss in liner´s integrity.

Therefore, the deformation behaviour of mineral sealing layers is a point of special interest when predicting and analyzing the overall behaviour of landfill systems. Fig. 1 gives an impression of this problem showing possible effects of uneven deformations acting on lining systems resulting into curvature.

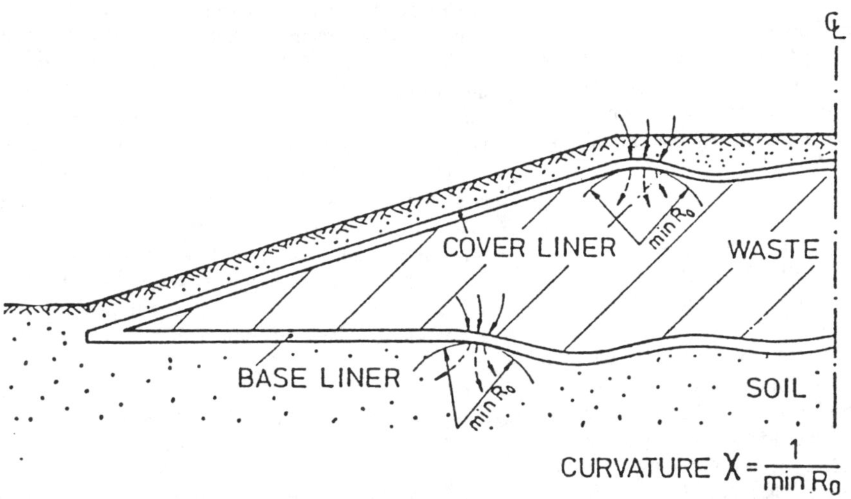

Fig. 1: Example for liner response to differential settlements in a waste landfill

Yet, very little is known about the influence of settlement-induced loading on the geotechnical behaviour of mineral sealing layers. This is opposite to the potential consequences for lining systems resulting from a damage caused by differential settlements. The geotechnical practice is sensibilized concerning this problem not for very long since only little attention is paid in previous publications. Daniel (1983), Gilbert and Murphy (1985) and Murphy and Gilbert (1987) point out geotechnical problems concerning the deformability of mineral cappings for hazardous waste landfills. They show that cracking of the cap might occur due to settlements caused by waste consolidation. The collapse of dumped waste could lead to uneven deformation with the possibility of opening voids in the cap, the worst case of deformation scenarios in landfill sites. Oweis and Khera (1990) developed a model where they compare the situation of the deformed cover liner with an elastic embedded beam using simple beam analysis. Assessing the deformability should be based on the comparison of ultimate tensile strains of liner materials. Larson and Keshian (1990) describe a similiar procedure to investigate the deformation potential inside a tailing pond by calculating tensile strains from settlement data and by comparing them with ultimate tensile strains of liner materials.

Realizing that data of geotechnical laboratory investigations in tensile zone are very rarely to be found since tensile properties are insignificant for the most applications, the few published investigations gain much attention. The investigations by Ajaz and Parry (1975) are the most convincing in this connexion since they compared properties of clayey soils in tension and compression and introduced sophisticated techniques for testing soil samples under tensile conditions.

The deficits in research concerning deformability problems of mineral layers is expressed by Crissman et al (1992), who stated that "the mechanism of cap deformation due to subsidence is not completely understood at the present time".

2 INVESTIGATIONS CONCERNING THE GEOTECHNICAL BEHAVIOUR OF DEFORMED MINERAL LAYERS

At the Institute of Soil Mechanics and Foundation Engineering of the Ruhr-University Bochum, Germany, an extensive research program funded by the Deutsche Forschungsgemeinschaft was carried out in 1988 to 1992. Provisional results from this research were published elsewhere (e.g. Jessberger et al. (1989), Stone et al. (1989), Scherbeck and Jessberger (1991)). A condensed presentaton of the final research results from this work will be given in this paper.

2.1 Application of centrifuge modelling

The main reason for using centrifuge model technique here is the possibility to simulate prototype stresses in a small scale model. In case of basal liners, prototype overburden depths of 30m waste or more are common, so that correct modelling of self weight stresses is essential. Previous studies by Stone and Wood (1987) showed from comparison between model tests under earth gravity (1·g) and increased gravity (n·g) that cracking is a condition heavily influenced by the presence of prototype stress levels. Since cracking is expected to be partly significant in this investigation, the centrifuge modelling principle is very applicable. Centrifuge modelling also provides excellent opportunities to change and vary boundary conditions during the model investigation.

The centrifuge model test reported here were performed on the Bochum 10m balanced beam Geotechnical Centrifuge under an acceleration field of 50 gravities (50 = scaling factor n). Details of operating characteristics and capabilities of the Bochum Geotechnical Centrifuge can be found in Jessberger and Güttler (1988).

Centrifuge scaling factors - major links between model and prototype behaviour - have been described extensively elsewhere, i.e. Jessberger and Güttler (1988). The geometry of a prototype lining situation (e.g. liner thickness, depth of overburden and water table etc.) is scaled down in centrifuge model by the factor of increased gravity level n. Other scaling relations between prototype and model do not match like this and have to be taken into account when analyzing test results. E.g., time t for consolidation is accelerated in centrifuge by a factor of n^2 relative to the corresponding prototype. On the other side, the coefficient of permeability k is scaled by n, see Tan and Scott (1983), meaning that the permeability in the model situation is n times the value of the prototype. When presenting model test results in this paper, values are standardized to prototype conditions regarding the influence of scaling laws mentioned above.

2.2 Model test package, material and test procedure

The centrifuge model tests were performed in a rectangular strong box representing a liner area of

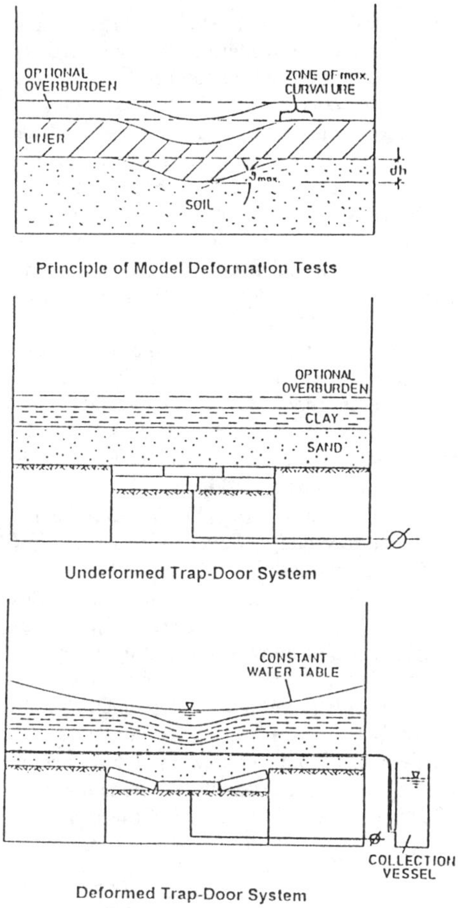

Fig. 2: Schematic drawing of centrifuge model package (cross section)

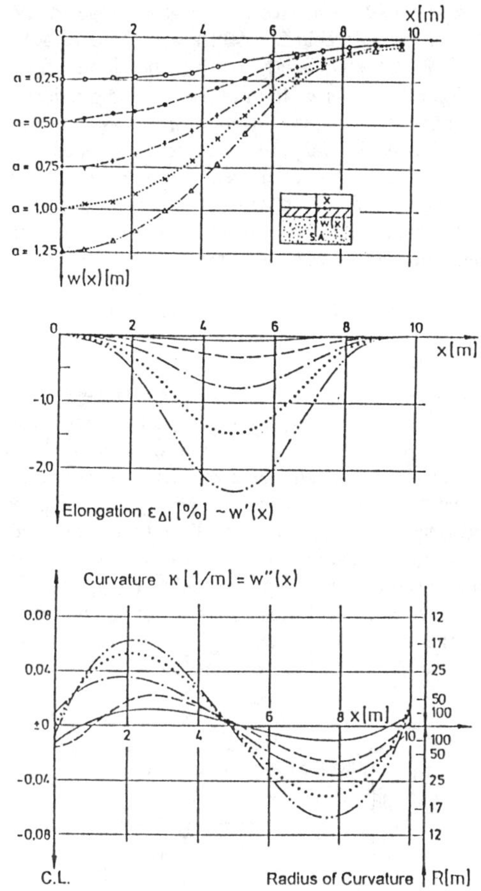

Fig.3: Deformation input on model liner (values in prototype scale)

650 m² in prototype scale. The test package involves a trap-door mechanism to initiate the differential deformation of the mineral liner under controlled conditions. The influence on the permeability behaviour is indicated by a water table on top of the layer permanently decreasing due to migration of water through the liner. Data sets obtained from model tests consists of flow rate data and optical records allowing to derive the actual deformation field of the liner in post-test analysis. Two CCD-cameras observe the liner surface to record the onset of liner deterioration. A more detailed description of the model package and experimental technique can be found in Stone et al. (1989) and Jessberger et al. (1989), respectively. Fig. 2 gives a sketch of the basic elements used in the centrifuge model.

Based on the results from several tests, Fig. 3 gives the quantity of differential deformation attainable with the centrifuge model package depending on the trap-door settlement, indicated by the value a. Mathematically derived from the shape of the deformed liner, curvature κ and elongation $\varepsilon_{\Delta l}$ are the two major components to quantify the deformation input.

The liner material preferentially tested in this study was a commercially available Kaolin clay of low plasticity (Liquid Limit = 44%, Plasticity Index = 16%) The Kaolin liners were pre-consolidated to produce an equivalent 95 % proctor density. After the liner material and an optional overburden load (e.g. sand load) were placed into the model container, centrifuge is accelerated to 50 gravities with fixed trap-door mechanism and the surface water level is raised to its designated value. Due to the circular water table in centrifuge the hydraulic gradient differs slightly along the longitudinal section of the liner with a mean hydraulic gradient i between 1 and 2 [-] in all tests. Consolidation of the liner to its equilibrium effective stress is - depending on liner thickness - achieved approximately after one hour and dropping of the trap-door starts at a predetermined rate of 1 mm/min until detection of failure or full trap-door settlement.

2.3 Test results

A test series with more than 30 single centrifuge tests was carried out varying mainly overburden load, liner´s consistency and type of liner material A typical result from the tests using Kaolin clay without overburden is shown in Fig. 4. The observed response of the cohesive liner to the induced curvature is characterised by two features, shear mechanism and cracking, depending on the degree of deformation, indicated by trap-door settlement a, and the material consistency I_C.

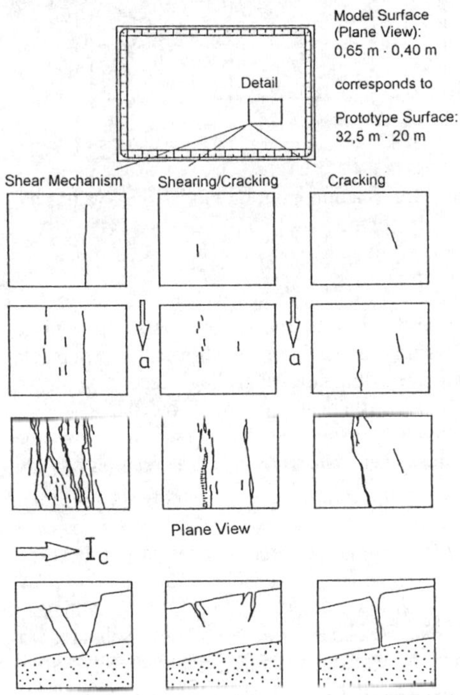

Fig. 4: Typical failure modes observed in centrifuge model tests

Shear mechanism was observed as the predominant failure in tests with an overburden load where shear bands were seen to extend the full depth of the liner with increasing curvature load. Tension cracking was the most significant failure mode in tests without overburden. The onset of tension cracking obviously seemed to be influenced by material plasticitity, consistency and liner thickness. Typical ultimate tension strains measured for Kaolin material at cracking ranged from 2% to 7 % beeing comparable to curvature loading with radius of curvature R less than 200 m.

The hydraulic performance was considerably different dependent on the mode of liner failure. Whereas shearing mode did not significantly influence the flow rate through the liner, tension cracking could lead to total failure of the liner indicated by sudden breakthrough, see Fig. 5.

When shearing mechanism is indicated to be the dominant failure mode, a double increase in flow rate and mean permeability k_m, respectively, was observed in the tests for maximum. Since this increase reflects the behaviour of the total liner surface aerea, that is in most parts unaffected by any deformation action, the results of permeability measurement have to be transfered geometrically to the deformed areas only, see Fig. 6. Following this procedure, a maximum increase in ruptured zone permeability k_1 by one order of magnitude could be identified in case of shear mechanism. Investigations

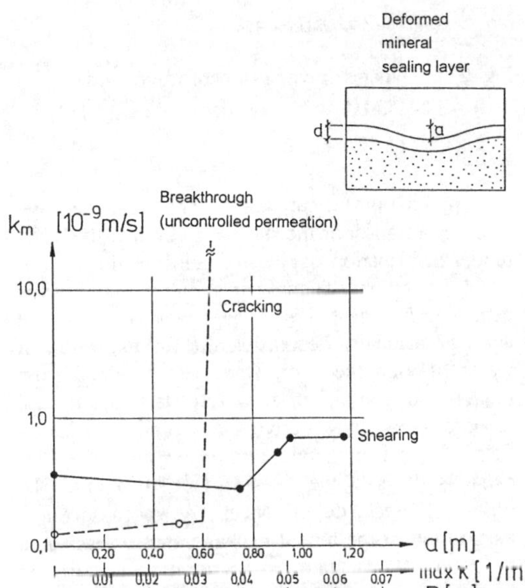

Fig. 5 Typical data record showing the relation between deformation and permeation

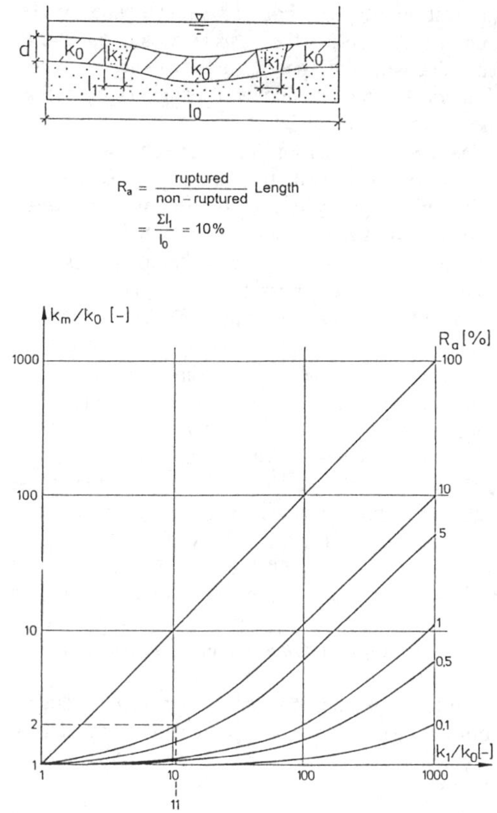

Fig.6: Derivation of permeability k_1 in ruptured areas of deformed liner

by Cressman et al. (1992) on the same topic show similiar results concerning the increase of permeability in shear zones caused by settlements.
These results underline the insignificance of shear mechanism and the importance of cracking mechanism when analyzing deformed mineral layers.

During variation of liner materials, the most significant changes in behaviour were observed, when using a granular material with a pore filling mixture of fine sand and bentonite. This material was developed by DYWIDAG AG, Essen, and has so far successfully been tested as mineral lining for landfills, cut off walls and embedding around leaking pipes. Major advantages in construction practice result from the ability of pre-mixing and from the low water content of 2 to 3%, allowing the use of vibro-compaction to attain high densities. This material with its granular behaviour seemed to be unaffected by the induced deformations and neither shear bands nor cracking mode could be obeserved. Mineral liner materials of low cohesion

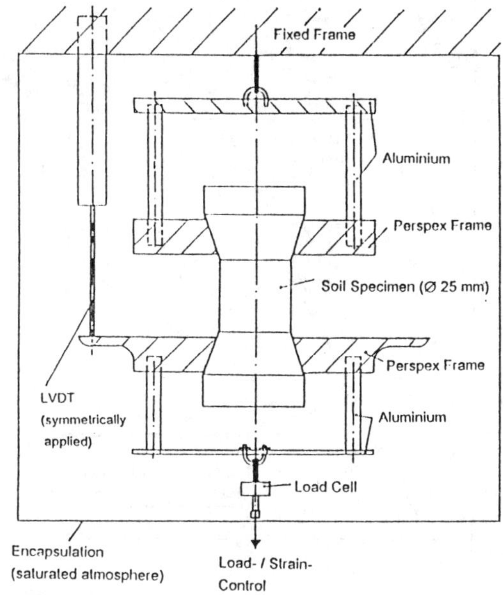

Fig. 7: Laboratory device for direct tension tests

are obviously not sensitive towards differential settlements.

In chapter 4, an assessment procedure for deformed mineral liners will be introduced that is based on the research work presented here. Therefore, more significant interpretations of the test results. could be found there e.g. the influence of overburden load.

3 INVESTIGATIONS ON TENSILE PROPERTIES

The results from centrifuge model tests clearly illustrate that curvature of compacted clay liners can lead to failure conditions resulting in an uncontrolled breakthrough of leachate. It was observed that this failure condition appears predominantly when the liner without overburden is able to form tension cracks. Therefore, the ability of cohesive material to fail in tension will be further analysed.

A direct tension test device was developed to determine the ultimate tensile strain properties of cohesive soils, especially those used in the centrifuge model tests. Fig. 7 shows a sketch of the laboratory device and Fig. 8 indicates a typical stress-strain relationship of a cohesive soil in tension.

Laboratory testing of tensile properties is significantly influenced by random effects, resulting into deviations of about 25 % concerning the ultimate

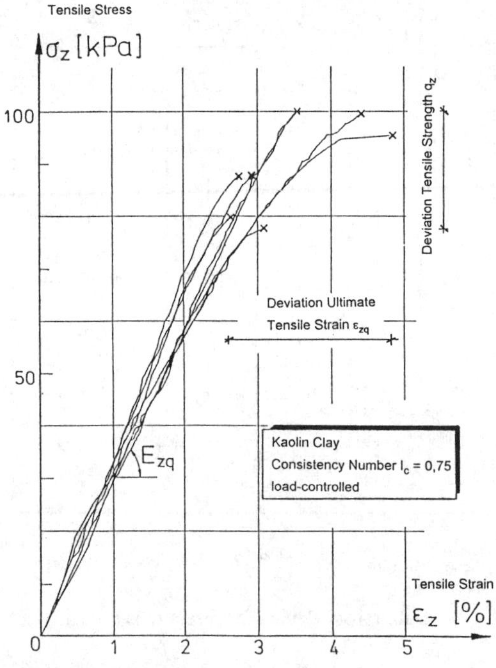

Fig. 8: Stress-strain curves for Kaolin in direct tension test

tensile strain ε_{zq} and about 10% concerning tensile strength q_z. Since the tensile strain is more important when interpreting deformability of mineral liners, an adequate number of single element test have to be carried out.

Meanwhile, numerous different cohesive soils were investigated with this device representing a wide range of plasticity and compaction conditions. An overview of these results is given in Tab. 1. The ultimate tensile strain ε_{zq} was computed as the mean value from minimum 4 single tests.

The most significant message deduced from these element tests concerns the quantity of ultimate tensile strain ε_{zq}. Typical tensile strain values for clays when compacted with standard and modified proctor energy can be found between 1,5 and 4,4%.

From other investigations, see Ajaz and Parry, 1975, and Scherbeck, 1992, it is known that ultimate tensile strain ε_{zq} will increase when material's consistency turns to plastic behaviour. This tendency is not prooved by all the results from element tests shown in Tab. 1, meaning that other conditions like compactive effort, plasticity, clay content, etc. can significantly influence the value of ultimate tensile strain.

The lower bound of tensile strain range deduced from the investigations shown in Tab. 1 fits well to

Table 1: Overview of tensile properties of different cohesive soils investigated with tensile testing apparatus shown in Fig. 6.

Subject	Clay A		Clay B		Clay C		Kaolin Clay	
Particle Content in [%]								
Clay	58		70		56		61	
Silt	37		25		41		39	
Sand	5		5		3		0	
Liquid Limit [%]	59		76		63		44	
Plasticity Index [%]	31		49		32		16	
Water content w [%], Consistency I_c [-]	w[%]	I_c[%]	w[%]	I_c[%]	w[%]	I_c[%]	w[%]	I_c[%]
100% ρ_{pr}	22	1,19	24	1,06	20	1,34	26	1,13
95% ρ_{pr}	29	0,97	33	0,88	26	1,16	30	0,88
100% ρ_{mpr}	16	1,39	19	1,16	18	1,41	21	1,44
95% ρ_{mpr}	21	1,23	24	1,06	24	1,22	25	1,19
Ultimate Tensile Strain	ε_{zq} [%]		ε_{zq} [%]		ε_{zq} [%]		ε_{zq} [%]	
100% ρ_{pr}	2,4		2,4		3,3		2,9	
95% ρ_{pr}	4,1		2,1		1,7		1,7	
100% ρ_{mpr}	1,5		1,6		4,4		4,2	
95% ρ_{mpr}	2,5		3,3		4,2		3,0	

Index: ρ_{pr} = standard Proctor density ρ_{mpr} = modified Proctor density other experimental results gained in this field, see Ajaz and Parry, 1975. For design purposes, a value of ultimate strain ε_{zq} = 1% seemed to be a conservative assumption when dealing with plastic clay materials of stiff consistency for sealing. In general, an experimental determination of tensile properties, varying boundary conditions like water content, compactive effort and material preparation is very much recommended.

4 RECOMMENDATIONS FOR ASSESSING DEFORMED MINERAL LINERS

The ETC 8 (European Technical Committee of the ISSMFE) publishes technical recommendations to standardize the works in the field of geotechnics of landfills and contaminated land, see GLC (1991). A new recommendation, R 2-11, is based on the research work described here and deals with the assessment of mineral sealing layer deformation. The main content of this procedure will be introduced here.

Input data for the following assessment approach are:

- the extent of deformation (e.g. uneven settlement of the landfill base or cap)
- the stress-strain behaviour of the sealing material, including its proporties under tension.

Firstly, the induced deformation is compared with the deformation capacity of the mineral sealing material. Secondly, the stability of an open crack and the influence of bending stresses is investigated. The concept of the validation procedure is shown in Tab. 2.

Table 2: Assessment procedure for deformability of mineral sealing layers (see GLC, R 2-11)

Content	Formula
STEP 1: Check deformation input and material properties	$\frac{\varepsilon_{zq}}{\varepsilon_{RF}} > \eta_V$ (>2,0) (see eq. 1 to 4)
STEP 2: a) Check vertical stresses σ_0 (overburden pressure)	$\sigma_0 > 2 \cdot c' \cdot \mathrm{TAN}\,(45° + \phi'/2)$ (see fig. 9)
b) Check deformation mechanism (shearing is acceptable)	req. c' > ex. c' (see fig. 10)

4.1 Deformability

The first step in assessing cohesive mineral sealing material is to compare the induced deformation and the deformation capacity of the mineral material in terms of tensile strains:

$$\varepsilon_{zq} \;/\; \varepsilon_{RF} > \eta_V \qquad \text{(eq. 1)}$$

The maximum deformation capacity of the mineral sealing material can be defined by the maximum tensile strain at the onset of cracking. This value is equivalent to the ultimate strain ε_{zq} measured in direct tension tests; alternatively the ultimate strain can be determined from bending tests, see Ajaz and Parry (1975).

The shape of the deformed mineral liner is given by the settlement curve w(x), from which the maximum edge fibre strain, ε_{RF}, can be specified by the geometrical values of elongation ($\varepsilon_{\Delta l}$) and curvature (κ) based on beam analysis. Shear stresses acting at the interface of mineral liner and landfill base lead to a displacement of the neutral axis in the cross section. This effect is considered by a factor 2/3 in eq. (4). The following relations between settlement curve w(x) and elongation and curvature, respectively, are fundamental:

$$\varepsilon_{\Delta l}\,(x) = w'\,(x) \,/\, [\mathrm{SIN}\,[\mathrm{ARCTAN}\; w'\,(x)] - 1] \qquad \text{(eq. 2)}$$

$$\kappa\,(x) = w''\,(x) \qquad \text{(eq. 3)}$$

$$\varepsilon_{RF} = 2/3 \cdot d \cdot \max \kappa + \varepsilon_{\Delta l} \qquad \text{(eq. 4)}$$

with:

$\varepsilon_{\Delta l}$	= maximum elongation [-]
w (x)	= shape of settlement curve [m]
w', w"	= first and second derivation from w (x)
κ	= curvature [1/m]
ε_{RF}	= maximum strain in edge fibre [-]
d	= thickness of mineral sealing layer [m]
max. κ	= maximum curvature (= max w" (x) = 1/R) [1/m]
R	= radius of curvature [m]

The safety factor of deformability, η_v is defined by the quotient of strains in eq. (1) and should exceed a recommended value of 2,0. If this is proved, the sealing material is considered capable of withstanding induced deformations without cracking and the barrier function of the mineral sealing is not significantly affected.

Overburden loads can be considered in this validation step by comparing the deformation induced edge fibre strain, ε_{RF}, with the overburden induced transverse strain. If the transverse strain exceeds the edge fibre strain, cracking of the mineral sealing layer is prevented.

4.2 *Stress*

If strain conditions satisfying eq. (1) are not fulfilled, stress conditions should be taken into consideration for the second validation step to determine the significant deformation mechanism. Local shear mechanisms can be accepted; on the other hand cracking will lead to a loss of liner integrity and has to be excluded.

4.2.1 *Required overburden load (excluding bending stresses)*

Overburden loads reduce the danger of cracking by suppressing the development of deformation induced tensile stresses. The required overburden load to avoid cracking can be deduced from the situation shown in Fig. 9. In this figure, the equation for calculation the height of a stable crack depth, z_R, is analysed in a diagram, showing that cohesive soils with effective friction angles φ' of about 20° resist cracking when the overburden load is minimum three times the cohesion value c'. The overburden load acting on the mineral sealing layer has to exceed the required load σ_0 from Fig. 9 by a safety factor as used in slope stability analysis (e.g. by 1,4).

4.2.2 *Influence of bending stresses*

The bending stress situation in critical liner cross section has to be analysed to determine if shearing or cracking is the system´s response on uneven deformation. In this step the existing material cohesion, ex. c', is compared with a computed cohesion, re. c', required for equilibrium at the sliding wedge at the moment before cracking, (see Fig. 10).

If the required cohesion is less than the existing cohesion, there is stable equilibrium in the system and further deformation would lead to cracking. On the other hand, if the relation ex. c' < req. c' is fulfilled, the shear resistance is actually too low to maintain equilibrium. At this stage shear mechanisms have already occured and hence the validation in the second step is fulfilled.

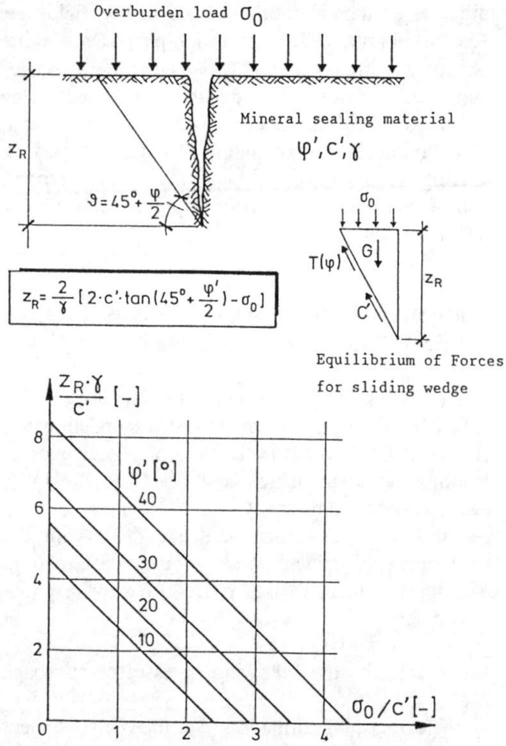

Fig. 9: Overburden load and cracking development

4.3 *Remarks on assessment practice*

The deformation assessment of mineral lining systems presented here is based on the knowledge of material properties as well as on deformation prognosis. From these two influences, the deformation prognosis is likely to be the most difficult for determination since in assessment phase only very little information about site conditons is available.

The deformation of base linings could be deduced from calculations using soil mechanical theories (e.g. elastic theory) refering to subsoil properties concerning stiffness. Realizing the stress situation in landfill bases, the deformation assessment given in Tab. 1, step 2, is fulfilled in most cases and therefore deformation analysis is not decisive. Special loading situations during construction phase have to be carefully regarded.

For capping, the situation is quite different, since the deformation behaviour of the waste body is often unknown in assessment phase and therefore deformation assessement is not applicable. It is recommended to strengthen efforts on detailed

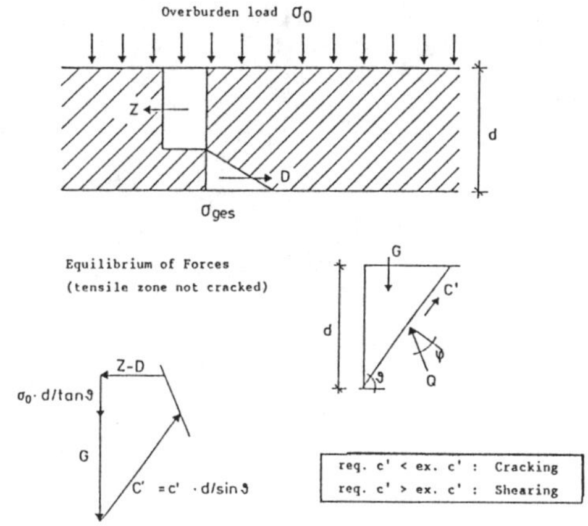

CASE a: **Overburden Pressure $\sigma_0 = 0$**

req. c' = $1/2 \cdot \gamma \cdot d \cdot (F_1 - 2/3 \cdot F_2) + q_z \cdot (2/3 - 1/12 \cdot R_E) \cdot F_2$

CASE b: **Overburden Pressure $\sigma_0 > 0$**

req. c' = $(1/2 \cdot \gamma \cdot d + \sigma_0) \cdot F_1 - \sigma_0 \cdot (1/3 + 1/12 \cdot R_E) \cdot F_2 + q_z \cdot (2/3 - 1/12 \cdot R_E) \cdot F_2$

with: γ = Specific Weight of Sealing Material [kN/m²]
d = Thickness of Mineral Layer [m]
σ_0 = Overburden Pressure [kN/m²]
q_z = Uniaxial Tensile Strength of Sealing Material [kN/m²]
R_E = Stiffness Ratio Compression / Tension [-] (e.g. Tangential Stiffness in Compression devided by Tangential Stiffness in Tension)
$F_{1,2}$ = Parameter from Diagram, see right [-]

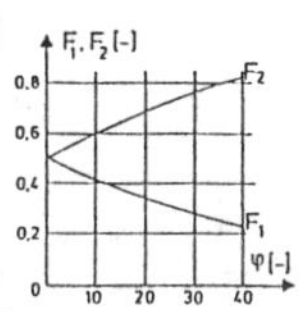

Fig. 10: Bending tensile stresses in the deformed sealing layer (tensile zone not cracked), see Scherbeck 1992.

deformation measurements after closure of the waste disposal site to be able to compare deformation input with material properties following the assessment scheme in tab. 1, step 1.

5 SUMMARY

The system behaviour of deformed mineral sealing layers was investigated in more than 30 centrifuge model tests. From these tests, it could be derived that deformed compacted mineral liners only stop to perform as effective hydraulic barriers when cracks occur. Depending on the plastic behaviour of the liner material tension cracks could appear, leading to an uncontrolled breakthrough; the presence of overburden loads will suppress this tendency. Instead, shearing mechanisms may occur and will lead to inner deformations but this response is unlikely to evoke a loss of integrity of the liner. Additional laboratory element tests quantified the influence of consistency, plasticity and compaction on the tensile strain behaviour of some cohesive soils, showing that they can typically attain ultimate tensile strains between 1.5 and 4.4%. For design purposes the ETC 8 will publish a recommendation about assessment of mineral sealing layer based on the results presented in this paper. This recommendation is refered to in this paper when interpreting test results from centrifuge modelling and element testing.

REFERENCES

Ajaz,A. und Parry,R.H.G. 1975. Stress-strain behaviour of two compacted clays in tension und compression. Géotechnique, Vol. 5, No. 3: 495-512.

Cressman,G.C., Cheng,S.C., Martin,J.P. Bredariol,A.S. 1992. Effects of subsidence distortion on integrity of landfill caps. In: Environmental Geotechnology, Editors Usmen and Acar, Balkema Publishers, Rotterdam: 229 - 235

Daniel,D.E. 1983. Shallow Land Burial of Low-Level Radioactive Waste. Journal of Geotechnical Engineering, ASCE, Vol. 109, No. 1, January 1983

Gilbert,P.A. and Murphy,W.L. 1985. Settlement and cover subsidence of hazardous waste landfills. Report No. EPA/600/2-85/035 Environmental Protection Agency, Cincinnati, Ohio, USA.

GLC - Geotechnics of Landfills and Contaminated Land 1991. Technical Recommendations "GLC". Edited by the German Geotechnical Society (DGEG) for the ISSMFE. Ernst & Sohn Publishers, Berlin, ISBN 3-433-01214-8, 76 pages (2nd revised and enlarged edition in preparation).

Jessberger,H.L. and Güttler,U. 1988. Bochum Geotechnical Centrifuge. In: Proceedings of the International Conference on Geotechnical Centrifuge Modelling - Centrifuge ´88, Paris, 1988, Editor J.F. Corté, Balkema Publishers, Rotterdam: 37 - 44

Jessberger,H.L. and Stone,K.J.L 1991. Subsidence effects on clay barriers. Géotechnique, Vol. 41, No.2: 185-194.

Jessberger,H.L., Güttler,U. and Stone,K.J.L. 1989. Centrifuge modelling of subsidence effects on clay barriers. Proc. 2nd Int. Landfill Symposium Sardinia ´89, Volume I, Porto Conte, Sardinia: XII1 - XII11

Larson,N.B. and Keshian,B. 1988. Prediction of strains in earthen covers. in: Hydraulic Fill Structures, Geotechnical Special Publication No.

21 of the ASCE, Fort Collins, Colorado, August 1988. Editors D.J.A. Van Zyl and S.G. Vick, New York, pp. 367-388.
Murphy,W.L. and Gilbert,P.A. 1987. Prediction/ Mitigation of subsidence damage to hazardous waste landfill covers. Report No. EPA/600/2-87/ 025. Environmental Protection Agency, Cincinnati, Ohio, USA.
Oweis,I.S. and Khera,R.P. 1990. Geotechnology of waste management. Publishers Butterworth & Co. Ltd.
Scherbeck,R. 1992. Geotechnisches Verhalten mineralischer Deponieabdichtungsschichten bei ungleichförmiger Verformungseinwirkung. Schriftenreihe des Institutes für Grundbau, Ruhr-Universität Bochum, Heft 16, 191 pages.
Scherbeck,R., Jessberger,H.L., Stone,K.J.L 1991. Mineral Liner Reaction from Settlement induced Deformation. In: Proceedings of the International Conference Centrifuge 1991, Boulder, Colorado, USA. Editors H.Y. Ko, F.G. McLean, Balkema Publishers, Rotterdam: 121-128
Stone,K.J.L and Wood,D.M. 1987. Model studies of soil deformation over a moving basement", Engineering Departement Cambridge University.
Stone,K.J.L., Güttler,U. and Jessberger,H.L. 1989. Some observations of the influence of deformation an a clay liner. Proc. 10 th National Conference Superfund ´89, Washington, D.C.: 537-542
Tan,T.S. and Scott,R.F. 1985. Centrifuge scaling considerations for fluid-particle systems. Geotechnique 37, No. 4: 131 - 133

Waste Disposal by Landfill – GREEN'93, Sarsby (ed.) © 1995 Balkema, Rotterdam, ISBN 90 5410 356 6

The use of waste paper sludge for landfill cover material

Thomas F. Zimmie, Horace Moo-Young & Keith LaPlante
Rensselaer Polytechnic Institute, Troy, N.Y., USA

ABSTRACT: A paper mill sludge from a paper manufacturing operation was substituted as the hydraulic barrier for landfill covers in Massachusetts. This study investigates the geotechnical properties of the paper sludge. Paper mill sludges have a high water content ranging from 150% to 250% and have a high degree of compressibility. Composed of 50% kaolinite and 50% organics, the sludge behaves like a highly organic soil. Paper fibers and tissues throughout the sludge created problems in testing, e.g. trimming and cutting. Atterberg limits were extremely high and were atypical for non--organic clays. Standard Proctor Compaction Tests were executed from the wet side due to the high initial water content. Since the sludge is stiff and unworkable at the optimum moisture content, a very wet water content is desirable for landfills. Consolidation tests revealed a large reduction in void ratio and high strain values which were expected due to the high compressibility. Triaxial shear strength tests were conducted on undisturbed samples. A large variation in the strength parameters resulted from the nonuniform mix of the sludge in the landfill. In general, laboratory permeability tests conducted on insitu specimens met the 1×10^{-7} cm/sec requirement for landfill covers. Freezing and thawing cycles increased the sludge permeability about one order of magnitude. Maximum permeability changes occurred within ten freeze/thaw cycles.

1 INTRODUCTION

The elevating cost of landfilling may be reduced by the use of unconventional material in the construction of landfills. The high price of disposal has sparked interest in the development of alternative uses for waste sludges (paper mill sludges and water treatment plant sludges). Paper mill sludges in spite of high water contents and low solids contents in comparison to clays, can be compacted to low permeabilities and can substitute for clays in landfill covers. Since 1975, paper mill sludges have been used to cap landfills in Wisconsin (Stoffel and Ham, 1979; Pepin, 1984).

The National Council of the Paper Industry for Air and Stream Improvement Inc. (NCASI) is conducting research on the use of paper mill sludges as a cover material for landfills (Swan, 1991). Early reports from four landfill cover test cells have indicated that the sludge may perform as an effective hydraulic barrier. A similar study is being conducted by the Erving Paper Mill in Massachusetts, in which several test cells were constructed to simulate a landfill cover insitu. The results of the Erving study support the findings of NCASI (Aloisi and Atkinson, 1991). Erving Paper Mill sludge is being used as the impermeable cover material on a portion of the Hubbardston Landfill in Massachusetts to examine the use of paper mill sludge as an alternative capping material.

This study investigated the geotechnical properties of Erving Paper Mill sludge. Atterberg limits, standard compaction tests, shear strength tests, field density, permeability, and consolidation tests were performed. The effects of freezing and thawing on permeability were also studied.

2 SOIL CLASSIFICATION

The paper mill sludge used in this investigation was obtained directly from the cover of the Hubbardston municipal landfill in Massachusetts. The sludge was originally transported to the landfill from the Erving Paper Mill in Erving, Massachusetts. The sludge is composed of approximately 50% kaolinite and 50% organics. The waste water treatment plant receives 96% of its effluent from the recycling paper mill and 4% from the Town of Erving.

2.1 Atterberg limits

The Atterberg limits of the sludge were determined by following the American Society for Testing and Materials (ASTM) procedure D4318 for liquid limit, plastic limit, and plasticity index of soils.

The sludge has a liquid limit, plastic limit, and plasticity index of 285%, 94%, and 191%, respectively. The specific gravity of the sludge solids was obtained by following the procedure of the American Society of Testing and Materials D854. The sludge solids have a specific gravity of 1.95.

The plastic limit was easily determined (i.e. behaves similar to clay); the liquid limit, on the other hand, was difficult to determine. Specimens were prepared from the wet side. When the paper sludge was passed through the number 4 sieve and air dried, the sludge formed flocs, was coarse in texture, and was not easily pulverized. Upon rewetting the dry sludge, large clods formed, and the sludge did not regain its initial plasticity, which made performing the Atterberg limit test from the dry side difficult. Paper fibers and tissues in the sludge caused problems when cutting the groove for the liquid limit test. The ASTM plastic groove tool has a tendency of catching the fibers and tissues and results in an unsatisfactory groove. The ASTM brass groove tool cuts a smoother groove and leads to more satisfactory tests results. Wang et al. (1991) reported similar results for a water treatment sludge.

2.2 Compaction test

A Proctor test was performed following the American Society of Testing and Materials (ASTM) procedure D698–78. Because of the high water content, the test was conducted from the wet side rather than from the dry side as recommended by ASTM. Furthermore, when water was added to dry sludge, large clods formed, the clods were difficult to break apart, and the sludge lost its initial plasticity.

During the drying process, the sludge was passed through the number 4 sieve and placed in a pan to air dry. Many trials were run to reach the optimum moisture content and density. The optimum dry density of 8.1 KN/m^3 was obtained at a moisture content of 49% as shown in Figure 1. The Proctor curve is skewed with only a small range of water contents on the dry of optimum side of the curve and with a wide range of moisture contents (50–250%) on the wet of optimum portion of the curve. At the optimum density and moisture content, the sludge is dry, stiff, and unworkable. A very high water content is desirable, if the sludge is to be used as a landfill capping material. These test results compare favorably to research conducted on water treatment plant sludges (Raghu et al., 1987; Environmental Technology Inc., 1989; Wang et al., 1991).

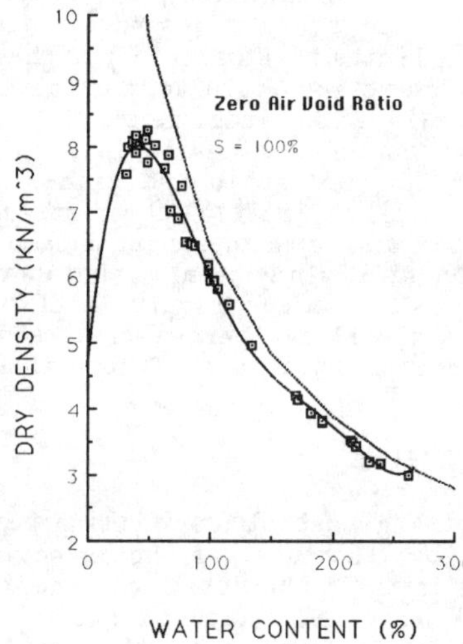

Figure 1 Proctor curve for Erving Paper Sludge

2.3 In–field Density

The density of the sludge was determined in the field using three methods: sand cone method following the procedure of ASTM D1556, the rubber balloon method following the procedure of ASTM D2167, and the nuclear density method. The nuclear density testing method was performed by Yankee Testing of Worcester, Massachusetts. Two sets of sand cone, rubber balloon, and the nuclear density tests were performed in July 1991 on the Hubbardston Landfill several weeks after construction. The sand cone method was performed in October 1991 on another section of the landfill several weeks after construction.

The results are shown in Table 1. Assuming that the sand cone test is the most accurate, the nuclear density device can be calibrated to the sand cone device.

Table 1 In place density determination at the Hubbardston Landfill

Method	Wet Unit Weight (KN/m^3)	Water Content (%)
Set 1 (July 1991)		
Sand cone	12.5	97.1
Balloon	13.7	145.3
Nuclear Device		
6 inch depth	13.3	115.6
12 inch depth	13.0	127.7
Set 2 (July 1991)		
Sand cone	11.4	159.6
Balloon	13.2	145.3
Nuclear Device		
6 inch depth	12.4	129.4
12 inch depth	12.8	115.3
Set 3 (October 1991)		
Sand cone	11.3	211
	11.4	221

2.4 Remolded permeability

Permeability tests were performed on Erving paper mill sludge at various water contents following the procedures of ASTM D5084 for measuring the hydraulic conductivity of saturated porous material using a flexible wall permeameter with backpressure. Results of the tests are shown in Figure 2. Near the optimum water content shown in Figure 1, the permeability increases for paper sludge. For clays, the minimum permeability typically occurs near the optimum water content. The minimum permeability for paper sludge occurs 50 percent wet of the optimum water content.

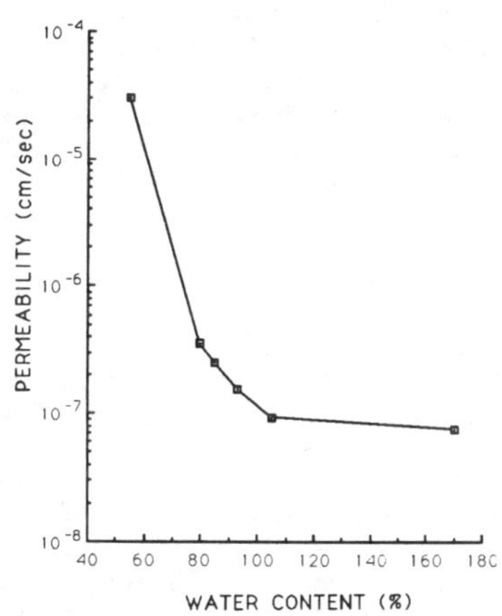

Figure 2 Permeability vs. water content

The construction of a landfill cover layer using paper mill sludge near the optimum moisture content is undesirable, due to the high permeability and the fact that the material is unworkable. A high water content is desirable for the construction of a landfill capping layer using paper mill sludge as the impermeable barrier layer.

3 CONSOLIDATION TEST

One-dimensional consolidation tests were conducted on samples taken from the landfill several weeks after construction and after nine months insitu following ASTM procedure D2435.

For the specimen taken from the landfill several weeks after construction, the initial water content was 206% as shown in Table 2. The void ratio was reduced by 2.986 from its initial value of 3.85. The compression index and coefficient of compressibility were calculated to be 1.24 and 1.30, respectively. Moreover, total strain values as high as 67% were encountered using a maximum load of 1.53×10^3 KPa, which was expected due to the highly compressible nature of the material.

For the specimen tested after nine months insitu, the initial moisture content was 106% as shown in Table 2. The final water content was 60%. The initial void ratio of 1.97 was reduced to 0.298. The compression index and coefficient of compressibility were calculated to be 0.71 and 0.74, respectively. At higher consolidation pressures, high strain values were measured. Low strain during the first increment suggests that the sludge was

Table 2 Summary of consolidation tests

Property	4–6 weeks	9 months
Initial water content	206	106
Final water content	91	60
Initial void ratio	3.85	1.970
Final void ratio	0.864	0.261
Compression index	1.24	0.71
Coefficient of Compression	1.30	0.74

consolidated insitu to larger effective stresses. These results compare favorably to consolidation test conducted on water treatment sludges (Wang et al., 1991; Raghu et al., 1987).

3.1 Settlement Calculations

Settlement values were estimated from the consolidation data. For the Hubbardston Landfill, the impervious layer was overlain by a vegetative support and drainage layer with a minimum thickness of eighteen inches: twelve inches consisted of top soil and six inches consisted of a drainage layer with a permeability equal to or greater than 1×10^{-3} cm/sec. The estimated load at the midpoint of the sludge is 13.8 KPa.

For the specimen taken several weeks after construction using a load of 13.8 KPa, the coefficient of consolidation and the strain were 3.4×10^{-4} cm²/s and 14%. For a 0.9 m layer, computations predict that the landfill cover will settle 12.8 cm. These results compare favorably with the observed settlement at the landfill. After one year, approximately 15.2 cm of settlement occurred at the Hubbardston landfill.

4 SHEAR STRENGTH TESTS

The shear strength behavior of the paper sludge was determined using consolidated undrained triaxial compression tests with pore pressure measurements following ASTM procedure D4767. Test series 1 used consolidation pressures of 34.5, 69, and 117 KPa, and Test series 2 used confining pressures of 34.5, 83 and 138 KPa.

For the six specimens sheared, the initial water contents ranged from 233% to 175%, and the final water contents varied from 128% to 80%. Pore pressure parameters A at failure (A_f) were 0.72 for Tests 1 and 0.64 for Tests 2. Moreover, the values of A_f indicate that the tested sludge is behaving in a similar manner to a normally consolidated clay (Lambe and Whitman, 1969). During the consolidation phase of the triaxial test, a large reduction in void ratio resulted due to the high compressibility of the sludge. This behavior is consistent with that observed during the consolidation testing reported previously.

For Test series 1, the effective angle of internal friction was 37 degrees with a cohesion of 2.8 KPa. For Test 2, the effective angle of internal friction was 25 degrees with a cohesion of 9.0 KPa. Failure is difficult to determine from the stress-strain curves, which are typical of soft compressible material in that they exhibit no sharp yield point. Failure has to be arbitrarily selected at some reasonable strain. For the purpose of this study, failure is defined at 10% strain. Obviously, if failure is defined at a different strain, the strength parameters would change. The variation in shear strength of Erving paper sludge may be attributed to the wide range of water contents used in the construction of the landfill cover and to the high organic content. Moreover, differences in the amount of fibers in the sludge matrix may alter the amount of cohesion measured in the paper mill sludge.

In studying the stabilization of a water plant sludge from Virginia for utilization as embankment material, Wang et al. (1991) concluded that both treated and untreated sludges were highly sensitive and thixotropic. They obtained similar results to that of Erving paper mill sludge using consolidated undrained triaxial tests, an effective angle of internal friction of 42.3 degrees and a cohesion of 6.9 KPa.

5 LABORATORY PERMEABILITY TESTING ON UNDISTURBED SAMPLES

Laboratory permeability tests were conducted on undisturbed samples taken from the Hubbardston Landfill on three occasions: July 1991, October 1991, and April 1992. Permeability tests were performed following the procedures of ASTM D5084 for measuring the hydraulic conductivity of saturated porous material using a flexible wall permeameter with backpressure. Samples were tested at a low confining stress of 34.5 KPa to simulate the worst case, that is the highest permeability. A differential pressure of 20.7 KPa was used to create a hydraulic gradient of 21 on the specimens.

5.1 Sampling Procedure

The best sampling procedure was discovered through trial and error using Shelby tubes. Slow static pressure (pushing the Shelby tube into the sludge layer with a constant vertical force) compressed the sludge during the sampling process and led to low recovery rates. A dynamic sampling process, like striking the Shelby tube with a hammer, resulted in high rates of recovery and minimal disturbance. Apparently, due to the fibers and tissues in the sludge matrix, a sharp blow was needed to cut through the sludge. The normal field procedure was to place the Shelby tube on the sludge, place a wood block on top of the Shelby tube, and strike the block with a hammer. This procedure resulted in the highest rates of recovery and the least disturbance.

5.2 Results

In general, the Shelby tube samples met the 1×10^{-7} cm/sec regulatory requirement for landfill impermeable barriers. Table 3 summarizes the permeabilities of the specimens. Examining Table 3, the water contents of the specimens taken from the landfill after construction varied from 146% to 210%. The first round of samples taken in July, 1991 were taken from sections of the landfill several weeks after construction. Specimen 3 with a permeability of 6.48×10^{-8} cm/sec met the regulatory requirements. For all practical purposes, specimens 1 and 2 with permeabilities of 1.49×10^{-7} and 1.04×10^{-7} cm/sec, respectively, met the regulatory requirements.

Seven specimens were obtained from the landfill in October, 1991. The samples were taken from sections of the landfill several weeks after construction. The seven specimens obtained yielded

Table 3 Summary of laboratory permeability test on insitu samples

SPECIMEN	PERMEABILITY	WATER CONTENT
	July 1991	
1	1.48×10^{-7}	160
2	1.04×10^{-7}	185.7
3	6.48×10^{-8}	165.5
	October 1991	
4	8.95×10^{-8}	183
5	1.12×10^{-8}	176
6	2.59×10^{-8}	173
7	1.95×10^{-8}	167
8	5.79×10^{-8}	171
9	4.00×10^{-8}	210
10	3.85×10^{-8}	206.4
	April 1992	
11*	9.47×10^{-8}	106
12	4.1×10^{-7}	189
13	7.44×10^{-7}	146
14	1.03×10^{-7}	172

*Nine month insitu permeability.

permeabilities ranging from 1.93 to 8.9×10^{-8} cm/sec, all meeting the regulatory requirement for permeability.

Four specimens were taken in April, 1992. Specimens 12, 13, and 14 were obtained from sections of the landfill several weeks after construction. For all practical purposes, specimen 14 with a permeability of 1.03×10^{-7} cm/sec met the regulatory requirements. Specimens 12 and 13 had slightly higher permeability values. Specimen 11 obtained in April, 1992 was taken from a section of the landfill that had been constructed nine months previously. Specimen 11, taken after 9 months, was dewatered and consolidated under an eighteen inch overburden. It was markedly stiffer and denser than samples obtained shortly after construction. The permeability for the sample met the regulatory requirements of 10^{-7} cm/sec.

6 EFFECTS OF FREEZE/THAW ON PERMEABILITY

One of the concerns in landfill design is the effect of freezing and thawing on the permeability of the cover material. A considerable amount of research has been conducted on the effects of freezing and thawing on compacted clays. The paper by Othman et al. (1993) summarizes the work on compacted clays. In general, the permeability of compacted clays increases from one to two orders of magnitude due to freezing and thawing. There is little or no information on the behavior of paper sludges due to freeze/thaw effects. Thus, a series of freeze/thaw tests were performed on the Erving Paper sludge.

Remolded specimens that were subjected to one dimensional freeze/thaw were made from the paper mill sludge. All samples were molded at a water content of 170% (wet of optimum), which approximately represents the water content of the sludge after placement in the landfill cover.

6.1 Procedure

Specimens were compacted into 10.2 cm high by 7.6 cm diameter PVC molds in three equal lifts, using six blows of a standard Proctor compaction hammer. For the mold used, it was empirically determined that six blows per layer would yield the proper density. For the given water content, the total weight of sludge required for each mold was approximated from the Proctor curve to be 540.9 grams. (Dry Unit Weight = 4.24 KN/m^3, Total Unit Weight = 11.5 KN/m^3). Specimens were sealed in plastic wrap and taped to prevent moisture loss.

The one dimensional freeze/thaw process used in this study has been previously described in detail by Zimmie and LaPlante (1990) and LaPlante and Thomas (1989). Permeability tests were performed when 0, 1, 3, 5, 10, 15, 20 and 25 freeze/thaw cycles were reached. At least two specimens were permeated at each of the cycles. Specimens were removed from the environmental room frozen to decrease disturbance, weighed, extruded, and placed in the triaxial permeameter.

Permeability testing followed ASTM procedure D5084 with backpressure application. Specimens were permeated under an effective stress of 34.5, 69 and 138 KPa. Due to the highly sensitive nature of the material, a low hydraulic gradient (21) was chosen. The permeation phase of the tests were generally conducted for two to three days.

6.2 Results and observations

Forty-eight permeability tests were performed on remolded Erving paper mill sludge samples that were subjected to a predetermined number of freeze/thaw cycles. Sixteen specimens were permeated at three different cell pressures to obtain the effective stress versus permeability relationship. The samples were highly compressible and decreased in moisture content from their initial value when placed under various cell pressures.

Figure 3 shows the results of the effects of freeze/thaw cycles on the permeability of paper sludge at various cell pressures: 34.5, 69 and 138 KPa. There is an increase in permeability due to freeze/thaw. Significant increases in permeability occur within ten cycles. At an effective stress of 34.5, 69 and 138 KPa, there is an increase of approximately one order of magnitude in permeability.

Figure 4 shows the relationship for Erving paper sludge between effective stress and permeability for different freeze/thaw cycles. Since permeability is dependent on effective stress and void ratio, the paper mill sludge was tested to evaluate the effects of effective stress as well as the effects of freeze/thaw on changes in permeability. In general, Figure 4 indicates that permeability changes at low effective stresses result in similar changes in permeability as at higher effective stresses. At higher and lower effective stresses, it appears that the change in permeability is about one order of magnitude. However, these results may not be conclusive. For example, freeze/thaw cycles 10 and 15 appear to be parallel with the zero freeze/thaw cycle line, which would indicate that the change in permeability at lower effective stresses is the same as for higher effective stresses. However, freeze/thaw cycles 1 and 3 exhibit about one order of magnitude change in hydraulic conductivity at lower effective stresses and tend to decrease in permeability by one half of an order of magnitude at higher effective stresses. Freeze/thaw cycles 1 and 3 display similar basic behavior to that of compacted clays. Obviously, further testing is required to establish more conclusive results.

With respect to the work of other researchers, Zimmie (1992) concluded that the largest changes of permeability occur at low effective stresses. His results illustrated a change in permeability of about two orders of magnitude at lower effective stresses and a change in permeability of one order of magnitude at

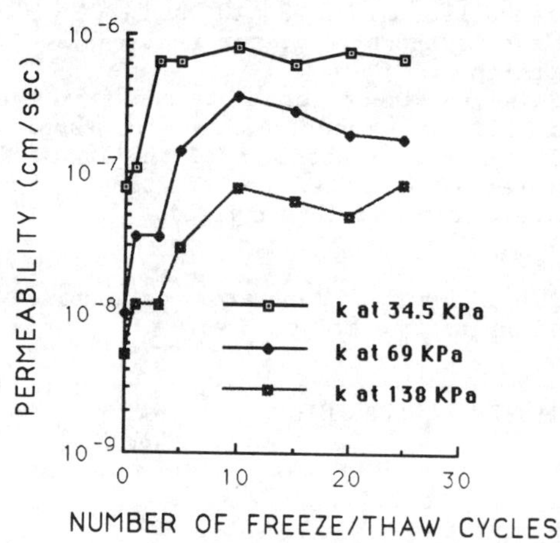

Figure 3 Summary of freeze/thaw effects on paper sludge

higher stresses for Niagara clay molded wet of optimum. Similar results were obtained by Othman and Benson (1991) on three Wisconsin clays.

Figure 5 shows the permeability versus void ratio relationship for frozen and unfrozen Erving paper sludge. The frozen specimens represent the twenty-fifth freeze/thaw cycle. At a low effective stress of 34.5 KPa, there is a 5% decrease in void ratio with an increase in

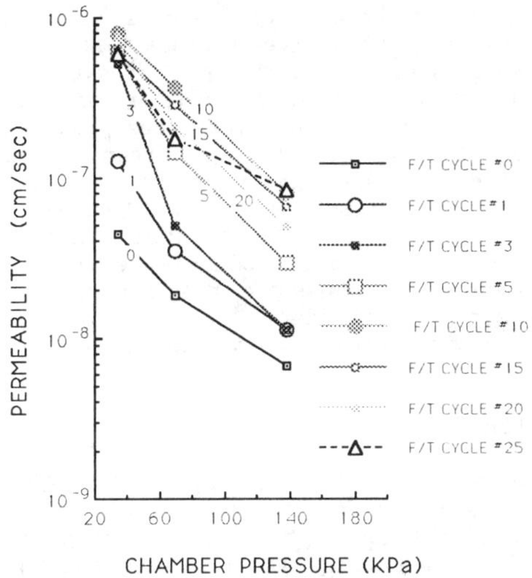

Figure 4 Effect of freeze/thaw on paper sludge in the permeability vs. chamber pressure plane

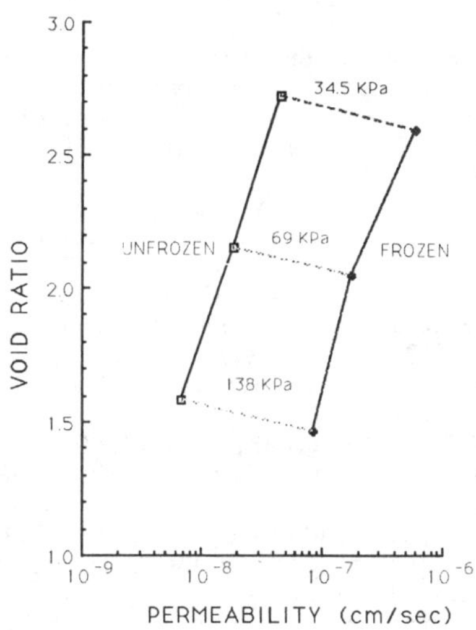

Figure 5 Effects of freeze/thaw on the permeability vs. void ratio relationship

permeability of one order of magnitude. The void ratio decreased by 5% at a confining stress of 69 KPa, while the permeability increased by one order of magnitude. At a high effective stress of 138 KPa, the void ratio decreased by 8%, but the increase in permeability was only one order of magnitude.

Chamberlain and Gow (1979) studied the permeability-void ratio relationship for fine grained dredged material subjected to freezing and thawing in a one dimensional consolidometer without backpressure. Their results exhibit similar basic behavior to paper sludge. The Ellsworth clay they tested showed considerable increases in permeability as well as a large reduction in void ratio after freezing and thawing. They proposed that structural changes and cracks created large increases in permeability. They were also of the opinion that high suction pressures developed in the freezing zone cause an increase in effective stress in the region below the freezing front. Thus, Chamberlain (1989) insists that the process significantly affects the engineering properties of soils that contain clay minerals. Since kaolinite comprises 50% of Erving paper mill sludge, it is not surprising that the engineering properties of the sludge are affected by freezing and thawing.

7 CONCLUSION

This study was performed to determine the geotechnical properties of Erving paper mill sludge, which has been used as the impermeable cover layer for a municipal landfill in Hubbardston, Massachusetts. Erving Paper mill sludge is characterized by a high water content, high compressibility, and high Atterberg limits and contains large amounts of fibers and tissues.

Some of the test procedures developed for clays were modified when applied to paper sludge. For example, the standard Proctor compaction tests and the Atterberg limits tests were conducted from the wet side. The high organic content appears to affect the plasticity of the sludge. The paper sludge behaves similarly to a highly organic clay. The Atterberg limits reported herein compare favorably with the research of others conducted on other water treatment plant sludges. As a result of the high initial water content, Standard Proctor Compaction tests were run from the wet side. A one hump curve was obtained. However, the Proctor curve is skewed with only a small range of water contents on the dry of optimum side. The wet of optimum portion of the curve extends over a wide range of moisture contents. Unlike a typical clay, the placement of the sludge at or near the optimum moisture content and density is unrealistic, since the sludge is dry and stiff near the optimum water content. When water was added to dry sludge, the sludge lost some of its plasticity, which is also typical for water treatment plant sludges.

A large decrease in void ratio as well as high strain values were encountered when consolidation testing was conducted on undisturbed samples. Settlement of 15.2 cm occurred at the landfill in the first year, which compares well with the predicted settlement calculated from the consolidation data. Consolidation test results compare favorably to previous research conducted on water treatment plant sludges.

The shear strength behavior of paper sludge is variable due to variations in the organic content, fiber and tissue content, and water content. The effective angle of internal friction varied from 25 to 37 degrees, while the cohesion ranged from 2.8 to 9.0 KPa.

In general, laboratory permeability tests conducted on insitu undisturbed samples met the maximum permeability requirements specified by the regulatory agency of 1×10^{-7} cm/s. However, several samples obtained shortly after construction had permeability values slightly higher than the regulatory requirements. These samples were tested at a low effective stress of 34.5 KPa. After the landfill cover consolidates, the water content will be greatly reduced and the permeability of the landfill cover should easily meet the regulatory requirements. After 9 months insitu, the laboratory permeability of the landfill cover was acceptable. The 9 month insitu specimen decreased in water content by 60% and had a permeability of 9.8×10^{-8} cm/sec.

One dimensional freeze/thaw tests were conducted on Erving paper mill sludge samples remolded at wet of the optimum water content (170%). An increase of about one order of magnitude in permeability occurred in the first ten freeze/thaw cycles. These results are similar to those studies conducted on compacted clays. The change in permeability of the paper sludge due to freezing and thawing at lower effective stresses appears to be similar to that for higher effective stresses. At higher and lower effective stresses, it appears that the change in permeability is about one order of magnitude. This observed behavior differs from that observed on compacted clays, where the change in permeability due to freezing and thawing is dependent on the effective stress (i.e. the largest changes of permeability occur at low effective stresses).

Observations of the Hubbardston landfill have indicated that paper mill sludge can provide an adequate hydraulic barrier. The substitution of paper mill sludge for clay can reduce the cost of disposal and is an excellent alternative in areas that do not have a local source of clay. However, the low permeability sludge barrier used for the landfill cover must be protected against the effects of freezing and thawing. The placement of a higher effective stress (larger overburden) on the landfill would further increase consolidation and reduce the natural permeability. Moreover, the placement of a larger overburden would increase protection against frost penetration.

REFERENCES

Aloisi, W. and Atkinson, D.S. (1990), "Evaluation of Paper Mill Sludge for Landfill Capping Material," Prepared for Town of Erving, MA. by Tighe and Bond Consulting Engineers, Westfield, MA.

Benson, C. and Othman, M. (1991), "Influence of Freeze/Thaw on the Hydraulic Conductivity of Compacted Clay," 14th Annual Madison Waste Conf., Sept. 25--26, Madison, WI, pp. 296--312.

Chamberlain, E.J. (1989), "Physical Changes in Clays due to Frost Action and Their Effect on Engineering Structure," in Proceedings for Int. Sym. on Frost in Geotechnical Engienering, Saariselka, Finland. Technical Center of Finland, pp. 863--893.

Chamberlain, E.J. and Gow, A.J. (1979), "Effect of Freezing and Thawing on Permeability and Structure of Soils," Engineering Geology, Vol. 13, pp. 73--92.

Environmental Engineering & Technology, Inc. (1989), "Water Plant Sludge Disposal in Landfills," Quarterly Report 1.

Lambe, T.W. and Whitman, R.V. (1969), Soil Mechanics, John Wiley and sons, New York, NY.

LaPlante, C.M. and Thomas, M.B. (1989), "The Effect of Freeze/Thaw Cycles on the Permeability of Niagara Clay," Masters of Engineering Thesis, Rensselaer Polytechnic Institute, Troy, NY.

Othman, M., Benson, C., Chamberlain, E.J. and Zimmie, T.F. (1993), "Laboratory Testing to Evaluate Changes in Hydraulic Conductivity of Compacted Clays Caused by Freeze--Thaw: State--of--the--Art," Hydraulic Conductivity and Waste Contaminent Transport in Soils, ASTM STP 1142, David E. Daniel and Stephen J. Trautwein, Eds., American Society for Testing and Materials, Philadelphia.

Pepin, R.G. (1984), "The Use of Paper Mill Sludge as a Landfill Cap," Proceedings of the 1983 NCASI Northeast Regional meeting, NCASI, New York, NY.

Raghu, D., Hsieh, H.N., Neilan, T., and Yih, C.T. (1987), "Water Treatment Plant Sludge as Landfill Liner," Geotechnical Practice for Waste Disposal 87. Geotechnical Special Publication No. 13, ASCE, pp. 744–757.

Stoffel, C.M., and Ham, R.K. (1979), "Testing of High Ash Paper Mill Sludge for Use in Sanitary Landfill Construction," Prepared for the City of Eau Claire, Wisconsin, by Owen Ayers and Associates, Inc. Eau Claire, WI.

Swann, C.E. (1991), "Study Indicates Sludge Could be Effective Landfill Cover Material," American Papermaker, pp. 34–36.

Wang, M.C., He, J.Q., and Jao, M. (1991), "Stabilization of Water Plant Sludge for Possible Utilization as Embankment Material," Report, Department of Civil Engineering, The Pennsylvania State University, PA.

Zimmie, T.F., and LaPlante, C.M. (1990), "The Effects of Freeze/Thaw Cycles on the Permeability of Fine Grained Soil," Proc. of the 22nd Mid–Atlantic Industrial Waste Conference. Drexel University, Philadelphia, PA, pp. 580–593.

Zimmie, T.F. (1992), "Freeze–Thaw Effects on the Permeability of Compacted Clay Liners and Covers," Geotechnical News, pp. 28–30.

Waste Disposal by Landfill – GREEN'93, Sarsby (ed.) © 1995 Balkema, Rotterdam, ISBN 90 5410 356 6

General report

M. Manassero
IG Ingegneria Geotecnica, Italy

M Manassero: The aspects of environmental geotechnics which are related to new landfill site construction can be listed as follows; mineral sealing layers and geosynthetic barriers, drainage layers using granular and geosynthetic materials, stability problems concerning the interface between geosynthetic and mineral layers, the mechanical and hydraulic behaviour of wastes (with by-product re-use being a major topic), risk analysis for the sub-soil around landfilled refuse. Session 4 is concerned with pollutant control which is the function of the first item on the preceding list.

Geotechnical engineering activities associated with mineral sealing barriers include; design (the selection and specification of materials to be used for lining), geometry (the overall shape), construction method (necessary to achieve an end result), performance (assessment of end result) control. Major specific Geotechnical considerations are suitability tests, construction, quality control, audit or quality assurance, long-term monitoring.

When considering a particular sealing material or technology it is necessary to deal with the main physical phenomena (and the related parameters) which play fundamental roles in pollutant migration control-hydraulic conductivity, diffusion, dispersion, sorbtion, degradation, coupled flow, suction and capillary action. Each of these phenomena can be characterised or quantified by a single parameter or group of determinants which in turn are influenced by many factors such as scale effect, environment (including chemical and physical environment), chemical compounds to be controlled/confined, boundary conditions etc. Some factors interact with each other to make the situation more complicated, eg sorbtion capacity can change long-term hydraulic conductivity, coupled flow phenomena such as osmosis, electrokinetics. Within the influencing factors it is worthwhile to consider scale effects. For example, the hydraulic conductivity which is the most familiar flow parameter for Geotechnical Engineers, can be influenced by physical/chemical phenomena related to the huge range of dimensions concerning engineered refuse containment - the containment liner may have a surface area of several hectares whereas control tests may involve a surface area of only several square meters and a volume of a few cubic meters (for infiltration tests, double-ring sealed infiltrometer tests etc.). Furthermore laboratory tests are conducted on small samples (a few centimetres long) where for considering clay grain structure size (for dealing with compatibility) and mineral structure effects the dimensions are in terms of microns and nanometers respectively. Size range effects means that geotechnical engineers have to work with a scale encompassing ten or eleven orders of magnitude relating to physical dimensions.

The papers presented to the session are concerned with pollutant control and the first group deal with the use of rather new materials for sealing layers.

The paper by **Baykal, Kavak and Alpatli** relates to material formed by mixing Kaolinite and Bentonite with rubber tyre chips. The main advantageous property of the tyre rubber is its ability to sorb organic contaminants and hydrocarbons and increase in volume at the same time. This volume increase causes a decrease in the voids of the mineral medium thereby keeping the hydraulic conductivity low. According to the Authors' data a Kaolinite - rubber mixture (with 10% tyre rubber by weight) has a hydraulic conductivity (with respect to gasoline as permeant) which is one-tenth of the plain Kaolinite. Problems can arise when the liquid in contact with the liner is not a pure hydrocarbon so that the volume increase of the rubber is insufficient to decrease significantly the hydraulic conductivity of the mixture so that the overall permeability may be higher than compacted clay alone.

Hall and Gilchrist present a summary paper about the use of geogrids to reinforce the soil cover of landfill. They also include a case history concerning the long-term stability behaviour of slopes.

Kokowski and Kokowska describe the use of a particular natural carbonate deposit to prepare a sealing mixture which can be used with both high water content (as a slurry for filling a cut-off wall) and low water content (to form a plastic mixture for compacted liners). The Authors suggest many potential applications, such as lining steep slopes, and data from many laboratory tests such as hydraulic conductivity are given, but unfortunately no field use of the material is reported.

The second group of papers are devoted to the more-or-less traditional use of natural clays for sealing layers.

Bouazza shows the approach to the characterisation of two Algerian clays for landfill linings. These clays exhibit typical behaviour in terms of hydraulic conductivity and classification indices. However in order to evaluate the Author's conclusions, such as the decrease of hydraulic conductivity with increasing hydraulic gradient, more details are needed of the actual laboratory test procedures.

Utilization of expansive soil in compacted clay liners is described by **El-Sohby, Elleboudy and El-Bakey.** With this topic it is important to recognise that the high shrinkage potential of expansive soils can be a problem when they are used to form a landfill liner. Moreover most expansive soils are highly active clays which leads to problems of compatibility with the leachate to be contained.

Jones, Murray, Rix and Humphrey suggest criteria for suitability testing of clays to be used in landfill liners. It has to be stressed that generally the sort of classification defined by the Authors can only be applied with any degree of confidence to deposits within a limited area, ie the criteria can only be used locally. Comparison with the many other criteria in existence would be useful.

The third group of papers relate to biological and chemical interaction between leachates and pollutant control barriers.

Fang describes the influence of biological factors such as attack by bacteria and tree roots, which are not generally considered in landfill design. Bacteria can change the characteristics of leachate, eg by converting it from a weak to a strong acid. Root penetration of a liner can increase leachate penetration via suction and electro-osmosis phenomena.

Dearlove presents detailed information about the interaction processes between organic-metallic leachate and clay liners in terms of ion exchange, physical sorbtion, bicolloidance and precipitation.

Manning, Owen and Hughes describe a very interesting idea for evaluating the compatibility of mineral barriers and chemical compounds. To do this they tried to measure the variation of silica content in the flow exiting from a barrier to judge the stability of the clay minerals forming the barrier. The data is very useful for examining the buffer capacity of clay liners.

Pamukcu, Newhart and Whittle describe electrokinetic technology for controlling the migration of heavy metals from landfill sites through containment barriers. The theory behind this technology is now rather well known but use of the technique is still in the experimental phase and there has been no practical application to the present time.

Within the final group of papers **Scherbeck and Jessberger** give a very clear framework to the settlement behaviour of mineral liners. So far it has proved very difficult to assess the settlement function in the proposed analytical procedure - much more experimental data are needed to give confidence to theoretical predictions. In addition the tensile behaviour of a clay is very much affected by its water content and it would have been useful to know the degree of saturation, or water content, of the clay which was tested.

The brief resume of the session papers was intended to stimulate deeper consideration of the articles. Specific topics which the papers suggest are worthy of further consideration and discussion are:

a) Geosynthetic clay liners versus compacted clay liners - this is an interesting and intensely practical topic.

b) Active/sorbant composite mineral liners where part of the liner has a sealing function and another part of it has a sorbing function.

c) Mineral mixtures (with particular reference to cementing materials).

d) Quality control of the performance of liners - this has not been considered in the session papers but it is one of the most difficult tasks when designing and constructing barriers for pollutant control.

M Manassero outlined a case history of the use of a cement-clay mixture to line the sides of a landfill site for disposal of toxic and noxious wastes. The side slopes needed to be as steep as possible to maximise the storage capacity in relation to the available surface area of the site. The natural slope comprised a gravelly sand, lightly cemented.

The original design employed a geosynthetic clay liner in a composite lining system covering the sides of the site. However this solution was not accepted by the National Authority who requested the use of a composite liner incorporating a compacted clay layer approximately one metre thick. Various alternatives were considered, eg flattening the side slopes to be able to place and compact lay in layers. Finally it was decided to place a cement-clay mixture which was prepared directly in-situ.

A 5% cement content was selected and the furnace slag cement used was spread directly, in-situ, onto the clay. The reduction of clay lump size and the mixing of the cement and clay was performed in-situ using a rotovator-type machine following immediately behind the cement spreader. The mixture was placed and compacted in horizontal lifts. To facilitate effective compaction the horizontal width of the placed clay was 3m which was subsequently trimmed back to give the final slope profile. The completed liner was covered with a polyethylene geomembrane.

In-situ quality control of the liner was provided by conducting a number of double-face infiltration tests with the Boatwell infiltrometer and taking sample cubes from the compacted layer for laboratory tests. Both laboratory and in-situ tests gave satisfactory results in terms of hydraulic conductivity and shear strength. Lack of homogeneity of the soil mixture was a major concern but close visual examination of the liner via trial pits and the extracted samples indicated that the mixture was surprisingly homogeneous. Laboratory and in-situ test results, in terms of hydraulic conductivity and strength, confirmed these direct observations.

As an extension to the preceding work a laboratory research programme has been undertaken at Ancona University to detect the influence of clay and cement type, clay/cement proportions, mixture preparation and pretreatment, compaction method and effort, initial water content and curing time and procedure, on hydraulic conductivity, strength and deformability.

As expected, increased curing time caused the shear strength to increase and the hydraulic conductivity to decrease. On the other hand, for a given curing time, very different shear strengths resulted from the different curing regimes. Samples subjected to a hydraulic gradient of about 30 immediately after compaction reached a significantly higher strength than the same mixture when simply cured in a water-tight bag. This if additional water is supplied to a mixture during the curing stage the chemical reactions within the cement are enhanced. This observation applies to peak shear strength - there is less difference at the residual strength stage of direct shear tests.

The pure clay exhibited a more or less constant hydraulic conductivity with time ie sustained permeation did not affect conductivity. There was a big decrease (about one order of magnitude) of hydraulic conductivity when a cement-clay sample was permeated immediately after compaction. If a cement-clay sample was cured in a sealed bag (for 30 days) and then permeated there was little reduction of hydraulic conductivity and no variation with time. Once again supplying water during curing improved the performance of the sealing mineral layers. The application of a hydraulic gradient during the curing stage not only enhances chemical reactions within the cement but facilitates re-arrangement of particles before hardening of the cement mixture and hence there is a reduction of hydraulic conductivity.

Bkaykal, Kavak and Alpatli
Rubber-Kaolinite and rubber bentonite liners.

Bouazza
Compacted clayey soils as cover materials in landfills.

Dearlove
Geochemical interaction processes between landfill clay liner materials and organo-metallic landfill leachate.

El-Sohby, Elleboudy and El-Bakey
Utilisation of expansive soils for lining of waste disposal landfills.

Fang
Bacteria and tree root attack on landfill liners.

Hall and Gilchrist
Steeply sloping lining system - stability considerations using reinforced soil veneers.

Jones, Murray, Rix and Humphrey
Selection of clays for use as landfill liners.

Kokowski and Kokowska
The use of lake sediment in sealing up building ground work as well as the bottom and the buttress of dumping ground.

Jones, Murray, Rix and Humphrey
Selection of clays for use as landfill liners.

Kokowski and Kokowska
The use of lake sediment in sealing up building ground work as well as the bottom and the buttress of dumping ground.

Manning, Owen and Hughes
Dissolved silica in landfill leachates and its possible significance for clay barrier stabilities.

Pamukcu, Newhart and Whittle
Use of electrokinetics to control the migration of heavy metals from landfill sites.

Scherbeck and Jessberger
Assessment of deformed mineral sealing layers.

Waste Disposal by Landfill – GREEN'93, Sarsby (ed.) © 1995 Balkema, Rotterdam, ISBN 90 5410 356 6

Discussion

I Peggs highlighted some problems related to the use of geosythetics in landfill covers. Settlements in caps can induce strains in the liners in excess of 1%. This strain will not be uniformly distributed, it will be higher at the edges than at the centre etc. Although there is a desire to have not more than 0.25% local strain within geomembranes, I Peggs considered this was not realistic in economic terms as the service strain in a geomembrane is likely to exceed this figure. Similarly the residual stresses in the seams will cause this value to be exceeded as will differential thermal effects, eg the result of welding together two geomembrane panels which have different temperatures. The temperature cycle to which the liner or cover is subjected, or cover settlements can easily generate strains in excess of 0.25%. Stress cracking is the 'Achilles heel' of HDPE geomembranes, particularly at their seams.

Problems with geomembranes may become evident virtually immediately or after a few years as illustrated by the following examples:

a) At a 100-hectare liquid impoundment site in Arizona, various problems became evident after only 9 months of service - there were air bubbles under the liner and a whole section was floating due to stress cracking which had enabled water to penetrate beneath the liner.

b) At an industrial plant in Southern Italy, a geomembrane beneath a waste cell suffered from cracking after 5 years. It was in a very cold environment, but the material had very poor resistance to stress cracking, and even the relatively small temperature variations occurring at this location were sufficient to cause stress cracking along the edge of a seam. In other locations the cracks propagated into the geomembrane.

c) At a liquid impoundment site in Canada an extruded fillet seam failed after 2 years as a stress crack propagated from the bottom to the top - the bottom of the seam had been overheated during liner installation.

d) The predominant site for stress cracking is at the edge of the extruded bead and the bottom geomembrane. The first stage of stress cracking is the generation of small crazes (unopened surface cracks) which eventually open (with no deformation of the membrane) and propagate to the bottom of the geomembrane.

e) Stress cracking was observed in the liquor pond of a pulp mill after only 6 months of service because of the presence of sharp wrinkles (where the geomembrane had been folded over) which had creased and cracked. There were also some very small waves present that had very little stress on them (compared to a wrinkle) but they had also stress cracked.

f) At another liquid impoundment stress cracking caused an exposed liner to shatter. A seam had been overheated and a stress crack was formed. Subsequently the temperature fell so that the liner became stressed and the crack ran from its original to a point at which it suddenly propagated in all directions so that the liner virtually exploded. The cracks went through seams and patchers about 70m along the side slope and went up into the anchor trench.

Tensile tests on HDPE samples shows that initially the material fails in a ductile manner then suddenly it fails in a brittle mode - this is where stress cracking occurs.

The stress rupture curve is obtained from a tensile test on a 'dog-bone' sample. The centre portion of the specimen contains a face razor notch which penetrates 10% of the thickness of the geomembrane in order to generate plane strain conditions in the bottom of the specimen (as per the field conditions). Different samples are loaded to different percentages of the material yield stress and the time to failure is then measured.

The initial reason for establishing stress rupture was to be able to extrapolate the data to the liner service life and thus obtain the permissible stress that could be applied to a HDPE geomembrane. Unfortunately with this kind of material behaviour the extrapolation cannot be performed. The transition time from ductile to brittle failure is very much a function of temperature - as temperature increases the time before the transition occurs decreases. For the different materials currently available failure time ranging from about 5 hours (the worst) to around 5000 hours (the best) have been measured. For covers, care must be taken either to select an HDPE with the best stress cracking resistance available or to move away from a semi-crystalline material and use a flexible, amorphous, low crystallinity material that does not have a yield point and has a high elongation at tensile failure.

D Daniel asked why engineers would want to use HDPE in covers for municipal waste landfills given that non-crystalline materials such as low density Polyethylene are available.

I Peggs stated that HDPE has good chemical resistance properties but the possession of a distinct high yield point does make it difficult to use because once this stress is exceeded the material becomes unstable and undergoes extremely large strains. Hence in design strains are limited to the order of 2-5%. Elimination of an abrupt yield point (as obtained with very low density Polyethylene or PVC) makes design much easier.

C Hall commented on the statement made by D Daniel in his keynote lecture that the provision of geotechnical stability of a liner is a straightforward consideration. C Hall agreed that it should be so but even the most elementary cover layer, comprising soil on geomembrane, has its uncertainties. At limiting equilibrium the essential factor of safety can be expressed as:

$$FoS = \frac{\alpha \tan \Phi}{\tan \beta}$$

where Φ is the friction angle for the soil
β is the slope angle
α is the coefficient of shearing interaction between the soil and geomembrane or other such seal.

There are several theoretical questions to be considered :-

1) If Φ is obtained from a shear box (presumably at low normal stress), which value is needed and which will be reported, ie Φ peak (since this is the test method norm) or Φ residual (since the soil cover layer will be in a loose, uncompacted state) or Φ other (for instance the value compatible with the low strain state which is to be aimed for in a geomembrane).

2) The test method for determining α and Φ is subject to many influences and also there are some potential flaws. For these reasons, it is frequently advised that reliance on published data, in final design, is inadvisable. Stability equations consider small differences in angles which dictate the disturbing and restoring forces. A few degrees, either way, will matter.

3) What is the definition of the inclination of a slope bearing in mind that;

   a) constructional variations and their effect on 2-D and 3-D stability analyses,

   b) differential settlement and the local steepening of slope gradient that this causes. (D Daniel mentioned a 500mm central depression of a 10m radius settlement with respect to a controlling strain of about 5%. This adds some 3° to the slope gradient and this may be sufficient to trigger instability.)

E Pasqualini added some comments on the influence of cement content and compaction water content on cement-clay mixtures. Hydraulic conductivity tests were conducted using flexible wall permeameters with uncured samples. There is a 'threshold' clay content for any significant effect, this was about 5%. The permeability of the mixture increases as the moulding water content decreases.

D Bishop stated that with regard to Geosynthetic Clay Liners (GCLs) one of the main purposes of the geotextile component was to keep the Bentonite in place, especially on sloping ground. However excavation of a variety of installed geosynthetics had indicated that if the geosynthetic's area weight was less than 300 g/m^2 then it would most probably be damaged at some stage in its life. Consequently it was queried whether the geotextile aspect of a Geosynthetic Clay Liner was important.

D Daniel replied that it can be of importance, particularly where there is a potential for penetration of the clay such as when granular drainage materials are placed in contact with the liner.

H Jessberger pointed out that in Germany there was no differentiation between final capping systems and liners, they are all classed as mineral sealing layers regardless of actual composition. With regard to the behaviour of GCLs several questions arise :

a) Do cycles of wetting and drying affect their integrity.

b) Is the self-healing and thixotropic nature of the Bentonite within a GCL affected by freeze-thaw cycles.

c) If the GCL has to span soft spots or voids then it will be stretched - is there any long term effect from localised thinning of the Bentonite or the tension in the geotextile.

d) A composite liner relies on there being good contact between the geomembrane and the clay layer - does the insertion of a geotextile between these layers affect detrimentally the performance of the system.

D Daniel responded by saying that, with regard to desiccation, large-scale laboratory tests will give some indication of performance but to evaluate the true behaviour field trials have to be conducted. In common with a mineral liner a GCL will crack upon drying out and the important question is not whether the material will swell upon re-wetting but what material (eg sand, silt, topsoil) falls into the cracks. Preliminary laboratory data on GCLs indicate no significant change in hydraulic conductivity due to freeze-thaw cycles. However more field data is needed, especially because the geosynthetic components will creep and it is unlikely that any gap will be maintained over a large period of time.

D Daniel concluded by saying that in his opinion the design of the mineral layer component of a landfill cap was much more difficult than the design of the equivalent layer for a basal liner.

R Jones referred to the paper by Hall and Gilchrist on steep-sided slope stability and enquired whether the geomembrane used was smooth or textured.

C Hall replied that in the case outlined it was a smooth geomembrane. This is the preferred situation because it means that the membrane is in a "benign state", ie free from significant strain, and it is only required to perform a separator function. If the surface were textured then the membrane would attract significant tensile and shear strain. Although the smooth geomembrane will have a low coefficient of frictional interaction stability should be achieved by a structural component within the system.

S Jefferis reiterated one of his previous points in that desiccation and freeze-thaw change the chemistry and chemical history of a clay. As a clay mineral freezes the pore water will freeze - initially it will be pure water which freezes until a eutectic mixture is achieved. Thus as the water freezes the clay is exposed to a eutectic solution and one should look at the pore water chemistry, particularly at the point of freezing, and when it begins to re-hydrate.

S Jefferis suspected that the Magnesium ion is the most significant in the swelling process. Above about 50 ppm the clay swelling is gently inhibited, certainly with Sodium-activated bentonites. So if there are concentrations well in excess of this value as the pore water begins to melt there may be some effects.

C Hall confirmed that the tilting table apparatus which was used to determine the shearing interaction behaviour (between soil and geotextile) for covers was only suitable for situations where there was a low normal stress across the sliding plane - as would occur for covers where perhaps the confining stress was due only to one metre of soil. But the test would also be appropriate to basal liner where the critical condition is for the soil cover layer in position prior to placement of waste. Hence it can apply to both, but one is a short term condition and one is long-term.

With regard to the damage of geomembranes **D Bishop** pointed out that one of the causes, ie stresses induced by installation and imposed weight of refuse, can be anticipated and measures can be taken to provide protection against these effects.

I Peggs agreed but emphasised the influence of cost and reiterated that by improving the damage resistance of the geomembrane itself the need for a heavy protective geotextile layer is eliminated.

5 Treatment of old landfill sites

Waste Disposal by Landfill – GREEN'93, Sarsby (ed.) © *1995 Balkema, Rotterdam, ISBN 90 5410 356 6*

Model studies of dynamic compaction

G.E.Barnes
Bolton Institute of Higher Education, UK

ABSTRACT: The technique of dynamic compaction consists of dropping a heavy weight from a considerable height to force particulate ground into a denser arrangement. The technique has been modelled in the laboratory using a cut-away box of sand to enable a visual observation of the phenomenon and provide a simple means for quantifying the effects. Some of the results are presented in this paper.

INTRODUCTION

A box with a stiff perspex side was filled with sand, placed in as loose a condition as possible. Density checks were carried out confirming that the packing was loose compared to the minimum density (D_r about 15%) and uniform across the box. At vertical intervals of about 100mm thin horizontal layers of dark brown sand were placed to act as marker bands.

Initially a steel mass, 50mm square, was used as the dropping weight but this produced an excessive mass/unit area in that the weight acted like a 'bullet' penetrating deeply into the sand, a process similar to that described by Mascardi in the paper by Greenwood and Kirsch (1983).

The mass/unit area was quadrupled by using a 100mm - square rigid steel block and the perspex front was stiffened to resist undesirable volume changes because of its flexibility. The effect of different energy levels and mass/unit area were investigated by dropping two masses free fall through the following heights.

Mass, kg	Height, m
6.7	0.5, 1.0, 1.5, 2.0
11.8	0.85, 1.5

After each drop a slide photograph was taken of the deformed sand for later appraisal and assessment.

Loose sand - phenomena

From the steps which appeared in the dark sand layers it was inferred that shearing occurred along planes emanating from the impact craters and distinct zones of shearing could be identified. Figures 1 - 5 show the interpreted zones of shearing which developed after 1, 5, 10 and 15 drops at each of the energy levels. By tracing the movement of the shear zones common features were noted at all energy levels, as follows:

a) 1 - 2 drops
Local vertical compression or compaction of the sand was observed immediately beneath the weight and the initiation of a conical shear zone occurred with a cone depth roughly equal to the width of the weight.

b) 5 drops
Several effects were noted:
i) The initial cone was more well-formed, denser and was being pushed down.
ii) Circumferential shear immediately beneath the weight (punching shear) developed into a deeper cone with a depth D now roughly equal to twice the width of the weight B. The smaller cone ($D \approx B$) tended to remain and progress inside the larger cone. It was noted that a lower mass/unit area produces inclined cone sides and the higher mass/unit area produces more vertical sides and

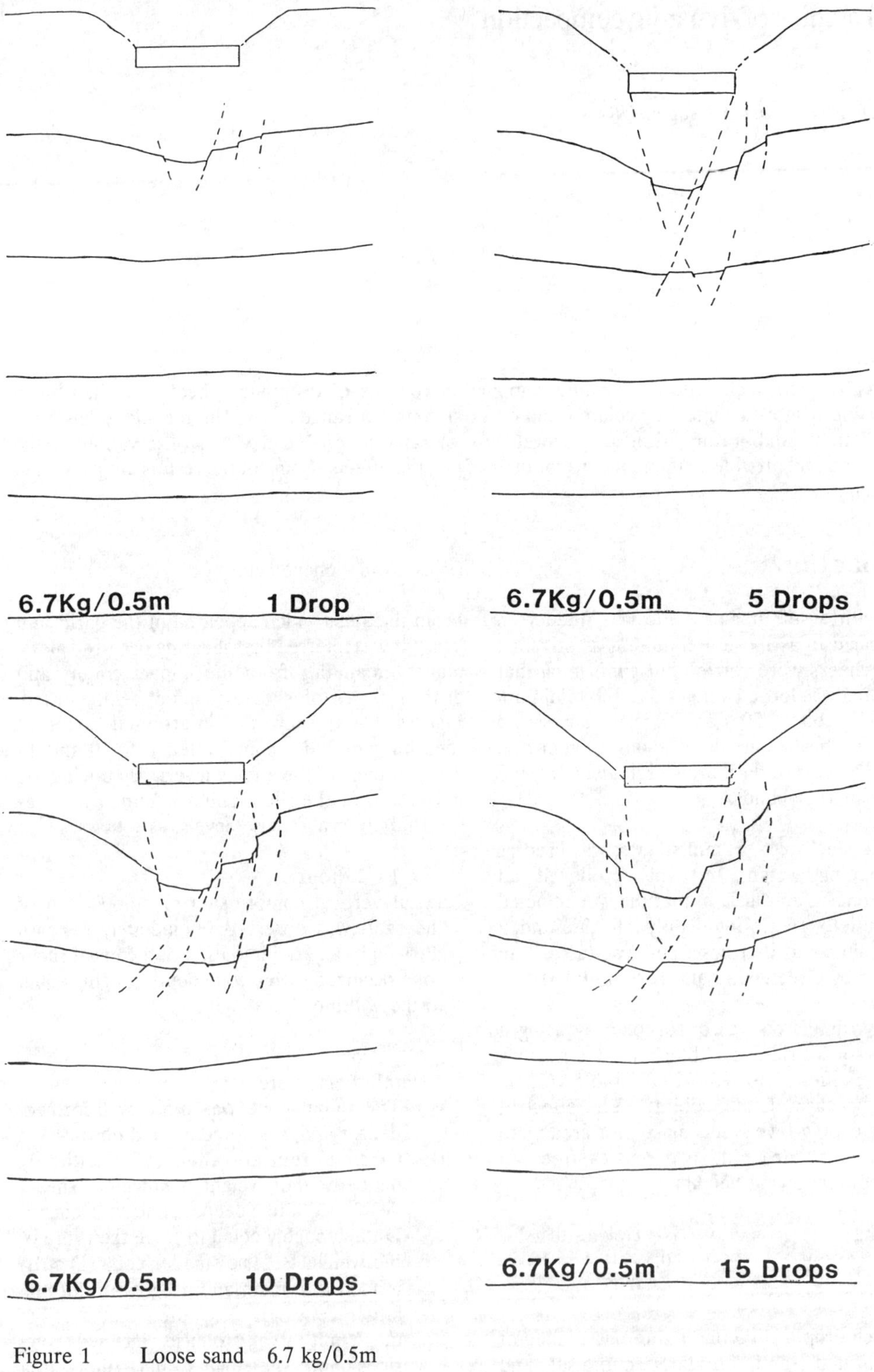

Figure 1 Loose sand 6.7 kg/0.5m

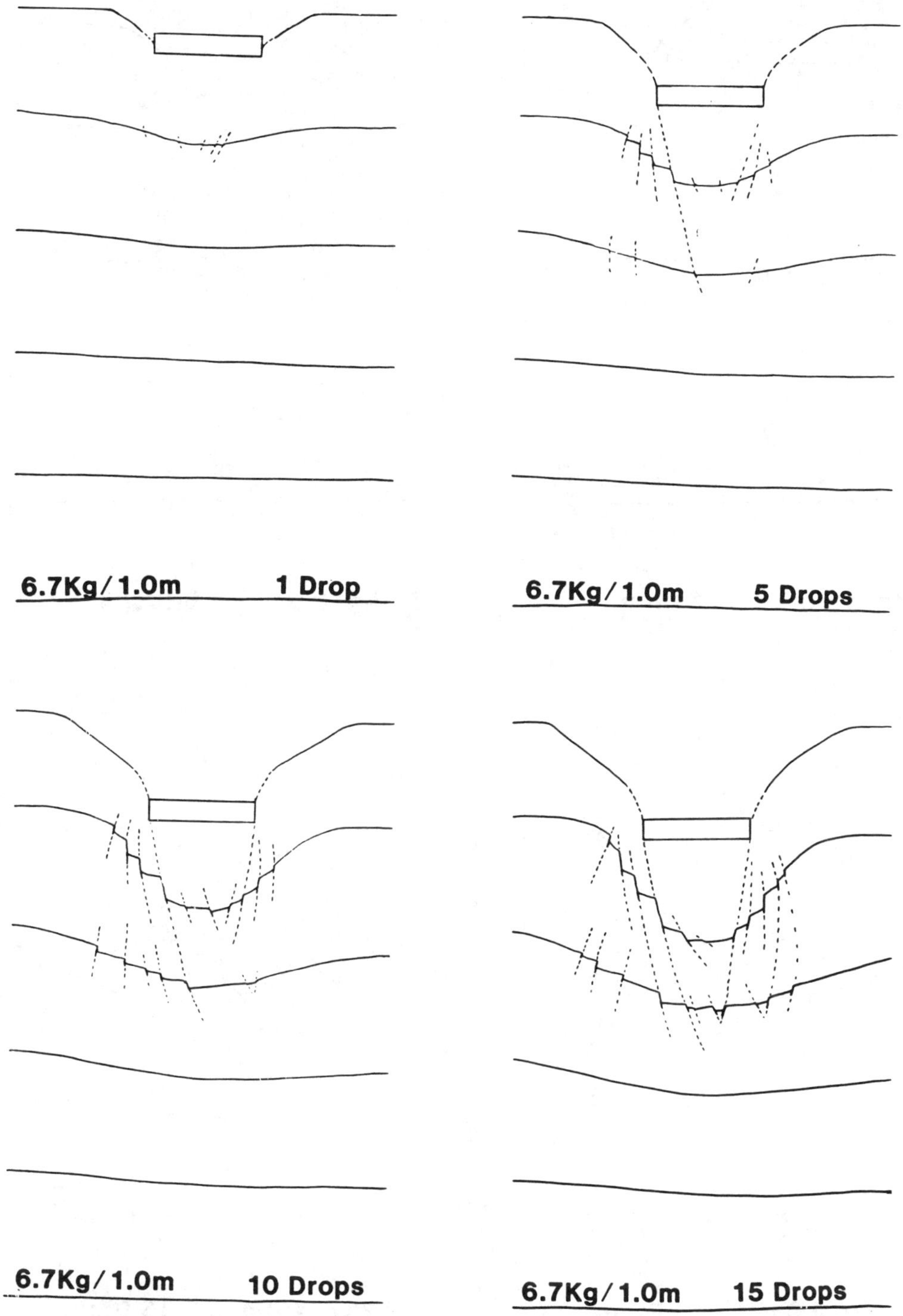

Figure 2 Loose sand 6.7 kg/1.0m

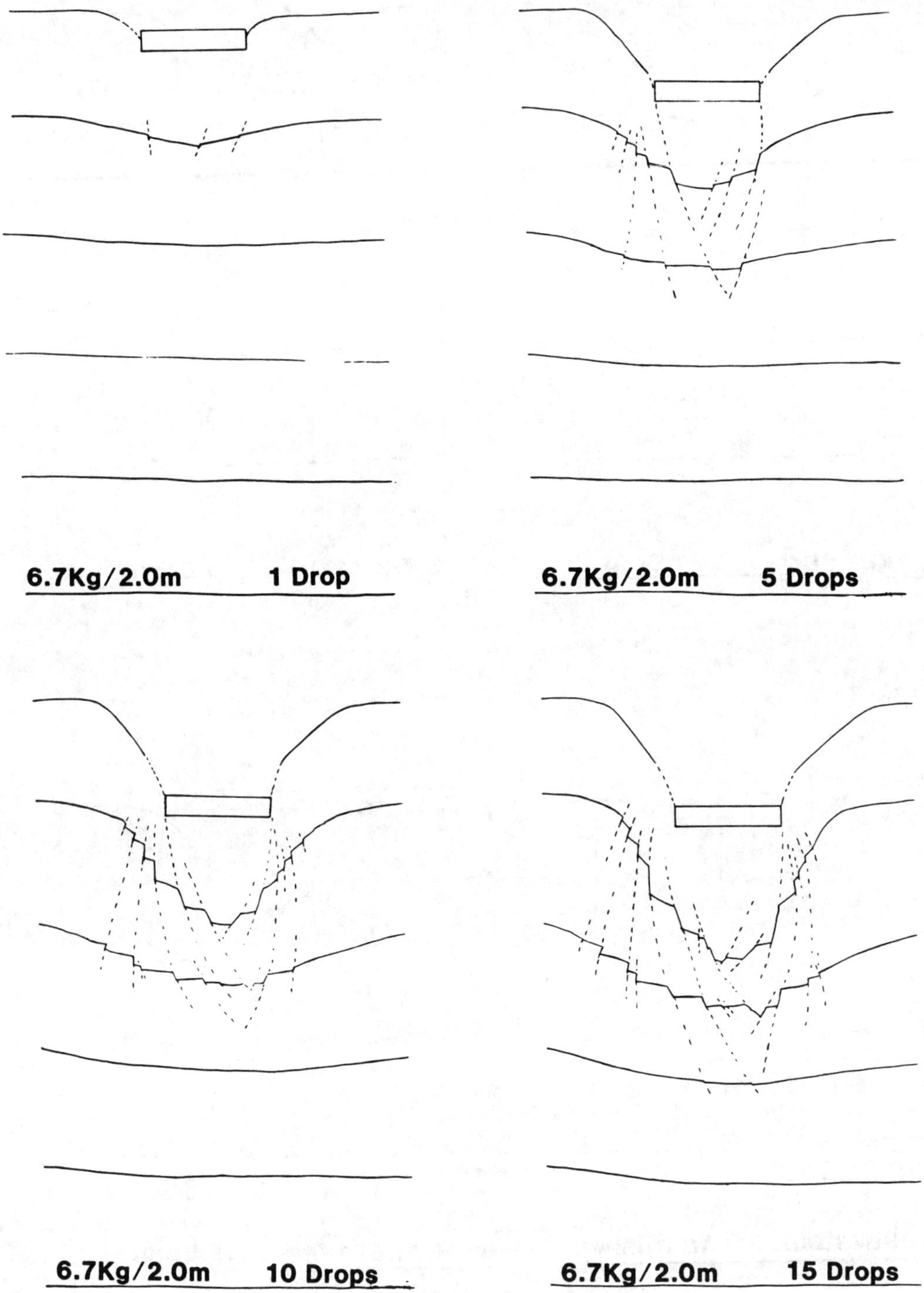

Figure 3 Loose sand 6.7 kg/2.0m

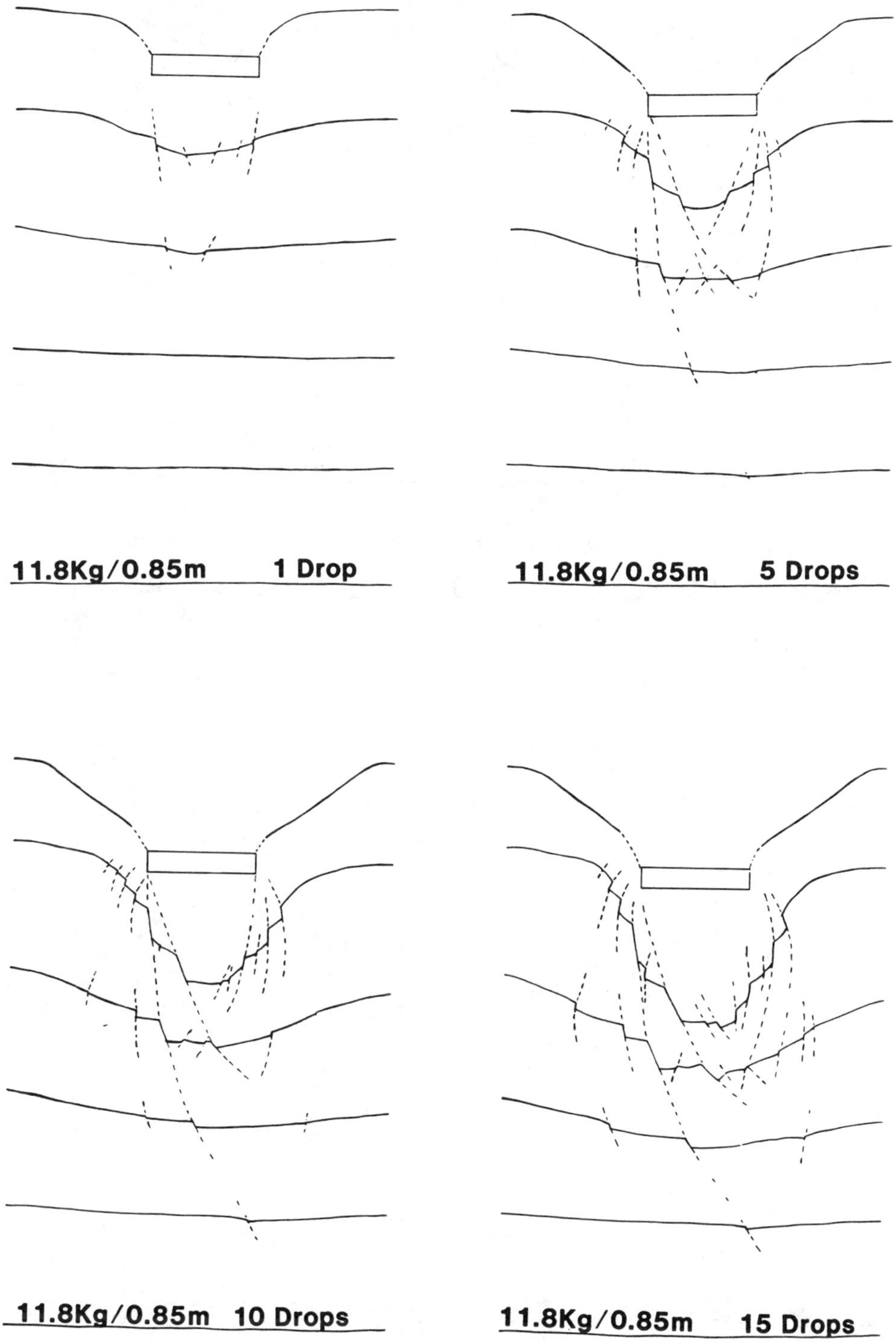

Figure 4 Loose sand 11.8 kg/0.85m

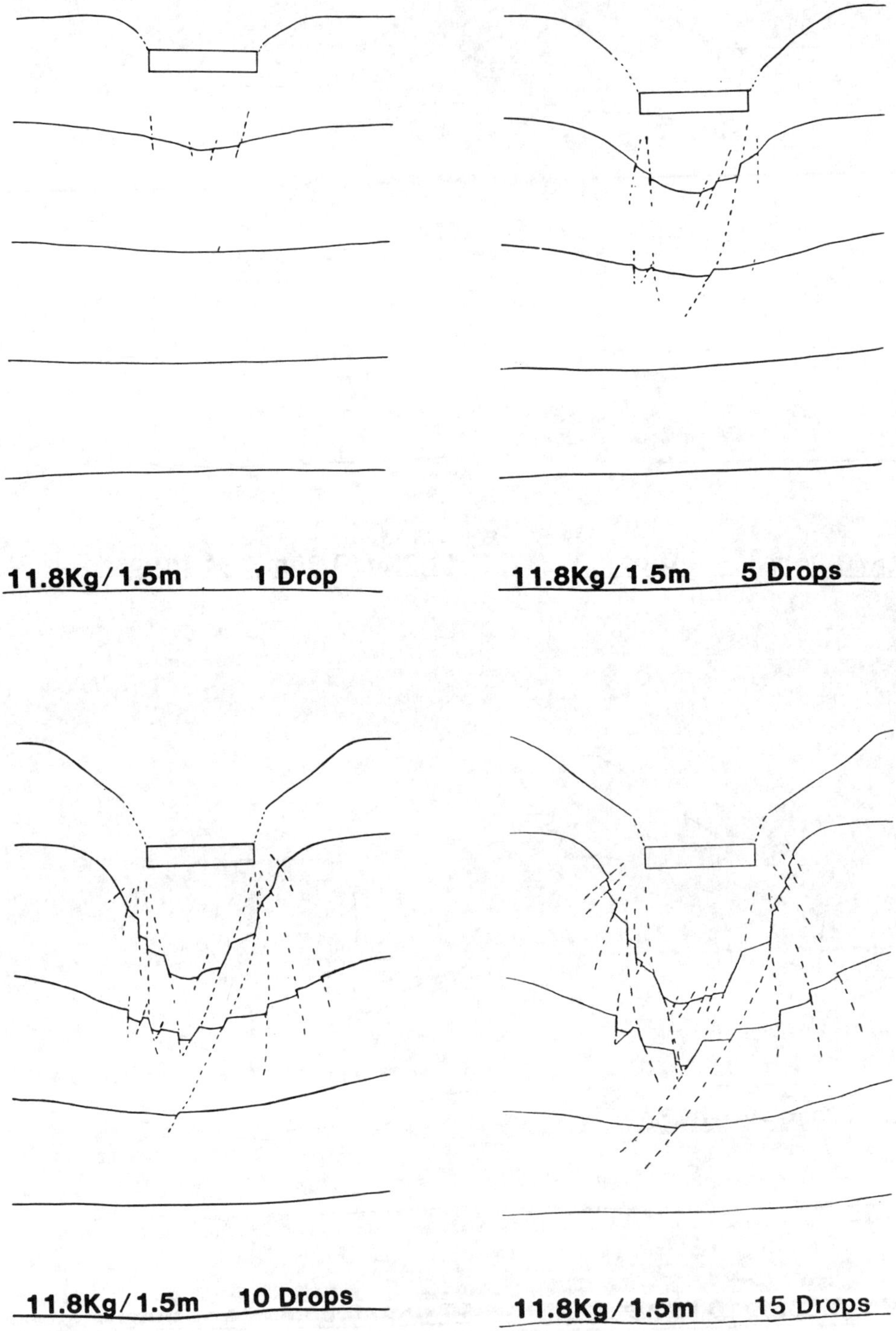

Figure 5 Loose sand 11.8 kg/1.5m

secondary cones or wedges alongside.

iii) Circumferential shear zones radiated laterally from the shear cone producing a wider punching shear effect, especially at the higher levels, although this could have been promoted by the low confining pressures around the edge of the impact crater.

iv) Radial compression occurred in the lower zones (between about D ≈ B and 2B) as a result of i) above.

v) Vertical compression with no distinct shearing beneath the deeper cone.

c) 10 drops

i) The deeper cone was pushed further down with more smaller cones produced within and outside. The higher mass/unit area tended to produce a deeper cone with depth D roughly equal to 3 - 4B.

ii) Circumferential shear zones were produced at deeper levels radiating up to 0.5B away from the edge of the weight.

iii) Vertical compression was noted up to B away from the weight edge and between B and 2B beneath the weight. This was probably associated with ii) but without the shear displacement discontinuities.

d) 15 drops

Further development from 10 drops occurred, particularly with internal and external cones/wedges produced.

Loose sand - depth of influence

The volume change of each soil band between the dark sand layers immediately beneath the weight was estimated (by determining areas) and plotted versus depth as shown on Figure 6. It was decided that an effective 'influence' occurred above a volume (or area) change of 10% so this would define the 'depth of influence'. These depths were plotted versus energy level (as $\sqrt{WH}$) on Figure 7.

This figure shows the same trend as found in practice but at a much higher level, probably as a result of the low confining stresses in the sand model and the smooth face provided by the perspex front. The phenomena described above are likely to be more inhibited in reality. The effect of mass/unit area did not seem to be

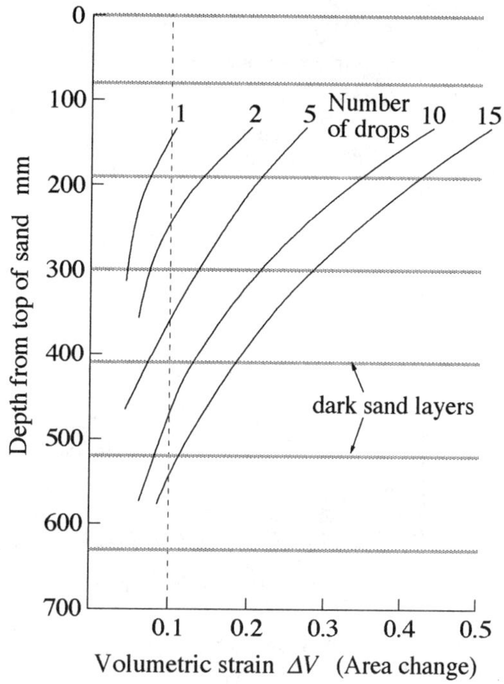

Figure 6 Criterion for depth of influence

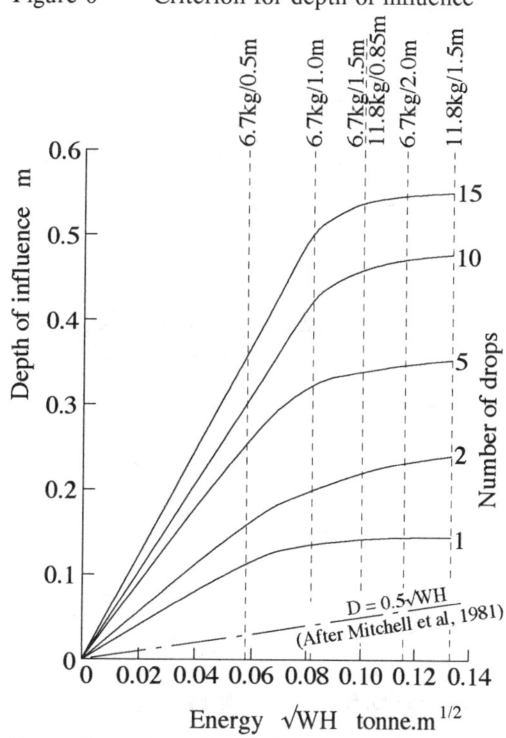

Figure 7 Depth of influence related to energy level

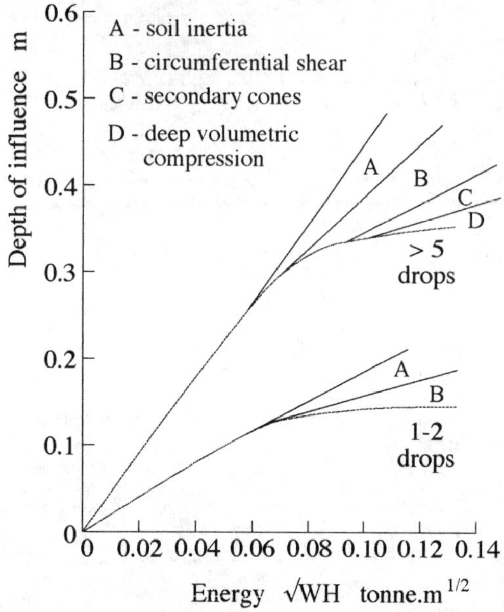

Figure 8 Dissipation of energy

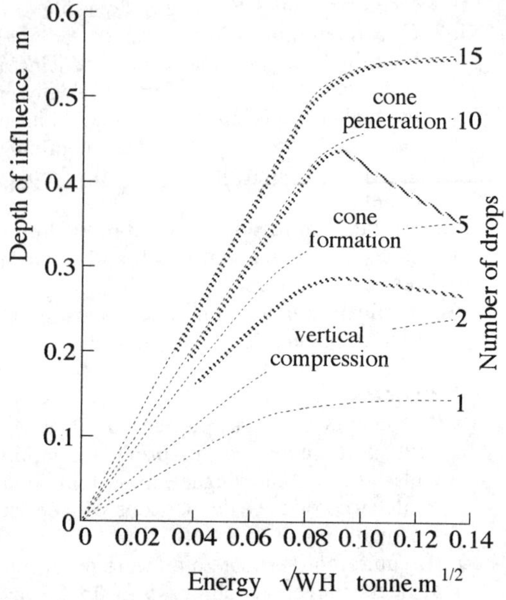

Figure 9 Three major phenomena

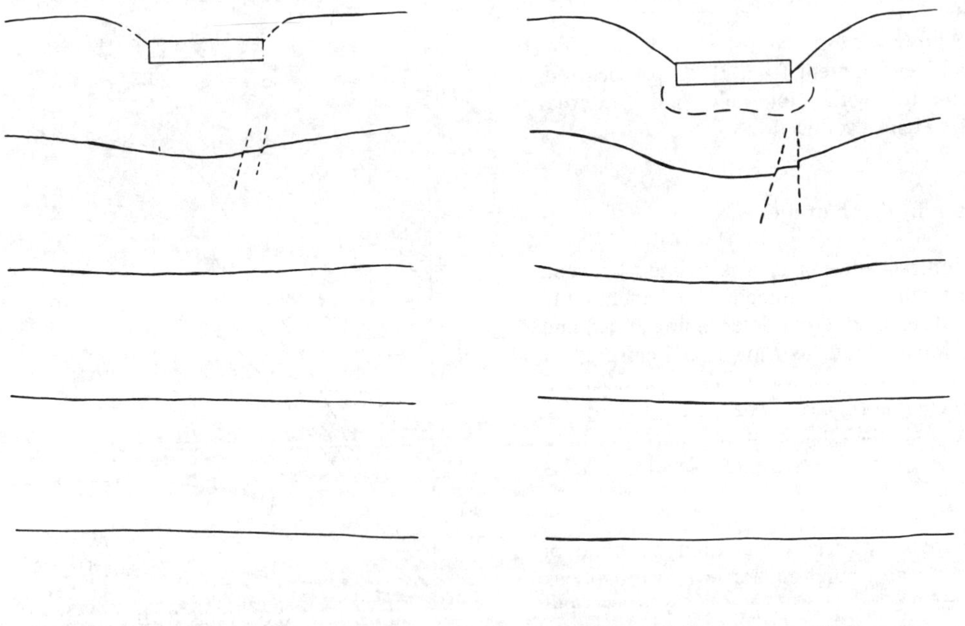

Figure 10 "Refuse" 6.7 kg/0.5m

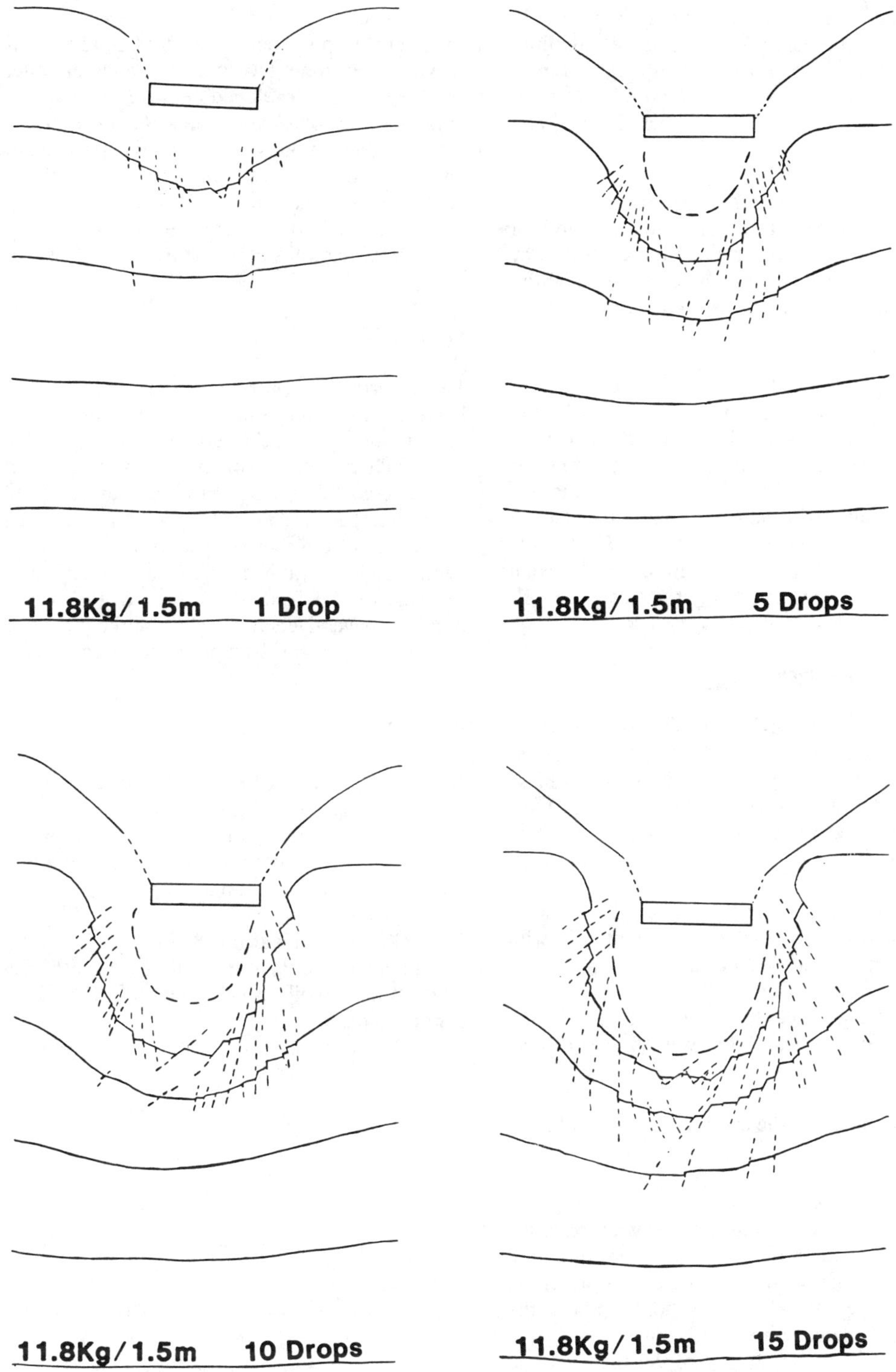

Figure 11 "Refuse" 11.8 kg/1.5m

significant. Also the concept of a threshold energy level (necessary to overcome elastic limit stresses) was not demonstrated although this may have been as a result of the high energy levels chosen and the effects of shock impact and vibration.

The most significant effect was the 'threshold depth of influence'. In other words the depth of influence per blow was not markedly affected by energy level above a certain value, although an increasing trend is noted with increasing number of drops.

It is suggested that this threshold depth of influence is a result of energy absorbed by the various phenomena as illustrated on Figure 8. Another way of illustrating this is to consider energy absorbed by three main phenomena, i.e. vertical compression, cone formation and cone penetration, as suggested on Figure 9. Energy dissipated in cone formation and cone penetration is considerable but not as effective.

"Refuse" - phenomena

In order to attempt to model the high compressibility and crushability of a domestic refuse material a proportion of dry Rice Crispies[R] was added to the loose sand during placing. Only two runs were carried out, at a low and a high energy level.

At the low energy level, Figure 10, the effect of a threshold energy level is more apparent with little effect even after 20 drops.

At the higher energy level, Figure 11, the development of shear zones was observed to be

a) 1 drop
A deep punching shear effect and a small local cone produced.

b) 5 drops
Punching shear develops deeper with extensive radiating circumferential shear zones. A 25mm thick layer of fine angular gravel was placed on the surface of the loose sand to model the granular blanket frequently placed on site. Because of the dry nature of the material and the deep craters produced much of this gravel fell to the bottom of the crater on withdrawal of the weight where it formed a gravel plug.

c) 10 drops
The gravel plug became larger producing a cavity expansion effect in the surrounding sand. A deeper cone developed beneath the plug. In practice, this 'plug' would be less likely to occur due to the capillary forces in the particulate medium and the tensile nature of many of the constituents, e.g. paper, plastic, cloth, etc.. In future studies this could be ameliorated by incorporating weak paper strips within the sand.

SUMMARY

The experimental procedure is far from perfect. Repeatability of sand density and area measurement were achieved. However, to enable more data collection, two runs were carried out in each box and although this does not appear to have affected the density for the second run the asymmetry of the shear cones does seem to have been produced by the relative incompressibility of the side of the box. Nevertheless, the results obtained have given an insight into the processes occurring during the dynamic compaction process.

REFERENCES

Greenwood, D.A. and Kirsch, K. (1983)
Specialist ground treatment by vibratory and dynamic methods. State of the Art paper, Piling and ground treatment for foundations, Thomas Telford, London.

Mitchell, J.K. and Katti, R.K. (1981)
Soil improvement - state of the art report, 10th Int. Conf. Soil Mech. and Found. Eng. Stockholm.

Waste Disposal by Landfill – GREEN'93, Sarsby (ed.) © 1995 Balkema, Rotterdam, ISBN 90 5410 356 6

Capping of an old dump containing industrial wastes

Reinhard A. Beine
Prof. Dr.-Ing. Jessberger + Partner Ltd, Bochum, Germany
Klaus Dahlmann
City Authorities, Bochum, Germany

ABSTRACT: A derelict sand pit was backfilled in the 70ies with a variety of wastes containing also production residues from the chemical industry. These residues represented a potential hazard to the groundwater. As the dump base is above groundwater level, a capping of the surface preventing the penetration of future rainfalls was sufficient.

Owing to the poor permeability of the underlying soil strata, a backwater level had been formed in the closed landfill. During an attempt to remove this backwater by pumping, only limited areas could be reached and small amounts of water collected. The pumped-off water was cleaned.

Soil air is extracted from beneath the surface capping for several reasons. The small amounts of methane determined in the landfill mean a potential hazard to the housing development in the immediate vicinity. In order to permit a longterm biological degradation of organic pollutants in an anaerobic athmosphere, a series of ventilation wells was arranged between the closed landfill site and the housing development, through which air is introduced into the landfill. Methane concentration in the soil air extracted could be reduced from initially 2 % to less than 0,2 %.

1 SITUATION

A sand pit in a Bochum suburb had been used in the 70ies to dump different wastes. They consisted mainly of slags, ashes and building rubble, only a small portion was domestic waste. Beyond that, also production residues from the chemical industry were deposited.

The base of the closed sand pit in up to 16 m depth consisted of silty and claycy sands which formed, so to say, a trough. The trough rim is at a level of approx. 7 m below the top of the site, it is covered by layers of fine to coarse sands. The low permeability of the subjacent sands caused the forming of a backwater level in the landfill which extended to overneath the trough rim. Rainfall water seeping into the landfill flowed over the trough rim into the sandy strata to the groundwater stratum. Vertical infiltration was considerably reduced because of the low permeability of the landfill base. The groundwater stratum consists of a fluviatile gravel layer the top of which is about 10 m below the landfill base. The groundwater level is also clearly below the landfill bottom. The groundwater in this aquifer shows no serious contaminations although dioxines, furanes and PCP's were detected in the landfill and in the backwater.

Methane concentrations in the landfill were low owing to the large portions of slag and building rubble, however, in some areas, not inconsiderable concentrations of high volatile halogen compounds were present.

After the closed sand pit had been backfilled, the landfill had been covered with a capping of at least 1,2 m thickness. The 20.000 m² surface of this abandoned landfill was used for agricultural purposes until after the investigation results became known, this exploitation was immediately stopped.

To valuate the situation, it is also important that adjacent to the border of the closed sand pit, residential houses had been built. The gardens and patio areas of these houses extended onto the abandoned landfill site.

2 REMEDIATION CONCEPT

The objective of the remedial action was to stop the movement of contaminated water and gas from the site. The remediation concept includes a providing containment with accompanying measures which, on the one hand, are required for the protection of groundwater and the residential area and on the other, are apt to reduce the hazard potential of this landfill site in the long run.

Individual elements of this remediation are the following:

- Surface sealing system with rainfall water

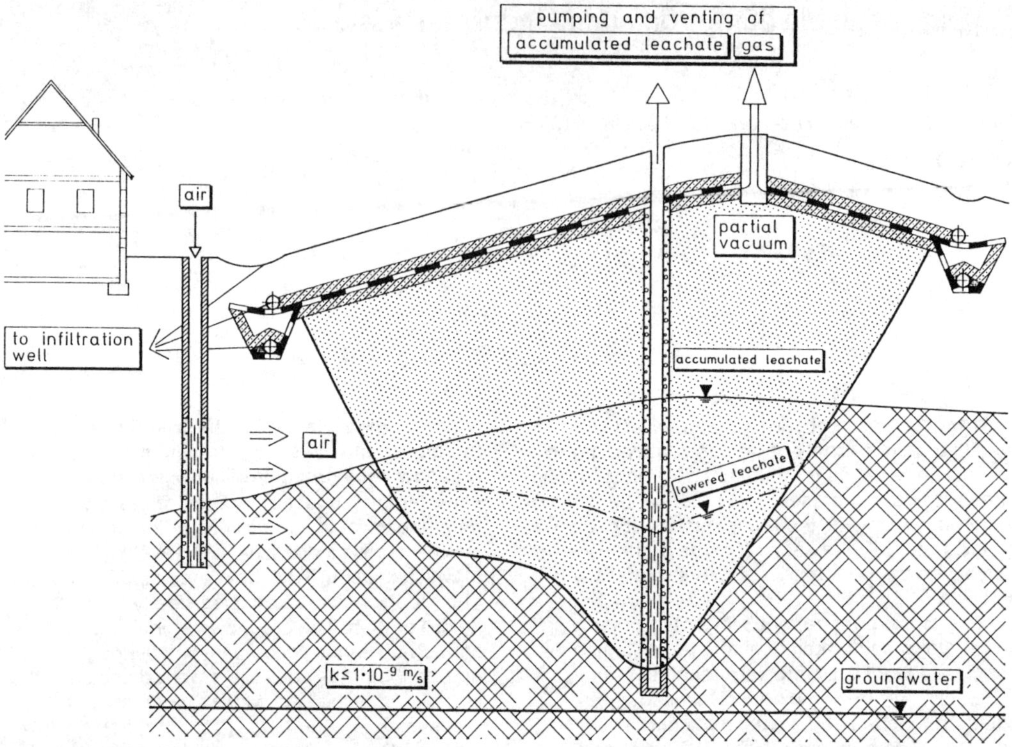

Fig. 1: Schematic diagram of the sealed waste pit

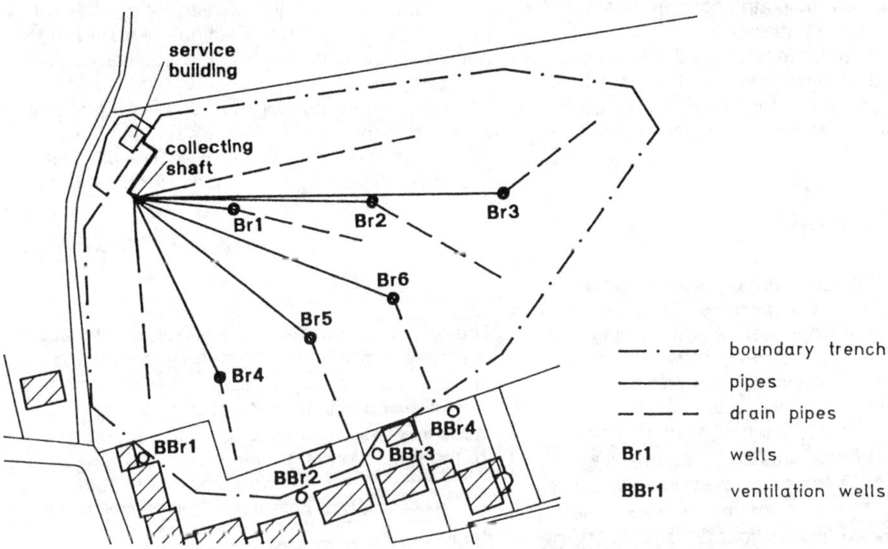

Fig. 2: Plan of the site

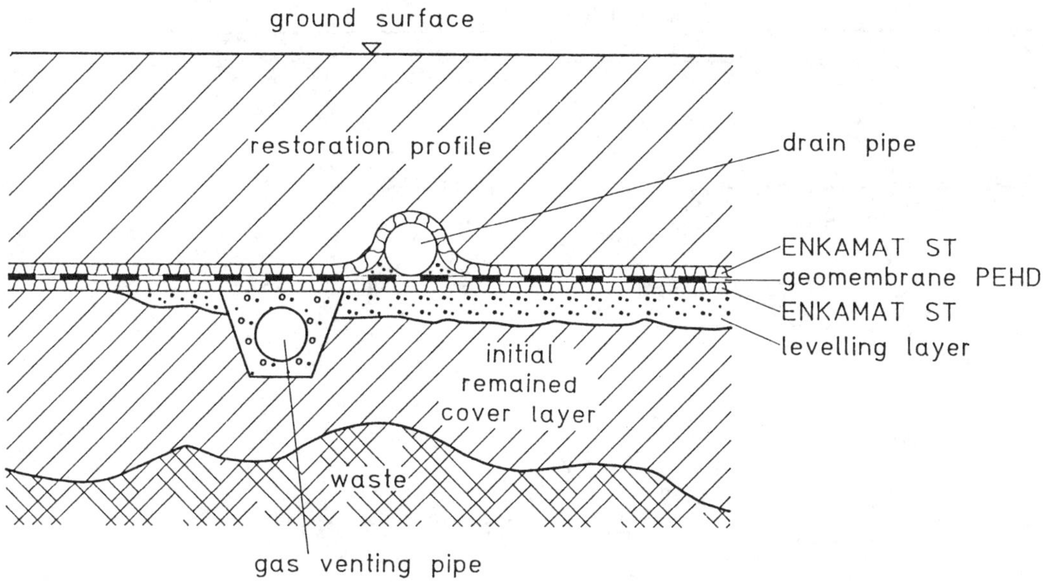

Fig. 3: Cross section of the capping system

drainage on the capping and soil air
drainage layer beneath the capping
- Pumping and treatment of backwater
- Extraction of soil air to prevent the migration of gas to the environment, particularly
to the residential buildings, while oxygen is simultaneously supplied to the landfill,
supporting the microbial degradation of organic components.

A schematic drawing of the remediation elements is contained in Fig. 1, a plan of the site is shown in Fig. 2.

3 Implementation

A decisive feature for selecting the sealing system was the condition that the site surface level should not be changed because of the adjacent residential housing development. In order to also comply with the demand for a frost-proof deep location ,
and because of the capping layer of merely 1,2 m thickness in certain parts, only a sealing system of low construction height would appropriate. The choice was in favour of a geomembrane with geocomposites as drainage layer each above and beneath (Fig. 3). Thus, the whole structure of the sealing system was only 5 cm thick, as against a corresponding structure of mineral material which would have had a thickness of at least 1 m. Around the landfill surface, the capping was shaped to form a boundary trench suited to drain the rainwater at three levels. A boundary tray is provided to drain surface water, beneath - at the level of the drainage layer - a drainage pipe is located, and another pipe has been installed approx. 3 m below the site in order to catch any rainfall water that may penetrate down to the bottom layers of the old landfill site (Fig. 4). Shafts have been provided at each of the deflection points of the boundary trench. The rainfall water catched is passed via a retaining basin to an infiltration well.

Distributed over the site surface, 6 wells were drilled into the landfill, through which backwater was to be pumped off and which, in the long run, should also serve for the extraction of soil air. The drainage layer beneath the geomembrane was additionally connected to the soil air extraction system.

To support oxygen supply to the landfill and for the simultaneous prevention of gas migrations, ventilation wells were provide outside the landfill. These wells go down to a depth of 15 m.

At the peak point of the landfill surface, provision has been made for a collecting shaft where the gas venting pipes from the different wells are converging. From the collecting shaft, the soil air is passed overneath the sealing to the service building which houses the blowers and adsorbent carbon filters to clean the soil air.

When building up the surface capping, the existing top soil layer on one half of the landfill surface was first removed to a depth of 80 cm and taken to an area outside the landfill surface and to the other half for intermediate storage. Subsequently, the wells were drilled and installed.

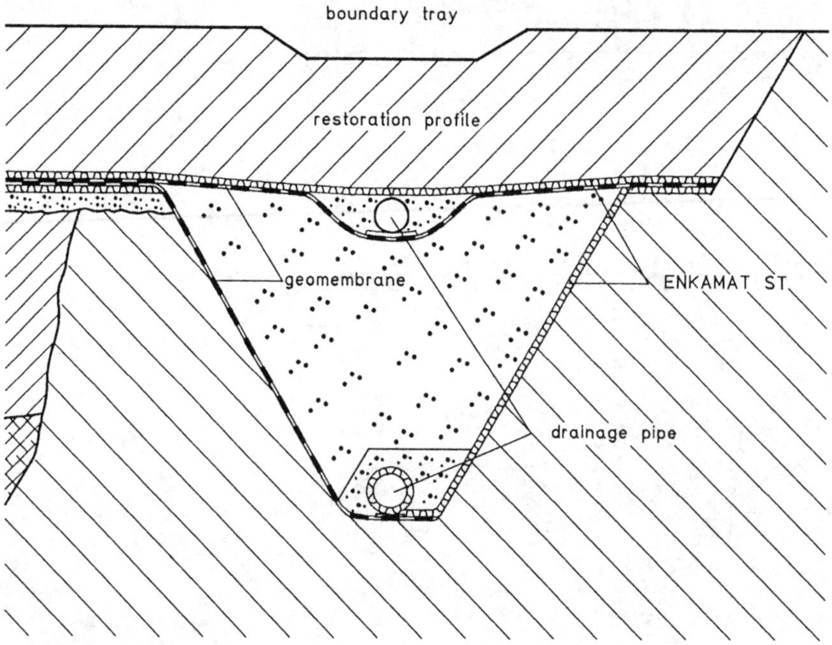

Fig. 4: Boundary trench

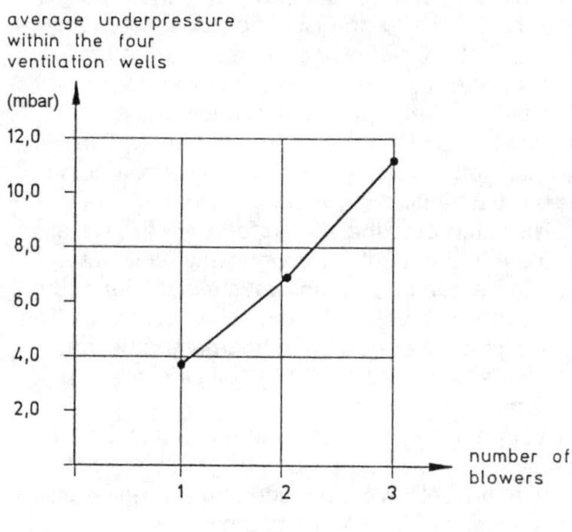

Fig. 5: Average underpressures within the ventilation wells related to number of blowers

In parallel, the boundary trench was prepared. The surface sealing was connected to it in small portions, which had to be done substantially by hand. No vehicles were allowed to drive on the geomembrane. Right after the capping including the drainage layers had been completed, the upper soil layer was replaced. During this work, too, no vehicles were allowed to enter the sealing.

The surface capping was connected to the previously prepared collecting shaft. The latter also was completely sealed off in its lower section. From the collecting shaft, three pipes were laid above the capping up to the service building located outside the landfill site.

Three blowers and two absorbent carbon filters each are housed in the service building. For backwater treatment, a feeder tank and an absorbent carbon filter were temporarily installed.

During the tests for the removal of backwater, it was found that backwater was present only in some areas and predominantly in certain strata. When drilling the six wells into the andfill, it became furthermore evident that considerably higher portions of silt were present than was originally known so that the quantity of detectable backwater was small. Removal of backwater was carried on until no water streamed into the wells any longer. Then, this measure was completed.

Soil air extraction is continued until no high volatile substances are obtained any longer and the organic landfill components have largely been degraded.

4 SOIL AIR EXTRACTION

After the surface sealing had been completed, the blowers for soil air extraction were put into operation. During the trial run, the plant constellation was fixed, i.e. the number of blowers

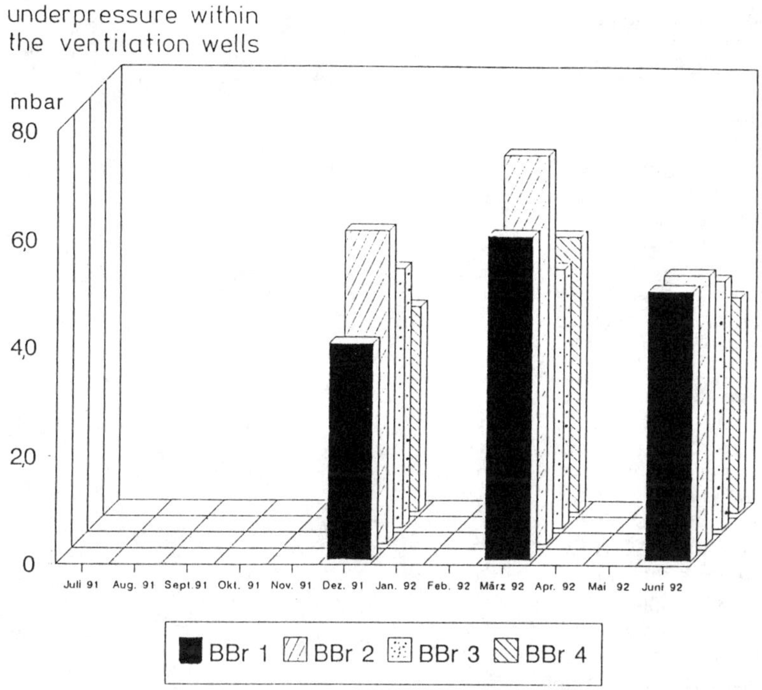

Fig. 6: Soil air components during a one year period

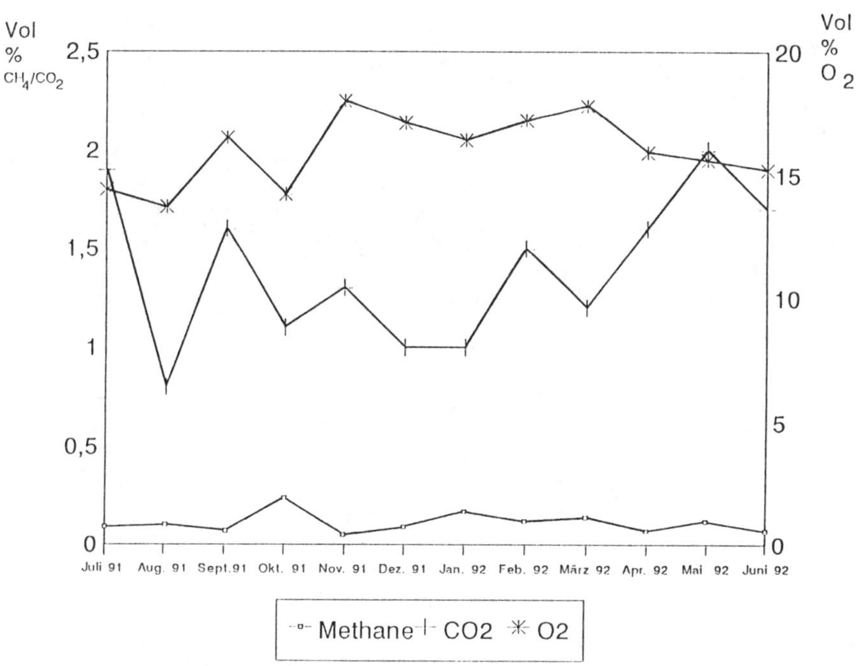

Fig. 7: Soil air components during a one year period

and the connection of pipe lines inside the gas collecting shaft. Each blower has a discharge capacity of approx. 300 m³/h. Depending on the number of blowers connected, different underpressures were measured in the ventilation wells outside the landfill (Fig. 5).

On basis of the results obtained, it was decided to run two blowers and to keep the third as standby unit. With this operation mode, an underpressure of 6,9 mbar was prevailing in the four ventilation wells. By further measurements during the operating phase, values of between 4 and 7 mbar were measured. These values are considered to be sufficient to guarantee an admission of air to the landfill (Fig. 6).

During trial operation in autumn 1990, a methane content of about 2 % was measured in the soil air extracted. This went down to approx. 0,2 % within one month and then remained unchanged in the observation period from mid 91 to mid 92. As can be seen from Fig. 7, CO_2 contents of round about 1,5 % and oxygen contents of 14-18 % were measured during the same period. No hydrogen sulphide was detected during the measurements made. The results were so interpreted that an anaerobic atmosphere continues to be present in small partial areas of the landfill where methane is produced. However, in the largest part of the landfill, an aerobic atmosphere has been created which is evidenced by the CO_2 content and the oxygen contents which are high for an old landfill.

Reference: Jessberger + Partner Consultants: Several reports and plans on the old landfill

Waste Disposal by Landfill – GREEN'93, Sarsby (ed.) © 1995 Balkema, Rotterdam, ISBN 90 5410 356 6

The possibility of biological reclamation of soda waste reservoirs 'White Seas' in Krakow, Poland

K.J. Boroń & H. Nagawiecka
Environmental Engineering and Geodesy Department, Agriculture University, Kraków, Poland

ABSTRACT: The soda waste reservoirs "White Seas" contain sediment waste from Kraków Soda Plant. It was stated that physical and chemical properties of soda sediment were unlike natural soil, but they provide proper water and air conditions for plant. Plant development was limited mainly by inadequate chemical properties of the sediment (high pH, lack of basic nutrients, toxic soluble salt concentration). The chemical composition and surface stability of sediment varies with age, the three year period since last innudation was the minimum necessity for biological reclamation. The improvement of physical and chemical properties of sediment by addition of peat or mineral soil could be useful for trees and bushes but was unecessary for grasses. The forest park type reclamation of soda waste reservoirs should be preferred.

1 INFORMATION ABOUT THE OBJECT

The Kraków Soda Plant is placed in south district of town Kraków, between railway and road line Kraków-Zakopane. The soda reservoirs exist in two compact complexes. The older part can be seen on a left side (reservoirs no 1-6) and younger one (res. no 7-19) on a right side of Wilga river bank. Total area under the waste reservoirs is about 70 ha. The reclamation of the oldest reservoirs no 1-6 and 19 was finished in 1990 and 1991 respectively. Soda production in Kraków Soda Plant has been stopped in 1990. The beginning of reclamation process started in reservoirs no 21-25 since that year. The reservoirs no 16-18 and 20 has been stayed out of reclamation up till now. The plan of soda waste reservoirs is shown in Fig. 1.

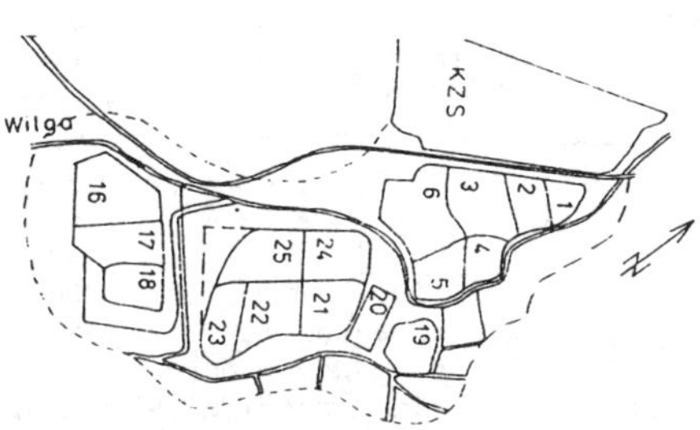

Figure 1. Plan of soda waste reservoirs of Kraków Soda Plant

2 PHYSICAL AND WATER PROPERTIES OF SODA SEDIMENT

The typical physical properties of soda waste sediment are presented in Tab. 1. The opening was placed in the reservoir whose period of the last innudation was two years.
In the studies carried out it was found that specific gravity of the soda waste material was 2.26-2.66 Mg m^{-3} which was

Table 1. Some physical properties of investigated opening.

Horizon cm	ρ_s	ρ_o	P_o	W_a	K_v
	Mg m^{-3}		vol. %		
0-15	2.48	0.33	86.89	78.40	-
15-25	2.50	0.32	87.32	79.70	-
25-50	2.50	0.64	74.40	62.40	62.02
50-75	2.66	0.43	83.76	76.00	43.02
75-100	2.66	0.40	85.00	80.30	54.89
100-125	2.26	0.43	81.02	74.00	45.40
125-150	2.26	0.41	81.90	81.20	38.28

Symbols:
ρ_s - specific gravity, ρ_o - bulk density,
P_o - total porosity, W_a - moisture content,
K_v - volumetric shrinkage

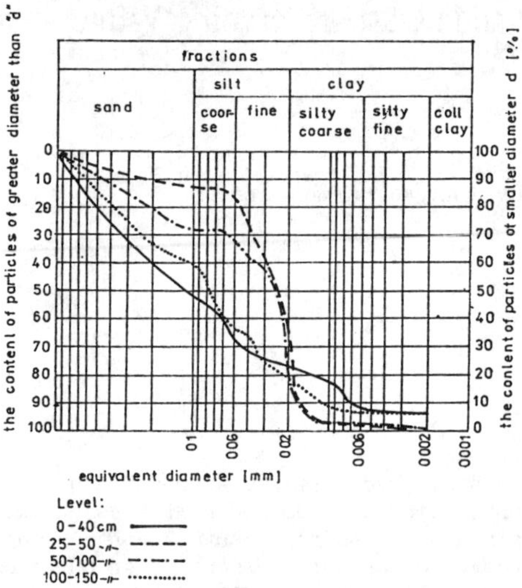

Figure 2. Particle size distribution.

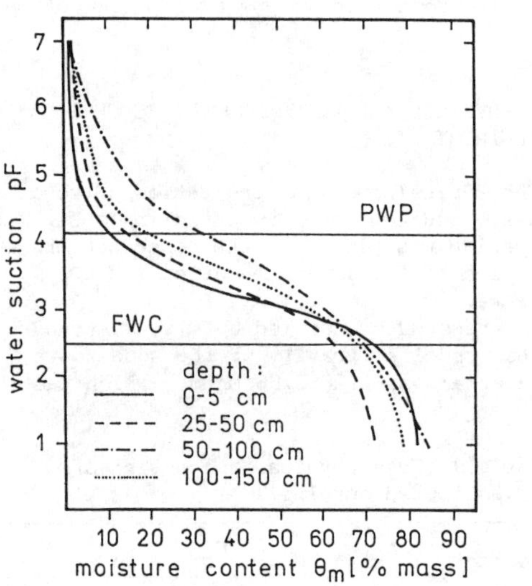

Figure 3. Water characteristic curves for soda waste substratum.

similar to the natural mineral soil. Bulk density was very low 0.32-0.64 $Mg\ m^{-3}$ and similar to organic soil. Total porosity was estimated on a base of specific gravity and bulk density. The data show that the values are rather high in comparison to mineral soil. Soda waste moisture content was 62.40-81.21 % in the profile. According to Nagawiecka and Klimek (1990) moisture content in soda waste in 20 m deep profile fluctuated between 31.25 % and 70.67 % . Volumetric shrinkage of soda waste material was very high 38.28 % - 62.02 % and was a cause of erosion phenomena occured in the waste reservoirs (Boroń, Ryczek, 1992).

Particle size distribution was illustrated by Fig.2. Texture of the profile was different for layers and independent on the depth. It may be inferred that the hydrotransported material was heterogenous.

The water characteristic curves for soda waste substratum (Fig.3) gives information about water ability for plant. The water availability for plant can be find as a result of moisture content corresponding Field Water Capacity (pF 2.5) and Permanent Wilting Point (pF 4.2) respectively. The values of available water > 50 % obtained for soda waste substratum were superior to typical mineral soil. It may suggest that water properties of soda waste substratum were good as a medium for plant development.

In the field experiment peat and mineral soil was added to 10 cm upper layer of soda waste profile. Some differences in physical and water properties of mixtures with comparison to clear waste were observed. The addition of mineral soil to soda waste caused 30 % - 50 % increase in bulk density. There was no difference in bulk density after peat addition but specific gravity of soda waste was lightened about 25 % . Field experiment supported the idea that mineral soil addition to soda waste substratum dimnished amount of water available for plant about 30 % .

3 CHEMICAL PROPERTIES OF SODA WASTE SEDIMENT.

The data of chemical properties of soda waste sediment were presented in the Table 2 and 3.

Presented in Table 2. data refer to the soda waste profile in which the period of the last innudation was three years. The values of available nutrients did not change along the length of the profile very much. There were small amounts of phosphorus and nitrogen but the great ones of potassium and magnesium with reference to typical minaral soil. The pH changes with depth was apparent but irregular. Fresh distilled soda waste was very alkaline (pH 10).

Based on the Table 3. it is easy to find that heavy metal concentration in soda waste profile was much below dangerous limits for all analized chemical elements. Very specific problem deals with the total soluble salt concentration which was rep-

Table 2. The distribution of available nutrients in soda waste profile

Depth cm	pH_{H_2O}	Available nutrients mg/100 g soil			N total
		P_2O_5	K_2O	MgO	%
10	6.0	0.50	34.80	70.50	0.006
20	6.7	0.60	34.80	70.00	0.006
30	7.7	0.75	37.00	68.00	0.006
40	8.1	0.65	37.00	66.00	0.005
50	9.7	0.50	37.00	69.50	0.005
80	7.2	0.50	37.00	69.50	0.006
100	7.6	0.25	26.00	68.50	0.003

Table 3. The distribution of some heavy metals in soda waste profile

Depth	Mn	Cu	Zn	Pb
cm	ppm			
10	15.00	2.20	7.30	4.50
20	12.50	3.10	5.55	3.95
30	30.00	3.10	6.75	3.50
40	15.70	3.35	8.40	4.75
50	23.75	3.00	7.90	4.30
80	15.75	2.85	7.80	4.95
100	19.75	3.15	8.15	4.90

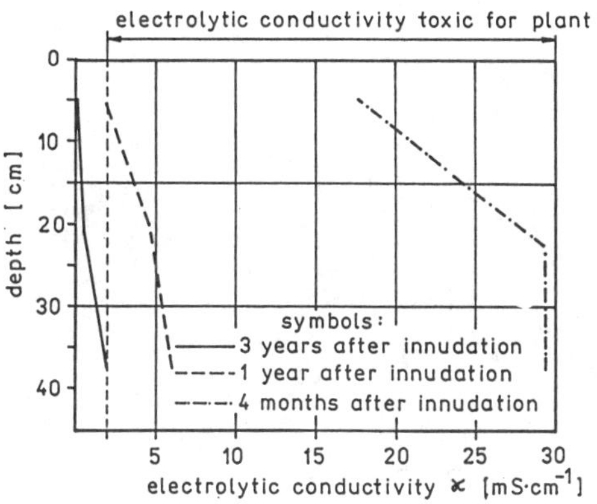

Figure 4. The electrolytic conductivity of soda waste.

resented by electrolytic conductivity (Fig.4).
Four months after innudation very high electrolytic conductivity was found in the profile. The values of the feature increased with depth in all presented cases. The longer period of the last innudation the smaller values of electrolytic conductivity was observed. Three year brake period was enough for electrolytic conductivity stabilization on the level non toxic for plant.

4 PLANT DEVELOPMENT

The field experiment was carried out on the waste reservoir in which the period of the last innudation was three years. Peat and mineral soil was added to upper layer of soda waste profile and holes for trees and bushes. Proper fertilization was applied for every species. The example of peat and mineral soil addition influence on green mass yield is shown in Table 4.

Table 4. The mean of total green mass of some grasses.

Grass	Soda waste	Soda waste with peat	Soda waste with soil
	T/ha		
fodder plant mixture	32.8	34.1	35.0
french rye grass	34.0	35.0	32.7

There were no visible influences of peat and soil addition on green mass of all tested grass species. The best grass species for biological reclamation were cocksfoot, meadow-grass, red fescue, white bent grass, french rye grass and fodder plant mixture. Among tested trees the best development was observed for such species as: birch, weeping willow and larch. The moderate development was observed for linden, maple tree. White poplar was the specie which occured spontaneously and developed very well on the soda reservoir. The species of bushes recommended for biological reclamation were: oleaster, sea-buckhorn, dogwood, gold currant bush, evonymus and caragana. Addition of peat and mineral soil caused intensive development of trees and bushes much better than clear soda waste substratum.

SUMMARY

From a practical perspective, the results indicate that forest-park type reclamation of soda waste reservoirs should be preferred. Peat and minaral soil addition can help the trees and bushes development.

Grasses developed well without any addition however mineral soil cover can stabilize and improve the mechanical resistance of the soda waste surface. Three year period after innudation was enough to eliminate unfavourable chemical properties of soda waste sediment like: high pH, toxic soluble salt concentration. This period made possible formation of superior water and physical properties of soda waste profile.

REFERENCES

Boroń, K., M. Ryczek 1992 Specificity of erosion on the decanters of soda waste of Cracow Soda Plant. Zesz. Nauk. A.R. Kraków 273 p.95-103

Nagawiecka, H., M. Klimek 1990. Possibilities of agricultural utilization of wastes from the post-sodium tank of the Kraków Sodium Plant. Zesz. Nauk. AGH 1186 p.105-116

Nagawiecka, H., K. Boroń, A. Gałka, K. Lipka 1975-1977. THe study of cultivation possibilities of soda waste at Kraków Soda Plant. Eng. Envir. Dept. Agric. Univ. in Kraków

Waste Disposal by Landfill – GREEN'93, Sarsby (ed.) © 1995 Balkema, Rotterdam, ISBN 90 5410 356 6

Successful development of closed landfill sites

R.Jeffrey Dunn
GeoSyntec Consultants, Walnut Creek, Calif., USA

ABSTRACT: Closed landfill sites are increasingly being utilized as development properties to construct and operate a variety of different facilities with many types of landuse. Successful project design and development requires careful and thorough evaluation of site conditions and redundant design features for buildings and infrastructure. Operations and scheduled maintenance must also be carefully planned and response actions considered.Major elements of post-closure landfill development require design considering: 1) landfill closure and waste containment; 2) control of landfill gas and leachate; 3) settlements; and 4) civil infrastructure constraints. Health and safety considerations must also incorporated into all aspects of the project. Civil infrastructure, settlement and foundations are discussed in this paper. Settlements of refuse materials are difficult to predict, but empirical and conservative evaluation seems to be the most suitable technique. Techniques are available to reduce settlements; however on most projects deep foundations, usually driven piles are used to support all lightly loaded structures. Civil infrastructure design generally follows a four step design process that starts with an analysis of historical records and geotechnical data. This is followed by definition of site constraints and boundary conditions. Civil works are then designed for predicted total and differential settlements. Finally future inspection and maintenance requirements are developed to deal with the uncertainties of actual performance.

1 INTRODUCTION

Recently there has been an increasing utilization of solid waste landfills for a variety of post closure land uses. Projects have ranged from parks or dedicated open space to commercial or industrial development, or in rare cases residential construction. A major driving force behind this utilization of what had often been thought of as derelict land, is that urban land within the United States has become increasingly scarce and expensive, thus making landfills and other sites which require additional development costs economically viable. An interesting benefit from this change in the value of these land parcels, is that development has often provided funds to properly close and maintain landfills where funds to do so were not formerly available.

In the United States, in particular the State of California, where the author practices as a consultant, most landfills are not more than sixty or perhaps seventy years old. In other parts of the United States, landfill sites may be older than this and in Europe it is not that uncommon to sometimes encounter waste deposits that are several hundred years old. The author was once told of the challenges and problems of dealing with medieval garbage in many areas of Europe. Conversely, on occasion developments are unknowingly constructed on waste deposits which are discovered either during or after construction. This situation provides a somewhat different range of problems and is not the direct topic of this paper.

2 THE BASIC DESIGN PROCESS

Successful design and development of

landfill site construction requires:
1. a high level of detailed site evaluation;
2. the incorporation of redundant project features;
3. careful construction with a higher than normal level of construction quality assurance; and
4. careful planning and implementation of inspections and maintenance, as well as potential response actions should problems occur.

Design must consider all of the following items:

1. site closure and maintenance of waste containment;
2. control of landfill gas and leachate:
3. total and differential settlements;
4. limitations on site infrastructure; and
5. health and safety issues during the investigation phases, site construction and site use or occupancy.

This paper deals with the three issues of settlements, foundations for structures, and infrastructure.

3 SETTLEMENTS

3.1 *Sources of settlement*

Settlements of landfilled solid waste materials result from a variety of sources including one or more of the following.

1. Due to the reduction of void spaces or compression of loose materials from the weight of the wastes themselves and any overlying materials such as covers or stockpiled soils or wastes.
2. Occasional movement of smaller particles into larger voids which form as the result of collapse, seepage, or vibrations such as from earthquake.
3. Dissolution of wastes by percolating water or leachate.
4. Volume reduction by biological decomposition or chemical reactions, which tend to be accelerated by high moisture content, warmer temperatures, poorer state of waste compaction and larger proportion of organic waste constituents. (Oweis and Khera, 1990)

Settlement can also occur in foundation soils and rocks underlying landfills, with soft soils such as clays typically exhibiting the highest foundation settlements.

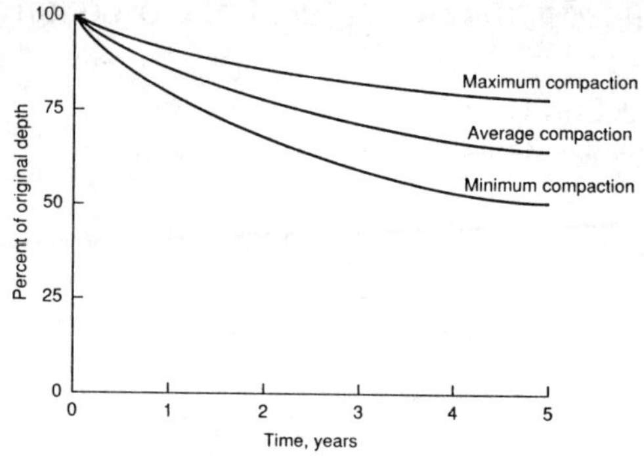

Fig. 1 Typical ranges of surface settlements of landfills containing compacted wastes

3.2 *Magnitude and rate of settlement*

Physical compression of landfilled wastes usually occurs relatively rapidly, typically within a matter of months of placement of the wastes or overlying loads. Decomposition processes on the other hand usually take years to reach completion, with the rate varying widely depending on the specific characteristics of a site and wastes contained in it. In total, settlements can be quite large, with total settlements often reaching or exceeding 25% of the initial waste thickness. Figure 1, from Tchobanoglous et al (1993) shows typical data generally applicable to a range of landfills and suitable as a starting point for evaluating settlements. As shown by this data settlements vary quite a bit in terms of the degree of compaction applied to the wastes at the time of placement.

Others have tried to quantify rates of settlement in landfill materials, by collecting settlement data in the field for a period of time. Sheurs and Khera (1980) present data from approximately two years of settlement monitoring of a waste site, which resemble those of peat or other highly organic soils. While peat may in some ways resemble solid wastes in that both contain organic materials. It is the authors opinion, that modeling solid waste as a peat material fails to acknowledge the occurrence of other processes resulting in waste settlement that do not occur with peat soils. It seems Geotechnical engineers often like to think of non-soil materials in soil like terms, and apply

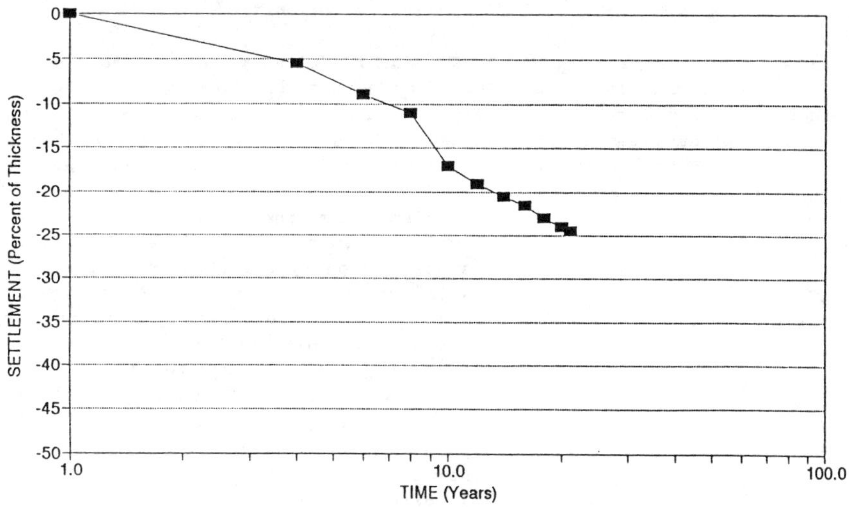

Fig. 2 Surface settlements of a landfill in Southern California, USA

soil models to them. However, at this time it is safe to say that solid wastes probably are not soil-like in their behavior, and thus should not be treated as such. Indeed other data in the literature such as that shown in Figure 2, (Coduto and Huitric, 1990) indicates that settlements may be distinctly non-soil-like when monitored over a longer period of time.

3.3 *Analysis of settlements*

Design of foundations for structures to be built on landfills requires that we analyze total and differential settlements in various areas of a subject site. Once depths of refuse are known over a landfill site, published values can then be used to provide initial estimates of settlements, suitable for conceptual design or selection of general foundation types. However, these values are not sufficient for actual design. Site specific data is required to provide a high enough degree of reliability for non-conservative design.

In rare occasions data over a number of years, such as that shown in Figure 2 is available for a particular landfill. More commonly, no data is available when a post closure development project is first proposed, and time is limited to collect the data. Thus it is recommended that one of the first activities initiated right away when considering post closure development is to undertake a program of settlement monitoring at a site. To the extent that plans are known, settlement should be concentrated in areas of planned site features such as buildings or roads. Settlement monitoring points should be laid out in sufficient number to provide data on differential and total settlements in various areas of a site. Relatively simple concrete monuments can be cheaply and easily installed at most sites, and then surveyed periodically to a set benchmark elevation by a licensed land surveyor, to provide surface settlement data. If settlement profiles with depth are desired, the use of more complex systems, such as Sondex tubes (Coduto and Huitric, 1990) are required. Initially surveys should be completed at fairly tight intervals of two weeks or a month, and then can be spread out after about six months, depending on the specific time requirements of any one project.

Once a series of settlement measurements are completed, it can often be plotted against the log of time with the results being a linear relationship between settlement and time. Then future settlements over a variety of time periods can then extrapolated linearly.

Failure to properly evaluate total and differential settlements can result in very poor performance of roads and buildings built on closed landfills and can also have drastic effects on underground utilities, particularly water or sewer lines that flow by gravity.

3.4 *Methods to reduce settlement*

If calculated total or differential settlements turn out to be too large a variety of techniques can and have been used successfully to reduce settlements and improve post closure performance. Among the techniques that have been used are:

1. allowing the wastes to age;
2. supplemental compaction;
3. surcharging;
4. grouting; and
5. dynamic compaction.

4 FOUNDATIONS

Settlements at landfill sites are nearly always large enough that only lightly loaded structures, two or perhaps three stories in height, can be supported on shallow foundations. Most other structures require deep foundations.

4.1 *Shallow foundations*

Shallow foundations typically include the following types:

1. conventional spread footings;
2. reinforced concrete mats; and
3. grid foundations consisting of column footings tied together with a system of grade beams and usually also including an integral reinforced concrete floor slab.

This last type is typically used on sites with large anticipated differential settlements. The additional stiffness of the grid system helps to distribute loads and even out settlements across the structure. Even when conventional spread footings are used it is still quite common to increase the amount of reinforming steel used thus stiffening the footings.

If shallow foundations are used generally conventional design techniques are employed. One difficult area is the analysis of waste strengths and thus the bearing capacity of the wastes. Most solid waste materials are difficult if not impossible to sample and test in a geotechnical laboratory. A limited amount of data on waste strengths is available in the literature, and at this time that seems to be what is used for shallow foundation design, with a higher than normal factor of safety applied in design to account for uncertainties about the waste properties.

4.2 *Deep foundations*

In California driven piles are the type of deep foundations most commonly utilized to support larger structures constructed on closed landfills. Drilled piles or piers are not typically utilized as they result in the removal of waste materials from the subsurface which then must be disposed of, often at great cost and difficulty. Piles are used to carry the structural loads to bearing strata below the waste materials, where they are supported by friction or end-bearing. The most typically used pile types are pre-cast prestressed concrete piles, steel H-piles, or steel pipe piles.

When designing piles for use at landfill sites the following items must be considered:

1. pile capacity;
2. pile downdrag due to waste settlement;
3. pile corrosion; and
4. maintenance of waste containment.

The later three items require a good deal of site investigation to develop a suitable picture of the landfill site, the subsurface environment, and the overall manner in which the pile foundations will be integrated in to the landfill closure.

Downdrag due to waste settlement can be quite large, and in extreme cases where waste depths approach 30 meters or more unmitigated downdrag can use up nearly the entire capacity of a driven pile of conventional size and type. Fortunately a number of techniques exist to mitigate downdrag. These include:

1. pre-drilling the piles;
2. driving through an outer casing;
3. drilling an oversize hole that is then filled with bentonite slurry; and
4. friction reducing coatings on piles.

Corrosion can also be problematic owing to the rather corrosive environment that decaying solid waste can produce. Acids, sulfates, and chlorides which develop in waste leachates can be quite corrosive to steel or concrete piles. Corrosion can be mitigated in a variety of ways including special cements, additives or coatings for concrete piles; and coatings or cathodic protection for steel piles.

Corrosion and downdrag issues have been recognized for some time. Within the last few years, concerns have now also focused on how the use of pile foundations may effect the integrity of waste containment. Concerns have focussed on both base or side liners either constructed or composed of natural materials, and the constructed landfill cover.

With regard to the base liner, either natural or constructed, project developers and their consultants must now develop sufficient information to demonstrate that use of pile foundations will not result in impact on ground water underlying a landfill. This requirement is typically satisfied by thorough investigation of the subsurface below the landfill to ascertain the existence of sealing layers that would be adversely impact by installing piles through them. It would be difficult if not impossible to seal piles which were driven through a geomembrane or even a thin clay liner, and in California this is expressly limited by government regulations. If a site is underlain by soft clays it can be shown relatively directly that a clay seal could be maintained around the pile during and after driving through the sealing layer. Alternatively, there are also sites where no natural or constructed liner exists below the wastes. In this case it is a matter of demonstrating that the use of driven piles will not promote the movement of leachate or contaminants downward into ground water.

In a similar manner the design must account for and demonstrate that foundations and structures are suitably tied into low permeability barrier layers that are incorporated into landfill covers to control escape of landfill gas and infiltration of precipitation. this is typically accomplished by careful detailing of the pile caps and the installation of a geomembrane above the pile caps, below the floor slab, and tying in the barrier layer surrounding the structure.

Pile construction methods must also be considered as part of the design process, both in terms of effect on pile capacity, but also with regard to environmental implications. Methods that push wastes downward may not be allowable as they may tend to breach the waste containment. Also methods that result in wastes being removed from the subsurface, such as predrilling, can be a problem in that these wastes must then be disposed of, usually at a location off-site. Should these excavated wastes contain hazardous materials, disposal can be quite costly.

5 Infrastructure

Design, construction and operations of civil infrastructure including roads and utilities requires careful consideration of the specific characteristics of a landfill site being developed. Numerous unsuccessful projects in the past have dealt successfully with the issues of foundation support, but failed to realize the importance of infrastructure issues. The result has been a need for continued rather expensive maintenance and retrofitting of many projects. In some extreme cases developed landfill sites have been abandoned as the costs to maintain or repair the sites become to high or the public perception of the site became too negative, so as render the project an economic failure.

Design of post closure infrastructure at landfill projects requires careful balancing of the costs to mitigate the effects of settlement initially, versus the costs of operations and maintenance during the project life. The design requires a close integration of civil, geotechnical and environmental engineers on the design team to meet the following design goals:

1. achieve acceptable performance as the site settles, where acceptable performance balances costs versus aesthetic acceptability;
2. avoid penetration of the landfill cover "barrier" layer; and
3. minimize future maintenance requirements.

Civil and geotechnical engineers practicing in California have developed a four step design process which works very successfully when properly applied.(Keech, 1992). These four steps are as follow.

1. Analyze historical documents and geotechnical reports.
2. Define site constraints or "boundary conditions"
3. Design for predicted total and differential settlements.
4. Define future inspection and maintenance requirements.

5.1 *Analysis of historical documents and geotechnical reports*

Historic documents can include original topographic maps, or grading plans of the site, if available. Landfill personnel can also be interviewed regarding specific items of landfill operations such as waste stream makeup, segregation of wastes, sequence of landfill operations, and existing landfill utilities and environmental control systems. This provides an initial model of the volume and thicknesses of waste and some initial ideas regarding potential orders of magnitude of settlement. Empirical data can then be used to develop preliminary settlement estimates. If site specific settlement monitoring has not already begun, it should be started immediately.

Geotechnical analysis can then be used to develop contour maps of the site showing items such as depth of wastes, depth of barrier layer, if any, thickness of refuse, depth of foundation material, and depth to bearing materials. Settlement estimates can then be refined and contours of predicted settlement calculated, both for existing conditions and for future fill conditions. In many cases a contour map of additional settlement per additional unit of additional fill thickness is also prepared. As appropriate, evaluation of possible settlement mitigation techniques can be initiated.

5.2 *Identification of site constraints*

The civil engineer then develops a picture of the site constraints or boundary conditions for the planned project. Typically these can be grouped into two differing boundary conditions. "Soft edge" conditions, usually associated with natural boundaries where differential settlements, although still possibly sizable are more gradual. "Hard edge" boundaries are those created by artificial boundaries, such as the edge of a pile supported structure, or deep elements such as a sewage pump station. Differential settlements at hard edge boundaries are typically abrupt vertical shear movements.

5.3 *Design for differential settlement*

Slope and grading design is developed to be compatible with the soft edge conditions. Slopes are designed to be suitable when differential settlements have reached their maximum estimated magnitudes. As necessary mitigation techniques such as surcharging, grouting, dynamic compaction, or even waste relocation may be used to meet project goals.

Building connections represent the most problematic hard edge issue to deal with. Access can be a particular problem and ultimate slopes must accommodate desired access to structures including vehicles, pedestrians, and the handicapped. Typically hinged slabs are used to for access connections, with the limiting design condition being angular rotation of the slab. Utilities must be built to remain serviceable while accommodating the differential settlements that can occur over a very short distance. Properly detailed flexible utility connections can mitigate the effects of the hard edge conditions.

Roads and pavements at closed landfill projects typically are constructed of flexible materials. Subgrade reinforcement such a geogrids or geotextiles which "bridge" softer or higher settlement areas can greatly improve service life of pavements, as can detailing of areas such as gutters and tie-ins to surrounding areas to reduce water infiltration which will weaken the pavement section. However, even with these mitigative design features, shorter pavement life can be expected from paved areas over landfills than from those on normal soil foundations.

5.4 *Define inspections and maintenance requirements*

Where economically possible it is important to design to minimize maintenance. However, some project owners will accept a higher level of required maintenance in the future to reduce initial construction costs. Inspection and monitoring must be considered part of on-going operations however. Typically the design team prepares an inspection and maintenance manual that stipulates procedures and schedules for these major inspection groups:

1. settlement monitoring;
2. slopes and pavements;
3. hinge slabs and access areas;
4. utility connections and vaults; and
5. utility piping.

Just as the design process must pay strict attention to details, so must inspections and maintenance.

6 Conclusions

Developments on closed landfill properties are not conventional projects. They represent unique economic opportunities, to improve and utilize what until recently was thought of as derelict land. However, successful development requires very careful design, construction and on-going future maintenance. The design team must work closely together and the process is typically one of evolution where different ideas are looked at in a variety of different ways, and the most appropriate solution is selected. The design process cannot be hurried and the costs are typically more than for conventual sites. In the end the result can be an aesthetically pleasing and successful post closure land use and enhanced environmental protection. The alternative to proper techniques is usually a project which is at least a failure aesthetically, and often an economic failure as well.

REFERENCES

Coduto, D.P. and Huitric, R. 1990. Monitoring landfill movements using precise instruments. Geotechnics of Waste Fills - Theory and Practice. American Society for Testing and Materials, STP 1070: 358-370.

Keech, M.A. 1992. Infrastructure and settlement design. Development of Closed Sanitary Landfills. California Integrated Waste Management Board.

Oweis,I.S. and Khera, R.P. 1990 Geotechnology of waste management. Butterworths.

Sheurs, R.E. and Khera, R.P. 1980. Stabilization of a sanitary landfill to support a highway. National Academy of Science. Transportation Research Record 754: 46-53

Tchobanoglous, G., Thiesen, H., and Vigil, S.A. 1993. Integrated solid waste management. McGraw Hill.

Waste Disposal by Landfill – GREEN'93, Sarsby (ed.) © 1995 Balkema, Rotterdam, ISBN 90 5410 356 6

Reclamation of industrial wasteland

D. D.Genske
Deutsche Montan Technologie, Baugrundinstitut, Essen, Germany
P.Noll
Entwicklungsagentur Östliches Ruhrgebiet, Bergkamen, Germany

ABSTRACT: Problems and strategies associated with the reclamation of derelict industrial sites and historic landfills are discussed. A surface sealing technique is introduced which in combination with hydraulic measures appears to be a cost effective remediation method. The sealing system chosen is a soil-geosynthetic compound structure with a geomembrane, a drain layer and a number of reinforcing geogrids. The efficiency of the sealing "sandwich" system is demonstrated in connection with a major remediation project in the German Ruhr District.

1 INTRODUCTION

The Ruhr District in Germany, being one of the oldest and largest mining regions in Europe, has been in a process of constant change over the last 40 years. Since the end of the second world war the coal industry has been declining. This in only partly due to the exploitation of the easy accessible coal resources. A major reason is the fall of the coal prices on the world market and the gradual increase of safety standards for coal mining in Germany. Besides the short but intensive coal boom during the oil crisis in the 70s it eventually became cheaper to buy coal from abroad rather than to mine it.

Because of this, the Ruhr District has had to cope with the economical problems associated with this much earlier than other regions in Germany. There has been a number of successful attempts to establish new industries on derelict mining sites, successful because of a close cooperation between the administrations involved, the participating municipalities, political parties, unions, scientists and economists. Successful because of the still intact infrastructure of the historically grown mining towns.

However, the initiation of new industries depends strongly on the availability of space. In former times the factories could expand without any significant restrictions. Today, on the other hand, the space available in the Ruhr District is rather limited. Since the cities have been growing together rapidly during the intensive phase of industrialization the Ruhr District has to be considered a megalopolis with an area of about 5000 km^2 and a population of some 5 million. Therefore, the one and only chance to gain space is to reclaim the areas formerly used by now obsolete industries. As a matter of fact in the Ruhr District every accessible site larger 10 acres is a derelict industrial site. 80% of this area used to be mining territory.

Yet, a high percentage of the reclaimable land must be considered as contaminated. A broad variety of contaminants leached into the ground, thus polluting the soil and groundwater. The remediation of the former mining sites in the Ruhr-

District and the reestablishment of alternative industries have now become both a challenge for city planners and a prestigious attribute for ambitious politicians. It has become the declared goal of the German government to convert the Ruhr District into the greenest industrial region in the world. One important piece in this campaign is to let the 'Internationale Bauaustellung' - an international city planning fair - take place in the Ruhr District, a plan which was settled in 1989. The first project will be presented to the public in 1994.

2 CONTAMINATION

The subsoil beneath a typical industrial wasteland in the Ruhr District can be divided into three main strata: an upper filling with foundation fragments, a middle stratum of quarternary sediments (mainly silty sands) and a fractured cretaceous marl as bedrock. The groundwater table is usually located at depths from 1 to 10 meters.

During the changing course of the history of the mining sites and former industrial complexes contaminants continuously leached into the ground. Some pollutants even migrated through the quarternary sediments into the fractured rock where they have been percolating through the joint network (fig 1). Furthermore, during the two world wars production of war equipment and missiles took place on the former mining sites. Beside the fact that a number of buildings were destroyed during the air raids, also documents concerning the function and location of facilities were lost. The destruction of the armament and mining facilities has certainly contributed to the contamination of the sites. Also, bombing craters as depicted in figure 2 were quickly filled with all kinds of material which were at hand at the moment. Nobody paid special attention to possible environmental hazard which might arise later from these fillings. Today we know that they rather often are serious sources of pollution. It is therefore fairly important to map all bomb impact craters and adjust the investigation program accordingly (DODT et al. 1993).

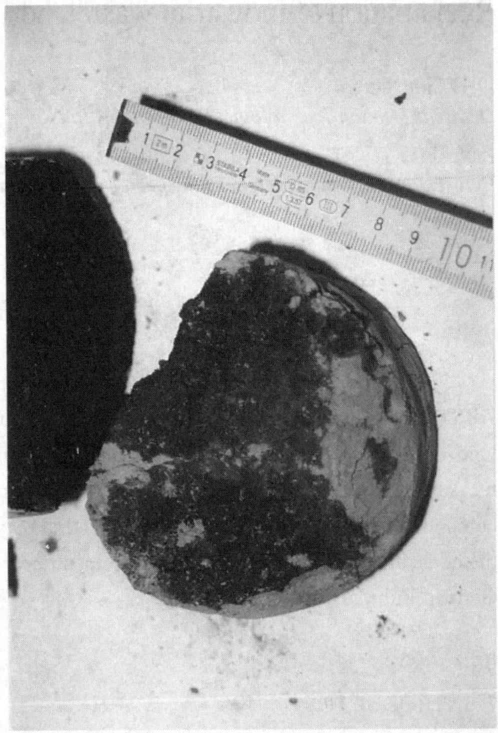

Fig. 1: Oil contamination on a joint in cretaceous marls 30 meters below the surface.

After the second world war the reconstruction of the German industry was pushed forward, especially in the Ruhr District, leaving no space for environmental thoughts whatsoever. More contaminants went into the ground making the situation worse. Today, city planners have to cope with serious pollution problems.

3 REMEDIATION

The key to an effective land re utilization plan is a harmonized management of ground investigation, risk assessment and clean-up strategies. A major

Fig. 2: An aerial photo from 1945 which shows the condition of a coal mine and coking plant shortly after an allied air raid (authorized research file, Ruhrkohle AG, Essen).

problem in this procedure is the processing of the vast amount of data: in the first step this refers to the historical material available such as former maps and aerial photographs of the factories established on the site under consideration. After the computer aided interpretation of this material a field campaign is laid out resulting in more data to be integrated into maps and 3D models. It is quite important that this presentation material must be understandable, to the point, and most of all: financiable. A strategy on how to synthesize the information available on a personal computer and produce appropriate presentation material in given for example in GENSKE et al (1992).

After the desk work and the field campaign a risk analysis concerning the contamination of the site is carried out. We will not go deeper into this point in this paper, we rather want to draw your attention to the reclamation of industrial wasteland such as derelict mining sites. A broad variety of remediation techniques have been proposed. It depends on a number of aspects which method is the most favorable one such as the future utilization of the site, the geological conditions, the grade and type of contamination, the time and money available to remediate the site. There are only a few methods, however, capable of both cleaning up the site and improving the ground conditions in a way that a redevelopment will be possible. Both aspects are of ample importance for the remediation activities in the Ruhr District as has been pointed out already in the introduction. The methods proposed so far can be divided into three main groups:

(1) replacement of the contaminated soil with clean coarse material,
(2) chemical, biological or thermal treatment on site or in situ (injections, mixing the soil with chemicals, etc.),
(3) sealing of the contaminated ground in combination with a soil improvement and hydraulic measures such as remediation wells.

Method 1 has the disadvantage that the contaminated soil will be difficult to find appropriate waste sites for. The second method has been proven to be quite expensive especially because the massive foundation fragments left in the ground obstruct tasks such as drilling, injecting or mixing (fig. 3). The third method is often favoured by environmental protection agencies because the excavated and contaminated soil is minimized. As a matter of fact, in Germany it has even become a legal guideline to avoid waste material as much as possible because of the lack of landfills available.

Fig 3: Massive foundation fragments and abandoned underground facilities are typical for industrial waste sites.

The surface sealing has to meet three tasks:

(1) it has to be waterproof to prevent the penetration of the precipitation into the contaminated ground,
(2) it has to be gasproof to stop the migration of toxic gas to the surface,
(3) it has to be stiff enough to allow a redevelopment, i.e. the construction of buildings and roads.

The third aspect accounts for the fairly inhomogeneous ground conditions. The massive fragments of the former foundations of the dismantled buildings have remained in the subground next to loose fillings, thus causing severe structural problems as

Fig. 4: The mine and coking plant "Graf Moltke" in 1926 with the shafts 3 and 4.

to the possible differential settlement of future structures.

4 THE "GRAF MOLTKE" SITE

The former mine and coking plant 'Graf Moltke 3/4' is located in the northwestern part of the Ruhr-District close to the city of Essen and covers an area of about 230.000 m^2. In 1873 the first shaft was excavated, three more followed within the next 30 years (fig. 4). A coking plant was built in 1903/04 followed by benzol and ammonia factories. Within the next 50 years additional industries were established, turning the site into a multiuse industrial complex. The coal production went up to 1 million tons per year in the 60s until the mine was closed down in 1971.

In the following years the Graf Moltke Mine became a typical industrial wasteland with all its negative attributes. The positive point, however, was the existing and still intact infrastructure of the surroundings and the immediate neighborhood of the site to the A2 Autobahn, one of the most frequented highways and an important east-west traffic connection which especially in the light of the German reunification has become a rather important aspect. It was therefore decided to remediate the site as a project of the 'Internationale Bauaustellung' and establish a high quality industrial park. The European Community provided the project with appropriate funding (European Fund for Regional Development EFRE) so that only about 50% of the remediation costs had to be supplied be the owners of the former mining site.

A serious hydrocarbon contamination has been detected in the vicinity of the former coking plant (fig. 5, upper graph). During the past decades the pollutants have migrated through the porous sediments into the fractured rock where they have been percolating through the joint network.

For the remediation of the "Graf Moltke" site it was required by the environmental protection agency to minimize all excavation. Therefore, the sealing and reinforcing technique was chosen for the heavily contaminated sectors (fig. 5, lower graph).

Fig. 6 depicts a composite soil - geosynthetic - system which was designed to satisfy all three objectives mentioned above. Basically, this sandwich system is composed of three elements:

(1) a lower supporting layer reinforced with geogrids,
(2) a drainage and sealing system, using geotextiles and geomembranes,
(3) and an upper reinforced foundation layer to account for structural loads.

Because the chosen technique is a rather innovating one it was decided to design and construct a 500 m^2 full scale test field on the site (fig 7). The aims of the field test were:

(1) to study the feasibility of using geosynthetics,
(2) to establish quality control tests and measures,
(3) to investigate possible damages during construction,
(4) and to create new design criteria for such structures.

The test field location was chosen at an area within the derelict mining site were the ground conditions were representative for the whole site. In order to monitor the stress-strain behavior of the two reinforced layers using geogrids as reinforcing elements

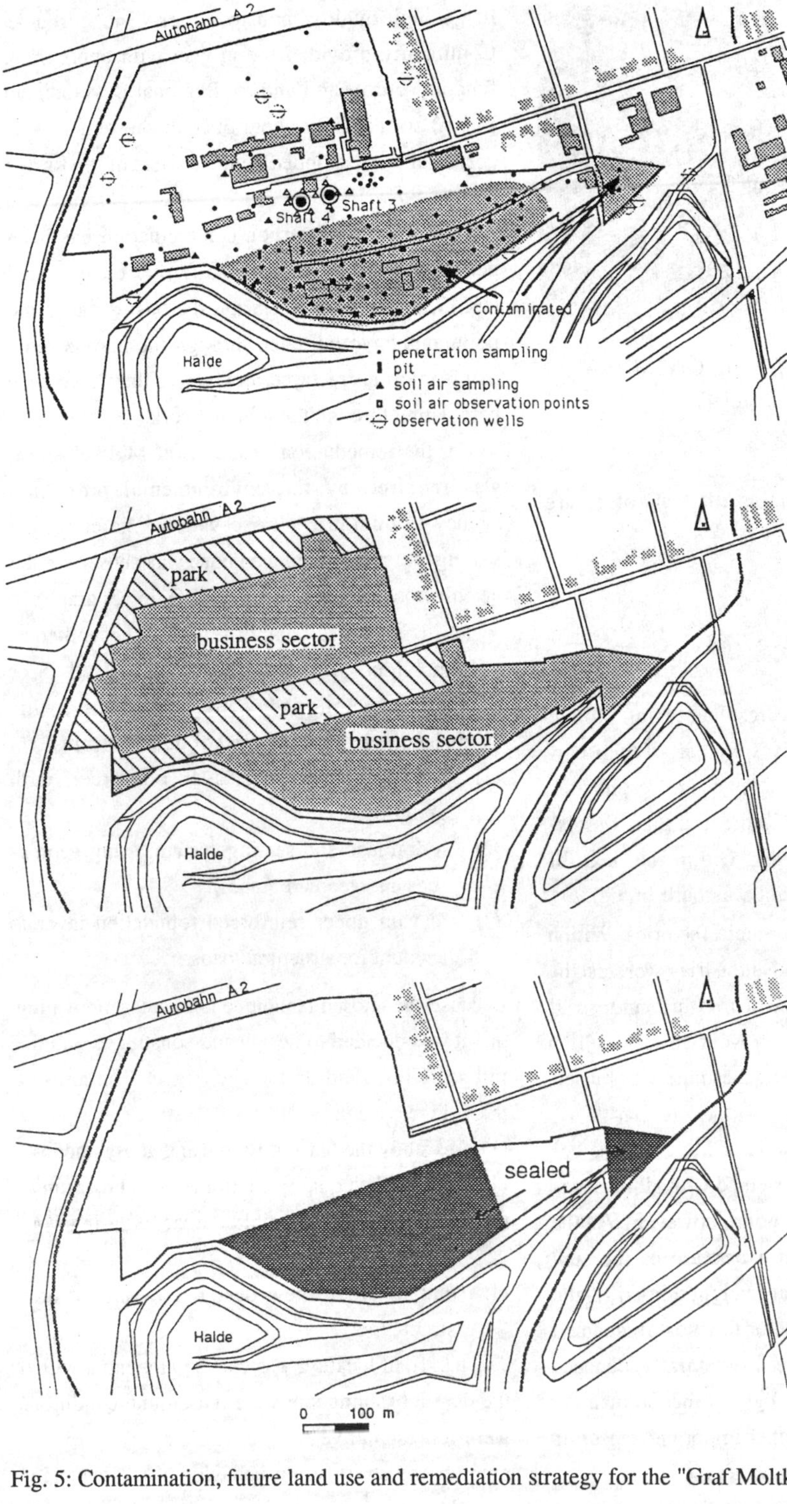

Fig. 5: Contamination, future land use and remediation strategy for the "Graf Moltke" site.

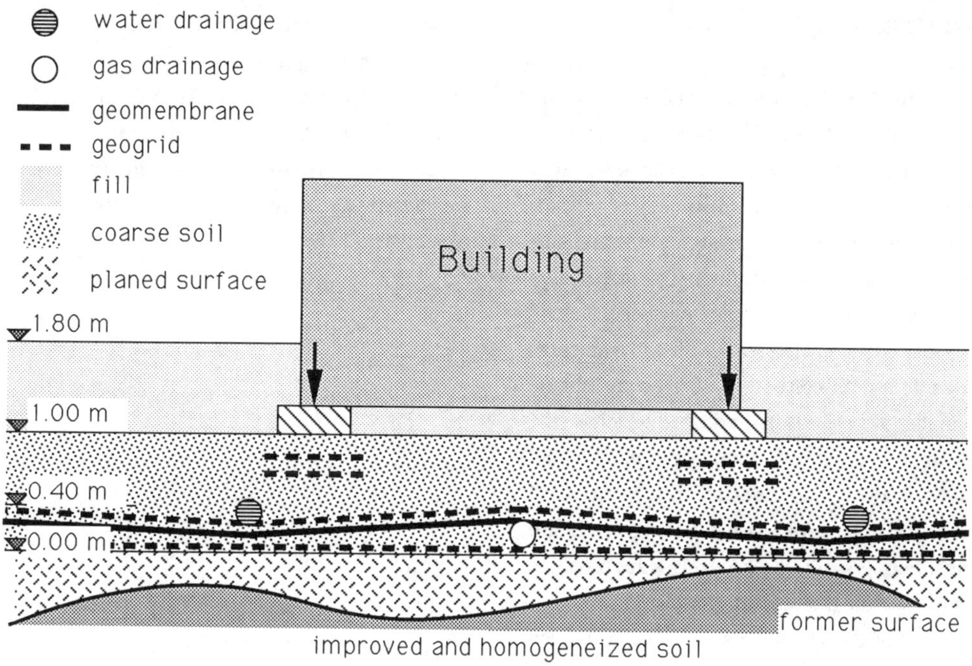

Fig. 6 A reinforced geotextile sandwich system. Besides the reinforcing geogrids a geomembrane is integrated.

Fig. 7: A test field to monitor the stress-strain behavior of the sandwich system.

Fig. 8: The loading test on the test site.

and the performance of the waterproofing geomembrane several measurement devices were placed within the test field. A loading test was carried out (as depicted fig. 8) to observe the stress-strain-behavior within the sandwich system with special regard to the deformation of the sealing geomembrane. Also, the absolute and relative settlement at the surface were measured. The results from this test field appear to be promising. They are commented in GENSKE et al. 1993a, 1993b).

In order to monitor the quality of the ground water a number of observation wells were installed, some of which may become recovery wells if a certain degree of contamination of the ground water is exceeded. The tolerable concentrations were negotiated with the environmental protection agency.

5 CONCLUSION

The reclamation of derelict industrial sites will be one of the most challenging tasks of the next decades. It is a problem where no general solution can be offered, no rule of thumb-strategy is at hand. It is a problem which calls for an interdisciplinary approach and for innovative ideas. This paper discusses one aspect of this complex business: a technique to seal contaminated sites with geosynthetics. A sandwich system consisting of a geomembrane and a number of geogrids is introduced which is stiff enough to compensate the inhomogeneities of the subsoil caused by foundation fragments and provides a sufficient foundation ground for future buildings. Furthermore, the geogrids help to reduce the strains within the sealing geomembrane. The effectiveness of the sandwich system was validated by means of a 500 m^2 test field on the site.

Since the sealing technique introduced minimizes the excavation of the contaminated soil and also allows a redevelopment of the site this remediation strategy appears to be both ecological compatible and cost effective.

REFERENCES

DODT, J., D. D. GENSKE, T. KAPPERNAGEL, P. NOLL 1993: Digital Evaluation of Contaminated Sites. *International Conference on Digital Image Processing: Techniques and Application in Civil Engineering*, February 28 - March 5, 1993, Kona, Hawaii.

GENSKE, D.D., H. KLAPPERICH, P. NOLL & B. THAMM 1993a. Surface Confinement Technique for Derelict Industrial Sites. *GEOCONFINE '93*, Montpellier 08-11.06.1993 (in press).

GENSKE D.D. , H. KLAPPERICH, P. NOLL & B 1993b. THAMM: Historic Landfills and Old Sites – Remediation and Design. *IV Int. Landfill Symposium*, S Margherita di Pula, Italy, 11-15 October 1993 (in press).

GENSKE D., GILLICH, M., H. KORIES & Ch. OLK 1992. Contaminated Sites- Data Processing, Visualization and Interpretation. *International Conference on Geotechnics and Computers*, 29.09.-01.10.1992, Paris.

Waste Disposal by Landfill – GREEN'93, Sarsby (ed.) © *1995 Balkema, Rotterdam, ISBN 90 5410 356 6*

Geotechnical characteristics of Zagreb waste disposal site and possibilities of its reclamation

D. Kovačić & D. Mayer
Faculty of Mining, Geology and Petroleum Engineering, Zagreb, Croatia

I. Muhovec
Faculty of Geotechnical Engineering, Varaždin, Croatia

ABSTRACT: The paper deals with the problem of Zagreb waste disposal site Jakuševac. For almost 30 years the uncontrolled open dump has been used. The composition and quantity of solid wastes are determined. The measures for the reclamation of the site are considered and the proposition for the construction of the sanitary landfill is given. The role of the slurry trench cut-off wall and the system of discharge wells is pointed out from the geotechnical point of view. The results of illustrative numerical analyses are presented.

1 INTRODUCTION

Zagreb has been faced with the problem of waste management for years. There is a number of uncontrolled waste disposal sites located at the edges of the city. All of them are typical unsanitary open dumps. The major site is Jakuševac. The site has been used since 1965 for disposal of municipal, hospital and partly industrial wastes.

Jakuševac is situated on the right bank of the Sava river in the south-east part of Zagreb. It occupies the area of about 600,000 m^2. The length of the site is cca 3 km and the width varies from one hundred meters to several hundred meters (fig. 1).

Fig. 1 Location of waste disposal site Jakuševac

The location of the site is extremely unfavourable as it intrudes into the city area. Besides it is located in the area which most probably has the influence to the existing and future well fields used by municipal water distribution system. Therefore it has been decided to stop the disposal of the wastes on the site and to start the remedial actions. These should minimize its negative influence to the environment and particularly to the groundwater.

2 COMPOSITION AND QUANTITY OF SOLID WASTES

The amounts of solid waste disposed daily at Jakuševac site is about 900 t. The composition of unprocessed waste is shown on Table 1. (Tušar et al. 1991).

Table 1. Composition of municipal solid waste in Zagreb (after Tušar et al. 1991)

Constituent	Composition (% by weight)
Construction and demolition debris	28.2
Organic waste	23.3
Paper	20.5
Glass	7.3
Soft plastics	6.6
Hard plastics	4.2
Textile	3.8
Iron	2.4
Leather and bones	2.2
Wood	1.1
Rubber	0.4

The chemical composition shows that the waste is made of 75.8% of mineral matter and 24.2% of organic matter. The chemical composition is particularly important due to its influence on the ground water. The amount of trace element and heavy metals in the deposited waste is shown on Table 2.

Table 2. Content of trace element and heavy metals in dry matter of municipal solid waste in Zagreb (after Tušar et al. 1991)

Element	Content in mg/kg dry matter
Manganese	678
Zinc	512
Lead	415
Cooper	251
Mercury	52.4
Cadmium	40
Arsenic	30
Chromium	23.1
Molybdenum	14.3
Selenium	10
Cobalt	7.9
Boron	2.8

3 HYDROGEOLOGICAL PROPERTIES OF THE SITE

Jakuševac waste disposal site is located on the area composed of Quaternary alluvial sediments forming the Zagreb aquifer. The thickness of the aquifer on the disposal site varies between 20 and 40 m. The value of the transmissivity coefficient varies between 1.5 x 10^{-1} to 3 x 10^{-1} m^2/s and the average coefficient of hydraulic conductivity is 1 x 10^{-2} m/s.

This is an unconfined aquifer, because a semipervious surface cover on the broader area was eroded by Sava river before its regulation, and it was partly eliminated due to the previous dredging of gravel (Mayer, 1980).

At the moment there is no reliable information about a confining lower bed. It can be said that an underlying impervious or semipervious stratum could be found at the depth of cca 30-50 m. The depth of this layer is not defined.

4 INFLUENCE TO THE ENVIRONMENT

The fact that dwelling houses are at the distance of not more than 200 m away from the site presents a local impact of the waste disposal to the environment. But on the regional scale, Jakuševac disposal site represents a serious source of pollution located in the area with especially inappropriate hydrogeologic characteristics considering groundwater protection.

However, negative influence of the waste disposal site up to now, on the groundwater quality are considerably smaller than expected. The in-situ tests proved a big groundwater pollution in the underlying strata of the site as well as on the area of Jakuševac village. There is considerably less pollution on the Mičevec area. In the zone of the future well field Črnkovec, about 2 km south-eastward from the site, there is no pollution evidence at all. The explanation of such condition can be found in the hydrogeological study (Mayer & Markovac, 1992), but the main raison appears to be rather low groundwater level in the area in relation to the bottom of the exiting dump (Pletikapić & Hubeny, 1991).

The danger is particularly increased by the planned hydrotechnical operations in the zone of Jakuševac. Namely, in the near future, the already mentioned well field Črnkovec with the capacity of several m^3/s is supposed to be put into operation. The construction of the hydroelectric power plant reservoir HE Drenje on the Sava river is also planned. These activities will essentially change the groundwater dynamics and they will disturb the processes which confine spreading of the groundwater pollution today.

It is therefore necessary, before starting to build the mentioned structures, to stop further disposal of waste on Jakuševac site and to perform its treatment.

5 GEOTECHNICAL ASPECTS OF THE RECLAMATION

5.1 *Site considerations*

In order to immobilize permanently existing pollution in the zone where it is present today, several options are discussed (Pletikapić & Hubeny, 1991; Lipić et al. 1991). They are classified within three main groups:

- the waste remain on the same location,
- incineration,
- removal of the waste to a new sanitary landfill.

In case the waste should remain on the actual site, remedy actions are a must. Technical solution of the problem basically consists of protection measures above the ground and protection measures below the ground level (fig. 2).

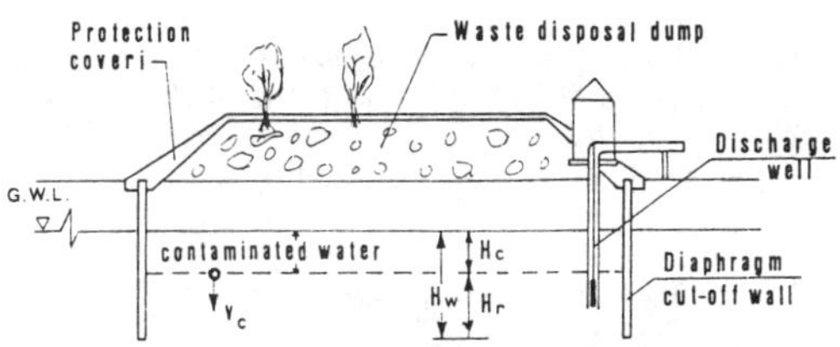

Fig. 2 Schematic cross section of sanitary landfill (not to scale)

The protection above the ground could be achieved by means of an impermeable soil cover. The surface layer should provide a suitable medium for plant growth.

In order to separate the waste from the groundwater an impermeable barrier should be constructed in the ground. As no bottom stratum of low permeability is available at reasonable depth, a feasible solution seems to be a combination of the slurry trench cut-off wall (diaphragm cut-off wall) and a system of discharge wells.

The system of wells should operate in the "protection groundwater stratum" $H_w = H_c + H_r$. As the "contaminated groundwater stratum" H_c will gradually increase due to the leachate production from the landfill, the depth of the "reserve stratum" H_r will decrease. The role of the discharge wells is the periodical pumping of the contaminated water which means the reduction of H_c and increase of H_r.

An automatic and continuous performance monitoring should be provided in order to control the position and quality of groundwater inside and outside the landfill.

5.2 *Cut-off techniques*

The above mentioned type of cut-off wall has no structural requirement but should be able to adjust itself to the settlements of surrounding ground. It can be constructed of unreinforced (plastic) concrete, i.e. a mixture of sand, gravel, cement and bentonite.

Diaphragm cut-off walls are usually built to prevent water movement, such as under embankment dams, along river banks, surrounding open pit mines or construction sites.

In the last twenty years slurry trench cut-off walls have become the principle type of barrier to mitigate pollution of groundwater, especially when a barrier is used as a remedial measure. Several situation in the USA where cut-off walls were used are listed by Ressi & Cavalli (1985).

The first experience in the design and construction of slurry trench cut-off walls in this part of Europe goes back to early 1970s. Four cut-off walls were made for the protection of the structures within the hydroelectric power plant HE SD2 on the Drava river near Ptuj, Slovenia (Muhovec, 1974). The components of the backfill were: 75 kg of cement, 1940 kg of gravel, 11.25 kg of bentonite and 240 l of water, for 1 m^3 of the mixture. The mechanical characteristics of the mixture were: Young modulus 3500 MN/m^2, uni-axial compression strength 1.4 MN/m^2, bending compression strength 240 kN/m^2. The mixture had coefficient of permeability of 3×10^{-9} m/s.

Similar backfill was used in the case of the cut-off wall for the protection of the excavation for the reactor of the nuclear power plant in Krško, Slovenia (Muhovec, 1975).

A new approach in cut-off technology took place in late 1960s. In this technique a self-hardening slurry is applied instead of concrete- bentonite mixture. The operations of excavation and backfilling are thus combined. The slurry usually consists of bentonite-cement with a retarding agent so that setting does not occur too quickly. The original development of the self-hardening slurries appears to have been made on tests carried out by Bush and Bares (1968) following the construction of the Arrow Dam on the Columbia River (Canada). It is interesting to point out that " ... the investigations were performed as personal projects rather than work assigned by the owners or his consultants ... ".

The first application of self-hardening backfill in this part of Europe is documented by Grubić et al. (1985). Slurry trench cut-off wall was constructed for the protection of an open pit mine at Drmno, near Kostolac (Serbia). An extensive laboratory testing was undertaken in order to find an optimal composition of cement-bentonite mixture.

The results of permeability tests gave a range of values for k between 8 x 10^{-8} m/s to 7 x 10^{-10} m/s.

The values of Young modulus obtained from a series of consolidated undrained triaxial tests were in the range between 15 MN/m^2 to 120 MN/m^2.

The appropriate characteristics of the specimen's compressive strength obtained from the same tests are presented on fig. 3. It can be seen that the decrease of the bentonite/cement ratio yields higher value of the vertical stress at failure σ_{1f}. Also, the influence of the confining pressure σ_3 on σ_{1f} is demonstrated.

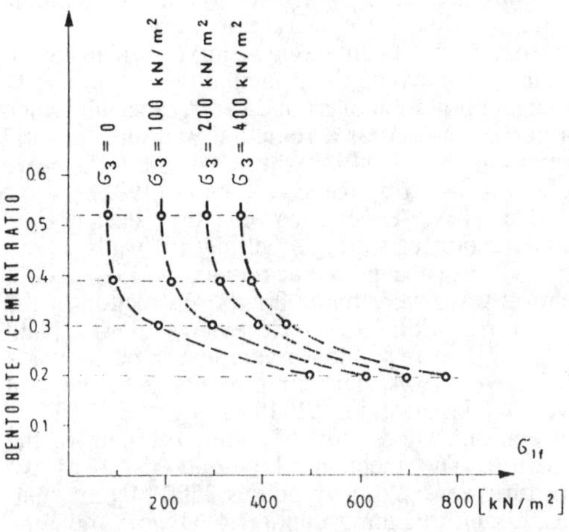

Fig. 3 Relationship between compressive strength and bentonite-cement ratio (after Grubić et al. 1985)

Brandl (1990) made a comparison between two techniques mentioned above. He also introduced the term "two-stages technique" for the method with classical backfill and "one-stage technique" for the method with self-hardening backfill.

In either case the design of the future cut-off wall on Jakuševac site should assess the effects of the existing contaminants in the ground on the properties of the wall. Laboratory experiments for determination of the behaviour of such walls under environmentally hostile conditions are described by Moura et al. (1991).

6 NUMERICAL MODELS

Two illustrative numerical analyses are presented. The aim of these analyses is to gain an understanding of the changes in porewater pressures as well as stresses in the soil and in the cut-off wall due to envisaged actions on Jakuševac site. Some of the results are presented hereafter.

6.1 *Steady state seepage analysis*

Two dimensional steady seepage conditions are assumed and finite element program FEDAR (Taylor & Brown 1967) is used. The following input data are assumed: the depth of the cut-off wall is 25 m and the capacity of the wells is 0.45 $m^3/s/m$. For the case of isotropic permeability the groundwater level inside the wall is lowered by the amount of cca 1 m (fig. 4)

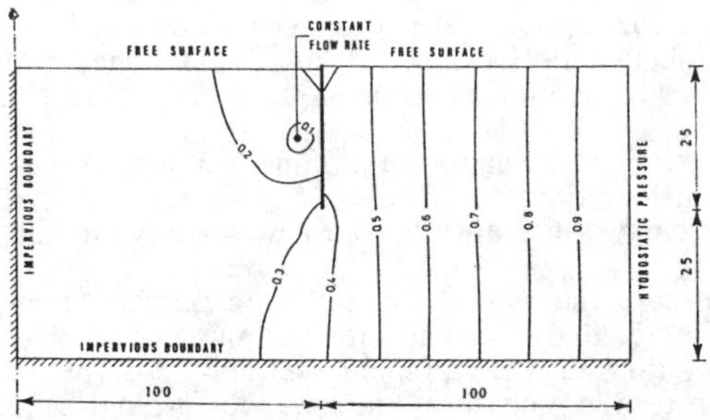

Fig. 4 Equipotential lines in the case of isotropic permeability

The influence of anisotropic permeability is modelled with a hypothetic ratio k_h/k_v = 10. Taking the same discharge ratio the groundwater level is lowered by cca 20 cm (fig. 5).

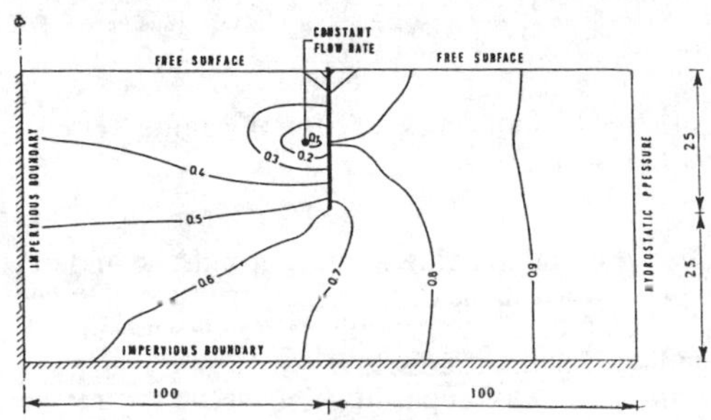

Fig. 5 Equipotential lines in the case of anisotropic permeability

6.2 *Stress-deformation analysis of the cut-off wall*

Plane strain, drained loading conditions are assumed and finite difference program FLAC (1991) is used. The following input data are assumed: the height of the landfill is 15 m, the density of the compacted waste is 700 kg/m^3, the thickness of the wall is 0.6 m, elastic constants of the soil are E_s = 30 MN/m^2 and ν_s = 0.3. Two types of cut-off technology are considered: classical backfill with typical value of E_b = 1000 MN/m^2 and self-hardening backfill with typical value of E_b = 100 MN/m^2.

The diagram of the horizontal displacements of the wall, due to surface pressure inside the landfill, is presented on fig. 6 a). The values are practically the same for both analyses. The distribution of bending moments for the case of self-hardening backfill is given on fig. 6 b). The same pattern but approximately ten times higher values are obtained for the case of classical backfill.

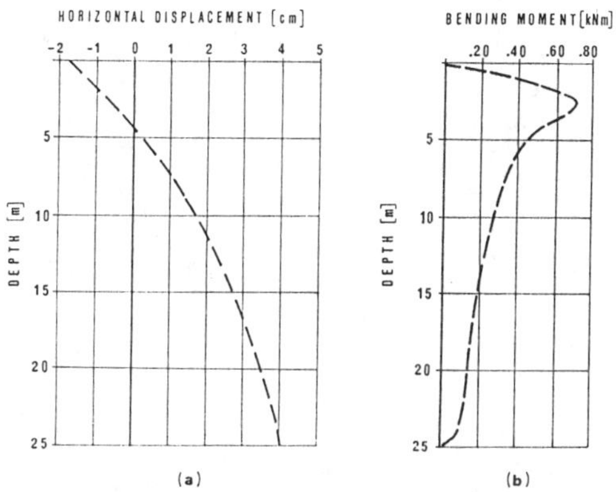

Fig. 6 Results of the stress-deformation analysis

In the case of classical backfill the influence of bending moment and axial force yields no tension stress and maximum compressive stress cca. 200 kN/m^2 in the critical cross section of the wall. In the case of self-hardening backfill the stress distribution in the same cross section is more favourable.

7 CONCLUSION

One of the most pressing problem facing Zagreb and its urban environment is the efficient and long term disposal of municipal solid waste. On the basis of data concerning composition and quantity of solid wastes disposed on the site, a number of solutions have been proposed in a recent study. A possible solution of the problem is the construction of sanitary landfill. The main components of the protection system are slurry trench cut-off wall and group of discharge wells. The role of the cut-off wall and the group of discharge wells is evaluated. The results of numerical analyses support the proposed technical solution.

REFERENCES

Brandl, H. 1990. Geomembranes for vertical waste containment sealing. In Den Hoedt (ed.), *Geotextiles, Geomembranes and Related Products,* p. 511-516. Rotterdam, Balkema.

Bush, E. & F. Bares 1968. Discussion on Proc. Paper 5231. *Proc. ASCE 94.* No. SM3: 578-581.

Grubić, N., S. Prokopović, M. Torbica, I. Sorić, D. Kovačić 1985. Primjena samoskrutive suspenzije za izvedbu vododržive dijafragme. *Saopćenja XVII kongresa JUDIMK.* Portorož.2: 243-252 (in Croatian).

Lipić, Z., Z. Pletikapić & G. Hubeny (1991). Pregled varijanata rješenja problema smetlišta Jakuševac. *Zagrebačka vodoprivreda* 10. No. 38: 38 - 42 (in Croatian).

Mayer, D. 1980. Mogućnost zagađivanja vodonosnih slojeva kao posljedica hidrodinamičkih značajki na području užeg porječja Save u SR Hrvatskoj. Disertacija. Sveučilište u Zagrebu (in Croatian).

Mayer, D. & Z. Markovac 1992. Hidrogeologija područja smetlišta Jakuševac (Zagreb), *Rudarsko-geološko-naftni zbornik* 4: 15-21. Rudarsko-geološko-naftni fakultet, Zagreb (in Croatian).

Moura, M.L., F.A. Weststrate & M. Loxham 1991. Effects of contaminants on the properties of cement-bentonite cut-off walls. *Proceedings of the International Conference on Environmental Pollution.* Lisbon. p. 322-325. Geneva. Interscience Enterprises Ltd.

Muhovec, I. 1974. HE SD2 - Dijafragme. Projektna dokumentacija. Institut Geoexpert, Zagreb (in Croatian, unpublished).

Muhovec, I. 1975. NE Krško - Zaštita građevne jame. Projektna dokumentacija. Institut Geoexpert, Zagreb (in Croatian, unpublished).

Pletikapić, Z. & G. Hubeny 1991. Pristup cjelovitom rješavanju zatvaranja smetlišta Jakuševac, *Zagrebačka vodoprivreda* 10. No. 37: 11- 13 (in Croatian)

Ressi, A. & N. Cavalli 1985. Bentonite slurry trenches. *Engineering Geology* 21: 333-339.

Taylor, R.C. & C.B. Brown 1967. Darcy flow solution with a free surface. *Proc. ASCE* 94. No. HY2: 25-33.

Tušar, B. et al. 1991. Studija utjecaja na okolinu sanitarnih deponija komunalnog otpada na širem području Zagreba. Arhiva građevinskog fakulteta, Zagreb (in Croatian).

- 1991. FLAC (Fast Lagrangian Analysis of Continua, Version 3.0, *User's Manual,* Itasca Consulting Group, Inc., Minneapolis, Minnesota.

Waste Disposal by Landfill – GREEN'93, Sarsby (ed.) © 1995 Balkema, Rotterdam, ISBN 90 5410 356 6

Geophysics and computer models in the evaluation of a landfill site in Blaby, Leicester

M.A. Meju
Department of Geology, University of Leicester, UK

ABSTRACT: Non-invasive geophysical (electrical and transient electromagnetic) methods have been deployed along and across the main axis of a backfilled excavation that now serves as a 100m × 200m recreational ground surrounded by residential houses on the outskirt of Leicester City in order to assess their usefulness in mapping such concealed features and any possible accumulations of polluted groundwater in the subsurface. The survey data show that the responses obtained in the filled areas are markedly different from those of the unfilled ground. Quantitative modelling of the processed data shows that the fill may be characterised by their resistivity structure and yielded estimates of fill thicknesses that are in good agreement with borehole information; and more importantly, identify a possible contaminated reservoir system not penetrated by the boreholes sunk in a previous investigation. The results demonstrate the need for pre-drilling geophysical surveys and computer modelling studies in landfill investigations.

1 INTRODUCTION

Geophysical imaging of the subsurface is a cost-effective way of obtaining very useful information required for engineering planning and remediation measures in contaminated land investigations. Of the several methods of geophysical exploration , the electrical and electromagnetic methods are the most popular owing to their inherent ability to detect changes related to fluid variations in the ground and the minimum capital and labour outlay required to use them in small scale surveys. The resistivity (or its reciprocal, conductivity) of a material is a measure of its ability to impede (or pass) the flow of electric currents through it. Earth materials have differing abilities to pass current flow and may thus be distinguished by measuring their resistivities. Note, however, that non-crystalline subsurface materials as in landfills are typically of various stages of aggregation and may thus have voids or pore spaces filled by fluids which may be of different elemental conce-

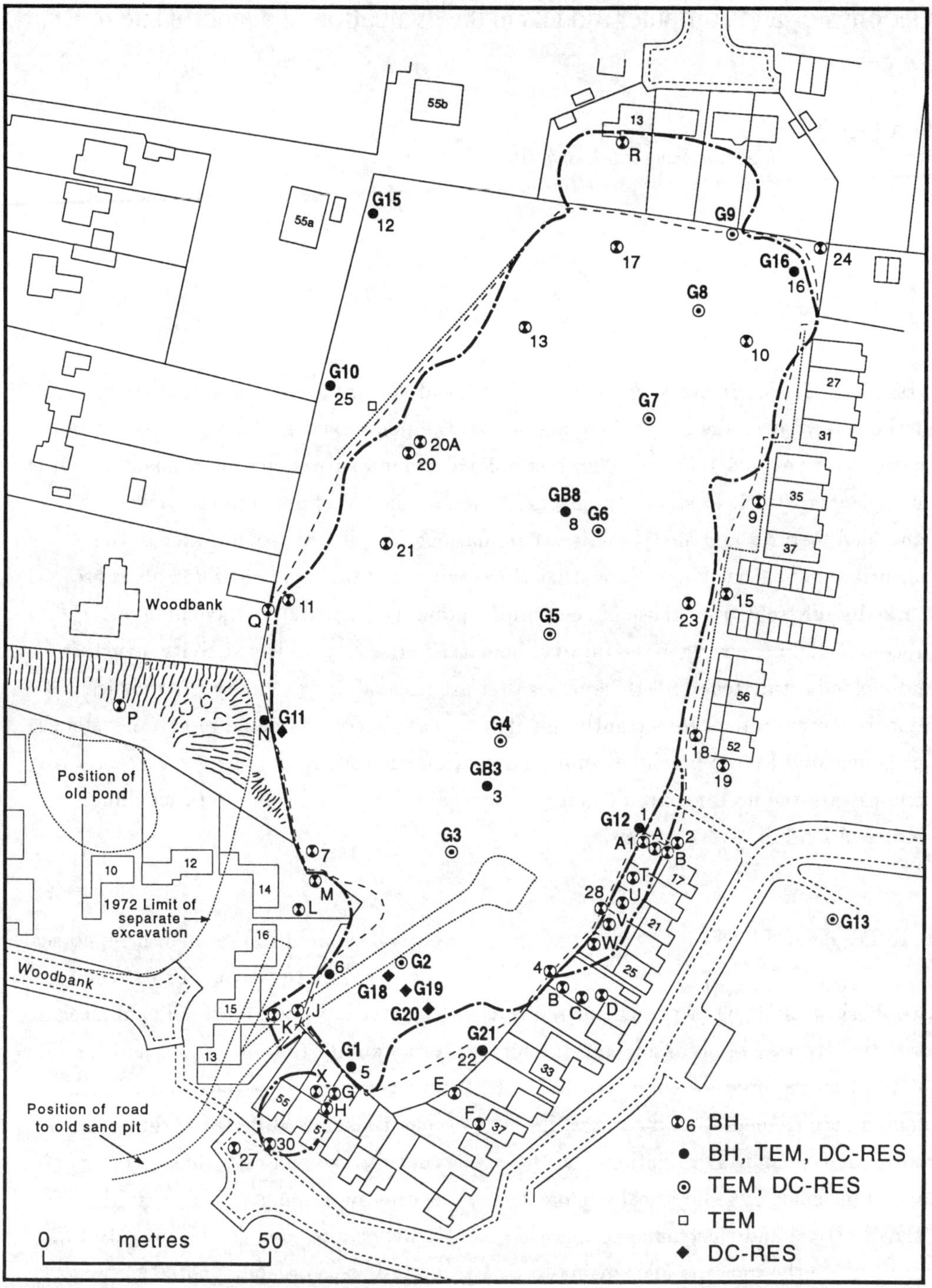

Fig. 1 Site Location Map

ntrations. Such pore spaces, when interconnected, often serve as the locale of groundwater and the material is said to be permeable to water flow and depending on its size constitutes an aquifer (i.e., a water storage system). An aquifer may contain fresh groundwater or saline fluids. The presence of saline fluids in the ground enhances its ability to conduct current. It is thus possible to locate any contaminant plume in the subsurface by measuring ground resistivity. It is stressed that a host of other factors (e.g., chemical composition and temperature) also affects the resistivity of a material and should be borne in mind when analysing field data.

The two principal ground resistivity measurement techniques employed in landfill investigations are the direct current (dc) resistivity and transient electromagnetic (TEM) methods and the underlying principles can be found in Parasnis (1986). For qualitative site characterisations and for modelling purposes, the field recordings are usually converted into apparent resistivities (since the ground is typically inhomogeneous and we can only measure directly the exact resistivity of a homogeneous ground). Resistivity data modelling involves finding a model (whose parameters are the resistivities and thicknesses of the constituent units) that explains the observed data and several techniques have been developed for this quantitative aspect of survey data analysis (e.g., Meju, 1992a,1993a,b). In this paper, it is shown that geophysical data can be used to adequately characterise a site by delineating the fill extent, its cover and groundwater conditions and the fill-induced resistivity alteration patterns of the subjacent geology. Since geophysical imaging is a cheap rapid process, it has economic implications and may be an alternative or complimentary method to time-consuming expensive borehole exploration in some geotechnical site investigations.

2 SITE DESCRIPTION, PREVIOUS WORK AND PROBLEM DEFINITION.

The landfill site is located about 12km southwest of the centre of Leicester City and is approximately 200m long in the north-south direction but the width varies from about 32m at the southern end to a maximum of approximately 100m at about 75m from the southern end (see Figure 1). The ground surface at this site is generally even but slopes towards the south. The regional geological stratification consists of Pleistocene glacial deposits (clays, sands and gravels) underlain by Triassic Mercia Mudstone deposits. From the available historical records, the glacial deposits have been exploited in the past and the site comprises several excavations that were backfilled between 1972 and 1982.
The site is surrounded by residential houses and serves as a recreational ground for the residents. One problem at this site is that significant levels of hazardous gases (methane and carbon dioxide) have been noted in some of the adjoining properties.

The Leicester City Council (LCC) promptly commissioned a gas monitoring investigation at the site in 1992 and the results were used to recommend remedial protective measures to be installed to stop any gas migration into the adjoining properties. Thirty-one 150mm diameter boreholes were drilled using light cable percussion techniques to a maximum depth of 11m (additional twenty-five 100mm diameter boreholes were drilled to a maximum depth of 2.3m using hand auger techniques) to monitor the gas levels in the ground by a firm of consulting engineers. The borehole distribution is also shown in Fig. 1. In general, the borehole data show that the fill materials vary over the site but are mostly top soil with brick and/or concrete fragments, plastic, paper, sacking, metal, rotting vegetation, bones, tyres and firm to soft clayey materials. Groundwater seepages were also observed in some of the boreholes (BH 3, 4, 8, 10, 13, 21, 22, and 23) in the main site. However, many of the hand dug (augered) holes were abandoned on obstruction so that some of them were too shallow to allow a detailed picture of the subsurface to be constructed. This is a major limitation of the drilling project. While boreholes provide excellent direct information about the ground, the resulting model of the subsurface may not be unique if the borehole data are insufficient and incomplete. Exploratory boreholes are not cheap and the need for a more cost-effective approach to contaminated land assessments leads one to ask if some comparable or even more complete information about the subsurface can be obtained from such constrained sites using cheap and efficient non-invasive means. The search for an answer to the above question forms the main theme of the present studies and a general solution may have implications for the optimum number of confirmatory boreholes and depth of investigation required to solve a given problem.

3 GEOPHYSICAL SITE SURVEYS

3.1 Grid layout

The grid layout for these surveys is shown in Fig. 1. The surveys consisted of a main north-south line running through the main axis of the playground with observational stations located every 25m starting from G1 (i.e., BH5) in the south to G9. Several other locations (including G13 outside the main site) were occupied using wider but irregular spacings allowing the construction of east-west profiles (G11-GB3-G12-G13 and G10-G5-G13) so as to obtain a better overall assessment of fill distribution as shown in Fig. 1.

3.2 Data acquisition and processing

The ABEM SAS300 Terrameter was employed in the dc resistivity surveys and the Offset-Wenner soundings utilised multicore cables (Barker, 1981). The expansion axis of the electrodes was appro-

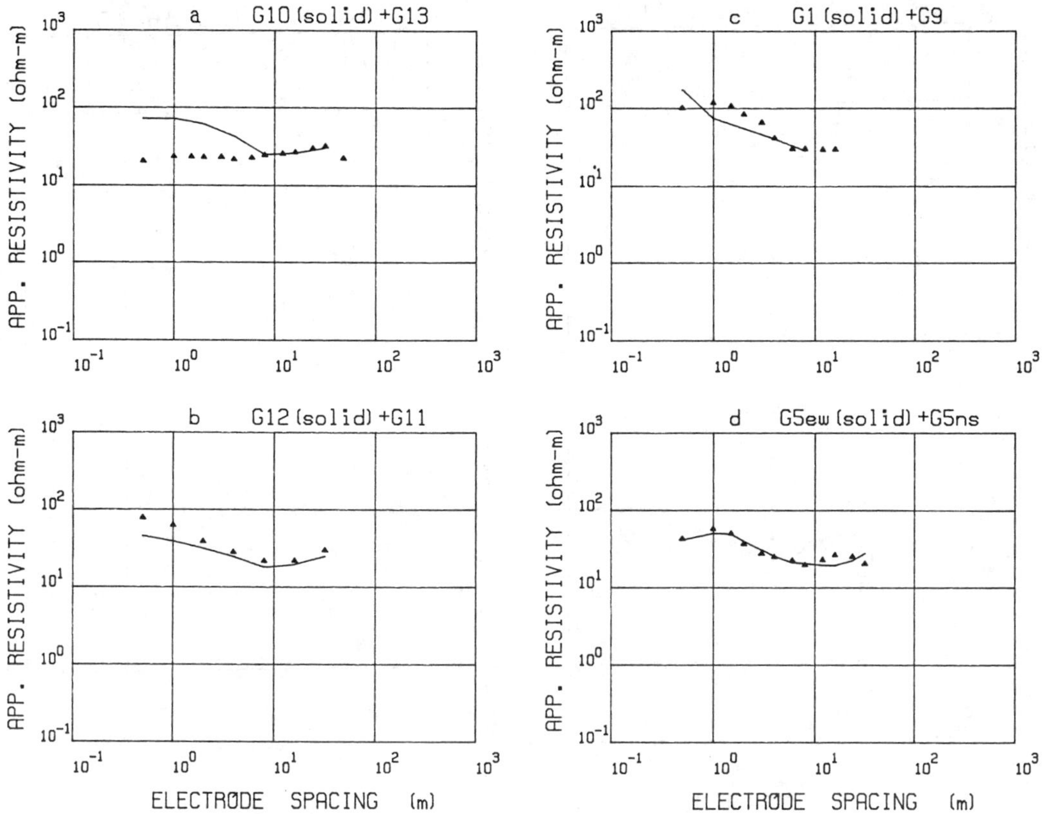

Fig. 2 Comparisons of offset Wenner sounding curves from the landfill site

ximately east-west for the main line and north-south for the cross-lines and the maximum spread length (and consequently, the depth of investigation) was constrained by the small size and shape of the site. Note that this limitation faced by the dc resistivity methods is not a major constraint for the TEM method.

The Sirotem Mk2SE system was used for the TEM surveys in the coincident loop geometry (i.e., with a square loop of 25m side-length serving both as the transmitter and receiver and centred on the dc resistivity central positions). The observational bandwidth was $0.0049msec$ to about $20msec$. The transmitter turn-off time was typically $54\mu sec$ for the $25m^2$ loops. This particular system images the deep section and was deployed to compliment the shallow dc images.

The various data sets were processed using standard techniques. The final results from these operations are apparent resistivities of the subsurface as functions of electrode spacing for the dc surveys and of time-windows for the TEM surveys.

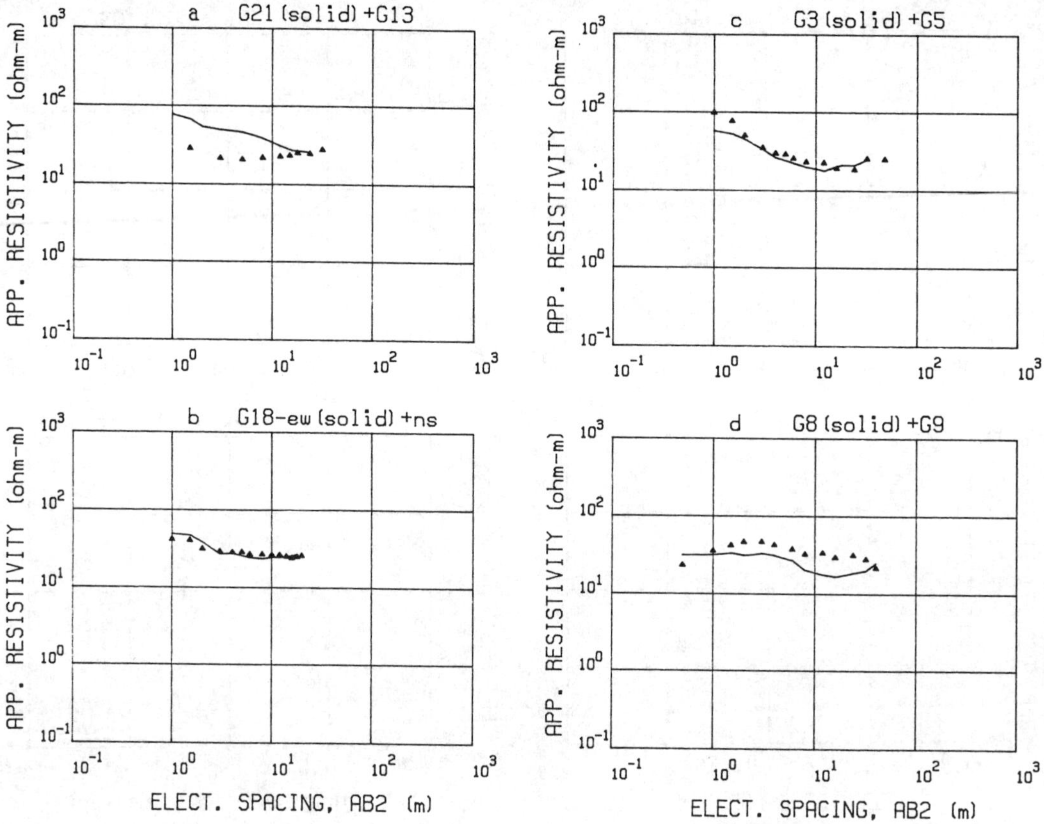

Fig. 3 Comparisons of Schlumberger sounding curves from the landfill site

4 QUALITATIVE SITE CHARACTERISATION

The usefulness of geophysical methods in landfill investigations may be gauged by their ability to discriminate between filled and unfilled ground. Sample results of the offset wenner, schlumberger and TEM surveys are shown respectively in Figs. 2, 3 and 4. Stations G10 and G21 are located outside the main areas of excavation and have only minor back-fill (as revealed by the borehole records provided by LCC) and their geophysical responses are shown together with those of G13 in Figs. 2a and 3a for comparison. The major differences at small electrode spacings (corresponding to shallow depths) and the concordance at large spacings are obvious. In Fig. 4a are shown the TEM responses for G13 and G5 (located near the centre of the main excavation). Note again the dramatic differences between the two sounding curves suggesting major resistivity alterations at depth underneath station G5. Overall, the above comparisons would seem to indicate that the backfilled excavation and the undisturbed land may be distinguished by their

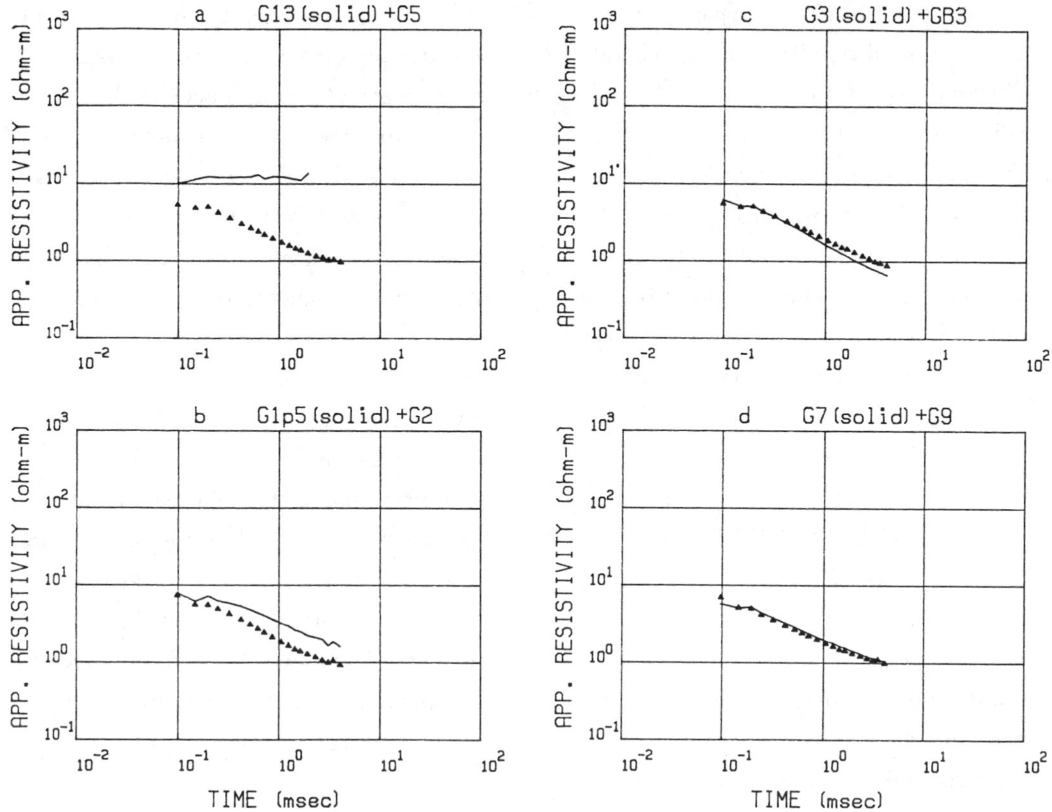

Fig. 4 Comparisons of TEM sounding curves from landfill site

resistivity characteristics. It is also obvious from Figs 4a to 4d that the whole area underneath the landfill site (and possibly beyond G9) have experienced significant resistivity alterations at depth. The degree of alteration appears to be highest underneath G3 (Fig. 4c) and noticeably less south of G2 (see Fig. 4b).

Stations G12 and G11 are located at the eastern and western margins of the excavation and the Wenner responses obtained are similar (see Fig. 2b). The responses at G1 and G9 (representing the south and north margins of the main site) are given in Fig. 2c; note the interesting structures at shallow levels which probably reflect differences in fill type and cover thickness in accord with borehole data. The Schlumberger data shown in Fig. 3c suggest similar ground conditions at G3 and G5 while it could be inferred that the ground is more disturbed at station G8 than at G9 (Fig. 3d). It is also possible to gauge the azimuthal variability of fill in the vicinity of any station by making measurements in more that one direction (e.g., north-south and east-west) as shown in Fig. 2d and Fig. 3b. These azimuthal soundings (cf.

Carpenter et al., 1991) appear to suggest similar ground conditions in the vicinity of the respective stations.

It is clear from the foregoing analysis that the landfill site can be characterised on the basis of resistivity response patterns. Let us now examine the problem of deriving useful quantitative information from the recorded responses since they may serve to guide any remediation work in such environments.

5 NUMERICAL MODELLING INVESTIGATIONS

The interpretation of resistivity depth-sounding data is beset by the problem of non-uniqueness since a number of completely different subsurface resistivity structures may adequately fit a given set of data. The usual approach to reducing non-uniqueness in resistivity data modelling is to incorporate *a priori* information in the solution process (e.g., Meju, 1993b). Typical *a priori* data may be those derived from existing boreholes or other previous investigations at the same site. In the absence of reliable *a priori* information, the common practice is to estimate the optimum values of the parameters of a hypothetical subsurface structural model that will satisfy the field observations. For a horizontally stratified ground, say, the optimum number of layers may be guessed from the shape of the sounding curve and the values of the layer parameters (i.e., thicknesses and resistivities) may then be sought using an iterative algorithm (e.g., Meju, 1992a).

A drawback of this approach is that there is no guarantee that the chosen number of parameters is truely representative of the subsurface resistivity structure and in recent times it has become fashionable to construct very smooth densely-parameterised models instead (see Meju, 1993c and references therein). An interesting development in resistivity data analysis is based on a simple datum-by-datum transformation of the field records into a resistivity-versus-depth image of the true subsurface model (e.g., Meju, 1992b). Sample applications of two of the data modelling techniques highlighted above to the apparent resistivity data from the landfill site will now be presented.

5.1 Application of direct-depth resistivity transformations and forward models

Stations G1, GB3 and GB8 are respectively centred on Boreholes 5, 3 and 8 (see Fig.1) and the results of depth transformation of the observed data at these stations are shown in Figs. 5, 6 and 7. In each figure, the resistivity-depth transformations are plotted (with error crosses) on the right-hand panel together with the borehole depth-locations of changes in fill types (denoted by numbers) and positions where groundwater (GW) seepages were encountered; the end of hole (EOH) positions are also indicated for comparison with the imaging depth attained after only a few

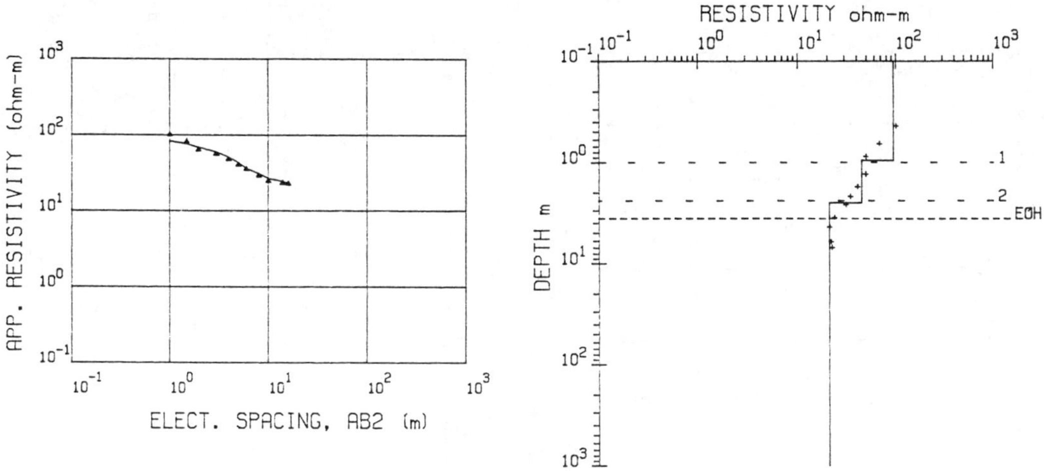

Fig. 5 1-D MODELS FOR SITE SLBH5

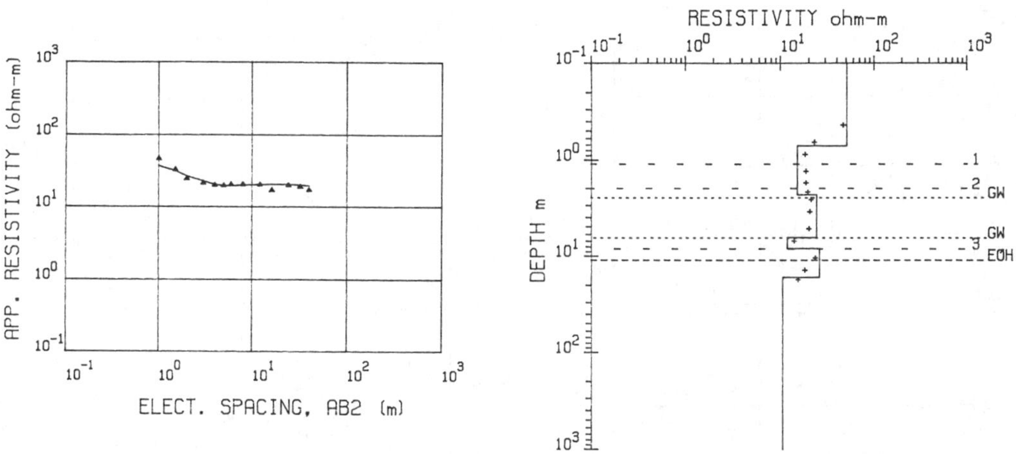

Fig. 6 1-D MODELS FOR SITE SLBH3

minutes of geophysical occupation. The computed theoretical responses of the interpretive layered-ground models for the imaging results are plotted together with the field data for comparison on the left-hand panels in all the figures in order that we can gauge the effectiveness of such data transformations. Notice the good agreement between the borehole information and the imaging results.

The general resistivity structure at this landfill site consists of a highly resistive cap whose thickness varies from 0.35m to 1m and an underlying less resistive fill that varies in thickness from 1.5m to 9m and whose basal part may be marked by a

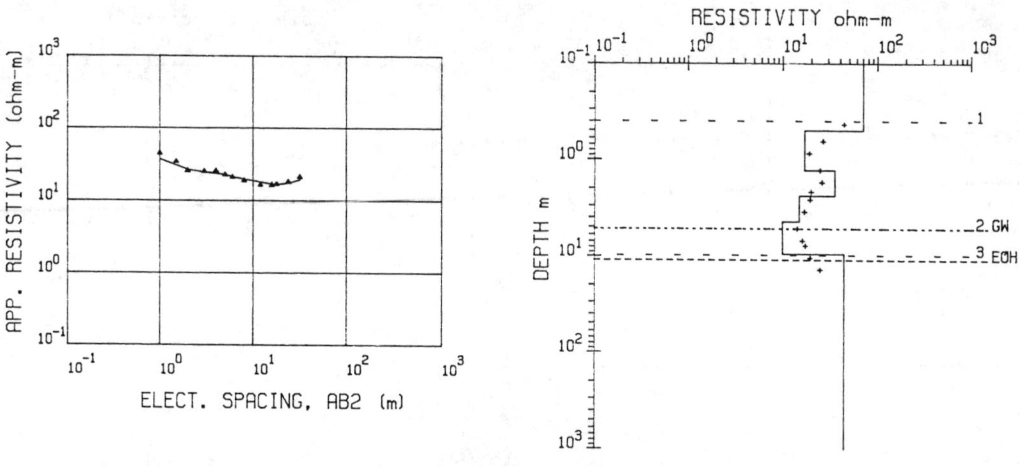

Fig. 7 1-D MODELS FOR SITE SLBH8

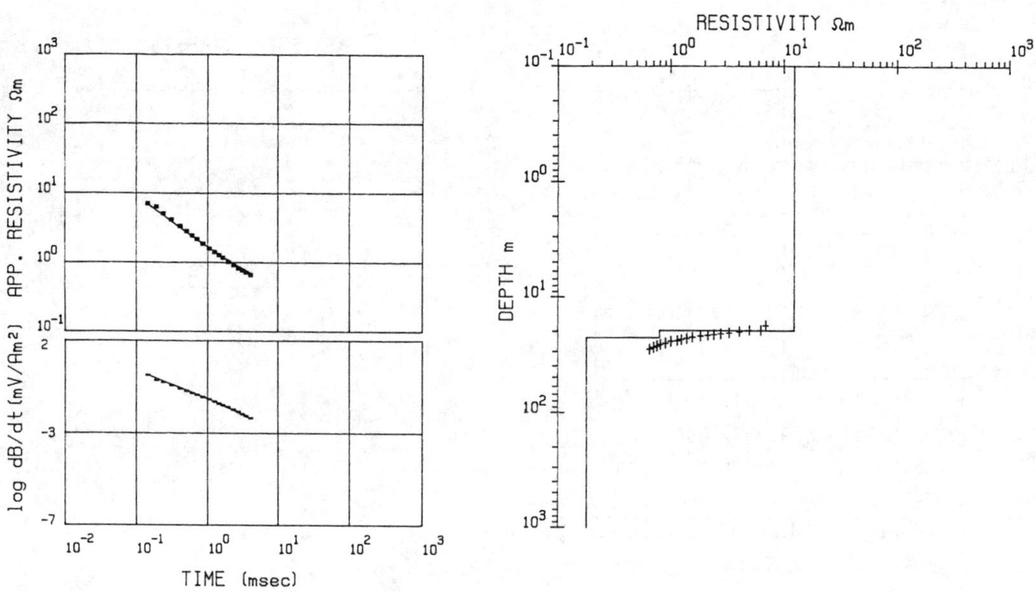

Fig. 8 1-D MODELS FOR SITE TEM-G3

zone with the lowest resistivity in the section. The subjacent geology is typically more resistive than the overlying backfilled zones as in GB8 and GB3 making it posible to delineate the vertical extent of the fill. It would also appear that the groundwater seepages encountered near the bottom of the fill materials have enhanced conductivities compared with those occuring at a higher position (2.4m) at GB3, say. The results also reveal a more complex fill-fluid structure at GB8

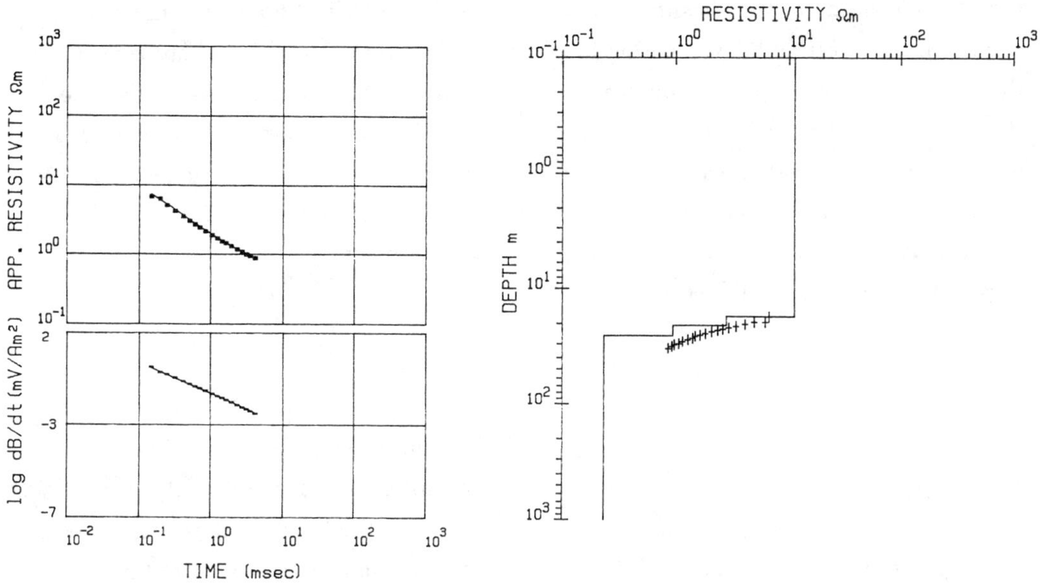

Fig. 9   1-D MODELS FOR SITE TEM-BH3

than is suggested by the borehole data (see Fig. 7); resistivity image may simply reflect changes in fluid saturation in the ground. Inspite of the above interesting results, one must bear in mind that there are geophysical methods have their limitations and are only effective where there are appreciable contrasts between the host geology and the fill materials. In resistivity sounding therefore, where the basal geological material is of similar or lower resistivity than the overlying fill as in G1 (Fig. 5), it becomes difficult to define the fill extent using only the suggested low resistivity basal marker as a guide. In such situations, one would require some extra information to be provided by either a borehole or an appropriate non-electrical geophysical method.

5.2 Quantitative data interpretation by nonlinear optimization

In Figs. 8 and 9, the resistivity-depth transformations of TEM data from stations G3 and GB3 are shown together with the optimal models derived by nonlinear regression analysis (Meju, 1992a) for comparison. Note the agreement between the two methods in the estimated depths to the zone of intense resistivity alteration (i.e., anomalously enhanced conductivity) at about 20m below the surface. It is also clear from these figures that the anomalous zone is relatively shallower and more conductive at G3 than at GB3. This may be related to the presence of a subterranean flow in the vicinity of G3 (it was gathered from historical records that there used to be a spring in this locality). In any case, a simple acceptable explanation for the observed TEM anomaly underneath the whole site is that it is simply due to percolating contaminated water from the fill being stored in a clayey very sandy material at depth. A deep borehole (possibly an injection well at G3) will suffice to clarify the true nature of the causative material.

6. CONCLUSIONS

The utility of non-invasive geophysical surveys in site investigations has been demonstrated using resistivity and transient electromagnetic methods of exploration. It has been shown that the resistivity patterns of undisturbed, partially disturbed and fully excavated grounds could be distinguished.

Numerical modelling of dc resistivity and TEM survey data acquired in the vicinity of boreholes has shown that the capping materials, the enclosed fill material and local groundwater levels may be distinguished from the basal geological materials. Also, the interpretive models have been shown to be in accord with borehole data suggesting that drilling costs may be minimized by having pre-drilling geophysical surveys which would then guide the siting of only a few exploratory and/or confirmatory boreholes. Importantly, the TEM models revealed the presence of a possible highly polluted groundwater reservoir system not penetrated by any of the previously drilled boreholes underlining the fact that a combined approach to landfill site assessments (involving the combined use of borehole and non-invasive geophysical probing techniques) is necessary to completely understand or solve a given problem.

ACKNOWLEDGEMENT

The author is grateful to the Pollution Control Unit of Leicester City Council for providing the borehole data and for permission to conduct the site investigations described in this paper.

REFERENCES

Barker, R.D. 1981. The offset system of electrical resistivity sounding and its use with a multicore cable. Geophysical Prospecting. 29: 128-143.

Carpenter, P.J., Calkin, S.F., & Kaufman, R.S., 1991. Assessing a fractured landfill cover using electrical resistivity and seismic refraction methods. Geophysics. 56:1896-1904.

Meju, M.A., 1993c. Inversion for smooth geoelectromagnetic models and the effective depth of inference: an interpretative analysis. Geophysics (in review).

Meju, M.A., 1993b. Biased estimation: a simple framework for parameter estimation and uncertainty analysis with prior information. Geophys. J. int. (in press).

Meju, M.A., 1993a. Geophysical Data Analysis: understanding inverse problem theory and applications. Continuing Education Course Notes, Society of Exploration Geophysicists Publ., Tulsa.

Meju, M.A., 1992b. Simple effective algorithms for in-field inversion of electrical and electromagnetic data. Paper presented at J.A.G. meeting on 'Developments in shallow electrical and EM surveys', Burlington House, Royal Astr. Soc., London. November 13, 1992.

Meju, M.A., 1992a. An effective ridge regression procedure for resistivity data inversion. Computers & Geosciences. 18:99-118.

Parasnis, D.S., 1986. Principles of applied geophysics. London: Chapman & Hall.

Waste Disposal by Landfill – GREEN'93, Sarsby (ed.) © 1995 Balkema, Rotterdam, ISBN 90 5410 356 6

Deep treatment of uncontrolled urban fills for the construction of a high capacity road system

C. Oteo & L. Sopeña
Laboratorio de Geotecnia Cedex, MOPT, Spain

ABSTRACT: The presence of uncontrolled and highly deformable anthropic fills (demolition rubble, dumped earth, etc.), which caused considerable settlement and collapse due to an increase in water content, was detected during the construction of a high capacity urban road network. This report describes a series of tests for improving the uncontrolled fill, using two types of treatment: construction of columns using the jet-grouting technique and gravel columns by water vibroreplacement. The results are analyzed, depending on the treatment control systems used (dynamic penetrometers, testification, etc.) for various positions and distances between the columns. A description is also given of the criteria adopted for the definitive treatment, together with the results obtained, highlighting the effectiveness of the methods and the conclusions eventually reached.

1. INTRODUCTION

When constructing new urban roads, it is often necessary to overcome problems related to the presence of considerable amounts of uncontrolled waste and fills of a variety of materials. Extensive and almost invariably unauthorized dumping grounds often exist in the outlying areas of large cities, consisting of thick layers of rubble (from demolition) and earth material (from the emptying of basements and urban excavations), including urban solid waste, and can be regarded as veritable rubbish dumps. Such kinds of anthropic deposits are potentially a source of many serious difficulties in road construction, either because they are often carefully hidden under new surfaces or because the problem has been underestimated by technicians.

One of the greatest repercussions is the intrinsic deformability of these fills and, above all, the effects of water which can reach the core, causing collapse through flooding, washing away, etc. Furthermore, these fills often lie in depressions, such as former stream beds, etc, as these were ideal places for dumping. Therefore, they lie in areas that are susceptible to infiltration, especially at the contact-points between the fills and the original land.

2. POSSIBLE SOLUTIONS

Although this article does not aim to be a state of the art report on the matter, we believe it is appropriate to point out the various possible solutions originally contemplated. In the face of the problems mentioned, there are "a priori" several types of activity, ranging from possible alterations to the route to the total removal and replacement of the unwanted material. However, in most cases neither of these options are totally feasible. It should be remembered that in urban areas road planning is largely conditioned by the presence of buildings, and that deep excavations could affect them. It is also extremely costly to use a large quality of replacement material, and an additional problem is created by having to find dumping grounds for the extracted material in other parts of the city. So, a combination of solutions must be taken, including treatment to improve ground conditions.

In choosing the most suitable system, numerous questions must be taken into account, these being dependent both upon the nature of the problem itself and environmental factors: proximity of buildings, availability of large amounts of water, economics, etc.

The following are among the deep ground improvement methods can be used in the case of uncontrolled urban fills:

- Preloading, with the drawback of the amount of time and the considerable volume of weight required. Furthermore, it may not be effectively applied to these urban fills.
- Dynamic compactation, of limited use owing to its effects on nearby buildings and installations, plus the fill thicknesses to be treated.
- Chemical and compactation injections and jet-grouting, quicker and easier to control.
- Gravel columns, well performed by tamping or vibroflotation, using both dry and wet methods.

3. SPECIFIC CASE

The case described in this report concerns a road in North-West Madrid, constructed upon fills dumped in a former stream bed. Surface earth fills concealed demolition remains, rubble, etc, dumped 10 to 12 years ago. A large capping layer as a sub-base for the pavement, together with a first layer of rolled asphalt, were constructed on top of these fills. Before placing the final pavement layer, considerable settling was observed, for several months (with and without rain), small 30 to 50 cm bumps forming with settlements of 30 to 60 cm in the centre.

It was decided to first carry out a geotechnical study (none existing before the work started), so several mechanical boreholes were dug, samples were extracted for inspection and a series of continuous dynamic penetration tests were performed; this procedure is usually very useful in the detection of uncontrolled fills.

4. GEOTECHNICAL CHARACTERISTICS OF THE SUBSOIL IN THE AREA

From a geological perspective, the fills in the study zone lie upon a former major stream bed, a confluence point for other secondary ones, cut into a Miocene substratum of coarse detrital facies, composed basically of sand and clay.

The lithogeotechnical column which represents the subsoil conditions in the area can be seen in Fig. 1. It also shows the grain-size chart of the earthy part of the fills and the corresponding Casagrande graph.

The fill thicknesses in the zone chosen to perform the tests, ranged from 8 to 10 m, and were made up of mixtures of different soils, basically granular, with abundant rubble from demolitions, urban remains, etc. The distribution of the layers where one or other of the materials predominates is variable, although the rubbish and elements most susceptible to deformation, tend to lie at a depth of 4 to 7 m.

Although the grain-size of the soils is more open (some undisturbed samples have been tested from the borings), it can be seen that the soils are sandy-silty and of low plasticity, mainly of the SC-type. The fines-content (% that passes through nº 200 ASTM seive) was from 20 to 40%, 80-96% of the material being smaller than 10 mm.

At the inspection time, rather than a general groundwater level being observed, this appeared in zones or at specific points. Where it was patent, it was located at the contact-point between the fill and natural ground, which is only to be expected at the bottom of a former stream bed.

Fig. 1, shows some of the S.P.T. values obtained from one of the borehole made prior to the tests; according to these values, the fills would be characterized by a standard resistance to dynamic sounding which on average was between N = 5-15 blows/30 cm. Between depths of 3 to 7 m, this parameter N, only gave values of 5 to 10 blows/30 cm.

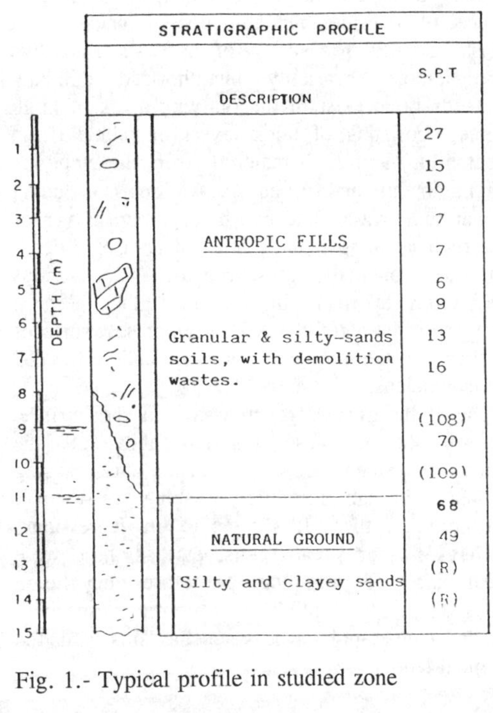

Fig. 1.- Typical profile in studied zone

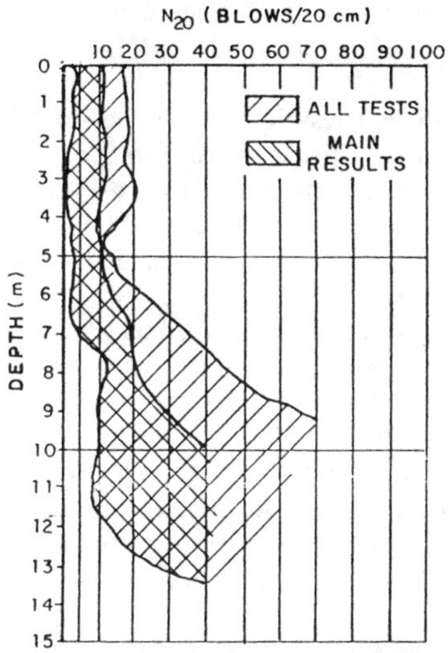

Fig. 2.- Envelopes of the dynamic penetration test results carried out before the treatments

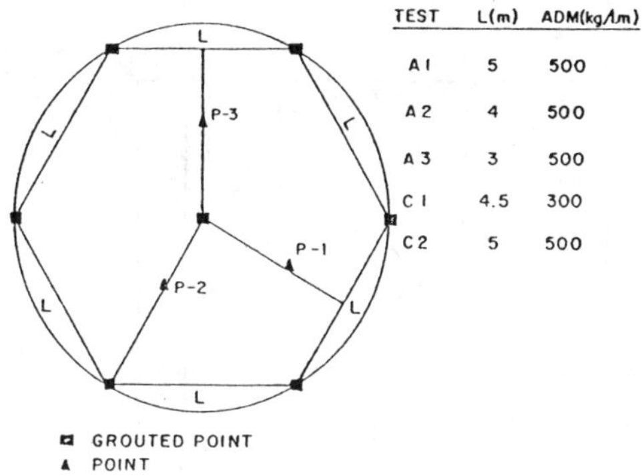

Fig. 3.- Characteristics of jet-grouting tests

Fig. 2, concerning the Borro-type continuous dynamic sounding tests, which represents the overall subsoil area tested, provides an even better idea of what happened. According to this data, the fill deformability could be represented by an apparent deformation modulus of about 50-75 Kp/cm^2, in the weakest zone. It could also suggest that the collapse value under water effect might have been about 5-8%.

In Fig. 2, it can be seen how there are considerable layers with a dynamic resistance below NB = 5 blows/20 cm, and there are certainly many with a resistance that is below NB = 10 blows/20 cm.

The average values of some parameters obtained from the undisturbed samples in the earth part, should also be mentioned: dry density 1.53-1.78 gr/cm^3; natural water content from 6-9% and 10-15%, depending on the layers. The following are basic values of the Normal Proctor compactation tests: optimum moisture 12.1% and maximum density 1.89 gr/cm^3, from which it can be deduced, by way of guidance, that the earth material from the fills would appear to be at degrees of equivalent compactation of approximately 75-85%, when comparing the natural density with the Proctor optimum. As is known, the sandy and silty-sandy formations of Madrid, are susceptible to collapse with these degrees of moisture and compactation, collapses occurring under vertical pressure of 200 kPa of about 3-5% (ESPINACE and OTEO, 1984). Furthermore, the presence of other material (rubble, plastic tubing, etc), gives rise to a large number of hollows which, on the whole, make the fill extremely deformable, and susceptible to collapse and erosion.

5. PRELIMINARY JET-GROUTING TESTS

Before starting the treatment of the area concerned, (fills over the Arroyo de la Veguilla), tests were conducted in two different types of zones: in one where the fill thicknesses were deepest (Zone 1), where they reached ten metres, and the other where they were much thinner (Zone 3), there being thicknesses of barely five metres. In each one of these zones some of the basic parameters governing the treatment were altered, such as the distance between treatment points and the envisaged mixture intake.

In all cases, and after a geometrical analysis of the question, an equilaterally triangular mesh, which is the one that gives greatest uniformity in the distribution of treatment points, was chosen. In both tests, a hexagon was chosen as the minimum test area, there being a hole at each corner point and another at the centre (Fig. 3).

The control and assessment system for the treatment consisted of carrying out a series of Borro-type dynamic sounding tests at strategically located points, both before and after the completion of the columns, so that they would, as a whole, be representative of all the different areas (intermediate, barycentre, etc). Likewise, a γ-γ testification was performed in the soundings, before and after the treatment (Fig. 4).

Figs. 4 and 5 shows some of the results obtained for the tests conducted in Zones 1 and 3, accor-

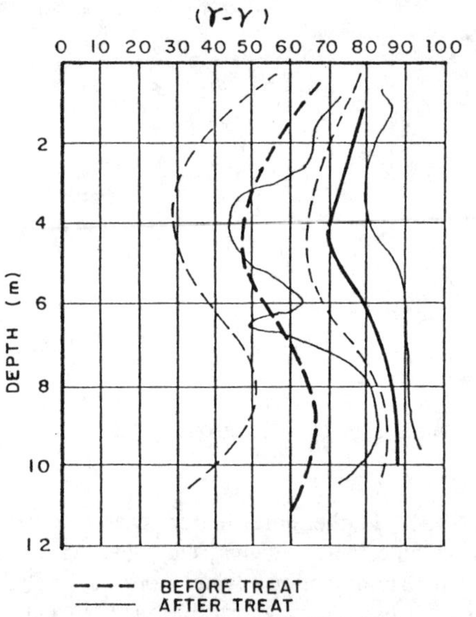

Fig. 4.- Control of the jet-grouting treatment by γ-γ testification (Tests of Zone 1).

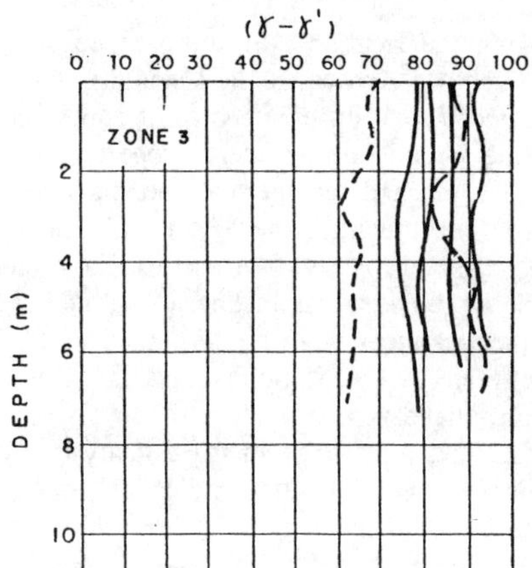

Fig. 5.- Control of the jet-grouting treatment by γ-γ testification (Tests of Zone 3)

ding to the strip that represents the testification of the control borings, which gives a relative value for the density of the ground. One of the conclusions drawn from these tests, was that the 5 m long mesh has a low effectivity, the maximum intake of cement being restricted to 500 kg/ml of column; it became clear that a finer mesh, with sides of 4 and 3 m, would be a good improvement.

However, the overall testification for the set of three tests meant a significant increase in the density indicators of the ground affected by the treatment, that shows an overall improvement in the fill which is more outstanding than that indicated in the sounding tests.

By combining a wider mesh with a greater intake, the difference between the two tests can hardly be appreciated, and this shows that the procedure could be efficient enough. It should be remembered that although the relative increase in density seems better, in the zones with thinner layers of fillings, the presence of rubble is of less importance, and the initial state is not as bad as in the rest of Zone 1.

In addition to these tests, the movements of the pavement, controlled by surveying, proved to be 2 to 5 mm. By later making sampling points, it was possible to observe that 50-60 cm diam soil-cement columns had formed, with pseudopodes 5-7 cm thick, full of mixture which, when separate from the column, extended 2-3 m from their axis.

6. TESTS WITH TAMPING GRAVEL COLUMNS

The column construction system used was the rolling procedure (wide-based Franki pile type), using the dry method.

The equipment used was a framework with a 55 cm diam steel tube, 15 m long. The tamping hammer weighed 2,75 T and the blow took place at a height of 4 m. Once the closed tubing had been driven until rejection (less than 1 cm of sounding per set of 10 blows), a start was made to pouring the gravel in, with successive quantities of 1.3 cm^3, poured every time an advance of 1.25 m was made. After this, the hammering began and the tubing was progressively removed.

A 2 m square mesh (see Fig. 6) was chosen for the "long" columns (driven until rejection), and then another "short" one, in the geometrical centre of the mesh. The order in which the columns were put in place is shown in Fig. 6, together with the order of the control tests (Borro-type dynamic sounding) programmed so as to obtain information regarding the effectiveness of both treatment stages.

The reason for using long columns (until rejection, i.e. 10 m in length) and short ones (6 m

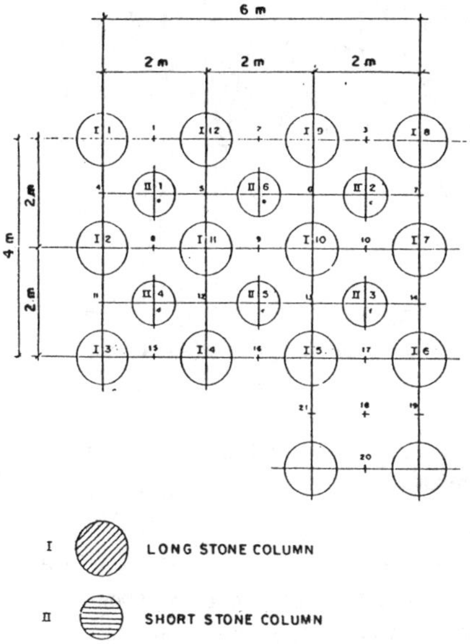

Fig. 6.- Plant of the stone column test mesh and control points

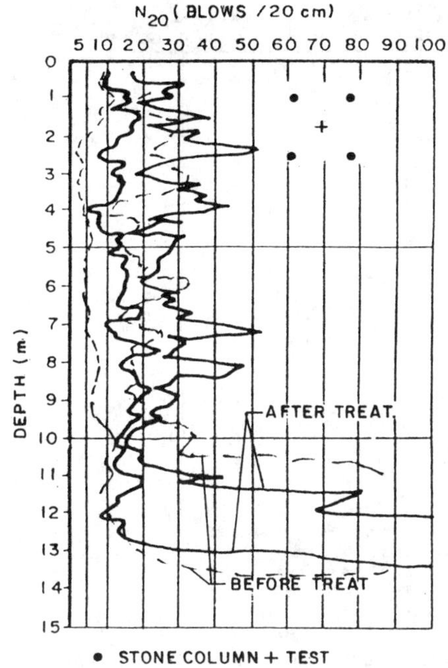

Fig. 7.- Envelopes of the dynamic penetration tests in square mesh of stone columns

long) arose from an attempt to obtain maximum effectiveness from the square mesh and a separation of 2 m, because from the onset it could be seen that the mesh was somewhat uncompacted at the centre of the square.

As far as the results were concerned, Fig. 7 shows the dynamic sounding test envelopes from before and after the long columns were put in place. Thus, these results refer to a time when only the long columns were present (corners of the squares).

There is a clear increase in sounding resistance. Starting from numerous layers with values of about 5 blows/20 cm, increases were obtained which corresponded to minimums of around 10 blows/cm in the first treatment, and of almost 15-20 blows/20 cm in the second one.

From a density viewpoint, once a specific relationship had been established with the sounding resistance in this type of test, considerable ground improvement was achieved. From this perspective, long column treatment can be considered reasonably acceptable for road networks, whereas the supplementary treatment with intermediate short columns is excessively intense (some experiences indicate a resistance of about NB = 10 blows/20 cm in this test, equivalent to the minimum recommended relative density for granular fills of this kind, in embankment cores).

Fig. 8 shows the horizontal movements brought about in a vertical section of the ground, when a long column is being positioned; these were measured by the installation of two inclinometers for auscultation purposes, situated one and two diameters away, respectively.

Not only can the considerable deformations caused by the movement of the ground be observed, but also how an increase in distance can reduce this effect. However, it should be remembered that, in the face of the major deformations induced by the rigidity of the response of the inclinometric tube, there is a tendency towards a distortion of the deformation.

A loss of effectiveness is particularly noteworthy along the layer of ground lying about 4 m from the surface. This phenomenon had already been observed, albeit less directly, when analyzing the sounding tests.

This occurrence is of considerable importance in practice, and generally affects this type of treatment, as will be seen later. It can be attributed, amongst other reasons, to a reduction in confined lateral pressure of the ground nearer the surface.

The columns put in place using this technique, had an average diameter ranging from 75 to 90

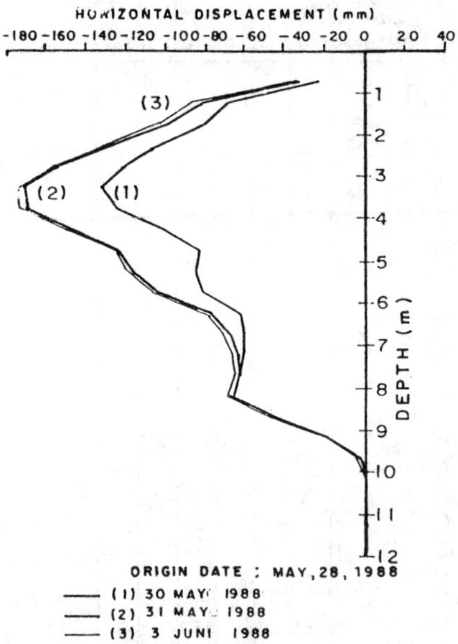

Fig. 8.- Horizontal displacements produced in a vertical 50 cm near a stone column (tamping method)

cm, with an average gravel intake of approximately 0.48 m^3/ml of column, for the kind we refer to as "light" (contribution of 0.120 m^3 of gravel every 0.25 m). They could also provide 0.96 m^3/ml of column (heavy kind) if around 0.120 m^3 of gravel every 0.125 m of column was provided. Virtually all the columns mentioned here were of the "light" kind.

7. TREATMENT PERFORMED AND RESULTS OBTAINED

In the light of the results obtained in the initial tests, a decision was taken to use stone columns, carried out with vibroflotation machinery, in the zones with a lesser water flow (so as not to reduce the permeability of the environment). The jet-grouting system was used at the remaining sites, to prevent damage to the existing pavement and to speed up the treatment process (Fig. 9).

The system used to place the gravel columns was based on use of a vibrator or torpedo similar to that used in the vibroflotation technique, driven down to the natural ground, with the usual water injection, after which the column was secured by adding gravel as the vibrator was removed progressively. This gravel was compacted by the

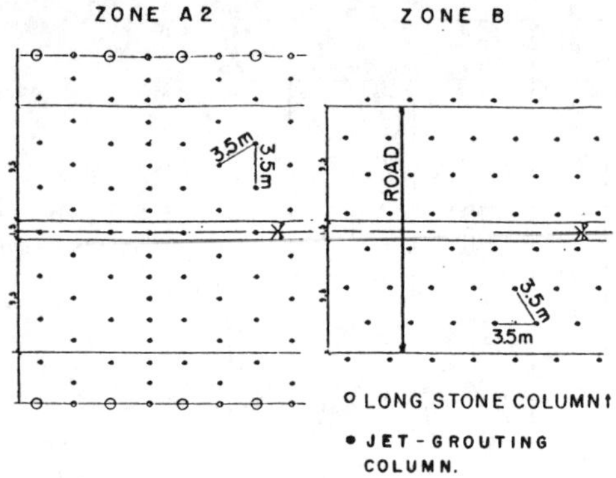

Fig. 9.- Mesh types used in massive field improvement

torpedo vibrations, as the device was removed and partially reinserted to increase effectiveness.

In classical terms, (the type of material with considerable fines and the absence of water), as in Mitchell's recommendations, notable differences were deduced with the vibroflotation itself. In fact, the injected water washed away a certain amount of the material, thus causing a partial ground replacement, i.e. it was not merely a question of providing gravel. So, the mixed treatment applied could be called vibroreplacement gravel columns by vibroflotation.

In short, the effect achieved, is not so much reinforcing the column itself, but rather an overall compacting of the subsoil mass. Furthermore, the purpose of using such a method was to get the large quantities of water employed to cause "a priori" the dreaded effects of water upon these types of fills (collapse, etc), when these effects were not yet harmful.

Each of the Bauer-type torpedos involved, consisted of an engine with eccentric, an antivibration device, an extension tube and head of suspension. The torpedo could function at a frequency of 3,000 r.p.m. and achieve a centrifugal force of 16 T or attain 3,500 r.p.m. with a centrifugal force of 25 T.

out, starting with one 3 x 3 m square, having long columns (down to the underlying natural ground) which, as will be seen, proved insufficient, after which a triangular one with sides of 3.5 m, was tried out, in which the specific volume allotted to each column is somewhat less, and this also proved to be unsuitable. The above-mentioned ineffectiveness in the upper layers of the ground was observed at depths of 4 to 6 m. So, finally, the use of short intermediate columns (6 m long)

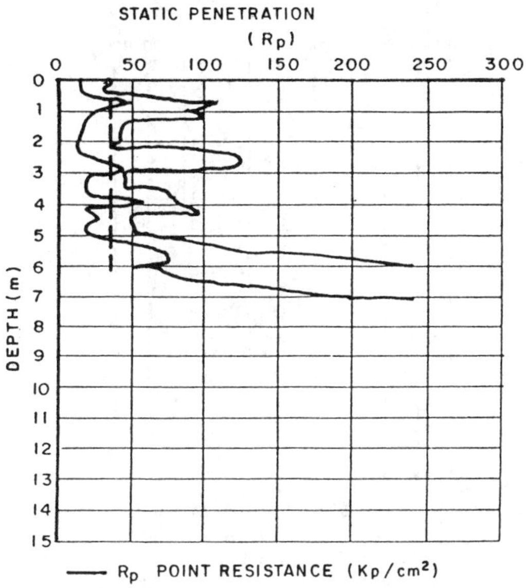

Fig. 10.- Envelope of the static penetration tests carried out between stone columns

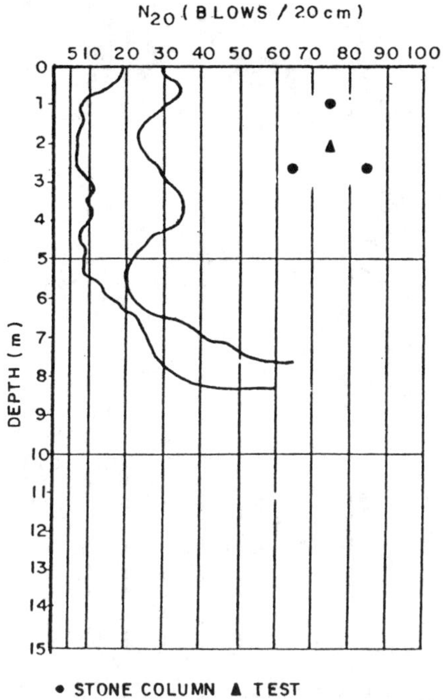

Fig. 11.- Envelope of the dynamic penetration tests carried out between stone columns disposed in triangular mesh (one month after)

was chosen. The mesh eventually selected as being most effective was of the triangular type, with sides of 3.5 and 3 m, and short (6 m) barycentric columns.

In these cases, the columns had an average diameter of 1.0 m and the gravel intake was approximately 0.8 m^3/ml per column.

In the face of the unsatisfactory results provided by the dynamic penetrometer control tests, a decision was taken to conduct static sounding tests, with a view to removing the possible influence of pore pressure that might have occurred during the test, this being a result of the water injected during the treatment.

Fig. 10 shows all the results for the square and triangular mesh zones, obtained from the static penetrometer tests performed after having placed the columns (test in the barycentres of the mesh).

It is worth mentioning that, after comparing the various tests of different types carried out during the trials, the conclusion was reached that there was an excellent correlation of the Rp = 4NB type, between both the dynamic and static resistance; Rp being the unit pressure at the tip and NB the blow of the Borro-type dynamic penetrometer.

With sounding resistance at the tip being clearly below 40-50 Kg/cm^2, the unsuitability of this mesh was obvious.

When analyzing these results, an attempt was made to take into account the effect of the time that had elapsed after the placing of the columns, owing to a possible delayed dissipation process of pore pressures, caused by the permeability differences of the different material involved, which might either have influenced an effectiveness assessment, or have caused a change in the cementing of the particles. (As Mitchell mentioned).

Fig. 11, shows the results for the triangular mesh with 3 m sides, and long columns, taken from tests subsequently conducted, after a post-construction period of one month.

The appreciable difference is significant, and although it cannot strictly be put down entirely to the time lapse (the ground layout is highly variable, and a comparison of tests from various zones is inevitable) it seems to give a good indication. The results for the triangular mesh with 3.5 m sides with a short (6 m) barycentric column, can be seen in Fig. 12; these results were forthcoming after the ground surface had been drained and a 1 m. thick first layer of embankment had been constructed, in order to "cap" the columns. There was a marked improvement, even in the shallowest layer of ground.

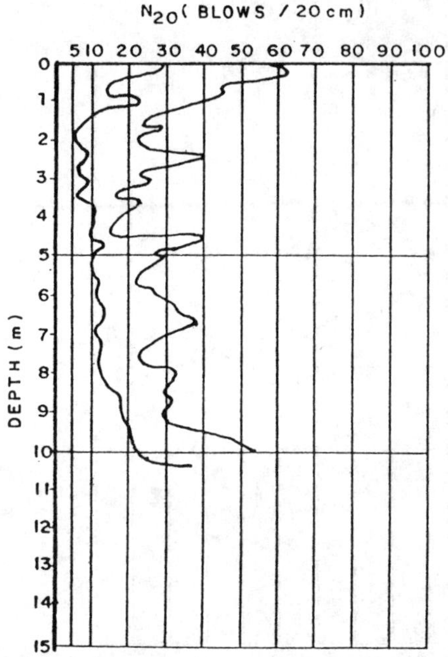

Fig. 12.- Envelope of the dynamic penetration tets after the end of the subgrade

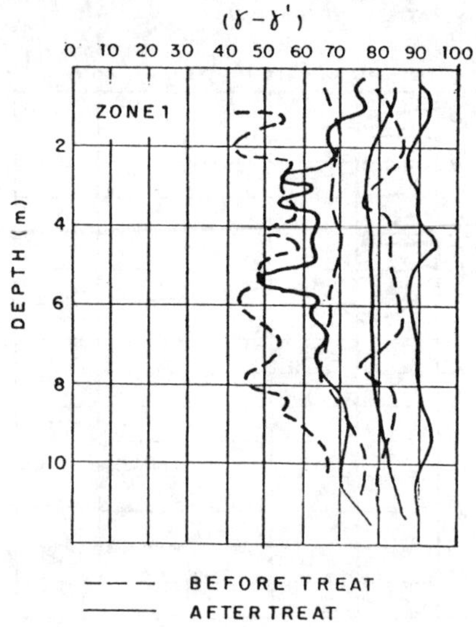

Fig. 13.- Global results of the control of the jet-grouting treatment by γ-γ testification (Zone 1)

In the zones which were considered to be free of water, the jet-grouting technique was applied (Fig. 9).

A particular treatment was applied depending both on the geotechnical characteristics of the subsoil in the area being dealt with and on the rest of the conditioning factors; jet-grouting columns were sometimes mixed with gravel columns. Fig. 11 shows two types of treatment used. At times, the general triangular mesh was supplemented by sealing with perimetrical columns, to increase the growth effect, thus limiting the volume affected by the injecting.

The jet-grouting zones were controlled using the same procedure as mentioned above, but this time the basic test system borings were tested by gamma ray absorption. To show the results obtained, Fig. 13 shows the general charts for Zone 1, before and after treatment. A growth in the values indicating the ground density can be observed for both minimum and average values. According to these, the 70% range, would increase to values of 80-90%.

It should also be pointed out that in the tests conducted after treatment, no hollows or cavities were detected in the vicinity of the corresponding borings. It seems plausible that the effect of hollow-filling achieved by jet-grouting, apart from giving a structural strengthening to the column, can also be detected from an average density increase in this testification, and not only from the absence of nearby hollows or cavities, but also by the presence of slabs, 3-7 cm thick, and up to 3-4 m long, full of the mixture, which serve as "flat jacks" inside the fill.

Likewise, mention must be made of the results of the direct visual observations from the sampling pits opened for that purpose in areas treated with jet-grouting columns. The injecting was aimed at the weakest nuclei, impelled by its kinetic energy, filling the hollows, cementing the loose mass of rubble, bricks, etc, going below the pavement-filling contact-point.

Reference should also be made to the results obtained from some of the tests performed using the Rayleigh-type surface waves technique (Cuéllar, 1992), the results of which are sufficiently conclusive, although it is a new technical skill not extensively applied as yet. The results would tend to suggest an improvement in subsoil compaction, indicated by an increase of 50-70% in the propagation of longitudinal waves, for the first 5 or 6 metres of fill (Fig. 14). Given that the average strain modulus depends on the square of the wave velocity, it could be said that the average deformability of the fill would have been reduced by 2.2 to 2.9 times after the treatment.

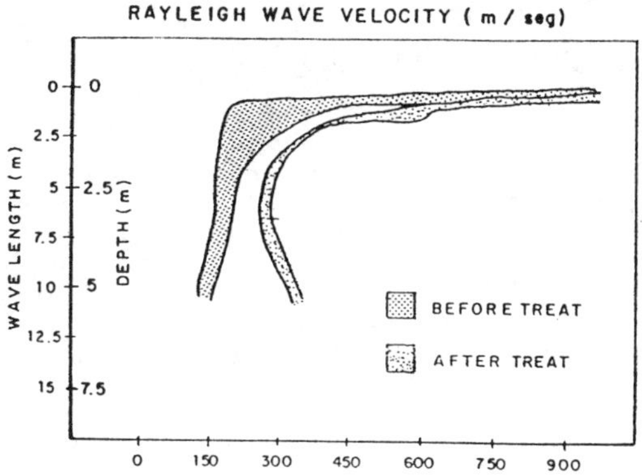

Fig. 14.- Results obtained by surface spectral Rayleigh wave control

A total of 23,000 m.l. of gravel columns were placed, plus 25,000 m.l. of jet-grouting columns. The cement consumption of the latter ranged between 459 Kg/m.l. (40% of the column total) to 700 Kg/col (10% of the total).

In view of the above, and taking into account the limitations of the test and control methods implicit in the jet-grouting system, it must be concluded that the treatment has achieved a notable improvement, which in overall terms can be regarded as satisfactory, as a result of the compaction it provided, and especially because it filled the hollows and, by cementation, strengthened the weak nuclei in the ground.

The settlement measured in the two years following the works, show no appreciable movements: from 5 to 10 mm at the most.

CONCLUSIONS

1) The two ground improvement systems used have given excellent results. The gravel columns have compacted the ground sufficiently and, furthermore, piezometric measurements have shown that the hydraulic conditions in the zone have undergone no changes.
2) The jet-grouting technique has made it possible to conserve the exisiting pavement, which has been restored to service with minimal repairs and a thin layer of coated asphalt. This has meant a great deal of financial saving at this stage of the works.
3) The mesh which proved effective was the triangular type, with sides not extending beyond 3-4 m.
4) The effects of jet-grouting on the ground, ought to be considered from three basic angles: creation of a resistant column of soil-cement; slight lateral compaction through displacement of the surrounding ground; and filling of hollows and weak zones, up to considerable distances from the drilling point.
5) With regard to gravel columns, water addition was considered appropriate, in order to accentuate and accelerate the collapse of the fills. The production performance obtained is notably better than with the dry method. Vibroflotation equipment can attain 160-200 m/day (columns 10 m long).
6) The presence of short gravel columns, about 6 m, at the barycentre of the primary triangles, proved to be highly effective in solving the problem of the area of ground that was looser near the surface. It was necessary to cap the columns with a 1 m thick layer of compacted earth at the surface.
7) As a control system, the application of the testification techniques in boreholes are regarded as generally acceptable as a means for overall assessment of the effect of jet-grouting and gravel column treatment and, in a supplementary capacity, the Rayleigh-type surface waves technique is also a valid and useful technique. In the case of gravel column treatment, the intensive use of continuous dynamic penetrometer tests, before and after improvement, are also an effective way of control.

ACKNOWLEDGEMENTS

The authors would like to express their thanks to the Roads Administration, who financed the work and research, and the engineers responsible (Mrs. Lazcano, Alberola and Guitart), the construction companies who carried it out (Terratest, S.A., Rodio and Geocisa) and the company in charge of quality control (Geoteyco, S.A.). We also wish to give our thanks to the laboratory staff at the Geotechnical Laboratory of the CEDEX who collaborated in this work, especially Mrs. Delgado, Nicolás and Quero, and Mr. Maldonado of the School of Mining Engineering.

BIBLIOGRAPHY

A.S.C.E. (1975). "Proc. of the Conf. on In-Situ Measurement of Soil Properties". North Carolina. U.S.A.

ESPINACE, R and Oteo, C. (1984). "Influencia del Grado de Compactación de Rellenos Arenosos en su Deformabilidad Bajo Cargas Estáticas y Cíclicas". Cuadernos de Investigación. C-10 CEDEX.

GREENWOOD, D.A. (1972). "Vibroflotation-Rationele for Design and Treatment". Symp. on Methods of Treatment of Unstable Ground. Sheffield.

HALKOLA, H.A. (1983). "In Situ Investigations of Deep Stabilized Soil". Proc. VIII E.C.S.M.F.E. Vol. 1. pp. 33-37. Helsinki.

MITCHELL, J.K. and Katti, R.K. (1981). "Soil Improvement. State of the Art Report". Proc. 10th. I.C.S.M.F.E. Stockholm.

MITCHELL, J.K. (1986). "Practical Problems from Surprising Soil Behaviour". Journal of Geotechnical Engineering, Vol. 112, N. 3, A.S.C.E.

MORGAN, J.G.D. and Thomson, G.H. (1983). "Instrumentation methods for Control of Ground Density in Deep Vibrocompaction". VIII E.C.S.M.F.E., Vol. 1, Session 1, pp. 59-65. Helsinki.

RATHGEB, E. and Kutzner, C. (1986). "Some Applications of the Vibroreplacement Process". Ground Treatment by Deep Compaction. I.C.E., London EC1V 9 Ad.

Waste Disposal by Landfill – GREEN'93, Sarsby (ed.) © 1995 Balkema, Rotterdam, ISBN 90 5410 356 6

Construction of mineral barriers with bentonite and fillers for sealing of waste landfills

S.D. Porphyris
Silver & Baryte Ores Mining Co. S.A., Athens, Greece

A. Weinmann
IBECO Bentonit-Technologie GmbH, Mannheim, Germany

ABSTRACT

Mineral barriers are used for base and cap seals of landfills, rectification of old dumping sites and containment of contaminants/toxic materials in the ground.
The paper describes the state of the art and common construction practice for construction of mineral barriers for landfills in Austria, Germany, Holland and Switzerland. Silts or fine clayey, silty sands can be enriched easily with bentonite (powderized high swelling clay) in order to achieve the desired $K_{(f)}$-values of 1×10^{-9} m/s or 1×10^{-10} m/s, as specified in various building codes, and EC-drafts.
The most common approach is to mix the bentonite into the soil with very powerfull soil milling machines (Mixed-in-place-Method).
This can be done either in-situ or on-site. In recent years more and more job sites were done using a Mixed-in-Plant-system, where a mixture of say one or two soils (base materials) are enriched with bentonite in a separate mixing plant (if more economical, also with fine fillers) to obtain a very homogeneous product.
In view of the increased practice of recycling materials and excavated soils from other sites for sealing of landfills, the Mixed-in-Plant-Method represents a very flexible and proven economical way of construction. In general the application of bentonite and fillers enables the use of most in-situ soils for the construction of mineral barriers.

1) INTRODUCTION

The protection of the biosphere and its media water, soil and air against contaminations is a common goal. Despite this "ideal" there were plenty of situations, procedures and incidents in the past decades which had not been considered carefully enough by man and there was an intensive increase of contamination in the above mentioned medias. The reasons were, partly, a purpose oriented optimism or lack of knowledge, but in some cases careless operation induced the incident - even though relevant experience would have been available from other sources. As a consequence of such development, our present society is forced to deal with remedial works on contaminated land. Further techniques and procedures have to be developed for the present operations in order to prevent continued or even greater damage of the environment and problems for the coming generations. A very essential item concerning this post and pre-care is the design and construction of mineral liners/ barriers for the containment of contaminated sites, upgrading of waste dumps and

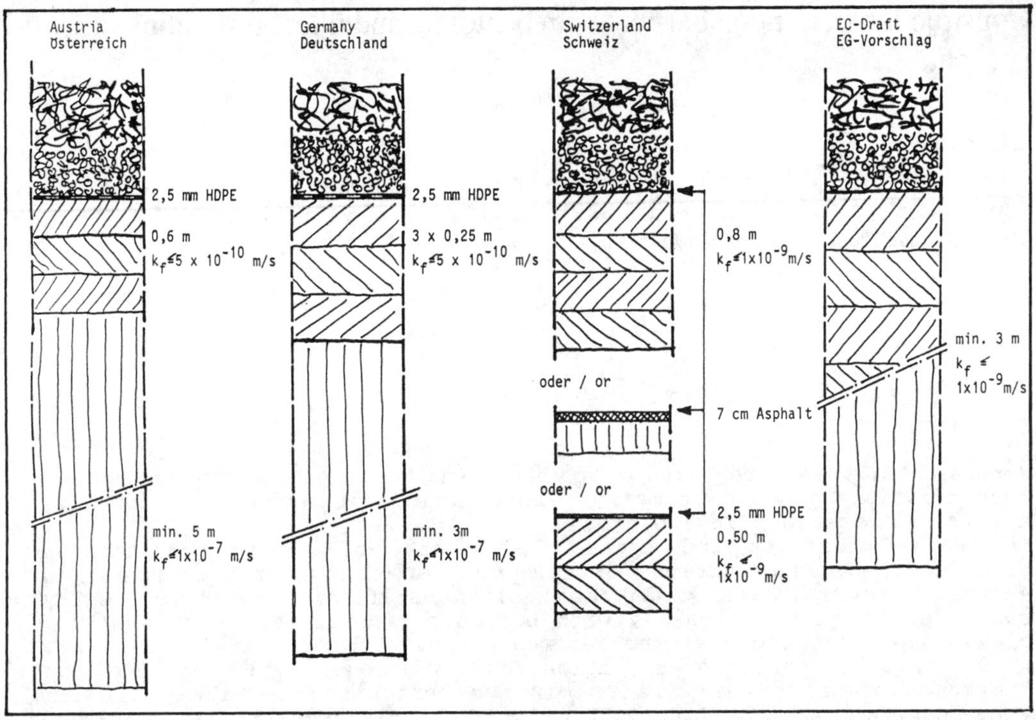

Fig. 1: Base sealing for domestic waste landfills (cross-section)

installation of new landfills. This paper describes first, briefly, the fundamental aspects and gives thereafter a more detailed description of the installation and application of mineral barriers.

2) NATIONAL REGULATIONS FOR LANDFILL LINERS IN AUSTRIA, GERMANY AND SWITZERLAND AND ACCORDING TO AN EC-DRAFT PAPER

The different soil and geological situation, the topography as well as distinguished national waste management concepts led to different liner constructions.

Figure 1 gives a comparative view of typical sections for the base sealing of a landfill for domestic waste in Austria, Germany and Switzerland. Please note the "or"-regulation for Switzerland. Furthermore, it displays the technical requirements as per EC-draft "Directive of the Council for landfills" (dated 22.05.91).

For cap sealing works, after completion of dumping, the regulations in the various countries are not, at this stage, defined in detail. In general, such barriers should prevent penetration of precipitation into the landfill, but also retain and respectively collect the generated gas. The common method of construction may comprise of a mineral liner, eventually combined with a HDPE-foil.

Intermediate liners in a landfill can be constructed like the base sealing.

Furthermore, such barriers may demand, in addition to the regular technical requirements, extraordinary provisions, like high plasticity and low cracking behaviour due to bending forces.

The above diagramm points out that in most cases the mineral liner constitutes the major sealing element of a landfill, especially in the aspect of long term stability. This development and the selection of that type of material may be supported by the

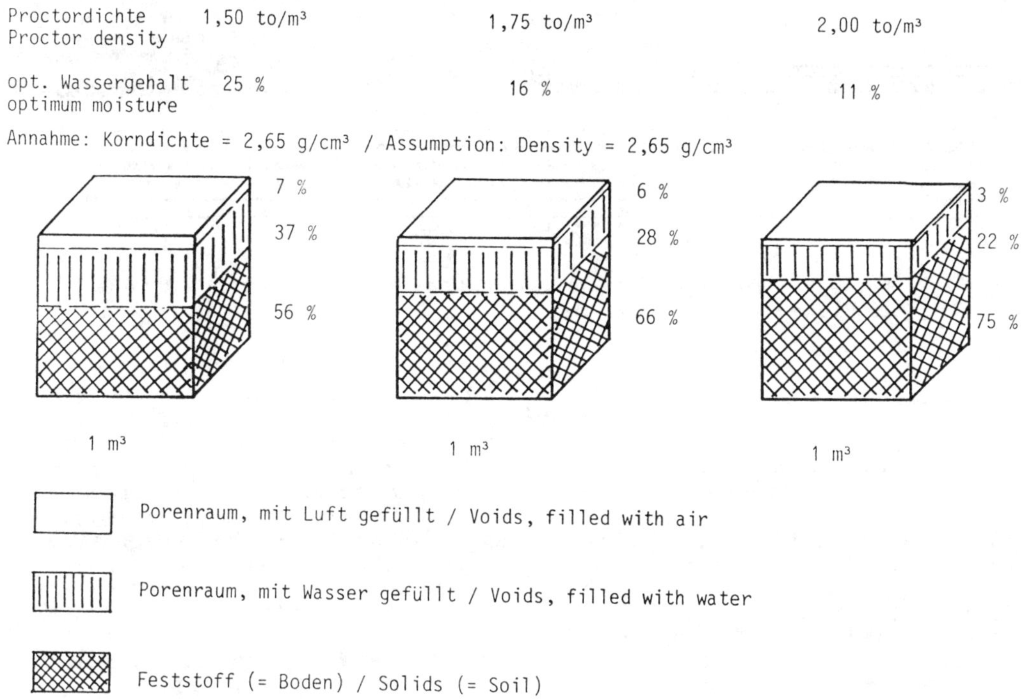

Fig. 2: Comparison of the percentage of solids, water and voids in various compacted soils.

proven function in the field - for example through the protective action of low permeable natural clay layers underneath or around hazardous old fashioned "toxic dumps" in regard to the rate of migration of contaminants, or to the retaining behaviour of subsurface layers, rich in clay minerals over geological ages.

3) TRANSPORT MECHANISM OF CONTAMINANTS THROUGH LINER SYSTEMS

a) Convection

The transport of contaminants by convection is connected with the movement of fluids due to high hydraulic gradients. More-over, the migration of gas has to be considered in cap sealings.

For the movement of fluids, voids, like very small channels, need to exist. A mineral barrier with a high percentage of fines, like natural clay, is one basic requirement for a retaining function. The very fine particles in clay slow down or even prevent completely the migration of contaminants. For stability reasons and proper functioning, a liner is compacted to about 100 % of the Proctor density at a given optimum of moisture, which means for a common clay a high humidity and due to this a "low" content of solids. A comparison of various soils (clay, silt and well-graded material) is shown in figure 2.

The figure shows the percentage of solids, water and voids related to 1 m³ of compacted barrier for a medium plastic clay, a common silty soil and a well-graded clay-silt-sand-material.

One has to conclude that the well-graded soil with some clay minerals (say 10-15 %) is a good prerequisit for a mineral liner due to its high content of solids and its low permeability value (K_f-value).

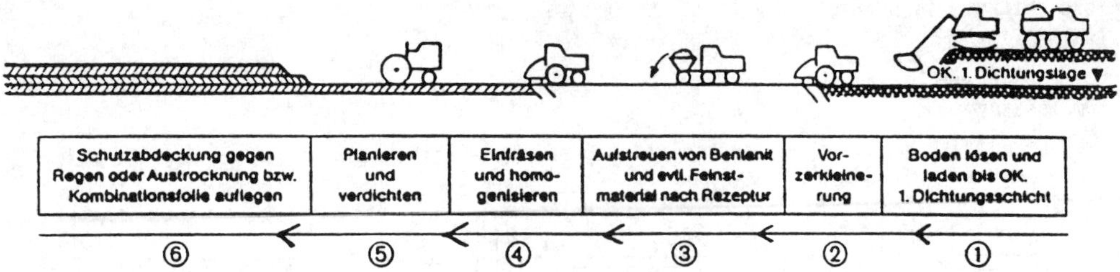

Fig. 3: Sequence of works for using bentonite with the Mixed-in-Place-method (in-situ)

b) Diffusion

Differential concentration of contaminants between two locations leads to diffusion. Molecules move towards the zone of lower concentration, which means for a landfill barrier from contaminated layers towards the clean soils. A detailed knowledge of why and how this happens in the soil-water-air-mixture of a mineral barrier is, up to now, not yet fully assessed. Therefore, one has to assume, that a high (ratio) content of solids in the barrier, in other words, a very dense soil packing, will minimize the migration of contaminants caused by diffusion.

A large number of lab tests and a few field investigations have already proved, that the sorption of clay minerals reduces the diffusion process. Bentonites, with the clay mineral montmorillonite as their major ingredient, are capable of adsorbing a number of contaminants and thus prevent their further migration.

4) GUIDELINES ON SOIL MECHANIC PROPERTIES FOR THE DESIGN AND SELECTION OF BARRIER MATERIALS

For the design of a landfill or a similar type of application the following basic rules should be noted:

* Constant well-graded grain size distribution in order to achieve a high bulk density (dry), in view of the good impact on characteristics concerning convection and diffusion.

* Some content of very fine particles including clay minerals, especially for provision of sorption capacity in the barrier.

* Low content of free lime and organic matter If the above guidelines are observed, the barrier made of soils thereof will feature here a very low permeability ($K_f < 10^{-10}$ m/s), a high shear strength and stability, and an appropriate plasticity combined with the sorption capacity for contaminants of the clay minerals.

5) IMPLEMENTATION OF SOIL-BENTONITE LINERS

a) Mixed-in-Place (in-situ)

A simple example of execution of a mineral liner is when the in-situ soil needs only to be enriched with a very low bentonite percentage in order to achieve the necessary values, especially its permeability value. This method, known in technical terms as Mixed-in-Place (in-situ), is presented in a diagramm on figure 3.

Very powerfull soil milling machines (usually with > 300 horsepower), are needed to disperse the bentonite in the soil mass. The workable thickness of the layers for this method lies between 0,25 and 0,3 m,and the daily output reaches up to 3.000 m².

b) Mixed-in-Place (on-site)

The preparation of mineral liner materials on stockpiles, also known as Mixed-in-Place (on-site), is used especially for

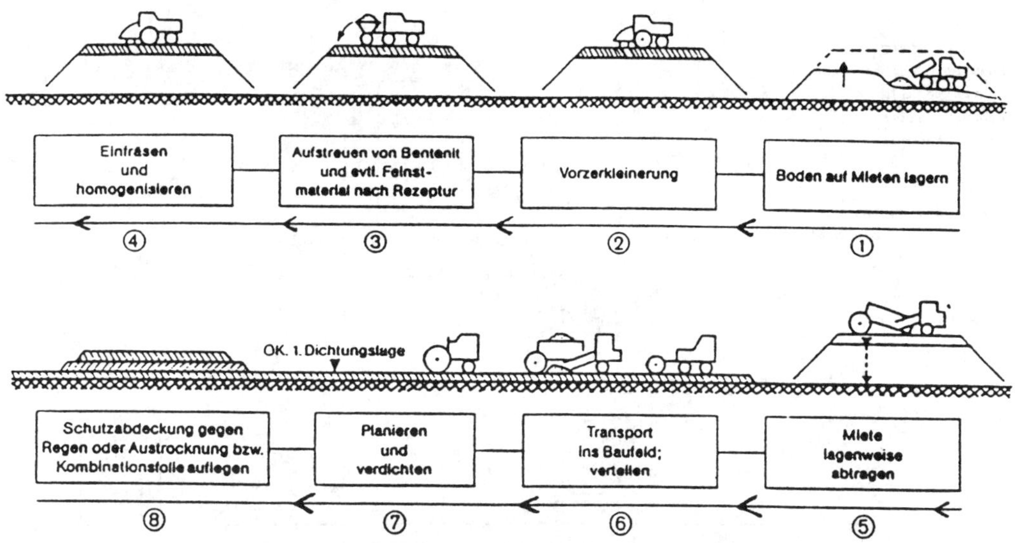

Fig. 4: Procedure of works for application of bentonite in the Mixed-in-Place-Process (on-site)

slopes, but also in congested sites, where the sealing procedure has to be executed. The usual procedure of this method is presented in figure 4. According to the conditions, the soil is brought from outside the job site and has, for this reason, to be loaded/unloaded and stockpiled.

It is furthermore possible, through careful execution, to produce a mixture of two kinds of soil as main components (for example sand and silt) and to enrich it with bentonite, after homogenizing. In some cases it is necessary to adjust the humidity. Thereafter, the mineral liner material will be taken from the stockpile to the final location, where it will be placed to profile and at the end compacted. The daily output of each machine work unit is comparable to the one described above under (a).

c) Mixed-in-Plant-Process (on-site/off-site)

About 15 years ago, the Mixed-in-Place-Process was established in landfill construction. The technical requirements and quality of works for landfills have been continuously increased since then, and the method described under the above items a) and b) has in some cases, reached its limits and capabilities. Similar to the production of plant-mixed construction materials (e.g. concrete) and to dam construction (e.g. clay cores), engineers and contractors started using modified batch plants for mixing of liner matetials. This process was called Mixed-in-Plant and offers to the par-ties involved, with a project, several advantages:

* Material preparation with little regard to weather conditions.
* Usage of one or two soils (as major components, depending on type of batch plant) along with milled clay minerals and fillers (e. g. bentonite and fly ash)
* Flexibility in case of changing soil conditions.
* Good monitoring and quality control
* Highly homogenous mixtures and good dispersement of fines
* Plant output, depending on soil type, up to 150 tons/hour.

Two types of batch plants which have been used on job sites in Germany and Switzerland are shown in figures 5 and 6.

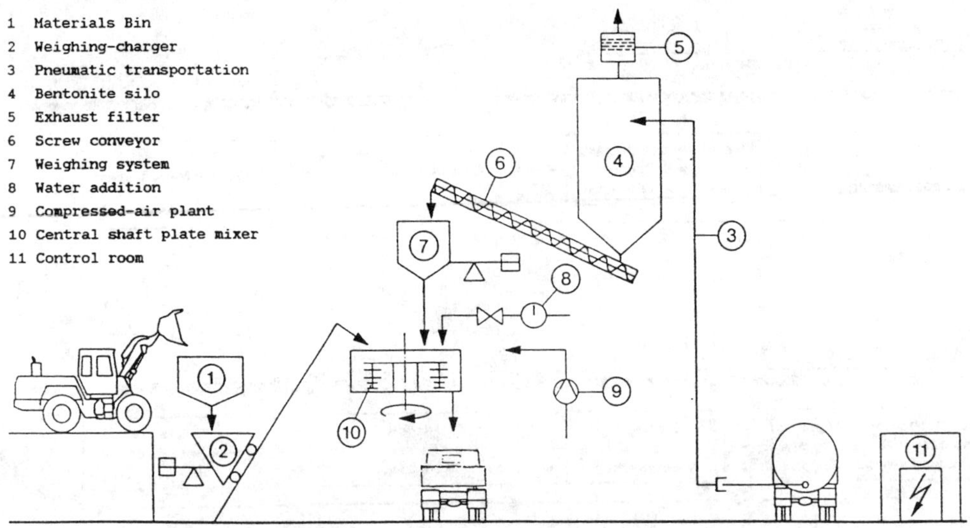

Verfahrens-Fließbild einer Mischanlage.

Fig. 5: Typical layout for discontinuous batch plant (stationary) for barrier construction

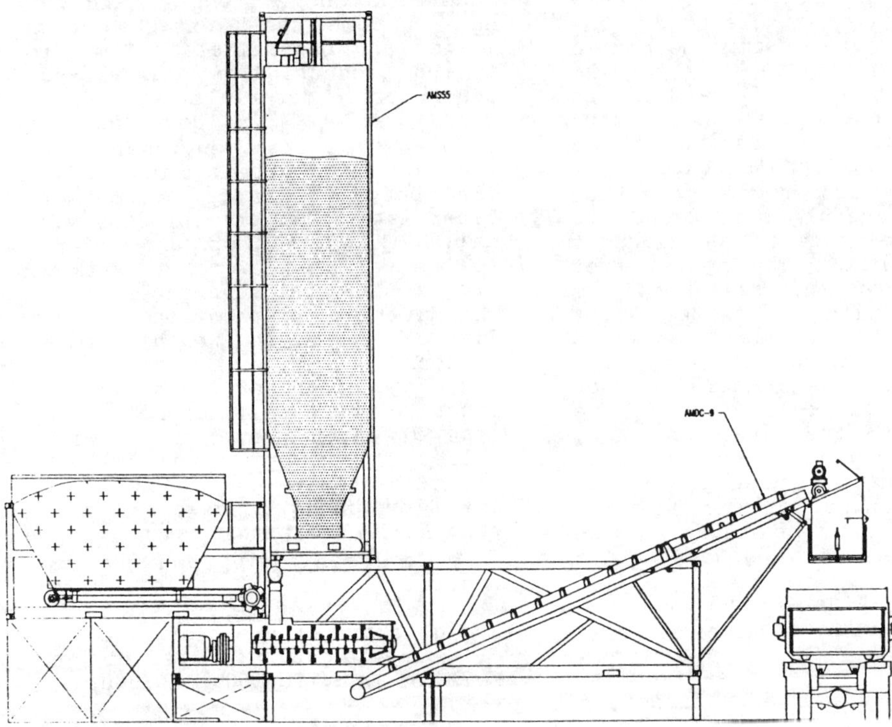

Fig. 6: Cross sectional view of mobile continuous batch plant

Discontinuous as well as continuous batch plants have been used in these applications. In Germany, about 30 landfill projects have been built with the Mixed-in-Plant-Process in the past few years. Therefore, the Mixed-in-Plant-Process is established and accepted in landfill construction as an equivalent to the process under a) and b).

The advantages of this process are obvious and the future perspectives of this process are the optimization of the barrier mix-tures according to local available materials, especially for large landfill projects (say > 50.000 m^2).

6) EVALUATION AND CONCLUSION

Barriers for environmental protection can be made of local soils with bentonite/filler addition. The necessary technology, plant and equipment is available in the market. Beside the Mixed-in-Place-Procedure, the materials can also be prepared in the sophisticated Mixed-in-Plant-Process. The function of clay minerals, especially bentonite, in the barrier can be best described as follows:

* Low permeability of the liner, due to the high swelling of bentonite
* Significant increase of plasticity
* Sorption of contaminants
* Blockage of voids in wide-graded types of soils

Finally it has to be concluded that, in regard to the numerous completed projects of landfills, mineral liners with bentonite additions constitute an established method of construction. By using natural building materials they contribute positively to the protection of the biosphere.

A substantial amount of the bentonite used for the above applications, especially in Central Europe, comes from the island of Milos (Melos), Greece, where some of the world`s largest and highest grade bentonite deposits are exploited. The bentonite is of the "calcium type" and is usually localy "activated" (sodium exchanged), crushed and dried and then shipped abroad for grinding and distribution.

Greece is the third bentonite producer in the world (after USA and ex-USSR) and the Greek bentonite has been used since many decades as a binder for foundry moulding sands, iron or pelletizing, drilling muds, civil engineering works like slurry-walls etc. and other applications.

REFERENCES:

Holzer, C., " Bestehende und geplante Regelungen betreffend die obertägige Abfalldeponierung in Österreich", in Deponietechnik und Entsorgungsbergbau; Fettweis, Golser & Hengerer (Publ.), 1992 Balkema, Rotterdam, Holland

Knight, L., "Landfill engineering advances in the UK", in Pollution Prevention; Issue October 1991

Kommission der Europäischen Gemeinschaften, "Vorschlag für eine Richtlinie des Rates über Abfalldeponien", 22. Mai 1991, Brüssel, Belgium

"Sechste Allgemeine Verwaltungsvorschrift zum Abfallgesetz: Technische Anleitung zur Vermeidung, Verwertung, Behandlung und sonstigen Entsorgung von Siedlungsabfällen (TA Siedlungs-abfall)", Bundesministerium für Umwelt, Entwurf vom 22.November 1991

"Technische Verordnung über Abfälle (TVA)", Der Schweizerische Bundesrat, 10.Dezember 1990

Weinmann, A., "Bentonit - ein Baustoff im Deponiebau", in proceedings to the seminar `Multi-Mineralische Deponie-Basisabdichtung`, 27. Nov. 1990, Bingen, Germany

Weinmann, A., "Mineralische Vergütung von Erdaushubmassen zum Bau von Deponiedichtungen" in proceedings to the seminar `Abfall-wirtschaft/Abfalltechnik`, 4.-5.11.92, Kaiserslautern, Germany

Waste Disposal by Landfill – GREEN'93, Sarsby (ed.) © 1995 Balkema, Rotterdam, ISBN 90 5410 356 6

Investigation of an eight year old slurry trench cut-off wall

P. Tedd & V. Paul
Building Research Establishment, Watford, UK

C. Lomax
Formerly MRM Partnership, Bristol, UK

ABSTRACT: An investigation of the behaviour of slurry trench cut-off walls used to contain contamination is described. An eight year old wall has been investigated and laboratory studies have included permeability, triaxial and chemical compatibility tests on samples from a number of walls.

1 INTRODUCTION

Slurry trench cut-off walls are increasingly being installed as barriers adjacent to landfill and contaminated land sites to control lateral migration of groundwater, leachate and gas. Much has been published about their construction and mix design (Chipp 1990, Jefferis, 1990, 1992). Their use in some recent projects is described by Barker et al (1992) and Ingle and Kavanagh (1992). The majority of the slurry trench walls constructed in the U.K. are less than 10m deep and are constructed using a continuous trench and a self hardening bentonite/cement slurry. The cementitious material is normally a blend of cement and ground granulated blastfurnace slag (ggbs). Pulverised fuel ash (pfa) has also been used.

The relatively recent use of slurry trench walls means that there is little information on their long term performance. The Building Research Establishment (BRE) is currently assessing techniques for examining the performance of these walls both in the short and long term. Prior to describing the findings of an investigation of an eight year old wall, the methods commonly used to test hardened slurry material to examine its compliance with typical design specifications are reviewed. The problems of testing a cementitious material at an acceptable early age to meet design specifications where the material properties continue to change for years are discussed. The difficulties and best practice of sampling and storage of both the liquid and hardened slurry are discussed.

The investigations carried out on the eight year old wall have revealed certain features of the wall. The methods used to investigate the wall are described and ideas on future testing are outlined. The results of laboratory testing, physical and chemical resistance and compatibility tests on samples from this wall and other walls are described.

2. REVIEW OF DESIGN SPECIFICATION AND TESTING METHODS

2.1 Design Specifications

Design specifications and testing of the hardened slurry material have been described in some detail by Jefferis (1990,1992). Tests on the hardened slurry are commonly carried out to measure the following properties:

a. Permeability
b. Strain at failure during drained triaxial compression
c. Unconfined compressive strength

For containment of aggressive contaminants chemical compatibility testing may also be required.

It is frequently specified that the coefficient of permeability is not larger than 10^{-9}m/s and the strain at failure is larger than 5%. The significance and relevance of these parameters and other tests are discussed in relation to the required performance of the wall. Most of the testing of the material is carried out using soil mechanics methods but the storage and timing of the testing tend to be based on concrete testing practice.

2.2 Sampling and storage

There is a contractual requirement to confirm that the hardened slurry material meets the design specification. To achieve this requirement, samples of the liquid slurry are usually taken from the trench at various depths. Slurry direct from the feed pipe will not contain material from the adjacent ground and therefore could be very different to that in the trench. The samples are then stored for testing at typically 28 and 90 days. Sampling the wall after the slurry has hardened is rarely undertaken. Methods of sampling the wall months or years after construction are referred to later.

To produce good quality samples, the material requires careful handling and storing. The recommendations made by Jefferis (1990) on the sampling and storage of the slurry material are worth restating. Some of these apply equally well to samples cored from a hardened wall. The need for particular care arises because the hardened slurry even when many years old is very susceptible to drying shrinkage and oxidation in air. Moisture contents are normally in excess of 150% and can be as high as 600%. At an early age the material is very weak and it becomes brittle with age. Knocking or dropping the samples can lead to cracking of the sample. The current best practice for storage and handling appears to be:

i. Plastic tubes should be used and are typically 100mm diameter and 600mm long. The cementitious material will cause corrosion of aluminium tubes. The use of demoulding oil is recommended prior to casting samples.

ii. Samples should be stored upright and once set they should be stored under water, but with the ends sealed with wax and capped. Experience has shown that where sample tubes have been capped and waxed, but not stored under water, cracking and oxidation have occurred such that some samples are unusable for testing.

iii. To reduce the risk of cracking, samples should be transported upright and only when they have attained some strength.

2.3 Permeability

The key parameter in the specification of slurry walls is the coefficient of permeability (k). By their very nature cut-off walls are usually installed in ground or fill of relatively high permeability and keyed into an aquiclude (an underlying stratum of low permeability). The effectiveness of the containment depends on the permeability and width of the wall (permitivity), the hydraulic gradient across the wall, the permeability of the aquiclude and on the depth of toe-in of the wall into the aquiclude. Most cut-off walls have been built 0.6m wide. The hardened slurry is specified by most designers to have a coefficient of permeability of 10^{-8} or 10^{-9}m/s, the lower value being increasingly required. This value is quoted in Waste Management Paper 26 (1986) and Waste Management Paper 27 (1989) where it is used to specify the properties of much thicker clay liners.

It is useful to quantify what a value of $k = 10^{-9}$ m/s means in terms of quantity of leachate through a wall and the breakthrough time for a particular hydraulic gradient. Assuming Darcy's law, the rate of flow, q, through a wall with a hydraulic gradient, i, of 2 can be calculated as follows.

$$q = kiA = 0.032 \times 2 \times A$$

$$= 0.064 \text{ m}^3 \text{ per year per m}^2$$

where
$k = 10^{-9}$m/s (0.032 m/year)
A is the cross-sectional area over which flow occurs

This flow rate per square metre is equivalent to the apparent velocity through the wall and gives a breakthrough time for a 0.6m wide wall of 9.4 years. The actual velocity assuming the material to have a porosity of 0.9 will not be significantly larger.

Permeability tests are normally performed in triaxial cells usually to BS 1377 although standards do not appear to be mentioned in the UK literature. Specified effective confining pressures range from 100 to 300 kPa and hydraulic gradients from 10 to 30.

Clearly laboratory permeability tests provide a valuable and necessary means of quality control on the hardened slurry to ensure it meets the required specification. However the following limitations exist:

i. Samples may not be representative of the actual material in the wall because of factors such as different stress conditions during setting and contamination from adjacent ground.

ii. The direction of the flow in the triaxial permeability test is vertical compared with the horizontal flow in the field.

iii. Only a small proportion of the actual slurry mix is tested.

2.4 Strain at failure

Another important property of the wall is its ability to sustain ground movement as any cracking of the wall could affect its overall permeability. The stress strain behaviour of the wall material is discussed in some detail by Jefferis (1990). A strain of 5% without failure is commonly quoted in specifications. According to Jefferis (1992) these values can only be achieved under drained triaxial conditions with the confining pressure larger than 50% of the unconfined compressive strength. Below this 50% confining pressure, brittle type failure at a strain of 0.2 to 2.0% can be expected. This implies that material less than 6m deep in a cut-off wall is unlikely to have sufficient confining pressure for it to behave in a non-brittle mode. The confining pressures in cut-off walls are difficult to estimate, but if the vertical stress is assumed to be equivalent to overburden pressure then taking a typical bulk unit weight of $12kN/m^3$ for the slurry material, the pressure at 6m would only be of the order of 72kPa. The unconfined compressive strength of slurry material is generally more than twice this value, see Section 2.5.

Some typical ranges of parameters used in drained triaxial tests are given in Table 1. This data was supplied by two test houses.

Table 1. Typical variables used in drained triaxial tests to obtain strain at failure

Age of sample, days	28 and 90
Confining pressure, kPa	50 to 300 (150)
Strain rate, % per hour	0.15 to 1.0

A confining pressure of 150kPa, which is most commonly used by one test house, is equivalent to a depth of 12.5m in the cut-off wall.

Although Jefferis (1992) describes the general behaviour of the material under varying degrees of confinement, little appears to have been published on the stress strain properties of the material.

The origins of the 5% strain at failure specification are unknown and according to Jefferis (1990) the value is arbitrary. Moreover the effective confining stresses required to achieve this value are in excess of the stresses likely to be found at depths at less than 6m. The effect of age of the sample does not appear to have been addressed. Even if the 5% strain at failure can be achieved with the necessary confining pressures at 28 and 90 days, it is questioned whether the same strain would be achieved on samples several years old, since as the material becomes stronger and stiffer its strain at failure will decrease.

2.5 Unconfined Compressive Strength (UCS)

Unconfined compression tests are carried out to confirm that the slurry is setting in a uniform way. Typical strengths required in specifications at 28 days are between 100 and 300 kPa. Schweitzer (1990) shows the development of UCS with age of a slurry mix, with mean strength of 350kPa at 28 days increasing to a mean of 1000kPa at four months. It appears that even after four months strengths are still increasing. The large scatter in the data is thought to result from the inhomogeneities in the mixing process and varying quantities of sand, presumably from the surrounding ground.

3 INVESTIGATION OF AN EIGHT YEAR OLD SLURRY TRENCH WALL

3.1 Background

The performance of the wall has been assessed by measuring the properties of samples taken from the wall, carrying out in-situ permeability tests within the wall and monitoring the ground water conditions adjacent to the wall. The study has involved a borehole investigation including installation of piezometers and wells on both sides of the wall, borehole sampling of the wall and installation of instruments in the wall for permeability measurement.

The wall was about eight years old at the time of the study and is adjacent to a landfill site which is being used for domestic refuse. The wall is 0.6m wide and approximately 7m deep at the location of the investigation, being keyed into an underlying stratum of clay. The top of the wall was capped with a layer of clay.

The mix used for the slurry was reported to be:

Bentonite	50 kg
Cement	63 kg
Slag	77 kg
Water	940 kg

Samples from the boreholes showed made ground between the surface and approximately 2m depth, peat between 2m and approximately 6m and below that a soft grey clay becoming stiff red brown with depth.

3.2 Piezometric levels and leachate analysis

Measurements from the well points and piezometers showed the maximum hydraulic gradient across the wall measured to date to be approximately two with the flow going away from the landfill.

Chemical analysis of the leachate showed that the risk of chemical attack of the cementitious material in the wall by the leachate is minimal as the pH was close to neutral and the sulphate concentrations in the leachate were low, falling into Class 1 as specified in BRE Digest 363 (1991). The chloride, magnesium and acid concentrations were also low. Only the ammonia concentrations could result in a moderately aggressive attack. It was interesting to note that close to the wall, the pH was nearly 12 indicating that the slurry wall was locally affecting the ground water.

3.3 Borehole sampling of the wall

A section of the top of the wall was exposed to determine its exact location and also to decide which method of drilling to use to sample the wall. Overspill from the wall had broken away parallel to the line of the wall. There was also evidence of a crack near the middle of the wall and running parallel to it. Samples of the overspill were removed to the laboratory for examination and testing, details of which are given later.

The surface of the top of the wall material was white, but immediately below the surface, it was the characteristic dark blue/green colour of a slurry mix containing ggbs. The change in colour is due to the ggbs in the material being oxidised from the dark blue/green sulphide to the white sulphate on the surface.

A preliminary examination of the top of the wall revealed it to be much harder and more brittle than expected. Therefore it was decided to use rotary coring with a diamond bit and water flush to obtain samples instead of the thin wall push sampling originally envisaged. Using this method the material was very easy to core. Coreline was used, giving 70mm diameter samples. The ends of the samples were waxed with a proprietary soft soil sealing wax and then capped leaving large air voids. A borehole on the centre line of the wall was drilled to a depth of 2.85m below the top.

The crack running along the middle of the top and parallel to the wall extended down to a depth of 0.95m. The material on both sides of the crack had oxidised to a depth of approximately 20mm. The wall material became weaker with depth and significant quantities of peat were found in the cores. In places the wall was very friable. The cores tended to break up into short lengths no greater than 300mm long. Pockets of peat were found even in the longer lengths of core.

The core and the overspill material have been used for permeability tests, triaxial tests and various exposure and compatibility tests. These are described in Section 4.

3.4. In-situ permeability tests

Various methods were considered to measure the in-situ permeability of the wall all of which involved drilling boreholes vertically on the centreline of the wall. A major concern with this approach is that at depth, it is difficult to be certain of the exact location of the borehole in relation to the sides of the wall.

There are few reported cases where in-situ permeability measurements have been made in cut-off walls. In one case (Bertero, 1983), piezometers were installed into the wall while the slurry was still fluid and permeability measurements were made some 3 months later. In this case the results were satisfactory giving permeabilities of less than 10^{-9}m/s.

One of the techniques considered was the BAT ground monitoring system (Torstensson, 1984). The system can be used for the following functions:

- permeability tests
- measurement of pore water pressure
- tracer tests
- sampling of ground water

An important aspect of the system is

that permeability tests can be carried out in a relatively short time.

The instrument consists of a sealed piezometer approximately 30mm diameter, into which access is gained for pressure measurement and sampling via a hypodermic needle puncturing a membrane. The piezometer part or probe is normally installed like a drive-in piezometer or into a pre-drilled borehole with the probe being pushed some distance below the bottom of the borehole to obtain a seal. Such a method of installation would be impossible in the hard brittle material found in cut-off walls. A method was tried in which a slightly undersize hole approximately 40mm diameter was drilled with a hand screw auger. The technique has proved satisfactory in another wall where BRE have been working, however it was not possible to obtain a satisfactory seal at this wall.

Another approach has been to drill an open borehole and carry out falling or rising head permeability tests. This approach has also been used by Schweitzer (1989). Using an analysis by Horn (1986), values of k have been determined for different hydraulic gradients as the water level in the borehole changes. It is considered by Schweitzer that this approach provides more representative values of k than laboratory tests as the volume of material tested is larger and it takes into account inhomogeneities in the wall. It has the advantage over a piezometer or BAT probe which only measure permeability for a small volume of material.

4. LABORATORY TESTING OF SAMPLES FROM THE EIGHT YEAR OLD WALL

4.1 Exposure and immersion tests

Samples from the top of the wall and from the borehole have been used to examine the behaviour of the hardened slurry in various environments.

1. Exposure to atmosphere
Part of the overspill material from the top of the wall and a core sample 2.5m below the top of the wall were stored in the laboratory at a temperature of 19°C and a RH of approximately 60%. Within hours the samples had oxidised causing the colour to change from the initial dark blue/green (sulphide) to white (sulphate). By 12 days large cracks had developed particularly in the sample from the top of the wall. By two months both samples had disintegrated into small pieces. The material became very weak, brittle and friable. As the core sample dried out, it revealed inclusions of peat in small lumps and many small fibres.

These tests demonstrate the importance of protecting the top of walls with a clay cap to prevent drying. The tests also demonstrate the care required in storing material even when many years old to prevent drying and oxidation. Most of the samples stored for any length of time in the coreline showed some evidence of oxidation despite being sealed with wax.

2. Immersion in water
A core sample approximately 2m below the top of the wall was immersed in water with a pH of 7.8. After two days the surface of the sample exposed to atmosphere had oxidised whereas the sample immersed in water had not. The surface of the sample immersed in water gradually oxidised after several weeks, but no cracking was observed even after two months. After two months the pH of the water had risen from 7.8 to 10.8.

3. Exposure to acetic acid
A core sample 70mm high from 1.8m below the top of wall was immersed in a solution of acetic acid with a pH of 2.9. The acetic acid was used to simulate a very aggressive landfill leachate. (The pH of the leachate at the site was approximately 7.) After two months the pH of the solution had increased to 5.8, but apart from the surface changing to near black, no other visible sign of deterioration had occurred. After six months the surface of the sample had softened considerably.

4.2 Permeability and chemical compatibility tests

These tests were carried out using a flexible wall permeameter in which aggressive permeants could be used as well as water, thus allowing the water permeability and chemical compatibility of the hardened slurry to aggressive leachates to be determined. The aggressive leachate used in these tests was acetic acid with a pH of 2.9, as used in the simple immersion tests.

Two tests were carried out on samples from the wall, one from the top and one from a core approximately 1.3m deep. The sample size was 70mm diameter by 70mm high. The effective confining pressure used was 100 kPa and the hydraulic gradients used varied from 3 to 210. Both

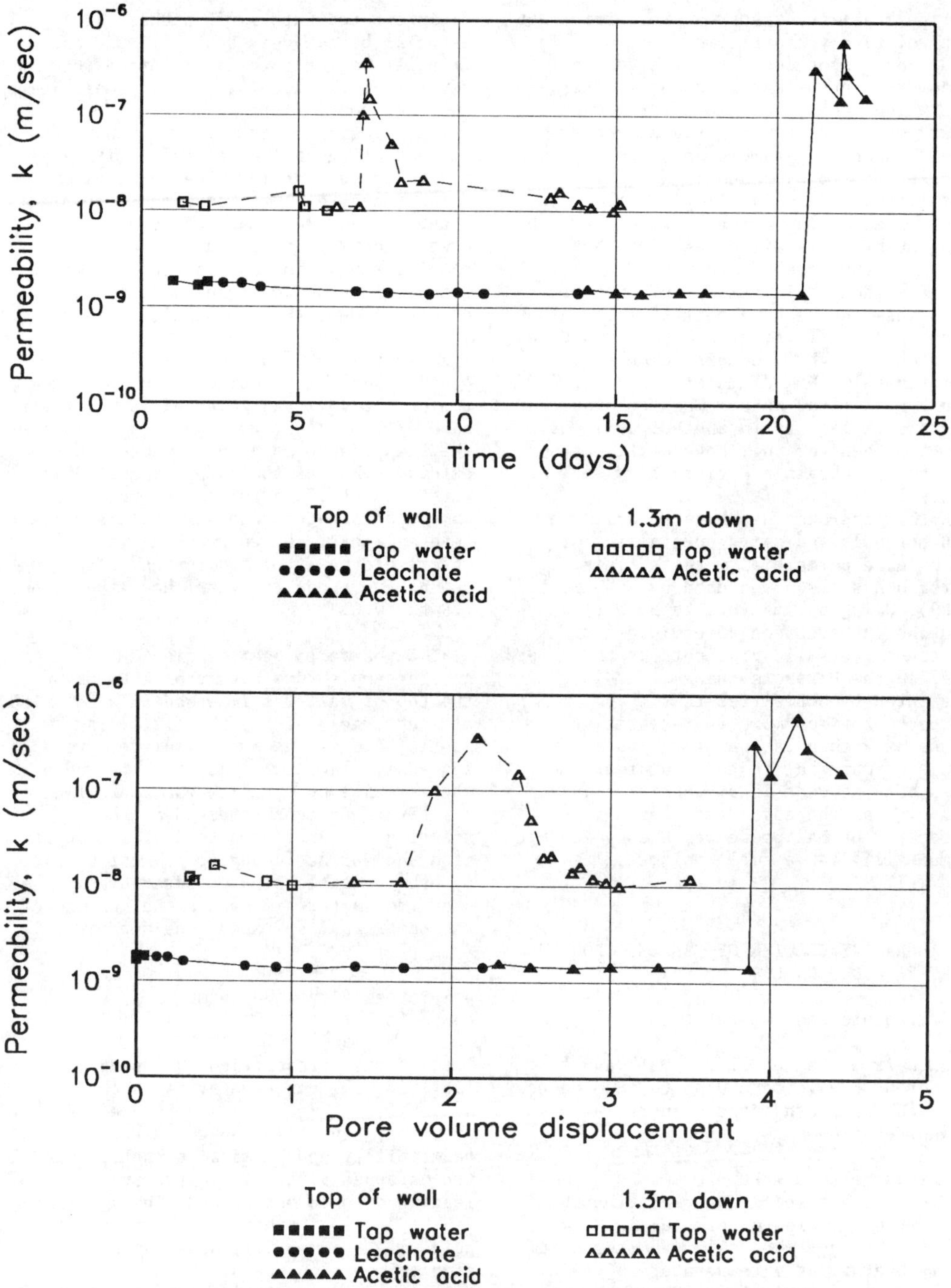

Fig. 1 Laboratory permeability results on samples from cut-off wall showing effect of using tap water, leachate and acetic acid, (above) plotted against time and (below) plotted against pore volume of sample

samples were saturated with tap water. The measured permeability for the two samples are plotted against time in Fig. 1a and pore volume in Fig. 1b, with variations in the permeant being indicated.

1. Sample 1, from top of wall. Permeabilities of nearly 10^{-9}m/s were measured using tap water or ground water. Increasing the hydraulic gradient from 3 to 100 to 210 did not affect its value. Changing the permeant to acetic acid resulted in an increase in permeability to between 10^{-6} and 10^{-7}m/s. This increase in permeability corresponded to a decrease in pH of the effluent. It is thought that this increase occurred because the acetic acid leached out the cementitious material from the sample making it more porous. The test was terminated before it was possible to establish if the permeability would have continued to increase or remain stable.

2. Sample 2, 1.3m below top of wall. Groundwater was not used in this test. It can be seen from Fig. 1 that using tap water, a permeability value of 10^{-8}m/s was obtained, nearly an order of magnitude larger than that in Sample 1. The larger value was probably due to the presence of peat in Sample 2 which was not present in Sample 1. Again changes in hydraulic gradient did not affect the permeability. Less permeation of acetic acid was required to cause an increase in the permeability, see Fig. 1b. The increase caused by the acetic acid was followed by a decrease to the original value of 10^{-8}m/s. The reason for this is not known.

4.3 Undrained triaxial compression tests

Table 2 summarises some engineering properties of hardened slurry taken from the old wall. The low unit weight compared to soil or concrete is reflected in the high water contents. The quick undrained triaxial tests were carried out with a cell pressure of 100kPa. The samples from the top of the wall were 38mm diameter by 76mm high and the strain rate 2.4% per hour. The core sample was 71mm diameter by 142mm high and the strain rate was 1.3% per hour. In each test failure occurred mainly by the formation of near vertical cracks. The strain at failure (maximum applied stress) in every case was significantly less than the 5% that can be achieved from a drained confined test on samples 28 days old. The stress/strain curve for the core sample is shown in Fig. 2 and is typical of the curves from the other tests. It can be seen in Table 2 that the sample from the core was significantly weaker and less stiff than the samples from the top of the wall.

Table 2. Properties of hardened slurry eight years old from quick undrained triaxial tests

Sample location	w %	γ kN/m^3	c_u kPa	E_u MPa	Strain at failure,%
Top 1	347	11.6	550	220	1.2
Top 2	412	11.3	330	165	0.6
Core	380	11.4	210	92	0.7

where
w is the moisture content
γ is the unit weight of material
c_u is undrained shear strength
E_u is the undrained modulus

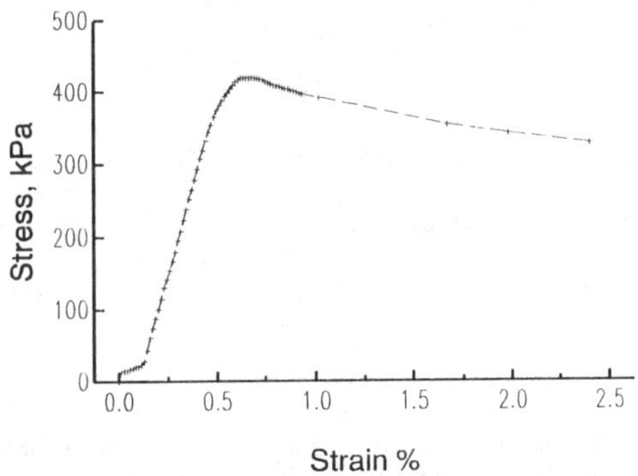

Fig. 2 Stress/strain behaviour of slurry sample from the eight year old wall

5 CHEMICAL RESISTANCE

The chemical resistance of hardened bentonite/cement slurries depends on the intrinsic resistance of both the cementitious material and the bentonite. Therefore when assessing the durability of slurry trench cut-off walls, consideration needs to be given to chemical attack on both components of the slurry mix. There is a substantial body of information on the effect of chemicals on both cement and concrete (ACI, 1987; Biczok, 1972, BRE, 1991) and clays (Acar et al, 1984, Alther et al., 1984, Madsen and Mitchell, 1989) and this can be used to identify potential problems. Care needs to be taken when using information on cement, concrete and clay durability to assess the likely performance of bentonite/cement slurries in aggressive environments. For example, Madsen and Mitchell (1989) note that the effect of chemicals on clays with a high water ratio, as in the bentonite in slurry cut-off walls, may be greater than on lower water content compacted clays such as those used as liners at landfill sites. Greater particle mobility and easier opportunity for fabric change are thought to be responsible for lower chemical resistance of the higher water content clays.

Two methods were used to assess the chemical resistance of hardened slurry from the eight year old wall, the immersion test and the permeation tests. In both cases acetic acid was used to simulate an aggressive landfill leachate. Further immersion tests have been carried out on samples from other walls using the following chemical solutions:

(a) sodium sulphate (28660 mg SO_4/l), sodium sulphide (320 mg S^{2-}/l), phenol (23 mg/l), pH 6.2.
(b) sodium sulphate (24930 mg SO_4/l), sodium chloride (490 mg S^{2-}/l), phenol (17 mg/l), pH 7.5;
(c) sulphuric acid (pH 2).
(d) sodium sulphate (3500 mg SO_4/l).

Solutions (a) and (b) were formulated using data from a number of site investigation reports on contaminated land obtained by BRE. Solution (a) reflected the ground chemistry of a highly contaminated gasworks site and solution (b) a chemical works site.

The samples were 100mm diameter by 150mm high. They came from a two year old bentonite/cement/pfa slurry wall and nine month old samples of a bentonite/cement/ggbs mix.

Samples were immersed in the solutions for six months. The pH of the solutions was monitored on a daily basis (except at weekends) and adjusted where necessary by the addition of concentrated sulphuric acid. In addition, the solutions were changed and renewed every month. This storage regime ensured that the samples were exposed to a constant chemical environment for the duration of the test. At each change of solution the chemical compatibility of each sample was assessed by visual examination and weight change.

Results from the immersion tests are shown in Table 3 and Fig. 3. They indicate that the samples containing pfa possessed a greater degree of chemical resistance than those containing ggbs under the conditions of the test. Two modes of failure, dissolution and cracking, were noted in the chemical immersion tests. The observed modes of failure were found to be dependent on both the chemical composition of the sample and the chemical solution in which the sample was immersed. Acid attack resulted in dissolution of both the pfa and ggbs samples. Pfa samples exposed to the sulphate solutions (a,b, and d) and the ggbs samples exposed to the weaker sulphate solution (d) showed cracking due to expansion reactions. The ggbs samples exposed to solutions (a) and (b) initially showed cracking due to expansion but ultimately disintegrated into a mush by dissolution.

It must be stressed that these immersion tests were very severe. They were also carried out on unconfined samples and may not reflect exactly the reactions that might occur in the ground where the hardened slurry is confined.

Table 3: Results of immersion tests for samples of bentonite/cement/ggbs mix

Solution	(a)	(b)	(c)	(d)
Percentage weight change at 60 days	disintegrated[3]	disintegrated[4]	-76.0[1]	+2.2[2]

Notes
-ive is loss of weight
1. Surface of sample rapidly changed colour from blue to orange
2. Sample showing slight cracking around edges
3. Sample suffered severe cracking after 1 week, disintegrated into mush by 15 days
4. Sample suffered severe cracking after 1 week, disintegrated into mush by 20 days

Solution	(a)	(b)	(c)	(d)
Percentage weight change at 150 days	disintegrated[1]	-5.6[2]	-48.0	-1.9

Notes
-ive is loss of weight
1. Sample lost 20% weight after 90 days and split into 2 pieces by 120 days
2. Sample showing signs of severe cracking around edges

Fig.3 Results of immersion tests on core samples from bentonite/cement/pfa slurry wall

Jefferis (1989, 1992) has carried out similar immersion tests in sulphate solutions and reports that whilst sulphate attack in unconfined conditions leads to expansive reactions and cracking of cement-bentonite, under confined conditions the reaction leads only to a slight softening of the material and no overall increase in permeability. Although severe, immersion tests provide a simple way of assessing the chemical resistance or durability of various mixes to a particular leachate. Barker et al (1992) have reported the use of immersion tests to modify mixes to improve their durability to a particular leachate.

6. CONCLUSIONS

There is little published information on the long term performance of slurry trench walls despite their use for more than 20 years. There are few reported cases of problems with cut-off walls.

Investigations at the eight year old wall have shown the hardened slurry to be a stiff brittle material that contains significant proportions of the adjacent ground. This contamination of the slurry has been observed in a number of walls and could affect the performance of the wall depending on the nature and proportion of the included ground. Exposure to atmosphere has demonstrated how susceptible the hardened slurry is to drying shrinkage even when many years old and the importance of protecting the top of the wall with an effective clay cap.

Rotary coring has proved to be a satisfactory method of obtaining samples from a slurry wall many years old. Thin wall or piston sampling may be suitable for a 28 day old wall but would not be possible when the wall has reached close to its full strength. Great care is required in the storing and handling of samples whether taken from liquid slurry or from cores of a wall.

A satisfactory inexpensive in-situ method of measuring permeability would appear very desirable. The main problem is the installation of an instrument centrally in the relatively narrow wall at depth. The use of small diameter hand augered open boreholes could provide a simple means of measuring permeability particularly near the top of the wall. Such a technique allows minimum disruption to the wall and overlying clay cap.

There appears to be a need for a standard approach to the testing of hardened slurry, particularly in relation to the control parameters used in triaxial tests to measure strain at failure.

Chemical immersion and permeation tests have shown that bentonite/cement slurries are susceptible to attack by strong acid and sulphate solutions. Initial results from the chemical immersion tests indicate that slurries containing pfa are more resistant to chemical attack than those containing ggbs.

FUTURE WORK

BRE is continuing its field and laboratory investigations into the behaviour of slurry trench cut-off walls. There is a need to have access to more walls particularly those which have been in the ground for a long time.

In collaboration with AMEC Piling, a trial length of wall has been built for purposes of research. Various in-situ permeability methods will be assessed and eventually part of the wall will be examined using a trial pit.

ACKNOWLEDGEMENTS

The co-operation of several site owners, consulting engineers and contractors is acknowledged. The work is funded by the Department of the Environment.

REFERENCES

Acar,Y.B., A.Hamidon, S.D.Field & L.Scott 1984. The effect of organic fluids on hydraulic conductivity of compacted Kaolinite. Hydraulic Barriers in Soil and Rock. Eds Johnson et al. ASTM STP 874. pp 171-187

ACI 1979. A guide to the use of water-proofing, damp proofing, protective and decorative barrier systems for concrete. Report by ACI 515. 1R-79.

Alther,G., J.C.Evans, H-Y.Fang & K.Witmer 1984. Influence of inorganic permeants upon the permeability of bentonite. Hydraulic Barriers in Soil and Rock. Eds Johnson et al. ASTM STP 874.

Barker,P.J., Godalming & A.Esnault 1992. Developments in Techniques of ground treatment for the protection and rehabilitation of the environment. Proc. of the 2nd Int Conf on Construction on Polluted and Marginal Land. pp389-394.

Bertero,M., P.Marcellino & A.Paviani 1983. Determination of in-situ permeability of slurry trench cut-off. Symp on recent developments in laboratory and field tests and analysis of geotechnical problems. Asian Institute of Technology, Bangkok, Thailand.

Biczok,I. 1972. Concrete corrosion and concrete protection. English edition. Hungarian Academy of Science. Budapest.

BRE Digest 363 1991. Sulphate and acid resistance of concrete in the ground. Building Research Establishment, Garston, Watford

Chipp,P.N. 1990. Geotechnical Containment Measures for Pollution Control. Int Conf. on Construction onPolluted and Marginal Land.Brunel pp103-116

DOE 1986. Waste Management Paper No 26. Landfilling Wastes, HMSO

DOE 1989. Waste Management Paper No 27. The Control of Landfill Gas, HMSO.

Horn,A. 1986. In-situ Prufung der Wasserdurchlassig von Dictwanden. Geotechnik,H.1.S.37

Ingle,J.L. & S.T.Kavanagh 1992. Pollution containment using cut-off barriers recent studies. Proc. of the 2nd Int Conf on Construction on Polluted and Marginal Land. pp389-394.

Jefferis,S.A. 1989. The design of cut-off walls for waste containment. Land Disposal of Hazardous Waste: Engineering and Environmental Issues. Eds Gronow J R, Schofield A N and Jain R K. Ellis Horwood Limited, Chichester.

Jefferis,S.A. 1990. Bentonite-cement cut-off walls for containment:from specification to in-situ performance. Symposium on Management and control of waste fill sites. Paper 5.

Jefferis,S.A. 1992. Containment - grout interaction, ASCE Speciality Conference. Grouting, Soil improvement and Geosynthetics, New Orleans, pp1393-1402

Madsen,F.T. & J.K. Mitchell. 1989. Chemical effects on clay hydraulic conductivity and their determination. Nr. 135. Mitteilungen des Institutes fur Grundbau und Bodenmechanik Eidgenoss-ische Technische Hochschule Zurich.

Schweitzer,F. 1989. Strength and permeability of single-phase diaphragm wall. Proc of XII ICSMFE. Rio de Janeiro. Vol 3. pp1515-1518

Torstensson,B.A. 1984. A new system for Ground water Monitoring. Ground Water Monitoring Review. pp131-138

Waste Disposal by Landfill – GREEN'93, Sarsby (ed.) © 1995 Balkema, Rotterdam, ISBN 90 5410 356 6

General report

C. Sagaseta
Division of Ground Engineering, University of Cantabria, Santander, Spain

SUMMARY: This Report covers the Papers presented to Session 5: "Old landfill sites treatment". The papers are not commented separately, but instead they are integrated into a general presentation of the topic. The Report is divided into three main sections: description of problems posed by old landfills, methods of evaluation of these problems and remedial measures.

1 INTRODUCTION

The theme of this Session covers a really broad scope, because all aspects of old landfills are included, from direct environmental impacts (contamination of aquifers, gas emissions, ...) to reclamation of old waste fill areas, including the improvement of their bearing capacity as foundation ground. It is not the aim of this Report to be a State-of-the-Art of such a wide scope, but only to present the body of related papers within a general framework.

A total of 9 Papers have been allocated to this Session. They differ not only in the problems addressed, but also in their scope, from chemical or biological to mechanical viewpoints. This variety of subjects and aims is one of the guiding ideas of the Symposium.

The Session papers, as well as relevant papers from other Sessions, are listed at the end of the References. They will be cited in the text underlined and followed by the paper number.

The Report is divided in three main chapters, devoted respectively to the presentation of problems, methods of evaluation and remedial measures. For each chapter, a general outline is presented, but only the particular topics addressed by the papers are detailed.

2 PROBLEMS ASSOCIATED TO OLD LANDFILLS

The problems posed by old landfills arise mainly due to the lack of control during their construction. They can be classified in two categories:

i) 'Direct' environmental impacts, i.e., effects produced by the landfill on its environment.
ii) 'Passive' effects, i.e., problems which appear when some action is taken in order to re-use the landfill for other purposes.

2.1 Active environmental hazards

The effects of uncontrolled old landfills on their environment are of variable nature. A comprehensive list, as given in the CIRIA Report (Leach and Goodger, 1991), can be summarized as follows.

- Physical effects:
  - Obstructions (old foundations, pipes)
  - Slope instability at the edges of the landfill
- Chemical and biochemical effects:
  - Health hazards, from substances affecting human, animal and plant life
  - Combustion. Fire and explosions
  - Gas and leachate emissions. Groundwater pollution
- Physico-chemical effects:
  - Expansive reactions of some slags, sulphate, pyrite, etc.

By far, the most important effect of old landfills on their environment is the danger of groundwater pollution through leachate infiltration. In absence of base seals, the permeability of the natural ground is the key parameter governing the problem. The paper by Kovačić et al. [5.4] presents an extreme example in Zagreb, of an old waste dump, used since 1965, covering an area of

600,000 m^2, located on the bank of Sava river, directly on the alluvial materials. The average permeability of the alluvium is 10^{-2} m/s, and no base seal has been installed. Several measures are being envisaged to avoid groundwater contamination. In contrast, in the case presented in the paper by Beine and Dahlmann [5.5], the ground permeability is of 10^{-9} m/s, which represents a natural barrier against leachate migration.

Slope stability problems in landfills have received considerable attention in the past. However, most of the reported cases of slope failures in domestic wastes landfills are associated to failure of the natural foundation ground (Sánchez-Alciturri et al. [3.6]), or along the base seal (Mitchell et al., 1990). The paper by Güler and Avci [6.3] is a remarkable exception to this rule, and a huge slide is reported in a domestic waste landfill. The instability seems to have been triggered by insufficient gas and leachate control.

2.2 Passive effects

The importance of passive environmental effects of old landfills depend mainly on the type of use planned.

In the paper by Boroń and Nagawiecka [5.6], the possibility of reclamation of soda waste areas in Krakow is analyzed, for the potential use for agricultural purposes. It is found that three years after inundation, the soil contents of toxic substances for plants has decreased to allowable values. Grasses develop well without any addition to the soil, whilst forest-type plantations require peat and mineral soil additions.

If the intended use of the landfill involves the construction of structures or transportation facilities, the lack of bearing capacity is the most important problem. Papers by Oteo and Sopeña [5.3] and Genske and Noll [5.7] deal with this topic. In close relation with the bearing capacity, self-induced settlements are one of the most relevant phenomena in old landfills. Several Symposium papers deal with this problem (Sánchez-Alciturri et al. [3.10], Dunn [5.9]). These topics which will be commented in more detail in the following sections devoted to evaluation and remedial measures.

3 EVALUATION OF PROBLEMS

3.1 Exploration methods

Quantitative evaluation of the problems presented in the preceding section requires a previous characterization of material properties, including laboratory and in situ testing. Most methods developed for soils have been used for wastes, although subjected to some restrictions.

Undisturbed samples are very difficult to obtain in most cases, and this fact enhances the relevance of in situ tests. However, the present experience with the interpretation of these tests in wastes is limited, and very often only qualitative.

Static and dynamic penetration tests have been extensively used, because they are an easy and inexpensive way to investigate the behaviour of wastes in situ and its variation with depth and time. From the available experience, it seems that domestic wastes behave in a way similar to loose fine granular soils (sands and silty sands), (Sánchez-Alciturri et al. [3.6]) (fig. 1).

Surface loading tests have also been used. However, the layered nature of landfills requires that the loaded area is large enough to involve a representative thickness of the landfill. On the other hand, the necessary loads are relatively low, and hence several low-cost systems have been developed, such as the use of skips (Charles and Driscoll, 1981), or conical heaps (Welsh, 1983; Schmertmann, 1993) (fig. 2). In the paper by Genske and Noll [5.7], reference is made to a large load test on an old landfill.

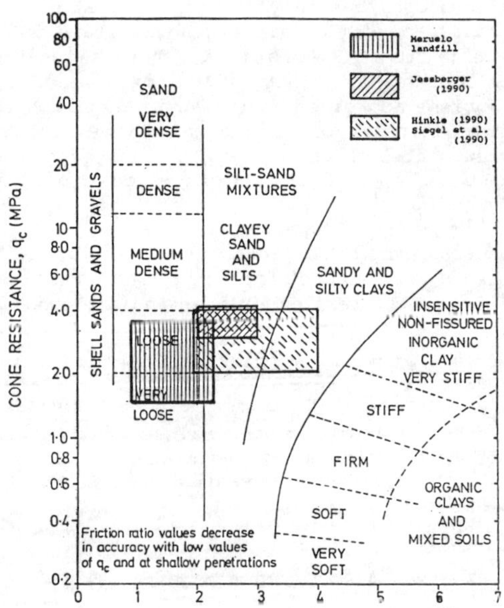

Fig. 1 Situation of CPT results in wastes in the Schmertmann's chart (Sánchez-Alciturri et al. [3.6])

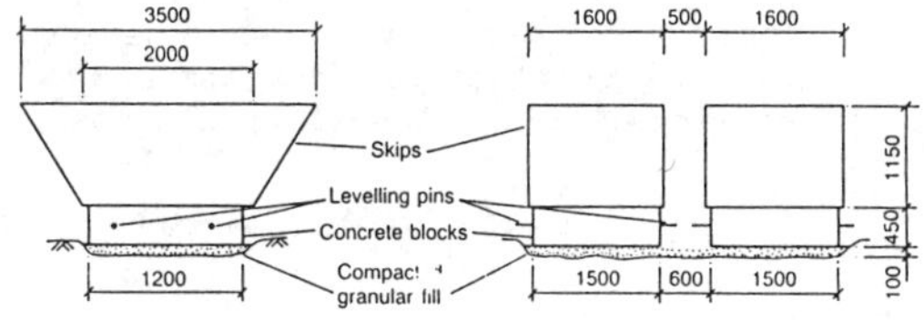

Skip test (Charles and Driscoll, 1981)

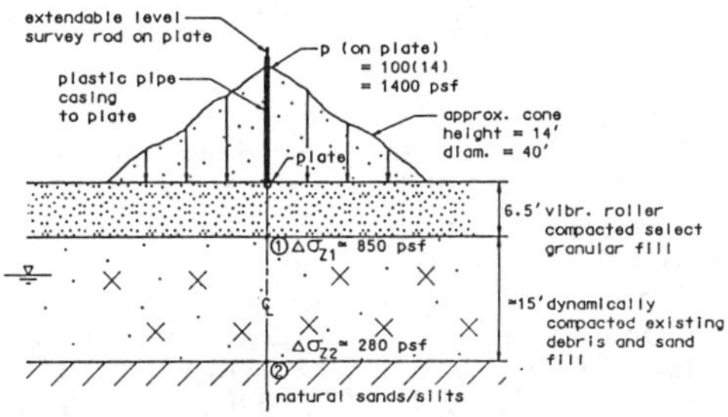

Conical load (Welsh,1983; Schmertmann,1993)

Fig. 2 Some tests arrangements for fills.

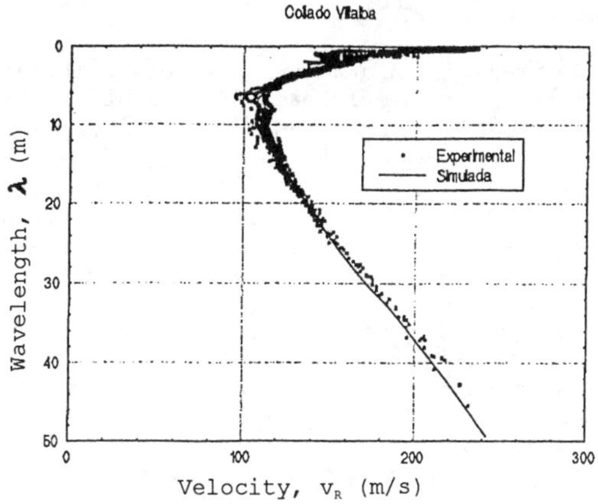

Fig. 3 Spectral Analysis of Surface Waves in landfills (Cuéllar et al., 1993).

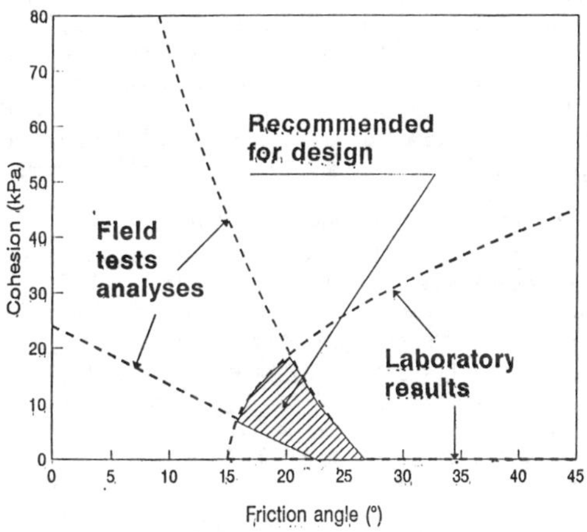

Fig. 4 Shear strength of wastes. Design values (Sánchez-Alciturri et al.[3.6])

Geophysical methods are very valuable for profiling landfills. Application of resistivity measurements are presented in the paper by Meju [5.2]. Seismic methods have the advantage that the measured parameter, the propagation velocity of seismic waves, is related to the material deformation modulus. Classical surface refraction techniques are not applicable, because of the presence of a stiffer surface crust. In the last years, the spectral analysis of surface waves (SASW) has been developed, providing an alternative way to overcome this difficulty (Abbiss, 1981). The method consists on measuring the velocity of propagation of Rayleigh waves of different frequencies. From the analysis of the 'dispersion curve' (wavelength vs. phase velocity), the profile of velocity with depth can be derived. Cuéllar et al. (1992) have applied this technique to landfills with satisfactory results (fig. 3).

3.2 Shear strength

Some wastes exhibit a shear strength behaviour similar to typical soils, and hence slope stability problems take place and can be analyzed accordingly. Some industrial and mining wastes and demolition debris belong to this category. In other cases, the waste behaviour is not 'soil-like', and hence slope stability problems are different from soils. Domestic wastes are an example of this, due to the combined presence of organic parts and paper and plastic strips. As a result, the material can sustain very large shear strains without signs of reaching failure (Jessberger, 1990). In these cases, strength parameters ('cohesion' and 'friction') are defined on the basis of allowable strains rather than a true strength. The papers by Sánchez-Alciturri et al. [3.6] and Jessberger and Kochel [3.3] deal with this point. It comes out from the available data that wastes develop a significant friction, behaving more as granular soil rather than as pure cohesive materials (fig. 4).

3.3 Compressibility

Compression of wastes by external forces, either structural loads or self weight, is in part immediate and in part delayed in time by creep. In sanitary landfills, internal biodegradation of wastes provides an additional component of differed settlement, which is added to the creep component of load-induced settlements.

For the analysis of immediate, load induced compressibility, methods derived from Soil Mechanics are followed, either using elastic approaches or e-log σ' laws (Sowers, 1973). Typical values of the compression index, C_c, for domestic wastes are in the range 1-4, corresponding to Young's moduli of 50-200 kPa.

For the analysis of delayed settlements, there are still few direct evidences of their final values, due to the lack of long enough observation periods and/or to uncertainties about the early stages of old landfills (precise dates and elevations, etc.). For self weight action only, several authors quote final settlements in the range 10-40% of the landfill thickness, depending on the type of wastes and degree of compaction achieved at the time of placement (Leach and Goodger, 1991).

With respect to the development of settlements with time, absolute settlements can not be measured if the early conditions of the landfill are not well defined. In this situation, only relative movements with respect to a given time, or, alternatively, the settlement rate (derivative of settlement with respect to time) can be measured.

Theoretical analysis of delayed settlements is difficult, due to the interaction of very different mechanisms, as stated above (creep, internal biodegradation), which follow different laws and are governed by different factors. Hence, only phenomenological approaches have been followed. A number of time-settlement laws have been proposed to match actual observations. Some desirable features of these fitting laws can be stated as follows:

a) To be dimensionally correct.
b) To be defined by a short number of parameters (1-3 is a usual range)
c) These parameters having a physical meaning, or being related to some known properties, so that reasonable ranges can be given to them.
d) To give reasonable predictions for long time intervals.
e) To separate the influence of as many factors as possible (for instance, landfill thickness, compacity, etc.).

Sowers (1973) proposed a logarithmic law, as used for secondary consolidation of soils. The settlements were assumed to be proportional to the landfill thickness, H:

$$s = H.\frac{C_\alpha}{1+e_0}.\log_{10}\frac{t}{t_1} \qquad (1)$$

The settlement rate then results:

$$v = \frac{ds}{dt} = 0.434.H.\frac{C_\alpha}{1+e_0}.\frac{1}{t} \qquad (2)$$

Eqs. (1) and (2) depend on two parameters, C_α and e_0. However, back analysis of recorded settlements in actual landfills presented by Sowers (1973) and supported by more recent observations, show that the coefficient C_α increases with void ratio, e_0, and hence, the ratio $C'_\alpha=C_\alpha/(1+e_0)$ can be taken as nearly constant. Values of C'_α are in the range 0.02-0.08, with most data in the lower range (0.02-0.05).

Yen and Scanlon (1975) presented measurements of settlement rates in two landfills along a period of more than ten years, and for landfill thickness from 12 m. to more that 30 m. They suggest a logarithmic variation of the settlement rate:

$$v = a - b.\log_{10}t \qquad (3)$$

The authors give values for the parameters "a" and "b". Both increase roughly linearly with the landfill thickness, H. Settlement rates from eq. (3) decrease with time, becoming zero for about 200-300 months. However, observations of older landfills show that settlements do continue for longer times (Coduto and Huitric, 1990; Dunn [5.9]). Similarly, measurements at earlier stages of landfills (Sánchez-Alciturri et al. [3.10])show that the settlement rate is not linear with log t when wider time periods are considered ($\Delta t/t \geq$ 5-10). On the other hand, for the time range covered by the Yen and Scanlon measurements (40 to 120 months), the settlement rates agree very well with the values predicted by the Sowers (1973) model (eq. 2), for C'_α of 0.03-0.05. The discrepancies claimed in the paper are due to a different definition of the settlement rate from eq. (2).

Among other laws which have been used, the linear viscoelastic model, proposed by Edil et al. (1990) and Mannassero has several merits and will be discussed here. This law comes from the parallel association of a linear spring and a dashpot and is expressed by:

$$s = s_f.(1 - e^{-\frac{t}{t_r}}) \qquad (4)$$

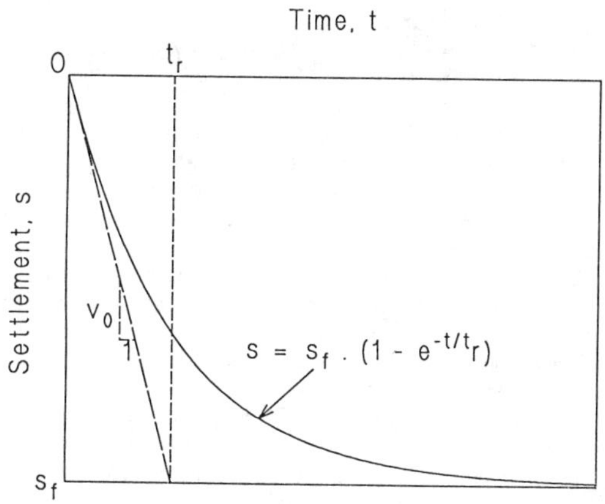

Fig. 5 Viscoelastic law for delayed settlements. Parameters

where s_f is the final settlement, and t_r is a 'reference time', defined as the time necessary to reach the final settlement, s_f, if the initial settlement rate, v_0, was kept constant (fig. 5).

The settlement rate is given by:

$$v = v_0 \cdot e^{-\frac{t}{t_r}} = \frac{s_f}{t_r} \cdot e^{-\frac{t}{t_r}} \qquad (5)$$

At the reference time, the settlement is 63% of the final value, s_f, and the settlement rate 37% of the initial rate, v_0.

This model is defined by two parameters, s_f and t_r (or v_0). Settlements are bounded by a final, finite value, reached for an infinite time. The expressions are well defined for any range of time. In the paper by Sánchez-Alciturri et al. [3.10] it is compared to the other proposed laws, showing reasonably good fitting to the observed settlement rates.

4 REMEDIAL MEASURES

In this Section, some possible solutions are discussed for old landfills. The solutions for each case depend primarily of the intended use of the landfill. If it is going to be reclaimed for recreational or agricultural purposes, then the only need is to isolate the landfill from its environment, in order to avoid contamination of adjacent soil and/or groundwater. However, if some superstructure is going to be built on it, then another solutions come into action.

4.1 Removal of wastes

Removal and replacement of wastes is always an alternative solution, but only in a few cases is possible. The main difficulty is the risk of contamination during the operations of excavating, handling and transporting the wastes. The difficulty of finding an alternative place for disposal of the removed wastes is usually a major problem.

4.2 Sealing

Two main topics are addressed by the Session Papers referring to old landfills sealing:

- Long-term effectiveness of old sealings, mainly of cut-off walls [5.1].
- Ways for providing isolation to unsealed old landfills [5.4, 5.5, 5.8].

The paper by Tedd et al. [5.1], presents a very interesting and well documented analysis of long-term performance of a bentonite-cement vertical cut-off wall, eight years old. The wall is studied by means of laboratory tests on undisturbed rotary samples of the wall and field observations. The evidence of cracking in the upper part, due to lack of confinement, is pointed out.

With respect to the isolation of unsealed old landfills, if the natural underlying ground is of low permeability (below 10^{-9} m/s), then it is only necessary to provide a top seal. The paper by Beine and Dahlmann [5.5] presents an actual case of this kind. The installed top seal has been carefully designed to provide adequate control of both leachate and gas emissions.

If the landfill is located on a permeable natural soil, then isolation to avoid groundwater pollution is more difficult, and involves lateral cut-off barriers together with internal pumping of leachate from relief wells. This kind of measures are proposed for a landfill in Zagreb in the paper by Kovacić et al. [5.4]. The paper by Porphyris and Weinmann [5.8] describes the available techniques for construction of bentonite barriers.

4.3 Improvement of bearing capacity

When the intended use of an old landfill involves the construction of some structures or facilities on it, it is usually necessary to improve the landfill bearing capacity. Deep compaction methods derived for soils are commonly used: preloading, dynamic compaction, stone columns, injections, jet-grouting, etc.

- Preloading

This technique has been applied in a number of cases (Charles et al., 1981; Oliden, 1987; Casanovas, 1989), when works schedule is compatible with the time needed for it. Usual surcharges are 2-6 m. of earth load, maintained for several months. Induced compression of the wastes follows the trends outlined in Section 3.2 above, and typically the settlement after one month of load application is 70 to 90 % of the settlement after one year.

- Stone columns

Stone columns formed by vibration were developped as an extension of vibro-compaction technique for soft clays and silts. It has been successfully applied to landfills in many cases, specially in the presence of groundwater, when other methods are less effective.

The paper by Oteo and Sopeña [5.3] describes a case in Madrid, where the construction of a highway on a landfill area required deep compaction treatment. The fill material was mainly composed of soils and construction debris, with only minor amounts of organic wastes. Dynamic compaction was discarded due to the proximity of buildings. Jet-grouting was used in dry zones, whilst stone columns were used in areas with significant groundwater flow in order to avoid alterations in the ground permeability. Static and dynamic penetration tests were used for control of results. Seismic Analysis of Surface Waves was also performed, showing a twofold increase of the propagation velocity before and after the treatment.

- Dynamic compaction

Although no paper to this Session deals directly with this method, it has been the subject of some lectures and discussions along the Symposium.

In this method, developped by Ménard (Ménard and Broise, 1975), a large weight (usually 10-20 t., with a base area of 3-4 m^2) is dropped from a great height (10-20 m.) on the ground surface, forming a crater. Usually, 5 to 10 blows are given at each point, on a square grid of 5-10 m. Several passes are given, some of them on a grid with points centered respect to the previous pass. A final 'ironing' pass is usually given. Generally, a layer of granular fill is extended first on the surface to improve traficability. After each pass, the surface is relevelled and craters are backfilled.

The depth of influence of the treatment is usually taken as proportional to the square root of the energy input in each drop:

$$h_i = \alpha \cdot \sqrt{\frac{W.H}{E_a}} \qquad (E_a = 1.0 \; tons.meter) \qquad (6)$$

Factor α in eq. (6) was taken as unity in early cases, but at present a value of 0.5 seems more realistic (Mayne et al., 1982).

The method is primarily adequate to compact loose granular materials, but it has also been applied for soft silts or silty clays, and in these cases the term 'dynamic consolidation' is often used.

In some cases, the method is applied in a different manner: the grids used in the successive passes are not shifted, and the drops are repeated in the same points, refilling the craters with granular material. In this way, the soft material is displaced downwards and sideways, and some kind of stone columns are formed. The terms 'tamped stone columns' or 'dynamic replacement' are sometimes used (Findlay and Sherwood, 1987).

Besides some references in the original reports on the method (Ménard and Broise, 1975), the first well documented applications of dynamic compaction to landfills are some cases in U.K. presented by Charles et al. (1981), and in America, the construction of a motorway in Arkansas (Welsh, 1983). Since then, it has been applied in a number of occasions. The general review of Mayne et al. (1984) includes some examples under the denomination of "miscellaneous fills". In Table 1 an updated summary is presented of cases of dynamic compaction of landfills. The first part of the table presents cases of domestic wastes. In the lower part, some cases of miscellaneous fills involving demolition refuse or industrial wastes have also been included. There are many other examples of compaction of loose clay or sand fills, as well as mine spoils, but they are out of the scope of the present report, and they have not been included in the Table.

Table 1 includes data about the site, intensity of treatment and results. The intensity of the treatment is usually defined by the specific energy applied:

$$E = \sum_{i=1}^{m} \frac{n_i \cdot W_i \cdot H_i}{e_i \cdot d_i} \qquad (7)$$

where "m" is the number of passes, and in each pass, "n" is the number of blows per point, W the weight of drop mass, H the free falling height, and "e" and "d" the grid spacing in the two directions.

Energy E is usually in the range 100-500 t.m/m^2 (fig. 6). The case presented by Welsh (1983), with 876 t.m/m^2, is still after ten years an upper bound of current practice.

The results of the treatment are usually given in different ways: increase of fill

Table 1. Cases of dynamic compaction of landfills

Site	Area (m^2)	Fill type	Depth (m)	Age (yrs)	W (t)	H (m)	A (m^2)	E (tm/m^2)	Settl. (m)	Reference
East London	(exp.)	D	6.5	40	14	14	4.0	260	0.58	Charles et al., 1981
Redditch	22,000	D	5-6	15	15	20	4.0	260	0.50	"
Hertfordshire	30,000	D	8	15-20	15	20	4.0	220	0.50	"
Indianapolis		DBE			15	21		440	1.00	Guyot et al., 1981 (*)
Baltimore		DBC	13	10	16	23		540	1.40	"
Arkansas US-71	18,000	D	6-12	5	18	28		876	1.6-2.5	Welsh, 1983
Pontevedra	34,000	D	8-10		12	20		212		Portilla, 1993
Barcelona B-10	24,000	D	6-8	10	15	20		225		Oliden, 1987
Haifa	14,000	D	25	20	11	15	4.0			Frydman and Baker, 1987
M25 Bell Lane	13,500	D	4.5	6-10	17	15	3.5	281	0.7-1.0	Perelberg et al., 1987
M25 Poyle		D	6	<10	15	20		150	0.60	Sherwood, 1987
Hull A-63	(exp.)	DI	12-17	20-30	15	15	4.9	675		Swain and Holt, 1987
Tulsa SH-11	90,000	DM	6		18	23	3.7	312	0.66	Snethen and Homan, 1991
Corby	(exp.)	E	24	>4	15	10	4.0	280	0.24	Charles et al., 1981
Rammey Marsh		I	5	10-20				220	0.45	Greenwood, 1982
Lonsdale		X						990	1.14	Dumas, 1982 (*)
Montreal		X						510	0.55	"
Barcelona B-10	16,000	EB	6-10	10	15	20		150		Oliden, 1987
M25 Bell Lane	5,500	E	4.5	6-10	17	10	3.5	170	0.2-0.4	Perelberg et al., 1987
"		IE	4.5	6-10	17	10	3.5	127	0.2-0.6	"
Queen's Dock	20,000	B	9	5-10	15	15		77	0.32	Reid and Buchanan, 1987
"	"	"	"	"	"	21		120	0.55	"
London Docks	67,000		9.5		15	22				Lord et al., 1987
Methil	8,500	EM	10		15	15		350	0.57	Slocombe & Moseley,1987
Port Clarence	15,000	EI	4		15	15		300	0.50	"
Aberdeen	8,200	EM	3-22		15	15		550	0.65	"
Gartsherrie	14,400	EMB	11		15	15		200	0.21	"
Briton Ferry	10,300	EIB	6.8		15	15		210	0.90	"
Princess Dock	70,900	EB	20		15	15		165	0.62	"
Illinois	(exp.)	M	4.5	> 10	16	20		300		Kousoftas & Kiefer,1990

KEY:
(exp.): experimental site
Wastes: D:domestic. M:mining. I:industrial. B:building. E:earth. X:miscellaneous.
Treatment: W/H/A: weight/height/base area of pounder E: applied unit energy
(*) Cited by Mayne et al., 1982.

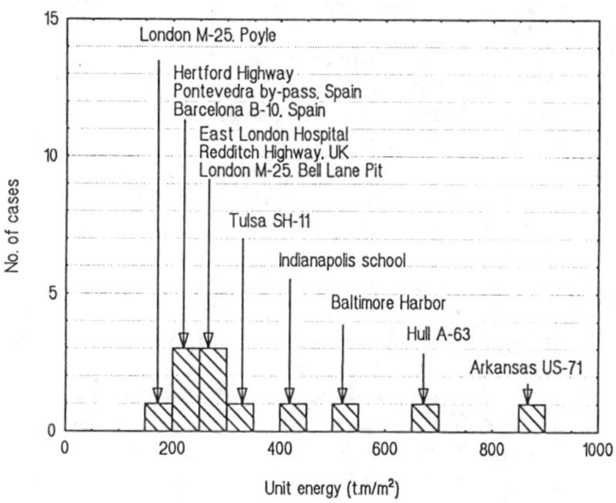

Fig. 6 Dynamic compaction of domestic wastes. Unit energy applied.

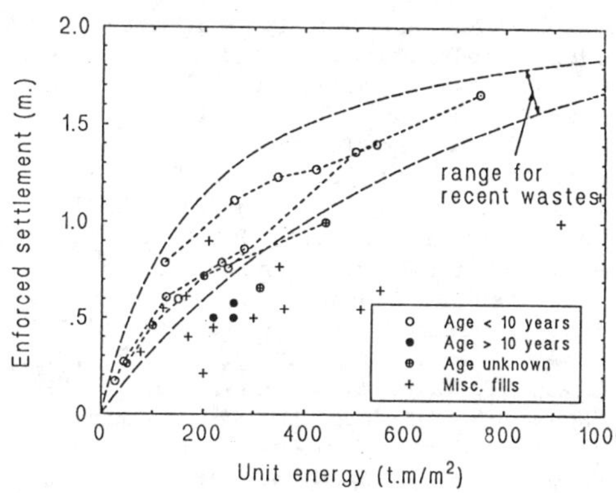

Fig. 7 Dynamic compaction of domestic wastes. Enforced settlement.

stiffness and strength, assessed by load tests or penetrometers (cf. Section 3.1 above), or performance records of the constructed facilities. An index of the degree of improvement is the enforced settlement of the landfill surface due to the dynamic compaction. This parameter has been measured in most of the cases presented in Table 1, and can be related to the unit energy, E, as suggested by Mayne et al. (1984) (fig. 7). As can be seen, recent domestic wastes are the most compressible cases, with settlements of 1.0-1.5 m. for energies of 200-400 t.m/m^2. By comparison, older sanitary landfills and miscellaneous fills have settlements of the order of half those values.

4.4 Special foundation techniques

As an alternative or complement to the above describe improvement methods, special foundation design is often used. The solutions depend on the features of each particular case, and range from piles (Dunn, [5.9]) to soil reinforcement, such as in the case presented by Genske and Noll [5.7], where a sandwich system is used, including improved soil, granular fill, geomembranes and geogrids, together with a careful drainage system both for gas and water.

REFERENCES

Abbiss, C.P. 1981. "Shear wave measurements of the elasticity of the ground". Géotechnique 31:1,91-104.

Cartier, G., Long, N.T., Pouget, P., Bargillat, R. and Cudennec, J.P. "Déchets urbains et pneumatiques usagés en Génie Civil. Proc. Xth ICSMFE, Stockholm. 3,603-606.

Casanovas, J. 1989. "Mejora de un relleno de vertidos de residuos urbanos". Rev. Obras Públicas. Dec. 1989, 909-916.

Charles, J.A., Burford, D. and Watts, K.S. 1981. "Field studies of the effectiveness of dynamic consolidation". Proc. Xth ICSMFE. Stockholm. 3,617-622.

Charles, J.A. and Driscoll, R. 1981. "A simple in situ load test for shallow fill". Ground Engineering 14:1,31-36.

Coduto, D.P and Huitric, R. 1990. "Monitoring landfill movements using precise instruments". Geotechnics of Waste Fills-Theory and Practice. (A. Landva and G.D. Knowles, eds.) ASTM, STP 1070,358-370.

Cuéllar, V., Valerio, J. and Muñoz, F. 1992. "Spanish experiences on the determination of the dynamic properties of soils through the analysis of surface waves". Invited lecture. Xth World Conf. Earthq. Eng. Madrid. (In press).

Dvirnoff, A.H. and Munion, D.W. 1986. "Stability failure of a sanitary landfill". Int. Symp. Env. Geot. (H.Y. Fang, ed.), Bethlehem, Pa, 25-35.

Edil, T.B., Ranguette, V.J. and Wuellner, W. W. 1990. "Settlement of municipal waste". Geotechnics of Waste Fills- Theory and Practice. (A. Landva and G.D. Knowles, eds.) ASTM, STP 1070, 225-239.

Findlay, J.D. and Sherwood, D.E. 1987. "Improvement of a hydraulic fill site in Bahrain using modified heavy tamping methods". Conf. on Building on Marginal and Derelict Land. I.C.E., Glasgow, 13-26.

Frydman, S. and Baker, R. 1987. "Construction of a bus parking station on a waste deposit site". Conf. on Building on Marginal and Derelict Land. I.C.E., Glasgow, 255-266.

Greenwood, J.R. 1982. "Dynamic compaction for the M25 Motorway at Rammey Marsh". Discussion on Ground Improvement. I.C.E. Meeting. Birmingham.

Hinkle, R.D. 1990. "Landfill site reclaimed for commercial use as container storage facility". Geotechnics of Waste Fills-Theory and Practice. (A. Landva and G.D. Knowles, eds.) ASTM, STP 1070,331-344.

Jessberger, H.L. 1990. "Stoffeigenschaften von Abfall im Hinblick auf Standsicherheitsuntersuchungen an Abfalldeponien". Neuzeitliche Deponietechnik. (H.L. Jessberger ed.) Bochum. 171-193.

Kousoftas, D.C. and Kiefer, M.L. 1990. "Improvement of mine spoils in Southern Illinois". Geotechnics of Waste Fills-Theory and Practice. (A. Landva and G.D. Knowles, eds.) ASTM, STP 1070,153-167.

Landva, A.O. and Clark, J.I. 1990. "Geotechnics of waste fill". Geotechnics of Waste Fills- Theory and Practice. (A. Landva and G.D. Knowles, eds.) ASTM, STP 1070,86-103.

Leach, B.A. and Goodger, H.K. 1990. "Building on derelict land". CIRIA Report. London.

Lord, A. 1991. "Options available for problem solving". Chap. 6 of 'Recycling derelict land' (G. Fleming, ed.). 145-195, T. Telford. London.

Lord, J.A., Mudd, I.G. and Williams, G.J. 1987. "Western Dock and Hermitage Basin reclamation, London Docks. Conf. on Building on Marginal and Derelict Land. I.C.E., Glasgow, 521-538.

Mayne, P.W., Jones, J.S. and Dumas, J.C. 1984. "Ground response to dynamic compaction". J. Geot. Eng. ASCE 110:6,757-774.

Ménard, L. and Broise, Y. 1975. "Theoretical and practical aspects of dynamic consolidation". Géotechnique 25:1,3-17.

Mitchell, J.K., Seed, R.B. and Seed, H.B., 1990. "Kettleman Hills waste landfill slope failure. Parts I and II". J. Geot. Eng. ASCE 116:4,647-691.

Oliden, J. 1987. "Vertederos de basuras y escombros. Problemas y tratamientos". Curso sobre Obras de Materiales Sueltos No Convencionales. CEDEX. MOPT. Madrid.
Oweiss, I.S. and Khera, K. 1990. "Geotechnology of waste management". Butterworths. London.
Perelberg, S., Boyd, P.J.H., Montague, K.N. and Greenwood, J.R. 1987. "M25 Bell Lane Pit: Ground improvement by dynamic compaction". Conf. on Building on Marginal and Derelict Land. ICE. Glasgow, 267-280.
Portilla, R. 1993. "Mejora del terreno por compactación dinámica en un tramo de autopista". EPTISA Techn. Note. Unpubl.
Reid, W.M. and Buchanan, N.W. 1987. "The Scottish Exhibition Center". Conf. on Building on Marginal and Derelict Land. I.C.E., Glasgow, 435-448.
Sánchez-Alciturri, J.M., Palma, J., Cañizal, J. and Sagaseta, C. 1993. "Post-closure bearing capacity of a sanitary landfill". Coll. Intern. Environnement et Géotechnique. Paris. 601-608.
Schmertmann, J.H., 1969. "Dutch friction cone penetrometer exploration of research area at Field 5, Elgin Air Force Base, Florida". Contr. Rep. S-69-4. U.S. Army Engrs. Wat. Exp. Stn. Vicksburg, Miss.
Schmertmann, J.H., 1993. "Conical test load measuring compressibility". J. Geot. Eng. ASCE 119:GT5,965-971.
Sherwood, D.E. 1987. Discussion Session 3. Conf. on Building on Marginal and Derelict Land. I.C.E., Glasgow, 407-411.
Siegel, R.A., Robertson, R.J. and Anderson, D.G. 1990. "Slope stability investigations at a landfill in Southern California". Geotechnics of Waste Fills-Theory and Practice. (A. Landva and G.D. Knowles, eds.) ASTM, STP 1070,259-284.
Singh, S. and Murphy, B. 1990. "Evaluation of the stability of sanitary landfills". Geotechnics of Waste Fills- Theory and Practice. (A. Landva and G.D. Knowles, eds.) ASTM, STP 1070,240-258.
Slocombe, B.C. and Moseley, M.P. 1987. "Experiences with dynamic compaction on derelict sites". Conf. on Building on Marginal and Derelict Land. I.C.E., Glasgow, 799-806.
Snethen, D.R. and Homan, M.H. 1991. "Dynamic compaction/stone columns. Test sections for construction control and performance evaluation at an uncontrolled landfill site". Deep Foundation Improvements: Design, Construction and Testing. (M.I. Esrig and R.C. Bachus, eds.). ASTM STP 1089, 199-211.
Sowers, G.F. 1968. "Foundation problems in sanitary landfills". J. San. Eng. Div. ASCE 94:SA1,103-116.
Sowers, G.F. 1973. "Settlement of waste disposal fills". Proc. 8th. ICSMFE. Moscow. 2,207-210.
Swain, A. and Holt, D.N. 1987 "Dynamic compaction of a refuse site for a road interchange". Conf. on Building on Marginal and Derelict Land. ICE. Glasgow,339-358.
Tonteri, K.J. and Lindroos, P.T. 1987. "Stability problems of landfill site founded on peat bog". Proc. Xth. ECSMFE. Dublin, Ireland, 1,337-342.
Welsh, J.P. 1983. "Dynamic deep compaction of sanitary landfill to support superhighway". Proc. 8th ECSMFE. Helsinki. 1,319-321.
Yen, B.C. and Scanlon, B. 1975. "Sanitary landfill settlement rates". J. Geot. Eng. Div. ASCE 101:GT5,475-487.

SIMPOSIUM PAPERS

Beine, R.A. and Dahlmann, K. "Capping of an old dump containing industrial wastes". [5.5].
Boroń, K.J. and Nagawiecka, H. "The possibility of biological reclamation of soda waste reservoirs "White Seas" in Krakow, Poland". [5.6].
Dunn, R.J. "Successful development of closed landfill sites". [5.9].
Genske, D.D. and Noll, P. "Reclamation of industrial wasteland". [5.7].
Güler, E and Avci, C.B. "Evaluation of the incident at the Ümraniye landfill, Istanbul, and proposed remedial investigations". [6.3].
Jessberger, H.L. and Kochel, R. "Determination and assessment of the mechanical properties of waste materials". [3.3].
Kovacić, D., Mayer, D. and Muhovec, I. "Geotechnical characteristics of Zagreb waste disposal site and possibilities of its reclamation". [5.4].
Mannassero, M. Contribution to Discussion Session 5.
Meju, M.A. "Geophysics and computer models in the evaluation of a land-fill site in Blaby, Leicester". [5.2].
Oteo, C. and Sopeña, L. "Deep treatment of uncontrolled urban fills for the construction of a high capacity road system". [5.3].
Porphyris, S.D. and Weinmann, A. "Construction of mineral barriers with bentonite and fillers for sealing of waste landfills". [5.8].
Sánchez-Alciturri, J.M., Palma, J., Sagaseta, C. and Cañizal, J. "Mechanical properties of wastes in a sanitary landfill". [3.6].
Sánchez-Alciturri, J.M., Palma, J., Sagaseta, C. and Cañizal, J. "Three years of deformation monitoring at Meruelo landfill". [3.10].
Tedd, P., Paul, V. and Lomax, C. "Investigation of an eight year old slurry trench cut-off wall". [5.1].

Waste Disposal by Landfill – GREEN'93, Sarsby (ed.) © 1995 Balkema, Rotterdam, ISBN 90 5410 356 6

Discussion

M Manassero referred to the concern about cement-bentonite slurry walls, in particular the deformation behaviour. Prior to the 1980's slurry walls were used primarily for cut-off walls beneath dams. In this case deformation of the wall is a major concern because of consolidation settlement of the surrounding soil (due to the self-weight of the dam) and the huge hydraulic gradient across the wall. For slurry walls the problems of containing landfill pollution is much smaller than for dams.

Available data about the deformation of cement-bentonite mixes invariably relate to undrained triaxial tests. Little data is available from drained tests. However triaxial tests conducted under drained conditions (after isotropic consolidation) give a much lower modulus of deformation than with consolidated-undrained tests. Also the undrained tests show very marked decrease in volume (almost a collapse behaviour), i.e. a contractive behaviour. This contraction determines the peak strength and the rather small failure strain of cement-bentonite mixes. If drainage is allowed, the pore pressure dissipates and completely different behaviour occurs - the strain rate becomes very important.

Another important aspect is the use of in-situ tests which can provide an insight into the mechanical behaviour of cement-bentonite mixes. Tests such as the piezo-cone penetrometer test can be used to assess the stress-strain behaviour and also the hydraulic conductivity. Recent work has involved assessment of the variation of hydraulic conductivity with penetration depth of a 4-month old slurry wall. The test is calibrated using both theoretical analysis and dissipation tests conducted in-situ.

S Jefferis said that this is a very important issue for the British community because they always perform their compression tests under drained conditions. Furthermore the crucial importance of the confining pressure, or confining pressure-to-strength ratio, on the behaviour of cement-bentonite mixes has to be pointed out.

M Manassero said that drained oedometer tests on cement-bentonite mixture exhibited a sort of yield, or destructive stress of about 2 bar - this is in the region where there is a change in behaviour, i.e. from dilational behaviour to contraction, with generation of a positive pore pressure.

S Jefferis pointed out that for the data quoted the confining pressure-to-strength ratio is about 1¼ and although it is feasible to have this value in laboratory tests the problem is to achieve this in the field.

M Manassero concurred with this point and said that the data quoted was for the first of a series of tests and it related to the situation at the bottom of an actual wall.

S Jefferis said that the problem is that in landfill situations the walls are shallow and confining stresses are low so the confined drained behaviour (of the laboratory) cannot be developed. Even with deep walls the top 5m will always suffer from low confining stresses.

M Manassero said that data showed that the top part of slurry walls has high hydraulic conductivity, believed to be due to a drying problem during curing of the top part of the wall. There is a need to develop a constitutive model for cement-bentonite mixes that covers the whole stress range.

S Jefferis's overall concern was to find a material that would behave as required. He was not sure that a cement-bentonite material has the perfect properties for a typical landfill application and if these properties are really needed from an engineering point of view then it may be necessary to include a synthetic membrane - this certainly may be the case for shallow walls.

M Meju commented, with regard to the use of geophysics for investigation, that the amount of money spent on exploratory drilling can be minimised by undertaking pre-drilling surveys using geophysical means such as resistivity sounding. M Meju pointed out that geophysics is able to detect boundaries very efficiently.

S Jefferis concurred but questioned whether boreholes were needed to calibrate the method.

M Meju said calibration was not needed, the method has been used to detect hydrocarbons (oil industry) and conductive materials (mineral industry) in the sub-surface, without calibration, hence it can be used readily with landfill sites.

S Waldron expressed an interest in settlements of landfill caps and the problems of deformation of the cap. She referred back to previous comments on the applicability, or otherwise, of geotechnical analyses to refuse. She felt that it was natural to extend the analytical treatment of peat to landfilled refuse since a peat deposit is a biological system similar to a landfill site which produces a lot of methane. Hence it seems a natural progression to look how peat settles and compare this to a landfill site. With a peat system there is a high water table - there is controversy over how much water is incorporated into a landfill site but methane will not be generated unless there is sufficient water for the bacteria and this seems to be quite a lot. In peat there seems to be dynamic settlement, i.e. fast, the bacteria degrade the organic matter and the older the matter is the more degradable it is so you have collapse from the bottom of the peat bog. The top material may not initially collapse if there is some support or structure and voids may suddenly appear when this top support is lost - S Waldron suspected that this behaviour also occurs within landfilled refuse. Studies need to be made of the way in which waste changes with time so that current geotechnical analysis methods can be extended and developed.

S Jefferis concurred that the material changes with time and the microbiological effects must be accounted for. However it must be remembered that peat is a very difficult material to deal with from a geotechnical point of view.

M Fairlie expressed concern over the use of stone columns in landfill primarily because of the potential lack of lateral support to the columns, particularly with time, due to degradation of the landfill. M Fairlie was seeking any information on the use of stone columns, successful or otherwise, in landfill.

C Sagaseta commented that he had no experience of using stone columns in landfill, and the only case history that he had seen relating to stone columns in fill had involved demolition waste and earth, i.e. not organic, not domestic waste - therefore it was inert.

Commenting on model tests of dynamic compaction (to simulate compaction of refuse) **H Jessberger** suggested that instead of using sand mixed with crushable flakes a paper material should be used with sand to simulate the bending and tensile strength of the refuse.

With regard to the use of geophysics to detect abandoned landfills then ETC8 (European Technical committee) contains advice on this.

H Jessberger asked for classification of the method of "gas sampling" referred to in the site investigation of landfills.

S Jefferis said that gas sampling, from probes or walk-over surveys, is an immensely powerful method of identifying chlorinated solvents, etc..

C Sagaseta said that geophysical methods of investigation have two ends -
a) identification of the problem i.e. extent and depth of the landfill - for this they are extremely useful,
b) determination of the mechanical properties of the fill - some methods can provide some guidance.

P Ansell directed a question towards P Tedd on the shear strength of slurry walls and the influence of inclusions on stiffness. P Tedd's accidental rebound curve seemed to indicate a stiffness very much the same as the as-cast samples. He attributed a reduced initial stiffness, of the as-sampled slurry, to inclusions but the rebound loop indicated a much higher stiffness. It therefore seems possible that the observed reduced stiffness was due to sampling effects rather than inclusions since the latter would have still been there when the rebound loop was performed, whereas sampling effects may have been reduced by the initial loading.

P Tedd replied that this was a possibility, bearing in mind that only two tests had been conducted, and that much more information is needed in the properties of old slurry walls.

S Jefferis said that fine material will certainly be suspended in the clay-cement system in normal situations. In the very bottom of the trench in coarse ground some grain-to-grain contact may occur. In a normal slurry contamination levels up to about 35% by volume would still permit the particles to float.

M Manassero asked for information on the relationship between the small-strain modulus, (which can be measured by geophysics), and the large-strain modulus of actual municipal waste (not a natural soil).

C Sagaseta replied that it has taken a long time to appreciate the difference between the low-strain modulus and the large-strain value for soils and we do not yet have this length of engineering experience with waste material. At the present time we are happy to have some relationship that can be used.

M Manassero added that geophysical measurements can be used to show directly a change in mechanical properties, e.g. before treatment and after treatment.

R Edwards noted that according to P Tedd a cement-bentonite slurry is very brittle and difficult to handle and suggested that perhaps test specimens are unavoidably damaged during preparation. This damage is highlighted by the change in stress-strain ratio. Perhaps samples should not be transported to the laboratory before 28 days or before they have had time to gain some minimum strength. The variation in stress-strain behaviour may indicate that even after 28 days (or even after 8 years in the case of the wall reported by P Tedd) sample preparation causes unavoidable damage to the specimens.

P Tedd agreed with the preceding points and added that it must be remembered that the material itself is very variable so unique properties will not be obtained - the mixing on site varies, the contamination varies, the preparation varies etc.. The specimens must be treated with a lot more care than soil samples, the material does break and oxidise very quickly and it has to be stored underwater or in some way so that air is excluded.

Regarding the use of the piezocone in investigating wall strength **P Tedd** felt that with british walls it would not be possible to push a piezocone into them.

S Jefferis observed that european walls appeared to have higher strengths than british ones and if high strength walls could be used in the UK then there would not be the problem with transporting the samples.

S Jefferis felt that oxidation did not damage the chemical structure of the cement-bentonite mix. At the end of a permeability test the sample is always fully oxidised because of the passage of oxidised water through it but if the samples are then left in an anaerobic environment they will return to their characteristic colour.

S Kavanagh endorsed previous speakers' comments about cut-off walls and felt that these highlighted the need for specifiers to think about what the wall is required to do and how it is expected to perform. The specification must be very specific over their requirements - a specification calling for a 60 or 100 year design life which does not include information on the constituents of the leachate or the hydraulic regime in the ground is not much use. Inclusion of a lot of natural material in the slurry always occurs when excavation is undertaken. It is very expensive to avoid such inclusion, by using super-dense slurries, but some contamination is inevitable. The specification for testing of the slurry should specify the sampling location, e.g. out of the discharge pipe, out of the mixer, from the top of the wall etc., because otherwise very different laboratory test results will be obtained according to the Contractor's sampling point.

6 Case histories

Waste Disposal by Landfill – GREEN'93, Sarsby (ed.) © 1995 Balkema, Rotterdam, ISBN 90 5410 356 6

A seepage problem associated with an old landfill in the greater Durban area

F.G. Bell & C.A. Jermy
Department of Geology and Applied Geology, University of Natal, Durban, South Africa

ABSTRACT: A waste disposal site in the greater Durban area had been used for codisposal of solid and liquid wastes and had been closed for a number of years. Leachate from the site was collected in a leachate pond located at the toe of the landfill along its eastern boundary. Little had been done since closure to ensure that leakage of leachate from the landfill or pond did not pollute the surrounding ground or surface waters and it was suspected that this was occurring. Small non-perennial and perennial streams occur in the area of the site and flow into a major river to the south east. The site itself is underlain by sandstones of the Natal Group and tillite of the Dwyka Formation. These rocks have been weathered to form a mantle of residual soil of variable thickness which is overlain by colluvium on the valley slopes and alluvium in the valley floor.

The investigation undertaken to determine the extent of the pollution included a number of inspection pits to allow the soil profiles to be examined and soil samples to be collected. Seepage water was also collected from the inspection pits for chemical analysis. The area downstream of the waste disposal site was geotechnically mapped.

Groundwater seepage in the inspection pits varied from slight to strong and occurred mainly at the interface between the more permeable sands overlying the residual soils. The results of the chemical analysis of water samples indicated that the groundwater and stream within the valley were being polluted by leachate seeping from the toe of the landfill and the leachate pond. The investigation revealed that the plume extended some 150 m downstream of the leachate pond.

Two main remedial measures were proposed. Firstly, a subsoil drain should be constructed immediately downstream of the landfill site. This should be located in bedrock, its purpose being to intercept seepage from the landfill. The drain should lead into a pump chamber that would remove polluted water which would then be conveyed to the sewer system. Secondly, to make doubly sure that the pollution was contained, a cement-bentonite cut-off trench, founded in bedrock, should be constructed 3 m downstream of the subsoil drain.

1 INTRODUCTION

Leachate is contaminated water which is produced when rainwater or groundwater flow through a landfill dissolving the soluble fraction of the waste. This occurs once the absorbent characteristics of the refuse are exceeded. Domestic refuse is a heterogeneous collection of many things (e.g. waste foods, garden rubbish, paper, cloth, ashes, building waste, plastic, metals etc), much of which is capable of reacting with water to give a liquid rich in organic matter, mineral salts and bacteria. At many landfill sites the production of leachate represents the major environment problem. This is because subsequent movement of leachate into the soil, groundwater, or surface water can cause considerable pollution. Severe leachate problems may occur in relation to operational or completed landfill sites and the most serious consequences result when the control measures are inadequate. It normally is easier and cheaper to implement leachate control measures before or during landfill operations than waiting for a leachate problem to arise after its completion.

Initially biological degradation of a landfill is aerobic but with the depletion of oxygen by microbial activity, anaerobic conditions quickly develop. The organic material biodegrades anaerobically and the leachate from a young landfill contains a high concentration of organic acids, ammonia and total dissolved solids. As a landfill ages and stabilizes, much of the readily biodegradable material breaks down and the organic content of the leachate decreases. Hence the biochemical oxygen demand (BOD) changes with time. It first builds to a peak as microbial activity in the landfill increases, this occurring after 6 months to 2.5 years after placement. After that the BOD concentration decreases until the landfill is stabilized some 6 to 15 years later. Also in a young landfill the amount of dissolved solids may exceed 10 000 mg/l with relatively high concentrations of chloride, sulphate, calcium, sodium or iron. As the landfill ages, the concentration of inorganic materials in the leachate tends to decrease (Table 1).

Leachate entering the ground mixes with the groundwater resulting in the formation of a leachate plume. Dilution often is effective in intergranular formations of high porosity and permeability where groundwater flow rates are high. Conversely, one of the worst situations occurs where groundwater movement is slow and takes place near the surface so that little dilution occurs. When leachate reaches the groundwater system its organic material tends to be biodegraded very slowly because the oxygen demand may rapidly deoxygenate the groundwater and there is no oxygen supply available to the groundwater system. If anaerobic conditions develop, then inorganic materials like iron and manganese may be dissolved in the groundwater.

However, as polluted water flows through the ground heavy metals may be adsorbed by cation exchange onto the surfaces of clay particles (the cation exchange capacity is not significant in clean sands). On the other hand the inorganic substances lead to mineralization of groundwater. Ions, especially chloride, may travel significant distances through soil without attenuation. Continuing input of inorganic ions due to seepage of leachate into groundwater can lead to it becoming unacceptable for drinking purposes.

A landfill site which had served the needs of an area within the Greater Durban region was closed in 1989. Unfortunately, however, seepage occurred from the toe of the landfill and it became necessary to carry out an investigation to determine the extent of the seepage, whether or not it posed a problem and, if so, what remedial measures should be taken.

The area surrounding the landfill site is undeveloped and consists of low rounded hills and small valleys. Small non-perennial and perennial streams flow into one of the notable rivers of the region. The area downstream of the toe of the landfill slopes gently eastwards, with a valley located immediately to the south east of the boundary of the landfill. A stream occupies this valley, commencing just downstream of the leachate pond and collection sump. A second stream occurs approximately 40 m to the north east of the previously mentioned

Table 1. Typical analyses of leachate from domestic landfill prior to dilution by surface runoff. (After Crawford and Smith, 1984).

	Age of refuse	
	<2 years old	>10 years old
pH	5.0-6.5	6.5-7.5
BOD_5	4 000-30 000	<100
COD	10 000-60 000	50-500
TOC	1 000-20 000	<100
Total solids	8 000-50 000	1000-3000
TSS	200-2 000	100-500
Total N (mainly NH_3-N or Org-N)	100-1 000	<100
Phosphate	5-100	<5
Chloride	500-2 000	100-500
Sulphate	50-1 000	<10
Iron	100-1 500	10-400
Sodium	500-3 000	<200
Potassium	200-1 000	50-400
Calcium	500-2 500	100-400

Notes:
(1) All units expressed as milligrams per litre, except pH
(2) Dilution by surface runoff from the landfill may substantially reduce these concentrations.

stream, and both flow into the river referred to.

The waste materials disposed of at the site consisted mainly of residential and commercial wastes. The leachate generated by the waste was collected in a leachate pond situated at the toe of the landfill adjacent to the eastern boundary of the site. A small bund surrounded the leachate pond on the downstream side. A trench around the perimeter of the landfill helped convey leachate seeping from the sides of the fill to the leachate pond. An attempt had been made at the head of the stream downstream of the leachate pond to form a leachate sump to contain leachate migrating beyond the pond and prevent pollution of the stream. Leachate was pumped from the sump periodically.

2 GENERAL GEOLOGY OF THE SITE

The solid geology of the area where the landfill is located consists of sandstones of the Natal Group and tillite of the Dwyka Formation (Fig. 1). These are overlain by a mantle of residual soils of variable thickness. The residual soils are, in turn, overlain by colluvial soils on the hillslopes and by alluvial soils in the valley floors.

The sandstones of the Natal Group are exposed in scattered outcrops across the site. It is light yellow in colour and varies from soft (where weathered) to medium hard rock. It is closely jointed. These sandstones occupy most of the area except to the south where they are downthrown by a fault against Dwyka Tillite. The sandstones have an overall dip to the south at angles ranging from 12° to 28°.

Where the tillite of the Dwyka Formation is exposed it is highly to extremely weathered. Closely spaced joints are still present in the weathered tillite, which varies from very soft to medium hard and grades from grey to light yellow in colour.

The residual soils associated with the sandstones are variable in character. They are generally dark reddish brown micaceous silty sands and sandy clay of soft to firm consistency. They range in thickness from 1.6 m to 2.1 m. However, in some places the residual soils are absent and here colluvial soils rest directly on sandstone. The particle size distribution of the residual soils is illustrated in Figure 2a.

The colluvial soils overlying the residual sands and clays commonly comprise dark brownish grey, loose to medium dense, fine to medium grained silty sand (Fig. 2b). The thickness of the colluvial soils ranges from

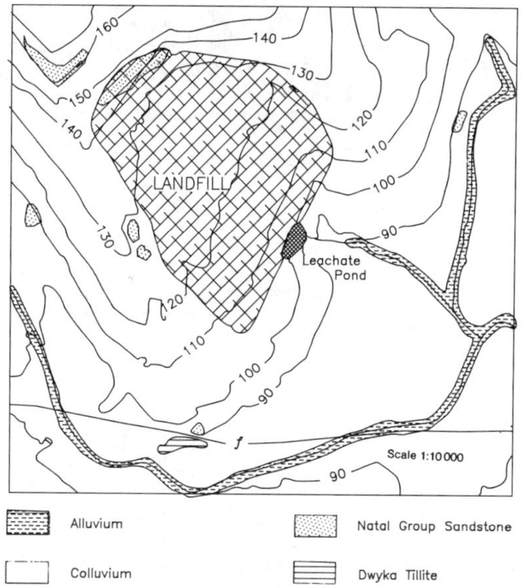

Fig. 1. Geology of the site.

0.3 m to 1.8 m. A pebble marker horizon usually occurs at the base of the colluvial soils. It is dark brownish grey, loose to medium dense, ferruginized sandy gravel. Most of the gravel-sized particles are composed of sandstone.

It is not easy to distinguish between the residual and colluvial soils overlying the tillite. These soils consist of dark brown, soft to firm, slightly sandy clay, which ranges in thickness from 0.3 m to 0.7 m. Again a pebble horizon occurs at the base of the soil. Patches of tillite gravel also overlie the tillite.

Alluvial soils primarily consist of light brown, very loose, fine to medium grained sand, clayey sands (Fig. 2c) and grey sandy clay of soft to firm consistency. They occur along the valley bottoms occupied by the streams and main river, and reach a maximum thickness of 2.5 m. In places, alluvial soils are overlain by colluvium.

3 INVESTIGATION OF THE SITE

The area around the site was geologically mapped and exposures were described in detail. Twenty two inspection pits were excavated (Fig. 3). The pits were positioned to give a general coverage of the area, and to enable a geological cross-section to be drawn of the subsoil conditions along the south east boundary of the landfill where seepage was occurring. The pits were

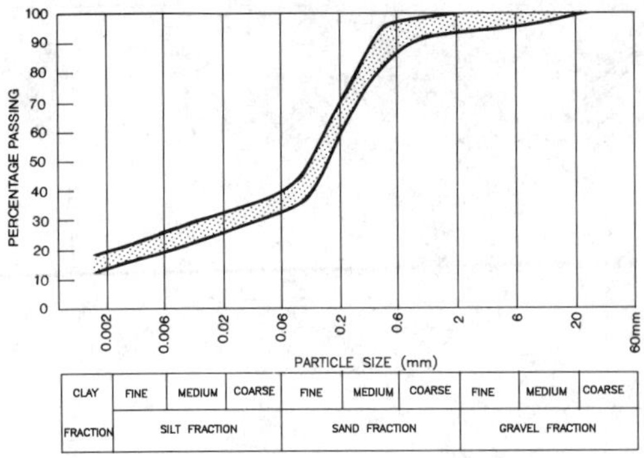

a

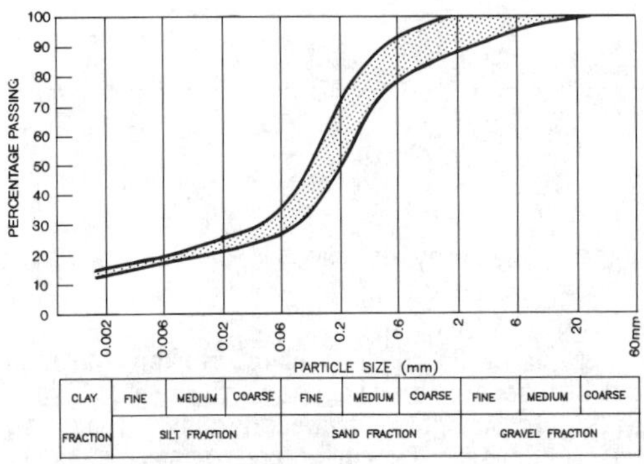

b

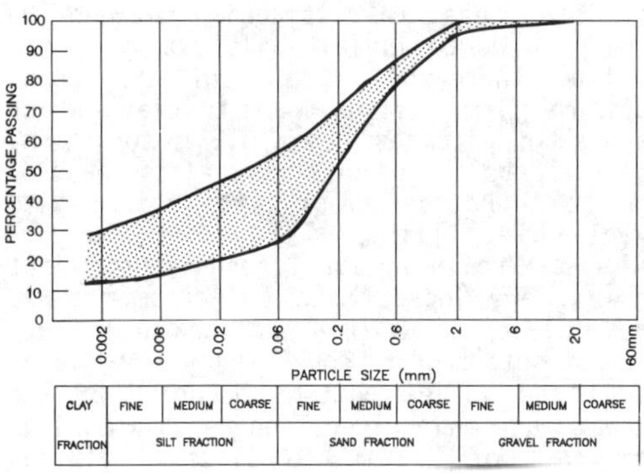

c

Fig. 2. Particle size distribution (a) residual soils (b) colluvial soils (c) alluvial soils.

excavated by a mechanical excavator except where soft wet marshy conditions existed immediately downstream of the toe of the landfill. Here the pits were excavated by hand. The latter were excavated to a metre depth whereas those which were mechanically excavated extended to 4.5 m below ground

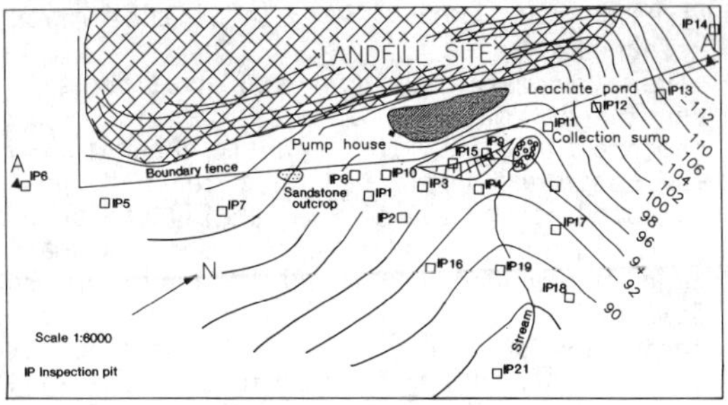

Fig. 3. Location of inspection pits. Contours given in metres.

surface. Profiles of the soils in each pit were made and soil samples were taken for testing. Water samples were also collected from the inspection pits. Dynamic cone penetrometer tests were carried out to assess the depth to bedrock and relative consistency of the subsoils in the valley floor immediately south east of the landfill. Bedrock occurred at between 3.5 m and 5.0 m below ground level.

Groundwater seepage in the inspection pits varied over the site from very slight to appreciable. It tended to occur within the upper 2 m of soil, principally at the interface between the more permeable colluvial sandy soils (estimated permeability, 10^{-3} - 10^{-5} m/s) and the underlying less permeable alluvial or residual soils (estimated permeability, 10^{-4} - 10^{-7} m/s). The water samples taken from the inspection pits were analyzed and examples of the results are given in Table 2. This should be compared with the figures noted in Table 1. Taken overall the results shown in Table 2, by comparison, suggest a relatively young landfill, although this is not the case. The explanation presumably is attributable, at least in part, to the sub-tropical climate with higher temperatures and rainfall.

The results of the investigation indicated that leachate is polluting the groundwater by seeping from the toe of the landfill and from the leachate pond, thereby demonstrating that the containment was inadequate. The concentration levels or isopleths of sulphides, chlorides, total dissolved solids (TDS), oxygen absorption, conductivity, chemical oxygen demand (COD) and pH downstream of the site are shown in Figure 4a-g. The maximum permissible concentration levels for each of these determinants are also shown in this Figure, as defined by the Water Act No 54, Section 21 of 1956 (as amended), "General Standards for Effluent". High levels of concentration

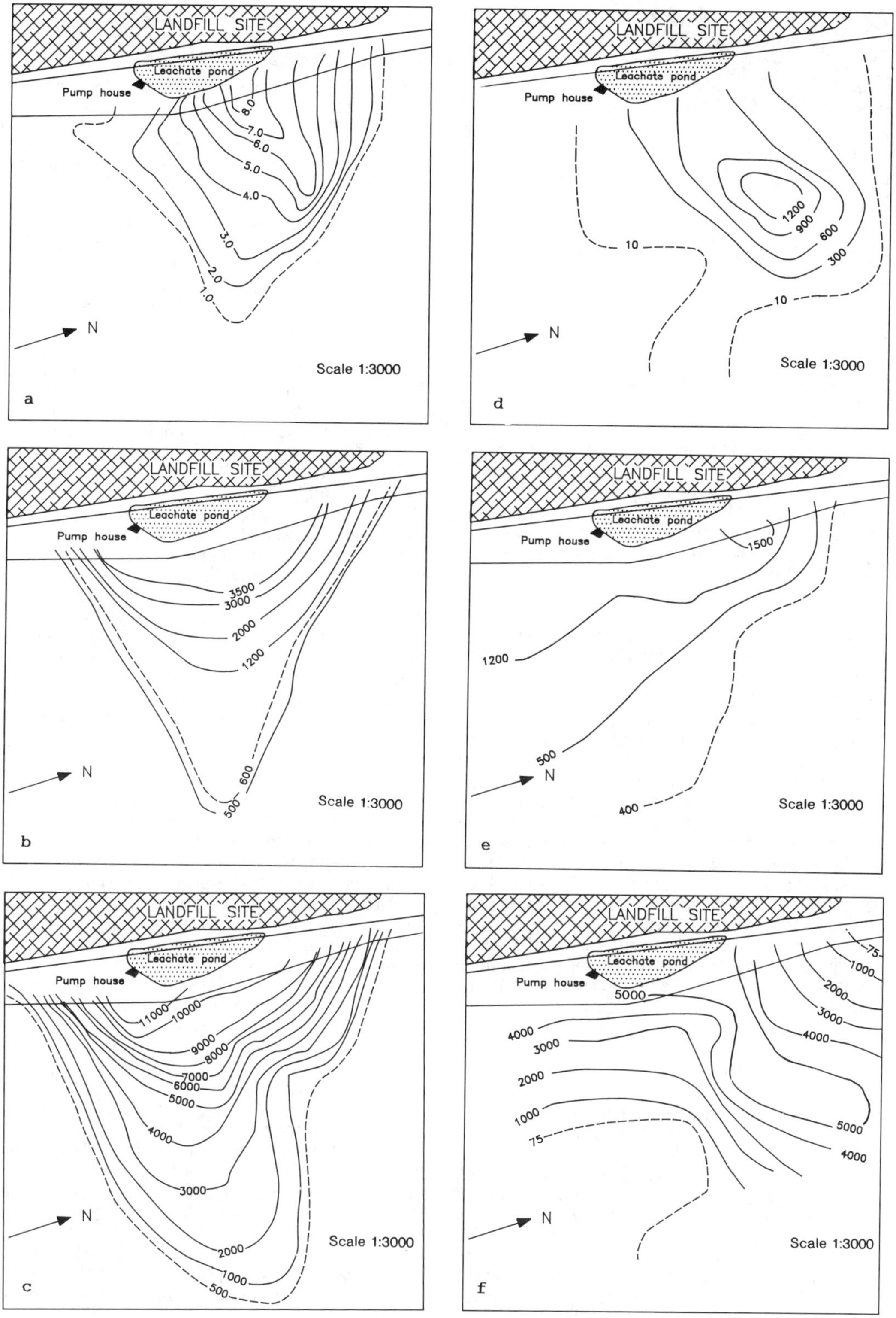
LANDFILL SITE
Leachate pond
Pump house
8.0
7.0
6.0
5.0
4.0
3.0
2.0
1.0
N
Scale 1:3000
a
LANDFILL SITE
Leachate pond
Pump house
3500
3000
2000
1200
600
500
N
Scale 1:3000
b
LANDFILL SITE
Leachate pond
Pump house
11000
10000
9000
8000
7000
6000
5000
4000
3000
2000
1000
500
N
Scale 1:3000
c
LANDFILL SITE
Leachate pond
Pump house
1200
900
600
300
10
10
N
Scale 1:3000
d
LANDFILL SITE
Leachate pond
Pump house
1500
1200
500
400
N
Scale 1:3000
e
LANDFILL SITE
Leachate pond
Pump house
75
1000
2000
3000
4000
5000
4000
3000
2000
1000
75
5000
4000
N
Scale 1:3000
f

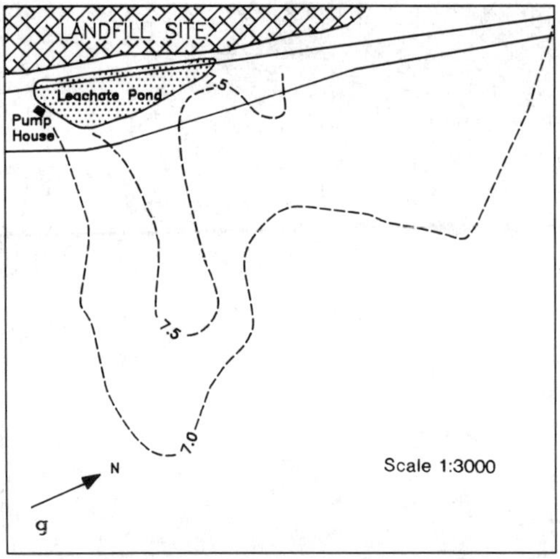

Fig. 4. Isopleths showing the distribution of (a) sulphides (b) chlorides (c) total dissolved solids (d) oxygen absorption (e) conductivity (f) chemical oxygen demand (g) and pH value in the groundwater. All in mg/l except conductivity, which is in mS/m, and pH value.

of pollutants were found which in most cases greatly exceeded the maximum permissible limits.

These pollution indicators show similar isopleth trends in that the plume originates from the leachate pond and extends downstream, the degree of pollution decreasing downstream of the landfill. However, Figure 4 indicates that sulphides decrease in concentration until acceptable levels are encountered beyond approximately 310 m downstream from the boundary i.e. beyond the 1.0 mg/l isopleth. On the other hand acceptable concentrations of chlorides are encountered some 180 m downstream from the boundary (i.e. beyond the 600 mg/l isochrone). The pH value of the groundwater downstream of the site remains around neutral.

4 REMEDIAL MEASURES

In order to avoid disturbing the landfill and to prevent further migration of seepage from it, it was decided to construct a cut-off barrier immediately downstream of the leachate pond, together with a subsoil drain and pump chamber. The principal qualities affecting the choice of a cut-off wall are its permeability, its resistance to pollutants, its strength and flexibility, and its cost. The maximum depth to bedrock at the proposed location of the cut-off barrier is around 5 m. Most of the soil present at this location is colluvial and residual sand, it reaching a maximum thickness of 4.5 m in the valley immediately downstream of the landfill, with alluvium beneath. In the centre of the valley groundwater is present at approximately 1.5 m below the ground surface.

It was suggested that the most suitable form of barrier would be a cement-bentonite cut-off wall. In a cement-bentonite cut-off, cement is added to fully hydrated bentonite slurry and backfilled into a slurry trench (Bell, 1993). The addition of the cement causes the slurry to harden, giving a strength comparable to that of stiff clay. Cement-bentonite walls are considered where suitable soil backfill is not readily available, the working area is restricted and some structural strength is required. A cement-bentonite wall generally contains 10 to 20% cement and 4 to 6% bentonite, the rest being water. The choice was economically attractive because its construction would involve limited use of materials and would be completed with a minimum sequence of operations. It was further suggested that if sand was used to replace some of the cement, about 50% by volume, then the cost of materials could be reduced significantly. A permeability of 10^{-9} m/s could still be achieved.

Impermeability is the most important property of a cut-off wall. There are three factors which account for the impermeability of a slurry trench, namely, the 'grouted' zone, the filter cake and the backfill. Bentonite initially may permeate the pores of the adjacent soil, depending upon the size distribution of the pores. As bentonite is thixotropic, when left undisturbed it gels and thereby seals the pores. The distance to which the soil is affected can range from almost zero in dense clays to a metre or so in loosely packed sands and gravels. A filter cake of loosely packed bentonite particles forms after a relatively short period on the side of the trench during excavation; this results from the filtrate as the fluid escapes into the surrounding soil. This filter is usually a few millimetres in thickness and has a permeability as low as 10^{-11} m/s. Hence, it acts as a watertight membrane.

The chief factors affecting the permeability of a cement-bentonite slurry are the cement-water ratio, the bentonite-water ratio and the mechanical procedure in making panel connections between the fresh cement-bentonite slurry and the set cement-bentonite. As noted, it is important that the bentonite be fully hydrated before any

Table 2. **Examples of chemical analyses carried out on groundwater obtained from inspection pits.**

Water Sample No. of inspection pit	Depth (metres)	pH	Chloride mg/l	Sulphide mg/l	Total dissolved solids mg/l	Conductivity mS/m	Oxygen absorption mg/l	Chemical oxygen demand mg/l
IP 4	0.9	7.45	3 620	5.10	9 003	1 254.6	619	2 960
IP 9	1.4	7.3	2 782	8.0	9 726	1 448.4	778	3 520
IP 10	1.7	6.9	3 938	1.4	11 047	1 397.4	74	656
IP 11	1.8	7.5	3 832	5.6	9 902	1 530.0	339	1 920
IP 17	0.4	6.9	76	6.1	557	85.7	1 577	6 240
IP 19	Surface	7.7	1 033	3.5	3 722	581.4	170	960
IP 20	0.6	6.9	504	0.48	2 166	300.9	40	272
IP 21	0.4	7.05	820	0.8	2 807	418.2	88	448
Spring S2	Surface	7.35	71	0.32	261	38.8	7	40
Maximum permissible range of limit*	-	5.5-9.5	600	1.0	500	400	10	75

*As laid down in the Water Act No 54, Section 21 of 1956 (as amended, "General Standards for Effluents").

cement is added. This means that the cement-water ratio is the factor which controls the ultimate permeability of the backfill (generally between 1:4 and 1:5). Activated sodium bentonites have large swelling potential and are stable with cement. Between 35 and 45 parts bentonite mixed with between 160 and 220 parts cement, and water gives a 'plastic' wall with an ultimate unconfined compressive strength of 1 MPa or over and a modulus of elasticity of 20 to 50 MPa. This allows a wall to deform under loading without cracks developing. In other words the cement-water ratio has a dramatic effect on deformability of cement-bentonite backfill. For instance, larger proportions of cement give higher strength to a wall and so it is less deformable. On the other hand, the higher the bentonite-water ratio, the more flexible is the wall.

Because it must have a low permeability, the continuity of a cement-bentonite wall is important. The depth of the slurry trench is controlled in many cases by the ground conditions, notably the depth of an impermeable formation, and the type of barrier to be constructed. The minimum depth of penetration into an impermeable formation at the base of the cut-off wall depends upon the nature of the formation. Only a very small penetration may prove satisfactory in competent impervious rock. The width is governed by the required permeability of the cut-off wall, the head of water across the wall, the size of the excavation equipment available, and the materials that form the wall. However, because the filter cake provides the principal barrier to groundwater movement, the width of the trench is not a major factor. The thickness of cement-bentonite cut-off walls normally ranges from 0.6 to 0.9 m. Such widths are satisfactory for depths of up to 30 m of hydrostatic head. The deviation from vertically is only important insofar as it affects the continuity and integrity of the wall. The permeability of a cement-bentonite wall may change with age especially near the surface where it is subjected to drying out and wetting. This can lead to the development of horizontal cracks. However, below the level of fluctuation in natural moisture content conditions are often almost ideal for curing. Any reduction which may occur in permeability with time must be balanced against the changing nature of leachate produced and the reducing concentration of pollutants within it.

It was proposed that the cement-bentonite wall and associated subsoil drain be constructed along the entire south eastern boundary of the landfill to a point 20 m south of the southern boundary and should be keyed into bedrock (i.e. the sandstones of the Natal Group). The addition of sand, obtained from the nearby stream, will mean that polymer treated sodium bentonite, Culseal, which is especially designed for containment of polluted water, will have to be used to ensure the durability and permeability of the cut-off wall. As the pH values of the water samples tested ranged from 7 to 7.5, the cement in the backfill should not be adversely affected.

It was suggested that the subsoil drain should be constructed before the slurry wall to allow the level of groundwater in the valley to be lowered. This will also entail constructing the pump chamber and linking the associated piping with the local sewer system.

The subsoil drain is to be laid in a series of stages because of the variation in elevation and soil type. It will drain under gravity flow into the pump chamber. It will consist of a 150 mm diameter pipe surrounded by stone aggregate. The latter will be overlain by sand which will extend to the surface. The aggregate will be enclosed in synthetic fabric to prevent clogging by fines, biological growths and the precipitation of inorganic substances. Blasting will be required to the south of the pump chamber location in order to obtain the desired gradient. It is important that the subsoil drain and pump chamber be sited at a lower level than the base of the slurry wall in order to permit leachate to flow back from the wall into the drain.

The pump chamber is to be constructed at the centre of the valley where bedrock is at the lowest elevation to allow gravity flow of the leachate from the collection pipes. It will be constructed with the aid of a caisson since the soil in this area is saturated and has low strength. The chamber will be sealed on the inside. The joints between the leachate lines and the pump chamber will be flexible and valves will be installed on the leachate lines so that they can be shut off during periods of maintenance. Automatic submersible pumps will be used in the pump chamber.

CONCLUSIONS

A landfill constructed in the Greater Durban area was located on sandstones of the Natal Group which are overlain by a veneer of residual, colluvial and alluvial soils. After its closure in 1989, seepage of leachate was noticed downstream of the landfill. This led to an investigation, which principally consisted of excavating a number of inspection pits in the suspect area. Water samples were taken from these pits and analysed. The analyses revealed that a leachate plume had developed downstream of the landfill and was polluting the groundwater.

The remedial measures proposed consist of a subsoil drain which will lead into a pump chamber. A cement-bentonite wall will be constructed immediately downstream of them to act as an impermeable barrier to any leachate seepage. The cut-off wall will be keyed into bedrock and will have a permeability of around 10^{-9} m/s. In this way it is hoped to prevent further pollution of groundwater occurring.

REFERENCES

Bell, F.G. 1993. Engineering Treatment of Soils. Spon, London.

Crawford, J.F. and Smith, P.G. 1984. Landfill Technology. Butterworths, London.

Waste Disposal by Landfill – GREEN'93, Sarsby (ed.) © 1995 Balkema, Rotterdam, ISBN 90 5410 356 6

Some experiences in environmental protection from the lignite ash impoundments waste water

Mladen Boreli & Dubravka Pokrajac
Faculty of Civil Engineering, Belgrade, Yugoslavia
Vladislav Halda
CIP, Belgrade, Yugoslavia

ABSTRACT: Problems of water protection from the pollution by waste waters from a two huge ash impoundments are considered. Ash is produced by lignite coal combustion. Annual ash production at the two plants is 6 - 7 million tons. It is transported from the plant to the impoundment in low-concentration suspension with ratio ash-water 1:10 to 1:13. About 2 m^3/s of water, needed for the transport at both plants, is pumped from the adjacent river Sava, and then discharged back as waste water. It is open, hydraulic ash transport system. Impoundments are situated on alluvial deposits, with semi- pervious surface layer and an underlying aquifer. The most significant pollutant is arsenic, which is to the great extent retained (temporarily) on the soil grains.

Usage of pollutant flux as a basic operational water protection criterion is suggested in the paper. By optimum use of natural (temporal) arsenic retaining, pollutant flux decreases, step by step. Possibilities to satisfy the adopted protection objectives within the frame of existing technology are analyzed as well as the optimum future modifications.

1. COAL AND ASH CHARACTERISTICS.

Problems of water pollution by ash impoundments and protection measures are considered. The coal is lignite, which is practically the basic thermal energetic resource of the country. The ash impoundments are in the vicinity of thermal power plants, on ground surface. Ash is transported in low concentration suspension, with the ratio ash-water from 1:10 to 1:13. The water for hydraulic transport is pumped from the river Sava and than returned to it, as waste water. (Open hydraulic transport of ash).

All the data in the paper correspond to the most exploited coal basin i the country- the Kolubara coal basin, particularly to the two main existing power plants: "Nikola Tesla-A" (abridged NT "A") and NT "B", with power 1650 and 1240 MW, in operation from 1970 and 1983, respectively. The yearly consumption of coal is 11 to 13 million tons; the resulting ash production is about 3 million tons, for each plant.

Coal characteristics. Lower heating value (with moisture) of the coal vary from 7.000 to 10.000 kJ/kg. It contains 48% of moisture, 20-25% of ash, and 0.5% of sulphur (0.3% remains in the ash, and the rest is combustible).

Ash characteristics. Ash is silicate, containing about 55% of SiO_2, and only 2.3% to 3% of CaO. No consolidation of ash in the depository is noticed, but in the transport tubes incrustation can not be neglected. The solubility of sulfate, as the most important macro-pollutant, is about 4 kg/t. Arsenic is the most important micropollutant in the ash with solubility of 0.4-0.8 g/t (of dry ash). For the existing hydraulic transport system ash in the depository is washed and has, for instance, only 0.03 g/t of arsenic. The difference in concentration between raw and washed ash is a source of water pollution.

Ash hydraulic conductivity is 10^{-4}-10^{-5} m/s in the "beach" zones (3 in fig. 1), and 10^{-5}-10^{-6} m/s in the lake zone of the impoundment (fig. 1). Peripheral levees are formed by the hydrocyclones, out of ash, with the conductivity of 10^{-4} m/s.

Relatively high hydraulic conductivity of ash is due to high volumetric porosity (about 0.7). The corresponding low bulk density of only 750 kg/m^3 makes ash deposits very sensitive to fluidization.

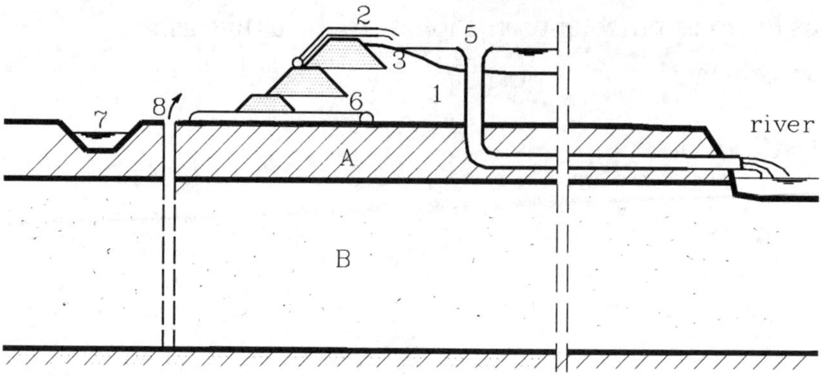

Figure 1: Typical cross section of an ash impoundment

Therefore, stability of impoundment slopes relies on water drainage. Drainage has also an important role in water protection.

2. TYPICAL LAYOUT OF THE ASH IMPOUNDMENT

NT ash impoundments have the areas of 350 and 450 ha, respectively. They are divided into three sections (cassettes): active cassette, standby casette and an inactive one, overgrown by grass to prevent air pollution. For NT "B" the cassette 3 will be used only if two new blocks are added to the plant, which would double its power.

A typical ash impoundment cross-section for the considered open ash hydraulic transport is given in figure 1. The ash-water mixture enters the ash impoundment through a system of short tubes (spigots), at the periphery of depository (denoted by 2. in fig. 1). The suspension then flows over the ash impoundment beach (3. in fig. 1), reaches the lake (4. in fig. 1), when practically all suspended ash has settled. Water is collected by an evacuator (5. in fig. 1) and returned into the river, as surface waste water.

While flowing over impoundment surface, through the "beaches" and the bottom of the lake, water infiltrates into the ash impoundment, leaving it in two ways: through a drainage system in the ash impoundment itself (6. in fig. 1), or through a system of drainage wells, drilled in the aquifer, around the ash impoundment (8 in fig. 1). A part of the polluted ground water is evacuated trough a drainage canal if it is in hydraulic connection with the aquifer, as in the case of NT "A" . The drainage channel also collects water from the ash impoundment drainage system, and drains the adjoining terrain as well.

Hydro-geological conditions. Both ash impoundments are formed on alluvial deposits (fig. 1). The hydro-geological profile can be simplified as two-layered medium. Upper layer (A in fig. 1) is of low permeability, 5-8 m tick (locally, at NT "A" far thinner), with conductivity of 10^{-7} m/s for NT "A", and $2x10^{-8}$ - $3x10^{-8}$ m/s for NT "B". Beneath the surface layer extends the aquifer (B in fig. 1) with transmissivity $10^{-2}x10^{-3}$ m^2/s.

3. POLLUTION MECHANISMS AND EFFECTS

Some information on water balance for the impoundment. Major part of water from hydraulic transport (about $1m^3/s$) infiltrates into the impoundment, either on the "beaches" or under the lake. For the plant NT "A", where the impoundment height is already 25m, almost all the water infiltrates. Ash is poorly compressible. Therefore, the amount of water which infiltrates into the upper layer is approximately proportional to the lake area and elevation above ground surface. On the contrary, compressibility of upper, less permeable layer is significant, reducing the rate of infiltration into underlying aquifer, as the impoundment grows. In the field conditions of NT "A" infiltration is dominant term of water balance, especially in summer, when energy production is reduced and evaporation from the lake increased. The power plant NT "B" is much lower, with less permeable upper layer than the one at NT "A". Nevertheless, the infiltration into the aquifer achieves 30% of total water amount. It can be concluded that the infiltration control is very important, for the slope stability, as well as for surface and ground water protection.

The presented qualitative elements of water balance influence the protection criteria and explain the proposed engineering solutions of the pollution problems using the existing ash deposing system.

Adsorption of pollutants on grain matrix. Practically there is no retaining of sulfate by the soil, at least in the range of field concentrations. Micro-pollutants, heavy metals, including the most important one, arsenic, are strongly retained by the soil, The ratio between concentrations of arsenic in solid and liquid phase has the values đ2đ:

* For sandy clay (surface layer)
  S/L = 78 mg/kg / mg/l
* For sand (upper part of aquifer)
  S/L = 5 mg/kg / mg/l

These values give only a very rough description of the exchange between water and soil grains, which is very complex and partly irreversible.

Radionuclides, with concentration in the waste water much lower than MCL (Maximum Contaminant Level), are strongly retained by the soil.

Degradation of ash impoundment pollutants. It can be accepted that all ash impoundment waste water pollutants are stable, both in the ground, and in the river water. The only exception is a non-typical pollutant, the mineral oil, which can be found temporarily in the waste water from the ash impoundments. It is not the result of coal combustion, but appears due to careless manipulation of the plant.

Users of ground and surface water endangered by the ash impoundment waste water. The types of the endangered consumers are:

* Thermal power plants themselves. They use ground water taken by the wells rather near the plants. Technical water is prepared by means of ion exchanger, whose efficiency is reduced in high degree as the total mineralization of captured water increases. The most important contribution to mineralization is given by sulfate.

* Sanitary water consumers in thermal power plants and the surrounding population. For the surrounding population a particular water supply system has been constructed, in order to avoid using ground water.

* Downstream water supply systems. The most important (but not the only) consumer of this type is Belgrade Water Supply System, whose principal supply zone is about 32 km downstream of the ash impoundment NT "A", and 50 km from the inflow of NT "B" effluents (the mouth of the creek Vukicevica)

Near and far field of the recipient (the river Sava) with respect to the ash impoundment waste water inflow. In the river Sava, rather complete mixing of the waste water with low pollutant concentration occurs about 16 km downstream from the waste water inlet. This length is found by detailed thermal field observations. Downstream of this limit, only the flux of pollutant is important, as the concentration field is homogeneous. Upstream, the concentration field is nonhomogeneous, depending, among other factors, on pollutant concentration at the waste water inlet.

4. POLLUTANTS CONCENTRATIONS

Concentration of some pollutants (arsenic, sulfates and total mineralization) in waste water at NT "A" and NT "B" thermal power plants are given in Table 1, as well as those in the recipients.

The following conclusions can be drown:

* The largest pollutant concentrations appear in surface (evacuator) water. In dry season, due to evaporation from the lake, pollutant concentration in the lake increases. It should be pointed out that the evacuator effluent represents only a fraction of the total pollutant outflow from the ash impoundment. The ratio between evacuator outflow and total outflow depends on the season of the year, electric production of thermal power plant, and, to the high degree, on the rate of infiltration into the ash impoundment. A rough estimation gives the value of this ratio of only several percents during the dry season for NT "A". (The evacuator spillway is in operation for only a very short time in this period). For NT "B" the ratio is noticeably higher, but not more than 25%.

* Ground water in the aquifer in piezometers, at some distance from the ash impoundment are polluted by sulfate (see piez. 56/3), but not yet by arsenic (C<MCL) which is still retained in the soil, and for NT "B" in the surface layer beneath the ash impoundment.

* Maximum arsenic concentrations reach 4 times average ones.

Table 1. Water quality near ash impoundments of NT thermal power plants

Power plant NT "A"	Arsenic mg/l			Sulphates mg/l			Mineral. mg/l		
1984-1990	C_{min}	C_{max}	$\bar{C}$	C_{min}	C_{max}	$\bar{C}$	C_{min}	C_{max}	$\bar{C}$
Evacuator	0.010	1.200	0.250	40	467	258	226	1272	810
Old rainage channel	0.000	0.050	0.031	37	434	274	270	1024	825
New drainage channel	0.000	0.120	0.037	49	424	284	171	1185	791
The river Sava upstr.	0.000	0.100	0.016	12	80	37	200	785	305

Power plant NT "A"	Arsenic mg/l			Sulphates mg/l			Mineral. mg/l		
1991	C_{min}	C_{max}	C	C_{min}	C_{max}	C	C_{min}	C_{max}	C
Evacuator	0.180	0.580	0.353	280	600	481	532	1165	789
Old rainage channel	0.008	0.021	0.015	374	574	487	275	995	601
New drainage channel	0.020	0.078	0.059	430	540	481	807	904	843
The river Sava upstr.	0.005	0.005	0.005	28	45	33	205	329	265
Sava downstr.	0.005	0.005	0.005	32	72	43	230	335	269
Piezometer P6/3	0.005	0.023	0.009	448	724	602	1063	1190	1128
Piezometar PP-5	0.005	0.009	0.006	34	100	68	414	662	594

Power plant NT "B"	Arsenic mg/l			Sulphates mg/l			Mineral. mg/l		
1984-1990	C_{min}	C_{max}	$\bar{C}$	C_{min}	C_{max}	$\bar{C}$	C_{min}	C_{max}	$\bar{C}$
Evacuator	0.000	1.870	0.300	190	525	249	369	1998	599
Drainage channel	0.000	0.300	0.052	100	508	256	369	1998	599
The river V. upstream	0.000	0.250	0.018	13	268	44	158	680	430
The river V. downstr.	0.000	0.500	0.158	79	462	243	251	1130	
Piezometer 23 (300 m)	0.000	0.005	0.003	5	105	37			

Power plant NT "B"	Arsenic mg/l			Sulphates mg/l			Mineral. mg/l		
1991	C_{min}	C_{max}	$\bar{C}$	C_{min}	C_{max}	$\bar{C}$	C_{min}	C_{max}	$\bar{C}$
Evacuator	0.188	0.252	0.211	294	525	402	427	760	584
Drainage channel	0.050	0.086	0.063	210	508	393	708	1698	995
The river V. upstream	0.005	0.098	0.041	34	116	82	342	573	471
The river V. downstr.	0.184	0.195	0.171	250	462	335	375	647	531
Piezometer 23 (300 m)	0.005	0.013	0.007	13	82	60	732	811	761

* The recipient itself (the river Sava) can contain noticeable concentrations of arsenic upstream the ash impoundments, which influences the quality of waste water directly, since the open hydraulic system of ash transport is used.

5. PROTECTION CRITERIA

The following types of criteria can be analyzed.

Global pollution criteria. Besides part of the pollutant which is tied to the soil particles by the irreversible adsorption, global influence of an ash impoundment is defined only by the total amount of ash and by pollutants solubility, regardless of ash deposition system. Possible effects of engineering solutions (except waste water treatment) are mainly spatial and temporal averaging of pollutant concentrations.

Criteria related to post-exploitation environmental pollution. If we analyze exploitation and post-exploitation period separately, the picture changes. Generally, the total amount of a pollutant in exploitation period remains same as in the previous case. The portion of this amount which remains on ash or soil particles depends on many factors (pollutant concentration, infiltration rate etc.)

By increasing of the pollutants retaining, better water protection is achieved during exploitation, but it is worse afterwards. The ratio of pollutant amount during exploitation and post-exploitation period is one of the most important and most delicate decisions which users of the plant, i. g. the whole society, must make. During the exploitation, protection of the environment is funded by the production, and afterwards by the global society. In the case of ash impoundments the exploitation period lasts about 30 to 40 years. Therefore, it is very difficult to make such an decision by standard optimization techniques, especially these which are based on economical analysis. It is rather a basic strategical choice of the society for the future. Part of the global pollution which is left to the oncoming generations depends to the great extent on the deposition system. An important advantage of the existing deposition system (hydraulic transport of low-concentration suspension) is that soluble pollutants in the ash are almost all washed out. All other deposition technologies are inferior in that sense.

Flux criteria. Total flux includes all the effluents which reach the river - either by surface or sub-surface flow. As it was stated, for the downstream consumers, in the far field zone, only total pollutant flux, which reaches the river, either in surface or in ground water effluents, is important. Consequently, flux criteria are very important for downstream river water users.

Flux criteria are very convenient for step by step water protection. Effects of each action on the ash impoundment are measured by the resulting change in pollutants flux.

Flux criterion is also specially convenient for systems of depositories in one coal basin, as is the case for the two plants in Table 1. Higher degree of protection at one plant can compensate lower one at the other plant from the same basin.

Despite their importance, the flux criteria have a disadvantage. It is their difficult control, requiring rather complex field investigations and computations.

Concentration criteria. They require that the concentration in all the effluent samples is lower than MCL all the time. The main advantage of these criteria is that they are easily controllable. Therefore, they are the base of protection laws. The fulfillment of these criteria guarantees that both surface and ground water is protected. They are very strict, because the are oriented to the worst conditions (evacuator water). In the case of ash depositories, the strategical orientation is that all the effluents are of the second class. Nevertheless, for the downstream consumers the flux criteria can be satisfactory.

Somewhat weaker concentration criteria state that MCL is exceeded in every 20th sample (95% reliability), with additional limit for maximum concentration C_{max}. This statement takes in the account both the effect of acute poisoning (C_{max}), and the cumulative toxic effects of pollutants.

Watershed oriented criteria. Generally speaking, surface water protection should be based on watersheds. This approach is often difficult to apply, specially for the most important, large rivers. The criteria of flux in one power production system are important step towards the water protection in the whole watersheds

Groundwater protection criteria. In the case of analyzed thermal power plants, part of the ground water (below ash depository and very close to it) has to be sacrificed, because the solution for their protection is much too difficult.

This inner zone of the depository influence should be under control of the power plant, and permitted concentrations in that zone may be much higher than those outside. The values of maximum concentrations in this zone are related to the technological problems (incrustation in the pipes etc.) and to the admissible risk of failure (see the following item).

Reliability criteria. The criterion of reliability should be stated for various protection problems, specially in the selection of technology. The basic principle is very simple: the bigger are effects of protection system failure, the higher reliability is needed. The closed circulation system, compared to the most commonly used (in Serbia) open system of hydraulic transport, produces much higher pollutant concentrations. Therefore, the degree of protection reliability should be much higher. The corresponding methodology is analysis of the uncertainties in the system. By the stochastic computations (Boreli, 1989) the influence of some parameters, assumed to be random is analyzed. The analysis includes all kinds of uncertainties, regardless of their origin. At the moment, only some aspects of the uncertainties are quantified, so the application of this criterion is still qualitative in general.

6. STRATEGY OF GROUND WATER PROTECTION

The final goal of water protection is second class of water for all the effluents, with low level of water pollution in post-exploitation phase. This strategical objective should be reached step by step, in most economical way, by optimization of deposition system changes and water protection. While approaching this objective, two aims are very important to achieve:

* reduction of pollutant flux as fundamental action
* protection of ground water supply zone.

The first step in water protection is obviously optimization of the existing deposition system operation in order to reduce impact. In order to improve surface water protection, increasing of infiltration is proposed. It is obtained by increasing of the lake level and some other actions. As a result, the total flux of heavy metals, particularly the most important - arsenic, is reduced by their sorption on clay, loamy and sandy grains. Ground water is protected by the system of drainage wells.

In order to increase pollutant sorption on ash and soil grains, and complete (temporarily) surface water protection, the closed hydraulic transport of ash is proposed as the second step. All the water collected by the evacuator and drained from the impoundment body is used for hydraulic transport again, i. g. recycled. The soil capacity for arsenic retention is such, that this step can last at least ten years.

The third step in water protection is the waste water treatment, which can be realised by purifying only 5% (maximum 10%) of the total amount now used for hydraulic transport. It was shown that this, rather expensive phase, can be postponed for some period by pre-treatment of coal. Loosing less than 10% of coal thermal potential, ash proportion can be reduced from 25% to 15%, and the pollutant flux to 70%. It seems to be a very attractive solution for further exploitation of the huge resources of low-caloric lignite.

7. CONDITIONS OF PROTECTIVE DRAINAGE WELLS OPERATION

Ash impoundments are surrounded by systems of drainage wells, drilled in the aquifer. The wells are generally used for two purposes: they mantain the stability of impoundment slopes and protect ground water in the aquifer.

Authors have made series of analyses concerning ground-water protection, using numerous field investigations and computations by the numerical model. The purpose of the computations was to determine the minimum discharge of the wells which satisfies both protection and stability criteria, for various seasons of the year. Besides other results, it was shown that in some period almost three times larger amount of water should be pumped, than is infiltrated from the impoundment, in order to capture all of it. (The protection criterion was that polluted water, infiltrated from the impoundment must not leave the area surrounded by the wells). The detailes of the computations are given elsewhere /2/.

As a result, the number, position and capacity of wells is proposed. The solution of this problem is not unique, so it should be found by opimization. It should be done according to these criteria:

* minimum number of wells
* minimum duration of wells exploitation (with full capacity, of a drainage well besides the depository is about 7 to 10 years)

* continuous well discharge, without quick changes (well adapted probes for automatic start-up of wells)

* optimum retardation of pollutants

* possibility of some wells' failure

* position of control piezometers, where mineralization is measured continuously, sulfates very often, and micro-pollutants couple of times a year.

REFERENCES

Boreli M.: Protection of Ground Water From the Impact of Ash Depositories, 4th International Mineral Water Association Congress, Ljubljana (Slovenia) - Portschach (Austria), Septembar 1991

Institute of Water Resources Research "Jaroslav Cerni", Institute of Hydraulic Engineering: Possibilities of Decreasingf of the Influence of Thermal Power Plants Ash Impoundments on the Ground and Surface Water Contamination, Technical report (in serbian), Belgrade, 1992.

Mining Institute, Belgrade: Coal, ash and environmental protection from the coal combustion, 1990.

Waste Disposal by Landfill – GREEN'93, Sarsby (ed.) © 1995 Balkema, Rotterdam, ISBN 90 5410 356 6

A case of environmental hazard caused by dumping in open ground Polychlorinated Biphenyls

G.Constantinou, G.Petrides & A.Charalambides
Cyprus Geological Survey, Cyprus

K.Kyrou
Water Development Department, Cyprus

ABSTRACT: Askarel oil, based on Polyclorinated Piphenyls (PCBs), a group of argenochlorine compounds, has been widely used in industry for its dielectric properties, thermal stability and fire resistance. PCBs are considered worldwide as a highly undesirable pollutant with low accute toxicity and moderate chronic toxicity. In 1987 a local scrap metals company imported from France a large number of old transformers for the purpose of reclaiming the scrap metal and it was reported that the askarel oil contained within them was dumped in a disused quarry. This paper describes the emergency measures taken to prevent further pollution and protect people and personnel involved with this incident, the field and laboratory investigations for the quantification of the pollution problem and the design and implementation of a secure landfill that ensured a safe disposal of contaminated material.

1 INTRODUCTION

At the beginning of 1987 a local Scrap Metals Firm imported in Cyprus a large number of transformers from France for the purpose of reclaiming the scrap metal within them. These transformers contained Askarel oil, the trade name of Polychlorinated Biphenyls. This particular oil contained Polychlorinated Biphenyls (PCBs) 60% and Trichlorobenzene 40%. As PCBs are considered worldwide to be highly undesirable pollutants it became apparent that the breaking up of these transformers could pose a serious threat to the environment. Uncontrolled dumping of Askarel oil in a disused quarry near the scrap yard in the outskirts of Limassol was reported and a technical committee under the chairmanship of the Director of Geological Survey Department was therefore set up to investigate the matter in detail and take all necessary measures to protect the environment, see FIG.1 for location. Members of the committee were representatives from the Ministries of Health, Commerce and Industry, Interior and Agriculture and Natural Resources.

At the beginning of the investigations it was established that a large number of transformers had already been dismantled and the oil was dumped in the quarry close to the scrap yard while a second batch of transformers with serious leakages, was stored at Limassol port awaiting customs clearance.

The amount of oil that was dumped in the quarry and which was estimated to be about 100 tons gave rise to several potential environmental and health problems.

2 POLYCHLORINATED BIPHENYLS (PCBs)

Polychlorinated Biphenyls (PCBs) is a group of argenochlorine compounds which are produced by the progressive chlorination of biphenyl. They are chemically and biochemically very stable compounds with low solubility in water but high solubility in organic solvents and polymers. Due to their high dielectric constant, fire resistance and thermal stability they find a wide application in industry.

PCBs have low acute toxicity but moderate chronic toxicity. Due to their high chemical and biochemical stability they persist in the environment and they accumulate in the fat tissues thus contaminating all forms of life. The highly chlorinated biphenyls undergo a very slow metabolism in the body and are consequently excreted in very low percentages. PCBs pose high occupational hazards and in people occupationally exposed, a broad spectrum of adeverse health effects have been reported ranging from skin and mucous membrane changes to malignant melanoma

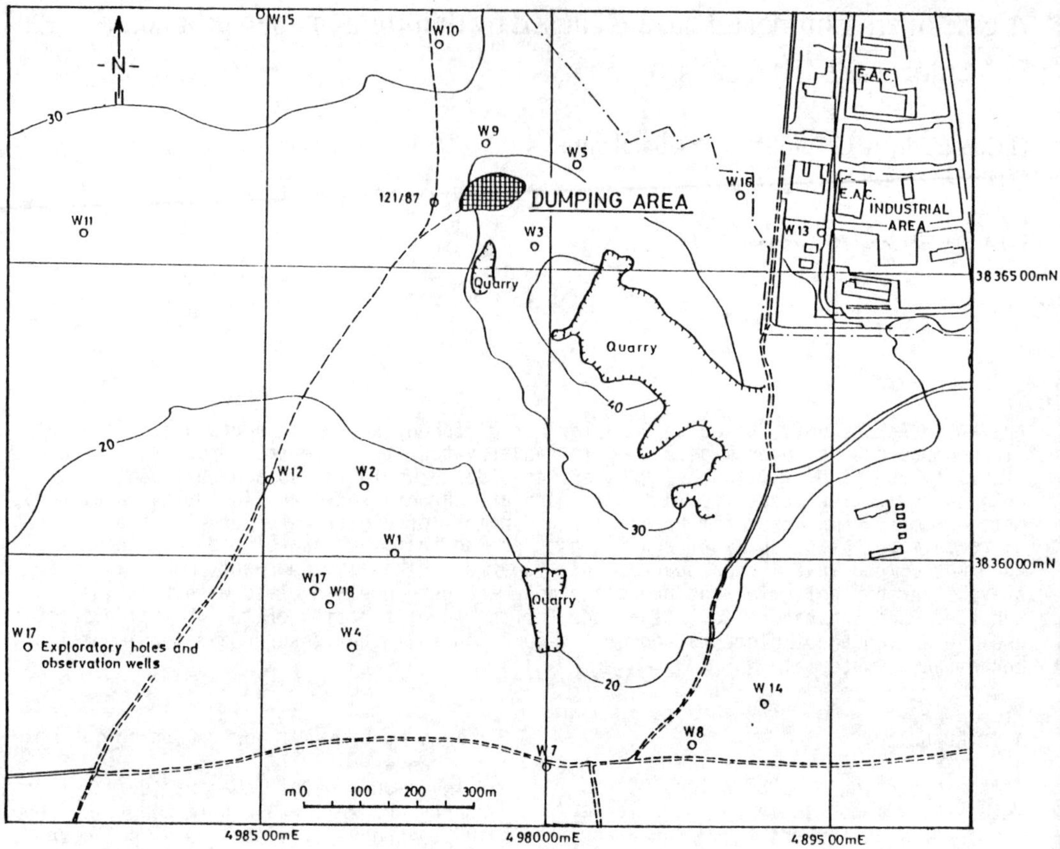

FIG 1: Location Map

and pancreas neoplasias.

As a result of all these effects to human health and the environment a code of practise exists in many countries and the use and manufacture of PCBs has been heavily restricted or banned in several countries.

3 EMERGENCY MEASURES

Before any long-term solutions to the problem were considered short term emergency measures had to be taken due to the large leakages that have occurred in order to contain them and to protect all personnel working in or near the contaminated areas.

In the case of the port area the oil from all leaking transformers was transferred to metal barrels taking all necessary safety precautions and the spillages which have occurred were treated with bentonite and sand and covered with heavy plastic sheeting to avoid further contamination and exposure of workers.

Similar measures were also taken at the scrap yard where all contaminated areas were closed to workers and contaminated materials such as wood and metal and barrels full of transformer oil, were isolated. Strict safety precautions were taken and workers coming in or near the contaminated areas were given safety equipment.

Meanwhile all necessary steps were taken for the exportation of all the transformers in the port area as well as all the contaminated material in the scrap yard and this was achieved shortly afterwards.

4 FIELD INVESTIGATIONS AND QUANTIFICATION OF THE PROBLEM

The immediate concern was to establish the geological and hydrogeological succession of the quarry area so as to determine the degree of danger of polluting the aquifers present in the more general area and at the same time establish the extend of the contamination in and around the quarry and

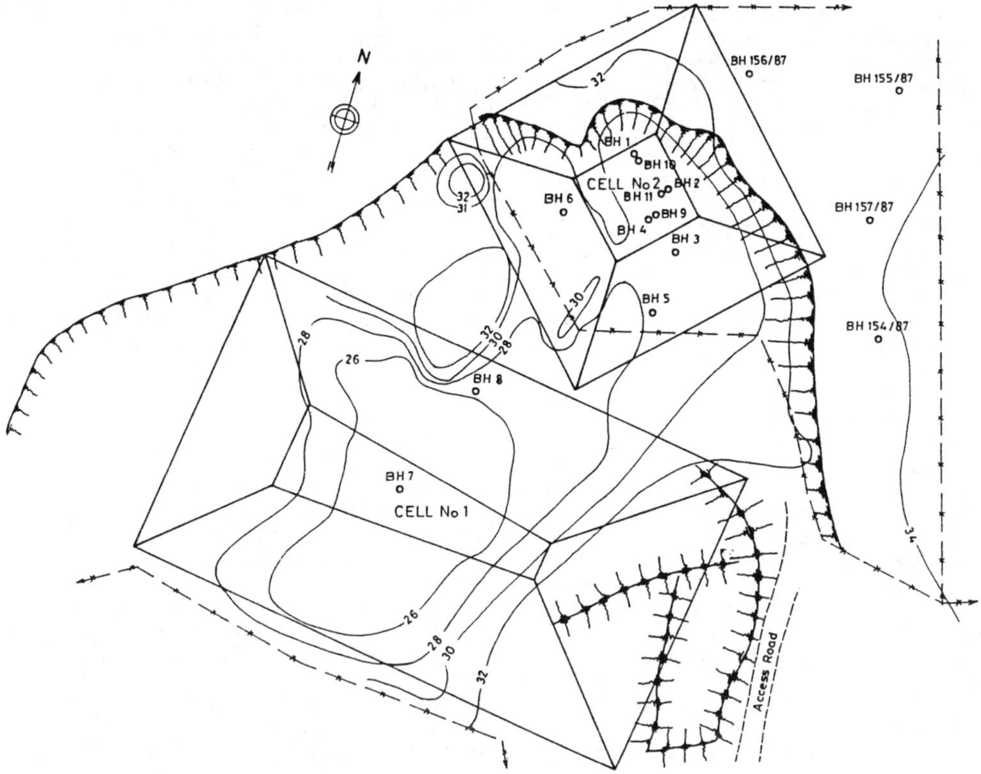

FIG 2: Details of site and layout of subsurface investigation

to establish the volume of polluted soil. The investigation started by selecting every available information including earlier hydrogeological and geophysical work in the vicinity of the quarry. This was followed by a geophysical survey, using the resistivity method. The investigations continued with the drilling of a total of 27 exploratory boreholes -FIGs 1 and 2. The drilling operations started at the quarry area where the Askarel was dumped the purpose being to determine the stratigraphical sequence below the quarry and obtain undisturbed soil and/or water samples for chemical analysis. Sixteen boreholes, FIG. 2, were drilled using a mobile auger rig. Bulk samples from the auger flights and undisturbed samples with shelby tubes were retrieved and sent for chemical analyses. The average depth of these boreholes was 22.0 metres except one borehole drilled (continuous core) down to a depth of 123 metres.

The second phase of drilling operations involved the drilling of eleven boreholes (rotary type). These boreholes were drilled around the quarry area. One line of boreholes along a circle with a radius of 100 metres from the quarry site and a second line of boreholes along an outer circle with a 300 to 500 metres radius from the quarry. The second group of boreholes had an average depth of 30 metres and they were converted into observation wells for future monitoring of the groundwater.

5 CHEMICAL INVESTIGATIONS

Extensive laboratory testing was carried out on a large number of soil samples retrieved from the boreholes in order to determine the width and depth of the contamination. Outside the quarry and in a radius of 1 Km, samples of surface soil and vegetation were collected as well as water samples from all drilling wells used for drinking supplies and agriculture in order to establish whether any contamination has occurred either in the groundwater through infiltration or in different agricultural products around the quarry as a result of air transport -see FIG. 1-.

The number of samples analyzed are summarized below:
- 17 water samples.
- 28 samples of vegetables, fruit and animal fodder.
- 51 soil samples from exploratory boreholes.
- 10 soil surface samples from spillage areas.

The chemical analyses were carried out by the Chemist of the Department in the GTZ pesticide residue analysis laboratory, West Germany as there were no facilities at the Geological Survey Department. All analyses were carried out using well established methods and standards by gasliquid chromatography using packed and capillary columns and ECD and Mass Spectrometric detectors.

The results of the analysis showed that all the ground water samples were uncontaminated thus proving as it was expected that no infiltration of PCBs had occurred in the aquifer. From the vegetable samples three samples, one of corn one of luzern and carob fruit were found to be contaminated by PCBs at a 20 ppb level, possibly due to contaminated dust transported by the prevailing winds. The subsoil analysis showed PCB contamination ranging from a few ppm in samples from depth to 300-350 ppm in near surface samples and to 3-4% in surface samples.

The results of the soil analyses showed a great variation in PCB concentrations, see Fig. 3. Having in mind all international bibliography on PCB contamination and the local conditions (precipitation, local geological conditions, land use etc) it was decided to consider as contaminated soil that had to be treated one way or the other any soil that contained more than 50 ppm PCBs. Having this criterion in mind the volume of contaminated soil in and around the quarry and down to depth of 5-7 metres was established to be of the order to 15000 m^3. Over and above this volume an additional volume of 3000 m^3 of contaminated soil using the some criteria, from the port area and the scrap yard were to be disposed in the same manner.

6 GEOLOGICAL SETTING - HYDROGEOLOGY

The survey showed that the dumping site is underlain by a sequence of Pliocene-Pleistocene deposits down to 95 m depth followed by Miocene gypsiferous marl and gypsum.

The uppermost part of the sequence is a 5-8 metres thick calcarenite horizon. The calcarenite is moderately to highly altered and it has been used in the past as fill material in the construction industry, hence the quarrying activity in the area. This horizon is followed by a 15-20 metres thick of yellowish silty marl which in places incorporates thin, irregular, calcarenitic bands. The yellowish colour is

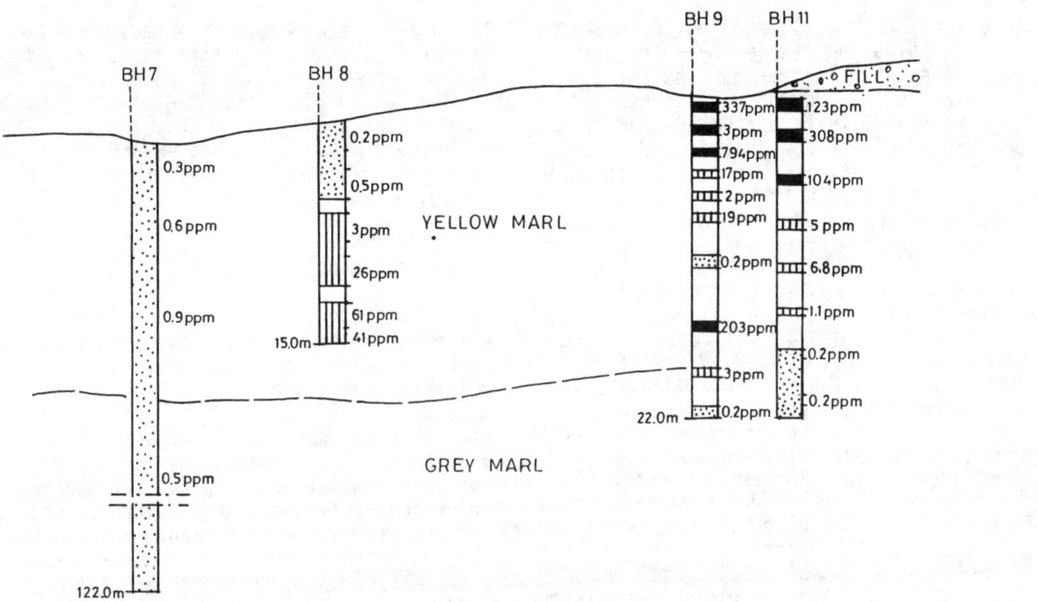

FIG 3: Geological Section. Showing extent of pollution and PCB concentration in ppm

attributed to weathering and alteration processes. Below the yellowish marl and down to depth of 65 metres there is a sequence of dark grey, overconsolidated silty sandstone and mudstones. At 95 m the sequence changes into Miocene gypsiferous marl and in turn, at 115 m depth, into gypsum. The whole sequence (drilled by a continuous cored hole) was aquiclude and no ground water was encountered -FIG. 3-.

Laterally the sequence shows marked variability which is partly attributed to facies variation and partly to tectonics which are dominated by north-south trending faults rendering the quarry site area into a horst like structure -FIG. 4-. East of the horst there is a thick accumulation of gravel and sand of Pleistocene - Holocene age which represents the main aquifer of the region. Tho the west of the site increasing occurrence of calcarenite in the marly sequence create favourable conditions for a local aquifer of rather limited yield.

At the quarry site, the uppermost calcarenite horizon has been removed by earlier exploitation and the floor of the quarry was at the time of Askarel disposal on the underlying silty marl.

The overconsolidated and fine grained fabric of the Pliocene Marl underlying the quarry site creates aquiclude conditions. Fissures and joints and the scattered occurrence of calcarenitic bands allow some ground water circulation especially within the yellowish (weathered) marl but this movement is estimated to be slow and limited.

7 COMBATING THE PROBLEMS

Quantification of the problem, mentioned above shows that the uncontrolled damping of Askarel in the quarry site, the opening up of transformers in the scrap yard and the leakages of transformers at the dock yard polluted a considerable quantity of soil in these areas, which considered hazardous and methods of disposing it safely had to be adopted.

Two methods of disposal were considered, that of incineration and that of secure landfill. The by far more efficient and effective disposal by incineration was not adopted because of the nature of pollution, involving soil masses of the order of 18000 m^3. Also incineration facilities were not available in the island. To set up such facilities would have involved very high costs and time. Alternatively, transporting the polluted soil abroad for incineration was also an impossibility involving high costs and additional pollution hazards during transportation.

The alternative left was secure landfill and every effort for speedy implementation of this solution was made.

8 SECURE LANDFILL

The first problem to be solved was to decide where the burial was to be made. The first option considered was to place the contaminated material in watertight cell(s) within the extensive bentonitic clay outcrops occurring near Pentakomo area some 22 Km from the quarry. The choice of bentonitic outcrops was obvious and offered a lot of advantages in terms of overall watertightness and the outcrops being generally of low agricultural value. To this end a number of sites were investigated, one with drilling. Eventually this option was rejected primarily because of fears of spreading the pollution during transportation from the quarry to the burial site (the trucks had to pass through Limassol town). A second and very important reason was that the ongoing site investigation at the quarry site had meanwhile proved that the geological sequence underlying the quarry itself, was watertight and aquiclude. It was thus decided to bury the polluted soil in specially constructed cells in the quarry area 80 metres southwest from the dumping point.

The burial cells were designed with the assistance of the Water Development Department which has long experience in the design and construction of irrigation ponds lined with impermeable membrane -see FIG. 2-.

The basic concept of the design was:

(a) To construct first a water tight cell 50x80x7 metres and place in it 13000 m^3 of heavily polluted material from the dumping area and
(b) To construct a second cell (40x45x7 m) in the depression that was to be created by the excavations and place in this second cell all other less polluted material.

Polluted material from the Limassol harbour dock-yard and the scrap yard was also to be imported and deposited in the cells. The cells were to be covered with unpolluted soil and the quarry would be reprofiled site so as to ensure good surface drainage conditions. Upon completion the area, nearly 10,000 m^2, was to be fenced.

The cells layout and design is given in Figs. 2 and 5. The membrane used was made of polythene, 1 mm thick sheets, joined together with the hot air method. Every effort was made to protect the membrane from being punctured by placing above and

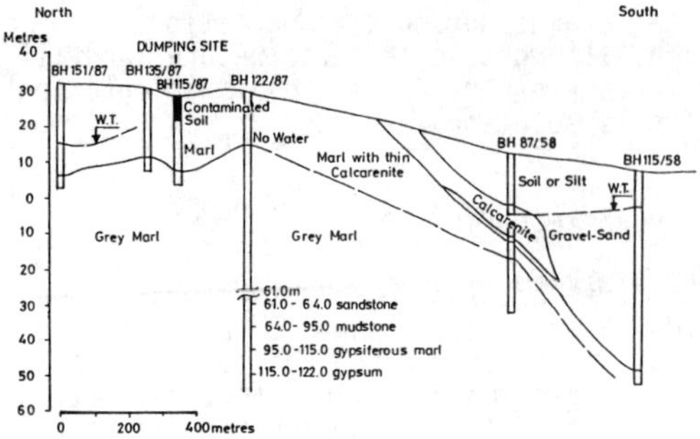

FIG 4: Hydrogeological Section

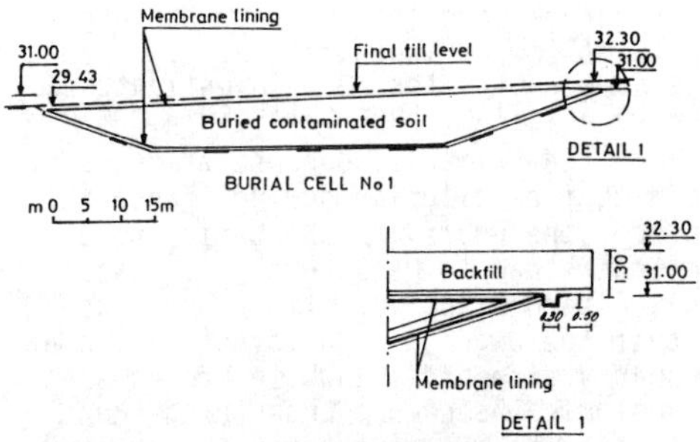

FIG 5: Section of cell and detail of the membrane emplacement

below it protective sandy layers. Also a 20 cm thick bentonitic layer was placed below the membrane as added security for watertightness -FIG. 5-.

9 IMPLEMENTATION

The burial procedure was carried out under the supervision of expertise from the Water Development Department and the Geological Survey Department and by subcontracting machinery from the Private Sector.

All necessary precautionary measures were taken in order to secure the safety of people involved in the construction and the nearby areas. All machinery were restricted to the site and only after completion of the works were allowed to leave after being properly decontaminated. People working at the site were provided with filter masks, heavy plastic uniforms, gloves and boots. At the end of each day the people changed clothes and went through a shower.

Operations started in early December 1987 and the construction and backfilling of the first cell was completed by the end of March 1988. The construction of the second cell was followed immediately and the whole operation was completed by June 1988. The quarry site is now fenced off and will stay out of reach until such time sciences will develop chemical or biological degradation procedures for PCBs.

The total cost of the works was CY£150,000 including the purchase of chemical laboratory equipment for follow up analyses.

10 CONCLUSIONS

The pollution problem caused by damping Askarel oil in a disused quarry near a populated area was fortunately, indentified early. The problem was quantified within a three month period of an intensive investigation programme. In parallel to the investigation various alternative solutions were considered and finally the secure landfill method was adopted. Implementation was carried out over a 7 month period but by the 4th month the heavily polluted part of soil was buried. Unavoidable delays were caused by the lack of proper chemical laboratory facilities and equipment (the analyses were made in Germany). Also the suitable for the purpose polythene membrane which had to be imported.

Despite the problems faced the time period for investigating, quantifying and solving the problem (8-10 months) is considered an achievement and despite the heavy local pollution caused, the spreading of contamination was arrested and the dangers of loosing the nearby aquifers with devastating consequences were averted.

The experience gained is considered invaluble. The chemical laboratory of the department has meanwhile been equipped with special gasliquid chromatographs with suitable detectors for the detection of PCBs and a variety of other organic pollutants. This together with the experience gained in the construction of the special cells for burial and the setting up of the needed infrastructure for the implementation of such works provide the conditions for pursuing in a more organized manner the search and eventual aversion of pollution from toxic wastes in the island.

The follow up of the whole operation involved periodic visits and visual inspection of the site. Also effort was made to keep the site free of any wild vegetation growth which in the dry summer months could become a fire hazard. Also survey checks determined any undue settlements. Groundwater samples from stategically located observation wells (borings) -FIG. 1, were collected in 1988 till 1993 and chemical analyses proved that no askarel permeated in the groundwater of the immediate vicinity.

Waste Disposal by Landfill – GREEN'93, Sarsby (ed.) © 1995 Balkema, Rotterdam, ISBN 90 5410 356 6

Innovative landfill design in a former sand quarry

H.L.J.Geusebroek & L.Luning
Grontmij Consulting Engineers, De Bilt, Netherlands

ABSTRACT: The landfill at Schinnen consists of two large sand and gravel quarries with steep slopes. New legislation demands waste to be dumped isolated from the surroundings. A subway was constructed in order to collect the leachate and to enable the ground water quality directly under the base liner to be checked. On the steep slopes a so-called "slope-creeping" base liner was made. In the landfill a gas extraction system is installed.

1 SITE DESCRIPTION

The landfill at Schinnen consist of two large sandpits with an area of 250,000 m^2 (phase 1, 2 and 3) and 180,000 m^2 (phase 4 and 5). The total capacity of Schinnen is 7.3 million m^3. In the proximity of Schinnen lies a third sand quarry Spaubeek (170,000 m^2 and a volume of over 3.3 million m^3).
The average rate of waste dumping is about 500,000 ton per annum. Phase 3 is presently used for landfilling. Phase 4 is already excavated. In the near future this quarry will be extended (phase 5). The two quarries are divided by a sunken road which is protected by law.
In both quarries sand and gravel are excavated till the water table is reached. The average depth of the site varies from 30 to 35 meters, with the water table at one meter below the bottom.

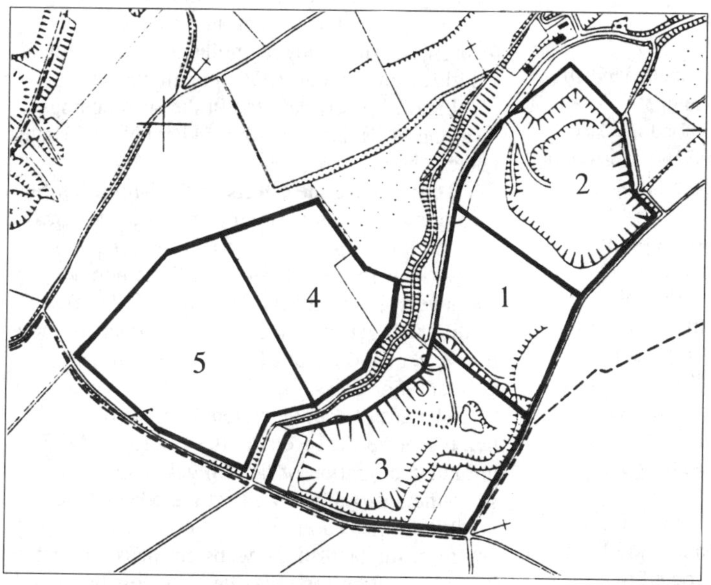

Fig. 1. map of the Schinnen quarries (scale 1:10.000)

The slopes of the quarry have gradients between 1:1 and 1:1,5. At the top of the slope no space is left in which the slope angle can be reduced. Until 1992 waste was dumped in phase 2. Neither phase 1 or 2 is provided with a base lining system.

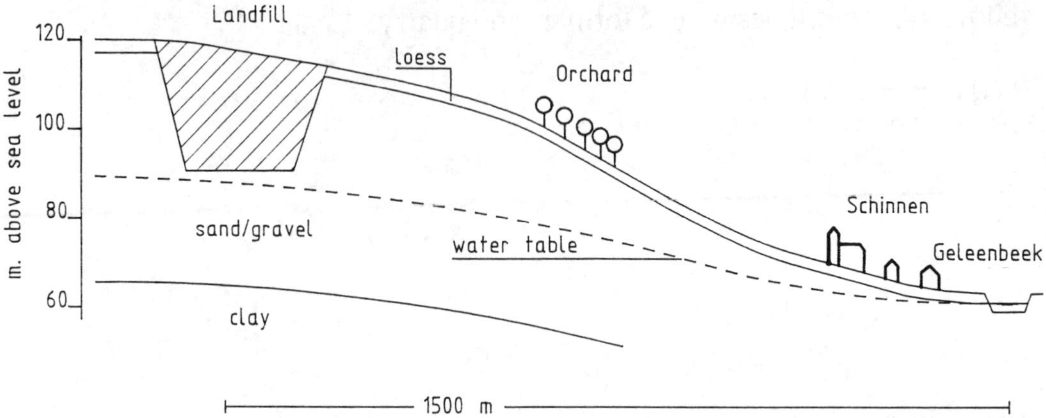

Fig. 2. Schematic cross-section of the landfill

2 GEOHYDROLOGICAL SITUATION

The quarries are situated on the plateau of Schimmert (50 km^2). The top of the plateau is covered with a layer of 4-6 meter of loess overlaying a layer of 40 to 60 m sand and gravel. Under this layer heavy clay deposits are situated. The ground water system is relatively small. The boundary of the catchment lies only 0.5 km south of the landfill. The ground water flows north to the Geleenbeek, which intercepts all the ground water. The distance to the stream is 1500 m. The velocity of the ground water is 90 m/yr.

3 DUTCH LANDFILL POLICY

According to Dutch legislation a modern landfill should be isolated from its surroundings (containment). Waste has to be dumped at least 0.70 m above the ground water level. The ground water has to be protected with a base lining system (composite).

Leachate must be collected and properly discharged. An on-site treatment could be necessary before discharging. Migration of landfill gas has to be prevented. After closing the landfill a top liner (composite system) has to be installed in order to prevent ground water pollution in the event of a failure of the base lining system. The effectiveness of the isolation has to be checked yearly (ground water quality and gas-emissions).

Since there is not much experience with landfills in quarries in the Netherlands, the Dutch legislation for gas emissions is restricted to regulations for prevention of stench and nuisance.

After closure the Regional Authority (several counties) will give the closed landfill to a so called after-care organisation. This public owned (Province and Counties) organisation has to check regularly ("till eternity") the landfill and its surroundings. If necessary this organisation has to restore the isolation. The necessary funds to operate will come from the former exploiter (after-care fund).

4 DESIGN ISOLATION PHASE 1 AND 2

Neither phase 1 or 2 is provided with a base lining system. Infiltrated rainwater becomes leachate and pollutes the ground water. In monitoring wells the following pollutants are detected: chloride, ammonium, COD and aromatics.

Landfill gas is detected in the surroundings, resulting in damage to crops and explosion danger under a house.

To minimize the effects of the landfill on the surroundings a composite top liner was designed together with a gas extraction system. The liner system consists of a 2 mm HDPE membrane overlayering a sand-bentonite layer (0.25 m). The landfill will get the old contours in order to restore the landscape. The top layer will get a vegetation of grass.

The gas extraction system consists of a number of vertical wells in the landfill body. To minimize migration of landfill gas to the surroundings, the gas wells at the edge of the landfill are placed close to each other. In the centre of the landfill the wells are placed to get enough capacity for gas collection. Just beneath the top liner a horizontal gas extraction system is installed in the top of the waste.

The extracted landfill gas is directed to a utilisation plant. Here the gas is upgraded to natural gas quality. This purified gas is sold and delivered to the Dutch gas distributing company. Remaining landfill gas (in case of overloading) is burnt in an advanced (high temperature) ground flare.

Phase 2 will undergo substantial settlement. To prevent damage to the top liner, a temporary liner will be made. Part of the temporary cover will consist of testing fields for new types of isolation materials. In this compartment gas will be extracted with the final system.

After most of the settlement has occurred this phase will get its final cover. Because of unequal settlement it could be necessary in the future to replace part of the gas extraction system.

5 DESIGN ISOLATION PHASE 3, 4 AND 5

To construct a modern landfill in a quarry with steep and long slopes gives various problems:

a) how to construct a reliable base (composite) liner for the slope. The composite layer consists of a synthetic layer on a mineral layer to provide optimum watertightness;

b) how to construct a leachate collection system, which has to be maintained and inspected 30 m below the surface;

c) how to check the ground water quality under the base liner.

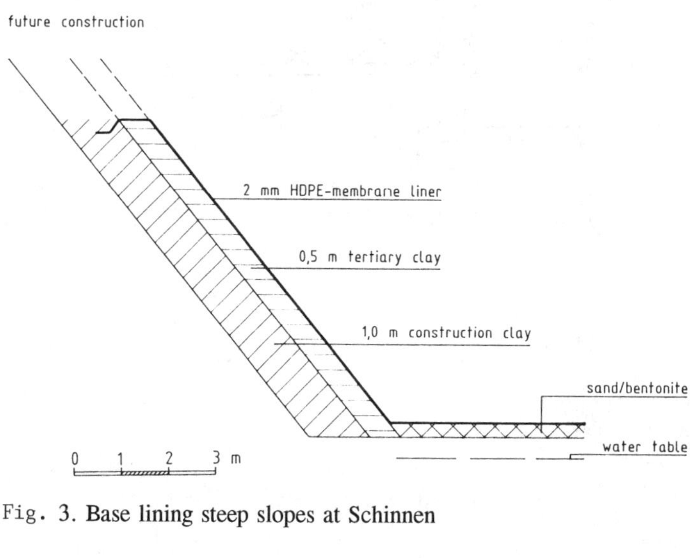

Fig. 3. Base lining steep slopes at Schinnen

To solve problem a) two solutions were taken into account: a "christmas tree" solution and a "slope-creeping" solution. The "slope-creeping" solution was chosen because of the "christmas-tree" uses more material and results in a greater loss of landfill volume.

No digging or anchoring is safe on the steep sand slope. Therefore a construction made of clay overlayering the sand slope was made in order to enable the attachment of the liner and to enable the compacting of a special so called tertiary clay with a low permeability.

The clay used for the construction is a type which is used for constructing dikes. On top of this construction clay a so called tertiary clay is used as water impermeable layer. A tertiary clay is a rare clay which is found in the province of Limburg. It contains smectites. When compacted this clay will have a very low permeability ($2*10^{-10}$ m/s) and can be used for constructing a combination liner. The compaction is done with a vibration-plate attached to a shovel which can reach about 10 m high. The HDPE-membrane is anchored in the tertiary and construction clay at a height 10 m high.

At the base of the landfill a sand/bentonite layer (0,25 m) is made to provide the mineral layer. If well compacted sand/bentonite (7 % bentonite) will have a very low permeability ($1*10^{-10}$ m/s). The base is also covered with a HDPE-membrane

To solve problem b) and c) the subway solution appeared to be the best solution. Negative aspects of the subway solution are the lack of experience and its relative high costs.

The subway enables maintenance of the leachate collection system and ground water sampling under the base liner. The landfill will have three cells. In the subway leachate from the three cells is collected separately to enable separate treatment. The leachate is treated in a UASB reactor (upflow anaerobic sludge blanket).

In the subway gas is separated from the leachate. This gas system is connected to the gas collection system with bore holes at level. After tipping in this compartment gas extraction wells will be placed and will be connected to the extraction system at level. In this part of the landfill the gas yield is supposed to be higher because no migration to the surroundings will occur (as a consequence of the base liner).

The ground water monitoring drains are individually connected to the subway in order to enable sampling and cleaning. All monitoring drains are connected together. In case of a serious leakage of the base liner it is possible to use the monitoring system as a ground water extraction system, thus preventing uncontrolled leakage to the surrounding aquifer.

The subway is two meter high to enable people to work in the subway. The subway is forced ventilated to prevent dangerous situations. The air quality is monitored constantly. In case of danger an alarm consisting of a flashing light signal and siren will alert the people inside.

All electric equipment is explosion proof. To clean the leachate collection drains a water supply system is installed. A special carriage was made for the water jetting equipment.

At the end of the subway a shaft is constructed. The shaft has four sumps to collect the leachate (3 sumps and one spare). From the sumps the leachate is pumped to the leachate treatment plant. At the other end of the subway a second shaft is built to enable the instream of fresh air and to provide an emergency-escape in case of danger.

6. ENVIRONMENTAL IMPACT ASSESSMENT STUDY

In the Netherlands activities with a possible high impact on the surroundings are obliged to make an Environmental Impact Assessment Study. All landfills with a capacity over 250,000 m^3 have to make such an EIA. In the EIA the activity (with all the designed constructions) and alternatives are compared in their impact on the environment. Authorities can use such an EIA to make their license for the landfill. In the case of Schinnen it appeared that because a significant proportion of the catchment would be isolated by the impermeable top and bottom linings, the ground water table would be lowered, resulting in some small but high quality streams drying up. To eliminate this effect all rainwater collected on the top cover will be infiltrated into the aquifer.

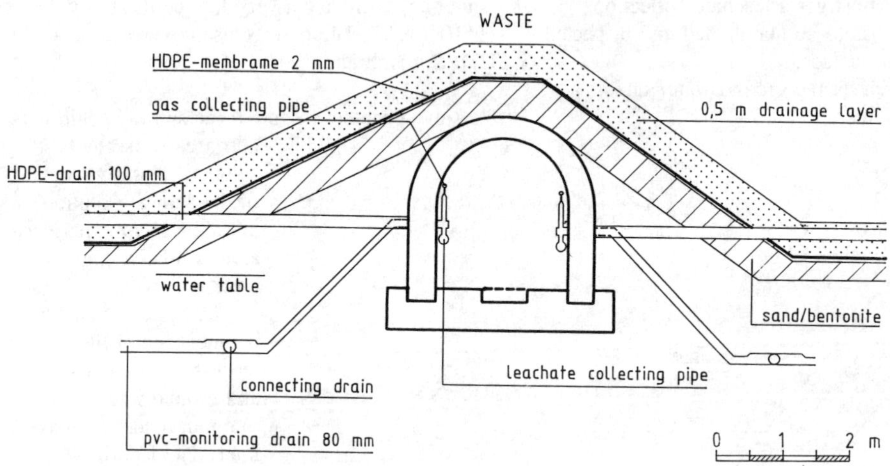

Fig. 4. Schematic cross-section of the subway

The subway was pre-fabricated in elements of 3 m length to enable a short construction time. The total length of the subway is 400 m. Because of the possible aggressive environment and the low thermal coefficient of expansion broken limestone was chosen as aggregate for the concrete.

7. PHOTO S OF THE CONSTRUCTION

In the following the construction of the modern landfill in phase 3 is shown.

photo 1: Compacting tertiary clay on a steep slope

photo 2: Subway, mineral base and slope liner

photo 3: The subway and the shaft

photo 4: Placing the membrane on the mineral layer

photo 5: The inside of the subway

photo 6:Leachate purification and gas utilisation plant and ground flare

Waste Disposal by Landfill – GREEN'93, Sarsby (ed.) © 1995 Balkema, Rotterdam, ISBN 90 5410 356 6

Evaluation of the incident at the Ümraniye landfill, Istanbul and proposed remedial investigations

Erol Güler & Cem B. Avcı
Civil Engineering Department, Boğaziçi University, Bebek, Istanbul, Turkey

ABSTRACT: A large waste landslide occurred at the Ümraniye municipal waste landfill located in İstanbul, Turkey on 28 April 1993. A biogas explosion is stipulated to have caused the waste landslide which travelled approximately 1 km down the slope of the valley where the landfill was located. The possibility of a slope stability failure due to increase pore water pressure was investigated. In addition, an assessment of the potential environmental problems which have arisen after the incident and the proposed remedial steps which must be taken to minimize the environmental impact of the waste landfill are presented.

1 INTRODUCTION

The present article provides an evaluation of the preliminary data obtained from site visits to the Ümraniye Landfill, İstanbul following the incident that took place on 28 April 1993. The incident consisted of a biogas explosion (as reported by newspapers and eyewitness accounts) which led to an entire section of the landfill side to slide down into a valley where residential housing was located in the path of the landslide.

The scope of the present investigation has been threefold:

• to assess the general site conditions based on the data obtained from site visits;

• to assess the potential problems created by the new site conditions following the landslide; these include landfill slope stability and the potential environmental impact of the landfill;

• to propose remedial steps that need to be taken immediately in order to minimize the environmental impact of the incident and the landfill.

2 SITE EVALUATION

The Ümraniye landfill located in İstanbul, Turkey covers an area of 16 ha and was used for disposing the municipal waste of 4.5 million inhabitants. The amount of waste disposed at the site was close to 2300 tons per day whereby the area represented the largest landfill to serve on the Anatolian side of Istanbul (Hürriyet, 1993). One estimate of the waste volume in the landfill that has accumulated after 20 years of operation stands at 5 million m^3.

The landfill area is located on top of a valley and is surrounded by hills which lead down along both sides of the landfill to the valley floor where a stream flowed perpendicular to the valley. A residential community is present at the base of the valley.

The landfill height is approximately 10-15 m above the surrounding topography. The top of the landfill is circular in shape and the side slopes leading down from the top of the landfill are close to 45^0. Gas venting systems and leachate collection systems have not been implemented during the active lifetime of the landfill.

It has been reported that a practice of installing an earth cover on top of waste was started only within the last few years in order to minimize odours; the previous practices apparently accumulated the waste and covered it with minimal usage of earth material.

The landfill stands on top of a hilly area which are approximately 100 m above the stream bed. Three sides of the landfill are surrounded by natural topography which undulates slowly away from the landfill; one side of the landfill stands on top of the former drainage basin created by two surrounding hills; this drainage pathway carried leachate from the landfill to the stream. The waste landslide took place along this drainage basin and after travelling an estimated distance of 1 km, ultimately came to rest at the bottom of the hills. The waste slide effectively blocked the path of the stream such that at the time of the site visits stream waters had created a small pool behind the waste.

3 ASSESSMENT OF POTENTIAL PROBLEMS

There are several potential problems which have been created from the waste landslide; these include problems related to the waste slope stability and the potential environmental impact of the new site conditions. A slope stability analysis was carried out using preliminary data obtained from the site visits and literature cited values of the strength parameters of municipal waste. These analyses were, at best, an attempt to establish sensitivity analysis of the conditions; the analysis were performed in order to investigate the conditions prior to the incident. Slope stability investigation must be carried out with site data if the landfill is expected to be used in the future for continued waste disposal.

The site conditions which have developed following the incident represent a much greater source of waste contamination than the previous conditions. The cause of the landslide was attributed to the explosion of biogas inside the landfill which subsequently led to the movement of the entire section of the landfill down the valley. The task of minimizing biogas accumulation should, therefore, be one of the main task of the remedial investigations. In addition, since the landfill waste is now spread on a much larger area, it is more likely that larger volumes of leachate will be generated than previously. Steps must therefore be taken to minimize the generation of leachate because the stream at the bottom of the valley discharges to the Bosphorus.

4 SLOPE STABILITY ANALYSIS

The length of the arc created on top of the landfill surface from the landslide is estimated at 200 m. Two observations were made during the site visits:

- a vertical delineation of past and present waste disposal practices were noted as follows; the bottom 20 meters of the exposed surface shows the presence of dark waste with minimal earth cover occurrence. The upper 20 meters of the cut, however, shows almost a predominance of earth material lying ontop of the waste. It had been reported that this material consisted mainly of construction debris and hauled earth.

The slopes of the landfill surfaces were noted to be quite variable; waste slopes as high as 60^0 were seen at certain locations and at other places, small slumps of earth cover and waste had occurred from the sides and rested at smaller degrees of inclination.

Landslide stability analysis was performed for the conditions prior to the incident. Although the exact topography of the area before the slide occurrence was not known, a rough estimate was be made based on local eyewitness accounts and interpreting the portions of the slope that remains intact. A limit equilibrium slope stability analysis was conducted using the following material properties. The properties of the natural soil is taken as follows based on experience with local soil conditions:

$\gamma = 20$ kN/m^3
$c = 100$ kN/m^2
$\phi = 18^\circ$

In general, it is difficult to determine the parameters for waste due to the heterogeneity, nature and size of different components of urban wastes. Cancelli et.al. (1987) proposes the use of the following parameters which were also adapted for the old waste material that is pure municipal waste:

$\gamma = 8$ kN/m^3
$c = 30$ kN/m^2
$\phi = 25^\circ$

The top layers of the fill were observed to have a great amount of earth material. The addition of earth is not expected to change the strength properties of the waste, however, the unit weight of the material is greatly increased due to the inclusion of large amounts of soil. So, for this layer, the parameters are taken as follows:

$\gamma = 15$ kN/m^3
$c = 30$ kN/m^2
$\phi = 25^\circ$

The pore pressure parameter r_u for the waste fill is taken to be 0.25; this assumes that the pore water pressure within the waste material is partially drained. For these conditions, the slope stability analysis gives a factor of safety of 1.3 which is adequate for such a landfill site (Figure 1). It is quite reasonable to assume that a large pore water pressure developed considering the fact that no drainage measures were built and that the topography causes water accumulation on the back of the slope. This assumption is also backed up by the fact that the month of April is the month with the highest level of precipitation and combined with the effect of melting snow causes the most severe condition in terms of ground water level. Analysis using undrained conditions for both levels of

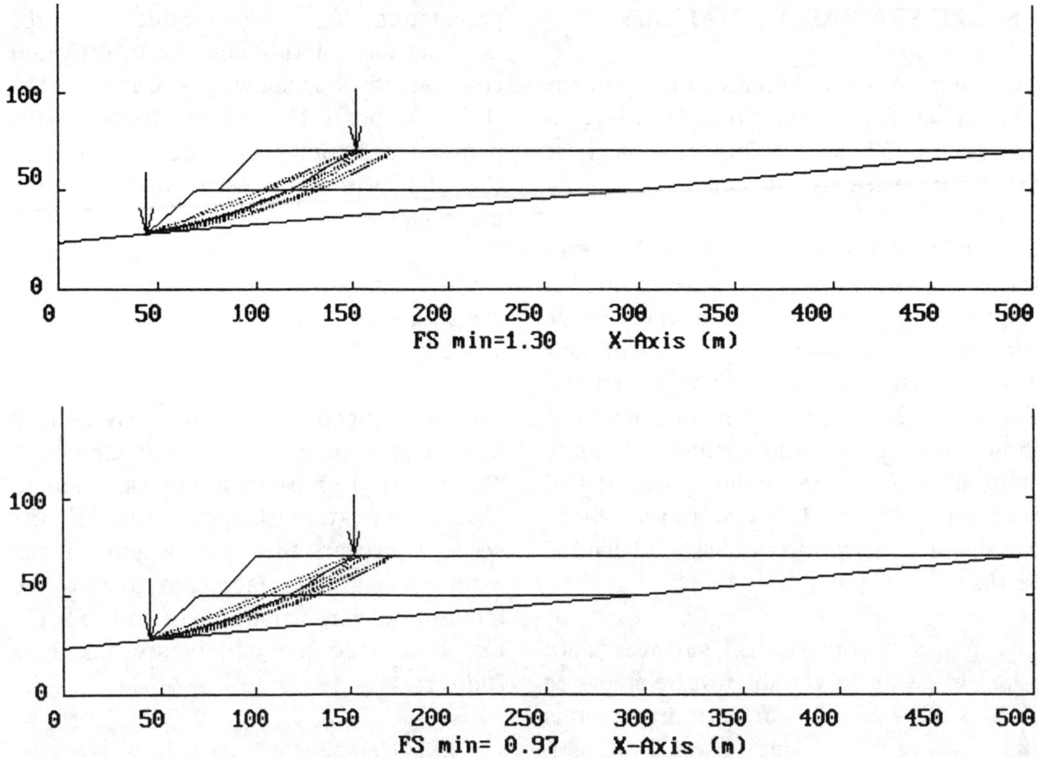

Figure 1. Factor of safety values and ten most critical potential failure surfaces for drained (top) and undrained (bottom) cases

waste gives a factor of safety of only 0.97 which indicates an imminent stability problem.

The result of the stability analysis combined with the fact that the failure occurred in the form of a flow indicate that this failure was mainly caused by the presence of excess pore water pressure, which almost liquified the waste mass.

5 REMEDIAL INVESTIGATIONS

Four remedial investigations should be undertaken at the landfill to minimize the environmental impact of the landfill: 1) investigation of the present stability of the waste mass and, if necessary, take the appropriate steps to ensure its stability 2) gas venting system installation and operation, 3) minimization of leachate generation and 4) collection of leachate. The present discussion outlines the methods which could be used to implement the above remedial steps. The implementation of these steps should only be performed following site data collection that must be collected to efficiently design the remedial steps.

5.1 Stability Investigation

Stability calculations should be performed after determining the present topography and material properties. It is very likely that some measures must be taken to provide sufficient stability given the preliminary observations. This can be achieved either by changing the topography of the present slope and providing drainage measures to relieve the pore water pressure.

5.2 Gas Venting

An active venting systems should be preferred over passive venting systems for minimizing the dangers of gas generation for the present site; an active gas venting systems should be installed consisting of a series of deep extraction wells connected by a header pipe to a blower that either delivers the gas for energy reuse purposes, or to an on-site burner or simply releases it to the atmosphere; the possibility of a reoccurrence of such an event should be minimized using the most effective means of gas venting systems.

The extraction well spacing, the header pipe selection, the blower and burner selections should all be made following investigations performed to collect the data for effective design of gas venting systems. It is also quite important that some kind of performance monitoring be implemented following the installation and operations of the gas venting systems.

5.3 Leachate Control

The wastes which now lie along the former drainage pathway that existed between the toe of the landfill and the former stream represent a large area of wastes exposed to the atmosphere. This exposed area represents a great potential for leachate generation due to rainfall; this leachate will ultimately descend into the stream at the bottom of the valley.

Two options exist for preventing leachate generation from this waste area: 1) remove the waste and transport it to some other location, or 2) leave the waste in place and construct an adequate cover to prevent rainfall infiltration into the waste layers. These steps, however, should be taken only after adequate information has been obtained regarding the stability of the remaining slopes of the landfill material.

The option of removing the wastes do not seem to represent an a viable solution given the fact that the landfill may be closed from this time on and that there are serious problems with respect to adequate waste disposal areas in Istanbul. The second method, which is to construct some kind of a cover, represents the quickest solution that is likely to be taken for this waste material.

Earth covers or geosynthetics could be used to cover the waste volume if the waste removal action is not selected. Adequate care should be taken to design an effective cover material.

6 CONCLUSIONS AND RECOMMENDATIONS

The observations made based upon site visits and preliminary analysis results indicate that several potential problems exist at the Ümraniye landfill which must be resolved as soon as possible to minimize the environmental impact posed by the landfill conditions. These potential problems and the proposed remedial steps are summarized as follows:

- Slope stability analysis of the conditions

existing prior to the incident indicate that the presence of high pore water pressure in the waste fill might have had a major role in the landslide occurrence. The data used in the analysis were based on very preliminary analysis and literature cited values for waste data. Site data must be collected and monitoring of the conditions at the site must be performed in order to establish and remediate the slope stability problems that might be present at the site.

• The fact that a biogas explosion is believed to have led to the waste landslide necessitates a preliminary site investigation for the design and subsequent implementation of an active gas venting system.

• A large loading of contaminants into the stream at the bottom of the landfill will likely occurs from the exposed waste surface area. Means of minimizing leachate generation and collection of generated leachate must be immediately implemented.

REFERENCES

Cancelli,A., DElia,B.,Sembenelli,P. (1987) "Groundwater problems in embankments, dams and natural slopes", Gen. Rep., IX Eur. Conf. Soil Mech. Found. Eng., Dublin,4, 1-30

Hürriyet (1993) Daily Turkish Nespaper

Waste Disposal by Landfill – GREEN'93, Sarsby (ed.) © 1995 Balkema, Rotterdam, ISBN 90 5410 356 6

Final capping and finishing of a landfill in Limburg, the Netherlands

Sjoerd E. Hoekstra
Akzo Industrial Systems bv, Arnhem, Netherlands

ABSTRACT: In early 1992, work to cap and finish the 22 ha. surface of the closed landfill at Horst was begun. Geocomposites are used to collect and discharge excess rain water and also leachate seeping from the inside towards the slope.

1 EXISTING SITUATION

Until the end of 1990, a 21-hectare site at Horst, Limburg, was used for storage of municipal and industrial wastes. Northern Limburg is a fairly thinly populated, largely agrarian region. As a result, and because of the proximity of a large vegetable and fruit auction mart, some 35% of the waste is composed of rapidly decomposing garden, vegetable, and fruit wastes. Sections predominantly filled with these wastes with a high content of organic material give rise to horizontal migration of leachate, causing wet spots in the slopes.

The landfill varies in height from 18 to 25 m; its base is located 0.5 m above the highest level of the water table. The slopes have an inclination varying from 1:3 to 1:5. It has been decided to maintain the existing relief. The site will be landscaped and planted so as to be an unobtrusive element in the local geography.

2 REMEDIATION PROGRAM

2.1 Area with a bottom seal

In 1987, the provincial government made renewal of a dumping permit conditional upon provision of the site extension with a bottom seal. This is why the first extension, sized roughly 4 hectares, was equipped with a 30 cm sand-bentonite layer. In 1989-90, a second 4-hectare extension was added with the same type of seal. The final capping of this part of the landfill could therefore be postponed until 1993 (see Figure 1).

2.2 Contaminated ground water

In 1991, a start was made on a clean-up of the ground water which was in part contaminated with toluene. By means of an array of pumps the ground water is being collected at the downstream edges of the site and pretreated prior to being piped to the regional water treatment plant.

2.3 Gas

By means of 19 gas extraction points sunk into the landfill, the methane formed in the waste will be collected and piped to a gas turbine to be burned for electricity.

3 CAPPING CONSTRUCTION

3.1 Provision of capping layers

Since no ground water protection had been provided for 13 hectares developed first for storage of wastes, steps to install a top seal upon this section were initiated in late 1991. The design of the final seal for this area is shown schematically in Figure 2.

Under the environmental rules applying in the Netherlands at the time, the top seal must not pass more than 20 mm precipitation per year at a hydraulic gradient of i = 5. This means that the k-value must not exceed $k = 1 \times 10^{-10}$ m/s.

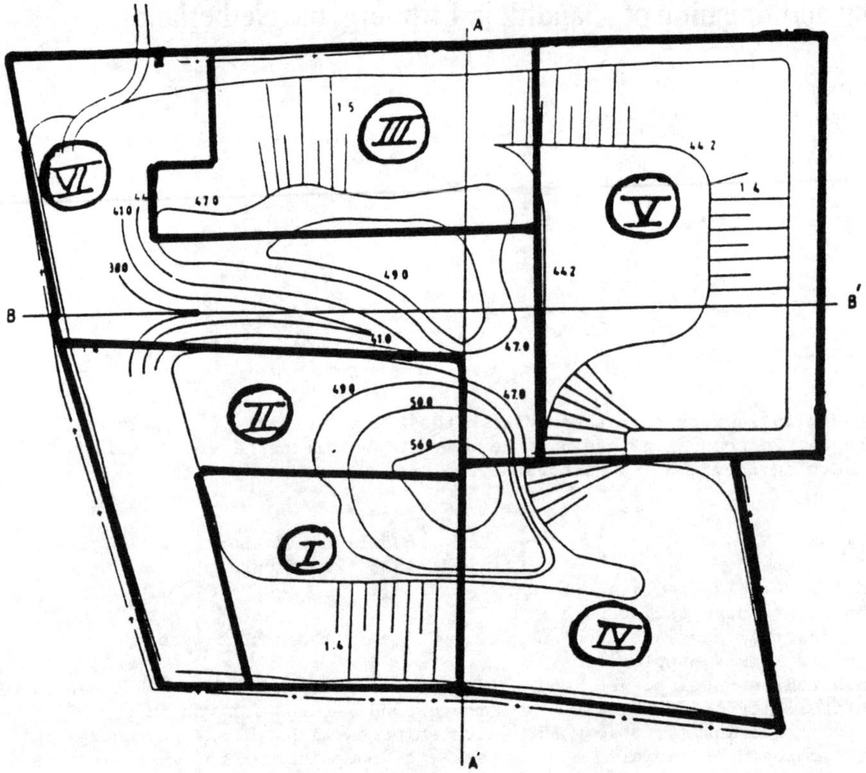

Figure 1. Overview of remediation measures in the landfill at Horst, Limburg.

	Bottom seal	Final capping,	Area
Phase I	none	constructed	
Phase II	none	in 1992	9 ha
Phase III	none		
Phase IV	completed 1987	scheduled for 1993	
	(0.3 m of sand-bentonite)		9 ha
Phase V	completed 1990		
	(0.3 m of sand-bentonite)	scheduled for 1993	
Phase VI	none	scheduled for 1994	4 ha

For this value to be achieved, the compaction of the selected sand-bentonite mixture needs to be at least 98% of the proctor density. From the pretrials it became clear that a compaction value of this magnitude would be only attainable if the support layer has a penetration resistance of not less than 2 N/m^2 (= 20 kN/m^2).

3.2 Wet spots

In those places in the slope where the covering soil is soggy due to leachate seeping out, special measures were considered necessary to obtain a stable, sufficiently watertight top seal. Without such measures, it was feared that the support layer consisting of 0.30 m of filling sand would be too wet to possess sufficient bearing capacity for satisfactory compaction of the superimposed 0.20-m-deep sand-bentonite layer.

A further design assumption was that the leachate will probably continue to flow to the wet spots in the slope for some years after placement of the top seal. Special drainage is therefore to be provided to safeguard integrity of the top seal. In these places, geosynthetic

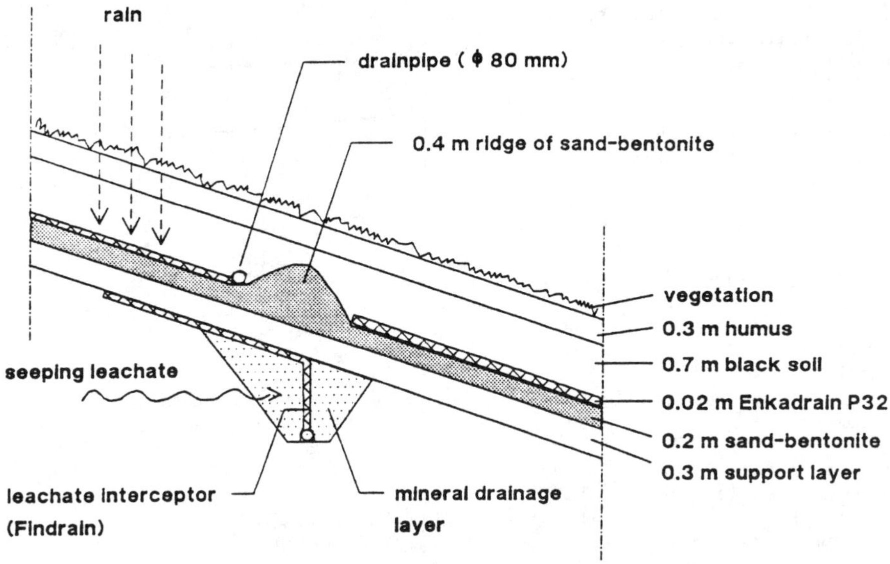

Figure 2. Diagram of capping construction

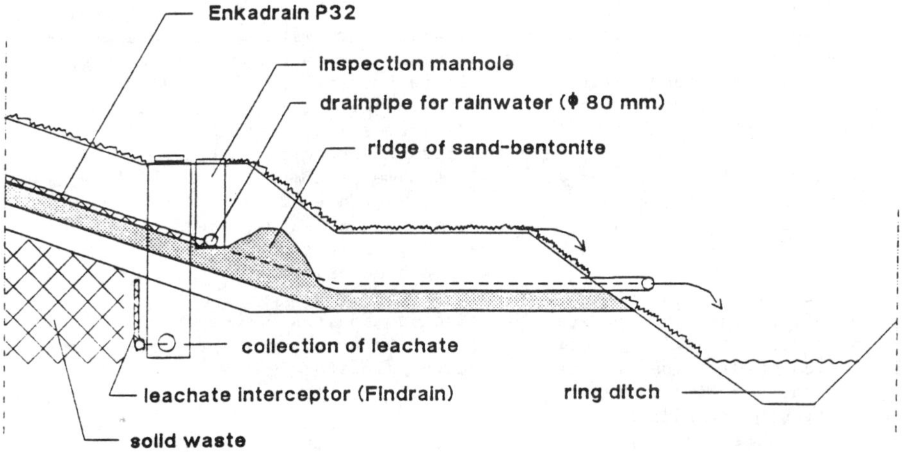

Figure 3. Diagram of construction at the foot of the slope

Findrain drainage strips therefore have been installed to ensure controlled moisture discharge.

The installation procedure was as follows: In the waste, 1-m-deep, vertical trenches were dug across the slope. With the Findrain interceptor and drain pipe in place, the trenches were refilled with mineral drainage material
(see Figure 2). The leachate is collected by the drain pipe and carried to the water treatment plant via the inspection manholes (see Figure 3).

Once the waste at the surface had dried sufficiently, the 0.30-m support layer with the requisite penetration resistance of at least 20 kN/m² could be placed.

4 INSTALLATION PROCEDURE

On an area of approx. 8 hectares of the section without bottom seal the top soil layer was first removed. Subsequently the support layer and on top of that the premixed sand-bentonite layer were placed.

The latter layer of 0.20 m thickness, containing 6-9% sodium bentonite, was compacted in situ to achieve a permeability coefficient of $k < 1 \times 10^{-10}$ m/s.

A separate protocol issued by the responsible authorities specifies procedural and quality control rules for this purpose and describes three phases in the quality assurance of the mineral seal:

a) Through preliminary investigations into the optimum moisture contents of various sand-bentonite mixtures, and establishment of the desired bentonite content, within a range of 6 to 9%.
b) Frequent checks by the contractor during placement.

c) Post-installation compliance tests according to a schedule fixed by the consultant and the client. Under this program, independent laboratories obtain readings for layer thickness, bentonite content, and degree of compaction.

Immediately after compaction and compliance testing, the sand-bentonite layer is covered by a 4m wide geocomposite drain and a topping layer of 0.70 m black soil and 0.30 m humus. This procedure ensures control of the moisture content of the sand-bentonite to the optimum value.

The finished sections are promptly sown with grass, in order to increase water absorption capacity and limit surface runoff. A well established plant cover reduces rainwater infiltration into the finishing layer by some 40%.

The rain water penetrating the soil layer is carried off by the geocomposite drain. Placed in sections each 50 m long, these drains have 80-mm-dia pipes at their foot to carry off the excess rain water from the section of slope immediately above. The water so collected is run to inspection manholes and then discharged to a ring ditch (Figure 3).

To prevent any uncontrolled flow of water along the slope, the sand-bentonite layer features 40-cm-high ridges. At these ridges, and at the very foot of the slope, the drain empties into a Findrain section with drain pipe which carries the water via inspection manholes to the ring ditch.

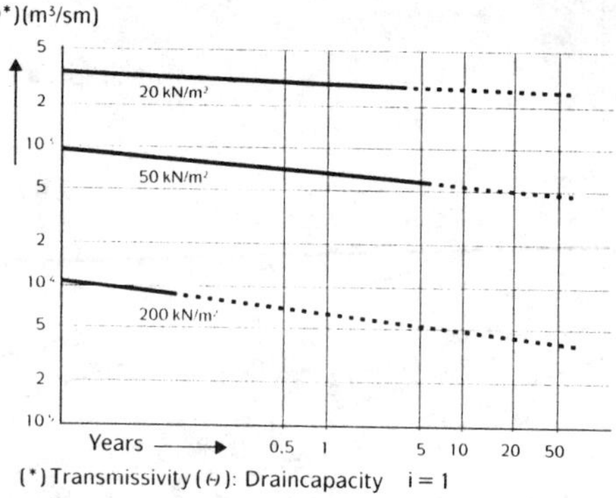

Figure 4. Drainage capacity of water-saturated Enkadrain ST under long-time pressure

5 GEOSYNTHETIC FINDRAIN AND SURFACE DRAINAGE LAYER

5.1 Geosynthetic filter-core composites

The selected Enkadrain-Findrains as described in 3.2 consist of a 1-m-wide polyamide core with an envelope of two thermally-bonded bicomponent nonwoven filters. Located between the core and the nonwoven filters is a drawstring which facilitates installation of the perforated PVC drain pipes.

The same geocomposite material is also used as surface drain for the removal of the excess rainwater. The drainage capacity of these geosynthetic composites has been described before (Hoekstra, et al., 1986, 1990).

Such composites both save valuable storage capacity and effectively remove excess rainwater.

5.2 Draincapacity

The high long-time drainage capacity of the geocomposite is due to the outstanding crush resistance of the core (geospacer), the large number of contact points of core and nonwoven filter, and the relatively high stiffness of the thermally bonded nonwoven.

The bicomponent filaments are fused together which makes this type of filter effectively resist being pressed into the geospacer by the weight of the covering soil. The open space providing a conduit for the water is therefore only slightly reduced by the pressure of the soil layer.

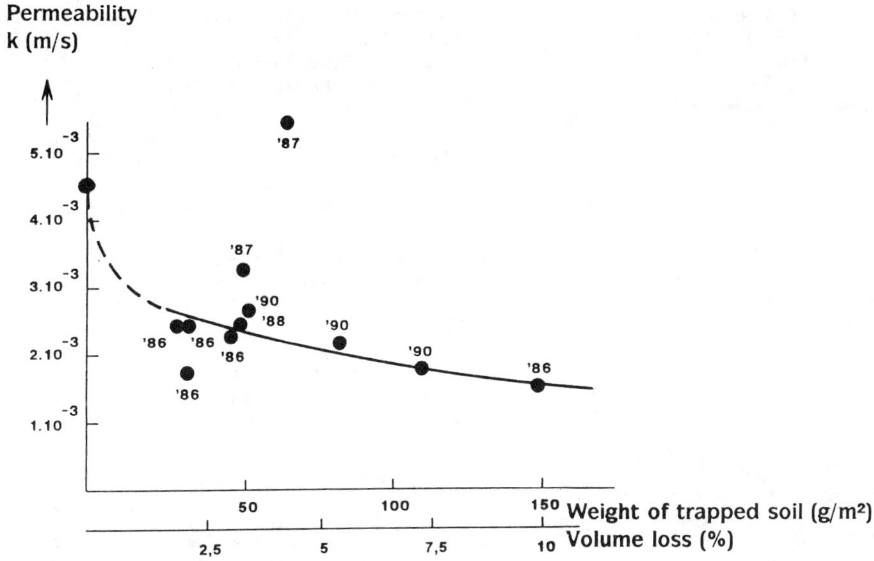

Figure 5 Relation between permeability k_f at i = 1 and soil uptake in a thermally-bonded Colback filter fabric (125g/m2) over a period of five years.

Figure 4 illustrates the long-time behavior over tens of years and the permanence of drainage capacity of this type of geocomposite in water-saturated condition under various loads.

5.3 Filterstability

Initially, such filters were mostly used in coastal protection constructions and shipping canals, Hence the primary research theme used to be the filter stability under dynamic waterflow conditions and heavy mechanical solicitations. Under these conditions thick, needle-punched nonwovens and heavy split-fibre fabrics are frequently used.

Rainwater seepage through the top cover of a landfill, however, is fundamentally different. Under these circumstances the advantage of the application of stiffer, thermally-bonded nonwovens as filter layer can be utilised.

The Staring Centre in Wageningen showed that under such conditions the soil structure has a decisive impact on the risk of clogging (Stuyt, L.C.P.M., 1987). A special study was made at the Deponie Gerolsheim (FRG) (Muth, W., 1988).

Over a five-year period the filter fabric and the soil on top of the filter were periodically sampled. For all filter samples both the weight increase due to soil particles that could not easily be washed off and the water permeability were measured (see Figure 5).

Table 1. Particle size distribution of (a) soil trapped in the Colback filter and (b) that of the covering soil.

Particle size (µm) between	Composition of soil trapped in the filter (%)	Composition of soil directly above the filter (%)
0 - 5	-	20
5 - 10	6.6	10
10 - 16	39.3	5
16 - 35	25.5	15
35 - 50	9.6	4
50 - 75	8.3	3
75 - 100	7.7	3
> 100	3	40

The results can be summarised as follows:

° Most soil particles are trapped in the first rainly month and add to the filter 40 - 100% of its own weight of 125 g/m².

° At about 1.10^{-3} m/s, the residual permeability over the five year period of the investigation remained more than

3.000 times higher than the permeability of the adjacent soil, given $k_f = 3.10^{-7}$ m/s.

° The final level or permeability is already reached after a couple of month.

Some of the filter fabric samples were further tested at the Staring Centre.

By means of high-frequency vibration, the trapped soil particles were removed from the filter, screened in a particle size analyser and counted by computer (see Table 1).

The results give rise to the following observations:

° Only some 3% of the particles in the filter are larger than 100 µm whereas the soil directly above the filter contains 40%.

° In the nonwoven no particles smaller than 5 µm are found although the soil directly above contains 20%.

Since these and other investigations showed that no soil particles remain in the drain core, this leads to this tentative conclusion:

° Only particles from a very thin layer of soil directly above the interface with the nonwoven are flushed out by water; particles up to 5 µm pass the filter, particles larger than 100 µm remain in the filter.

Further research on the structure of the soil layer above the filter has been carried out at the Staring Centre.

A soil sample has been analysed using a computer-assisted tomography X-ray scanner (Stuyt, L.C.P.M., 1987). The results further clarify the movement of only a small amount of the soil particles by water seeping through a soil cover. No clogging of the filter and no changes in the soil structure above the filter were found.

6 CONCLUSIONS

° The landfill at Horst has been in operation from 1987 till December 1990 under the Dutch "IBC" criteria then applying.
In early 1992, work started on the final capping construction, that will permanently limit the ingress of rainwater to the required level of less than 20 mm per year.

° The use of geocomposite drains permitted the application of a construction method which combines economic with technical attractiveness. The result is a stable cover, which from the ecological and engineering angles is an appropriate element in the landscape.

° For in-plane water transport geotextile composites with a separate filter and drainage function are preferred.

° Thermally-bonded nonwovens posses a relative high stiffness. This makes them attractive as filterlayer on top of the draincore under the load of a soil layer.

REFERENCES

Hoekstra, S.E. and Berkhout, H.C. (1986) "Geotextile/"Geospacer Composites for Environmental projects", 3rd Int. Conf. on Geotextiles, Vienna, Austria , 343-347.

Hoekstra, S.E. and Berkhout, H.C. (1990) "Geosynthetics for safe waste disposal", 4th Int. Conf. on Geotextiles, Geomembranes and Related Products, The Hague, the Netherlands, 487-492.

Muth, W., Versuchsanstalt für Wasserbau, FHS Karlsruhe, (1988), "Filterstabilität von Verbundmatten Enkadrain, Mülldeponie Gerolsheim", Karlsruhe, Fed. Rep. of Germany.

Stuyt, L.C.P.M. and Oosten, A.J., (1987) "Mineral and Ochre Clogging of Subsurface Land Drainage Systems in The Netherlands", Geotextiles and Geomembranes 5, 123-140, Elsevier Applied Science Publishers, Barking, England.

Waste Disposal by Landfill – GREEN'93, Sarsby (ed.) © 1995 Balkema, Rotterdam, ISBN 90 5410 356 6

Land disposal of dredged mud

A.M.O.Mohamed, R.N.Yong & R.Galvez-Cloutier
Geotechnical Research Centre, McGill University, Montreal, Que., Canada

ABSTRACT: This study assesses the feasibility of landfilling contaminated sediments by evaluating its compatibility with a natural clay barrier. A clay barrier is overlain by a contaminated sediment and subjected to leaching with an acidic solution of pH = 3. Physical and chemical properties of a natural Quebec clay sample were determined. Chemical analysis of the pore fluid of various ions as a function of time were conducted and the results are discussed. Diffusion parameters were calculated and compared with other published results obtained under similar conditions.

1. INTRODUCTION

Dredged mud generally constitutes fine matter with a dry matter content of 2-50%, depending on the dredging method. Grain size distribution analysis reveals that the material fraction < $16 \mu m$ may vary between 10-70%, and the fraction > $210 \mu m$ is usually not higher than a few percent. Common contaminants found in sediments include compounds of heavy metals (Zn, Cd, Cr, Cu, Pb, Hg), as well as organic compounds (PCB's, HAP, MAH, organic chloride compound (EOCl)). Other metals may be present as sulphides and carbonates, or complexed to the high organic matter content, or bound to the minerals.

Traditional disposal alternatives of dredged mud include: (1) subaqueous (open water) disposal; (2) on land disposal; (3) coastal marine disposal in capped mound deposits above the prevailing see floor, and (4) disposal in subaqueous depressions and capping deposits in depressions. These categories differ primarily in the physicochemical conditions and transport processes potentially capable of removing contaminants from dredged material at the disposal site.

Land disposal of dredged mud is mainly concerned with: (1) change of pH; (2) change of redox conditions; (3) formation of soluble, complexing organic compounds, and (4) microbial interactions. These factors affect the chemical stability of sediments and may weaken the bonding strength of heavy metals in sediment particles and thus increase their mobility. In order to minimize potential adverse effects upon disposal of dredged material, assessment of metal mobility in dredged material and its interactions with clay barrier system are of great importance.

In this study, leaching column tests were used to investigate the compatibility of the leachate generated from the dredged mud and the proposed clay barrier system. In order to simulate the behaviour of metal in the dredged mud under typical environmental conditions, a solvent leaching technique was used. Samples were collected after water passage of different pore volumes. Chemical analysis were performed on the pore water solution and on the solid matrix. Migration profiles of various metal ions were studied. Also, the diffusion parameter of various metal ions was calculated via (1) experimental migration profiles; (2) analytical solution of the advection-diffusion equation, and (3) root square method.

2. MATERIALS AND METHODS

2.1 Sediment properties

The chemical characteristics of the sediment used in this study is shown in Table 1. In particular, the total carbon content, PCB's and Poly aromatic hydrocarbons concentrations in the

Table 1. Sediment characteristics

Parameter	
pH	6.3
Temperature °C	18.5
Water Content (% by weight)	120.8
Conductivity (μ mho/cm)	3.05
Total Organic Matter%	5.83
Sodium Na^+ (mg/kg)	278.0
Potassium K^+ (mg/kg)	1836
Calcium Ca^{2+} (mg/kg)	2426
Magnesium Mg^{2+} (mg/kg)	6880
PCB's	< 0.02
Total organics	3.15
Monoaromatic hydrocarbons	ND
Polyaromatic hydrocarbons	17.5
Total oil and grease	5.6
Cadmium Cd^{2+} (mg/kg)	65
Chromium Cr^{3+} (mg/kg)	12
Mercury Hg^{2+} (mg/kg)	413.2
Lead Pb^{2+} (mg/kg)	103.2
Zinc Zn^{2+} (mg/kg)	1891.3

sediment are high. Heavy metals such as lead, mercury and zinc present in concentration higher than all limiting criteria. Cadmium and chromium concentrations are to be considered a serious problem.

Table 2 Clay Properties

Property	
Physical	
Liquid Limit %	60
Plastic limit %	19
Plasticity index	41
Natural water content%	46
Specific gravity	2.74
Void ratio	0.6
Porosity	0.37
Maximum dry density (Mg/m^3)	1.67
Optimum moisture content %	24
Wet density (Mg/m^3)	2.04
Permeability (m/sec)	
@ 3 psi	2E-10
@ 18 psi	4E-10
Chemical	
Specific surface area (m^2/g)	107
Cation exchange capacity (mg/100g)	33.6
Pore fluid analysis	
Sodium, Na^+, mg/l	480
potassium, k^+ mg/l	55
Calcium, Ca^{2+} mg/l	100
Magnesium, Mg^{2+} mg/l	43
Chloride, Cl^-	25
Bicarbonate, HCO_3^- mg/l	6
pH	7.8
Conductivity (μ *mho/cm*)	2.59

2.2 Soil Properties

The physical and chemical properties of Quebec natural clay which used in this study are shown Table 2.

2.3 Test Procedure

The test procedure for leaching tests consisted of using leaching columns of 0.1 m diameter by 0.128 m high. The soil used in the leaching cells were prepared in the following manner. A 2.5 Kg sample of fine (passing sieve No. 8), air dried soil was mixed with an appropriate amount of distilled water to obtain a 25% moisture content which was slightly higher than the optimum water content. The soil was allowed to cure overnight in a humid room. It was subsequently compacted statically in two layers. The total height of the compacted clay was 0.05 m. Four replicate cells were structured for testing. After passage of one pore volume of distilled water under a hydraulic head of 12.7 m, the contaminated sediment was placed on the top of the compacted clay in the leaching cells. The height of the sediment in the leaching cells was 0.078 m. In the leaching cell reservoir, which is located at the top of the sediment, a sulphuric acid solution of pH3 was added in order to simulate acid rain condition and accelerate the mobility of heavy metals from the contaminated sediments. The sulphuric acid solutions then leached through the multi-layer system of sediment and compacted clay. After leaching a specific pore volume of the compacted clay, the test specimen was vertically sectioned into 5 layers of 10 mm each. The prescribed pore volumes passage of the sulphuric acid were one, three, and five respectively. Each section and the collected effluent were analyzed for soluble ions. The soluble cations, anions (Cl^-, H_2CO_3) and heavy metals (lead, copper, zinc and chromium) in the free liquid were determined from pulverized samples.

3. RESULTS AND DISCUSSION

3.1 Migration of heavy metals:

Passing an acidic solution through the leaching cells would lead to an increase in the hydrogen ion concentration in the pore fluid. Since hydrogen ion has a greater affinity for the clay particle than most other cations at lower pH, one expects a greater concentration of exchangeable cations to be desorbed from the outer layers of the clay particles. Since heavy metals interact with clay through the exchange of cations, one expects some heavy metals to be desorbed from the sediment. To test this hypothesis, the pore fluid of the sediment was analyzed in solutions of pH 1.5 and pH 3.0. The results are shown in Table 3.

The pH of the pore fluid of the sediment analyzed in the acidic solution of pH 3 after shaking for several hours was 7.0. This shows that the sediment has a buffering capacity. The buffering capacity of the sediment is significant in that it reduces the ability of the increased hydrogen ion concentration to release heavy metals. This is evident in the analyses of the heavy metals in the clay and in the effluent.

Copper nor chromium were detectable in the pore fluid of the sediment analyzed in the pH 3 solution. Since chromium is trivalent, it is expected to be strongly bound to the sediment. At pH 1.5, copper was detected in the sediment pore fluid. Lead was not detectable in any of the clay columns nor in any of the effluent samples. A possible explanation for this is that the soluble lead measured in the pH3 solution has been adsorbed onto the top layer of the clay. By the middle of the first layer, complete adsorption of the small amount of soluble lead could have taken place. Trace amounts of zinc was detected in the pore fluid of the sediment analyzed in the acidic solution of pH 3.0. Initially, 2ppm zinc were detected in the clay layer prior to passing the acidic solution through the cell. After passing the acidic solution through the cell, 36 ppm zinc were detected in the clay column after one pore volume of effluent had been collected. In addition, 16 ppm zinc were collected in the one pore volume of the effluent. These results suggest that zinc is more mobile than any of the other heavy metals. This was expected since zinc has a smaller ionic radius and a greater ionic potential than any of the other heavy metals.

On the contrary, lead has a large ionic radius,

Table 3 heavy metal concentrations in sediment pore fluid

Concentration (ppm)	Cu	Pb	Zn	Cr
pH 3	nd	7	18	nd
pH 1.5	40	36	500	nd

nd = non detected

and a small ionic potential relative to the other heavy metals (Yong, Mohamed and Warkentin, 1992). Since lead and zinc both have the same valence, but lead has a greater ionic radius, the electron cloud of lead would emit a weaker electric field and is more likely to precipitate from solution than zinc, which would emit a stronger electric field. Also, since lead has a larger ionic radius it is more likely to polarize and form covalent bonds, whereas zinc is more likely to form ionic bonds. Based on this information, one would expect the following affinity series for the heavy metals which is similar to that reported by Mohamed et al. (1992)

$$Cr^{3+} > Pb^{2+} > Cu^{2+} > Zn^{2+}$$

3.2 Migration of Cations

The concentrations of various ions in the sediment as a function of pore volumes are shown in Fig. 1. Since the water content of the sediment was measured to be 128%, the soluble cations in the sediments are considered to move freely through the sediment. As a function of time, the concentrations of soluble cations and anions were decreased with time; hence the input concentrations at the clay/sediment interface decreased as a function of number of collected pore volumes. The monovalent cations are more concentrated in the top layers of the clay than in the sediments as shown in Figs.2 and 3 which depicts the sodium and potassium migration profiles as a function of depth and pore volumes. As the collected pore volumes increased, the concentrations of both sodium and potassium increased. Similar results however obtained for magnesium as shown in Fig. 5. However, for calcium migration profiles, the concentration of calcium as a function of depth was decreased as shown in Fig. 4. This could be attributed to

Table 4 - Calculated Diffusion Parameter x 1.0E-04 m^2/day

Ion	Na^+	K^+	Ca^{2+}	Mg^{2+}	Cl^-
This study	2.501	2.241	1.959	2.004	----
Mohamed and Yong (1993)	3.002	7.878	7.573	7.878	7.573
Yong and Warith (1990)	---	0.4	0.12	0.14	0.33

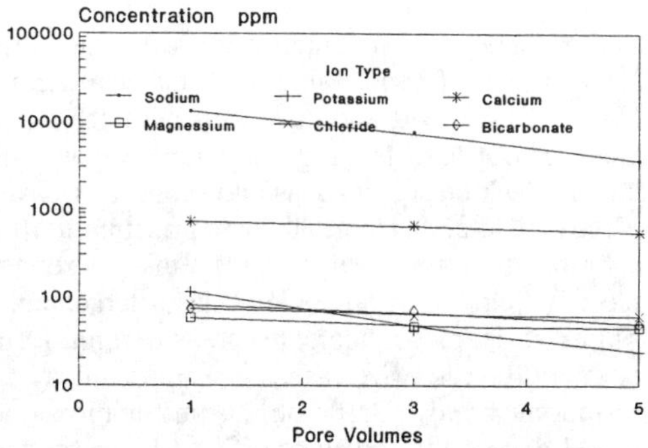

Fig. 1 Input concentrations at clay/sediment interface.

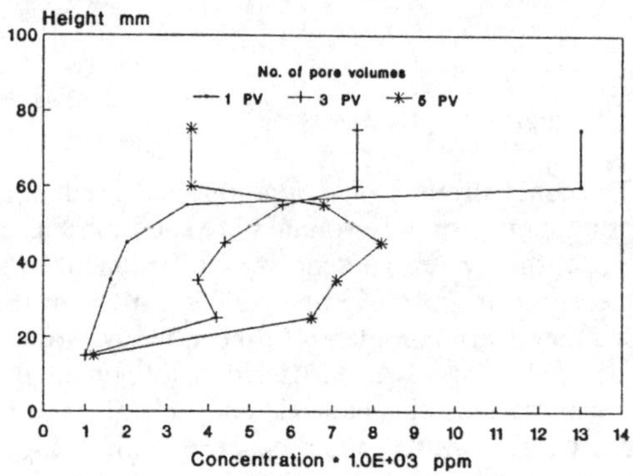

Fig. 2 Sodium migration profile

adsorption of calcium by the clay particles and calcium exchange with monovalent cations.

3.3 Migration of Anions

One may expect the clay particles to maintain charge balance through incorporation of anions into the outer layers. However this is known not to be the case since the clay particles are negatively charged. Rather, as mentioned earlier, charge balance is maintained through the adsorption and desorption of cations, Also, substitution of oxygen or hydroxide by chloride within the clay structure is not feasible since chloride has a much larger ionic radius. Consequently, one expects little or no reaction of chloride and bicarbonate with clay particle and that these anions are mobile and have high concentrations within the pore fluid of the clay. The anion concentrations in the sediment as a function of collected number of pore volumes are shown in Fig.1. It can be seen that as a function of pore volumes, the anions concentration decreased. The migration profiles for both chloride and bicarbonate ions are shown in Figs. 6 and 7. The migration profiles show that H_2CO_3 ion moves faster that Cl^- ion which suggests that Cl^- could be adsorbed in the diffuse layer of the clay while H_2CO_3 is desorbed from the clay. Also, the variation in anion concentration could be the result of speciation of these anions within the pore fluid. Since a detailed chemical analysis of the pore fluid, speciation of these anions is only a speculative hypothesis.

4. DETERMINATION OF THE DIFFUSION PARAMETER

To determine the diffusion coefficient through the clay barrier, the root time method was used. (Yong, Mohamed, and Warkentin, 1992). The diffusion coefficients were calculated by using: (1) diffusion-advection equation; (2) nondimensional analysis; (3) Fourier series analysis; and (4) root time technique. The diffusion coefficients are then calculated from the following equation (Mohamed and Yong, 1993):

$$D=\frac{0.2435L^2}{t_{90}} \quad (1)$$

where: D = diffusion parameter of the specific ion; L = depth of the clay layer, and t_{90} = time

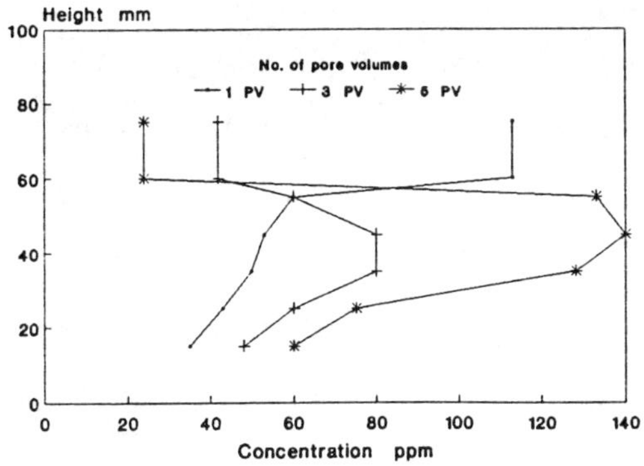

Fig. 3 Potassium migration profile

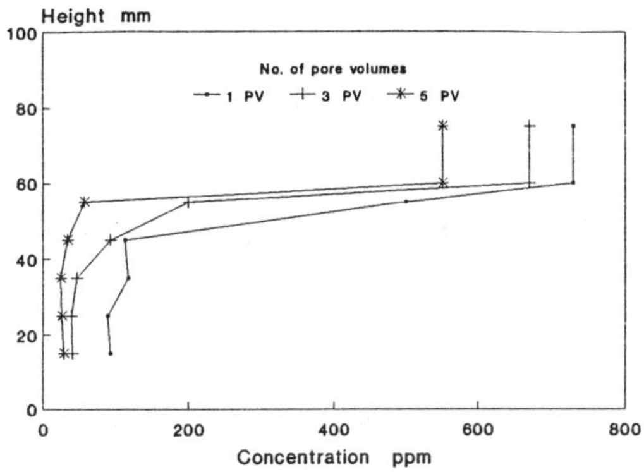

Fig. 4 Calcium migration profile

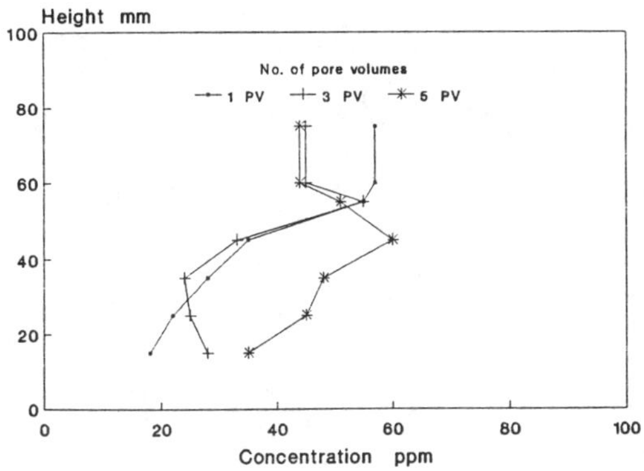

Fig. 5 Magnesium migration profile

which corresponds to 90 percent equilibrium time of ion concentration gradient at steady state. The calculated diffusion parameters for each cation is shown in Table 4.

These diffusion parameters are compared with those calculated by Yong and Warith (1990) and Mohamed and Yong (1993). The calculated values determined in this study are of the same order of magnitude but less than those determined in a similar study by Mohamed and Yong (1993). In Mohamed and Yong's study a highly acidic influent of pH 1.3 was passed through a 120 mm thick of natural clay layer under a hydraulic head of 10.6 m. The clay used in that study had a CEC of 14 meq/100g and a permeability of 2.3 x 10^{-9} m/sec. The diffusion parameters determined in this study are less since the clay has a lower permeability (4.0 x 10^{-10} m/sec). The diffusion parameters determined by Yong and Warith (1990) are an order of magnitude less than the values determined in this study. The clay used by Yong and Warith (1990) has a CEC=60 meq/100g, a permeability of 7 x 10^{-9} m/sec and is subject to a hydraulic head of 2.5 m. The low hydraulic head relative to the head used in this study (12.7 m) could explain the difference in the calculated and reported diffusion parameters.

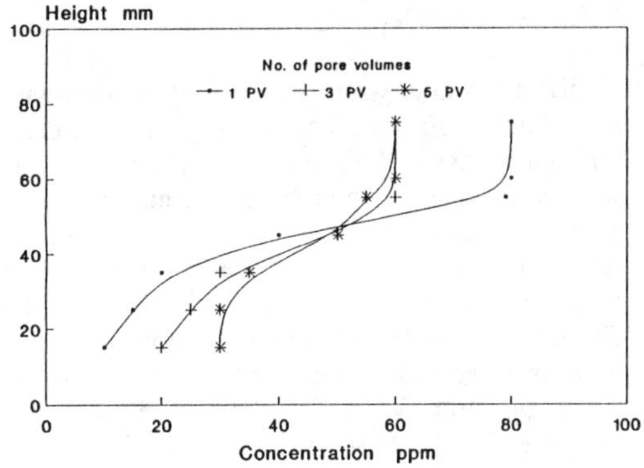

Fig. 6 Chloride migration profiles

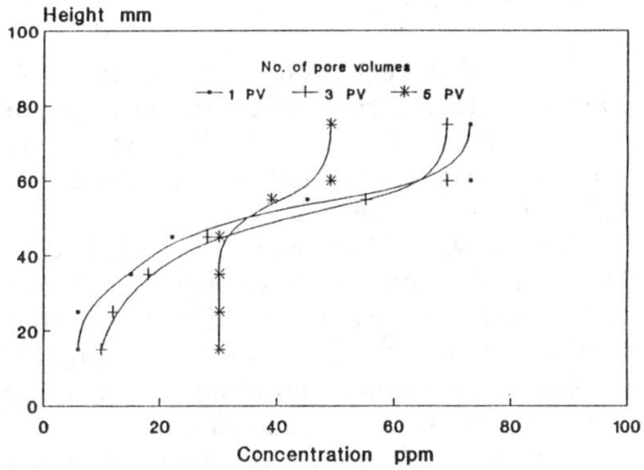

Fig. 7 Bicarbonate migration profiles

5. CONCLUSION

This study addressed the migration of heavy metals and metal ions contained in the sediment through a layer of Quebec natural clay. The study has demonstrated that the sediment possess a high buffering capacity to resist pH change. Hence, heavy metals are not detected due to the high pH of the sediment. The calculated diffusion parameters for various cations were one order of magnitude different form those reported by Yong and Warith (1990) and in close agreement with those calculated by Mohamed and Yong (1993).

6. REFERENCES

Mohamed, A.M.O., Yong, R.N. and Tan, B.K., 1992. Mitigation of acidic mine drainage: barriers for reactive tailings. ASCE, Proceeding Environmental engineering Water Forum '92, Baltimore, pp. 457-462.

Mohamed, A.M.O. and Yong, R.N 1993. Diffusion of contaminants through clay barrier under acidic condition. In the Engineering Geology of Waste Storage and Disposal, 29th Annual Conference, Engineering Group of the Geological Society of London, Sept. 6-9, 1993.

Yong,R.N., and Warith, M. 1990. Contaminant migration effect on dispersion coefficients. ASTM, STP 1095, Hoddinott, K.B. and Lamb, R.O. (Eds.), pp. 69-80.

Yong,R.N., Mohamed, A.M.O., and Warkentin, B.P. 1992. Principles of contaminant transport in soils. Elsevier, Amsterdam, 327 p.

Waste Disposal by Landfill – GREEN'93, Sarsby (ed.) © *1995 Balkema, Rotterdam, ISBN 90 5410 356 6*

Quality of landfill leachates and their effect on groundwater

M. Niininen & P. Kalliokoski
Department of Environmental Sciences, University of Kuopio, Kuopio, Finland
E. Pärjälä
Environmental Protection Office, City of Kuopio, Finland

ABSTRACT: Quality of landfill leachates and their effects on groundwater were studied at 12 landfill sites in Finland. Seven sites located in important groundwater areas and four of them even in eskar areas, where groundwater is easily contaminated because of permeable soil. The rest of the sites were selected because they have recieved considerable amounts of sludges and industrial wastes. Some effects of leachate on groundwater were found at all sites studied, though groundwater quality varied considerably in the surrounding areas of the landfill sites. In general, the effects of landfill activity appeared most clearly as high concentrations of organic matter (COD, TOC), manganese, iron, ammonium and chloride, and as high conductivity. Heavy metal concentrations were low in groundwater. However, cyanide concentrations were quite high. Chlorophenols were detected in small amounts.

1 INTRODUCTION

Disposal of the large volume of waste produced in our modern society is a significant problem. Much of our solid waste continues to be placed in municipal landfills. In the past, quite little attention was given to proper site selection and operation of landfills. Several cases of surface and groundwater pollution caused by landfills have been reported around the world, and thus toxic leachates belong to the major risks for groundwater quality. Previously, landfill sites were often improperly located in Finland. For example, as many as 63 landfills were located in important groundwater areas used for municipal water supply. Hazardous waste was often buried into landfills in the past. Landfills were sometimes located even in eskar areas, which are often significant groundwater areas and where groundwater is also easily contaminated because of permeable soil. Hence, contamination of groundwater caused by landfills was considered to be important to study in Finland.

2 MATERIALS AND METHODS

2.1 Site selection

Leachates from 12 Finnish sanitary landfills and quality of groundwater in landfill areas were studied in 1984-1986. Landfill sites were located in Joutsa, Joroinen, Kannonkoski, Kontiolahti, Kouvola, Kuopio, Lappeenranta, Lempäälä, Leppävirta, Pieksämäki, Tampere and Turku (Fig. 1). The age of studied landfills varied from 9 to 40 years. A special interest was given to landfills located in eskar areas (Joutsa, Joroinen, Kannonkoski and Kontiolahti). Sewage sludge has been disposed to several Finnish landfills and industrial wastes to some landfills. Considerable amount of industrial wastes have been placed in the landfills of Kouvola, Kuopio, Lappeenranta, Leppävirta, Tampere and Turku. Hazardous waste was suspected to be buried in the landfills of Tampere and Lempäälä. Four landfills (Joutsa, Joroinen, Tampere and Turku) were already abandoned.

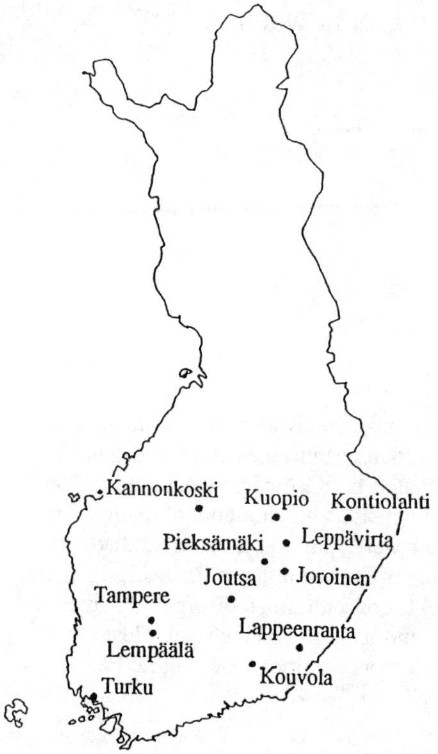

Fig. 1. Locations of studied landfills in Finland.

2.2 Sample collection and analysis

Leachate samples were taken with decanters from flowing water, when possible. Samples from groundwater monitoring wells were collected with a submercible pump or groundwater sampler. Chemical and physical analyzes belonging to the general water quality control (COD, BOD, nitrogen, phosphorus, pH, conductivity and total solids) were determined according to Finnish standard methods and methods described by The Finnish National Board of Waters and the Environment. TOC was analyzed with a carbon analyzator Beckman 915A. Additional analyzes of metal, cyanide, PCB and chlorophenol concentrations were also done. Metals and cyanides were determined according to Finnish standards 3044 and 3039, respectively. PCB and chlorophenols were identified with gas chromatography Perkin-Elmer Sigma 2000.

3 RESULTS

3.1 Leachate samples

Composition of leachates determined in this study as well as in a previous study of Finnish landfills is presented in Table 1. There are also corresponding results from studies made in England and Germany. In this study, quality of leachate varied a lot at different landfills and also at different parts of one landfill. Great amount of organic material (COD, TOC) and high chloride, iron and ammonium concentrations were typical for leachates. BOD levels were generally low. Heavy metal concentrations were low on the average although high maximum concentrations were occasionally found.

Cyanide was detected in all leachates. Average cyanide concentration was 62 µg/l (variation range <5-650 µg/l).

PCBs were analyzed from 14 leachate samples collected in five landfills. PCB concentrations varied between <0.05 and 5.00 µg/l.

Chlorophenols were detected in leachates of all five studied landfills. Two landfills had low levels of chlorophenols, but in three landfills concentrations were clearly elevated. Especially 3,4-dichlorophenol, 2,4,6-trichlorophenol and 2,3,4,6-tetrachlorophenol levels were high (Table 2). In one landfill 2,4,5-trichlorophenol was the main component.

3.2 Groundwater samples

Characteristics of groundwater in vicinity of landfills are presented in Table 3. Some contamination of groundwater was observed in every landfill area. Effect of landfill on groundwater quality was seen as high concentrations of organic matter (COD, TOC), manganese, iron, ammonium and chloride and as high conductivity.

Heavy metal concentrations were low in groundwater compared to EEC maximum admissible concentration levels in drinking water (Table 3). However, cyanide concentrations were high in several cases.

Table 1. Leachate characteristics in Finland and in Central Europe.

parameter	England, 15 landfills (Robinson *et.al.*, 1982)	Germany, 20 landfills (Ehrig, 1983)	Finland, 26 landfills (Ettala, 1988) *)	Finland, 12 landfills **)
total solids, mg/l	-	-	99 (76)	80 (1-520)
conductivity, mS/m	-	-	240 (170)	214 (5-580)
pH	6.2-7.4	6.1-8.0	7.1 (0.5)	7.1 (3-8)
COD_{Mn}, mg/l	-	-	99 (77)	230 (7-1200)
COD_{Cr}, mg/l	66-11 600	3 000-22 000	-	600 (52-5 200)
TOC, mg/l	21-4 400	-	-	379 (17-1 900)
BOD, mg/l	<2-8 000	180-13 000	300 (520)	270 (1-3 900)
NH_4-N, mg/l	5-730	741	53 (35)	88 (1-480)
total-P, mg/l	<0.02-3.4	5.7	1.4 (3.0)	0.5 (<0.02-3.9)
Cl, mg/l	70-2 777	2 119	250 (170)	170 (1-600)
Fe, mg/l	0.1-380	15-925	42 (67)	7.7 (<0.1-692)
Mn, mg/l	0.3-26.5	0.7-24	2.5 (1.7)	1.9 (0.1-7.4)
Mg, mg/l	12-480	250-600	-	41.8 (23.8-59.8)
Ca, mg/l	165-1 150	80-1 300	170 (79)	67 (65-69)
Zn, mg/l	<0.1-1.0	0.6-5.6	0.5 (1.2)	0.2 (<0.1-1.4)
Cd, µg/l	<5-10	5.2	5.4 (8.8)	0.6 (<0.1-4.7)
Cr, µg/l	<50-160	275	33 (32)	9 (<1-134)
Ni, µg/l	50-160	166	41 (19)	98 (<3-394)
Pb, µg/l	50-220	87	29 (35)	14 (<1-42)
Cu, µg/l	10-150	65	31 (34)	4 (<1-56)

*) standard deviation in parenthesis
**) varation in parenthesis

International restrictions of different chlorophenol compounds in drinking water are under progress. Canadian national limit values in drinking water are 900 µg/l for 2,4-dichlorophenol, 5 µg/l for 2,4,6-tricholorophenol and 60 µg/l for pentachlorophenol. Compared to these concentrations, chlorophenol levels were low in studied groundwater samples (Table 2.).

4 DISCUSSION

Great amount of organic material (COD, TOC) and high chloride, iron and ammonium concentrations were typical for leachates in this study as well as in other published researches. Typically low BOD levels indicate that the organic matter is no more easily biodegradable. On the other hand, BOD test may be inhibited by toxic compounds of leachate while these compounds can more efficiently be detected with COD or TOC analyses.

High cyanide concentrations observed nearly in all landfill leachates may partly be of microbiological origin. Both fungi and bacteria are known to be able to produce cyanide from different compounds (Timonin 1941; Turner 1961; Walker 1954).

Low concentrations of PCBs in leachates may be caused by very low solubility of PCB to water. Thus, PCB compounds are more likely to be found in solid organic matter or oil fraction.

Chlorophenol concentrations were relatively high in leachates. In most cases, the concentra-

Table 2. Chlorophenol concentrations (µg/l) in leachate and groundwater samples.

	leachate	groundwater
2,3-dichlorophenol	nd	nd
2,4-dichlorophenol	nd-12.82	nd-1.50
2,6-dichlorophenol	nd-0.61	nd-0.17
3,4-dichlorophenol	nd-22.0	nd-0.59
2,3,4-trichlorophenol	nd-1.95	nd-0.21
2,4,5-trichlorophenol	nd-88.0	nd-2.61
2,4,6-trichlorophenol	nd-9.38	nd-0.15
3,4,5-trichlorophenol	nd-5.65	nd-0.55
2,3,4,6-tetrachlorophenol	0.09-20.4	nd-1.48
pentachlorophenol	0.02-4.44	nd-0.23

nd = not detected (detection limit for tri-, tetra- and pentachlorophenols is 0.005 µg/l and for dichlorophenols 0.02 µg/l)

Table 3. Characteristics of groundwater in landfill areas and EEC guide levels and maximum levels for drinking water.

parameter	average (variation range)	EEC limits, GL/MAC*
conductivity, mS/m	23.7 (4.2-120)	40/-
pH	6.5 (4.8-7.9)	6.5-8.5/-
COD_{Mn}, mg/l	13.3 (1-44)	5/2
COD_{Cr}, mg/l	30.6 (<10-73)	-
TOC, mg/l	12.6 (1-92)	3.5/-
BOD_7, mg/l	25.6(<2-190)	-
NH_4-N, mg/l	2.52 (<0.01-28)	0.04/0.4
tot.-N, mg/l	2.81 (<0.1-23)	-/1
Cl, mg/l	12.1 (<1.0-106)	25/-
Fe, mg/l	2.7 (<0.1-26)	0.05/0.2
Mn, mg/l	0.56 (<0.1-2.5)	0.02/0.05
As, µg/l	4 (<1-19)	-/50
Cd, µg/l	0.2 (<0.1-1.3)	-/5
Cr, µg/l	1 (<1-13)	-/50
Cu, µg/l	3 (<1-33)	100/3 000
Ni, µg/l	24 (<3-190)	-/50
Pb, µg/l	3 (<1-25)	-/50
Zn, µg/l	138 (<10-1 000)	100/5 000
CN^-, µg/l	57 (<5-781)	-/50

* GL = guide level, MAC = maximum admissible concentration

tion ratios between various chlorophenol compounds suggested their origin to be a wood preserving formulation previously widely used in saw milling industry. According to these results chlorophenol concentration in landfill leachate can be in the same magnitude as in run-off waters from saw milling areas (Valo *et.al.* 1984). In one landfill, 2,4,5-trichlorophenol was the main component which may indicate traces of herbicides.

Some effects of landfill areas on groundwater quality were detected at all sites studied. Groundwater quality varied considerably in the surrounding areas of the landfills indicating the complexity of the hydrogeological and physicochemical conditions.

In general, the effects of landfill activity appeared most clearly as high concentrations of organic matter (COD, TOC), manganese, iron, ammonium and chloride and as high conductivity. However, low oxygen level can lead to high concentrations of manganese and iron also in unpolluted groundwater, depending on the soil type of the area. Therefore, manganese and iron are not as reliable indicators of contamination in groundwater as in surface water.

Cyanides, which have relatively high mobility in soil, were found in groundwater samples. Concentrations of chlorophenols in leachates were in some cases qiute high, but in groundwater samples concentrations were low.

The effects on groundwater quality were most clearly observed at large landfills which may simply be result of larger waste volume. Clear correlation between groundwater quality and disposal of industrial wastes was not observed.

REFERENCES

Ehrig, H.J. 1983. Quality and quantity of sanitary landfill leachate. *Waste Management & Research* :53-68.

Robinson, H.C., Barber, C. & Maris, P.J. 1982. Generation and treatment of leachate from domestic wastes in landfills. *Wat. Pollut. Control* 54:465-478.

Timonin, M.I. 1941. The interaction of higher plants and soil microorganisms. III Effect of by-products of plant growthon activity of fungi and actinomycetes. *Soil Science* 52:395-413.

Turner, E.M.C. 1961. An enzymatic basis for pathogenic specifity in Ophiobolus graminis. *Journal of Experimental botany* 12:169-175.

Valo, R.,Kitunen, V., Salkinoja-Salonen, M. & Räsänen, S. 1984. Chlorineted phenols as contaminants of soil and water in vicinity of two Finnish sawmills. *Chemosphere* 13:835-844.

Walker, J.B. 1954. Inorganic micronutrients of Chlorella. III Quantitative requirements for iron, manganese and zink. *Archives of Biochemica Biophysica* 53:1-8.

Waste Disposal by Landfill – GREEN'93, Sarsby (ed.) © 1995 Balkema, Rotterdam, ISBN 90 5410 356 6

Development of a landfill cell with artesian groundwater conditions

R.W. Sarsby
Bolton Institute of Higher Education, UK

ABSTRACT: The paper relates to the development of a landfill cell at a location where there was a groundwater pressure in underlying shales and sandstones.

INTRODUCTION

At this site there is a high groundwater pressure in the rocks underlying a thin cover of boulder clay. Originally it was a water reservoir which has been decommissioned because of the decline in local manufacturing industries. A plan view of the site is given in figure 1.

Prior to the site being sold, by the Water Authority, to a developer, the earth dam across the valley was breached. The site developer, had no expertise in engineering or waste dispersal, and at that time (mid 1980's) definitive technical guidelines were not in existence and the engineering of waste disposal sites was evolving rapidly.

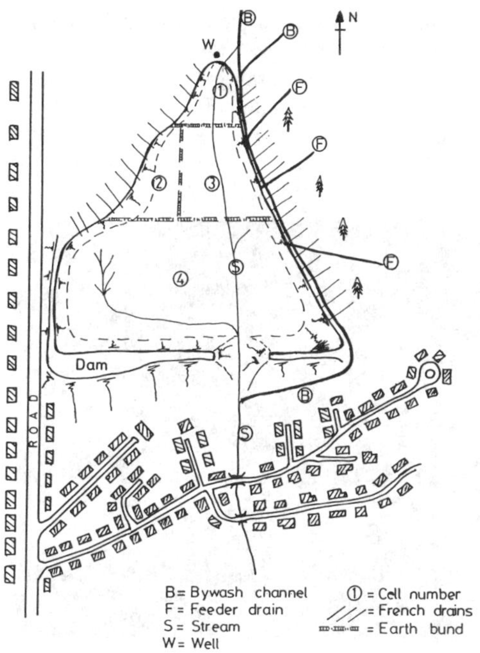

Figure 1 Site Plan

The base of the reservoir had a thin covering (from 0.5m to 3m) of low permeability boulder clay underlain by shattered shales and fissured sandstone (see Plate 1). These rocks formed the

Plate 1 Exposed underlying rock with springs

bed of the river upstream of the reservoir and approximately 5km away they outcropped in the local hills, at a significant height above the reservoir. These shales and sandstones were being charged with water in the hills and were the source of natural springs in the vicinity of the reservoir. When the site developer started to prepare the reservoir for tipping he destroyed and removed the thin clay cover in places (by uncoordinated excavation of clay, using unsuitable plant and methods etc.) so that springs were developed in the base of the reservoir. Furthermore the Water Authorities were aware that the original basal seal of the reservoir was defective because they had never been able to fill the reservoir to the design level.

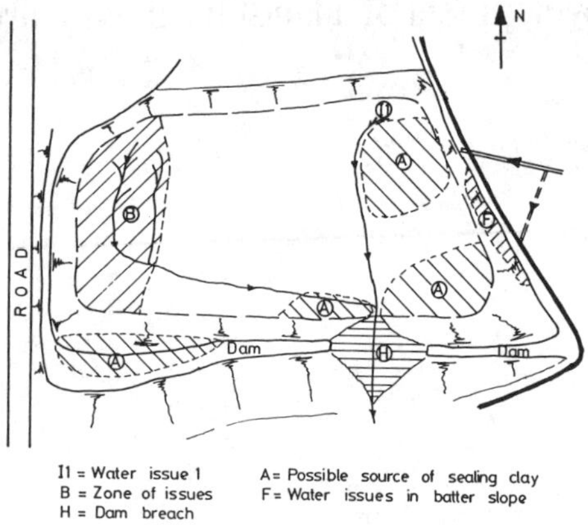

Figure 2 The final cell

SITE DEVELOPMENT

The Water Authorities required that the refuse should be underlain by a low permeability (10^{-6} mm/sec maximum) layer of compacted clay to prevent contamination of the groundwater. The site was developed progressively in self-contained cells. The final cell to be developed was number 4 (see figure 2) and it posed several major problems:-

1) There was insufficient suitable clay on the base of the site to form the compacted clay liner and fill the breach in the dam. The unsupervised early work of the site developer had turned a lot of suitable clay into unusable slurry.
2) The base of the cell was always very wet due to water issues and springs from the underlying rocks.
3) Even if an acceptable clay liner could be placed in the wet condition it would be lifted due to pressure build-up in the groundwater. When the liners for the first cells were constructed standpipes were installed in the underlying rocks. Using the water levels from these standpipes, the water level achieved in the reservoir when it was used and the local topography it was estimated that the static head of water in the underlying rocks would be approximately 8m above the top of the finished liner.
4) The Water Authorities would not allow a permanent under drainage system to be installed beneath the waste. The disposal site had to be entirely self-contained and measures to reduce the uplift from the groundwater could only be temporary.

SOLUTION

The solution which was proposed, and accepted by the Water Authorities was:

a) Install relief drains to collect all groundwater issues. (see plate 2)
b) Lead all relief drains to a central point, i.e. a vertical manhole founded on an impermeable base in the breach in the dam.

Plate 2 Relief drain from one issue

c) Connect the relief drains to a vertical pipe, with outlet pipes at different levels, located inside the manhole.

d) Place a compacted clay liner over the relief drains and areas where there is insufficient existing clay cover - at this stage the clean water from the relief drains is allowed to discharge freely from the manhole so that no uplift is applied to the clay liner.

e) Raise the level of the outlet in the manhole so that the head in the groundwater is one metre above the liner. Leave the liner under this head for a specified time (one month) to check for leaks.

f) Partially fill the breach in the dam and encase the manhole with low permeability clay compacted to the same standard as the liner - see figure 3 and plate 3.

g) Install a leachate drainage collection system and central extraction point. (Plate 4 shows the extraction chimney and the dam breach).

h) Place and compact the waste in layers and progressively fill the breach so that no run-off from the waste escapes via the breach. At the same time progressively raise the level of the relief drain outlet in the manhole so that the ground water head rises but not sufficiently high to lift the liner and waste overburden.

i) Continue the preceding operations until water ceases to flow from the outlet in the manhole. The steady-state groundwater head has now been developed and so the outlet can be closed, i.e. there is no longer an underdrain operating beneath the waste.

The proposal was accepted by the authorities for several reasons:

1) The integrity of the compacted clay liner is checked via a pressure test.

Plate 3 Relief drain manhole

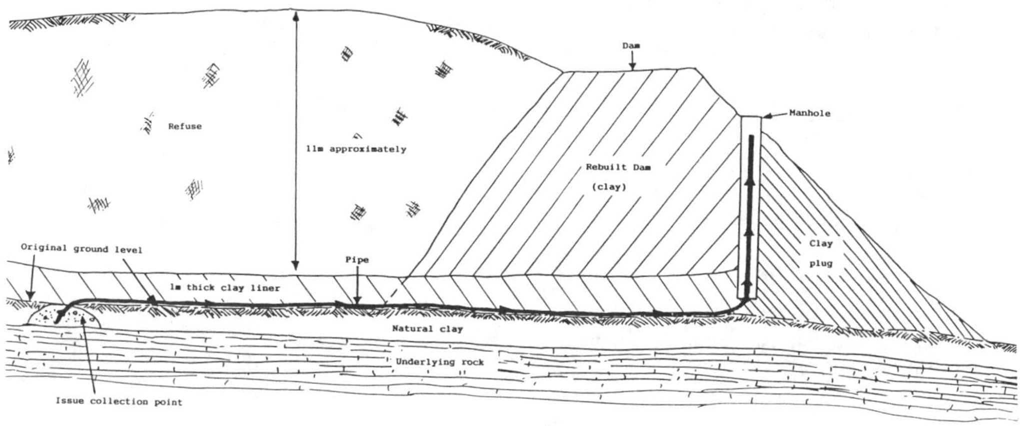

Figure 3 Arrangement of temporary relief drainage

Plate 4 Cell 4 - looking towards the dam breach

Plate 5 Top of the filled site

2) The temporary relief drains eliminate the possibility of liner rupture due to uplift.
3) There is no permanent underdrain operating beneath the waste.
4) During and after filling of the site the purity of the groundwater is checked at the outlet pipe - if the water were contaminated it would be piped away for treatment.
5) If groundwater pollution is detected at any time the pollution can be 'captured' by reducing the head at the outlet pipe so that the groundwater flows towards this low pressure area.
6) Once the site has been filled and capped and leachate is extracted there will be a positive hydraulic gradient into the cell rather than out of it.

CONCLUDING REMARKS

The site has now been filled and completed - the top part of the relief drain manhole is visible to the right of the centre of plate 5.

No pollution has been detected of any water taken from the outlet pipe. The outlet pipe discharges into the bywash channel (shown in figure 1) to the south of the original earth dam. Whilst the water from beneath the waste disposal site is clean the surface of the water in the channel carries a white scum due to pollution of the stream from a farm.

The water level in the manhole became static at about 7m above the liner and so discharge from the relief drains ceased and the site is thus self-contained.

The waste has now been shaped for capping. Although the cap is required (by the local planning officer) to have a final slope of 1 vertical to 40 vertical, the officer has not defined how long after capping this slope should be achieved. Estimates have been made (using modified geotechnical procedures) of the compression of the waste to try to ensure a final cap slope of not less than 1 in 40 - time will show the accuracy of the analysis.

Waste Disposal by Landfill – GREEN'93, Sarsby (ed.) © 1995 Balkema, Rotterdam, ISBN 90 5410 356 6

Stable isotope analysis as a means of identifying the source of methane

S. Waldron & A. J. Hall
Glasgow University, UK

A. E. Fallick
Scottish Universities Research & Reactor Centre, East Kilbride, UK

R. A. Gilmour & M. MacDonald
University of Strathclyde, Glasgow, UK

ABSTRACT: The application of stable isotope analysis in the identification of methane is demonstrated with five case studies whose combined $\delta^{13}C$ and δD values ranged from -66 to -44‰ and -352 to -263‰ respectively. This is characteristic of CH_4 produced by bacteria and so suggested the source of the gas was landfill. In this paper the theoretical background behind stable isotope analysis will be explained, the cryogenic separation method and CH_4 analysis method will be outlined and the advantages of using stable isotope analysis and problems that may occur when interpreting the results will be discussed.

INTRODUCTION

At 6.30am on 24 March 1986 a methane gas explosion completely destroyed a bungalow in Loscoe Derbyshire, badly injuring the three occupants (Williams & Aitkenhead, 1991). Derbyshire County Council initiated a non-statutory inquiry to identify the source of the gas and its migration pathway, the outcome of which showed the gas to have migrated from a filled and capped landfill site, situated 70m from the site of the explosion.

As early as 1983, there was evidence of gas migration from the landfill site, which had not been in use for a year, into the residential area. A resident, experiencing problems, dug into his lawn to discover why it had died. A 0.5m deep hole exposed muddy water bubbling with gas, a percentage of the ebullitive flux composing of methane. Despite the absence of carbon monoxide, the occupier assumed that the flux was from underlying coal seams and contacted British Coal who installed a stand-pipe with flame trap to allow the soil gas to vent harmlessly to the atmosphere. Analysis of the gas by British Coal (although by what method is not stated in the relevant literature, Williams & Aitkenhead, 1991) suggest that the gas was not associated with coal seams, but rather had a composition similar to that produced from rotting material. Natural gas was supplied to the houses, but investigations several months later by East Midlands Gas Board suggested that the leaking gas was not mains gas.

In the case of Loscoe, if the full significance of the distressed vegetation and 'hot spots' had been realised when first observed some two years prior to the explosion, it is possible that the authorities may have been able to prevent continued gas seepage and so the explosion. Early sampling with appropriate analysis could have revealed quite clearly the origin of gas and a proper investigation subsequently undertaken to trace this source.

Where there is more than one potential source of origin a method is needed that can distinguish between the different sources. Stable isotopic analysis of methane is one such tool that can be used for this differentiation. In this paper, using case studies, this use of stable isotopes in identifying the source of unknown methane will be outlined. It can be definitive by itself but there are however occasions when this technique may need to be complemented by other forms of geochemical analyses. The drawbacks to the technique and a comparison of other techniques is also discussed.

PRINCIPLES OF STABLE ISOTOPE ANALYSIS

Hydrogen and carbon, elements intimately associated with the biosphere, hydrosphere and lithosphere, are amongst those especially susceptible to natural isotope fractionation i.e. the selective partitioning of one isotope into a compound. Approximately 98.89% of all carbon in nature is ^{12}C (carbon of mass number 12), and 1.11% of all carbon is ^{13}C. Hydrogen has two stable isotopes whose abundances are ^{1}H= 99.985% and ^{2}H = 0.015%. The ratio of these two isotopes for each element in natural materials varies slightly around these average values as a result of isotopic fractionation during physical, chemical and biological processes. However differences between materials in the range of several parts per thousand can be significant.

Isotope fractionation is a consequence of the fact that certain thermodynamic properties of molecules depend on the masses of the atoms of which they are composed. Bonds formed by the lighter of two isotopes are weaker and therefore more easily broken, making the molecule with the lighter isotope more reactive than a similar molecule containing the heavier isotope. In methane formation, bacterially controlled fractionation processes such as bacteriogenic production from landfill waste or from fermentation in an animal's rumen, result in methane richer in light isotopes than methane formed from an organic substrate as a more direct result of elevated temperatures and pressures, such as occurs in the formation of fossil fuels. Different sources therefore have different isotopic signatures so measurements of the ratios of $^{13}C/^{12}C$ and $^{2}H/^{1}H$ in the methane molecule allow the characterisation of sources.

A relative difference function, the δ value, is used for reporting stable isotope abundances and variations and is the quantity actually measured on isotope ratio mass spectrometers. If the absolute ratios of the standards employed are known, the absolute ratio of any sample is readily calculated from its δ value. The δ value is defined as follows:

$$\delta_x = ((R_x - R_{standard})/R_{standard})10^3$$

where, R is an atomic ratio and by convention is always written as the ratio of the heavy to light isotope. $R_{standard}$ is the corresponding ratio in a standard and in the case of methane $R_x = (^2H/^1H)_x$ or $(^{13}C/^{12}C)_x$. The reference standard for carbon is the Pee Dee Belemnite (PDB) (Craig, 1957), the standard for hydrogen is Standard Mean Oceanic Water (SMOW). The δ value, then, is the difference in isotopic ratio between a sample and a standard, expressed in parts per thousand, or per mil (‰). Negative values indicate that the sample is depleted by that amount relative to the standard, while positive values indicate the standard is enriched in that amount relative to the standard. Fig.1, a plot of $\delta^{13}C$ (‰) against δD (‰), illustrates the suggested boundaries for bacteriogenic methane and thermogenic methane. Further information about the theory behind stable isotope analysis can be gained from O' Neil in Valley et al., 1986.

EXPERIMENTAL TECHNIQUE

Gas samples were collected in one litre Tedlar gas sample bags using a vacuum method designed by Mr. L. Thomson of Strathclyde University. Unfortunately it was later discovered that such bags are not suitable for long term storage of methane samples due to diffusion of the gas through the Tedlar membrane. However in terms of reference to these results methane stored in a bag for fifteen days became 5.5‰ isotopically heavier in $\delta^{13}C$ and 23‰ in δD. Samples were analysed soon after collecting and for these particular sites such change in isotopic composition would make little difference to the origin of gas attributed for each site. Gas samples are now collected in evacuated glass bottles.

Gas samples are prepared for mass spectrometry at the Scottish Universities Research & Reactor Centre, East Kilbride in a specially constructed and dedicated vacuum line similar to that designed by Stevens of the Argonne National Laboratory, Illinois (Stevens & Rust, 1982).

Before analysis begins the diffusion pumped vacuum line is degassed while open to high vacuum by flaming with a gas torch. The sample gas is then introduced by a syringe from the sampling vessel via a septum into an evacuated area of known volume. This is connected to a fixed volume capacitance manometer so that the amount of gas introduced can be measured quantitatively.

Liquid nitrogen is then placed over the first trap. This will cryogenically remove any atmospheric carbon dioxide, water, hydrogen sulphide and higher weight hydrocarbons while allowing the methane and other non-condensable gases (non-condensable at -196°C i.e. nitrogen, hydrogen, helium, argon, carbon monoxide) to pass through. The sample is then allowed to expand into the vacuum line passing through another liquid nitrogen trap as it expands. This is a safety precaution in the unlikely event that there are traces of condensable gases still present.

The methane is then combusted by passing through an electrically heated quartz furnace containing copper oxide at approximately 750°C. Oxygen produced at this temperature by the copper oxide is the oxidant, and carbon dioxide, water and oxides of nitrogen are formed. Carbon monoxide may also be present in the reaction by-products. With the exception of carbon monoxide all are frozen into a liquid nitrogen trap placed on the line after the furnace. Products forming as a result of incomplete oxidation if substantial in volume e.g. carbon monoxide, OH, nitrogen oxide can be converted by passing through a trap containing Schutze reagent (iodine pentoxide on silica wool), a powerful oxidising agent, and similarly collected at liquid nitrogen temperature. However Stevens (pers. comm.) has shown that if not done this has minimal influence on the isotopic value and so is probably unnecessary. With all the desired products of the oxidation of methane collected in a liquid nitrogen trap the line can then be opened to the vacuum pump and any undesired non-condensable gases pumped away.

After the oxidation of the methane sample to water and carbon dioxide the traps containing the gases are warmed to -78°C with an acetone-dry ice mixture, allowing the carbon dioxide and any oxides of nitrogen to sublimate. These then pass over a copper furnace at a temperature of 650°C. Nitrous oxides are unstable at this temperature and will become reduced to leave free nitrogen, whereas the carbon dioxide remains stable and can be trapped at liquid nitrogen temperature. The carbon dioxide can then be measured quantitatively and collected for analysis on the VG Sira 10 triple collector mass spectrometer.

Hydrogen is extracted from the gas samples as follows. The purified water is allowed to warm to room temperature and the line heated so that the water evaporates. It is then passed through uranium turnings at ~750°C. This reduces the water to hydrogen which is pumped by a Toepler pump into a chamber of known volume.

Once most of the hydrogen has been pumped away the trap and surrounding piping are flamed to release adsorbed water. Any water that passes through the furnace (usually only a fraction of a mM) is condensed into a second liquid nitrogen trap and then recycled through the uranium furnace to ensure quantitative conversion. The total yield of hydrogen is then measured with a fixed volume capacitance manometer. Using the Toepler the sample is transferred to an evacuated sample tube for transfer to the mass spectrometer.

The hydrogen is analysed using a VG Micromass 602B with a modified inlet system. This inlet system makes use of mercury pistons which enables very small samples of gas (at the present moment <4μM) to be introduced into the mass spectrometer. The $\delta^{13}C$ permil value for carbon dioxide is measured in a similar manner using a VG Sira 10 Mass Spectrometer. The condensable nature of carbon dioxide allows the use of an inbuilt cold finger to make the isotopic ratio of very small samples possible. As with hydrogen the reference gas is calibrated by running against standards.

At present analysis are quoted to an accuracy for $\delta^{13}C$ of ±2‰, for δD of ±5‰.

CASE STUDIES

The isotopic composition of methane measured from each case study is given in Table 1 while their composition in relation to other sources of methane can be seen from Figure 1.

Case A : This is probably the simplest of the case studies. The site was documented as containing 'inert' industrial and commercial waste since opening in 1981, but has periodically been giving methane concentrations of 30%. It is situated in Carboniferous age Westphalian Middle and Lower Coal Measures, with the cyclic sequence of mudstones, siltstones, sandstones, coals and seatclays overlain by thin drift deposits of pebbly silts and clays. From Fig. 1 (symbol A), it can be seen that the carbon and hydrogen isotope ratios indicate that the methane that is present has the signature

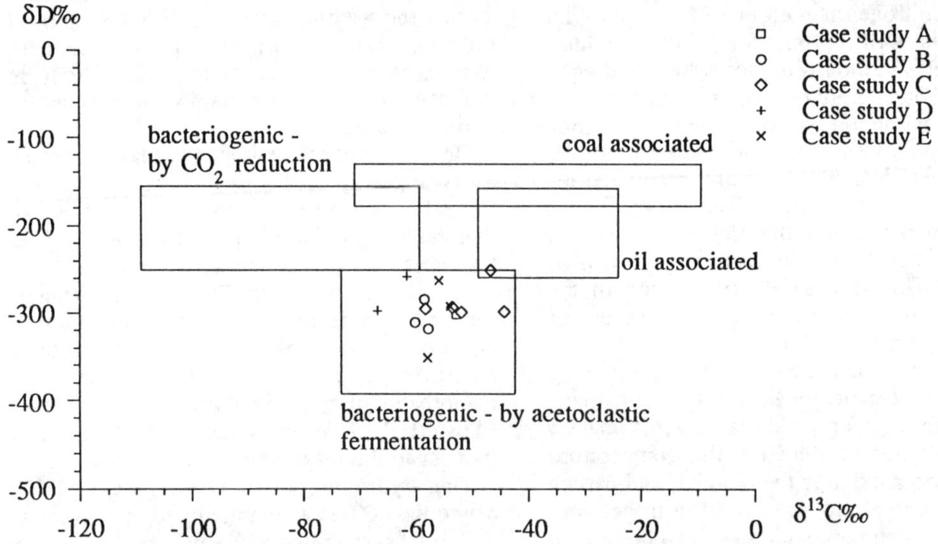

Figure 1. Plot outlining boundaries of δ13C‰ and δD‰ in naturally occurring methanes. Adapted from Schoell, 1988.

characteristic of methane produced from landfill. Further investigation revealed that in one area of the site where methane concentrations in excess of 50% have been recorded, domestic waste was dumped during the first year of the site being operational.

Table 1: Analysis results
* signifies data from sample collected outwith landfill site.

Case study	δ13C ‰	δD ‰
A	-53	-302
B	-58	-318
	-60	-310
	*-59	*-284
C	-54	-294
	-52	-300
	-58	-296
	-44	-299
	*-47	*-251
D	-62	-258
	-66	-298
E	-58	-352
	*-54	*-293
	*-56	*-263
Mains gas	-45	-197

Case B : This landfill site utilises a disused sand and gravel quarry. Infill started in 1952, and the bottom is neither graded nor lined. The waste, mainly loose with a small percentage baled is composed of 90% industrial, 7% commercial/household and 3% from demolition. 15,000 tonnes is estimated to be received weekly. A cell system is employed and the intermediate cover material is discarded sand/silt from a neighbouring quarry.

Site B is situated adjacent to a residential area, with the nearest house being only a distance of 15m from the site boundary. Despite having boreholes within the site to vent the gas build-up, boreholes placed adjacent to the edge of a completed and capped part of the site were found to be venting methane up to concentrations of 41%. The proximity of the houses and uncertainty to the extent and pathway of methane migration meant that an analysis technique was needed that could help in defining the source of the gas as quickly as possible.

Samples were taken from within the site and from the external borehole for comparison. The results are plotted on Fig.1 (symbol B). The composition of the gas found outwith the site is virtually identical to the gas found within the site. It can therefore be assumed that the landfill site is indeed the source for the gas found in the external boreholes and remedial action be taken identify the extent of the migration pathway and/or curtail the leakage.

Case C : This landfill site utilises a disused opencast coal mine. Infill started in 1984 and the bottom is lined with high density polyethylene. The waste is baled with to high density capacity of 1.15 tonnes/m^3 and in composition is 45% household waste, 13% commercial waste, 7% street litter, 29% subsoil and 6% demolition waste. A cell system is employed and the intermediate cover material is subsoil. The site is drained with the runoff pumped to a treatment plant for purification. There are plans underway to utilise the gas to power this purification plant, but at present the gas is flared when the pressure builds up.

However, part of this site is a result of unsupervised dumping pre-1986 and therefore is not so well planned. At the moment methane vents at concentrations of up to 70%. As with case B, monitoring boreholes positioned just outside the site, next to the edge of a river and by a public walkway, were found to have concentrations of up to 29%. Stable isotope analysis of methane was employed in this case to ensure that the source of the methane was from the landfill site and not from thermogenic methane associated with the coal measures from the old colliery. Samples from gas vented from boreholes within the old site and from those adjacent to the periphery of the site were analysed and the results are shown on Fig1.

Methane found outwith the site has an isotopic signature attributable to production by bacterial fermentation (i.e. normal landfill gas) and is similar to the signature of the methane produced within the site. The sample outwith the site is isotopically heavier, but this may be attributed to partial oxidation of the gas by bacteria during migration, a phenomenon which is documented as leaving the resultant methane isotopically heavier (Coleman, Risatti & Schoell, 1981). The samples analysed however are quite distinct from the isotopic signature of methane associate with coal and so the coal measures can be disregarded as the likely source of methane found in the monitoring boreholes.

Case D : Explorative boreholes put down during drilling to estimate the extent of a landfill site believed to be the source of migratory methane intersected underground mine workings. As with case C, stable isotope analysis of the gas was undertaken to identify the source of the gas, eliminating either the coal measures or the landfill site. From Fig.1 it can be seen that the isotopic composition of the sample collected is representative of methane produced by bacteria by fermentation, such as the process that happens in a landfill site. The underlying coal measures can therefore be disregarded as the source of methane.

Case E : This landfill partially utilises the space left by quarrying of a quartz-dolerite sill for roadstone but also extends into coal measures. There is, or was, a disused mine entrance within the site (it may now be covered) and in coal seams can still be seen to outcrop in one cell. In fact, during sampling, people have been seen pushing coal filled wheelbarrows from the site.

Three boreholes, placed just outside the edge of this landfill site, have consistently shown methane to be present up to concentrations of 70%, with an isotopic signature representative of bacterially produced methane. The positioning of these boreholes is of interest for, from the geological map, they appear to be situated directly up dip of the outcropping coal seams. No methane has been found present at any of the other boreholes outwith the site which may imply that discontinuity surfaces e.g. bedding planes, within the sedimentary sequence are acting as a conduits along which methane, if produced by the coal, may migrate.

Isotope analyses again suggests that the source of the methane is landfill, although the isotopically heavier values of the gas in the boreholes suggests that during migration it has undergone bacterial oxidation.

DISCUSSION :

Schoell, (1988) has compiled data from recorded variations in the isotopic composition of bacteriogenic methane to outline the zones shown on Fig. 1. which represent methane produced by acetoclastic methanogenesis or by the reduction of carbon dioxide. $\delta^{13}C$ for the former ranges from -73 to -42‰, δD ranges from -390 to -250‰, while for methane produced by the reduction of carbon dioxide $\delta^{13}C$ ranges from -110 to -60‰ and δD from -250 to -150‰.

Unfortunately there is very little data available that combines both typical $\delta^{13}C$ and δD values for methane associated with oil and gas, however $\delta^{13}C$ composition alone is well documented. Thermogenic methane (including deposits associated with crude oil) shows $\delta^{13}C$ ranges from -50 to -25‰ (Stevens, 1988). δD typically has a range from -260 to -150‰ (Schoell, 1980). $\delta^{13}C$ values of methane absorbed in coals show a wide range, -12 to -71‰ and are independent of the degree of coalification of the coal from which the gas was released (Deines, 1990). The lower values of this range overlap with the attributed ranges for methane produced by bacteria, however less than 10% of all coal methane samples measured fall in the range -55 to -70‰ while 50% of measured $\delta^{13}C$ for coal methane samples fall between -25 and -40‰. Furthermore δD of methane associated with coal is isotopically heavier than δD of bacteriogenic samples with typical values ranging from -180 to 163‰ and is therefore quite isotopically distinct.

In all the aforementioned case studies the isotopic signature of each sample analysed was distinctive of methane of bacteriogenic origin and therefore attributed to the landfill. These results were used to eliminate the sources of methane that would be likely to give another signature, and combined with further information about each site to pin-point the likeliest source of gas, which for each of these cases this was the landfill. Although some samples did plot in the area of overlap between bacterially produced methane and thermogenic produced methane associated with crude oil, the proximity of the gas sampled to landfill and the characteristic bacteriogenic signature suggested that the origin of the gas was from landfill.

Stable isotope analysis of methane can give a signature which is indicative of the process by which the methane formed, i.e. thermogenic or biogenic, however there is one drawback in that within each range there are several sources which may give similar signatures. Methane produced by a ruminant, from a sewerage digester or from anaerobic, organic rich sediment will all have an isotopic signature that is similar to methane produced in a landfill from decomposing rubbish. Likewise, the two main thermogenic sources, coal gas and natural gas, will have a similar signature.

Bacteriogenic methane that has undergone partial oxidation by bacteria during migration can have a resultant isotopic signature that is similar to methane of thermogenic origin, as the bacteria responsible for the oxidation preferentially consume the lighter isotopes leaving the gas isotopically heavier (Coleman et al, 1980). The resultant isotopic composition of a landfill gas that may has been fractionated as a result of oxidation is shown on Fig. 1, Case C. However research carried out by King et al, 1989 which is consistent with Coleman et al., 1980, show that the change in δD value of methane which has been partially oxidised by bacteria is 8-14 times greater than the change in $\delta^{13}C$ value. It should therefore be possible to recognise methane which has been subjected to partial oxidation by bacteria during migration by using the combined carbon and hydrogen isotopic analysis technique, for although the $\delta^{13}C$ value for the gas may be consistent with a thermogenic signature, the δD value may be outside the field normally accepted for thermogenic gas. If bacterial oxidation of landfill gas is suspected then characterisation of samples from mains supply gas or coal related methane in the specific area for comparison would aid identification of the source of the unknown methane.

Another problem that may lead to confusion in identifying the source of methane is where the mixing of two isotopically distinct gases occurs e.g. the mixing of a landfill gas and gas from a coal seam. For example, if the landfill gas started with a composition of $\delta^{13}C$ = -65‰, δD = -300‰, and the coal-derived methane started with an isotopic composition of $\delta^{13}C$ = -29‰, δD = -150‰, with both gases present in equal volumes, then by simple mass balance calculations the resultant isotopic composition of the methane would be $\delta^{13}C$ = -37‰, δD = -225‰. It can be noted that this resulting composition may be similar to the isotopic composition of methane that has been fractionated as a result of oxidation during migration.

In situations such as these, most other conventional analysis techniques such as determination of the chemical composition of the problem gas and age dating by ^{14}C determinations (refer to Williams & Aitkenhead for a further explanation) may also give rise to ambiguous results. If used in conjunction with stable isotope analysis they should be capable of defining the source. In comparison to these techniques, particularly age dating by ^{14}C determinations, stable isotopic analysis of methane is a quick and inexpensive way of determining the likely source of an unknown methane sample. If there is confusion from the initial analysis the results will serve to eliminate certain sources and will be useful in suggesting which alternative, complementary technique should be used. Furthermore, the volume of gas that is required for stable isotope analysis (1cc of 100% CH_4, or the proportionally appropriate volume of a diluted sample e.g. 10cc at 10% CH_4 concentration) is much smaller than that required for ^{14}C age dating. There may be instances where the volume of gas available is not sufficient for ^{14}C determinations, whereas in the majority of cases, unless the sample size is very small and at very low, in the order of L.E.L. (lower explosive limit), concentrations, there should be sufficient gas for stable isotope analysis.

CONCLUSIONS

The use of stable isotope analysis of methane to identify the source of unknown methane is viable because different processes of methane generation give rise to large differences in isotopic signatures. It is a relatively quick and inexpensive technique, particularly useful because it utilises small volumes of gas. Alone, it can often give a definitive answer but in more complex environmental/geological situations can be complemented by other available geochemical analysis techniques or information to provide the solution.

REFERENCES:

Coleman, D.D., Risatti, J.B. and Schoell, M. 1981. Fractionation of carbon and hydrogen isotopes by methane-oxidising bacteria. *Geochim. Cosmochim. Acta* 45: 1033-1037.

Craig, H. (1957). Isotopic standards for mass-spectrometric analysis for carbon dioxide. *Geochim. Cosmochim. Acta* 12: 133-149.

Craig, H. (1961). Standard for reporting concentrations of deuterium and oxygen-18 in natural waters. *Science* 133: 1833-1834.

Deines, P. (1980). The isotopic composition of reduced organic matter. *Handbook of Environmental Geochemistry*, Vol 1, The Terrestrial Environment, A. Published by Elsevier.

King, S.L., Quay, P.D. and Lansdown, J.M. (1989). The $^{13}C/^{12}C$ kinetic isotope effect for soil oxidation of methane at ambient atmospheric concentrations. *Journ. Geophys. Res.* 94: D15, 18273-18277

O'Neil, J.R. (1986) Theoretical and experimental aspects of isotope fractionation, in Stable Isotopes in High Temperature Geological Processes, *Reviews in Mineralogy*, 16: 1-40. Published by BookCrafters Inc.

Schoell, M. (1980). The hydrogen and carbon isotopic composition of methane from natural gases of various origins. *Geochimica et Cosmochimica Acta*, 44, 649-661.

Stevens, C.M. and Rust, F.E. (1982). The carbon isotopic composition of atmospheric methane. *Journ. Geophys. Res.* 87, C7: 4879-4882.

Williams, G.M. & Aitkenhead, N. (1991). Lessons from Loscoe: the uncontrolled migration of landfill gas. *Quarterly Journal of Engineering Geology*, 24, 191-207.

Waste Disposal by Landfill – GREEN'93, Sarsby (ed.) © 1995 Balkema, Rotterdam, ISBN 90 5410 356 6

General report

N. Kalteziotis
Public Works Research Centre, Athens, Greece

N Kalteziotis: The presentation of fully-documented case histories is a vital part of the process of developing and improving any engineering process. Case histories should; highlight previous mistakes and misconceptions (so that they can be avoided in future practice), declare successes and good practice (to provide a firm foundation of practice which can be improved further) and, point out topics or areas which require further research and investigation. Not unnaturally engineers are reluctant to display their failures in public and in addition case histories of successes often do not contain all the necessary details for various reasons, e.g., commercial confidentiality, lack of experience of preparing articles for publication. It is therefore pleasing to have presented within this session several detailed case histories which relate to successes and failures.

The general goal of the paper by **Hoekstra** is to show the application of geosynthetic materials to the storage of solid wastes and the benefits of such materials. With geosynthetics it is possible to select or develop a composite that meets the requirements of a specific problem, and so achieve any economic benefit and a technically-satisfactory design - Hoekstra's paper illustrates such a situation.

Leachate exiting from the waste slope made construction of a sand-bentonite sealing layer impossible because of the lack of a good supporting layer. By using fin drains in slits cut into the waste the leachate could be removed. With the increasing variety of geosynthetics available on the market it is important to maintain awareness of the essential properties of the products selected for a certain function (such as filtration and drainage in this case history.

Waldron, Hall, Fallick, Gilmour and MacDonald describe a method to help determine when a methane supply is from a natural source or from decomposing organic matter.

Methane and carbon dioxide are the main gases produced by the anaerobic decomposition of the biodegradable organic waste components in Municipal Solid Waste. In the initial phase of anaerobic decomposition the carbon dioxide predominates. With time it decreases and the methane content increases and appears to stabilise after approximately four years. The proportion of methane in landfill gas can reach 50-60% and a landfill may produce methane at this rate for ten or more years. in the presence of oxygen a methane content in air of between 5 and 15% by volume is explosive. Carbon dioxide alters these limiting proportions for explosion.

Geological or geothermic gasses, ie mine gas, natural gas, are in general characterised by a low carbon dioxide content and a higher proportion of the higher hydrocarbon gases (ethane, propane). Biogenic gases (which arise from naturally-occurring biodegradable deposits such as peat) and landfill gas, are characterised by a high proportion of ethane and propane.

Distinction between landfill and other biogenic bases may be made on the basis of age or the presence of traces of modern synthetic compounds such as CFCs (which until their invention in the late 1930s were not found in nature). Methane from ground strata can be oxidised by the aerobic bacteria in soil to produce carbon dioxide. Hence some landfills show little or no methane but high carbon dioxide levels. Similarly ethane and longer-chain hydrocarbons are the most easily digested by bacteria and so the lack of ethane is no absolute guarantee that a gas derives from natural sources. However the presence of methane and ethane in a ratio of about 20 to 1 makes it fairly certain that natural gas is present. But at high ratios it is still possible that the source of the methane is landfill. Alternatively, the gas may be a mixture from both sources.

However it is sometimes necessary to distinguish between different potential sources of gas. Williams and Aitkenhead (1991) summarised a case history of a methane explosion which took place in March 1986 at Loscoe (Derbyshire) and which destroyed a bungalow. Two people were injured but fortunately there were no fatalities. Although a coal seam was initially contemplated as the source of the methane, investigation defined the source as a landfill from which gas was migrating via both coal and sandstone beds to the bungalow (even though it was further away from the site than other houses).

The method outlined by Waldron et al involves isotope content, to help determine the origin of a methane supply, ie whether it is from natural sources of from decomposing organic matter.

Methane may be characterised isotopically by its radio-carbon (Carbon 14) or by the ratio between the stable isotopes of carbon (Carbon 13 to Carbon 12) and the stable isotopes of Hydrogen (Hydrogen 2 to Hydrogen 1). Stable isotope analysis is a method which uses natural isotopic distributions to characterise different sources of methane. If the distribution of C13 : C12 is plotted against H2 : H1, different types of methane exhibit different values and this allows a distinction to be made between methane made thermogenically and that formed biogenically.

Waldron et al worked in conjunction with Strathclyde University Water and Environmental Management Unit to identify the source of methane found venting outside (but adjacent to) landfill sites, using stable isotopic analysis. In the most densely populated region of Scotland the Midland valley, the infilling of quarries with refuse is problematic because most of the disused quarries are situated in coal-bearing strata which have their own problems with methane emissions. Determination of the origin of a pollutant, eg methane, is important because of the "polluter pays" principle.

In the case studies outlined by Waldron et al stable isotopic analysis showed the gases to be of bacteriogenic origin and it was therefore attributed to migration from the landfill waste. Waldron et al believe that stable isotope analysis is a better method than the conventional Carbon 14 identification techniques because the former method only requires small volumes of the gas, in low concentrations. As with other geotechnical techniques there are situations where interpretation can be more complicated such as microbial oxidation of gas and mixing with methane from another source.

Güler and Avci reported an evaluation of preliminary data obtained from site visits to a large waste landslide which took place in Istanbul in April 1993. The landfill had no leachate collection system, gas venting or properly-designed capping system. The cause of the landslide was attributed to the explosion of biogas inside the landfill. The waste subsequently travelled approximately one kilometre down the slope of the valley within which the landfill was located.

The Authors examined the possibility of a slope stability failure due to development of large pore pressures caused by the lack of drainage and a local topography which encourages water to accumulate at the back of the landfill slopes. The Authors cover remedial measures which must be taken and the problems which have to be resolved to minimise the environmental impact posed by the condition of this landfill.

Constantinou, Petrides, Charalambides and Kyrou present a case of environmental hazard concerning soils contaminated with Polychlorinated Biphenyls (PCB's). PCB's are considered as a highly undesirable pollutant, having low acute toxicity but moderate chronic toxicity, which need special care. These compounds have the general formula; Carbon-12, Hydrogen - X, Chlorine - Y, where X equals 0 to 9 and Y equals (10 - X). There are 219 such compounds commercially available in various mixes. They are very stable, both chemically and biochemically, with low water-solubility but highly soluble in organic solvents and polymers. Due to their high dielectric constant, fire resistance and thermal stability these compounds have wide application in industry, but their chemical and biochemical stability means that they persist in the environment. PCBs accumulate in the fatty tissues of living organisms this contaminating all forms of life.

Constantinou et al describe the pollution problem caused in Cyprus in 1987 by dumping askarel oil (based on PCB's) in a disused quarry adjacent to a populated area. The paper describes the emergency measures taken to prevent further pollution and to protect people and personnel involved with this incident and the field and laboratory investigations undertaken to quantify the pollution. In parallel with their investigation, alternative methods of disposal were considered and finally a secure landfill method was adopted. The implementation of the latter method involved burying the polluted soil in specially constructed watertight cells in the quarry area.

Large volumes of dredged material have to be placed in confined disposal areas because of environmental restrictions or economic factors. About 20% of dredged material is landfilled by hydraulic placement as a slurry in dyke containment areas. Common contaminants found in these dredgings include compounds of heavy metals and organic compounds. **Mohamed, Yong and Galvez-Cloutier** present a comprehensive study of metal mobility within dredged material and contaminant interaction with a dam barrier system.

The Authors carried out leaching-column tests to investigate the compatibility of leachate generated from the dredged mud and the proposed clay barrier. They simulated the metal behaviour under typical environmental conditions using a solvent leaching technique. The study demonstrated that the dredged mud possessed a high buffering capacity to pH change and hence heavy metals are not detected due to the high pH of the sediment. The calculated diffusion parameters for movement of various cations through the clay barrier were of similar magnitude to other published values.

Bell and Jermy report a case of ground water pollution caused by leachate seeping downwards from the toe of a closed landfill and its leachate pond. The landfill is located on sandstones which are overlain by a veneer of residual colluvial and alluvial soils. The investigation to determine the extent of the pollution consisted primarily of excavating a number of inspection pits for soil profiling sampling and for collection of water specimens for chemical analysis. The remedial measures consisted of; a subsoil drain (set into the bedrock) located immediately downstream of the landfill site and leading into a pump chamber, and a sand-bentonite wall keyed into bedrock which is constructed 3m downstream of the subsoil drain.

The most common, and perhaps most obvious, location for an engineered domestic and trade waste disposal site is an abandoned quarry or opencast coal site if the floor area is sufficient to produce a reasonable into life and particularly if there is sufficient overburden clay or similar material in the immediate vicinity for using as a seal. The quarry face or faces form one or more of the containment boundaries. Construction of a suitable and effective containment barrier against the quarry face and around the sides of the site is difficult because the steepness of the boundaries and the desire to keep the barrier as thin as possible makes it difficult for construction plant to operate. **Geusebroek and Luning** present a case history concerning the construction of a modern landfill in a quarry with steep, high slopes.

Three problems had to be addressed; construction of a suitable barrier against a steep quarry face, formation of a reliable leachate system (to which there had to be access), provision of a facility to check the quality of the groundwater beneath the liner. The first problem was solved by forming a composite liner using an HDPE membrane and a 0.5m layer of compacted clay. To compact the clay at a steep angle a heavy vibrating plate compactor was mounted on a hydraulic boom which could reach a height of about 10m. The second and third requirements were satisfied by constructing a passage beneath the base liner. This 'subway' is 2m high to enable personnel to work inside and pipes pass from it to the groundwater beneath the base liner. The strength of this subway system is that should serious leakage of the basal liner be detected by groundwater monitoring then the polluted groundwater can be extracted via the monitoring system thus preventing spread of the pollution away from the site.

Niininen, Kalliokoski and Parjala investigated the quality of sanitary landfill leachates and their effects on groundwater. Their studies pertain to twelve sites of various ages in Finland, some of which were even located in esker areas where groundwater is easily contaminated because of the permeable soil.

Leachate from decomposing landfill contains various concentrations of organic and inorganic chemicals. These chemicals can have a serious impact on the groundwater quality beneath and beyond a landfill. The concentrations of the various constituents vary considerably according to the age and character of the landfill and the events preceding the time of sampling. Leachate generated from the disposal of hazardous waste may contain elevated amounts of heavy metals eg mercury, lead, cadmium and toxic substances such as arsenic and organic compounds. Various mechanisms operate in the attenuation of the constituents found in the leachate as it migrates through the subsurface soil. In general heavy metals are removed by ion exchange reactions as leachate travels through soil while trace organics are removed by adsorption.

On the basis of chemical and physical analysis of leachate and groundwater samples Niininen et al concluded that groundwater had been affected by the leachate to differing extents according to the hydrogeological and physio-chemical conditions of the area surrounding the landfills. In the groundwater samples quite high concentrations of cyanides were found whereas chlorophenols and heavy metals were present in low concentrations. Unfortunately the Authors do not provide details of the form of construction of these waste disposal sites.

Boreli, Pokrajac and Halda give a general description of a global approach to the disposal of an ash which is the tailings from lignite. They indicate the use of pollutant flux as an operational criterion for water protection. The sorption properties of the natural sub-soil are emphasised as a contributor to the containment of pollution migration.

Bell and Jermy
A seepage problem associated with an old landfill in the greater Durban area.

Boreli, Pokrajac and Halda
Some experiences in environmental protection from the lignite ash impoundments waste water.

Constantinou, Petrides, Charalambides and Kyrou
A case of environmental hazard caused by dumping in open ground polychlorinated biphenyls.

Geusebroek and Luning
Innovative landfill design in a former sand quarry.

Güler and Avci
Evaluation of the incident at the Umraniye landfill, Istanbul and proposed remedial investigations.

Hoekstra
Final capping and finishing of a landfill in Limburg, the Netherlands.

Mohamed, Yong and Galvez-Cloutier
Land disposal of dredged mud.

Niininen, Kalliokoski and Parjala
Quality of landfill leachates and their effect on groundwater.

Waldron, Hall, Fallick, Gilmour and MacDonald
Stable isotope analysis as a means of identifying the source of methane.

Waste Disposal by Landfill – GREEN'93, Sarsby (ed.) © 1995 Balkema, Rotterdam, ISBN 90 5410 356 6

Discussion

A Malone : My office is particularly concerned with the stability of landfills, undertaking an independent review of the design of all new landfills in Hong Kong. The geotechnically-related topics covered include;

a) the link between the economy and solid waste production,
b) landfill provision now and in the future,
c) geotechnical aspects of certain particular classes of landfills,
d) specific issues relating to these sites.

Hong Kong is a small territory, being overall about 60 km square, at the mouth of the Pearl river on the south coast of China. It consists of a large piece of the mainland and two islands. It has a population of 6 million yet its land area is rather less than that of London contained within the M25 orbital motorway.

Hong Kong's economy is very strong and this has a major impact on solid waste production and over the last 15 years the annual amount of landfill has risen to 6/7 million tonnes. Construction waste is a major problem in Hong Kong, e.g. in 1990 of a total of 5 million tonnes of refuse, 2.9 million tonnes were construction wastes (at landfill).

In Hong Kong there are four operating landfills at the present time with a further life of 4/5 years at most - the long-term plan is to introduce three very large landfills with 20-25 million cubic metres capacity. After collection the refuse is taken to transfer stations where it is containerised and then taken by sea/road to the landfills. These landfills will operate for 10/15 years and after that incineration will probably have to be used. The main concern, in terms of water contamination, is not for surface water (Hong Kong is supplied from China and catchments away from the landfills) but contamination of enclosed sea bodies. So the design of landfill sites is mainly concerned with prevention of pollution of water courses.

Now to geotechnical aspects of landfills in Hong Kong. There are principally two settings - hillside valleys and coastal reclamation.

Hong Kong suffers very high annual rainfall (about 2.2m of rain per year) and watercourses carry very large quantities of water over short time periods. Rainfall is highly seasonal, ie virtually all between May and September. So the principal problem for a landfill is how to cope with the surface water flowing in the valley containing the waste. Then there is the problem of stability - in the bottom of the valleys there is an accumulation of hill slope debris material (colluvium) and this material is very difficult to engineer embankments on. In the lower parts of the slopes there are high groundwater tables (within 5/6m of the surface). The other problem always present in hilly areas in Hong Kong is boulder fall - the hillslope landfill is 'fir-tree' construction, ie like a slimes tailings dam - if a trapezoidal form of slope face were used there would be no room for the waste.

For coastal sites the typical depth of soft to very soft post-glacial clays are 5 to 20m. Consequently there are the normal problems of marine embankments on soft clays. Some years ago there was a major failure of a landfill on soft clay involving about 300,000 cubic metres of material. However these situations are covered by conventional geotechnical analyses and there are no particular difficulties if the correct techniques are used. Marine pollution is the major problem because many of these waters are used for mariculture (firm farming) and fishing is a very important industry for Hong Kong. There is also the question of conservation - corals exist in the area and the marine biota is very rich (particularly in the Eastern part of Hong Kong).

Highly complicated, high technology engineered landfills are designed for materials other than construction waste. Traditionally construction waste has been used in Hong Kong to form land. To promote this activity a 'public dump' is established to which people have free access for the disposal of inert material within a certain specification and traditionally this would be the disposal point for construction waste. However the rapid programme of building in the late 1980s onwards meant that insufficient 'dump' space was available and there is the problem is what to do with the excess. One possibility would be to take it to an engineered waste disposal site but the landfill charge/levy creates the difficulty of enforcement of fly-tipping, ie illegal tipping. Hence the current strategy is to provide 'protective public dumps' to facilitate taking the refuse to the dump rather than the engineered landfill site. Thus at the present time marine reclamation is being promoted to create new land solely for public dumping.

The urban landfills are located in hills very close to developments. Hence the major problems involved in their restoration are gas and leachate control.

Contaminated mud - not a classical landfill problem but important in Hong Kong. The marine reclamation for the development of the airport involves the excavation by hydraulic-dredging techniques and grab-dredging of around 25 million cubic metres of contaminated mud - the principal contaminant is heavy metals. The strategy for disposal (which is actually a complex package of measures with the principal one being to discourage excavation) still means that at least 20 million cubic metres of contaminated refuse will be disposed of in a sea-bed landfill, ie a sea-bed borrow pit. This will be achieved by performing dredging and dumping into a sea-bed pit followed by a cap of 2m of sand and 1m of clean mud. The geotechnical problems involve; ensuring that the cap is stable under tractions from current and wave loading, and ensuring that the sand can be placed to form a stable blanket.

A van Autenboir: Limburg University's involvement with waste disposal is 'upstream' of the engineering works, ie its work is to evaluate sites, assess their impact on groundwater and thereby formulate general principles of rehabilitation/remedration/protection. The actual engineering work is done by others although it may be subsequently inspected by the Author.

Most of the work (site investigations, monitoring of groundwater quality, etc.) is undertaken on a contract basis and eight people are involved permanently on the work. Recently the group produced a vulnerability map of groundwater (this really should have been produced 150 years earlier before major industrialisation, in Europe).

The flemish part of Belgium is very flat with several aquifers (for drinking water) which are threatened by diffuse pollution, eg sulphates, nitrates from agriculture. The quality of surface waters is poor/very poor and the area is overpopulated so that the NIMBY syndrome can be very effective especially when there is a symbiosis between the local 'green' committees and the politicians. Such a liaison caused one of the best landfill sites (from the geological point of view) to be abandoned with the refuse going to other places where the groundwater was much more vulnerable. The most recent census indicated approximately 1800 contaminated sites in Flanders along.

A detailed examination of five landfills has shown; no problems with gas, very limited influence on the groundwater near to the landfills, and no need for extensive intervention. The proposal is therefore that funding available for reclamation/rehabilitation should go to other areas.

One site investigation concerned an old zinc

smelter occupying an area of about 47 hectares with the groundwater on top of a gravel layer which was itself on top of a sandy layer. The permeability of the ground varied between 360m/day to 3000m/day. On the site processes had included roasting of the ore, production of sulphuric acid, reduction by a thermal process etc. and there was about 500,000 m^3 of contaminated soil at the site.

The groundwater was contaminated on the site and in a narrow area extending some 2/3 km from the plant. However the groundwater was only contaminated in the gravel part of the aquifer. In the underlying sands the water was clean - this is believed to be due to the high hydraulic conductivity of the gravel so that the hydraulic head in the sand exceeded that in the gravel. The remedial action proposed comprised a fully-engineered local landfill with a composite liner with nothing being done about the groundwater.

Flemish coal tips have also been investigated because of a problem of increased sulphate and chloride content in some water wells, and also because of observed high concentrations of heavy metals (eg arsenic, beryllium) close to the tips. The main reason for this occurrence seems to be oxidation of the pyrites content and the pumping of the deep ground water which is salty. In the region there are seven deep mines, seven sites each with tips containing about 30M m^3 each. Most of the sites are not directly on sandy aquifers.

The treatment which has been advised is to leave the tips undisturbed because they consist mainly of shales. Thus weathering (particularly alternating wetting/drying cycles) very rapidly produces an impermeable crust on the tip thereby preventing air and water entry thus halting oxidation of the pyrites.

Finally a problem with sludges arising from the zinc industry and the fertiliser industry. The sludge is pumped into lagoons, about 8m deep, and excess water is pumped back to the processing area. A very thin crust (about 1m) forms on top of the sludge but beneath it is completely liquid. The lagoons have very poor quality liners and some leakage is occurring.

The intention is to dewater the sludge in-situ by compressing it. This will provide extra storage space but it is also hoped that it will cause plugging of the leaks and also increase the stability of the lagoons so that the sides can be raised. The sludge has a permeability of the order of a silty clay. For remedial works to be viable de-watering of the sludge has to be; rapid (2/3 years), undertaken in-situ and cheap - work has just started to find a solution.

Closing session

Waste Disposal by Landfill – GREEN'93, Sarsby (ed.) © 1995 Balkema, Rotterdam, ISBN 90 5410 356 6

Concluding remarks

Session chairs/general reporters were invited to present their views of the salient points to come out of their session:

Y Wasti: Session one covered general aspects including legislation, regulation, current practice, national and international dimensions. The papers touched on a variety of subjects but the one which generated greatest interest, and spilled over into other sessions, was landfill lining systems, especially for final covering.

One of the specific points raised was that in order to render waste environmentally safe in a short time rainwater must be allowed to infiltrate the waste and thereby promote degradation. People enquired whether this presented a conflict with current approach to landfill cover technology which aims to provide encapsulation of the waste - there was no consensus of agreement on this point. Some contributors opinioned that even if the waste is left uncovered water will not penetrate uniformly and will produce concentrated, isolated flow parts so that the anticipated uniform degradation is not achieved. Professor Daniel suggested leaving the waste open for a certain time (a 'grace period') and then cover. The time period will depend on each specific site.

Another issue addressed was the use of geomembranes in lining systems. It was emphasised that Construction Quality Assurance requires an understanding of polymer chemistry as well as engineering requirements. The major current usage of HDPE was highlighted as was the necessity for trying, developing and testing of alternative materials. A further issue concerned protection of geomembrane liners (both during their installation and due to placement of a drainage layer). Other items mentioned were asphalt lining layers and factors affecting hydraulic conductivity values.

The highlight of the session was undoubtedly the keynote lecture by Professor Jessberger with its reference to the recommendations of European Technical Committee 8 on landfill design and remedial works. The five critical issues covered by the recommendations which are intended to form an authorative guidance document are; mineral liner and cover deformations, design of leachate drainage systems, certified and non-certified geotextiles and geomembranes, cut-off walls, contaminant migration assessment.

M de Silva: The session which was entitled 'Basic Principles' was a combination of aspects of; landfill design, biological processes occurring within landfill, leachate chemistry and use of new materials in waste encapsulation.

In his keynote lecture on risk assessment Dr P McKendry detailed an engineered containment approach to landfill which allowed a more rigorous engineering treatment of design and construction which in turn has enabled the methodology of quantitative risk assessment to be applied.

However it has become clear that more materials research needs to be conducted in order to be certain about some risk assessment figures particularly in terms of the long-term behaviour of some materials being used. The composition and characterisation of leachate is very important in terms of its effect on materials. However this clearly needs to be done on an individual site basis because of the influence of refuse composition, 'cocktail-effects', etc. Also the increased amount of recyclable waste being diverted from the refuse stream into landfill will cause significant changes in leachate chemistry, so ongoing investigation of leachate-liner interaction is required.

Clearly a lot of work is being done on rather elaborate composite liner systems for environmental protection but given that their long-term behaviour cannot be predicted it would seem sensible to design landfills to stabilise waste more rapidly. This would tend to favour a containment-type system with controlled liquid inputs particularly when dealing with more reactive refuse such as Municipal Solid Waste. This does not completely rule out dry entombment (which is probably still applicable to contaminated soils) but it is not really appropriate for biologically-reactive wastes. Furthermore dry entombment would not be consistent with the british concept of certificates of completion, (which signify a stabilised waste deposit) for finished disposal sites.

Biological control systems, particularly methane oxidation, were also discussed within the session. My experience is that it is not a particularly effective way of reducing methane (levels of oxidation of methane of about 10% have been achieved) although there is scope for further development of the method. However the current design of landfills does not lend itself to effective methane oxidation as a way of controlling methane emissions. However given the potential for energy production combined with the possibility for rapid stabilisation a way forward for the design of new sites is to consider them as energy generators - bearing this in mind there is a need to re-evaluate the potential gas yields.

With the greenhouse effect of methane, in comparison to Carbon Dioxide, more effective controls are clearly required. This problem was recently considered in the British parliament and it was stated that UK methane emissions were 1.4% of the global value with most of it originating from landfill sites.

There is further scope for investigating low-cost options for liners, e.g. use of Pulverised Fuel Ash, particularly if rapid refuse stabilisation is chosen so that the active life of the decomposing matter is less than the break-through time for leachate. If this is combined with good leachate management (which is analogous to leachate treatment) it could lead to effective generation of landfill gas particularly since it must be remembered that leachates themselves have a gas-generating capacity, so the treatment could involve anaerobic digestion of leachate.

Other treatments of leachate were discussed such as the use of cellulose skeleton ion exchangers which are able to capture contaminants on their surface and electrokinetic remediation which is the use of electric current to attract ions to an electrode from which they can then be removed.

Looking at the future, I believe that there will be much more waste minimization and recycling (this will affect the economics of waste - figures of 40p to £4 per household per annum have been quoted to separate waste materials in England). Waste minimization and recycling will reduce the amount of landfilled refuse. If the waste is adequately separated at source then the fraction going into the landfill can be excavated (after it has stabilised) and be used as a soil conditioner (the residue is an anaerobically-derived compost). Using fabricated anaerobic digestion technology and process controls the waste can be rapidly digested together with maximum gas production. The excavated cells can be refilled and used in rotation - this is the concept of the sustainable landfill.

H Y Fang: The topic which has to be addressed is the definition of refuse/garbage and identification of its engineering properties - this is one of the great unknowns. However several things are known about the engineering behaviour of refuse. Firstly, we cannot solve the problem by simply using mathematics. Secondly the behaviour of garbage is not described by a Mechanical energy approach, it fits within thermal/electrical/magnetic processes. Refuse is three-dimensional and random, it is not a linear, uniform material. Consequently we cannot apply Darcy's law or Terzaghi's theory of consolidation and we have to develop new analyses.

In general the conference has approached refuse analysis in the conventional way - discussion has centred around 'how' rather than 'why'. To solve the problem we need to start from basics, e.g.

why are we undertaking a particular process, why does a particular liner work/not work. The United States Environmental Protection Agency initially only required a clay liner at landfill sites, then two years later they specified double liners and then another two years later triple liners are required - this is because they really do not know what they are doing. Delegates to the conference should concentrate on identifying the fundamental engineering issues specifically to refuse/garbage.

Terzaghi's work on the engineering behaviour of soils was undertaken at a time when the behaviour of waste was not a problem. Now it is a concern and we must develop specific theoretical treatments rather than adapt theories which were never intended to represent the behaviour of garbage.

D Daniel: We tend to solve the easiest problems first and leave the most difficult until last. In my talk I spoke about landfill covers and some of the challenges I perceived which exist in the design of landfill covers - for instance, I opinioned that clay liners placed over freely-deposited Municipal Solid Waste would not survive. On the other hand Professor Jessberger believed that the situation was not this bleak. But overall it must be agreed that covers over freely-deposited waste are still an area with many questions and few answers. The problem may be solved in the near future but I am not optimistic because it is such a difficult challenge.

There has been a significant amount of debate over when the final cover should be placed over refuse. Philosophically it may seem desirable to cover waste immediately (apparently this does happen in Switzerland) but this is almost never done in the United States - this may be another example of solving the problems that are easiest to solve.

I am enthusiastic about geosynthetic clay liners and it was surprising to observe the lack of interest amongst delegates because they are generating a lot of enthusiasm in the United States. It is clear that there are dramatic differences in perception of landfill problems between Europe and America. For example in the northern part of England the boulder clay is frequently used as a natural clay liner without engineering controls and yet a decade ago there was intense debate in America over the need for clay liners to be engineered.

Furthermore there is the fundamental problem that our engineering controls for landfills seem to bear no relationship (in terms of regulation) to natural geological conditions. The requirements for liners are not keyed, in any way, to the natural hydrogeological vulnerability of a site. To make progress we need to develop criteria for relating the engineering controls to factors such as climatic conditions, hydrogeological vulnerability of the site, etc. The permeability criteria for liners (such as a maximum value of 10^{-6} mm/sec) appear to have been arbitrarily defined - how can engineers design landfill covers when the allowable amount of water passing through the cover has not been specified.

Several speakers in Session 4 referred to "geochemistry". This concern for detailed knowledge about geochemical reactions is interesting because existing regulations do not mention geochemistry. At most there may be some vague comment about compatibility between the liner and the leachate.

One of the clear areas in which progress is being made in pollution prevention is in innovative use of materials, particularly those which are themselves waste products.

In closing I would say that there still are many areas of confusion/uncertainty but that we have moved from a position of thinking that we know all the answers to one of addressing our lack of knowledge.

S Jefferis: The importance of site investigation, in the context of old landfills, is the ability to locate accurately the landfill. The feeling from the session was that geophysical methods of site investigation are very good at finding these old sites. In most cases there is very substantial contrast between the landfill and the surrounding ground and boreholes are not needed to calibrate the geophysical investigation techniques - this is not always true but the technique is very powerful. Electrical resistivity, ground-probing radar and gas monitoring (using mobile monitors of inserting probes into the ground and extracting vapour for measurement) are all very useful techniques. It may be possible to obtain a small-strain soil modulus from small-strain seismic work, however this modulus may not be much use for predicting settlements.

It does not appear logical to apply Terzaghi's consolidation theory to the behaviour of landfilled refuse since the model is not the correct one. Consolidation theory is about the diffusion of water out of materials of relatively low permeability where effective stress rules apply whereas these wastes are soft materials with a wide range of sizes and an enormous range of permeability. A peat related model may convey some ideas of very rapid, and perhaps dynamic, settlement. But peat has a very substantial tensile strength in the horizontal plane and it is believed that this tensile strength is not exhibited by landfill materials which break down to smaller fibre lengths. There are fundamental issues of total settlement, which is controlled by finance because of the extra waste which can be deposited as the landfill settles, and differential settlement, which is the key parameter for cap design. There are problems at the edge of the cap where large differential settlements are bound to occur because of the rapid variation in fill depth.

The bearing capacity of old refuse may be improved using stone columns and jet grouting, although there was a suggestion that these methods should only be used in non-organic wastes where lateral support will exist. Although dynamic compaction appeared to be very promising for bearing capacity improvement there is the problem of re-initiation of methane production due to aeration and this may seriously delay a project.

Other topics within Session 5 related to gas and leachate control and slurry walls. In the UK the specification for slurry wall barriers was produced to suit earth dams and it is not really appropriate to the low confining stress level which occurs in landfills. Since the cut-off material cannot be changed dramatically it is necessary to change the design approach and recognise the range of material properties that can be achieved. Thus the stress-strain behaviour is a fundamental issue.

Chemical interaction between clay and leachate does not seem to be a major concern (at the present time scale) and it is suspected that the interaction will not give more than an order increase in permeability. It does not seem to be the problem that it appeared to be ten years ago when chemistry was one of the overriding issues and engineering and physics tended to get forgotten. We have been getting the engineering and the physics wrong that we need to concentrate more on these areas. On the other hand if the concern is not about barriers but about drainage then the chemistry may be much more important. If leachate is to be extracted from landfills the clogging of drains due to inorganic deposition of carbonates and sulphides and organic bio-slimes from bacterial growths may be very significant so that the permeability drops from very high values (for open pipework) to effectively zero. The drains must be kept open if recycling of leachate is to be used within landfills.

The issue of thermal cracking should have been recognised years ago because it is accepted that landfills are hot. There will be moisture cycling in clay liners and natural materials which leads to cracking and greater knowledge is needed about the healing of cracks, e.g. what happens when they re-hydrate. Work on slurry walls has shown that after cracking they will often seal again completely as a result of flow through them.

In the future there will be fewer, but larger, landfills. The volume to surface area ratio will be higher which means that there is less periphery for leakage and if the main cost of developing a site is the work on the periphery then the unit cost of landfill will decrease. But the move towards deeper landfills will mean that stresses are much higher. We will have substantial engineering structures and it will be necessary to consider the stiffness of the base, the effect on mineral barriers etc.. The high stress level may be beneficial for mineral barriers in that it may prevent cracking, any chemically-induced shrinkage will be countered by strains produced by the high stresses. The effect on geosynthetics is uncertain but it is likely that a lot more buffer layers will be needed to protect them. There will be more gas and leachate and if it escapes there will be a larger pollution plume, it will progress further, and its chemistry may be significantly different from the leachate of today so that leachate-clay interaction may become significant.

Since there will be fewer and larger landfills they should be constructed in appropriate geological conditions and monitoring must be undertaken since they will be engineering structures on a grand-scale. Without monitoring data it will not be possible to assess whether the waste has become stabilised or not. Finally, it should be remembered that today's waste sites could be tomorrow's problem and so we must get the engineering right.

N Kalteziotis: Most of my points have been covered in preceding reports and I will add only a few comments.

Only a limited number of case histories were presented and much more field data is needed relating to fill settlement, slope stability, mechanical properties of refuse etc.. The presentation of fully-documented case histories is always a problem because of confidentiality aspects and a natural unwillingness to exhibit one's failures. However it is through case histories that confidence is developed for advancement of an engineering subject. To protect the environment from the pollution arising from landfilled refuse engineers need to be able to assimilate elements of chemistry, physics and microbiology and there is an urgent need for the development of models for waste containment sites.

I hope that by the time the next GREEN conference is held on engineered waste disposal that the preceding points will have been addressed.

Following the comments by Session chairs/general reporters comments from the conference delegates were invited:

B Leach: My overall feeling from the meeting is that there are a considerable number of areas where not only is further research and thought necessary but there is a need to go back to basic principles and ask the fundamental questions.

J Khaffaf: The important issue of the influence of earthquake/seismic activity on landfills does not seem to have been addressed - what are the consequences and what precautions need to be taken.

T Zimmie: The effect of seismic activity on interface friction between geosynthetics was mentioned briefly and in future in the United States it will be mandatory to undertake a seismic analysis of a landfill. Landfills appear to be relatively unaffected by earthquakes in that it is "like shaking a bowl of jelly", i.e. the contents of the bowl may move around but they are contained within the bowl. But certainly more information is needed, e.g. on dynamic friction angles.

F Y Fang: Clay liners have a high concentration of acid, ie a low pH. Under static loading, e.g. unconfined compression, direct shear, triaxial test, the acid helps to increase the shear strength of the clay liner. Under dynamic loading the opposite is true because the interparticle energy is destroyed by the excitation - this effect of pH value on dynamic shear modulus was reported in 1986. So landfills cannot be designed for dynamic conditions using a static approach.

F Crozier: The issues of earthquakes was raised at a recent public enquiry regarding a landfill site in North Wales. Whilst the United Kingdom is not regarded as an area of seismic activity North Wales is susceptible to earthquakes and recently suffered one of magnitude 5.2 on the Richter Scale.

A Malone: I would like to know where the forums area in which technical points relating to landfill design are being discussed in the United Kingdom. Also there appear to be no journals concentrating on this topic and publication of articles is diffuse and widely scattered.

D Daniel: Currently in America about 50 - 70% of activity in geotechnical engineering community is related to environmental problems. Yet a suggestion to change the title of an ASCE journal from "Geotechnical Engineering" to "Geotechnical and Geo-environmental Engineering" was turned down.

H Y Fang: Leachate is a multi-phase solution, and to study it technically certain additional parameters need to be considered such as, surface tension, dielectric constant. The conventional geotechnical approach is not correct and therefore the wrong parameters are selected. Only sand is considered susceptible to liquefaction because with clays there are additional forces relating to heat of working and kinetic dispersal. In terms of pollution of water from clays there are many complex variables. For instance sorption itself can be divided into a part which is a physical phenomenon and a part which is a physical-chemical-biological phenomenon. Soils engineers know that water is absorbed by soil but they do not understand the mechanisms.

S Clam: In the United Kingdom the construction of landfills below the water table is generally discouraged. Properly-constructed landfills of this form could give excellent containment if underdrainage and in-cell drainage are included so that the waste is subject to an external positive water head as described in Professor Sarsby's talk. There is also the benefit that any mineral liner would remain saturated. The principal contaminant losses will then be limited to diffusion.

D Daniel: In America there are certain states which have had no choice but to allow landfills below the water-table e.g. Wisconsin, Louisiana. They are frequently called "zone-of-saturation landfills" and the earliest ones which were built around 1980 did experience some problems. This was because of inadequate impermeability of linings and leachate collection systems being unable to cope with the groundwater flowing into the landfill. In recent years below the water-table landfills have generally worked well. For a recent below-water-table landfill in Texas a groundwater pressure release system comprising a geonet/geotextile combination was installed below the clay liner and once the landfilled refuse produced sufficient ballast loading the pressure relief would be switched off. The natural soil at the site was relatively clayey so that the inflow of fresh water into the landfill is not expected to be a problem. If zone-of-saturation landfills are operated and designed correctly there is no reason why they cannot function in many ways better than an above-the-water table landfill in the short term. In the long-term there is a question of chemical diffusion of leachate for all landfills.

S Jefferis: There may also be some geotechnical issues associated with below-the-water table landfills, e.g. it may not be possible to achieve steep side slopes. In Canada this form of construction is encouraged because it is possible to counterbalance outward diffusion with inward advection.

H Y Fang: There is a lot of concern about healing of cracks in liners but little attention has been paid to the reasons for cracking. There are four reasons for cracking; moisture content decrease, shrinkage, temperature change, water content increase beyond the saturation value. This latter type of cracking is the worst case because air (in the form of bubbles) is carried in the water. Diffusion is the movement of ions from one location to another and this movement is covered by Frick's law. However when the concentration of ions becomes very high this law ceases to be applicable.

S Waldron: If one is concerned about the possibility of air bubbles in groundwater coming into a landfill site it should be remembered that there will already be gas, in solution, in the leachate.

The closing address of the symposium was given by the chairman of the organising committee:

R W Sarsby: The intended coverage of the GREEN symposia is contained in the name of the series i.e. **G**eotechnics **RE**lated to the **EN**vironment. Waste Disposal by Landfill was perceived as the major current topic within the general area of environmental geotechnology and hence its selection as the theme for this first conference.

During the period of the meeting there has been an exchange of information, ideas and questions on all facets of engineered waste disposal from global issues to consideration of the behaviour of individual discrete components. Having listened to discussion of the various topics I would hope that by the time of the next GREEN conference on Waste Disposal by Landfill that solutions, consensus of opinion or significant performance data will be available for the following topics :-

a) The important global considerations.
   For example, what are the functions of this engineered landfill over the whole of its life cycle, what are we trying to design - tomb or cooking pot?

b) The internal design of the landfill.
   Controlled influx and distribution of water and efficient extraction of leachate and gas is needed if the waste is to be decomposed thoroughly and at a controlled rate so that the site can be stabilised rapidly or operated as a bio-reactor. To complete this design (which will mean that subsequently the waste will have to be placed in a controlled engineering manner) we need to have 'models' for the chemical, biological, mechanical etc. processes occurring within the landfill mass. Definition of the internal behaviour will permit the effects at the landfill boundaries to be predicted, e.g. cap deformation, leachate production rate.

c) Liquids and gaseous collection systems around the waste.
   These systems are a vital part of the containment provision for the decomposing waste. If the drains around the waste are fully effective then pressure differentials will not exist to drive pollutants through the active barrier layers. Pollutant movement by diffusion and other mechanisms can be restricted by periodic 'flushing' of the drainage system and recirculation of leachate can aid waste decomposition and leachate neutralisation. Hence it is vital to be able to design drainage systems that will function for significant times under highly adverse conditions. We need to be sure that not only will the drain itself not clog but that liquid and/or gas can pass from the waste into the drain.

d) Whole-life-cycle behaviour.
   Each waste disposal cell undergoes significant changes during its life-cycle of construction, filling, sealing, stabilisation and rehabilitation. Furthermore the nature of the cells which are developed will change with time. The composition of the leachate will change as the refuse in a landfill ages and also as the composition of the waste-stream itself changes. The rate, and quantity, of leachate and landfill gas production will be affected by waste capping practice and control of inflex and eflux of liquids and gases.
   The long-term effects of leachate and landfill gas on mineral and synthetic layers are still unknown. There are a multitude of factors to be considered such as chemical aggression, biological attack, temperature changes and gradients, stress cracking, etc.. There is still considerable discussion about the length of time for which a landfill containment system must function - design lives of a hundred or more years are quoted. However if the drains function properly, and the waste degrades so that its liquid storage capacity increases and the viscosity of the leachate increases, the time-period for which the pollutants are mobile may be relatively short.

e) Performance monitoring of landfilled refuse.
   To design effective containment sites parameters are needed that describe the properties of the waste during the life-time of a site. Such information is obviously needed for the proper design of the cap of the waste but it is also vital for designing systems to regulate decomposition of the refuse if we move towards the bio-reactor approach. Since the waste is not homogeneous and it decomposes with time long-term field monitoring of actual landfills needs to be undertaken. Performance monitoring is also necessary for the development of analytical methods and to assess the stability of the construction materials in aggressive conditions.

f) Rehabilitation of old sites.
   Sites operated as bio-reactors will have to be rehabilitated at the end of their

working life. Simple containment sites will also need to be reclaimed at some stage because there is only a finite amount of land for habitation and, even with the promotion of recycling, large volumes of refuse will continue to be produced each year for the foreseeable future. Rehabilitation and redevelopment processes for these old sites must be efficient, effective and environmentally-friendly and the development of this facet of engineered waste disposal is a difficult and interesting challenge for geotechnical engineers.

I thank all those who have attended this first GREEN symposium for their contribution to this successful meeting.

Prof. R. W. Sarsby
Bolton, October 1994

Waste Disposal by Landfill – GREEN'93, Sarsby (ed.) © 1995 Balkema, Rotterdam, ISBN 90 5410 356 6

Author index